Australian Moths: A Natural History

Edited by

Roger L. Kitching

Royal Entomological Society

CABI

CABI is a trading name of CAB International

CABI
Nosworthy Way
Wallingford
Oxfordshire OX10 8DE
UK

Tel: +44 (0)1491 832111
E-mail: info@cabi.org
Website: www.cabi.org

CABI
200 Portland Street
Boston
MA 02114
USA

Tel: +1 (617)682-9015
E-mail: cabi-nao@cabi.org

A catalogue record for this book is available from the British Library, London, UK.

ISBN-13: 9781800627666 (hardback)
9781800627673 (ePDF)
9781800627680 (ePub)

DOI: 10.1079/9781800627680.0000

Commissioning Editor: Ward Cooper
Editorial Assistant: Theresa Regueira
Production Editor: Rosie Hayden
Typeset by Straive, Pondicherry, India
Printed and bound in the UK by Halstan & Co. Ltd, Amersham

Contents

This book is dedicated in friendship,
respect and admiration
to

TED EDWARDS
(1945–2023)

Lepidopterist Extraordinaire

Contributors

Michael J. Andren
Biodiversity Conservation and Science, NSW department of Climate Change, Energy, the Environment and Water, Coffs Harbour, New South Wales 2450, Australia
Mick.Andren@environment.nsw.gov.au
Louise A. Ashton
School of Biological Sciences, University of Hong Kong, Hong Kong SAR, China
lashton@hku.hk
Andrea Battisti
Department of Agronomy, Food, Natural Resources, Animals and Environment, University of Padova, Legnaro, Padova, Italy
Ethan P. Beaver
Research School of Biology, Australian National University, Acton, Australian Capital Territory 2601, Australia, and Australian National Insect Collection, CSIRO, Black Mountain, Australian Capital Territory 2601, Australia
Georgina E. Binns
School of Natural Sciences, Macquarie University, Macquarie Park, New South Wales 2109, Australia
Michael F. Braby
Australian National Insect Collection, CSIRO, Canberra, Australian Capital Territory 2601, Australia, and Division of Ecology and Evolution, Research School of Biology, Australian National University, Acton, Australian Capital Territory 2601, Australia
michael.braby@anu.edu.au
David R. Britton
Elizabeth Macarthur Agricultural Institute, NSW Department of Primary Industries and Regional Development, Menangle, New South Wales 2568, Australia
david.britton@dpi.nsw.gov.au
Matthew Brookhouse
Fenner School of Environment and Society, Australian National University, Canberra, Australian Capital Territory 2600, Australia
matthew.brookhouse@anu.edu.au
Catherine J. Byrne
Tasmanian Museum and Art Gallery, Department of State Growth, Hobart, Tasmania 7000, Australia

Lyn Cook
School of the Environment, University of Queensland, Brisbane, Queensland 4072, Australia
Julia Cooke
School of Environment, Earth and Ecosystem Sciences, The Open University, Milton Keynes, United Kingdom
julia.cooke@open.ac.uk
V. Alistair Drake
School of Science UNSW Canberra, University of New South Wales, Canberra, Australian Capital Territory 2617, Australia, and Institute for Applied Ecology, University of Canberra, Canberra, Australian Capital Territory 2617, Australia
a.drake@unsw.edu.au
Eleanor Drinkwater
School of Science, Western Sydney University, Penrith, New South Wales 2751, Australia
Mark A. Elgar
School of BioSciences, University of Melbourne, Parkville, Victoria 3010, Australia (deceased)
Mike Elias
Australian National Insect Collection, CSIRO, Black Mountain, Australian Capital Territory 2601, Australia
Marc E. Epstein
California Department of Food and Agriculture, Sacramento, California 95832-1448, USA
marc.epstein@cdfa.ca.gov
Juliane Farrell
School of the Environment, University of Queensland, Brisbane, Queensland 4072, Australia
Gary P. Fitt
CSIRO Health and Biosecurity, Brisbane, Queensland 4102, Australia
Michael J. Furlong
School of the Environment, University of Queensland, Brisbane, Queensland 4072, Australia
m.furlong@uq.edu.au
Peter C. Gregg
School of Environmental and Rural Sciences, University of New England, Armidale, New South Wales 2350, Australia
Kai Hart
School of the Environment, University of Queensland, Brisbane, Queensland 4072, Australia
Peter Hendry
Sheldon, Redland City, Queensland 4157, Australia
bevyjoy93@gmail.com
Robert J.B. Hoare
New Zealand Arthropod Collection (NZAC), Bioeconomy Science Institute (Manaaki Whenua Landcare Research Group), Auckland 1142, New Zealand
hoarer@landcareresearch.co.nz
Axel Kallies
University of Melbourne, Parkville, Victoria 3010, Australia
Beverley M. Kitching
Independent Scholar, Glenning Valley, New South Wales 2261, Australia
Roger L. Kitching
School of Environment and Science, Griffith University, Nathan, Queensland 4111, Australia
r.kitching@griffith.edu.au
Yu-Chi Lin
Department of Life Science, National Taiwan Normal University, Taipei 116059, Taiwan
Rose M. Lownds
School of Science, Western Sydney University, Penrith, New South Wales 2751, Australia

Ying Y. Luo
yingyluo@outlook.com
Connor Marsland
School of Science, Western Sydney University, Penrith, New South Wales 2751, Australia
Ian McMillan
Imbil, Queensland 4570, Australia
icmcmi@gmail.com
Nicole McMullen
School of the Environment, University of Queensland, Brisbane, Queensland 4072, Australia
Kathryn B. McNamara
School of BioSciences, University of Melbourne, Parkville, Victoria 3010, Australia
Peter B. McQuillan
School of Geography, Planning and Spatial Sciences, University of Tasmania, Hobart, Tasmania 7000, Australia
Liz Milla
Centre for Australian National Biodiversity Research, CSIRO, Black Mountain, Canberra, Australian Capital Territory 2601, Australia
liz.milla@csiro.au
Scott E. Miller
National Museum of Natural History, Smithsonian Institution, Washington, DC 20013-7012, USA
MillerS@si.edu
Michael Moore
South Australian Museum, Adelaide, South Australia 5000, Australia
michael.moore@samuseum.sa.gov.au
Helen F. Nahrung
Forest Research Institute, University of the Sunshine Coast, Sippy Downs, Queensland 4557, Australia
Tim R. New
Department of Ecological, Plant and Animal Sciences, La Trobe University, Melbourne, Victoria 3086, Australia
Erica H. Odell
RMIT University, Bundoora, Victoria 3083, Australia
Lynda Perkins
School of the Environment, University of Queensland, Brisbane, Queensland 4072, Australia
David Rees
P.O. Box 406, Gungahlin, Australian Capital Territory 2912, Australia
Donald P.A. Sands
CSIRO Health and Biosecurity, Brisbane, Queensland 4102, Australia, and School of the Environment, University of Queensland, St Lucia, Queensland 4072, Australia
Mark Schutze
Biosecurity Queensland, Queensland Department of Primary Industries, Brisbane, Queensland 4001, Australia
Jessa H. Thurman
CSIRO, Agriculture & Food, Dutton Park, Queensland 4102, Australia
Mizuki Uemura
School of the Environment, University of Queensland, Brisbane, Queensland 4072, Australia
Kate D.L. Umbers
School of Science, Western Sydney University, Penrith, New South Wales 2751, Australia, and Invertebrates Australia, Osborne Park, Western Australia 6017, Australia
Andrew Walker
Institute of Molecular Biology, University of Queensland, Brisbane, Queensland 4072, Australia

Joshua Waters
Indigenous Knowledge Systems Laboratory, Deakin University, Burwood, Victoria 3125, Australia
Joshua.waters@deakin.edu.au.
Allan J. Wills
Biodiversity and Conservation Science, Department of Biodiversity, Conservation and Attractions, Manjimup, Western Australia 6258, Australia
David A. Young
D'Estrees Entomology & Science Services, MacGillivray, South Australia 5223, Australia
Myron P. Zalucki
School of the Environment, University of Queensland, Brisbane, Queensland 4072, Australia

Preface

This book had its inception in 1999 when, after interminable author-driven delays, the volume *Biology of Australian Butterflies* appeared. This was number 6 in the *Monographs on Australian Lepidoptera* series and was a multi-authored volume edited by me, Rhondda Jones, Naomi Pierce and Elly Scheermeyer. The idea of a complementary volume on the Australian moths was always in my mind but the many demands on the time of a senior academic continually pushed back the priority the work demanded. I was periodically encouraged by others, particularly Scott Miller and Myron Zalucki, to pursue the idea. Retirement has finally allowed the project to emerge, accompanied by a resolution that it would be conceived, produced, edited and published within a 2-year period. Once the work became a priority, I drafted an outline and circulated it to senior lepidopterists of my acquaintance: Michael Braby, Peter McQuillan, Scott Miller and Myron Zalucki. These workers were both encouraging and helpful at that embryonic stage. The general idea developed further during the inaugural meeting of the Moths and Butterflies Australasia Association held in Canberra in November 2023.

So what happened next? In general, my ambition has been to recruit authors to report on their specialties, summarizing research published principally but not exclusively since the appearance of Ian Common's *Moths of Australia* in 1990. The chapter on the history of study of Australian moths (Chapter 2) clearly has had to be an exception to this guideline. Nevertheless, in the 35 years since 1990, several revolutionary shifts in our approaches to understanding moth biology have occurred. Not least has been the maturing of molecular approaches to biosystematics and phylogenetics, giving much more certainty on evolutionary pathways with sometimes dramatic consequences for taxonomy and classification. Further, there has been a blossoming of interest in the Australian moths by professional and amateur entomologists alike driven in no small part by Common's book.

What should a book of this kind contain? An early decision was that it could not cover all groups of moths in detail. We await another scientist with Ian Common's breadth of knowledge and experience to produce a modern single volume that achieves that goal. Not only has research interest across the fauna been heterogeneous, with many groups awaiting much-needed attention, but potential authors had to be willing and able to produce an account of their topics within my imposed and demanding time frame. An impressive set has risen to the challenge and the resulting volume covers a wide swathe of the moth fauna.

I recruited authors under three general rubrics. Some I asked to produce overviews of particular aspects of the biology of moths, focusing on but not necessarily limited to the Australian fauna. Topics such as overall evolution and diversity, biogeography, life histories, natural enemies, migration and dispersal, community ecology and conservation make up the set. No attempt was made to review moth comparative anatomy or physiology. Second, I recruited a set of authors who would produce accounts of the Australian fauna within particular superfamilies, families or subfamilies. This is the section of the work where significant gaps were left in coverage but a 'complete' account would have expanded the product to several (probably unpublishable) volumes! The third group of authors was asked to produce accounts of particular species (or small groups of species) that have attracted particular research attention. I have divided this last section into two parts: the first on well-worked pest species, the second on 'other' prominent species. Again, these sections could have comprised dozens of chapters: the selection included is representative and novel. A final chapter summarizes the outcomes from a questionnaire circulated among authors late in the production process asking them to identify gaps in knowledge and priorities for future work on the Australian moth fauna. This important chapter is intended as a guide to future workers as they make choices about where to place their research efforts.

Research interest in the Australian moths is a burgeoning area for both amateurs and professionals. New insights are published every year and many others remain within the collective minds of the coterie of moth enthusiasts. There can never be, and should never be, a last word on any aspect of our understanding of the evolution, systematics, ecology, behaviour, or whatever, of the moths or any other segment of biodiversity. Too often, though, in my professional life as an entomologist I have watched immense knowledge and wisdom disappear with the passing of particular individuals – I'm sure we can all bring to mind poignant examples. This is always regrettable and always unavoidable. One goal of this work, then, has been to capture some of this wisdom and these insights. I am sure it contains errors (although I have striven, as editor, to minimize these) and it certainly isn't the last word on any part of its subject matter. If parts of its audience rise up aghast and set out to rectify any shortcomings (as they see them), it will have succeeded.

On behalf of all authors I remember and honour Professor Mark Elgar, an ex-student, friend and author in this volume, who passed away in August 2025. His chapter on the gumleaf skeletonizer, with Kathryn McNamara, stands as his swansong.

Finally, I pay homage to the late Ted Edwards (1945–2023), to whom this book is dedicated. It is probably fair to say that no-one knew more about Australian Lepidoptera than Ted. He worked within the Australian National Insect Collection (ANIC) for over 50 years, first as assistant ('Experimental Officer') to a sequence of curators of the Lepidoptera – Ian Common, Ebbe Nielson, Marianne Horak – later, after formal retirement in 2001, as a CSIRO Honorary Fellow. Ted was a meticulous nomenclaturist and bibliographer. His best known and most used contribution is undoubtedly the many sections of the Australian Lepidoptera checklist (Nielsen *et al.*, 1996) for which he was responsible – the Checklist is perhaps the most widely consulted of any publication on the Australian Lepidoptera. There can be very few among the community of lepidopterists interested in the Australian fauna, including most of the authors in this volume, who do not owe Ted for willing and cheerful assistance spiced with matchless experience and, sometimes pithy, wisdom.

Vale, Ted, te salutamus!

Reference

Nielsen, E.S., Edwards, E.D. and Rangsi, T.V. (1996) *Checklist of the Lepidoptera of Australia*. Monographs on Australian Lepidoptera 4. CSIRO Publishing, Collingwood, Victoria.

Acknowledgements

Individual acknowledgements of assistance appear in the appropriate chapters, as do thanks for the use of photographs and other illustrations. I pay special homage to those who nurtured and mentored my interest in the Australian Lepidoptera since my arrival on the continent in 1971: Ian Common, Ebbe Schmidt Nielsen, Marianne Horak and, of course, Ted Edwards. For the development of the work as a whole, I am particularly grateful to Scott Miller, Myron Zalucki, Peter McQuillan and Michael Braby for their critiques of the initial proposal and ongoing support during the production process.

All chapters have been refereed and have benefited from these critiques. Any errors of omission or commission, of course, remain with the chapter authors. Referees have included most of the authors of chapters in the book as well as John Grehan, the late Jeremy Holloway, Peter Mackay, Akihiro Nakamura, Erik van Nieukerken, Peter Ridland, Mark Sperling, Terry Whitaker and Andy Young. I thank them collectively for this assistance.

I am grateful to Albert Orr who kindly supplied the artwork which graces the book's cover. Ward Cooper, Lucy Pritchard, Theresa Regueira, Rosie Hayden, and Holly Regan-Jones have seen the book through the labyrinthine processes of assessment, production and publication for which I thank them sincerely.

Finally, I thank my partner Beverley who, as well as being a chapter author, has also put up with my 'moth book paranoia' and my absence from social life during the production process.

Roger L. Kitching
Glenning Valley, NSW,
January 2026

Foreword

This book provides the much needed new foundation for understanding the fascinating moth fauna of Australia. The chapters here review and synthesize the last 35 years of research on these moths, fill some of the gaps in knowledge, and provide an expert assessment of the most important topics for future research.

Because of my own work on systematics and ecology of moths of New Guinea, I have followed that on Australian moths since my first trip to Papua New Guinea in 1983. At that time, well before the internet, there was no cohesive way to access the fragmented knowledge of Australian moths other than visiting a major collection of well-curated physical specimens, and the best overview of the fauna was Ian Common's chapter in the original *Insects of Australia* (1970).

The publication of Common's masterful *Moths of Australia* in 1990 was a sea change in synthesis and accessibility of knowledge, followed shortly by the comprehensive *Checklist of the Lepidoptera of Australia* (1996), and a series of other now classic publications in the *Monographs on Australian Lepidoptera* series. For my own work on the New Guinea fauna, these publications collectively allowed me to start to understand the fauna, and provided entry points for exploring the overlaps in the Australian and New Guinea faunas.

In parallel, a series of publications opened up the South-east Asian fauna, including Henry Barlow's *An Introduction to the Moths of South East Asia* (1982), Gaden Robinson and others' *A Field Guide to the Smaller Moths of South-East Asia* (1994), Jeremy Holloway's series *Moths of Borneo* (1984–2011), and Holloway and others *The Families of Malesian Moths and Butterflies* (2001). The advent of the internet allowed the production of the many online resources that exist today, which go much further in making information available to the public.

As those publications opened up the fauna, many researchers responded with diverse products over the years. Thus, it is timely that the present volume synthesizes the body of knowledge that has accumulated since the last comprehensive volume on Australian moths in 1990. The present volume is, however, different in its nature from the 1990 volume. There is no single person with Ian Common's breadth of knowledge of the fauna, so individual authors have treated the range of individual topics. While the coverage is not consistent across all taxa and topics, this approach also allows more depth in some topics that will be of special interest to readers.

Just as the volumes before it did, I expect this volume will catalyse renewed interest in the moths of Australia, not just by professionals and students but also by the public. The final chapter offers many insights into the opportunities, and I hope that students think of it as a 'shopping list' of ideas for dissertation topics. Emerging technologies in genetics, genomics, digital imaging and artificial intelligence promise to combine with the best of traditional natural history and taxonomy to drive new discoveries that will inform future management of challenges in conservation, agriculture and forestry.

I am personally grateful to Ian Common, Ebbe Nielson, Ted Edwards and Marianne Horak for building the modern foundation of our knowledge in publications, databases and collections, and to Roger Kitching for his vision and leadership in pushing that foundation to the next level through this volume.

Scott E. Miller
Smithsonian Institution
Washington, USA

1 Moths and The Australian Fauna

Roger L. Kitching
School of Environment and Science, Griffith University, Nathan, Queensland, Australia

Abstract

The insect order, Lepidoptera, comprises almost 160,000 described species with probably at least as many again awaiting description and discovery. About 88% of the known species are moths. The earliest moths may have been bryophyte feeders as larvae adapting to the greenery of the angiosperms with which they co-developed through the Cretaceous and Palaeogene. The modern moth fauna retains a full range of plant- (and fungus-) feeding habits as well as excursions into detritivory and coprophagy, predation and, even, ectoparasitism. Adults', habits range from non-feeding through to blood sucking. Interactions with plants have allowed the development of antipredator devices such as distastefulness and aposematism. Other antipredator devices include camouflage and protective resemblance to a range of other organisms as well as behavioural responses from flash colouration to antisonar jamming. The Australian moth fauna is estimated at over 10,500 described species organized into about 85 families, five of which are endemic to the continent. Of global families, the Oecophoridae dominate and Australia may contain a large majority of the world fauna. Following this introduction, this volume is organized into four major parts: eight general chapters; 13 dealing with selected subfamilies, families or superfamilies; 10 with pest species; and a concluding five describing species of conservation and cultural importance. A final chapter reviews knowledge gaps and research opportunities to guide future moth studies in Australia.

1.1 Introduction

A recent publication referred to the moths as the ‘Other Lepidoptera’. This was clearly in contrast to the butterflies which have a wider and more favourable presence in the public eye. Moths, in contrast, are perceived either as pests of the cupboard and pantry or as creatures of the dark. Certainly, the butterflies have a diurnal edge and because of their daytime habits are, as adults, often brightly coloured and highly visible. Their very name, in English, is redolent of summer days and a frivolous lifestyle. The central role that butterflies have played as evolutionary models, of course, undermines this assumption of frivolity.

Nevertheless, simply in terms of species richness within the order Lepidoptera the moths outscore the butterflies by at least an order of magnitude. A recent family-by-family estimate suggests there are about 19,000 species of butterflies globally (van Nieukerken *et al.*, 2011; James and Lohmann, 2024). An attempt by the same authors to estimate the numbers of known and named moth species arrived at a figure just under 139,000 species (Lees and Zilli, 2019). For both of these estimates, we must add a guesstimate

Corresponding author: r.kitching@griffith.edu.au

DOI: 10.1079/9781800627680.0001

of the number of species awaiting description and/or discovery. Here we enter the realms of speculation – some would say, fantasy. There are many complications about any extrapolation of this kind, not least because different researchers have different concepts of how to define a 'species'. The surge in molecular profiling as an integral part of lepidopteran systematics has complicated these matters of opinion even further with lively, ongoing and unresolved (even unresolvable) discussions about the extent and nature of 'cryptic' species within the order. These debates are, of course, entirely healthy as part of science.

Whatever species model one espouses, however, there is a general consensus that our knowledge of the butterflies is much more mature than that of most moth families. Generously, we may guess that as many as a further 10–15% of butterfly species remain to be recognized. For many of the moth families, even the optimists would settle for a figure of 50%. For some lesser known groups of the microlepidoptera, the global figure may be much higher. In the most recent assessment, Wagner (2025) says '[m]ore than 160,000 species [of Lepidoptera] have been described, with likely two to three times that number yet to be named...'

The sheer size of the moth families means their members, variously, exhibit an immense range of ecologies and behaviours. Although fundamentally herbivorous and, it is hypothesized, primitively bryophyte feeders, some moth larvae have deviated from this original green theme in almost as many ways as can be imagined. Those that have retained the plant-feeding habit – the vast majority, it must be said – have continued to exploit the bryophytes, gymnosperms and angiosperms. All plant parts are exploited variously by moth larvae, from roots to seeds. Indeed, it is supposed that throughout the Paleogene the moths and flowering plants co-radiated, each providing services to the other, food and shelter on the one hand, pollination and nutrient cycling on the other. Plant-associated feeding behaviours go far beyond the simple free-living mode. Many larvae have developed leaf tying and rolling. Others spin vast webs and live gregariously. Yet others are associated with galls. Whole clades burrow in wood. Some lineages have deviated into feeding on moribund, even dead, leaves and leaf litter. From detritivory, it was a small jump to coprophagy: some moth larvae feed on dung, from koala scats in Australia to sloth droppings in the Neotropics. Others have evolved associations with living animals. These range from keratin-feeding horn-moths in Africa to carapace-feeding parasites of tortoises in the Americas (Lees and Zilli, 2019). A few have become genuine free-living predators, either 'grazing' on scale insects or, in the case of some extraordinary Hawaiian eupitheciines, acting as sit-and-wait predators striking at and feeding on passing adult insects. Other highly specialized families are kleptoparasites in ant nests or, even, genuine ectoparasites feeding on the haemocoel of leaf hoppers.

Although most studies on species–species interactions involving moths include one species – usually the moth – exploiting another, there are a few cases of closer relationships, mutualisms or apparent mutualisms (Pierce and Dankowicz, 2022). Many moth larvae associate with ants. Pierce and Dankowicz (2022) record this phenomenon in one form or another globally across 13 moth families. Often this involves larvae feeding within ecological spaces protected by ants as they tend Homoptera or simply defend foraging territories or nests. The ants simply ignore the larvae for whatever reason. In Australia, this association has evolved into a more exploitative one on the part of the moth larvae. Species of Psychidae, Cyclotornidae, Pyralidae, Crambidae and (perhaps) Oecophoridae and Erebidae have obligate relationships with ants, feeding either on debris or ant brood within ant nests. These sorts of relationships clearly involve a cost to the ants either by the loss of predatory opportunity or, more directly, by brood predation on the part of the moth larvae.

A more convincing case of a true mutualism has evolved between heliozelid moths and their *Boronia* host plants in Western Australia. The moths pollinate the plants and their larvae feed on the consequent developing seeds. This brood pollination mutualism is discussed in Chapter 11.

Adult moths are basically flower feeders, graduating from primitive pollen chewing to nectar feeding as their characteristic proboscides evolved. Adult sloth moths live within the pelage of their hosts, feeding on skin products and algae. Further variations on this theme have appeared, with a few species feeding on mammal tears and a small number in one Asian lineage have even become blood suckers (Bänziker, 2021). Others have lost the adult feeding habit altogether.

Intra- and interspecific communication has evolved via olfaction, sight and hearing. The chemical ecology and evolution of the moths has been critical in developing the entire field of chemical biology and is a vast subject. Lees and Zilli (2019) provide an excellent general introduction. Externally active hormones – 'pheromones' – play important roles in moth biology. Both females and, less commonly, males use chemical signals to attract mates. These pheromones can act over long distances and receptor organs may detect these chemical signals in minute quantities. Pheromones may also lead to lekking behaviour in which aggregations of males form, to which mate-seeking females may be attracted visually and chemically. In some cases males have evolved a variety of organs which enhance the dispersion of pheromones. The Australian arctiine, *Creatonotus gangis*, is an extreme case, extruding disproportionately large coremata from its abdomen to enhance lekking. Chemical signalling, directly and indirectly, also occurs interspecifically.

Distastefulness of potential prey species may come about through sequestration of toxic chemicals from larval feeding habits and has frequently led to the evolution of warning colouration. Uniformly bright colour in moths, especially when there is no sexual dimorphism, is commonly taken as an indication of such warning colouration. It depends, of course, on visual signalling through which potential vertebrate predators learn to associate distinctive patterns and colours with distastefulness. The immense subfamily of the tiger moths and their allies, the Arctiinae (Erebidae), may well be the largest clade of aposematic species in the animal kingdom (Wagner, 2025; see also Chapter 21).

Such antipredator visual signalling may take other forms in moths, including daytime camouflage whereby forewing colouration matches backgrounds or caterpillars resemble (often uncannily accurately) twigs or other plant parts. Other visual signalling may occur where larvae and/or pupae have evolved to resemble other organisms ranging from ants to, it is supposed, the heads of snakes and even apes. One form of this antipredator visual signalling is so-called flash colouration where the hindwings may contain strong, often 'eyed' patterning, which when revealed by the resting adult moth may startle the predator into moving on. Visual signalling within moth species presumably occurs when sexual dimorphism is seen in adults although chemical cueing may also be involved.

Communication by sound is also commonplace in moths and several highly successful families have evolved distinctive organs dedicated to both producing and receiving signals from conspecifics. These probably predate the evolution of those quintessential moth predators, the bats (Kawahara *et al.*, 2019), and possibly were originally involved in mate finding. Subsequently, however, a secondary adaptation of the sound system aids in confusing the sonar mechanisms of the micro-bats in a classic 'arms race'.

This ecological range and versatility has resulted in characteristic assemblages of moth species on virtually every 'green' continent where they occur from the seashore to mountain tops. It has also meant that a subset of species has switched their trophic preferences to human society. The number of moth species that have actually become pests of human existence represent a very small proportion of overall moth diversity. These few, however, have a massive impact. A wide range of species attack growing crops, others infest stored products, yet others destroy woollen clothing. A small number balance the ledger. Silk moths have been exploited for millennia and remain the base of a global industry. Others have been introduced as biocontrol agents, occasionally with spectacular success. Finally, like their more meretricious cousins the butterflies, moths have been effective experimental models in scientific developments in population ecology, evolutionary biology, chemical communication and physiology.

1.2 The Australian Fauna

The global moth fauna is currently organized into about 130 families grouped into 43 superfamilies based on increasingly authoritative phylogenetic analyses. Twenty-six of these superfamilies contain just single highly distinctive families: others contain more than ten families, with the Gelechioidea topping the table with 21 families within the superfamily. The original qualifier 'about' is and will remain necessary as opinions differ among systematists as to the exact status of many groups.

The numbers quoted here are those used in Table 1.1 and are drawn from van Nieukerken *et al.* (2011) in their assessments of global species richness within the Lepidoptera. Table 1.1 lists

Table 1.1. The moths: global and Australian species richness. Asterisked families are the focus of other chapters in this volume. Table author's own.

TAXON	GLOBAL SPECIES RICHNESS (Van Nieukerken *et al.* 2011)	AUSTRALIAN SPECIES RICHNESS		TAXON	GLOBAL SPECIES RICHNESS (Van Nieukerken *et al.* 2011)	AUSTRALIAN SPECIES RICHNESS		TAXON	GLOBAL SPECIES RICHNESS (Van Nieukerken *et al.* 2011)	AUSTRALIAN SPECIES RICHNESS	
		CHECKLIST (1996)	MOST RECENT ESTIMATE			CHECKLIST (1996)	MOST RECENT ESTIMATE			CHECKLIST (1996)	MOST RECENT ESTIMATE
MICROPTERIGOIDEA				NEPTICULOIDEA				Bucculatricidae	297	14	26++
Micropterigidae*	160	3	12	Nepticulidae	819	14	280+	Gracillariidae	1866	198	300+
AGATHIPHAGOIDEA				Opostegidae	192	18	'100s'	YPONOMEUTOIDEA			
Agathiphagidae*	2	1	2+	ANDESIANOIDEA				Yponomeutidae	363	56	?
HETEROBATHMIOIDEA				Andesianidae	3	0	0	Argyresthiidae	157	1	1
Heterobathmiidae	3	0	0	ADELOIDEA				Plutellidae	150	25	25?
ERIOCRANOIDEA				Heliozelidae*	123	37	300+	Glyphipterygidae	535	57	57?
Eriocraniidae	29	0	0	Adelidae	294	11	?	Ypsolophidae	163	0	0
ACANTHOPTEROCTOIDEA				Incurvariidae	51	13	100+	Attevidae	52	0	0
Acanthopterocteridae	5	0	0	Cecidosidae	16	0	0	Praydidae	47	0	0
Aenigmatineidae*	1	0	1	Prodoxidae	98	0	0	Heliodinidae	69	4	4
LOPHOCORONOIDEA				PALAEOPHATOIDEA				Bedelliidae	16	2	2
Lophocoronidae	6	3	6	Palaeophatidae	57	11	11	Lyonetiidae	204	40	40?
NEOPSEUSTOIDEA				TISCHERIOIDEA				UNASSIGNED			
Neopseustidae	14	0	0	Tischeriidae	110	0	0	Prodidactidae	1	0	0
MNESARCHAEOIDEA				UNASSIGNED				Douglasiidae	29	1	1
Mnesarchaeidae	7	0	0	Millieriidae	4	0	0	SIMAETHISTOIDEA			
HEPIALOIDEA				TINEOIDEA				Simaethistidae	4	0	0
Palaosetidae*	9	1	1	Eriocottidae	80	1	1	GELECHIOIDEA			
Prototheoridae	12	0	0	Psychidae*	1350	188	300*	Autostichidae	638	13	13+
Neotheoridae	1	0	0	Tineidae	2393	187	370+	Lecithoceridae	1200	47	?
Anomosetidae*	1	1	1	GRACILLARIOIDEA				Xylorictidae	524	275	275+
Hepialidae*	606	119	164	Roeslerstammiidae	53	23	23	Blastobasidae	377	13	?
								Oecophoridae*	3308	1504	5000+

Schistonoeidae	1	0	0
Lypusidae	21	0	0
Chimabachidae	6	0	0
Peleopodidae	28	0	0
Elachistidae[1]	3201	197	250+
Syringopaidae	1	0	0
Coelopoetidae	3	0	0
Stathmopodidae	408	104	106
Epimarptidae	4	0	0
Batrachedridae	99	30	?
Coleophoridae	1386	6	13
Momphidae	115	1	?
Pterolonchidae	8	0	0
Scythrididae	669	24	?
Cosmopterigidae	1792	386	'many'
Gelechiidae	4700	453	1600+
ALUCITOIDEA			
Tineodidae	19	12	16+
Alucitidae	216	6	'more'
PTEROPHOROIDEA			
Pterophoridae	1318	42	50?
CARPOSINOIDEA			
Copromorphidae	43	7	?
Carposinidae	283	40	?
SCHRECKENSTEINIOIDEA			
Schreckensteiniidae	8	0	0
EPERMENIOIDEA			
Epermeniidae	126	21	?
URODOIDEA			
Urodidae	66	0	0
IMMOIDEA			
Immidae	245	13	15?
CHOREUTOIDEA			
Choreutidae	406	15	15

GALACTICOIDEA			
Galacticidae	19	6	?
TORTRICOIDEA			
Tortricidae	10387	632	1200+
COSSOIDEA*			
Brachodidae*	137	23	44
Cossidae*	971	91	186+
Dudgeonidae*	57	3	10+
Metarbelidae*	196	0	0
Ratardidae*	10	0	0
Castniidae*	113	24	42
Sesiidae*	1397	16	21
ZYGAENOIDEA			
Epipyropidae	32	4	6?
Cyclotornidae	5	5	12+
Heterogynidae	10	0	0
Lacturidae	120	28	?
Phaudidae	15	0	0
Dalceridae	80	0	0
Limacodidae*	1672	70	70 + a 'few'
Megalopygidae	232	0	0
Aididae	6	0	0
Somabrachyidae	8	0	0
Himanopteridae	80	0	0
Zygaenidae	1036	21	29
WHALLEYANOIDEA			
Whaleyanidae	2	0	0
THYRIDOIDEA			
Thyrididae	940	56	60?
HYBLAEOIDEA			
Hyblaeidae	18	4	4
CALLIDULOIDEA			
Callidulidae	49	0	0

PYRALOIDEA*			
Pyralidae*	5921	472	ca 600
Crambidae*	9655	738	ca 1100
MIMALLONOIDEA			
Mimallonidae	194	0	0
DREPANOIDEA			
Cimeliidae	6	0	0
Doidae	6	0	0
Drepanidae	660	10	13?
LASIOCAMPOIDEA			
Lasiocampidae	1952	73	82
BOMBYCOIDEA			
Apatelodidae	145	0	0
Eupterotidae	339	6	7
Brahmaeidae	65	0	0
Anthelidae	94	74	'many'
Carthaeidae	1	1	1
Endromidae	59	0	0
Bombycidae	185	3	3
Saturniidae	2349	14	20?
Sphingidae*	1463	64	87
GEOMETROIDEA			
Epicopeiidae	20	0	0
Sematuridae	40	0	0
Uraniidae	686	36	40
Geometridae*	23002	1200+	2000+
NOCTUOIDEA*			
Scrancidae*	100*	1	1
Oenosandridae*	8	8	8
Notodontidae*	3700*	79	80?
Erebidae*	24569	890	1300++
Euteliidae*	520	35	35
Nolidae*	1738	167	180+
Noctuidae*	11772	577	1000+

[1]Note van Nieukerken's concept of the Elachistidae (s.l.) includes the Depresariidae, Hypertrophidae, Ethmiidae, Agonoxenidae and Stenomatidae - all regarded either as separate families or as subfamilies of the Oecophoridae (s.l.) in the Australian Checklist.

all families together with the species richness counts given by van Nieukerken and his colleagues, including those which do not occur in Australia. I have used the family combinations as adopted by van Nieukerken *et al.* even when these are at odds with the 'usual' Australian usage (as in, for example, the Elachistidae). The second column adjacent to the family names in the table contains the number of species recorded from Australia generally as listed in the authoritative checklist of Nielsen *et al.* (1996). I have attempted to match the numbers listed in that checklist with those given by van Nierkerken *et al.* (2011) even where the latter authors have assigned different status to particular taxa. In most cases this was relatively straightforward as subfamilies became recognized as families or vice versa. For the mighty Noctuoidea, this reassignment was more challenging given the more fundamental and dramatic changes which have happened within the superfamily since the checklist appeared (see Chapter 20).

Of course, the number of named entities ('species') within a list is not an estimate of the entire Australian fauna. Many species lie recognized but unnamed, even sorted, in museum drawers, yet others await discovery in the field. Peter McQuillan refers to these as the 'known, unknowns' and the 'unknown, unknowns' respectively. Where authors have published recent credible estimates of the entire fauna, or of the known but undescribed 'species', I have included these in Table 1.1. These estimates are sourced from a variety of places. In some cases expert sources have 'guesstimated' the likely total size of particular families. In many cases these have been non-numerical or only quasi-numerical estimates – 'a few', 'many', '100s', etc. In a few cases I have translated such guesses into numbers – by increasing described species by a 'sensible' fraction. For smaller families the BOLD database includes known but undescribed species, allowing a more realistic estimate of the 'known' fauna. Finally, for several families there is no reasonable way of estimating the size of the Australian fauna. In these cases the Estimate column simply includes a '?'. Overall, these estimates should not be taken too seriously, simply indicating the depth of our ignorance in some cases.

The named Australian fauna falls into 85 families organized into 34 superfamilies. In general, the relative richness of each family reflects that seen in the global fauna (correlation across all families excluding Australian zeroes, 81%) but there are important deviations from this pattern. Five families are wholly endemic to Australia, a further one is endemic to Australia and New Guinea. Three of these are monospecific: the 'primitive' Aenigmatineidae, the hepialoid Anomosetidae and the bombycoid Cartheidae. The curious ant predators, the Cyclotornidae, are represented by five described species and the noctuoid Oenosandridae by a further eight. The bombycoid Anthelidae, however, are considerably more diverse. According to Zwick (2008), Australia is home to 74 species in eight genera, with New Guinea adding a further 20 species in two genera. Only one genus (*Anthela*) is shared between the two land masses. Known larval foodplants include several characteristically Australian genera such as *Eucalyptus*, *Acacia* and *Casuarina*.

For non-endemic families, assessing the relative importance (in species richness terms) of other families in the Australian fauna is challenging given the patchiness of knowledge both here and elsewhere. The information presented in Table 1.1 suggests that *described* species of Xyloricitidae and Stathmopodidae make up more than 25% of the *known* global named fauna. The emphases on 'described' and 'known' are deliberate and make such assessments very preliminary. Other families such as the Hepialidae, Nepticulidae, Opostegidae and even Heliozelidae are also likely to have very large Australian faunas should the anticipated sizes of the taxa ever be matched by firm knowledge. The real stand-out, though, is the Oecophoridae. Even considering named species alone, the Australian fauna already makes up more than half of the known global fauna. The group has evolved strong links with the equally successful Myrtaceae in Australia. Sensible expert estimates suggest Australia may house as many as 4000–5000 species of oecophorids. The current world fauna totals 3308 species. It must be re-emphasized, however, that compared with other southern continents (Africa and South America), our fauna is relatively well known – although we may not perceive it as such – and any cross-regional comparisons can only be extremely tentative.

1.3 The Book in Hand

This work rides on the shoulders of its many predecessors (Kitching and Edwards, 2020;

Chapter 2, this volume). Its inception and history are described in the Preface. Fundamentally, the 36 chapters that appear between this Introduction and the concluding chapter fall into four sections, each of which provides different sorts of information for students of the Australian moths.

Chapters 2–9 (Part 1) are general overviews of key aspects of moth biology. Chapter 2 examines the history of the accumulation of knowledge of the Australian moth fauna, focusing on the scientific approaches which have accrued since the 17th century, yet mindful of the vast store of age-old Indigenous knowledge. The following seven chapters describe aspects and patterns in moth evolution, ecology, behaviour and conservation. These are described generally in a 'top-down' fashion starting with global issues of definition and scientific development and gradually focusing down on key Australian examples and data. These opening eight chapters together make up a near complete introduction to the general biology of the moths in their Australian context, omitting only accounts of their comparative morphology and physiology.

The next 13 chapters (Part 2: Chapters 10–22) are taxonomically focused accounts of selected taxa at the supraspecies level: superfamilies, families and subfamilies. Although, inevitably, only a selection of these categories within the lepidopteran phylogeny, they range from basal groups such as the Hepialidae through to the highly derived Sphingidae and Noctuidae. The selection of these taxa reflects areas of active research in Australia and beyond as well as the availability of authors.

The remaining two parts of the work comprise accounts of individual species, or small groups of species, which have brought them to special attention for one reason or another. In both cases, an introductory chapter (23 and 33 respectively) examines the broader general contexts of 'pest species' and 'other significant species'. So Chapters 23–32 (Part 3) focus on a range of pest species, both introduced and indigenous, which have been the focus of attention by economic entomologists over long periods of time. Chapters 33–37 (Part 4) focus on that set of species which have come to human attention other than as pests, either because they are of particular conservation concern or they have a presence in the human consciousness for other reasons.

1.4 Conclusion

Research interest in the Australian moths is a burgeoning area for both amateurs and professionals. New insights are published every year and many, many others remain within the collective minds of the coterie of moth enthusiasts. There can never be, and should never be, a last word on any aspect of our understanding of the evolution, systematics, ecology, behaviour or whatever of the moths or any other segment of biodiversity. Too often in my professional life as an entomologist, I have watched immense knowledge and wisdom fade with the passing of particular individuals – I'm sure we can all bring to mind poignant examples. This is always regrettable and always unavoidable. One goal of this work, then, has been to capture some of this wisdom and these insights. I am sure it contains errors (although I have striven, as editor, to minimize these) and it certainly isn't the last word on any part of its subject matter. If parts of its audience rise up aghast and set out to rectify any shortcomings (as they see them), it will have succeeded.

References

Bänziger, H. (2021) *Vampire Moths: Behaviour, Ecology and Taxonomy of the Blood-Sucking Calyptra*. Natural History Publications, Kota Kinabalu, Malaysia.

James, D.G. and Lohmann, D.J. (2024) *The Lives of Butterflies: A Natural History of our Planet's Butterfly Life*. Princeton University Press, Princeton, N.J.

Kawahara, A.Y., Plotkin, D., Espeland, M., Meusemann, K., Toussaint, E.F.A. *et al.* (2019) Phylogenomics reveals the evolutionary timing and pattern of butterflies and moths. *Proceedings of the National Academy of Sciences* 116, 22657–22663.

Kitching, R.L. and Edwards, E.D. (2020) Books about Australian moths 1805–2020: An annotated bibliography. *Australian Entomologist* 47, 119–154.

Lees, D.C. and Zilli, A. (2019) *Moths: Their Biology, Diversity and Evolution*. Natural History Museum, London.
Nielsen, E.S., Edwards, E.D. and Rangsi, T.V. (eds.) (1996) *Checklist of the Lepidoptera of Australia, Monographs on Australian Lepidoptera 4*. CSIRO Publishing, Collingwood, Victoria.
Pierce, N.E. and Dankowicz, E. (2022) Behavioral, ecological and evolutionary mechanisms underlying caterpillar-ant symbioses. *Current Opinion in Insect Science* 52, 100898.
van Nieukerken, E.J., Kaila, L., Kitching, I., Kristensen, N.P., Lees, D.C. *et al.* (2011) Order Lepidoptera Linnaeus, 1758. In: Zhang Z.Q. (ed.) Animal biodiversity: An outline of higher-level classification and survey of taxonomic richness. *Zootaxa* 3148, 212–221.
Wagner, D.L. (2025) *Moths of the World: A Natural History*. Princeton University Press, Princeton, N.J.
Zwick, A. (2008) Molecular phylogeny of Anthelidae and other bombycoid taxa (Lepidoptera: Bombycoidea). *Systematic Entomology* 33, 190–209.

2 The History of Moth Study in Australia

Beverley M. Kitching[1] and Roger L. Kitching[2]*
[1]*Independent Scholar;* [2]*School of Environment and Science, Griffith University, Nathan, Queensland, Australia*

Abstract

The accumulation of knowledge on the Australian moths began with the First Nations people informing them both domestically and spiritually over many millennia. The discovery, mapping and colonization of Australia by Western nations brought the fauna to the attention of naturalists. Elements of the Lepidoptera were among collections made by exploring fleets and first descriptions were by European lepidopterists. The Scott family and John Lewin were resident naturalists who began the process of bringing the fauna to a wider audience in the early days of the colony, complementing the efforts of European-based workers. Over the cusp of the 19th and 20th centuries, many thousands of species were collected and described by a large number of authors dominated by Alfred Jefferis Turner, Edward Meyrick, Francis Walker and Oswald Lower. Most were amateurs. The mid-20th century saw professional lepidopterists take on a leading role led by Ian Common and Ebbe Nielsen with crucial support from Ted Edwards. Throughout this period of gradual development, museums, societies and journals were established, as was the Australian National Insect Collection (ANIC) held by the CSIRO. A healthy set of state and national collections are now held in the ANIC and in state and territory museums. A thriving set of journals persist. Technological advances and a community of amateur and professional naturalists promise an exciting and productive future for moth studies in Australia.

2.1 Introduction

Modern studies of moths are based on a deep Western tradition of natural history dating back, perhaps, to Pliny the Elder and beyond (see, for example, Chapter 19). As Western science emerged from the European Renaissance, so names began to be associated with particular fields and subfields. For the Lepidoptera we honour 17th- and 18th-century savants: Thomas Moffat, James Pettiver, Benjamin Wilkes, Moses Harris, Maria Sybilla Meriam and others (Salmon, 2000). A whole subculture of dilletante and other amateur lepidopterists was established in Western Europe preceding, stimulating and, ultimately, invigorated by the Linnean revolution in the mid-18th century. The wide-ranging voyages of exploration of the time brought back natural treasures from around the globe, giving home-based lepidopterists material to incorporate into their scientific schemes and to induce wanderlust in the more restless of them.

A little later, in the first half of the 19th century, two young naturalists gave in to this wanderlust. A young Cambridge graduate, admittedly more interested in rocks and beetles than

*Corresponding author: r.kitching@griffith.edu.au

DOI: 10.1079/9781800627680.0002

in moths and butterflies, and a dissatisfied surveyor who was deeply interested in the Lepidoptera set off on voyages of discovery motivated by their passions for natural history. Between them, Charles Darwin and Alfred Russel Wallace initiated a scientific revolution which changed science and much else. The rest, as we say, is history!

These exclusively European developments in natural science were progressing coincidentally with the period during which the 'new' land of Australia was discovered, explored, mapped (in part) and, ultimately, colonized. A select group of naturalists, both in Europe and later in Australia itself, took particular interest in the hitherto isolated and certainly novel biota of Australia. These investigators, variously, brought with them the weight of European developments in natural history and began applying these ideas to the *tabula rasa* which they saw as the Australian flora and fauna.

What was rarely appreciated among these colonial savants was that this was no more a *tabula rasa* awaiting their endeavours than had been the flora and fauna of Europe before, say, the Renaissance. Indigenous Australians had lived with and by the biota for 60,000 years before the 'light' of Western inquiry was directed onto their environment. During this time Australia's inhabitants had developed an acute, detailed and perceptive knowledge of their environment and its animals and plants and had incorporated this knowledge into their everyday lives, from the domestic to the spiritual. This deep knowledge is embedded in language and culture and only in recent times has become appreciated by Australia's natural scientists. And only in recent times, as is evident in several chapters in this volume, has any of this knowledge been integrated with mainstream scientific endeavours.

2.2 First Nations Knowledge

The First Nations people of Australia undoubtedly recognized the vital role that lepidopteran communities could play in the 60,000+ years of their occupation of Australia. Much of their knowledge has been lost since the arrival of Europeans in the 18th century (Clarke, 2026). Even to access the knowledge of surviving groups, ethno-entomologists face a number of problems in researching the traditional knowledge base of what were non-literate cultures. Access to such traditional knowledge can be controlled by age, status or gender. Information may be transmitted in storytelling, song and dance. Insects feature in mythology, place names, personal names, seasonal calendars and cultural practices. They were exploited for food, medicine, pigment and poisons, among other things (Si and Turpin, 2015), and served as ecological indicators of resources and natural phenomena. Moths feature culturally from the Caterpillar Dreaming of the Arrentye people of central Australia to the vital role of the bogong moth to seasonally alpine peoples (Flood, 1980) and the importance of witchetty grubs in the diet of many groups such as the Pitjantjatjara (Tindale, 1953, 1966). The powerful ancestral beings of the creation story of Caterpillar Dreaming are the larvae of four species of hawkmoths: *Hyles livornicoides*, *Hippotion celerio*, *Coenotes eremophilae* and *Agrius convolvuli*.

Although, as Si and Turpin (2015) noted, there has been little focused research on First Nations knowledge and use of insects, their survey of the work of linguists in the production of dictionaries, for instance, provides useful information especially where the proliferation of vocabulary is associated with orders of insects including the Lepidoptera. Traditional Indigenous Australians understand the intimate relationships among moths and their host plants. This is often reflected in their naming of edible insects to include the name of the host plant from which the edible grub was collected (Yen, 2005, 2010). For example, the witchetty grub of the *Endoxyla leucomochla* moth (Cossidae) is one of the best known edible grubs consumed by First Nations people. It is called witjuti by the Adnyamathanha people of South Australia's Central Desert and is named so for its host plant the witchetty (witjuti) shrub, *Acacia kempeana* (Faast and Weinstein, 2020).

Wood-boring caterpillars of cossid moths, like the witchetty grub, feed within the stems, branches, bark and trunks of a wide variety of host plants (see Chapters 15 and 26). The majority of edible wood-boring grubs used by First Nations people are found in *Acacia* species (Faast and Weinstein, 2020). This suggests an understanding of the processes and dynamics governing edible insect distributions and factors affecting their palatability. The bogong moth (*Agrotis infusa*) is another popular and nutritious food source (see Chapter 37). Their immense aestivating populations not only provided an incoming feast and opportunity for intertribal meetings, but also came to be markers of a shift in the season and availability of associated resources.

Other behaviourally prominent moths including day-flying species, web builders, processionary caterpillars, stinging cup-moth larvae and many others were known and named by First Nations people. Documentation of much of this knowledge would no doubt preserve the knowledge itself but would also inform modern scientific inquiry.

2.3 Explorers and Collectors

The first Western explorer and collector of Australian insects was Sir Joseph Banks (1743–1820). He was renowned in the scientific community of his day as a naturalist, botanist and patron of natural sciences. Banks was made president of the Royal Society in 1778, a position he held for 40 years. The Banks collection of some 40,000 specimens collected during his life is held by the Natural History Museum in London. It contains the first entomological specimens, including moths, from Australia which he collected as a member of Cook's voyage in 1768–1771. Banks was invited to join the voyage and he funded eight others to join him including the Swedish naturalist Daniel Solander, the Finnish naturalist Herman Sporing and two artists, Sydney Parkinson and Alexander Buchan.

The first to examine Banks' Australian material was the Danish entomologist Johan Christian Fabricius (1745–1808) who described more than a dozen new species from Australia in publications between 1775 and 1794 (Fabricius, 1775, 1792–1798) including '*Pyralis*' (now *Rhadninoscops*) *koenigiana* (Tortricidae, 1775), *Maruca vitrata* (Crambidae, 1787), *Hellula undalis* (Pyralidae, 1781), *Hippotion velox* (Sphingidae, 1793) and '*Tinea*' (now *Achroia*) *grisella* (Pyralidae, 1794). The earliest of all of these (by page priority), however, was '*Bombyx*' (now *Donuca*) *castalia* (Erebidae, 1775). Fabricius, a pupil of Linnaeus himself, was probably the leading entomologist of his era: he travelled to England while professor at the University of Kiel in order to study the collections in the British Museum, including those of Banks.

Of course, other exploratory expeditions sampled the Australian fauna in addition to Cook. The Australian circumnavigation of Matthew Flinders (1801–1803) certainly had a scientific focus but was undoubtedly botanically biased. A number of French expeditions such as those of Jean-François de Galaup La Perouse (1788), Nicolas Baudin (1800–1803; see below) and Louis de Freycinet (1817–1820) collected significant material. They all carried '*naturalistes*' and made extensive collections. Much of this material was destined for the Musée d'Histoire Naturelle in Paris where entomologists such as Pierre Latreille and Achille Guenée added to our described fauna.

2.4 Collectors and Describers

Many authors have been responsible for describing and naming the Australian moth fauna and a full account would fill many volumes. Some species, of course, were cosmopolitan and had been described from non-Australian material by Linnaeus himself. These include the two 'Australian' sphingids, *Agrius convolvuli* and *Hippotion celerio*, named in the 10th edition of his *Systema Naturae* (Linnaeus, 1758) more than 10 years before Captain Cook visited Australia in 1770.

2.4.1 Donovan and Lewin: the first Australian moth books

European settlement in Australia began in 1788 and insect collections made by some residents soon followed. The earliest known descriptions and illustrations of Australian moths from this time were published in Europe. Not all of the early describers visited Australia. Edward Donovan (1768–1837) and John William Lewin (1770–1819) both published books in 1805 (Donovan 1805, Lewin 1805) with beautiful hand-coloured illustrations of Lepidoptera which included the first descriptions of several Australian moths. Donovan apparently never left London during his working life and based his descriptions on specimens collected by Joseph Banks during the Cook expeditions but Lewin based his descriptions on direct observation. In 1800 Lewin, in contrast, became a free settler in New South Wales and remained there until his death in 1819.

Lewin (1805) was the first author to describe the life-history – larvae, pupae and foodplant – of any Australian moth. Due to publishing delays and supposed lack of dating, several of Lewin's descriptions and names were subsumed by those of Donovan (Kitching and Edwards, 2020). Other publications containing original descriptions followed over the next 40 years; William S. Macleay (1826), Jean Baptiste Boisduval (1832) and George French Angas (1847). By the mid-19th century, British museums were extending their holdings and as the various Australian states were established as British colonies, new descriptions of moths mostly appeared in scientific publications in London. As royal societies and Linnaean societies were established in Australia, so publication habits changed and many descriptions appeared locally. It remained noticeable, however, that what authors considered more significant discoveries were reserved for the parent journals of Britain: a habit which, some might claim, still persists!

2.4.2 The Scott family: naturalists and illustrators

Following the Lewin and Donovan volumes, Alexander Walker Scott (1800–1883) and his daughters Harriet (1830–1907) (Fig. 2.1) and Helena (1832–1910) (Fig. 2.2) produced two lavishly illustrated volumes entitled *Australian Lepidoptera and their Transformations*. The first of these appeared in 1864 and the second in four parts from 1890 to 1893 after Alexander's death (Scott, 1864, 1890-98). These volumes were principally based on observations and paintings made during their residence at Ash Island on the Hunter River. Alexander was also assisted by his stepdaughter Mary Ann Calcott who kept his notebooks and assisted her step-father and her sisters with their entomological work; she may also have been responsible for the detailed anatomical line drawings which are appended to the painted plates.

Fig. 2.1. Harriet Scott, ca. 1863, attributed to Helena Scott. Reproduced with permission, State Library of New South Wales.

This work is outstanding for several reasons, not least because it was based on the rearing of species from their early stages, a real focus on observation and natural history. This was unmatched until the much overlooked work of Dumigan (1934). The work of the Scotts has been extensively noted, reviewed and variously reproduced (Ord, 1988; Braby and Olsen, 2011; Finney, 2018). It is noteworthy that the Scotts' volumes were the first appearance of female authors in the Lepidoptera literature in Australia. As far as we can tell, more than a century was to pass before the publication of *A Guide to the Common Moths of the Adelaide Region* by P.B. McQuillan and Jan Forrest in 1985. Jan Forrest was the Collection Manager for the South Australian Museum.

2.4.3 The heyday of moth descriptions 1860–1920: Turner, Meyrick, Walker and Lower

In 1866 William J. Macleay (later Sir William) published his description of the sphingid *Darapsa*

Fig. 2.2. Helena Scott ca. 1864. Photo: Public Domain work of art, Wikimedia.

moorei (later synonymized with Cramer's *Daphnis hypothous*) in Volume 1 of the *Transactions of the Entomological Society of New South Wales* (Macleay, 1866). This was the first Australian entomological society and journal (Moulds, 1999; Hale and Parsons, 2024) which Sir William himself had founded in 1861. Sir William was one of four (Alexander, George, William Sharp and William John) who, together with collector and curator George Masters, conjointly accumulated the insect collection which later became the basis of the Macleay Museum housed at the University of Sydney (see below). The Macleays were among the first of a long line of amateur collectors which over the years has made such a vital contribution to our knowledge of the Australian Lepidoptera. George Masters, initially an employee of Alexander Macleay and curator of the Macleay collections until his death in 1912, may lay claim to have been Australia's first professional entomologist.

From these early efforts, substantial attention was being paid to the Australian moth fauna both at home and abroad. Many thousands of species were described and named. *The Checklist of the Lepidoptera of Australia* (Nielsen *et al.*, 1996) lists the 10,583 species named to the present together with their authors, dates of description and synonyms.

In order to identify the principal authors involved in describing the Australian fauna, we selected a random 32 of the 314 pages of the *Checklist*. For that 10% of the list of the Australian fauna, we have abstracted the names of authors and dates of publication of the moths listed as currently recognized. We have not included information on synonyms or other historical names that are not currently valid. In this fashion we encountered 1084 'valid' species, just over 10% of the entire fauna.

Sixty-eight authors or pairs of authors were responsible for the current names of these 1084 species. The large majority of these (57 in fact) described fewer than 10 of these species as now recognized. Four authors alone account for over 75% of all the names we encountered. Alfred Jefferis Turner (1861–1947) tops the list, having been responsible, alone, for 344 (31.7%) of the 1084 names we sampled. He is said to have described 3500 species of moths in his lifetime with publications stretching from 1894 to 1947 (Monteith and Mather, 1986). English entomologist Edward Meyrick (1854–1938) featured second in our samples with responsibility for 275 or 25.4% of the names in articles published from 1878 to 1937. Francis Walker (1809–1874), prodigiously productive and controversial entomologist of the British Natural History Museum, achieves third place having named 136 (12.5%) of our sample of 1084 species in publications from 1854 to 1869. Finally, South Australian entomologist Oswald Bertram Lower (1863–1925) described 63 (5.8%) of our sample of 1084 species over the period 1892–1923. No other authors came close to these four in our sample in terms of sheer productivity. It is noteworthy that three of the four were officially amateurs with more or less distinguished careers in other professions. Only Walker was actually employed as an entomologist. We return to them shortly.

Our subsampling of the *Checklist* also highlighted the heyday of descriptions of the Australian moth fauna. Again we speak of the 1084 species in our sampling of the entire fauna. Species in that list were formally described from

1758 (the Linnaean species *Hippotion celerio*) to 1995 (the saturniid, *Opodiphthera excavus* described by David Lane). Both these dates, however, represent extreme outliers. Sixty-eight percent of the 1084 species we examined were described between 1861 and 1920, with just under 30% named between 1901 and 1920. Only seven species in our sample were described since 1961. In no small way, this distribution of descriptions reflects the activities of Turner, Meyrick, Walker and Lower mentioned above. Each of these entomologists thus deserves special attention in any account of the history of study of the Australian moths.

Alfred Jefferis Turner (1861–1947) (Fig. 2.3) was born in Guangzhou, China, on 3 October 1861. He was the eldest of six children born to Frederick Storrs Turner and Sophia Mary Harmer. He was educated in England at Amersham Hall and the City of London School before studying medicine at University College, London, from which he graduated with First Class Honours in 1886. In 1888 he moved to Brisbane, Australia, to become the first resident surgeon at the Hospital for Sick Children in 1889. Even after he went into private practice in 1893, he remained the honorary visiting physician to the hospital. In 1901 he qualified for the Cambridge Diploma in Public Health and in 1904 he became the visiting medical officer to the Diamantina Hospital for Chronic Diseases. In 1916 he felt obliged to contribute to the war effort and went to England to join the Royal Army Medical Corps. After his return to Brisbane, he became director of the Central Tuberculosis Clinic in 1927 and part-time director of infant welfare until his retirement from clinical practice in 1937. During this period infant mortality dropped in Queensland from 50.7 to 35.6 per thousand registered births. He died in December 1947. His entry in the *Australian Dictionary of Biography* (Thearle, 1990) focuses primarily on his medical contributions but Turner was indeed one of the most influential lepidopterists in Australia. He became President of the Entomological Society of Queensland in 1930. He named 450 new genera and four new families (Lambkin and Daniels, 2023). He published 240 articles in both entomological and medical journals. His collection of more than 50,000 moths was bequeathed and ultimately transferred to the Australian National Insect Collection (ANIC) (Upton, 1997), where it formed an early backbone of the Lepidoptera holdings.

Fig. 2.3. Alfred Jefferis Turner. Photo: Public Domain work of art, Wikimedia.

Edward Meyrick (1854–1938) (Fig. 2.4) was born in Wiltshire, England, and educated at Marlborough College where he joined the Natural History Society. His passion for Microlepidoptera was remarked on by the teacher who founded the society. He enrolled at Trinity College, Cambridge, in 1873 where he obtained First Class Honours in the Classical Tripos in 1877. In 1880 he became a Fellow of the Royal Entomological Society of London. He then moved to Australia, teaching Classics at Sydney Grammar School before moving to New Zealand in 1881. There he taught at the Cathedral Choir School in Christchurch from 1882 to 1883. After leaving the school, he remained in New Zealand to study the Microlepidoptera and published a monograph entitled *New Zealand Geometrina* in the *Transactions of the New Zealand Institute* (Meyrick, 1884). He then returned to Australia to teach at the King's School, Paramatta, NSW, until his return to England in 1887. He returned to Marlborough College in 1887 as Classics master where he remained until his retirement in 1914. For the last 15 years of his career he served as President of the Marlborough College Natural History Society.

During his time in Australia, he went on collecting expeditions to South Australia, Tasmania, Mt Kosciuszko, NSW, and to Western Australia in 1886–1887. In his collections from Perth and Albany were many new species and genera, especially of microlepidoptera. By 1890 he had published some 55 articles in Australian and international journals. Eight years after his return to England, he published his *Descriptions of Australian Microlepidoptera* (published in Sydney, 1904) (Meyrick, 1904) and in collaboration with his colleague, O.B. Lower (1863–1925), he published a *Revision of the Australian Psychidae* (Meyrick and Lower, 1907). Kaila (2011) credits Meyrick as the first author of his day to give a comprehensive account of any of the Australian micromoths.

Meyrick based his taxonomic identification on external characteristics, primarily wing venation. He refused, vigorously, to countenance the use of characters of genitalia as diagnostic tools.

By the end of his career he had authored more than 400 publications, an astonishing output for an amateur. In addition to his work on Australian and New Zealand microlepidoptera, he also published extensively on micros from Britain, the Pacific Islands and India.

In 1902 Meyrick became a Corresponding Member of the Linnean Society of New South Wales and in 1904 a Fellow of the Royal Society, London. He donated a significant portion of his collection of Lepidoptera, comprising more than 100,000 specimens, to the British Museum of Natural History.

Francis Walker (1809-1874) (Fig. 2.5) was born in London into a wealthy and educated Quaker family. His father John was a fellow of the Royal Society, the Linnean Society and the Royal Horticultural Society. As a boy aged 7–11 Walker lived with his family in Switzerland (1816–1820) and in 1818 he and his brother Henry made a collection of butterflies on the slopes of Mt Pilatus so at the age of nine he was a budding entomologist. In 1834 he became the

Fig. 2.4. Edward Meyrick. Photo: Creative Commons, Wikipedia.

Fig. 2.5. Francis Walker. Photo: Public Domain work of art, Wikimedia.

unpaid editor of the *Entomological Magazine* for a year and in 1837 he began working on a contract basis for the British Museum of Natural History (BMNH). In 1844 he became a curator at the BMNH and in 1848 was contracted to catalogue all of their entomological collections except the Coleoptera. He was paid 1 shilling for each new species and 1 pound for each new genus. He was, in addition, funded by his family. This work continued until 1873 and was published at the insistence of the Director, John Edward Gray.

It is Walker's work on the Lepidoptera collections which gives him such an important position among the early describers of the Australian moth fauna. He described numerous species belonging to a variety of families including the Nolidae, Erebidae, Sphingidae, Geometridae, Noctuidae, Xyloryctidae and Crambidae. In the five years between 1859 and 1864 alone he described more than 130 species of Australian moths. Walker was a prolific entomological author but remains controversial. He was described as a 'careless' taxonomist who often described the same species more than once under different specific names, which left large numbers of synonyms for his successors to sort out. After his death in 1874 one obituary in *The Entomologists Monthly Magazine* described him as 'having done an amount of injury almost inconceivable in its immensity' (Carrington, 1884) whereas Newman (1874) described him as 'the most voluminous and industrious writer on Entomology this country has ever produced'.

Oswald Bertram Lower (1864–1925) (Fig. 2.6) was born in Adelaide in 1864, the third son of Samuel Lower who had migrated from the UK to Australia in 1854. He studied pharmacology and by 1884 was a practising pharmaceutical chemist. In 1888 he moved to Broken Hill and with his younger brother, Sydney, set up a successful business, the Barrier Ranges Apothecaries. He was a man of wide interests, a keen cricketer, a notable tenor and a prominent freemason. He had a keen interest in philately and was said to have owned the finest collection of stamps in Australia. He was also a founding member of the Barrier Field Naturalists Club which was founded in 1920. In the early 1890s Lower was living part of the year at Parkside in Adelaide and after his retirement the family moved there. Lower was 45 when he married Eva Davies in 1909 and they had two sons. Even after moving permanently to Adelaide, Lower retained an interest in the Broken Hill community and still worked occasionally in the pharmacy until 1924, the year before his death (McQuillan *et al.*, 2023).

Fig. 2.6. Oswald Bertram Lower. Reproduced with permission, Bonita Coventry.

Lower collected mainly around Adelaide and Broken Hill and although his first taxonomic paper dealt with specimens collected in South Australia (Lower, 1892), he soon described specimens sourced from other areas of Australia through a wide range of contacts, family members and fellow entomologists. He also bought specimens from dealers. He produced a catalogue of Victorian Heterocera (Lower, 1893) and by 1894 he was describing new species of moths from central and northern Queensland. He was the first entomologist to study the moths of the mallee and mulga woodlands of southern Australia. He discovered hundreds of new species particularly from the semi-arid zone of South Australia. Between 1892 and 1923 he named around 1000 species of both moths and butterflies. His collection of 40,000 specimens became the nucleus of the Lepidoptera collection of the South Australian Museum. He produced

45 solo publications plus two co-authored with Edward Meyrick.

Lower was elected a member of the Royal Society of South Australia in 1888 and in 1892 the first of his many papers in their Proceedings was published. He was elected as a member of the Linnean Society of New South Wales in 1901. Recognition from overseas included membership of the Entomological Society of London in 1892 and of the Zoological Society of London in 1907. Lower was one of the first Australian-born entomologists to be recognized as a credible lepidopterist. He contributed to the knowledge of some 43 families of Australian Lepidoptera and described around 9% of the currently named species, which is the most described by any Australian-born author (McQuillan *et al.*, 2023).

2.5 Professionals and Advisors

Lower's publications spanned a period of 31 years from 1892 to 1923 which fits almost exactly within the time period which Ian Mackerras (1948) called the 'period of the amateurs' (with apologies to George Masters). As the 20th century progressed, however, the number of institutions which employed entomologists increased so that the contribution of professionals increasingly outweighed that of the amateurs. It must be added that few of these professionals were lepidopterists, and amateurs and 'part-timers' continued to play critical roles, as they do to this day.

Professional entomologists can be classified in a variety of ways. They may be taxonomists, ecologists, behaviourists, physiologists or what have you – that is, defined by what disciplinary subfield they bring to their study of insects. Indeed, entomology has itself been regarded as a subfield of wider sciences: of zoology on the one hand, of agricultural science on the other. Quite separately, entomologists may be driven by fundamental or applied problem sets which may define their primary places of work – predominantly universities versus state organizations of one kind or another. Finally, they may be researchers or teachers. Most working professionals will map onto these contrasts in complex ways; seldom is it a case of either/or, rather a melange of activities driven by interests, employers and, not least, available funding.

So, focusing on the Lepidoptera, a majority of those employed to study moths have been applied entomologists tasked with understanding particular pest species and the development of mechanisms for their management. Many of these contribute to the species accounts later in this volume. Whether or not these workers would describe themselves as 'lepidopterists' is probably open to question; almost certainly their formal job descriptions would not use the term. Curiously, there are several who would describe themselves as ecologists or ethologists on a 'professional' basis yet as lepidopterists when wearing 'amateur' garb. This often reflects early passions for butterflies and moths later tempered by the need to earn a living!

Turning then to the few professionals who might see themselves as lepidopterists above all else, a very few names stand out. Since the early 1950s the ANIC has had a curator of Lepidoptera, which position has been generally perceived as the senior appointment in the field (by the profession at large, that is, if not by the office holders themselves). Since its inception only three persons have held this position. The earliest holder of the position was, of course, Ian Common, succeeded in 1982 by Ebbe Schmidt Nielsen. Both scientists made fundamental contributions to their fields, both played wider roles within and beyond their employing institutions, and both saw part of their role as being to encourage and facilitate the many other workers on the Lepidoptera as amateurs or part-timers. Both were supported by their research associate E.D. ('Ted') Edwards (1945–2023). Nielsen was succeeded as curator by Marianne Horak who continued (and continues) the fundamental and essential work of the ANIC both as a researcher and focus of Lepidoptera studies as doyenne of Australian lepidopterists.

We focus here on Common, Nielsen and Edwards, deliberately excluding surviving practitioners from this historical account.

2.5.1 Ian Francis Bell Common (1917–2006)

Ian Common (Fig. 2.7) was born in Alford, Lincolnshire, England. The family migrated to Australia and Ian was brought up in Toowoomba, Queensland, and developed an early interest

Fig. 2.7. Ian Francis Bell Common. Reproduced with permission, © CSIRO.

in Lepidoptera while still at school. He was educated at Toowoomba Grammar School and then at the University of Queensland where he gained Bachelor's and Master's degrees in philosophy and agricultural science. He was awarded a Doctor of Agricultural Science by the University of Queensland in 1969. In 1938 he joined the Entomological Society of Queensland, serving as Secretary in 1938–40. At the same time as he was studying, in 1944, he was appointed as Research Officer by the Queensland Department of Agriculture and Stock studying moth infestations in wool stores. In December 1944 he went first to Biloela and then from 1945 to 1948 to Rockhampton to work on pests of cotton and tomatoes and then on yellow-winged locust. In 1948 he was appointed to the CSIR Division of Economic Entomology as Technical Secretary to the Chief of Division, A.J. Nicholson. He was, however, able to work half time on the taxonomy of Lepidoptera. In 1951 he transferred to work full time on Lepidoptera, focusing initially on pest species.

With his appointment as curator of Lepidoptera for the nascent Australian National Insect Collection within the now CSIRO Division of Entomology in Canberra, the study of Australian Lepidoptera underwent a dramatic change. Over the course of his career, Common published some 100 papers and seven books on the Lepidoptera. Summarizing the full oeuvre here is both impossible and inappropriate but key highlights are of historical importance.

During the 1950s, Common undertook a series of revisions of pest moth groups in Australia, including *Heliothis*, *Agrotis*, *Persectania*, *Pectinophora*, *Scirpophaga* and *Epiphyas*, and published his classic work on the migration and dynamics of the bogong moth (Common, 1954). Establishing the exact identity of these major pest species underpinned all subsequent applied ecological work of vital importance to Australia's agriculture.

During his time as curator, Common built the ANIC Lepidoptera collection into an immense resource in size and geographical location noted for its high curatorial standards. In order to expand the extent and coverage of the collection, he and Murray Upton undertook a series of highly organized field trips, the results of which highlighted how little was known about the moth fauna of Australia.

As Common's research matured, he paid more and more attention to the higher systematics of the Lepidoptera as a whole, defining or redefining families, subfamilies and tribes as well as essential generic and specific descriptions. In the wider global community of lepidopterists, he is probably best known for his overviews of the entire Order. These began with the seminal chapter on the Lepidoptera in the first edition of *Insects of Australia* (Common, 1970) subsequently updated in collaboration with Nielsen in 1991 (Nielsen and Common, 1991; see also Common, 1975). These chapters provided the basis for the book length account that became *Moths of Australia* (1990).

Common strongly believed that his duties should include the production of popular and semi-popular guides to the Australian Lepidoptera. In 1963 he published *The Australian Moths* in the Jacaranda Pocket Guides series, (Common, 1963) the first general guide to the moths of the continent.

Common retired in 1982 and entered perhaps the most significant period of scientific productivity in his already stellar career. As already mentioned, he published his massive and internationally significant volume *Moths of*

Australia in 1990. This work was globally reviewed and received many accolades including the Whitley Medal of the Zoological Society of New South Wales. The importance of this volume to the study of Australian moths cannot be overstated (Kitching and Edwards, 2020). The work remains the go-to volume for all Australian lepidopterists interested in the moth fauna and is a key work in the global entomological literature.

Common then went on to publish three major volumes on the *Genera of the Australian Oecophorinae* (Common, 1994, 1997, 2000), establishing a generic framework for the future study of this richest of all Australian moth families (see Chapter 12).

In 1996 he was awarded the Karl Jordan Medal of the Lepidopterists Society of the USA and in 2001 was made an Officer of the Order of Australia for services to entomology and to community education. In 2003 he was given the Jacob Hübner Award for Lepidoptera Systematics, by the Association for Tropical Lepidoptera.

Ian Common is without any doubt the modern cornerstone of Australian studies of Lepidoptera. He not only contributed significant, often seminal, works across the Lepidoptera but also curated the Lepidoptera collection of the ANIC as a model of how such major collections should be established, organized and maintained. He encouraged many younger lepidopterists, international visitors and, not least, his assistants and successors.

2.5.2 Ebbe Schmidt Nielsen (1950–2001)

Ebbe Schmidt Nielsen (Fig. 2.8) was appointed as curator of Lepidoptera within the ANIC in December 1982 after Ian Common retired. Nielsen was born into a farming family near Silkeborg in central Jutland, Denmark. His parents instilled his early interest in natural history. Inspired by reading about collecting and identifying moths and butterflies, he began collecting specimens at the age of 14 and later joined an entomology club in Aarhus. After completing his National Service, in 1971 he enrolled as a biology student at Aarhus University while also working part-time for the Aarhus Natural History Museum. He completed his studies at the University of Copenhagen. After graduation he worked in the Entomology Department of the Zoological Museum of Copenhagen University, joining an in-depth study of a beech stand ecosystem and obtaining his MSc 1976. He then became the first PhD student of distinguished lepidopterist Niels P. Kristensen, being awarded his doctorate in 1980.

Fig. 2.8. Ebbe Schmidt Nielsen. Reproduced with permission, © CSIRO.

Nielsen began publishing short papers on ant-lions, lacewings, moths and local faunistics from 1970 (Halliday and Edwards, 2003). His first major work was the *Checklist of Danish Lepidoptera* (Karsholt and Nielsen, 1976) followed by a monograph on the Elachistidae of Scandinavia (Traugott-Olsen and Nielsen, 1977). In 1978 he was invited to participate in a 6-month expedition to South America. This experience provided him with much material for his PhD, brought him into contact with a range of 'basal' families of the Lepidoptera, and triggered an enduring interest in the fauna of the southern hemisphere. Many publications reflect this period of research, including the definitive monograph on South American hepialids (Nielsen and Robinson, 1983) and the discovery of a new basal family, the Heterobathmiidae (Kristensen and Nielsen, 1979).

Nielsen had visited the ANIC at the invitation of Ian Common in 1981 and in 1982 was appointed by CSIRO as Common's replacement. His remit was to work primarily on primitive Lepidoptera and other Lepidoptera inventories as well as overseeing the maintenance and curation of the ever growing collection. Initially Nielsen worked on two taxa of primitive Lepidoptera: the family Lophocoronidae (Nielsen and Kristensen, 1989a) and the genus *Fraus* in the family Hepialidae (Nielsen and Kristensen, 1989b). Early in his CSIRO career, Nielsen saw the need for an annotated checklist of the Australian Lepidoptera as a fundamental tool for any and all studies on the Australian fauna. The *magnum opus* was finally published in 1996 (Nielsen *et al.*, 1996).

Nielsen was deeply engaged with CSIRO Publishing. He helped found the journal *Invertebrate Taxonomy* in 1996 (retitled *Invertebrate Systematics* in 2002). He was chair of the editorial committee and a frequent contributor. From a more focused Lepidoptera perspective, his establishment of the ongoing series of *Monographs on Australian Lepidoptera* was of particular significance. These stemmed from his conviction that Lepidoptera (and other) studies demanded authoritative monographs rather than piecemeal short papers (but see discussion in Chapter 38). He served as chief editor of the series until his untimely death in 2001. He was also co-author of three of the volumes (Nielsen and Kristensen, 1989; Robinson and Nielsen, 1993; Nielsen *et al.*, 1996) in the series (which now stands at 13 volumes with more in preparation). His full publication output is listed by Halliday and Edwards (2003) and, as well as his many taxonomic contributions, includes works on general systematics, biological inventories, biodiversity and informatics. He became probably the most successful global advocate for the science of systematics, especially insect taxonomy.

During the 1990s, Nielsen became less involved with fundamental research and more concerned with science policy. With his understanding of the concept of biodiversity, he also understood that it was vital to develop knowledge of species and their characteristics and that therefore biological collections such as the ANIC were important sources of information. The growth of the world's collections, however, was greatly outpacing the ability of scientists to handle the immense amount of data contained within them. Nielsen became a champion for the Global Biodiversity Information Facility (GBIF), an initiative of the OECD which came into being in March 2001 (Edwards, 2003). Nielsen was on his way to attend the inaugural session of the governing board in Montreal when he died of a heart attack on March 7, 2001.

During that inaugural meeting, the GBIF Ebbe Nielsen Prize was created to honour his memory, later revamped as the GBIF Ebbe Nielsen Challenge. In 2003 a special issue of *Invertebrate Systematics* was published as a tribute to the man and his work (Horak and Halliday, 2003).

Nielsen was awarded the Karl Jordan Medal for Lepidopterology by the Lepidopterists' Society (1990) and the David Rivett Medal from the Australian Academy of Science (1990). He was awarded the Ian Mackerras Medal by the Australian Entomological Society (1992). He was a Foreign Associate of the United States National Academy of Science and a Foreign Fellow of the Royal Danish Academy of Sciences and Letters.

2.5.3 Edward David ('Ted') Edwards (1945–2023)

Ted Edwards (Fig. 2.9) probably knew more about the Australian Lepidoptera, on the tarsus, so to speak, than anyone in the modern era. Ted was inspired in childhood by his father, another keen amateur lepidopterist. Following graduation from the University of Sydney with a degree in agricultural science and a short stint as a research agronomist at the Yanco Agricultural Station, he was appointed to CSIRO as Experimental Officer under Ian Common. So began 50 years of association with the ANIC. He worked collaboratively with all three distinguished curators of the Lepidoptera collections – Ian Common, Ebbe Nielson and Marianne Horak. After his formal retirement in 2001 he continued association with the ANIC as a CSIRO Honorary Fellow.

However, Ted was far more than an 'assistant'. While the more senior curators focused on sorting out key families, writing global overviews

Fig. 2.9. Edward David ('Ted') Edwards. Reproduced with permission, © CSIRO.

or driving international taxonomic change (see, for example, Nielsen and Kristensen, 1989b; Common, 1990; Horak, 2006), Ted familiarized himself with the breadth and minutiae of the ever burgeoning ANIC collection of Lepidoptera. Ultimately, he was the one (perhaps the only one) who could look at virtually any newly arrived anomalous moth, stalk the precipitous stacks of cabinets, pull out a drawer and tell you 'it's in here somewhere'. Alternatively, he would peer at your tentatively named specimen, shake his head and skip to some quite different family, then point at a particular unit tray so you realized you had, once again, been misled by that noctuid-looking geometrid or geometrid-looking noctuid (or whatever). And this help was given (almost) always with good humour and a wry smile.

This is not to say that Ted's only contribution was as willing helpmeet to others. Ted was a meticulous nomenclaturist and bibliographer. His best known and most used contributions are undoubtedly the many sections of the Australian Lepidoptera checklist (Nielsen *et al.*, 1996) for which he was responsible. His more narrowly focused treatment of the butterflies alone, published as part of the *Zoological Catalogue of Australia* by the Australian Biological Resources Survey (ABRS) (Edwards *et al.*, 2001), is a true masterpiece of detail, reliability and completeness. This work untangles the early history of butterfly taxonomy in Australia as well as the sometimes tortuous paths the subject has followed since those early days.

Ted Edwards made many excursions into more traditional taxonomy often in conjunction with other lepidopterists, describing new species of lycaenids, hesperiids, hepialids and castniids, even co-authoring the description of a new family of Australian moths (Kristensen *et al.*, 2015). He continued work on a proposed monograph on the Australian Castniidae, sadly incomplete at the time of his death.

Perhaps Edwards' most used popular contributions are as co-author in general guidebooks, most notably with Paul Zborowski in *A Guide to Australian Moths* (Zborowski and Edwards, 2007) and with Glenn Cocking and Suzi Bond in *Moths in the A.C.T.* (Cocking *et al.*, 2022). He also played a role in several of the excellent *Moths of Victoria* series led by Peter Marriott (Hewish *et al.*, 2014, 2016; Marriott *et al.*, 2017).

The true breadth of Ted's influence, however, is best seen in the Acknowledgments sections of virtually every publication on Australian Lepidoptera over many years. They almost all acknowledge, often humbly, the key role played by Ted in accessing not only the riches of the ANIC, but also the further riches of his encyclopaedic knowledge of the Australian fauna and his willingness to share this knowledge with all and sundry.

Ted received several well-deserved honours. *A Guide to Moths of Australia* received a Whitley Award in 2007, he was made a member of the Order of Australia in 2012, and, most significant of all, he was awarded the Karl Jordan medal in 2015.

2.6 Museums, Societies, Journals and the National Collection

As the study of entomology grew in Australia before, during and after colonization, so individuals, as we have described, grew to prominence. Their efforts were contemporaneous with the gradual establishment of supporting 'infrastructure'. Indeed, many of those we have discussed in terms of their contributions to growing knowledge of the Australian Lepidoptera were also closely engaged with the various colonial museums and the founding of specialist societies and journals. It is in the museums that their collections, ultimately, could find permanent homes. The societies brought together those of common interest and could, potentially, provide mentors to burgeoning talents, and the associated journals provided outlets for records and results of these efforts. The history of development of state museums and the rise and sometimes demise of local scientific societies and journals has been well documented and is noted here only briefly. We also comment upon the national insect collection (the ANIC) which, perhaps uniquely, resides not in a museum but within the CSIRO, the national scientific organization.

2.6.1 Museums

From the late 19th century to the present day, museums have played a vital role in acquiring and curating entomological collections, including, of course, moths. Sir Joseph Banks (1743–1820) not only went on collecting expeditions in his younger days but also served as a trustee of the British Museum for 42 years. His collection in the Natural History Museum, London, is maintained as a separate entity.

The British, of course, were not the only nation involved in voyages of exploration which resulted in vast collections of entomological specimens. Nicholas Baudin, for example, sent back more than 100,000 zoological specimens including large numbers of moths. Baudin himself did not survive the voyage home and died of tuberculosis in Mauritius in September 1803. The collection did eventually reach Paris but not all of it went to the Musée National d'Histoire Naturelle (MNHM). As a result of the Napoleonic Wars, much of the wider collection was destroyed by invading troops or sold off and dispersed to other collections (Case, 2022). The MNHM, however, still houses a significant collection of thousands of specimens from Baudin's voyage.

There was pressure from the earliest days of colonial settlement that museums be established in Australia. From the earliest establishment of the state museum in New South Wales in 1827 followed in short order by Tasmania (1846), Victoria (1854), South Australia (1859), Queensland (1862) and Western Australia (1897), naturalists in all states were well served (in principle at least) by supporting institutions. A museum was established in the Northern Territory in 1966.

2.6.1.1 Australian Museum

Originally known as the Colonial Museum or the Sydney Museum, the Australian Museum is the oldest natural history museum in Australia and the fifth oldest in the world (Strahan, 1979). The establishment of a museum had first been planned by the Philosophical Society of Australasia in 1821. Specimens were collected but the Society folded in 1822 (Thomsett, 1993). When Alexander Macleay arrived in Sydney in 1826 to become New South Wales Colonial Secretary, he lobbied strongly for the foundation of a museum. The Colonial Museum was founded in 1827 by Earl Bathurst, Secretary of State for the Colonies, who provided £200 annually for its upkeep (Watson, 2012). It was administered directly by the Colonial Government until 1836 when it was renamed the Australian Museum and a Committee of Superintendence of the Australian Museum and Botanical Garden was established. Alexander Macleay's son, William Sharp Macleay, served as Chair until 1835 when that committee was abolished and the Australian Museum Act was enacted (Docker, 2012).

During the 1860s hundreds of moths and butterflies were donated to the Museum by Alexander Walker Scott and his daughters Helena and Harriet. These had been collected during their research for their volumes on the Australian Lepidoptera (Scott, 1864, 1890–1898) (as discussed earlier). Of particular importance is the type specimen collection, some of them described by A.W. Scott himself, others by Edward Meyrick and Arthur Sidney Olliff, the first Australian Museum entomologist appointed in 1885. The Scotts' donations were dispersed throughout the Museum's collection but were finally brought together (after 150 years) in 2011 for an exhibition of the art of the Scott sisters.

The museum's collections grew substantially as a result of enthusiastic field studies in the 1880s and 1890s. Through the 20th century numerous collections were donated by both amateur and professional entomologists, including the G.A. Waterhouse collection of Australian butterflies. Within the entomology collection of some 1.8 million pinned specimens, the Museum counts the Lepidoptera as one of its particular strengths.

2.6.1.2 Macleay collection

The Macleay collection (Stanbury and Holland, 1988) was reputably the largest insect collection of its time, containing a trove of marked and unmarked types as well as many specimens considered 'significant' in global terms. Horning (1988) notes that some 60% of the Museum's insect holdings were non-Australian and included material from many notable early collectors and travellers. These had been amassed by Alexander Macleay and his son, William Scott Macleay. At the time of W.S. Macleay's death in 1865, the collection comprised 250 drawers of insects. Alexander's nephew, William John Macleay, created an additional collection of Australian and exotic insects which was combined with the earlier Macleay collections. W.J. Macleay inherited the entire collection in 1865 and subsequently donated it to the University of Sydney. The original Macleay Museum was built in 1886–7 within the grounds of the University. The gift to the university came with the condition that its curator at the time, George Masters, continue in that position, and Masters remained in the post until his death in 1912. Subsequently the collection was curated only intermittently.

Elizabeth Hahn was curator from 1958 to 1962 and, among other things created a catalogue of the holdings of type specimens. In 1969 concern was raised about the security of the many type specimens and other 'significant' material held in the neglected collection. After what appears to have been a complex and vituperative negotiation, more than 9000 specimens were transferred on 'permanent loan' to the ANIC although, as Ken Key suggested (reported by Upton, 1997), the remaining collection remained (and remains) of major entomological importance. The University appointed Woody Horning as curator in 1982 and he remained in place until retirement in 1994. Subsequently the curation again lapsed and access was limited (Upton, 1997). The museum itself was finally dissolved in 2016 and its collections were amalgamated into the Chau Chak Wing Museum in 2020 and are now part of the Sydney University Museum.

2.6.1.3 Tasmanian Museum and Art Gallery

In 1838 the Mechanics Institute of Hobart made an agreement with the Government for a room in the Customs House to be used as a museum and in 1843 Lieutenant-Governor Eardley-William also attempted to set up a museum. The Tasmanian Museum and Art Gallery (TMAG) was founded by the Royal Society of Tasmania in 1846 and officially established in 1848. The museum moved to Harrington Street in 1852 until, in 1861, the government funded a new building which was completed in 1862. Until 1885 the museum remained part of the Royal Society when a Board of Trustees was set up under the state government to manage both the museum and the Royal Botanical Gardens. It was not until 1952 that the boards of the museum and gardens were separated and the museum became known as the TMAG.

Although collectors and entomologists made notable contributions to the understanding of Tasmanian moths during the 19th century, most of the resulting specimens were deposited in museums in Victoria and New South Wales. The early 20th century did see a significant expansion of the TMAG's insect collection, an expansion which has continued into the present century. In 2013 the Tasmanian Forest Insect Collection was transferred to the TMAG, a collection which has contributed to the documenting and preserving of the moth diversity of Tasmania's forests.

2.6.1.4 Queen Victoria Museum and Art Gallery, Launceston

The Queen Victoria Museum and Art Gallery, Launceston had its beginnings in 1842 with its original collections coming from those of the Royal Society of Tasmania and the Launceston Mechanics Institute. It was originally called the Victoria Museum and Art Gallery and was officially opened in 1891. The first curator, Herbert Hadley Scott, was appointed in 1897. The natural history collections increased to such an extent that in 1907 an extension was built to house a zoological gallery. The name of the museum was amended to the Queen Victoria Museum and Art Gallery in 1926 to avoid confusion with the state museums of Victoria. As the size of collections grew, more extensions were built in the late 1920s.

Tasmania is home to a considerable moth fauna, with more than 1200 described species across some 60 families, with the Oecophoridae, Geometridae and Tortricidae being particularly diverse. There are many undescribed species in the QVMAG and elsewhere, particularly among the Microlepidoptera. The museum is actively engaged in building a comprehensive collection of Tasmanian moths aimed at filling geographical and taxonomic gaps in knowledge of the local fauna. The museum also mounts associated public events, including holding a Lepidoptera Week in 2022.

2.6.1.5 Queensland Museum

The Queensland Museum was founded on 20 January 1862 by the Queensland Philosophical Society (QPS). The QPS was given the use of rooms in the historic windmill on Wickham Terrace, one of them being set aside for the classification and arrangement of specimens donated to the QPS. The Society continued to operate the Museum until 1871 when it became the responsibility of the Queensland Government. Charles Coxon of the QPS was appointed Honorary Curator and, in 1873, Karl Steiger was appointed Custodian. In 1875, Henry Tryon (1865–1943) was the first entomologist to be appointed by the Museum. He left the Museum in 1893 upon appointment as Government Entomologist.

The museum continued to grow during the 1880s and early 1890s and staff numbers reached 15 but the depression of the 1890s cut this back to only five. When the Museum moved into its new home in the Exhibition Building in 1899 and then reopened to the public in 1901, there were only four staff members (Mather, 1986). The decline that followed led to a major review and in 1910, Ronald Hamlyn-Harris was appointed as curator reporting directly to the Premier. The museum was expanded, its displays revitalized, honorary scientists were appointed, an education programme put in place and a library established. Harris appointed Henry Hacker as Museum Entomologist in 1911. His adventurous field trips first by bicycle then motorcycle gained him quite a reputation. Even after his transfer to the Agriculture Department in 1923, he remained in charge of the Museum's insect collection until 1943. In 1911, the first woman was appointed to the permanent staff. Eileen Murphy, initially a stenographer, went on to register specimens, write labels and compile catalogues. She occupied her post for 42 years.

The holdings of Lepidoptera were substantial. A.J Turner was associated with the Museum in an honorary capacity and described numerous species from the collection of his collecting colleague, Wilfred Bourne Barnard. Barnard had begun collecting moths in 1929 and built a substantial and important collection including some 750 of Turner's type specimens. The Barnard collection was bequeathed to the growing ANIC in 1940 but they rejected the bequest and the collection went to the Queensland Museum. The ANIC decision was subsequently regretted and described as 'the one that was given away' (Upton, 1997).

The Barnard collection made a substantial contribution to the Museum's holdings of Lepidoptera. The Museum also received the celebrated Dodd collection in 1987. This comprised 40 cases of tropical insects, mostly Lepidoptera, assembled by Frederick Parkhurst Dodd (1861–1937) and his son Alan Parkhurst Dodd (1896–1981) (Monteith, 1991). Alan Dodd had added substantially to the collection in his later years after his work on the introduction of *Cactoblastis* moth as a control for prickly pear. Much of this material was from New Guinea. An additional significant accession containing large numbers of moths (and other insects) was the University of Queensland Insect Collection which was donated in 2011 and amalgamated with the Museum's collection.

The current collection of insects at the Queensland Museum numbers more than 3.9 million specimens, including more than 38,000 types, and is the largest collection of Queensland insects in existence.

2.6.1.6 National Museum of Victoria

In 1854 the Victorian Government Assay Office brought its natural history collections together as the Museum of Natural History and put some of them on public display. In 1856 Frederick McCoy, Professor of Natural Sciences at the University of Melbourne, transferred these collections to the university. The official starting date for the National Museum of Victoria was 1858 and McCoy was appointed as Director. In 1899 the Museum moved to premises in Russell Street where it remained for the next 100 years. In 1983 it was merged with other museums as Museums Victoria. Over the years the collections grew through the work of staff, donations from amateur naturalists and purchase. Of particular note is the collection of Lepidoptera donated to the Museum by George Lyell (1866–1951) in 1932 to which he added further specimens over the next two decades, donating this later collection to the Museum in 1946. The collection contains around 12,000 butterflies and 40,000 moths. Lyell said his favourite species was *Dudgeonea actinias*, the Queensland wood moth, a rare species of which seven specimens are in the Lyell collection (Coleman, 2021). At the beginning of 2019 a multidisciplinary team from the Museums Victoria Archives began a McCoy Seed Fund project focused on the Lyell collection.

2.6.1.7 South Australian Museum

In the first half of the 19th century there had been attempts to set up mechanics' institutes in South Australia but it was not until 1856 that the colonial government provided the first government-funded, purpose-built cultural institution building (Hancock, 2021). That year the South Australian Legislative Council passed the South Australian Institute Act which incorporated the South Australian Institute under the control of a Board of Governors and provided for a library and a museum as part of the new organization. It was only in 1939 that the South Australian Institute Museum became, officially, the South Australian Museum.

Frederick George Waterhouse was appointed as first curator in 1856. In 1883 Johann Gottlieb Otto Tepper (1841–1923) was appointed natural history collector and from 1888 to 1911 he held the position of entomologist, the first appointed by the museum. Tepper described the giant wood moth (Cossidae) *Endoxyla cinereus*, and also a number of other macromoths including the impressive *Hylaeora caustopis* (Notodontidae) and the lovely *Grammodes ocellata* (Erebidae). His popular book *Insects of South Australia* (Tepper, 1890) included valid descriptions of these and other moths. His entomological specimens are held by the museum.

In 1919 Norman Tindale (1900–1993) was appointed to the museum as entomological assistant to Arthur Mills Lea (1868–1932) who had been invited to apply for the position of entomologist in 1911. Tindale, primarily known for his (later) anthropological work, made a particular study of the Hepialidae and Lea, whose primary enthusiasm was beetles, did work on *Cydia pomonella*, the codling moth, and on *Levuana iridescens*, the coconut moth. The nucleus of the Museum's Lepidoptera collection is the Oswald Bertram Lower Collection of some 40,000 specimens.

2.6.1.8 West Australian Museum

The original West Australian Museum began life as the Geological Museum in 1891 in the Old Perth Gaol. It became the Perth Museum in 1892. The Old Gaol was demolished in 1899 to make way for the Jubilee Building which housed both the Museum and Art Gallery until 1972 when a new Francis Street building was opened. This building was closed in 2011 due to asbestos problems, all collections were put into storage until the newly constructed Western Australian Museum reopened in 2019 with more than 3 million specimens located in state-of-the-art storage in the Harry Butler Research Centre (museum.wa.gov.au, 2025).

Widespread collecting of insects including moths began in the 19th century in Western Australia but most material was lodged elsewhere, often in eastern museums. There does not appear to have been any single 'monumental' collection such as emerged in some other states. The present-day Museum's Entomology collection comprises over 500,000 pinned specimens and 1000 microslides. The collection also holds around 800 holotype specimens.

2.6.1.9 *Museum and Art Gallery of the Northern Territory*

The Museum and Art Gallery of the Northern Territory (MAGNT) houses collections that are representative of the art, natural history and culture of the region. The MAGNT was established in 1966 with Colin Jack-Hinton as its first Director. It was first housed in the renovated Old Town Hall but this was destroyed by Cyclone Tracy in 1974. Much of the MAGNT collection was lost or damaged. For the next few years both the staff and collection were housed in multiple buildings around Darwin. Eventually a purpose-built museum building was constructed and opened in September 1981. The MAGNT now includes the Museum of Central Australia in Alice Springs and several other centres and historical buildings.

Although insect collections were made from the 19th century on as part of the exploration of the vast and remote areas of northern Australia, this material became part of collections elsewhere such as the Dodd, Macleay and Scott collections as well as overseas repositories. Collection and study of Northern Territory moths increased significantly during the 20th and 21st centuries with research focusing on diversity and geographical distribution and the role of moths within ecosystems. The principal Northern Territory Insect Collection, however, is not housed at the Museum but at the Berrimah Farm Science Precinct of the Department of Agriculture and Fisheries, in Darwin. It includes more than 56,000 specimens including Lepidoptera, collected over a 50-year period.

2.6.2 Societies and journals

Lepidopterists, especially those based in Australia, needed local outlets for their work and these were promptly provided by local learned and enthusiasts' societies. A summary of the exclusively entomological journals that have appeared in Australia is presented in Table 2.1. The earliest of these was yet another Macleay product, the *Transactions of the Entomological Society of New South Wales* which appeared in two volumes between 1862 and 1873. This was shortlived

Table 2.1. Australian entomological journals, past and present. Titles in bold are current.

1862–1873 *Transactions of the Entomological Society of New South Wales*. Published by the Entomological Society of New South Wales (1st establishment). Two volumes appeared.

1949–1952 *The Australasian Entomologist*. Published by the Amateur Entomologist's Society, Australian Branch.

1952 *The Australian Entomologist*. Published by the Entomological Society of NSW (2nd establishment).

1964–1976 *Journal of the Entomological Society of Australia (NSW)*. Published by the Entomological Society of Australia (NSW).

1978–present ***General and Applied Entomology***. Published by the Entomological Society of NSW (3rd establishment) (volume numbers continued from *Journal of the Entomological Society of Australia (NSW)*).

1971–present ***Victorian Entomologist***. Published by the Entomological Society of Victoria.

1972–1992 *Australian Entomological Magazine*. Published privately 1972–1986 then by the Entomological Society of Queensland, 1987–1992.

1993–present ***The Australian Entomologist***. Published by the Entomological Society of Queensland (volume numbers continued from the *Australian Entomological Magazine*).

1962–1966 *Journal of the Entomological Society of Queensland*. Published by the Entomological Society of Queensland (continued as the *Journal of the Australian Entomological Society* with uninterrupted volume numbers).

1967–1995 *Journal of the Australian Entomological Society*. Published by the Australian Entomological Society (continued as the *Australian Journal of Entomology* with uninterrupted volume numbers).

1996–2013 *Australian Journal of Entomology*. Published by the Australian Entomological Society (continued as *Austral Entomology* with uninterrupted volume numbers).

2014–present ***Austral Entomology.*** Published by the Australian Entomological Society (volume numbers continued from *Australian Journal of Entomology*).

2008–2023 *Metamorphosis Australia*. Published by the Butterfly and Other Invertebrates Club, Brisbane.

and transmuted in 1874 into the *Proceedings of the Linnean Society of New South Wales* (PLSNSW), again under Macleay patronage, with a more general natural historical remit. The journal has thrived and still appears. Many early moth descriptions by Turner and others appeared in the pages of the PLSNSW.

It was not until the mid-20th century that further explicitly entomological journals appeared. Entomological societies in New South Wales, Queensland and Victoria were established and all produced journals of varying degrees of longevity.

The saga (and no other word is appropriate) of the various entomological societies in New South Wales has been chronicled in detail by Hales and Parsons (2024). These societies underwent occasional name changes as did the journals that they produced, finally settling to the current journal of the Entomological Society of New South Wales, *General and Applied Entomology* (1978–present). The ambitions of some that the NSW journal (under an earlier name) should transmute into a national journal were frustrated when this transmutation actually occurred in Queensland.

The Queensland Entomological Society was established in 1923 and began publication of the *Journal of the Entomological Society of Queensland* in 1962. The Queensland Society had always seen itself as the precursor of a national entomological society and this came to pass in 1965 (although the Queensland society persisted). The Queensland journal was renamed as the *Journal of the Australian Entomological Society* but with a simple continuation of journal numbers (Wylie, 2023). The national journal changed names first to the *Australian Journal of Entomology* (1996–2013) and then to its current label, *Austral Entomology* (2014–present) as fashions and remits evolved.

Queensland was also the home of *Metamorphosis Australia* (2008–2023) published by the Brisbane-based Butterflies and Other Invertebrates Club which perished for lack of support in 2023.

Meanwhile the *Victorian Entomologist* (1971–present) was established and continues publication by the Entomological Society of Victoria which also produces occasional publications, including the excellent series *Moths of Victoria*. This was begun in 2008 by Peter Marriott and, to date, comprises nine booklets with associated computer discs expanding on the content of each.

New South Wales was also the birthplace of the privately published *Australian Entomological Magazine* founded by Max Moulds in 1972 and modelled on the British *Entomologist's Monthly Magazine*. This was passed to the Entomological Society of Queensland in 1987 and became *The Australian Entomologist* in 1993. It remains a place of choice for the publication of many studies on the Lepidoptera (Monteith and Wright, 2023).

In addition to these explicitly entomological outlets, much work on the Australian moths has appeared in more general journals produced by the royal societies, museums and state naturalists' organizations in each state, as well as the aforementioned Linnean Society of NSW.

Finally, we note that CSIRO Publishing has produced national (becoming international) journals in which many larger works on the Australian and global Lepidoptera have appeared. The *Australian Journal of Zoology* was first published in 1953 and *Invertebrate Systematics* (previously *Invertebrate Taxonomy*) began publication in 1987.

2.6.3 The Australian National Insect Collection

The Australian National Insect Collection (ANIC) is without any doubt the pre-eminent repository of preserved insects (and other terrestrial arthropods) in Australia. It houses over 12 million specimens with, probably, well over a million specimens of Lepidoptera and has been curated to a high level. No study of Australian moths is possible without reference to the ANIC and its superb quality and essential nature are recognized worldwide. The history of the national collection is intricately bound up with the genesis and development of the erstwhile Division of Entomology of the CSIRO and its predecessor the Division of Economic Entomology of the CSIR. This history up to 1991 has been chronicled in fine detail by Upton (1997) in his book *A Rich and Diverse Fauna* which describes the gradually evolving focus on cutting-edge taxonomy which has become the hallmark of the ANIC and its associated curators and other research staff.

Once the importance of a general insect collection and a reliable taxonomy of the Australian fauna had overcome the initial overwhelmingly utilitarian inclinations of the organization's early political promoters, the national collection built rapidly through donations of private holdings and targeted collecting by specialist curators. It was finally gazetted as a resource of national importance in 1962.

For the Lepidoptera, this was overwhelmingly due to the energy and scholarship of the first curator of Lepidoptera, Ian Francis Bell Common. As we discussed earlier, Common had joined CSIRO as 'Technical Secretary' in 1948. Fifty percent of his initial role in the organization was to research the ecology of the Lepidoptera. This focus on the Lepidoptera became a full-time job in 1951.

The moth holdings of the national collection received a huge boost in 1948 when the 208 drawers of the Turner collection were transferred to Canberra. Many other major donations followed: the Brandt New Guinea collection in 1955, the Mosse–Robinson collection in 1956, the Robinson collection of micromoths in 1983, the McQuillan collection in 1986, and many others. As we noted above, many types were transferred on permanent loan from the Macleay Museum in the mid-1970s.

The holdings of Lepidoptera were also substantially increased as a result of a series of major remote-area field trips chronicled in detail by Upton (1977) as well as the continuous running of light traps in bush behind the CSIRO buildings in Canberra.

The ANIC is now part of CSIRO's National Research Collections Australia, based, as of 2024, in new purpose-built premises.

2.7 A Promising Future

This brief history of moth studies and the associated institutions could have filled an entire book. Restrictions of space and time have obliged us to focus on only a few central characters and institutions. What is clear is that Australia's 'rich and diverse fauna' has attracted amateur and professional attention since it first came to scientific notice.

Over the 250 years since Johan Fabricius' first descriptions of Australian moths, much has changed technically, reflecting both technological advances and scientific fashion. Many of these changes have been alluded to in this chapter, even more are revisited in Chapter 38. Simple changes have made huge differences to efficiency and effectiveness. These range from the almost universal switch from lacquered corrosible pins to standard stainless steel, from open drawers to unit trays, from wooden to steel cabinets. Of more strictly scientific concern has been the incorporation of genitalia studies in species descriptions through to the barcode revolution of the early 2000s.

Above the species level, developments in cladistics and phylogenomics have revolutionized attempts to reconstruct the evolutionary history of the Lepidoptera, as is clearly evident in many of the chapters in this volume. Associated developments in 'museomics' and informatics have occurred in parallel with those in the study of phylogenetics: indeed, the one has often informed the other in a hugely constructive way.

Almost all of these developments have occurred in the lifetimes of currently active lepidopterists. All have had their immediately enthusiastic adherents while slowly overcoming (mostly) entrenched traditionalists.

One obvious theme runs through the history of moth studies in Australia: the initial, continuing and future role of amateur scientists. These amateurs may be enthusiastic nature photographers, collection-driven hobbyists, serious backyard and backroom scientists, or professional biologists maintaining a 'sideline' in their youthful passion for the Lepidoptera. The vast majority of Australia's moth fauna does not impact on agriculture though those that do will always receive a lion's share of professional engagement. A second growing concern of 'professional' naturalists, however, is more all-encompassing. Human wellbeing depends on a healthy environment and moths, and other insects, play vital roles in maintaining ecosystem health. These ecosystem services provided as free goods by nature are now receiving increasing attention from ecologists generally assisted by amateur volunteers. Such 'citizen scientists' are a vital part of present and future endeavours in the study of moths and their roles in natural and modified ecosystems.

It can reasonably be claimed that Australian moth studies have never been in better shape, in no small part as a result of those who preceded us as scientists, scientific administrators and (even) visionary politicians; we do indeed stand on the shoulders of giants! We have a network of first-rate museums and national collections, we have a scattering of professional scientists in universities and research institutes, and we have technological tools which could only have been dreamed of a generation ago. And we have a growing army of amateur naturalists with a passion for understanding, recording and preserving our natural heritage – of which the moths are a vital part.

Acknowledgements

We thank Peter McQuillan and Erica Odell for providing information and for constructive comments. We are also grateful to Andrea Wild and Rob Birtles (CSIRO) who helped source photographs of Ian Common, Ebbe Nielsen and Ted Edwards. We thank Bonita Coventry, Oswald Lower's descendant, for the image which is Figure 2.6.

References

Angas, G.F. (1847) *South Australia Illustrated*. Thomas McLean, London.

Boisduval, J.B.A.D. (1832) *Voyage de Découvertes de L'astrolabe exécuté Par Order du Roi, Pendant les Années 1826–1827–1828–1829, Sous le Commandement de M.J. Dupont D'Urville. Faune Entomologique de L'océan Pacifiques Avec L'illustration Des Insectes Nouveaux Recuillis Pendant le Voyage. Partie 1. Lépidoptères*. J. Tastu, Paris.

Braby, M. and Olsen, P. (2011) *A Flutter of Butterflies*. National Library of Australia, Canberra, ACT.

Carrington, J.T. (1884). Francis Walker obituary. *Entomologist's Monthly Magazine* 11, 140–141.

Clarke, P.A. (2026) *Aboriginal People and Terrestrial Invertebrates in Australia: Historical and Cultural Relationships*. CSIRO Publishing, Melbourne, Victoria.

Cocking, G., Bond, S. and Edwards, E.D. (2022) *Moths in the A.C.T.* Privately published, Canberra, ACT.

Common, I.F.B. (1954) A study of the ecology of the adult bogong moth, *Agrotis infusa* (Boisd.) (Lepidoptera: Noctuidae), with special reference to its behaviour during migration and aestivation. *Australian Journal of Zoology* 2, 223–263.

Common, I.F.B. (1963) *Australian Moths*. Jacaranda Press, Brisbane, Queensland.

Common, I.F.B. (1970) Lepidoptera (moths and butterflies). In: Mackerras, I.M. (ed.) *Insects of Australia*. Melbourne University Press, Collingwood, Victoria, pp. 765–866.

Common, I.F.B. (1975) Evolution and classification of the Lepidoptera. *Annual Review of Entomology* 20, 183–203.

Common, I.F.B. (1990) *Moths of Australia*. Melbourne University Press, Collingwood, Victoria.

Common, I.F.B. (1994) *Oecophorine Genera of Australia I. The* Wingia *Group (Lepidoptera: Oecophoridae). Monographs on Australian Lepidoptera* **3.** CSIRO Publishing, Collingwood, Victoria.

Common, I.F.B. (1997) *Oecophorine Genera of Australia II. The Chezala*, Philobota *and* Eulechria *groups (Lepidoptera: Oecophoridae). Monographs on Australian Lepidoptera* **5**. CSIRO Publishing, Collingwood, Victoria.

Common, I.F.B. (2000) *Oecophorine Genera of Australia III. The* Barea *Group and Unplaced Genera. Monographs on Australian* Lepidoptera **8**. CSIRO Publishing, Collingwood, Victoria.

Docker, R. (2012) The Museum's Early Days. Available at: https://australian.museum/about/history/the-museums-early-days/ (accessed 28 September 2025).

Donovan, E. (1805) *An Epitome of the Insects of New Holland, New Zealand, New Guinea, Otaheiti and Other Islands in the Indian, Southern and Pacific Oceans Including the Figures and Descriptions of One Hundred and Fifty-Three Species of the More. Splendid, Beautiful and Interesting Insects Hitherto Discovered in those Countries, and which for the Most Part Have Not Appeared in the Works of any Preceding Author*. Privately Published, London.

Dumigan, E.J. (published anonymously) (1934) *Our Insects*. Queensland Department of Education, Brisbane, Queensland.

Edwards, E.D., Newland, J. and Regan, L. (2001) Lepidoptera: Hesperioidea, Papilionoidea. In: Wells, A. and Houston, W.W.K. (eds) *Zoological Catalogue of Australia 31.6*. CSIRO Publishing, Collingwood, Victoria.

Edwards, J.L. (2003) Ebbe Nielsen's bioinformatics legacy. *Invertebrate Systematics* 17, 11–13.

Faast, R. and Weinstein, P. (2020) Plant-derived medicinal emtomochemicals: An integrated approach to biodiscovery in Australia. *Austral Entomology* 59, 3–15.

Fabricius, J.C. (1775) *Systema Entomologiae: Sistens Insectorum Classes, Ordines, Genera, Species, Adiectis Synonymis, Locis, Descriptionibus, Observationibus*. Korte & Flensberg, Leipzig.

Fabricius, J.C. (1792–1798) *Entomologica Systematica emendata et aucta, I–III*. C.G. Proft, Copenhagen, Denmark.

Finney, V. (2018) *Transformations: Harriet and Helena Scott, Colonial Sydney's Finest Natural History Painters*. New South Publishing, Sydney, NSW.

Flood, J. (1980) *The Moth Hunters: Aboriginal Prehistory of the Australian Alps*. Australian Institute of Aboriginal Studies, Canberra, ACT.

Hales, D. and Parsons, R. (2024) History of the entomological societies of New South Wales. *General and Applied Entomology* 52, 41–69.

Halliday, R.B. and Edwards, E.D. (2003) Bibliography of Ebbe Schmidt Nielsen. *Invertebrate Systematics* 17, 157–160.

Hewish, M., Marriott, P., Edwards, T., Kallies, A. and Williams, S.C. (2014) *Moths of Victoria. Part 5. Satin Moths and their Allies, Geometroidea (A)*. Entomological Society of Victoria, Melbourne, Victoria.

Hewish, M., Marriott, P., Edwards, T., Kallies, A., Williams, S. and Byrne, C. (2016) *Moths of Victoria. Part 7. Bark Moths and their Allies, Geometroidea (D)*. Entomological Society of Victoria, Melbourne, Victoria.

Horak, M. (2006) *Olethreutine Moths of Australia (Lepidoptera: Tortricidae), Monographs on Australian Lepidoptera 10*. CSIRO Publishing, Collingwood, Victoria.

Horak, M. and Halliday, R.B. (2003) Lepidoptera systematics and biology: A tribute to Ebbe S. Nielsen. *Invertebrate Systematics* 17, 1–160.

Horning, D.S. (1988) Entomology. In: Stanbury, P. and Holland, J. (eds) *Mr Macleay's Celebrated Cabinet*. Macleay Museum, University of Sydney, Sydney, NSW, pp. 126–130.

Kaila, L. (2011) *Elachistine Moths of Australia (Lepidoptera: Gelechioidea: Elachistidae), Monographs on Australian Lepidoptera 11*. CSIRO Publishing, Collingwood, Victoria.

Karsholt, O. and Nielsen, E.S. (1976) *Systematik Fortegnelse over Danmarks Sommerfugle – Catalogue of the Lepidoptera of Denmark*. Scandinavian Science Press, Klampenborg, Denmark.

Kitching, R.L. and Edwards, E.D. (2020) Books about Australian moths 1805–2020: An annotated bibliography. *Australian Entomologist* 47, 119–154.

Kristensen, N.P. and Nielsen, E.S. (1979) A new subfamily of micropterigid moths from South America. A contribution to the morphology and phylogeny of the Micropterigidae, with a generic catalogue of the family (Lepidoptera: Zeugloptera). *Steenstrupia* 5, 69–147.

Kristensen, N.P., Hilton, D.J., Kallies, A., Mila, L., Rota, J. *et al.* (2015) A new extant family of primitive moths from Kangaroo Island, Australia, and its significance for understanding early Lepidoptera evolution. *Systematic Entomology* 40, 5–16.

Lambkin, C.L. and Daniels, G. (2023) A history of insects collected, collectors, and collecting in Queensland. *Australian Entomologist* 50, 261–294.

Lane, D.A. (1995) A new species of *Opodiphthera* Wallengren (Lepidoptera: Saturniidae) from northern Australia. *Australian Entomologist* 22, 115–122.

Lewin, J.W. (1805) *Prodromus Entomology. Natural History of Lepidopterous Insects of New South Wales, Collected Engraved and Faithfully Painted after Nature*. Thomas Lewin, London.

Linnaeus, C. (1758) *Systema Natura*, 10th edn, vol. 1. Laurentius Salvius, Stockholm.

Lower, O.B. (1892) Descriptions of new south Australian Lepidoptera. *Transactions of the Royal Society of South Australia* 15, 5–17.

Lower, O.B. (1893) A catalogue of Victorian Heterocera. *Victorian Naturalist* 10, 12–15, 31–34, 51–54, 65–68, 81–84, 97–100, 113–116.

Mackerras, I.M. (1948) Alfred Jefferis Turner and amateur entomology in Australia. *Proceedings of the Royal Society of Queensland* 60, 69–88.

Macleay, W.S. (1826) Annulosa. Catalogue of insects collected by Captain King R.N. In: King, P.P. (ed) *Narrative of a Survey of the Intertropical and Western Coasts of Australia Performed between the Years 1818 and 1822*, vol. 2. John Murray, London, pp. 438–469.

Macleay, W.J. (1866) No title. *Transactions of the Entomological Society of New South Wales* 1, 53–55.

Marriott, P., Edwards, T., Kallies, A., Hewish, M. and Williams, S. (2017) *Moths of Victoria. Part 8. Night Moths and their Allies, Noctuoidea (B)*. Entomological Society of Victoria, Melbourne, Victoria.

Mather, P. (1986) *A Time for a Museum: The History of the Queensland Museum 1862–1986.* Queensland Museum, Brisbane, Queensland.

McQuillan, P.B. and Forrest, J.A. (1985) *Common Moths of the Adelaide Region*. South Australian Museum, Adelaide, South Australia.

McQuillan, P.B., Edwards, E.D. and Camilleri, J. (2023) Oswald Bertram Lower (1864–1925): A south Australian pioneer in the discovery of Australia's biodiversity. *Historical Records of Australian Science* 35, 1–15.

Meyrick, E. (1884) A monograph of the New Zealand Geometrina. *Transactions and Proceedings of the New Zealand Institute* 16, 49–113.

Meyrick, E. (1904) Descriptions of Australian microlepidoptera. XVIII. Gelechiadae. *Proceedings of the Linnean Society of New South Wales* 29, 255–441.

Meyrick, E. and Lower, O.B. (1907) Revision of the Australian Psychidae. *Transactions of the Royal Society of South Australia* 31, 192–208.

Monteith, G.B. (1991) *The Butterfly Man of Kuranda: Frederick Parkhurst Dodd*. Queensland Museum, South Brisbane, Queensland.

Monteith, G.B. and Mather, P. (1986) Single cells to spiny shells. Invertebrate zoology. In: Mather, P. (ed) *A Time for a Museum: The History of the Queensland Museum 1862–1986*. Queensland Museum, Brisbane, Queensland, pp. 172–197.

Monteith, G.B. and Wright, S. (2023) Origin and history of the scientific journal *The Australian Entomologist*. *Australian Entomologist* 50, 385–402.

Moulds, M.S. (1999) The history of Australian butterfly research and collecting. In: Kitching, R.L., Scheermeyer, E., Jones, R.E. and Pierce, N.E. (eds) *Biology of Australian Butterflies, Monographs on Australian Lepidoptera 6*. CSIRO Publishing, Collingwood, Victoria, pp. 1–24.

Newman, E. (1874) Death of Mr Walker. *Entomologist* 7, 260–264.

Nielsen, E.S. and Common, I.F.B. (1991) Lepidoptera (moths and butterflies). In: Naumann, I. (ed) *Insects of Australia*. 2nd edn. Melbourne University Press, Collingwood, Victoria, pp. 817–915.

Nielsen, E.S. and Kristensen, N.P. (1989a) The Australian moth family Lophocoronidae and the basal phylogeny of the Lepidoptera-Glossata. *Invertebrate Taxonomy* 10, 1199–1302.

Nielsen, E.S. and Kristensen, N.P. (1989b) *Primitive Ghost Moths: Morphology and Taxonomy of the Australian Genus Fraus Walker (Lepidoptera: Hepialidae S. Lat.)*. Monographs on Australian Lepidoptera 1, CSIRO Publishing, Collingwood, Victoria.

Nielsen, E.S. and Robinson, G.S. (1983) Ghost moths of southern South America (Lepidoptera: Hepialidae). *Entomon* 4, 1–192.

Nielsen, E.S., Edwards, E.D. and Rangsi, T.V. (1996) *Checklist of the Lepidoptera of Australia, Monographs on Australian Lepidoptera 4*. CSIRO Publishing, Collingwood, Victoria.

Ord, M. (1988) *Ash Island Series Volume 1. Historical Drawings of Moths and Butterflies*. Craftsman House, Roseville, NSW.

Robinson, G.S. and Nielsen, E.S. (1993) *Tineid Genera of Australia. Monographs on Australian Lepidoptera 2*. CSIRO Publishing, Collingwood, Victoria.

Salmon, M.A. (2000) *The Aurelian Legacy: British Butterflies and their Collectors*. University of California Press Berkeley, California.

Scott, A.W. (1864) *Australian Lepidoptera and their Transformations Drawn from the Life by Harriet and Helena Scott: With Descriptions General and Systematic*, vol. 1. Jan van Voorst, London.

Scott, A.W. (1890–98) *Australian Lepidoptera and their Transformations by the Late Alexander Walker Scott with Illustrations Drawn from the Life by his Daughters, Harriet Morgan and Helena Forde*, vol. 3. Australian Museum, Sydney, NSW.

Si, A. and Turpin, M. (2015) The importance of insects in Australian aboriginal society: A dictionary survey. *Ethnobiology Letters* 6, 175–182.

Stanbury, P. and Holland, J. (eds) (1988) *Mr Macleay's Celebrated Cabinet*. Macleay Museum, University of Sydney, Sydney, NSW.

Strahan, R. (1979) *Rare and Curious Specimens: An Illustrated History of the Australian Museum 1827–1979*. Australian Museum, Sydney, NSW.

Tepper, J.G.O. (1890) *Common Native Insects of South Australia Part II. Lepidoptera, or Butterflies and Moths*. E.S. Wigg & Son, Adelaide, South Australia.

Thearle, M.J. (1990) Alfred Jefferis Turner (1861–1947). *Australian Dictionary of Biography* 12, 13.

Thomsett, S. (1993) A history of the Pacific collections in the Australian Museum, Sydney. *Pacific Arts* 7, 12–19.

Tindale, N.B. (1953) On some Australian Cossidae including the moth of the witjuti (witchety) grub. *Transactions of the Royal Society of South Australia* 76, 56–65.

Tindale, N.B. (1966) Insects as food for Australian aborigines. *Australian Natural History* 15, 179–183.

Traugott-Olsen, E. and Nielsen, E.S. (1977) The Elachistidae (Lepidoptera) of Fennoscandia and Denmark. *Fauna Entomologica Scandinavica* 6, 1–299.

Upton, M.S. (1997) *A Rich and Diverse Fauna: The History of the Australian National Insect Collection 1926–1991*. CSIRO Publishing, Collingwood, Victoria.

Watson, F. (2012) Bathurst's letter. *Historical Records of Australia Series* 1(14), 210.

Wylie, R. (2023) Symposium welcome address and a history of the entomological Society of Queensland. *Australian Entomologist* 50, 146–155.

Yen, A.L. (2005) Insect and other invertebrate foods of the Australian aborigines. In: Paoletti, M.G. (ed) *Ecological Implications of Minilivestock: Potential of Insects, Rodents, Frogs and Snails*. Routledge, Abingdon, UK, pp. 367–387.

Yen, A.L. (2010) Edible insects and other invertebrates in Australia: Future prospects. In: Durst, A.B., Johnson, D.V., Leslie, R.N. and Shono, K. (eds) *Forest Insects as Food: Humans Bite Back*. FAO, Bangkok, pp. 65–84.

Zborowski, P. and Edwards, E.D. (2007) *A Guide to Australian Moths*. CSIRO Publishing, Collingwood, Victoria.

3 Origins and Evolution

Roger L. Kitching[1]* and Peter B. McQuillan[2]

[1]*School of Environment and Science, Griffith University, Nathan, Queensland, Australia;* [2]*School of Geography, Planning and Spatial Sciences, University of Tasmania, Hobart, Tasmania*

Abstract

This chapter presents an overview of the evolution of the Lepidoptera, with special focus on groups with extant Australian representatives. Current views on the deep time origin of the Lepidoptera in the Carboniferous are discussed with a brief account of early fossils, as well as information from dated, recent phylogenies. The extant mandibulate moths are described with focus on the Micropterigidae and Agathiphagidae. Basal families within the proboscis-bearing Glossata are reviewed, focusing on the Australian families Aenigmatineidae, Lophocoronidae and the non-hepialid Exoporia. The major developments within the vast group of 'higher' Lepidoptera, the Heteroneura, follow, again with a focus on some Australian representatives. Consequences for approaches to classification and a brief overview of the evolution of feeding strategies conclude the account.

3.1 Introduction

Three recent lines of discovery illuminate our current understanding of the origins and subsequent evolution of the Lepidoptera. The combination of authoritative phylogenies based on multigene molecular studies, the continued discovery of early fossils and the careful study of surviving families that derive from some of the earliest divergences within the order combine to provide an increasingly comprehensive understanding of the origins and radiations of the Lepidoptera. This understanding has had major impacts on the higher classification of the order.

A sequence of major phylogenomic studies and reviews of the phylogeny of the Lepidoptera have been published over the last few years (Regier *et al.*, 2013, 2015; Mitter *et al.*, 2017; Kawahara *et al.*, 2019) with others focusing on key families or segments of the overall phylogeny of the order. The size and coverage of the datasets used together with increasingly powerful analytical tools have improved clarity and understanding of the evolution of the order through time. The most recent comprehensive phylogeny published by Akito Kawahara and colleagues (Kawahara *et al.*, 2019), summarized here as Fig. 3.1, is the latest and most comprehensive phylogeny and interpretation and provides the essential backbone to this chapter (with additional comments on the recently described Australian family, Aenigmatineidae, inserted as appropriate). The first Lepidoptera volume of the reissued *Handbook of Zoology* (Kristensen, 2003) provides detailed morphological information on all superfamilies and cross-references to selected chapters of that work are also provided here where appropriate.

*Corresponding author: r.kitching@griffith.edu.au

DOI: 10.1079/9781800627680.0003

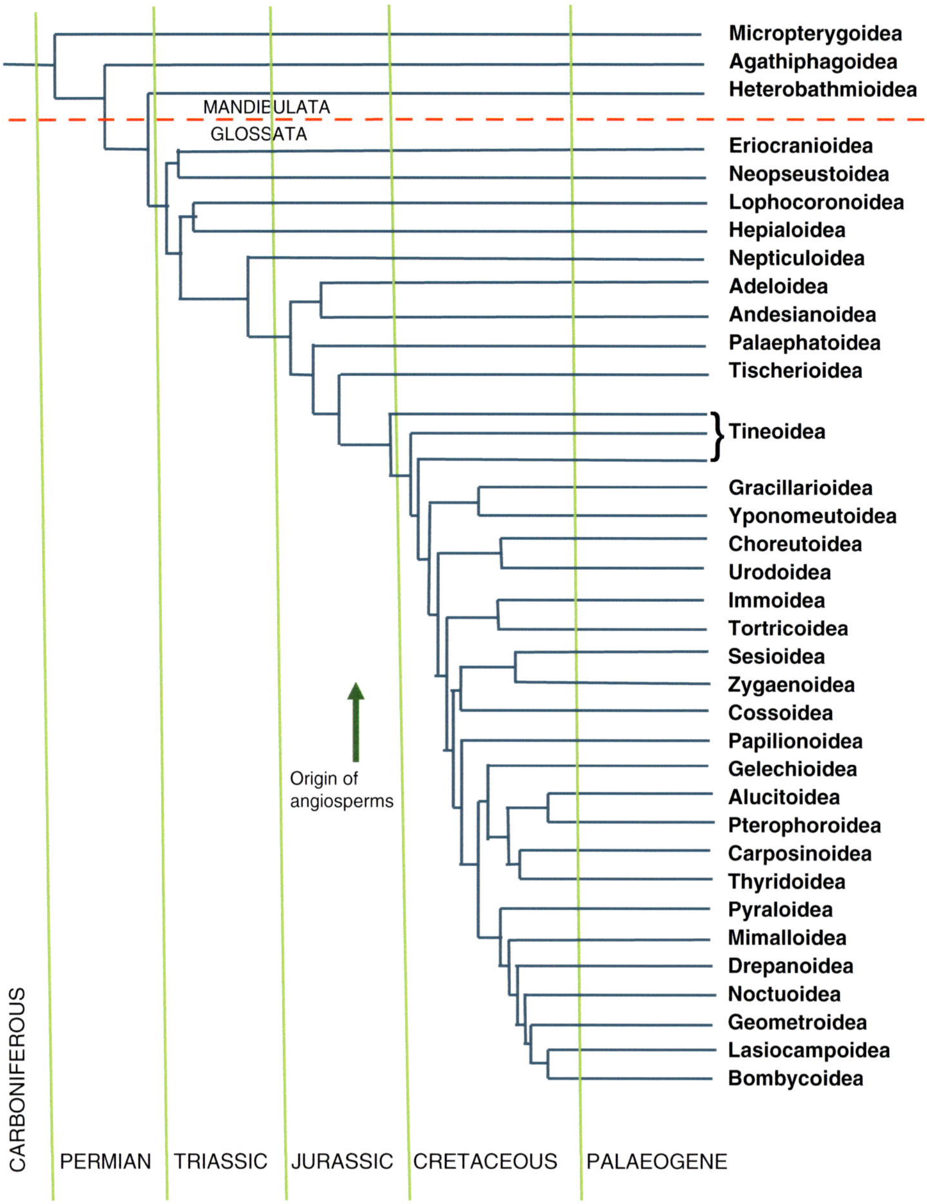

Fig. 3.1. Phylogenetic hypothesis for the Lepidoptera superfamilies, modified and simplified from Kawahara *et al*. (2019). Note some minor superfamilies are omitted. Relative size of the superfamilies is shown in Table 1.1.

3.2 Deep Time Origins

It is generally supposed that the Lepidoptera and Trichoptera form a super-order, the Amphiesmenoptera, together with a recently described wholly extinct order, the Tarachoptera (Mey *et al.*, 2017). The Amphiesmenoptera are postulated to have arisen some 350–250 mya in the Carboniferous, with the paraphyletic group Necrotaulidae probably containing the common ancestor of the modern members of the superorder (that is, the Trichoptera + Lepidoptera). Ivanov and Sukatsheva (2002) provide a reconstruction of a stem amphiesmenopteran (reproduced in Grimaldi and Engel, 2005). Common features linking both Trichoptera and Lepidoptera include, but are not restricted to, female heterogamy (that is, the females contain XY sex chromosomes, not the males), wings covered in setae and looped anal veins on the forewings (Grimaldi and Engel, 2005). The monophyly of both orders is well established based on both molecular and morphological evidence.

So, turning to the Lepidoptera, when did the earliest members of the order arise? This remains a matter for debate although modern analyses have gradually pushed the origin further and further into the past. The earliest fossils, although of great interest, shed little light on the earliest lepidopterans. The earliest known lepidopteran fossils are scales found in German drilling cores dated at the Triassic/Jurassic boundary (*ca* 200 mya) (van Eldijk *et al.*, 2018). These scales are hollow and are presumed to come from moths with proboscides (the Glossata): that is, they are relatively derived compared with what we know of the very earliest Lepidoptera. Next in line are the fossilized wings of *Archeolepis mane* from the early Jurassic of southern England (Whalley, 1985). A re-examination of these fossils confirms the presence of both scales and a looped anal vein (cited in Grimaldi and Engel, 2005).

As recently as 2019, Lees and Zilli asserted that 'the origin and early diversification of Lepidoptera would have occurred in the early Jurassic period about 225–240 million years ago' (Lees and Zilli, 2019). This is about the same time frame to which the fossil glossatan scales mentioned earlier have been ascribed and, by which time on that evidence, major lepidopteran radiations had already taken place. So in the absence (and intrinsic unlikelihood) of future key stem fossils, recent authors have turned to extrapolation (back in time) from more and more authoritative phylogenomic reconstructions using molecular clock assumptions. The most comprehensive of these is that of Kawahara *et al.* (2019) based on an analysis of a massive 2098 protein-coding genes from 186 species across the gamut of the Lepidoptera. These authors confidently assert that the earliest Lepidoptera were mandibulate, bryophyte-feeding leaf miners in the late Carboniferous (*ca* 300 mya). Not only is this considerably earlier than the conclusions of previous authors, it is 100 million years earlier than the oldest known fossils. The late Carboniferous predates the final break-up of the Pangaea supercontinent (and, hence, the origin of Gondwana) so exactly where the first Lepidoptera appeared remains unknown (and, perhaps, unknowable). The European locations of the earliest fossils are much too late in the phylogeny of the order to inform us in this respect.

3.3 The Mandibulate Families

The molecular analyses of Kawahara *et al.* (2019) confirm the conclusions of earlier workers with respect to the major innovations along the lepidopteran tree. The modern representatives of the very earliest of these branches are all mandibulate and their current biological interactions allow some inferences about the habits of the earliest Lepidoptera.

Three extant families of mandibulate moths are recognized globally: the Micropterygidae, Agathiphagidae and Heterobathmiidae. Each of these taxa is also generally accorded superfamily and suborder status consigning all other Lepidoptera into the immense, fourth suborder Glossata which alone contains probably 99% of the entire order. Kawahara *et al.*'s phylogeny places the Micropterigidae as sister to the other mandibulate families (and hence to all other Lepidoptera), with the Agathiphagidae as a separate clade, sister to the Heterobathmiidae which, in turn, is sister to all later Lepidoptera. This resolves earlier, weaker interpretations which had a Micropterigidae + Agathipagidae clade as the most basal. The Heterobathmiidae are confidently identified as the first angiosperm feeders (as larvae), sister to the proboscis-bearing Glossata with which they form the clade designated by Regier *et al.* (2015) as the Angispermivora. This clade encompasses the massive and consequential parallel evolutions of the Lepidoptera and the angiospermophyte plants.

3.3.1 Micropterigidae

At the end of the 19th century, renowned British/Australian entomologist Alfred Jefferis

Turner (1861–1947) visited Mt Tamborine just south of Brisbane and collected a number of diurnal micromoths which he sent to British schoolmaster and entomologist Edward Meyrick (1854–1938) in England. Meyrick had spent time in Australia and New Zealand (see Chapter 2) and had encountered mandibulate micropterigids in New Zealand. Turner's material from Mt Tamborine, however, was the first of the family he had encountered from Australia. In 1902 Meyrick published a short paper in the *Entomologists' Monthly Magazine* describing *Palaeomicra* (now *Tasmantrix*) *calliplaca*, the first of these super-primitive moths to be described from Australia (Fig. 3.2).

Following Kawahara *et al.*'s phylogenomic analysis, the extant Micropterigidae (sometimes termed the suborder Zeugloptera) appear to be the sister group to all the rest of the surviving Lepidoptera, having originated in the early Permian about 280 mya. Distinctive fossils occur in amber from the Baltic to Burma, leading Gibbs (2014) to suggest that the family reached its heyday in the Jurassic (*ca* 140 mya). Nevertheless, modern representatives of this radiation remain extant, with more than 160 named species globally organised as 20 genera. Many others are known but await naming.

Fig. 3.2. Micropterigidae: *Tasmantrix thula*, North Queensland. Photo: Bron King.

A phylogeny of the family has been developed by New Zealand entomologist George Gibbs, British phylogeneticist David Lees and colleagues, which illuminates the evolutionary history of the family. Gibbs devotes an extended section of the much expanded second edition of his book *Ghosts of Gondwana* (2016) to the Micropterigidae, principally because of its role in understanding elements of New Zealand's faunal development. The biogeography of the family is discussed in Chapter 4.

A preliminary phylogeny was presented by Gibbs at the International Congress of Entomology in Brisbane in 2004 in collaboration with Japanese colleagues. Expanded versions were published in subsequent works, culminating in the summarizing diagram in Gibbs' (2016) book. Figure 3.3 is a simplified version of the latest published phylogenies. The left-hand side of the figure presents the entire family phylogeny and the right-hand side shows that of the southern hemisphere *Sabatinca* lineage. These phylogenies are based on an analysis of sections of the mitochondrial 16S rRNA gene only (unlike more recent phylogenies) but appear to be well supported. Gibbs (2016) suggests that critical parts of this RNA-based phylogeny are supported by more recent (but unpublished) work incorporating further genes. Work to date supports the monophyly of the family and places the Palaearctic genus *Micropteryx* within other micropterigid clades. This appears to resolve the question reiterated by Scoble (1992) as to whether the Micropterigidae are indeed monophyletic or should be regarded as two separate families. This dichotomy had been suggested by Sonnenchein and Hauser (1990) based on the heterogeneous cellular anatomy of spermatozoa they observed within the family (that is, in *Epimartyria* and *Micropteryx* respectively).

Systematically, then, the known Australian micropterigids and their kin elsewhere in the world have received, and continue to receive, considerable attention. When we turn to natural history and ecology, we are far less well informed. What little we know is summarized on iNaturalist and Don Herbison-Evans' Lepidoptera webpage (https://lepidoptera.butterflyhouse.com.au).

The 12 known Australian micropterigid species have all been found in wet forests. Nine named

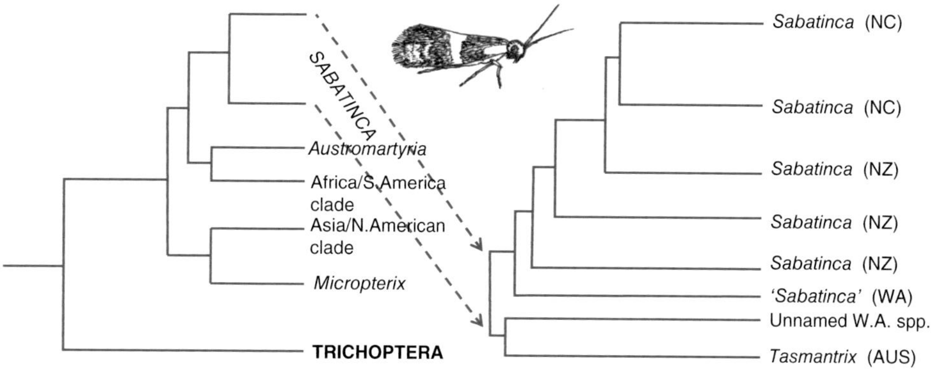

Fig. 3.3. A phylogenetic hypothesis for the Micropterigidae. Redrawn and modified after Gibbs (2016).

species occur along the east coast of Australia, four in rainforest, the remainder in moist eucalypt forests. The three as yet unnamed species from Western Australia again complicate the picture. One species was discovered in eucalypt forest, a second in montane heathland and a third is known only from a single road drain (Moir and Young, 2023). Larval habits are known for *Austromartyria porphyrodes*, the larvae of which have been found in rotting logs, presumably feeding as either detritivores or fungivores. The larval habits of the remaining species are unknown. The New Zealand clade of true *Sabatinca* are known to be leaf miners in bryophytes and this may be a hint as to where Australian larvae should be sought. It has been suggested that one of the 'new' Western Australian species follows this habit. The other two, it is suggested, are leaf litter feeders presumably targetting fungi or directly feeding on detritus (Moir and Young, 2023). There is a little more information about adults which, on general principles, are assumed to feed on pollen or fern spores. Aggregations of adults of *A. porphyrodes* have been found clustered beneath the fronds of *Blechnum* ferns and *Tasmantrix thula* on the flowers of *Guoia lasioneura* (Sapindaceae). *Tasmantrix tasmaniensis* has been observed perched in sunshine on trackside blackberry bushes fringing *Nothofagus cunninghamii*-dominated rainforest.

3.3.2 Agathiphagidae

In February 1947 New Zealand botanist Bayard Parham (1902–1987) was employed by the Fijian Department of Agriculture. While in Fiji, he noted that a large proportion of seeds of the commercially important Pacific kauri pine (*Agathis vitiensis*, Araucariaceae) was infested with an insect larva living within them. He sent a sample of these to New Zealand entomologist Lionel Dumbleton (1905–1976) of the DSIR in Nelson, who reared them to the pupal stage. This allowed him to claim them as members of the Micropterigidae. A similar species was known from the Queensland kauri (*A. robusta*) and Queensland entomologist Alf Brimblecombe (1909–2001) was able to rear adults from these 'after many years of effort'. So, in 1952, Dumbleton described both species based on Brimblecombe's Queensland material and adult material extracted from within pupae of the Fijian species. He noted that these did not fit with any described micropterigid genus, so placed the species, *queenslandensis* and *vitiensis* in a new genus that, aptly, he called *Agathiphaga* (Fig. 3.4).

These species sat in an almost forgotten corner of lepidopterology until they came under the eye of one of the most brilliant lepidopterists of our time, Nils Peder Kristensen (1943–2014). He recognized them as a new basal family, the Agathiphagidae, distinguished from the Micropterigidae by adult venation, abdominal gland structure and a range of larval features (Kristensen, 1967). Initially Kristensen placed his new family within the glossate infraorder Dacnonypha based on the structure of the larval mandibles but subsequent analyses, including recent exhaustive molecular studies, place it as a

Fig. 3.4. Agathiphagidae: *Agathiphaga* sp., Speewah, Kuranda, Queensland. Photo: Paul Zborowski.

separate grouping in deep time. Traditionally this has been classed as the suborder Aglossata, sister to the entire Angiospermivora. The unique presence of an M4 vein in the forewings of Agathiphagidae links it with two mandibulate families from Jurassic amber, the Mesokristenseniidae and the Ascololepidopterigidae, as well as the Trichoptera (Mey *et al.*, 2021).

The Queensland species is also remarkable for its extended pupal diapause which has frustrated most attempts to rear out adult moths. Upton (1997) records this diapause as persisting up to 12 years. Dr Don Sands monitored a group of old kauri trees at Rainbow Beach, Queensland, over several years, recovering infested seeds from undisturbed leaf litter at the base of the remaining trees. Again, these proved very difficult to rear. From 70 seeds with larvae, he managed to rear two adults which only partially eclosed. Sands' opinion, based on emergence times, is that adult agathiphagids probably feed on male spores high in the canopy of mature trees. He suggests that mating probably occurs in the canopy followed by oviposition into early developing seeds within the female cones (D.R.P. Sands, pers. comm.).

The range of the Pacific species, *A. vitiensis*, has now been extended to include Vanuatu and the Solomon Islands. *A. queenslandica* occurs in two disjunct populations reflecting the remaining stands of its host tree. It occurs from Cooloola to Gympie in south-east Queensland, and from Ingham to the Atherton Tablelands in the tropical north. Stands of kauri have been extensively logged, so the tree and, probably in consequence, the moth are rare. At one stage, indeed, it was feared to be extinct following forestry-driven devastation of mature host plants. We now know that this is not the case (D.P.A. Sands, pers. comm.) but clearly the species' ongoing existence depends upon conservation of mature host trees. As far as we can detect, only two photographs of adults of the Queensland species have been published – a black and white image in Common (1990) and a colour version in Zborowski and Edwards (2007).

Neither the Micropterigidae nor the Agathiphagidae have larval connections with the flowering plants (the angiosperms) and we may assume that adult feeding on angiosperm pollen is a derived (that is, relatively recent) behaviour. The third extant mandibulate family, the southern South American Heterobathmiidae, however, occur as leaf-mining larvae in species of *Nothofagus* (Nothofagaceae). The adults also are presumed to feed on the pollen of their host plants (Kristensen and Neilsen, 1983). The Heterobathmiidae are well established as the sister group to the Glossata; that is, all the rest of the Lepidoptera.

3.4 The Basal Glossata

The evolution of proboscides ('glossae' or 'tongues') in the Lepidoptera, which probably occurred some time in the mid-Triassic, was almost certainly the single most consequential innovation to arise within the order (Krenn, 2019). It permitted efficient access to liquids, probably initially water or plant sap, and then nectar produced by flowering plants explicitly to attract potential pollinators. In glossatans the mandibles become non-functional in the adult moths and the galeae of the maxillae form a tube-like structure. The molecular analyses by Regier *et al.* (2015), confirmed and strengthened following the discovery of the enigma moth (Family: Aenigmatineidae) endemic to Kangaroo Island (see below) (Kristensen *et al.*, 2015), suggest that proboscideal musculature has likely evolved twice, once within the Exoporia and again in the Heteroneura. This paraphyly of the Myoglossata is confirmed by Kawahara *et al.*

(2019). Kawahara *et al.* (2019) recognize three basal clades within the Glossata. The first two of these are homoneurous, having similar venation on both fore- and hindwings. The third clade has heterogeneous venation across fore- and hindwings and is designated, accordingly, the Heteroneura; it contains all the remaining 21 or so moth superfamilies.

The most basal of the two homoneurous clades contains the two superfamilies Eriocranioidea and Neopseustoidea and is proposed by Kawahara *et al.* (2019) as the sister group to all other Glossata. In this respect they agreed with the analyses of Bazinet *et al.* (2017) but differed from Regier *et al.* (2015), who placed the single family Eriocraniidae in this role, and Kristensen *et al.* (2015), who placed the Lophocoronoidea/Exoporia as the basal group.

The Neopseustoidea contains three families: the Acanthopteroctetidae, the Neopseustidae and the newly discovered Aenigmatineidae. The Aenigmatineidae did not feature in the Kawahara, Bazinet or Regier analyses but is clearly established by Kristensen *et al.* (2015) as a relatively recent sister to the Neopseustidae which the previous authors did include. Regier *et al.* (2015) and Bazinet *et al.* (2017) clearly linked Acanthopteroctenidae and Neopseustidae as a clade. Kristensen *et al.* (2015) also linked these families together along with Aenigmatineidae. These last authors listed subtle but derived skeletal and antennal characters linking these three families.

Kawahara *et al.*'s (2019) second clade comprises the Lophocoronoidea and the Exoporia which are placed as sister to the Heteroneura. This confirms the arrangement presented by Bazinet *et al.* (2017). Other analyses have placed the Neopseustoidea in this role (Kristensen *et al.*, 2015; Regier *et al.*, 2015). Traditionally the Exoporia has contained two superfamilies: the single-family Mnesarchioidea (endemic to New Zealand) and the Hepialoidea. On the basis of their molecular analyses, Regier *et al.* (2015) strongly recommend combining these two entities into a single superfamily, the Hepialoidea. This would make the name Exoporia redundant. All members of this grouping share important female genitalia characteristics intermediate between the simple single opening Monotrysia and the heteroneuran group Ditrysia in which mating and ovipositing openings are separate (Dugdale, 1974) (see below).

The Hepialoidea (without the Mnesarchioidea), notionally, contains five families, three of which were included in the molecular analyses of Regier *et al.* (2015). Simonsen (2018) suggests that Regier *et al.* (2015) 'decided to follow Scoble (1992) and synonymised the four smaller families Neotheoridae, Anomosetidae, Palaeosetidae and Prototheoridae with Hepialidae'. Now Regier and his colleagues discuss the Hepialoidea at length but, in a very close reading, we can find no evidence for this supposed synonymization. Indeed, only three putative hepialoid families were included in their analysis – the Prototheoridae, the Palaeosetidae and the Hepialidae (*s.l.*). In addition, strictly speaking, Scoble did not formally synonymize these groups, simply stating 'in the present work they are all treated as Hepialidae (*sensu lato*)'. So the status of the four 'minor' families remains uncertain as does that of the set of genera sometimes referred to as the 'primitive' Hepialidae. All authors agree, however, that the Hepialidae (*sensu stricto*) – that is, without the 'primitive' genera – is a monophyletic family in good standing. These issues are discussed further in Chapter 10.

In summary, then, the first glossatans, in addition to having mouthparts modified as proboscides (sometimes much reduced), were both homoneurous and monotrysian. Modern survivors of these early glossatans form four superfamilies.

The Australian fauna contains members of two of these three basal clades. The Eriocranioidea (family Eriocraniidae) are an exclusively Holarctic group. Within the Neopseustoidea, the Acanthopterocteridae are restricted to western North America, the Crimea and Peru and the Neopseustidae to southern South America and south-east Asia. The Aenigmatineidae, however, are exclusively Australian. The second clade contains the exclusively Australian Lophocoronoidea plus a well-developed fauna of Hepialoidea. This includes members of the families Palaeosetidae and Anomosetidae, the genus *Fraus* (one of the 'primitive' hepialid genera), and of the Hepialidae (*s.s.*). Finally the Heteroneura contains, as elsewhere, the greater part of the lepidopteran fauna (see below).

3.4.1 The Aenigmatineidae

In October 2009 South Australian entomologist Richard Glatz collected two adult female moths while sweeping vegetation in and around native cypress trees (*Callitris gracilis*) on Kangaroo Island, South Australia. Further captures followed later in the month. Subsequent searches in 2013 found larvae and pupae beneath the bark of *Callitris* twigs. The adults collected ultimately formed the type series of a new species, *Aenigmatinea glatzella* (Fig. 3.5). The specific name pays homage to the species' discoverer and to the tonsured nature of the scale patterns on the adult heads (thus demonstrating the existence of a sense of, albeit subtle and polyglot, humour among insect taxonomists). These were and remain the only species known so far in the family Aenigmatineidae (Kristensen *et al.*, 2015). The species received the common name, 'enigma' moths.

The species is, of course, homoneurous and monotrysian and has much reduced mouthparts in the adults. Other anatomical features suggest this reduction is a secondary feature and that the family belongs in the glossatan clade. Molecular analysis strongly confirms this and places them into the clade made up, hitherto, of Acanthopteroctenidae and Neopseustidae. Adding the Aenigmatineidae to this clade substantially strengthens its identity compared with Regier *et al.*'s (2015) analysis. As already indicated, the three-family clade is now identified as sister to the Lophocoronoidea/Exoporia clade (Kawahara *et al.*, 2019).

Enigma moths are closely associated with *Callitris gracilis* at most suitable locations on Kangaroo Island (A. Young, pers. comm.). *C. gracilis* (formerly known as *C. preissii*) forms a low woodland with *Eucalyptus diversifolia* and other species. This vegetation type occurs on deep alkaline sands mainly on stranded beach lines in the eastern part of the island (Ball and Carruthers, 1998). *C. gracilis* is widely distributed over much of southern Australia but targeted searching has failed to find any other locations for the Aenigmatineidae.

Fig. 3.5. Aenigmatineidae: *Aenigmtinea glatzella*, female ovipositing on *Callitris*, Kangaroo Island. Photo: Andy Young.

The association with a gymnosperm is noteworthy in a stem lineage of Lepidoptera (*cf.* the Agathiphagidae). Fossil conifers in Australia date back to the early Cretaceous although related species elsewhere go back to the Jurassic. Fossils of Cupressaceae are more recent, with diverse Tasmanian fossils from the early Oligocene *ca* 32 mya (Paull and Hill, 2010). Although derived from rainforest ancestors, modern *Callitris* species became largely adapted to fire-prone biomes, including in southern Australia, during the later Neogene (Crisp *et al.*, 2019). Kristensen *et al.*'s (2015) phylogeny, which includes the Aenigmatineidae, does not indicate divergence dates. The position of the family in the wider phylogeny of the Angiospermivora suggests that its adoption of a coniferous host plant is either a retention of or a reversion to an ancestral habit.

Adult enigma moths are day-flying and appear, it seems, in the early spring (late September/October). They are closely attached, even as adults, to their host plants. The absence of a fully developed proboscis does not prevent uptake of liquids, as Kristensen *et al.* (2015) infer from the presence of a strongly developed sucking pump. Eggs are inserted beneath bracts on the growing tips of *Callitris*. Larvae establish chambers beneath the bark and feed on adjacent plant structures (phloem and callus tissues). Larvae do not appear to make tunnels although, presumably, larval chambers increase in size as larvae mature. Kristensen *et al.* (2015) observed mixed ages of larvae co-occurring and inferred a multiseason life cycle accordingly.

3.4.2 The Lophocoronidae

Dryland woodlands dominated by mallee eucalypts are spread over much of the drier parts of Western Australia, South Australia and Victoria (with minor incursions into other states and territories). Dominated by a variety of species of multistemmed eucalypts and with a sparse understorey often of chenopod saltbush, these ecosystems have developed unique biotas. It was in the mallee, along and beyond the Nullarbor Plain in the late 1960s, that Ian Common and Murray Upton encountered what turned out to be two species of tiny, primitive moths that led to the erection of the family Lophocoronidae by Common (1973) (Fig. 3.6). A third specimen collected much earlier (in 1949 in fact) in the Australian Capital Territory added a third species to the family. Follow-up fieldwork by Ebbe Nielsen (1950–2001) and Ted Edwards (1945–2023) in the early 1980s scoured the mallee zone for more material which ultimately led to the description of three more species as well as the first female of the family (Nielsen and Kristensen, 1996). Initially placed by Common with the Eriocraniidae and Acanthopterocteridae in the now discarded grouping Dacnonypha, more morphological and anatomical studies placed the Lophocoronidae in its own superfamily, sister to the Exoporia. This has been subsequently confirmed by extensive molecular evidence.

Fig. 3.6. Lophocoronidae: *Lophocorona pediasia*, nr. Eucla, South Australia. Photo: Doug Hilton.

Kawahara *et al.*'s (2019) phylogeny puts the Lophocoronoidea/Exoporia divergence somewhere close to the Triassic/Jurassic boundary (*ca* 200 mya). The extant lophocoronids have been associated with a group of six extinct species, assigned to the genus *Acanthocorona*, from Burmese amber dating to the late Cretaceous (*ca.* 99 mya) (Mey *et al.*, 2021).

The six species of Lophocoronidae are exclusively from the drier parts of southern Australia, almost always associated with mallee ecosystems. The external and internal anatomy of adults have been well described (especially of the species *pediasia*) and the phylogenetic position and evolutionary importance of the family have been established. No immature stages are known and no life cycle information is available for any of the six described species. Based on the behaviours of members of adjacent families, Nielsen and Kristensen (1996) speculate that larvae are likely to be internal feeders. These authors also suggest, on anatomical grounds, that lophocoronid pupae will be adecticous (that is, with non-functional mandibles). These remain speculations.

3.4.3 The Non-hepialid Exoporia

The unresolved nature of the status of the various minor goups of Hepialoidea that occur in Australia has already been discussed. A full account of the Hepialidae themselves is given in Chapter 10. Here we briefly examine the two remaining putative families which occur in Australia. Both families were described by A.J. Turner (1922) based principally on wing venation and the presence or absence of tibial spurs. Turner recognized their close affinities with the Hepialidae and each other. Indeed, he suggested that the Palaeosetidae might well be derived from the Anomosetidae 'by reduction'.

The Palaeosetidae were discovered by Turner in Lamington National Park, south-east Queensland, in 1921. They were free flying in primary subtropical rainforests associated with ferns and mosses. He described them in the genus *Palaeoses*. Ian Common describes encountering an additional species of *Palaeoses* from the

Atherton Tablelands; this remains undescribed. Other authors have added further genera to this grouping from Asia and South America, most recently from western China (Liao *et al.*, 2021). Although details of adult biology for some of the non-Australian species have been reported (including male 'lekking'), no further natural historical details for the Australian species have been reported.

The Anomosetidae were collected, like the original palaeosetids, in Lamington National Park (Fig. 3.7). They were described as *Anomoses hylecoetes* by Turner (1916) (being elevated to family status by him in 1922). Further specimens have been found in subtropical rainforest in north-eastern New South Wales. Common (1990) speculates 'it is possible that the larvae feed in decaying logs or on detritus on the rain forest floor'. The very brief account by Nielsen in the Australian checklist (Nielsen, 1996a) appears to confirm this, reporting the late Murdoch DeBaar's observation that 'they have been reared from larvae feeding in a rotten log'.

3.5 The Heteroneura: An Explosive Radiation

Once moths became heteroneurous and adopted angiosperm foodplants, their diversity expanded in step with that of the flowering plants. This is discussed at length by Regier *et al.* (2015). According to Kawahara *et al.* (2019), this almost explosive radiation began in the Triassic but accelerated massively in the Cretaceous. It represents what Conrad Labandeira calls the fourth phase of plant–arthropod association in deep time. As in previous, pre-lepidopteran, phases major lags were involved in the availability of resources such as particular plant tissues and any subsequent fossil evidence of the widespread exploitation of these resources (Labandeira, 2006).

Fig. 3.7. Anomosetidae: *Anomoses hylecoetes*, Korora, NSW. Photo: Nick Lambert.

We associate the availability of a proboscis and a sucking mechanism with the lepidopteran flower association and yet a convincing fossil proboscis exists from the late Jurassic/early Cretaceous boundary at least 15 mya before evidence of the earliest angiosperms (Labandeira, 1997, 2006). Nevertheless, by the K-T event that marked the end of the Cretaceous, fossil evidence suggests that the full range of modern moth families were in place. This does not mean that the mass extinction event did not have major impacts on insect (including lepidopteran) diversity: it did. Labandeira *et al.* (2002) suggest that the diversity of the late Cretaceous was not matched until the Palaeocene/Eocene boundary, a lag of at least 10 million years. In addition to his many more technical articles, Labandeira (2005) presents a highly readable summary of the palaeoecology of insect herbivores.

This co-expansion and interweaving of plant and angiosperm diversification is just that. It is not what has become known as co-evolution and should not be referred to as such. No doubt tight co-evolutionary pairwise interactions – so-called 'evolutionary arms races' – did evolve but the bigger picture was the co-existence of opportunity – for Lepidoptera to exploit plants and for plants to exploit Lepidoptera. Nevertheless, as the Tertiary era developed the Lepidoptera were established as the primary group of insect herbivores with all of their associated biological splendour. Few if any angiosperms escape exploitation by lepidopteran larvae. This in turn has led to the emergence of anti-feedant plant chemicals, to the use of those poisons by insects, to warning colouration, mimicry and other fascinating interactions. Of course, other agents, principally insectivorous vertebrates such as birds and bats, were involved either directly or through the cascade of opportunity and interaction associated with lepidopteran/plant associations (see, for example, Regier *et al.*, 2015).

In summary, then, the Heteroneura evolved to become the dominant superclade within the so-called Angiospermivora. It is not our purpose here to deal with all of the many superfamilies recognized within this radiation. It can be summarized as four clades each more diverse and evolutionarily more 'successful' than the preceding one. These are the non-ditrysian Heteroneura, the basal Ditrysia, the Neoditrysia and the Obtectomera. The base of each of these clades is associated with a major innovation within the order Lepidoptera.

The Monotrysian Heteroneura comprise a small group of ten families globally, organized into five superfamilies. Davis (2003) gives an excellent extended account of the group which he describes as 'questionably monophyletic'. Recent molecular studies confirm this assertion. Following Kawahara *et al.* (2019), the Nepticuliodea form a basal clade comprising the Nepticulidae and Opostegidae. The remaining three subfamilies form three clades culminating in the Palaophatoidea which are confirmed as sister to the remaining Lepidoptera – the Ditrysia. The Incurvarioidea (called Adeloidea by Kawahara *et al.*, 2019) contains six families and forms a clade with the newly discovered South American Andesianoidea. This clade is sister to the Tischerioidea and the Palaephatoidea.

The Australian fauna comprises six families: Nepticulidae and Opostegidae (Nepticulioidea); Heliozelidae, Adelidae and Incurvariidae (Incurvarioidea); and, Palaephatidae (Palaephatoidea). The Nepticuloidea have been catalogued recently (van Nieukerken *et al.*, 2016) and, for Australia, these authors list 58 species in four genera from the Nepticulidae and 24 species in four genera for the Opostegidae. A recent phylogeny of the global Heliozelidae discloses a very diverse Australian clade associated with the plant families Rhamnaceae, Rutaceae and Sapindaceae (Milla *et al.*, 2018). The origin of the family is also confidently estimated to be in late Cretaceous Australia (Milla *et al.*, 2020). Chapter 11 in this volume presents an extended account of the Heliozelidae. The Adelidae and Incurvariidae of Australia await recent taxonomic and phylogenomic treatments although the jarrah leaf miner (*Perthida glyphopa*) has received much attention as a pest species (see Chapter 24).

The Palaephatidae are a small Gondwanan assemblage of moths initially described from South America by Don Davis (Fig. 3.8). Two genera of Australian species, *Azaleodes* and *Ptyssoptera*, were reassigned to this family by Nielsen (1987). He also suggested that the palaephatid genera fell into two groups – one wholly Australian, the other containing genera from both continents. The thorough molecular analysis of Bazinet *et al.* (2017) finally confirmed this but went further in demonstrating that the 'family' was in fact paraphyletic. They suggest that the South American nominate genus *Palaephatus* is a separate clade, sister of the Ditrysia. A mixed Australian/South American group of genera, including *Azaleodes*, show close affinities to the Tischerioidea (a grouping which does not otherwise occur in Australia). They advocate wider sampling and concurrent morphological analyses to strengthen this hypothesis.

Fig. 3.8. Palaephatidae: *Ptyssoptera* sp., alpine zone, Tasmania. Photo: Kristi Ellingsen.

Many of these basal heteroneurans are leaf miners at some stage in their larval lives, others become case bearers later, and at least one incurvariid is a gall former. Several families or subfamilies have particular associations with Australian plant families, particularly the Myrtaceae and the Proteaceae.

According to the Kawahara *et al.* (2019) phylogeny, the Ditrysia arose in the late Jurassic, radiating subsequently mostly in the Cretaceous, with the angiosperms, to produce the massive diversity we now observe in the order Lepidoptera. The 22 or so superfamilies can be divided into four groupings.

The basal Ditrysia are the set of four superfamilies that, evolutionarily speaking, precede the monophyletic Apoditrysia. The Tineoidea contains about six families, four of which occur in Australia. The Gracillarioidea contains four families, all represented in Australia. The Yponomeutoidea are a set of seven families with six occurring in Australia. Finally, the Gelechioidea subsumes 17 families, 13 of which occur in Australia. Recent phylogenies, including that of Kawahara *et al.* (2019), suggest the Tineidae may not be monophyletic.

Some of these basal families have radiated strongly in Australia. Within the Tineoidea, the Psychidae may contain well over 300 species (see Chapter 14), the Tineidae about 370 species (Robinson and Nielsen, 1993). Within the Gracillarioidea, the Gracillariidae has over 450 species. The Yponomeutoidea contains the monocot-feeding Glyphipterygidae with over 240 species. Many families in these superfamilies remain poorly worked in Australia and diversity estimates are, at best, approximate. By way of example, the monograph on the Elachistinae by Kaila (2011) took the Australian fauna from 19 (Nielsen, 1996b) to 148 species, with 128 being described as new. This even takes into account the fact that some of the original 19 were either synonymized or reassigned elsewhere.

Extended general articles on these superfamilies in the *Handbook of Zoology* are Davis and Robinson (2003) (Tineoidea and Gracillarioidea), Dugdale *et al.* (2003a) (Yponomeutoidea) and Hodges (2003) (Gelechioidea).

The Apoditrysia is a huge monophyletic clade which encompasses all the rest of the Lepidoptera. Scoble (1992) suggests that the unifying derived morphological character (synapomorphy) for this enormous group is the possession of a sternum (ventral plate) on the second abdominal segment which has its anterior corners elongated into extended processes ('a tortricid-type abdominal sternum 2'). This monophyly is confirmed by recent molecular phylogenies (Regier *et al.*, 2013; Mitter *et al.*, 2017; Kawahara *et al.*, 2019).

The first few superfamilies within the Apoditrysia form a group distinguished by having pupae with mobile anterior segments differentiating them from the more derived Obtectomera (see below). Again the morphological distinction of monophyly of the Obtectomera is confirmed by subsequent molecular studies. The non-Obtectomera Apoditrysia comprises the large superfamilies Tortricoidea, Zygaenoidea, Cossoidea and Sesioidea with a few much smaller additions basally.

The Kawahara *et al.* (2019) phylogeny places the two-family clade [Choreutoidea + Urodoidea] as basal to all other Apoditrysia, followed by the [Immoidea +Tortricoidea] pairing. The remaining non-Obtectomera form a single monophyletic superclade encompassing the Sesioidea, Cossoidea and Zygaenoidea. Each of these superfamilies is hypothesized to be monophyletic.

The Australian fauna within this group comprises 16 families grouped into nine superfamilies. Numerically speaking, these are dominated by the Tortricidae (Tortricoidea) with, by estimate, well over 1000 species. Only the Cossidae, with an estimated 200+ species, show any other exceptional richness (estimates based on Nielsen and Common, 1991). Some families, notably the Zygaenidae, are exceptional by their modest diversity in Australia.

General accounts of these superfamilies appeared in the first *Handbook of Zoology* Lepidoptera volume (Edwards *et al.*, 2003; Epstein *et al.*, 2003; Horak, 2003). Monographs on the Australian fauna in particular families or subfamilies are Tarmann (2004) and Horak (2006). Accounts in this volume include Chapters 15 (Cossoidea) and 16 (Limacodidae).

Regarding the Obtectomera, the semi-mobile pupae of the less derived clades were superseded in the mid-Cretaceous by a set of superfamilies in which pupae became 'obtect' – that is, immobile. Named the Obtectomera, this group survives as about 17 highly successful superfamilies which include the most familiar and diverse of families within the Lepidoptera. The monophyly of this huge clade is confirmed in recent phylogenies in spite of the fact that some species in the highly derived superfamily, Bombycoidea, appear to have regained (or perhaps never lost) some pupal mobility (Brock, 1990).

As well as the radiation in parallel to the angiosperms already discussed, it is in groups such as the Obtectomera that some workers have suggested that an 'arms race' between the larger moths and the insectivorous bats has driven both diversification and behaviour in the Lepidoptera (Corcoran and Conner, 2012; ter Hofstede and Ratcliffe, 2016). In very brief summary, this idea suggests that once bats evolved echo location then the moths responded by developing sensitive hearing structures. Some have even suggested that this was the reason (or one reason, at least) why the vast clade of butterflies became almost wholly diurnal. More recent analyses in parallel with more and more sophisticated, dated, phylogenies undermine this idea. Hearing organs (generically 'typana') have evolved at least six times within the Lepidoptera and several of these developments predate the radiation of insectivorous bats (see discussion in Kawahara *et al.*, 2019). Tympanal organs are by no means homologous structures across the taxa which exhibit them. This does not, of course, exclude individual species and groups of taxa of moths from having responded to the 'bat challenge' by evolving decoy sound-producing mechanisms (Kawahara and Barber, 2015), more sophisticated hearing mechanisms and other avoidance behaviours but, as a major driving force of diversification, this theory seems unlikely.

Within the Obtectomera, Kawahara *et al.* (2019) identify four monophyletic clades. The first of these comprise the butterflies which, together, comprise the sister clade to all the remaining Obtectomera. Next, the [Gelechioidea + Alucitoidea + Pterophoroidea + Calliduloidea + Thyridoidea] appear as a monophyletic grouping sister to the distinctive Pyraloidea. The small (and non-Australian) superfamily, the Mimallonoidea, then form a monophyletic clade with a massive group informally named the Macroheterocera (see below).

The Australian fauna includes all superfamilies of the Obtectomera except the Whalleyanoidea (Madagascar), Mimallonoidea (the Americas) and Calliduloidea (Madagascar, South-east Asia, Solomons). Excluding for the moment the 'Macroheterocera', by far the most diverse superfamilies are the Pyraloidea and the Gelechioidea. The Australian Pyraloidea comprise the two families Pyralidae and Crambidae and together contain well over 1200 described species. The two largest subfamilies, Pyralidae: Phycitinae and Crambidae: Spilomelinae, are known to contain a wealth of species awaiting description, while others are sorely in need of revision (see Chapter 17). It is in the Gelechioidea, however, that truly massive diversity is seen. The Cosmopterygidae is estimated at 850 species, the Gelechiidae at 1580. It is in the Oecophoridae, though, that we see the single largest radiation of Australian moths with an estimate of more than 5000 species (Common, 1994) (see Chapter 12).

Handbook accounts of these moth superfamilies are Dugdale *et al.* (2003b), Minet (2003) and Munro and Solis (2003).

The final monophyletic grouping of moth superfamilies identified within the Ditrysia: Obtectomera clade in the Kawahara *et al.* (2019) phylogeny have been termed, informally, the Macroheterocera (or, simply, 'macromoths'). Five clearly defined, monophyletic clades appear under this designation in that phylogeny: the superfamilies Drepanoidea, Noctuoidea, Geometroidea, Lasiocampoidea and Bombycoidea becoming increasingly 'derived' in that sequence. All are well represented in the Australian fauna.

The Drepanoidea are represented in Australia by the single family Drepanidae. The second drepanoid family, as recognized by Minet and Scoble (2003), the Epicopeiidae, is restricted to eastern Asia. Ten species of drepanids are described from Australia, mostly rainforest insects in the north east but inhabiting dry woodlands in the south west (Edwards, 1996a; Zborowski and Edwards, 2007). No life histories are known.

The Noctuoidea is the largest superfamily in terms of species globally. In Australia, with close to 2500 named species, it ranks second only to the Gelechioidea. It is also the superfamily in which the greatest modifications have been made to its internal arrangements since the 'traditional' listings of Common (1990), Nielsen and Common (1991), Edwards (1996c) and Kitching (2003). This reflects, for the most part, the thorough molecular analyses of Zahiri *et al.* (2011) and subsequent follow-up work. These authors recognized six families to which has recently been added the Scrancidae (St Laurent *et al.*, 2024). Based on wing venation, these group into three 'trifid' families – the Scrancidae, Notodontidae and Oenosandridae – and four 'quadrifid' families – the Erebidae, Euteliidae, Nolidae and Noctuidae. Each of these groups is a monophyletic clade within the superfamily. Many previously familiar 'families' such as the Arctiidae and Lymantriidae were reduced accordingly to subfamily status in this major revision. These novel family designations have been reinforced by further phylogenomic analyses (Regier *et al.*, 2017) and are now uniformly considered to be monophyletic. The separation of the Scrancidae from the Notodontidae resolved the erstwhile paraphyletic nature of the

Notodontidae (*s. l.*). Various within-family phylogenies are now being worked on but, according to St Laurent *et al.* (2023), issues remain to be resolved (see Chapter 20 for further discussion of this issue).

All seven noctuoid families are represented in Australia (including Scrancidae – the genus *Lasioceros*); indeed, the Oenosandridae are endemic to the continent and comprise eight species organized into four genera (Edwards, 1996b). The Erebidae with up to 16 recognized subfamilies in Australia, and the Noctuidae with ten, are by far the largest of these. These are discussed further in Chapters 20, 21 and 22. The Handbook chapter is Kitching and Rawlins (2003).

The Geometroidea as currently recognized contains five families globally. These five form a monophyletic clade in recent phylogenies (Regier *et al.*, 2013; Rajaei *et al.*, 2015; Kawahara *et al.*, 2019; Ding *et al.*, 2023). In this sense it subsumes the sometimes recognized Uranioidea (*sensu* Edwards, 1996d). Of the five geometroid families, the Geometridae dominate in terms of numbers of species with over 23,000 described worldwide (Scoble, 1999). The Uraniidae are a poor second in this regard with about 700 species globally. The remaining families are small and local: the Epicopeidae (Asia, 25 species), Sematuridae (*ca.* 42 species principally in South America) and Pseudobistonidae (south-east Asia, two species). The Australian fauna comprises about 2500 named species of Geometridae with a comparable number known but awaiting taxonomic attention. The Uraniidae are represented by 33 named species (Edwards, 1996d). Within-family phylogenies have been developed for the Geometridae (Abraham *et al.*, 2001; Murillo-Ramos *et al.*, 2019). The family is discussed at length in Chapter 18. Minet and Scoble (2003) discuss the 'drepanoid/geometroid assemblage' in Chapter 17 of the *Handbook of Zoology*.

The Lasiocampoidea as currently viewed contains but a single family, the Lasiocampidae. The superfamily was initially established as separate from the Bombycoidea to contain the Lasiocampidae and Anthelidae based on an earlier morphological (principally wing venation) designation of the two families as sister taxa (Minet, 1991, 1994). This combination is retained in Lemaire and Minet's (2003) account in the *Handbook of Zoology*. Later extensive molecular work returned the Anthelidae to the Bombycoidea (see below). The Lasiocampoidea do, however, sit in a monophyletic higher clade with the Bombycoidea.

Globally the Lasiocampidae are represented by about 2000 named species divided into about 220 genera in five subfamilies (van Nieukerken *et al.*, 2011). The Australian fauna comprises 73 named species organized into 12 genera principally in the nominate subfamily Lasiocampinae. The most recent species description was in 1942 (https://lepidoptera.butterflyhouse.com.au/lasi/lasiocampidae.html). Edwards (1996b) notes 'there are considerable numbers of well recognised undescribed species' and unresolved cryptic species in some genera such as *Entometa*.

The Bombycoidea as a superfamily has benefited from a sequence of increasingly authorative accounts based on more and more sophisticated molecular analyses (Regier *et al.*, 2008; Zwick, 2008; Zwick *et al.*, 2013). In a closely argued account, Zwick *et al.* (2013) propose that ten families be recognized within the superfamily. They reduce the nominate family Bombycidae to two subfamilies (the Old World Bombycinae and the New World Epiinae). They also expand the Endromidae to include the erstwhile family Mirinidae and two tribes previously included in the Bombycidae. The subfamily Phiditinae is elevated to family level. Previously recognized families – Saturniidae, Sphingidae, Carthaeidae, Anthelidae, Brahmaeidae, Eupterotidae and Apatelodidae – remain unchanged in status.

Globally, the superfamily contains just over 6000 named species organized into 520 genera. By far the most diverse in terms of described species are the Saturniidae (3454 species) followed by the Sphingidae (1602 species). These exact counts are taken from Kitching *et al.* (2018). The Saturniidae and Sphingidae, of course, also happen to be amongst the largest, most spectacular and hence most studied groups of moths. The Australian bombycoid fauna contains about 180 described species with others known and awaiting description and wider revisions. Six of the ten bombycoid families are represented, two of which are Australasian endemics. The Carthaeidae comprises a single south-west Australian species and is associated with Proteaceae. The more widespread Anthelidae comprises 74 Australian species and a further group in New Guinea. Many others await description. The 87 known Australian Sphingidae have recently been extensively revised and monographed (Moulds *et al.*, 2020) (see Chapter 19).

3.6 Classification Consequences

The long-accepted rule of thumb in post-Darwinian biology is that classification *should* reflect phylogeny. On the other hand, we have inherited from Linnaeus the top-down kingdom (perhaps, now, domain) to species hierarchy. The conundrum facing modern taxonomists, therefore, is how to impose the analogue nature of Darwinian phylogenies into the digital categories of the Linnaean hierarchy. The addition of formal and semi-formal qualifier categories – sub-, super-, infra-, etc. – just complicates this situation.

This led Carter and Kristensen (2003) to advocate abandoning all formal labels for groupings within the Lepidoptera above the level of superfamily, simply referring to any such grouping as a clade. This would mean that everything from suborders to infraorders to grades (and other such designations) would be condemned to the museum of lepidopterological history. These authors did not suggest that actually naming some of these higher categories was necessarily inappropriate – Glossata, Ditrysia, Exoporia, Neolepidoptera, Heteroneura, Obtectomera and so forth may well be useful descriptors as long as they are not forced into categories which imply generality across orders, classes and so on. If we adopt this viewpoint, we are taking a strictly evolutionary stance and circumventing the analogue/digital conundrum.

Of course, to make this free-form system work for any large taxon, comprehensive, high-quality, authoritative phylogenies are required. For the Lepidoptera, now, we have such phylogenies. They may not agree perfectly but they do, for the most part, converge on a consensus view of the major (and many minor) monophyletic clades. Some other orders of insects still await such consensus. Moreover, they serve as working hypotheses which can be tested as new constellations of data emerge, especially from morphological and biological studies of the poorly known immature stages.

Useful and evolutionarily justified supraspecies categories recognized by the International Code of Zoological Nomeclature, then, are the superfamily (-oidea), the family (-idae), the tribe (-ini) and the genus (no particular ending). These too are best defined following intensive lower level molecular/morphological phylogenies. Following the revolutionary rearrangement of the Noctuoidea (Zahiri *et al.*, 2012), other superfamilies are, progressively, receiving similar treatment.

Acknowledgements

We are grateful to Doug Hilton, Bron King, Nick Lambert, Andy Young and Paul Zborowski for permisssion to reproduce photos. We thank George Gibbs for comments on the Micropterigidae.

References

Abraham, D., Ryrholm, N., Wittzell, H., Holloway, J.D., Scoble, M.J. and Löfstedt, C. (2001) Molecular phylogeny of the subfamilies in Geometridae (Geometroidea: Lepidoptera). *Molecular Phylogenetics and Evolution* 20, 65–77.

Ball, D. and Carruthers, S. (1998) *Kangaroo Island Vegetation Mapping*. Planning South Australia Technical Report, Adelaide.

Bazinet, A.L., Mitter, K.T., Davis, D.R., van Nieukerken, E.J., Cummings, M.P. and Mitter, C. (2017) Phylotranscriptomics resolves ancient divergences in the Lepidoptera. *Systematic Entomology* 42, 305–316.

Brock, J.P. (1990) Pupal protrusion in some bombycoid moths (Lepidoptera). *Entomologist's Gazette* 41, 91–97.

Carter, D.J. and Kristensen, N.P. (2003) Classification and keys to higher taxa. In: Kristensen, N.P. (ed.) *Lepidoptera, Butterflies and Moths,* 4(1) *Handbook of Zoology.* De Gruyter, Berlin & New York, pp. 27–40.

Common, I.F.B. (1973) A new family of Dacnonypha (Lepidoptera) based on three new species from southern Australia, with notes on the Agathiphagoidea. *Journal of the Australian Entomological Society* 12, 11–23.

Common, I.F.B. (1990) *Moths of Australia*. Melbourne University Press, Carlton, Victoria.

Common, I.F.B. (1994) *Oecophorine Genera of Australia 1. The Wingia Group (Lepidoptera: Oecophoridae)*. Monographs on Australian Lepidoptera 3. CSIRO Publishing, Collingwood, Victoria.

Corcoran, A.J. and Conner, W.E. (2012) Sound strategies: The 65-million-year-old battle between bats and insects. *Annual Review of Entomology* 57, 21–39.

Crisp, M.D., Cook, L.G., Bowman, D.M.J.S., Cosgrove, M., Isagi, Y. and Sakaguchi, S. (2019) Turnover of southern cypresses in the post-Gondwanan world: Extinction, transoceanic dispersal, adaptation and rediversification. *New Phytologist* 221, 2308–2319.

Davis, D.R. (2003) The Monotrysian Heteroneura. In: Kristensen, N.P. (ed.) *Lepidoptera, Butterflies and Moths,* 4(1) *Handbook of Zoology.* De Gruyter, Berlin & New York, pp. 65–90.

Davis, D.R. and Robinson, G.S. (2003) The Tineoidea and Gracillarioidea. In: Kristensen, N.P. (ed.) *Lepidoptera, Butterflies and Moths,* 4(1) *Handbook of Zoology.* De Gruyter, Berlin & New York, pp. 91–117.

Ding, W., Xu, H., Wu, Z., Hu, L., Huang, L. *et al.* (2023) The mitochondrial genomes of the Geometroidea (Lepidoptera) and their phylogenetic implications. *Ecology and Evolution* 2023, 13e9813.

Dugdale, J.S. (1974) Female genital configuration in the classification of the Lepidoptera. *New Zealand Journal of Zoology* 1, 127–246.

Dugdale, J.S., Kristensen, N.P., Robinson, G.S. and Scoble, M.J. (2003a) The Yponomeutoidea. In: Kristensen, N.P. (ed.) *Lepidoptera, Butterflies and Moths,* 4(1) *Handbook of Zoology.* De Gruyter, Berlin & New York, pp. 119–130.

Dugdale, J.S., Kristensen, N.P., Robinson, G.S. and Scoble, M.J. (2003b) The smaller microlepidoptera-grade superfamilies. In: Kristensen, N.P. (ed.) *Lepidoptera, Butterflies and Moths,* 4(1) *Handbook of Zoology.* De Gruyter, Berlin & New York, pp. 217–232.

Dumbleton, L.J. (1952) A new genus of seed-infesting micropterygid moths. *Pacific Science* 6, 17–29.

Edwards, E.D. (1996a) Drepanidae. In: Nielsen, E.S., Edwards, E.D. and Rangsi, T.V. (eds) *Checklist of the Lepidoptera of Australia*, Monographs on Australian Lepidoptera. 4, CSIRO Publishing, Collingwood, Victoria, p. 229.

Edwards, E.D. (1996b) Lasiocampidae. In: Nielsen, E.S., Edwards, E.D. and Rangsi, T.V. (eds) *Checklist of the Lepidoptera of Australia*, Monographs on Australian Lepidoptera. 4, CSIRO Publishing, Collingwood, Victoria, pp. 255–257.

Edwards, E.D. (1996c) Noctuoidea. In: Nielsen, E.S., Edwards, E.D. and Rangsi, T.V. (eds) *Checklist of the Lepidoptera of Australia*, Monographs on Australian Lepidoptera. 4, CSIRO Publishing, Collingwood, Victoria, pp. 271–333.

Edwards, E.D. (1996d) Uraniidae. In: Nielsen, E.S., Edwards, E.D. and Rangsi, T.V. (eds) *Checklist of the Lepidoptera of Australia*, Monographs on Australian Lepidoptera. 4, CSIRO Publishing, Collingwood, Victoria, pp. 230–231.

Edwards, E.D., Gentili, P., Horak, M., Kristensen, N.P. and Nielsen, E.S. (2003) The cossoid/sesioid assemblage. In: Kristensen, N.P. (ed.) *Lepidoptera, Butterflies and Moths,* 4(1) *Handbook of Zoology.* De Gruyter, Berlin & New York, pp. 181–197.

Epstein, M., Geertsema, H., Naumann, C.M. and Tarmann, G.H. (2003) The Zygaenoidea. In: Kristensen, N.P. (ed.) *Lepidoptera, Butterflies and Moths,* 4(1) *Handbook of Zoology.* De Gruyter, Berlin & New York, pp. 159–180.

Gibbs, G.W. (2014) *Micropterigidae (Insecta: Lepidoptera). Fauna of New Zealand*. Manaaki Whenua Press, New Zealand.

Gibbs, G.W. (2016) *Ghosts of Gondwana: The History of Life in New Zealand*, 2nd edn. Potton & Burton, Nelson, New Zealand.

Grimaldi, D. and Engel, M.S. (2005) *Evolution of the Insects*. Cambridge University Press, Cambridge, MA.

Hodges, R.W. (2003) The Gelechioidea. In: Kristensen, N.P. (ed.) *Lepidoptera, Butterflies and Moths, vol. 4(1) Handbook of Zoology.* De Gruyter, Berlin & New York, pp. 131–158.

Hofstede, H.M. and Radcliffe, J.M. (2016) Evolutionary escalation: The bat-moth arms race. *Journal of Experimental Biology* 219, 1589–1602.

Horak, M. (2003) The Tortricoidea. In: Kristensen, N.P. (ed.) *Lepidoptera, Butterflies and Moths,* 4(1) *Handbook of Zoology.* De Gruyter, Berlin & New York, pp. 199–215.

Horak, M. (2006) *Olethreutine Moths of Australia (Lepidoptera: Tortricidae)*. Monographs on Australia Lepidoptera, 10, CSIRO Publishing, Collingwood, Victoria.

Ivanov, V.D. and Sukatsheva, I.D. (2002) Order Trichoptera Kirby 1813. The caddisflies. In: Rasnitsyn, A.P. and Quicke, D.L.J. (eds) *History of Insects*. Kluwer, Dordrecht, pp. 199–219.

Kaila, L. (2011) *Elachistine Moths of Australia (Lepidoptera: Gelechioidea: Elachistidae)*. Monographs on Australian Lepidoptera, 11. CSIRO Publishing, Collingwood, Victoria.

Kawahara, A.Y. and Barber, J.R. (2015) Tempo and mode of anti-bat ultrasound and jamming in the diverse hawkmoth radiation. *Proceedings of the National Academy of Sciences* 112, 6407–6412.

Kawahara, A.Y., Plotkin, D., Espeland, M., Meusemann, K., Toussaint, E.F.A. *et al.* (2019) Phylogenomics reveals the evolutionary timing and pattern of butterflies and moths. *Proceedings of the National Academy of Sciences* 116, 22657–22663.

Kitching, I.J. and Rawlins, J.E. (2003) The Noctuoidea. In: Kristensen, N.P. (ed.) *Lepidoptera, Butterflies and Moths,* 4(1) *Handbook of Zoology.* De Gruyter, Berlin & New York, pp. 355–401.

Kitching, I.J., Rougerie, R., Zwick, A., Hamilton, C.A., St Laurent, R.A. *et al.* (2018) A global checklist of the Bombycoidea (Insecta: Lepidoptera). *Biodiversity Data Journal* 6, e222236.

Krenn, H.W. (2019) Fluid-feeding mouthparts. In: Krenn, H.W. (ed.) *Insect Mouthparts—Form, Function, Development and Performance*. Springer International Publishing, Dordrecht, pp. 47–100.

Kristensen, N.P. (1967) Erection of a new family in the lepidopterous suborder Dacnonypha. *Entomologiske Meddelelser* 35, 341–345.

Kristensen, N.P. (ed.) (2003a) *Lepidoptera, Butterflies and Moths,* 4(1) *Handbook of Zoology.* De Gruyter, Berlin & New York.

Kristensen, N.P. and Neilsen, E.S. (1983) The *Heterobathmia* life history elucidated: immature stages contradict assignment to sub-order Zeugloptera (Insecta: Lepidoptera). *Zeitschrift für Zoologische Systematik und Evolutionforschung* 21, 101–124.

Kristensen, N.P., Hilton, D.J.H., Kallies, A., Milla, L., Rota, J. *et al.* (2015) A new extant family of primitive moths from Kangaroo Island, Australia, and its significance for understanding early Lepidoptera evolution. *Systematic Entomology* 40, 5–16.

Labandeira, C.C. (1997) Insect mouthparts: Ascertaining the paleobiology of insect feeding strategies. *Annual Review of Ecology & Systematics* 28, 153–193.

Labandeira, C.C. (2005) The fossil record of insect extinction: New approaches and future directions. *American Entomologist* 51, 14–29.

Labandeira, C.C. (2006) The four phases of plant–arthropod associations in deep time. *Geologica Acta* 4, 409–438.

Labandeira, C.C., Johnson, K.R. and Wilf, P. (2002) Impact of the terminal Cretaceous event on plant–insect associations. *Proceedings of the National Academy of Sciences* 99, 2061–2066.

Lees, D.C. and Zilli, A. (2019) *Moths: Their Biology, Diversity and Evolution*. Natural History Museum, London.

Lemaire, C. and Minet, J. (2003) The Bombycoidea and their relatives. In: Kristensen, N.P. (ed.) *Lepidoptera, Butterflies and Moths,* 4(1) *Handbook of Zoology.* De Gruyter, Berlin & New York, pp. 321–353.

Liao, C.-Q., Hirowatari, T. and Huang, G.-H. (2021) First report of the family Palaeosetidae (Lepidoptera: Hepialoidea) from mainland China, with a description of a new *Ogygioses* species and its phylogenetic position within the genus. *Journal of Asia-Pacific Entomology* 24, 1040–1049.

Mey, W., Wichard, W., Müller, P. and Wang, B. (2017) The blueprint of the Amphimenoptera – Tarachoptera, a new Order of insects from Burmese amber (insecta: Amphiesmenoptera). *Fossil Record* 20, 129–145.

Mey, W., Léger, T. and Lien V.V. (2021) New taxa of extant and fossil primitive moths from South-East Asia and their biogeographic significance (Lepidoptera, Micropterigidae, Agathiphagidae, Lophocoronidae). *Nota Lepidopterologia* 44, 29–56.

Meyrick, E. (1902) An Australian species of Micropterygidae. *Entomologists' Monthly Magazine* 38, 60–61.

Milla, L., van Nieukerken, E.J., Vijverberg, R., Doorenweerd, C., Wilcox, S.A. *et al.* (2018) A preliminary molecular phylogeny of shield-bearer moths (Lepidoptera: Adeloidea: Heliozelidae) highlights rich undescribed diversity. *Molecular Phylogenetics and Evolution* 120, 129–143.

Milla, L., Moussalli, A., Wilcox, S.A., van Nieukerken, E.J., Young, D.A. *et al.* (2020) Phylotranscriptomics resolves phylogeny of the Heliozelidae (Adeloidea: Lepidoptera) and suggests a Late Cretaceous origin in Australia. *Systematic Entomology* 45, 128–143.

Minet, J. (1991) Tentative reconstruction of the ditrysian phylogeny (Lepidoptera: Glossata). *Entomologica Scandinavica* 22, 69–95.

Minet, J. (1994) The Bombycoidea: Phylogeny and higher classification (Lepidoptera: Glossata). *Entomologica Scandinavica* 25, 63–88.

Minet, J. (2003) The Axoidea and Calliduloidea. In: Kristensen, N.P. (ed.) *Lepidoptera, Butterflies and Moths,* 4(1) *Handbook of Zoology.* De Gruyter, Berlin & New York, pp. 257–261.

Minet, J. and Scoble, M.J. (2003) The drepanoid/geometroid assemblage. In: Kristensen, N.P. (ed.) *Lepidoptera, Butterflies and Moths,* 4(1) *Handbook of Zoology.* De Gruyter, Berlin & New York, pp. 301–320.

Mitter, C., Davis, D.R. and Cummings, M.P. (2017) Phylogeny and evolution of Lepidoptera. *Annual Review of Entomology* 62, 265–283.

Moir, M.L. and Young, D.A. (2023) Insects from the southwest Australia biodiversity hotspot: A barometer of diversity and threat status of nine host-dependent families across three orders. *Journal of Insect Conservation* 27, 3–18.

Moulds, M.S., Tuttle, J.P. and Lane, D.A. (2020) *Hawkmoths of Australia: Identification, Biology and Distribution*, Monographs on Australian Lepidoptera 13. CSIRO Publishing, Collingwood, Victoria.

Munro, M. and Solis, M.A. (2003) The Pyraloidea. In: Kristensen, N.P. (ed.) *Lepidoptera, Butterflies and Moths,* 4(1) *Handbook of Zoology.* De Gruyter, Berlin & New York, pp. 233–256.

Murillo-Ramos, L., Brehm, G., Sihvonen, P., Hausmann, A., Holm, S. *et al.* (2019) A comprehensive molecular phylogeny of Geometridae (Lepidoptera) with a focus on enigmatic small subfamilies. *PeerJ* 7, e7386.

Nielsen, E.S. (1987) The recently discovered primitive (non-Ditrysian) family Palaeophatidae (Lepidoptera) in Australia. *Invertebrate Taxonomy* 1, 201–229.

Nielsen, E.S. (1996a) Anomosetidae. In: Nielsen, E.S., Edwards, E.D. and Rangsi, T.V. (eds) *Checklist of the Lepidoptera of Australia*. Monographs on Australian Lepidoptera, 4, CSIRO Publishing, Collingwood, Victoria, p. 23.

Nielsen, E.S. (1996b) Elachistidae. In: Nielsen, E.S., Edwards, E.D. and Rangsi, T.V. (eds) *Checklist of the Lepidoptera of Australia*. Monographs on Australian Lepidoptera, 4, CSIRO Publishing, Collingwood, Victoria, p. 96.

Nielsen, E.S. and Common, I.F.B. (1991) Lepidoptera (moths and butterflies). In: Naumann, I.D. (ed.) *Insects of Australia*, 2nd edn, Melbourne University Press, Carlton, Victoria, pp. 817–915.

Nielsen, E.S. and Kristensen, N.P. (1996) The Australian moth family Lophocoronidae and the basal phylogeny of the Lepidoptera-Glossata. *Invertebrate Taxonomy* 10, 1199–1302.

Paull, R. and Hill, R.S. (2010) Early Oligocene *Callitris* and *Fitzroya* (Cupressaceae) from Tasmania. *American Journal of Botany* 97, 809–820.

Rajaei, H., Greve, C., Letsch, H., Stüning, D., Wahlberg, N. *et al.* (2015) Advances in Geometroidea phylogeny, with characterization of a new family based on *Pseudobiston pinratanai* (Lepidoptera: Glossata). *Zoologica Scripta* 44, 418–436.

Regier, J.C., Cook, C.P., Mitter, C. and Hussey, A. (2008) A phylogenetic study of the 'bombycoid' complex (Lepidoptera) using five protein-coding nuclear genes, with comments on the problem of macrolepidopteran phylogeny. *Systematic Entomology* 33, 175–189.

Regier, J.C., Mitter, C., Zwick, A., Bazinet, A.L., Cummings, M.P. *et al.* (2013) A large-scale, higher-level, molecular phylogenetic study of the insect order Lepidoptera (moths and butterflies). *PLOS One* 8, e58568.

Regier, J.C., Mitter, C., Kristensen, N.P., Davis, D.R., van Nieukerken, E. *et al.* (2015) A molecular phylogeny for the oldest (nonditrysian) lineages of extant Lepidoptera, with implications for classification, comparative morphlogy and life-history evolution. *Systematic Entomology* 40, 671–704.

Regier, J.C., Mitter, C., Mitter, K., Cummings, M.P., Bazinet, A.L. *et al.* (2017) Further progress on the phylogeny of Noctuoidea (insecta: Lepidoptera) using an expanded gene sample. *Systematic Entomology* 42, 82–93.

Robinson, G.S. and Nielsen, E.S. (1993) *Tineid Genera of Australia (Lepidoptera)*. Monographs on Australian Lepidoptera, 2, CSIRO Publishing, Collingwood, Victoria.

Scoble, M.J. (1992) *Lepidoptera: Form, Function and Diversity*. Oxford University Press, Oxford, and The Natural History Museum, London.

Scoble, M.J. (ed.) (1999) *Geometrid Moths of the World: A Catalogue (Lepidoptera: Geometridae).* CSIRO Publishing, Collingwood, Victoria.

Simonsen, T.J. (2018) *Splendid Ghost Moths and their Allies. A Revision of Australian* Abantiades, Oncopera, Aenetus, Archaeoaenetus *and* Zelotypia *(Hepialidae).* Monographs on Australian Lepidoptera, 12, CSIRO Publishing, Collingwood, Victoria.

Sonnenchein, M. and Hauser, C.L. (1990) Presence of only epipyrene spermatozoa in adult males of the genus *Micropterix* Hübner and its phylogenetic significance (Lepidoptera: Zeugloptera, Micropterigidae). *International Journal of Insect Morphology and Embryology* 19, 269–276.

St Laurent, R.A., Goldstein, P.Z., Miller, S.E. and Robbins, R.K. (2024) Hiding in plain sight: Phylogenomics reveals a new branch in the Noctuoidea tree of life. *Systematic Entomology* 49, 258–278.

Tarmann, G.M. (2004) *Zygaenid Moths of Australia: A Revision of the Australian Zygaenidae (Procridinae: Artonini).* Monographs on Australian Lepidoptera, vol. 9. CSIRO Publishing, Collingwood, Victoria.

Turner, A.J. (1916) A new micropterygid from Australia. *Transactions of the Entomological Society of London* 48, 391–393.
Turner, A.J. (1922) Observations on the structure of some Australian Lepidoptera Homoneura including the diagnosis of two new families. *Transactions of the Entomological Society of London* 1921, 592–604.
van Eldijk, T.J.B. Wappler, T., Strother, P.K., Weijst, C.M.H. van der, Rajaei, H. *et al.* (2018) A Triassic-Jurassic window into the evolution of Lepidoptera. *Science Advances* 4, e1701568.
van Nieukerken, E.J., Kaila, L., Kitching, I., Kristensen, N.P., Lees, D.C. *et al.* (2011) Order Lepidoptera Linnaeus, 1758. In: Zhang, Z.Q. (ed.) Animal biodiversity: An outline of higher-level classification and survey of taxonomic richness. *Zootaxa*, 3148, 212–221.
van Nieukerken, E.J., Doorenweerd, C., Hoare, R.J.B. and Davis, D.R. (2016) Revised classification and catalogue of global Nepticulidae and Opostegidae (Lepidoptera, Nepticulioidea). *Zookeys* 628, 65–246.
Whalley, P. (1985) The systematics and palaeogeography of the Lower Jurassic insects of Dorset, England. *Bulletin of the British Museum (Natural History)(Geology)* 39, 107–189.
Zahiri, R., Kitching, I.J., Lafontaine, J.D., Mutanen, M., Kaila, L., Holloway, J.D. and Wahlberg, N. (2012) A new molecular phylogeny offers hope for a stable family level classification of the Noctuoidea (Lepidoptera). *Zoologica Scripta* 40, 158–173.
Zborowski, P. and Edwards, E.D. (2007) *A Guide to Australian Moths*. CSIRO Publishing, Collingwood, Victoria.
Zwick, A. (2008) Molecular phylogeny of Anthelidae and other bombycoid taxa (Lepidoptera: Bombycoidea). *Systematic Entomology* 33, 190–209.
Zwick, A., Regier, J.C., Mitter, C. and Cummings, M.P. (2013) Increased gene sampling yields robust support for higher level clades within Bombycoidea (Lepidoptera). *Systematic Entomology* 36, 31–43.

4 Biogeography

Roger L. Kitching
School of Environment and Science, Griffith University, Nathan, Queensland, Australia

Abstract
The subscience of biogeography attempts to explain the spatial distributions of animals and plants from the global to the local. Key driving forces are plate tectonics, adaptive radiation, dispersal and human activities. For the Australian moths, insightful analyses are restricted to those groups for which reliable information on within-continent distributions is available. This chapter provides a general background to biogeography and in-depth discussions of the likely driving forces across selected taxa, viz. the Micropterigidae, Oecophoridae, Zygaenidae, Sphingidae and Noctuidae: Heliothinae.

4.1 Introduction

The science of biogeography attempts to describe and explain the spatial patterns of life we observe across the face of the Earth. As such, it is interested in both the patterns themselves and the processes that have produced and continue to influence those patterns across scales from the local to the global. Inevitably this focus on the *spatial* involves delving into the *temporal* – the history of change up to and beyond the here and now. Alfred Russel Wallace is widely seen as the father of the science with his massive compendium *The Geographical Distribution of Animals* (Wallace, 1876). That seminal work has the informative subtitle '..with a study of the relations of living and extinct faunas as elucidating the past changes of the Earth's surface'. In other words, how do past events produce the patterns we observe currently? Wallace's narrative largely concerned vertebrates and was descriptive rather than analytical, but it interwove a set of scientific threads that have engaged biologists ever since.

The incorporation of insects and entomologists into the field was a later development reflecting the sheer diversity involved, the lack of credible phylogenies and, perhaps most telling, inadequate information on distributions at the species level. We revisit this issue below with special reference to the moths. When insects did begin to impact on biogeographical ideas, it was largely a result of entomologists working on faunas of the Pacific region including Australia (e.g. Darlington, 1965; Munroe, 1965; Gressitt, 1974). Perhaps inspired by Wallace, island faunas figured large in this renaissance (Holloway, 1977, 1979, 1986–2011).

Three recent works are the intellectual foundation for this chapter. The account of Peter

Corresponding author: r.kitching@griffith.edu.au

DOI: 10.1079/9781800627680.0004

Cranston and Ian Naumann in the second edition of *Insects of Australia* (1991) is essentially a primer of biogeographical ideas up to that date. The contribution of Jeremy Holloway and Ebbe Nielsen to the first Lepidoptera volume of the *Handbook of Zoology* (Holloway and Nielsen, 2003) reviews some of these ideas with special reference to the Lepidoptera on a global scale. Finally the much expanded second edition of George Gibbs' work *Ghosts of Gondwana* (2016) is probably the most up-to-date and thoughtful reflection on the state of historical biogeography published in recent years. Do not be deceived into thinking it is only about and relevant to New Zealand (although, certainly, it is that as well). This is a work to be read and reflected upon by all those interested in biogeography. I shall not rehearse the ideas of these authors in full but some preliminary generalizations are essential in setting the scene for any biogeographic discussion.

Cranston and Naumann (1991) suggest there are three fundamental branches to the science of biogeography, as follows.

Ecological biogeography attempts to explain the presence, absence or relative abundance of particular species, and consequently of sets of insect species, based on the prevailing environment in any particular place. In their account, Cranston and Naumann focus on the 'physiological envelope' of tolerance to climatic conditions exhibited by particular species. I add to this notion of environment the biotic variables that impinge upon the ecological 'health' of particular species: from the availability of key resources such as food plants or nesting sites, to the degree of impact of local predators, pathogens and competitors.

Dispersal biogeography is caricatured as 'spread from a point of origin' paradigm and depends substantially on ideas on the vagility (dispersive ability) of particular species (or their ancestors). When dispersing organisms overcome some physical or ecological barrier, they become isolated and speciation events will occur over time. Movement of 'founding' organisms is facilitated by habitat corridors, filtering by 'stepping stones' such as islands or habitat patches, or may be entirely due to chance as in the case of storm- or current-driven trans-oceanic movements. The familiar subfield of *island biogeography* relies heavily on dispersalist ideas with more or less random arrivals being balanced against ongoing on-island extinctions (Macarthur and Wilson, 1967).

Vicariance biogeography effectively suggests that species are more or less static and it is the Earth that moves them through geological processes such as plate tectonics, the rise and fall of land masses, even ideas of an expanding Earth. The somewhat eccentric biogeographer Leon Croizat developed this idea (1958) into what he called *panbiogeography*, constructing 'tracks' across the globe based on perceived (i.e. non-cladistic) phylogenetic relationships among taxa. With the growth of Hennigian cladistic phylogenetics (Hennig, 1950), more precise geographical hypotheses could be erected based on more and more confident understanding of relationships of descent among species. The combination of modern phylogenetics with vicariance ideas has become known as *cladistic vicariance biogeography* which remains a dominant paradigm within the field.

Of course, it seems obvious that these three approaches to understanding the spatial distributions of individual species and identifiable assemblages of species can and must be complementary. A *synthetic biogeography* will draw on all three general ideas and others, and the interactions among them.

4.2 The Dimensions of Biogeography

Underlying any approach to explanation in biogeography are four classes of processes. Again these will interact both spatially and temporally. They are seldom alternatives.

4.2.1 Plate tectonics, sea levels and time

Continents and parts of continents move. Convection currents within the molten magma core of the Earth shift the cooled crust which sits atop them in predictable and measurable ways. If we consider the configuration of land masses since the Carboniferous (>299 mya) (a credible starting point for the Lepidoptera – see Chapter 3) then this is the time when all the continents were coming together to form the single supercontinent

Pangaea which persisted through the Permian (299–252 mya) and Triassic (252–201 mya) periods. Pangaea in due course began to split, first into two supercontinents, Laurasia and Gondwana, in the Jurassic (201–145 mya). This process continued into the Cretaceous (145–66 mya) during which period India, Madagascar and Africa sequentially broke away from Gondwana. Australia remained connected to Antarctica, New Zealand and South America. The Australian part of this landmass remained considerably to the south of its present position with a commensurately cooler climate. Following the dramatic K-T event, the Palaeogene period began (66–23 mya) and Australia continued to drift northward, severing its connections with other landmasses in consequence. From the end of the Palaeogene, land masses were more or less as we know them.

Other geological events were overlaid on these tectonic shifts. Cycles of cold, polar dominance imposed ice ages upon the world with associated sea-level changes opening up and closing off terrestrial connections among major land masses (John, 1979). Such connections, of course, permit mixing of faunas across otherwise insurmountable gaps. The glacial events that could have impinged upon the history of the Lepidoptera began in the Permian when the 'Late Palaeozoic Icehouse' event occurred (360–260 mya). Of much more consequence for the Lepidoptera are the Quatenary glaciations which have occurred cyclically over the last 2.5 million years. Most recently the Riss (180–130 kya) and Wurm (70–10 kya) glaciations created connections across shallow seas, allowing dispersal both from and to the Australian land mass. Of most use as an explanatory tool is the existence of the 'recent' Australia/New Guinea land bridge, the so-called Sahul shelf, which existed from about 18 kya to 10 kya. This connection created the Pleistocene Australia/New Guinea land mass sometimes called Sahul. Over the same time period, to the north and east, the great islands of Java, Sumatra and Borneo were connected with each other and the Malayan Peninsula, forming a second land mass, Sundaland.

Finally the recent renewed interest in the idea of a 'Zealandia' continental mass to Australia's east linking the current land masses of New Zealand, Norfolk Island, Lord Howe Island and New Caledonia could be important biogeographically (see, for example, Gibbs, 2006, 2016). This land mass is postulated to have broken away from the Gondwana supercontinent some 83–79 mya, gradually sinking subsequently until, by about 23 mya, only the few emergent remnants remained, sometimes having persisted with volcanic assistance. Over the tens of millions of years of its postulated existence, such a large land mass housed evolutionary radiations of many kinds and its position not so very far from the Australian mainland logically would allow for faunal spillovers or exchanges.

4.2.2 Evolutionary radiation

Fragmenting land masses may well have carried ancient species or sets of species away from their original points of origin and close relatives. Inexorable evolutionary processes on these drifting land masses would lead to adaptive radiation and diversification. This is not the place to deal with the vast literature on speciation and radiation, but suffice it to mention two or three dimensions of particular importance to the Lepidoptera.

Perhaps the key dimension in any evolutionary radiation event is *isolation*. Basically this is any process which produces premating isolation through time, cutting of one segment of a founding population from others. We generally think of this as geographical isolation driven by water barriers, mountain ranges or dry corridors in the case of Australia. However, other kinds of isolation are equally, sometimes more, important. Behavioural patterns may lead to reproductive isolation. Even apparent trivialities like timing of flight activities may lead, first, to cohort partitioning and, ultimately, to taxonomic separation – speciation.

Of special importance for an overwhelmingly herbivorous group like the Lepidoptera is the development of more and more specialized *host plant relationships*. We observe many moth clades which have adopted, or been adopted by, particular plant clades. Some of the better known of these have been with chemically distinctive plant families such as the Moraceae, Apocynaceae, Proteaceae, Myrtaceae and Euphorbiaceae. Although each of these families has developed particular antiherbivore devices, so in turn each has come to host a suite of co-evolved lepidopteran species. As the plant clade diversifies, so can the moths. Of course, in one sense, there is a price to be paid: close dependence on such plants makes breaking away from this chemical dependance evolutionarily challenging. This is the burden of *contingency*. In other words, future evolutionary options are frequently constrained by past evolutionary change.

This also raises the sometimes vexed question of specialization. In its most primitive form, this involves simply counting up the number of plant taxa (I usually use genera) that a herbivore species is known to utilize. Recent discussions have advocated and developed methods to incorporate the degree of similarity among the known host plants. This can be 'simple' phylogenetic relatedness or any other measure of affinity – physical, chemical, locational or what have you (Jorge *et al.*, 2014). Other authors have suggested that the set of food plants of a herbivore is better expressed as its *host repertoire* rather than breadth or width or even range to capture the multidimensionality of host plant choice (Braga and Janz, 2021). Of course, both of these undoubtedly useful ideas demand more and more data on the herbivore species – which is seldom available especially in large, diverse faunas.

Poisot *et al.* (2011) built a useful conceptual model of how ecological specialization – the key to evolutionary radiation – may arise (Fig. 4.1). Basically environmental predictability and phylogenetically programmed preferences will tend to promote the evolution of specialization whereas dispersal and longevity may be countereffects. The biological factors involved all act under historical genetic constraints – another form of contingency.

4.2.3 Migrants and adventives

Species are mobile. In particular, pterygote insects such as the Lepidoptera can fly. We know that many species of moths can cover large distances either through their own powers of flight or more or less aided by pressure fronts and the associated winds (see Chapter 7). From a biogeographical viewpoint this means that the composition and integrity of faunas, at whatever spatial scale, will be continually challenged by adventive individuals from other regions without human assistance. Where we have records, we know that in any year there are a set of vagrant arrivals from nearby land masses – in Australia's case, the tropical islands to our north. The big question then is whether such adventives can form viable populations. The 'one pregnant female' trope is

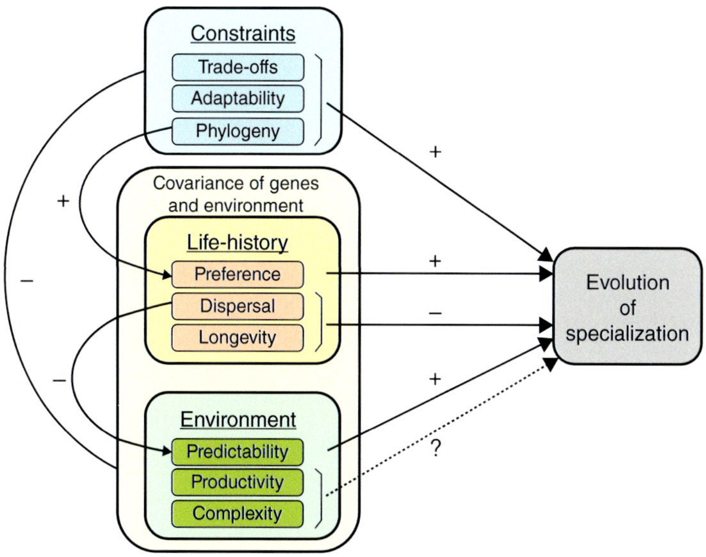

Fig. 4.1. Conceptual model of the evolution of specialization (after Poisot *et al.* 2011, from *Ecology Letters* 14). Figure used with permission from John Wiley and Sons Ltd.

just that but we do know, mainly from European butterfly studies, that in some years substantial numbers of vagrant arrivals may occur including, from time to time, extraordinary rarities (Eeles, 2023). On occasion these will establish short-term populations or, even more rarely, persistent presences on their 'new' land masses. Such vagrants can impact existing populations through predation, competition, introduction of pathogens, even the introduction of further, linked adventive species (Dufour *et al.*, 2024).

These 'rare' events become almost commonplace when one imagines the available time frames over which such chance events may be accumulated. Add to that the changing nature of barriers as geological events change the very topography of land masses and intervening water barriers together with changing climates and the ongoing arrival of candidate species becomes not a lottery but a certainty.

Of course, it is establishment rather than arrival that is the key to changing biotas. Establishment depends not only on intrinsic properties of the vagrants but on the state of the 'receiving' ecosystem. Clearly the physicochemical properties of that environment must fall within the fundamental, physiologically determined tolerances of the arrival but there must also be available 'space' within the biological community into which it is inserted. In that respect, some arrival times may be more auspicious than others. Newly available 'spaces' created by changes in sea levels, recent invasions by other species opening up interactive opportunities or, even, unoccupied resource space within existing communities all provide opportunities for new arrivals to establish themselves.

Such vacant potential niche space may arise through earlier extinction events (Levinton, 2013). Since their putative origin close to the Carboniferous/Permian transition (Kawahara *et al.*, 2019), the Lepidoptera have persisted through three such events usually labelled as major 'mass extinctions'. The best known of these were at the end of the Permian 'when life nearly died' (Benton, 2003), the Triassic/Jurassic event (Blackburn *et al.*, 2013) and the Cretaceous/Palaeogene (often referred to as the K-T event) event marking the Cretaceous/Tertiary boundary (Schulte *et al.*, 2010). These are three of the 'big five' extinction events referred to by palaeoecologists, the other two being before the Lepidoptera existed. In addition, several more geographically or taxonomically restricted events have also been identified as causing mass extinctions in at least some major taxa. There were four of these in the 50 million years separating the Permian and Triassic events, four to five between the Triassic and K-T boundary and at least two in the subsequent period (at the Eocene/Oligocene boundary and in the middle Miocene). The current so-called anthropogenic 'sixth extinction' is discussed in section 4.2.4.

Mass extinctions are almost always discussed in terms of the biota lost: the major taxa of plants, marine life, mega-amphibians or dinosaurs. Terrestrial insects are seldom mentioned except for a few archaic orders which disappeared from the fossil record at the end of the Permian. One of these orders, the Palaeodictyopterida, is described by Grimaldi and Engel (2005) as 'primary herbivores (that) probably caused much of the plant-tissue damage in the Permian and later Carboniferous'. That their disappearance at the end of the Permian left vacant potential niche space for terrestrial insect herbivores seems likely.

Later mass extinctions are frequently cited as having led to substantial reorganizations of floras so must have had substantial impact on insect faunas dependent upon them (like the Lepidoptera). Nevertheless, several studies suggest only minimal impact on insects. A recent important study of insect–plant interactions across the Triassic/Jurassic boundary by Xu *et al.* (2024) reported almost no changes in frequency of occurrence of a wide range of herbivorous feeding strategies across a selection of major plant groups. The plant assemblages turned over across the extinction event but the herbivorous insects appear to have been able to switch to alternative, often related, host plants over the mass extinction period. Only skeletonizers of dipterodacean ferns appear to have disappeared at the end of the Triassic. For the K-T event, Grimaldi and Engel (2005) assert that there is no evidence of major impact on insect faunas *at the family level* and suggest that insects in general may be more resistant than most other animals to natural mass extinctions.

4.2.4 The hand of humans

The final major mechanism that may be invoked in explaining the current distribution of organisms is that they have been moved either accidentally or deliberately by human activity.

Homo sapiens is generally supposed to have evolved some time before 160,000 years ago in Africa where the earliest known fossils originate. From Africa our species spread to other habitable continents, reaching Asia about 70 kya and Australia not long afterwards (perhaps 50–65 kya). Clearly, the arrival of such a versatile, co-operative and intelligent species would have impacted upon the existing fauna of the day. Waterhouse (1991) suggests the only insects that accompanied the earliest Australian arrivals were probably species of ectoparasites. This does not mean, of course, that human establishment of the continent did not have early biogeographical impacts, just that these were likely to have been on the existing fauna rather than through the advent of exotic species. Two processes may have been involved: the decline of the vertebrate megafauna and the frequency of fire. For the Lepidoptera, any such impacts would have been processed through changes in the flora. Where megafauna still exist, we know that they are largely responsible for maintaining the dynamic balance between forest and savannah (Cardosa *et al.*, 2019; Gordon *et al.*, 2023), introducing a dynamic heterogeneity to the landscape which can only lead to richer insect assemblages.

The extensive Australian megafauna included heavyweight herbivores which undoubtedly would have modified the vegetation around them much as modern-day elephants do in Africa and Asia. Towards the mid-Pleistocene many of the megafauna appear to have become extinct coincident with climate change associated with the penultimate glacial maximum ending about 135 kya. Some survivors, however, overlapped with the establishment of humans on the continent for at least 17,000 years (Westaway *et al.*, 2016). The extent to which humans contributed directly to the demise and ultimate extinction of the megafauna is still debated but, for whatever reason, there is no evidence for their persistence after 43 kya. Rule *et al.* (2012) suggest that the absence of the megafauna led to a shift in vegetation from mixed rainforest to sclerophyllous vegetation and a sustained increase in the frequency of fire. Westaway *et al.* (2016) suggest that evidence of genetic bottlenecks in the flora supports the idea that human-modified fire regimes may have impacted on the flora at large but, they suggest, these are not evident until much later (*ca.* 10 kya), long after the last evidence of the megafauna. Whatever the immediate reason, these vegetation changes would inevitably have had significant impacts on the insect fauna and, in particular, herbivorous groups such as the Lepiodoptera.

With the arrival of Europeans, first as explorers and then as colonists, a Pandora's box of exotic biota began to infiltrate the native flora and fauna. Precise records for that period are few especially as almost all insect arrivals were accidental. Exotic Lepidoptera would have arrived in stored foodstuffs, in clothing and other cloth goods, in seed stocks and as incidental passengers on ships. As agricultural industries became established so other pests arose – some native and some introduced.

With respect to the Lepidoptera involved, Nielsen and Common (1991) suggest most pest species in Australia are actually native. Where exotics have become established, they suggest, it is as pests of stored foodstuffs and fibres, orchard and vegetable crops. In biogeographic terms anything which changes the structure of biotas may be important. In this respect native pest species may be more noteworthy than exotic ones. Commercial forestry using native timber species, for example, will change vegetation structure and composition with cascading effects on the native fauna whether they are designated 'pests' or not. These issues are discussed at greater length in Chapters 23 and 28.

I cannot conclude this section without mention of the so-called '6th Extinction' (Leakey and Lewin, 1995; Kolbert, 2014). The general proposition that humans are driving a sixth major mass extinction was, again, based on well-known vertebrate and plant groups, with a few insects added in for inclusiveness. It was not until 2014 that Rodolfo Dirzo and his colleagues published their 'Defaunation in the Anthropocene' in *Science*, suggesting that invertebrates (particularly insects) were implicated in global faunal declines (Dirzo *et al.*, 2014). This theme was expanded upon by Wagner (2020) focusing on 12 well-monitored assemblages of insects in Western Europe and North America showing dramatic declines in abundance in recent years. Wagner synthesized the driving forces involved from deforestation, changed fire frequency, encroaching urban and agricultural development, to widespread use of new-generation pesticides (among others). The overarching impacts of a warming climate act as an intensifier for all these drivers (see also Chapter 8). There remains an open question as to whether comparable declines are occurring beyond the highly developed industrial agricultural regions of the economic North. Tropical regions of known high diversity have few datasets that can be brought to bear on the question (Boyle *et al.*, 2025). Targeted resampling is in progress in several regions.

4.3 Patterns, Hypotheses and Data Demands

I have demonstrated in section 4.2 that there is a wide range of explanatory tools available to the biogeographer to account for observed distributions of animals and plants, including the Lepidoptera. These span time scales from the geological to the immediate, and spatial scales from the global to the local.

Any further biogeographical analysis requires more or less accurate knowledge of current distributions and here the biogeographer of Australian moths must address a major hurdle. There are few taxa at the family or subfamily level where there are recent taxonomic analyses, food plant data and reliable, fairly complete distribution maps. This is in contrast to the butterflies and it is not surprising that recent biogeographical analyses have focused on the Papilionoidea (Kitching, 1981; Kitching and Dunn, 1999). Two general sources provide some exceptions to this rule. First, the series of *Monographs on Australian Lepidoptera* under the current general editorship of Marianne Horak provides excellent accounts of a number of families with either graphical or descriptive distribution information. Useful accounts, in this respect, are available for the Hepialidae, Tineidae, Oecophorinae, Heliothinae, Zygaenidae, Olethreutinae, Elachistinae and Sphingidae. These volumes contain distribution information at the genus or species level from which, with care, overarching patterns can be collated. Second, there is the extensive database represented by the *Atlas of Living Australia* (Belbin *et al.*, 2021) which collates point data for the entire Australian biota and which can be 'mined' for specific distribution data at any given taxonomic level. These two sources used in conjunction are essential tools in any approach to the geography of Australian moths.

Here I focus on five selected taxa in an attempt to synthesize the various forces which have shaped our moth fauna. Taxon-specific accounts elsewhere in the book describe biogeographic scenarios for additional groups (e.g. Heliozelidae, Chapter 11; Geometridae, Chapter 18).

4.4 Australian Case Histories

4.4.1 The Micropterigidae

The evolutionary placement, diversity and ecology of the Micropterigidae are discussed in Chapter 3. Of particular relevance to their biogeography is Fig. 3.2 which presents the best currently available phylogeny of the family (Gibbs, 2016).

Gibbs' analyses recognize five major clades within the family. His hypothesis suggests an origin sometime close to the Permian/Triassic boundary. Kawahara *et al.*'s (2019) more general phylogeny suggests a rather earlier origin within the Permian. Gibbs goes on to suggest that the micropterigid line split into two toward the end of the Jurassic (*ca.* 150 mya) coincident with the break-up of Pangaea and the establishment of southern and northern supercontinents. These splits persist in modern micropterigids as southern and northern hemisphere clades. The northern hemisphere microterigids split into what he calls the 'Asia/N American' clade and the 'European *Micropterix*' clade. It is, however, the three southern hemisphere clades that are of most interest in the context of this account.

A so-called 'southern sabatincoid' clade includes single species from Australia, sister to a group of species from South Africa, Madagascar and southern South America with an outlier species as far north as Costa Rica. The Australian member of this clade, *porphyrodes*, was removed from the catch-all genus *Sabatinca* (*s.l.*) by Gibbs (2010) and placed in a new genus, *Austromartyria*, of which it is currently the sole described member.

The remaining two southern hemisphere clades now contain a total of four named genera:

Aureopteryx, *Tasmantrix*, *Zealandopteryx* and *Sabatinca* (*s.s.*). The first three of these genera (split from *Sabatinca* (*s.l.*) again by Gibbs in 2010) form the so-called 'Australian' group (even though it has a New Zealand member). The now restricted genus *Sabatinca* (*s.s.*) forms a group with members in New Caledonia, New Zealand and Western Australia. Seven species, including *calliplaca*, were placed in the Australian endemic genus *Tasmantrix*. The Australian species *sterops* was joined with a New Caledonian species in *Aureopteryx*. The remaining two new genera, *Zealandopteryx* and *Nannopteryx*, principally contained species from New Caledonia (with one New Zealand species in *Zealandopteryx*). Mey *et al.* (2021) have recently described an additional species of *Aureopteryx* from Vietnam, breaking the Australasian endemism of the genus (and demanding further analysis).

Erecting a satisfactory biogeographically-based explanation for the evolution of these groupings was complicated significantly by the discovery, in 2012, of three new species from Western Australia (the first micropterigids known from that state). They probably represent at least two new genera, one comprising two species which belong to the 'Australian' group as defined by Gibbs (2010) and the other grouped with the *Sabatinca* group. All three are local endemics, two extremely so. The possibility of further species being discovered within the undersampled south-western bioregion has been suggested.

The species associated with the *Sabatinca* group are of particular interest. Hitherto this group had been particularly associated with New Zealand and New Caledonia, remaining elements of the proposed Zealandia supercontinent (see below). It was assumed that the origin of the *Sabatinca* group probably lay in the early existence of that supercontinent. The discovery of the Western Australian member throws considerable doubt on this as preliminary molecular analyses suggest it is the most basal of the group.

4.4.1.1 The biogeographic scenario

The Micropterigidae are of special interest as their phylogeny suggests a principal driving role of plate tectonics and the post-Triassic break-up of Pangaea – a truly vicariant foundation. Other lepidopteran families are associated with the southern supercontinent of Gondwana but only the microptergids appear to have ridden the division of Pangaea into northern and southern supercontinents with deep-time consequences for their within-family phylogeny.

The northern hemisphere isolate diversified to become the speciose genus *Micropteryx* in the Palaearctic and a sister group of genera in Asia and North America.

The fortunes of the southern hemisphere group of genera effectively reflect the gradual break-up of Gondwana itself. As noted above, this group had generally been assumed to divide into an Australian lineage and a group restricted to New Zealand and New Caledonia, in other words, radiations associated with continental Australia and the erstwhile eastern 'continent' Zealandia. This neat and convincing model had to be set aside with the discovery of a true Western Australian species of *Sabatinca* in 2012.

According to Gibbs (2016), the putative continent of Zealandia split from what is now the Australian land mass in the early Cretaceous. If the recently discovered West Australian *Sabatinca* is indeed ancestral to the clade now principally occupying New Zealand and New Caledonia, then ancestral *Sabatinca* must have preceded (or perhaps coincided with) the excision of Zealandia from eastern Gondwana. Once isolated, they radiated as Zealandia subsequently submerged, leaving the terrestrial fragments we identify today (New Zealand, New Caledonia, Norfolk Island) producing in due course the exceptional (in microptergid terms) diversity we observe currently in New Zealand and New Caledonia.

The sister group of sabatincoid genera meanwhile underwent parallel radiation in Australia, producing the species now assigned to *Aureopteryx* and *Tasmantrix*. Other inconspicuous species may well await discovery.

4.4.2 The Oecophoridae

No account of the biogeography of Australian moths could be considered complete, or even adequate, without a section on the Oecophoridae, in particular the subfamily Oecophorinae. Here I add some biogeographical speculations to the otherwise comprehensive account by Robert Hoare which is Chapter 12 of this volume.

The massive radiations of the Oeccophoridae in Australia place it as probably the most species-rich animal taxon on the continent, possibly by as much as an order of magnitude. There are no definitive, species-level distribution maps (and those on iNaturalist are certainly incomplete and, given inherent challenges in oecophorid identification, possibly unreliable anyway). There is a useful recent phylogeny by Kim *et al.* (2016) using mitochondrial CO1, nuclear ribosomal DNA and several nuclear protein-coding genes from a total of 43 'core' oecophorids and a similar number of other species from closely associated taxa. The study focused on East Asian and European taxa but did include a few Australasian examples. No material from South America or Africa was included. Finally, and of greatest moment, are the three volumes of Ian Common's *magnum opus* (Common, 1994, 1997, 2000) revising the generic structure of the Australian Oecophorinae, listing described and known species for every genus. In addition larval host plants are listed wherever known.

So why has this family succeeded in such an extraordinary fashion in Australia? Two ecological factors seem to be the key. First and foremost, the larvae of most Australian oecophorids (as far as is known) have overcome the chemical defences of the plant family Myrtaceae and have to a large extent specialized on this widely available food resource. As Hoare suggests, this must involve a highly specialized gut biota perhaps sequestered from the associated leaf litter. A microbiological study of the oecophorid gut would be most instructive.

So, in an after-the-fact fashion, we can erect hypotheses as to why the oecophorids have been so successful in Australia. It is more challenging to attempt to explain *how* this has come about, especially as such host plant records as exist do not suggest very tight relationships within the Myrtaceae at the species level. Speculating about the history of such a radiation requires repeated assumptions of isolation acting to separate perhaps small populations so that the accrual of mutations can produce genetic isolation. The general small size and, indeed, often the wing shape of oecophorids make an assumption of low levels of vagility a reasonable generalization. What is then needed is some idea of changes in the patchiness of suitable habitat through time; a combination of spatial environmental heterogeneity and low moth mobility would set the scene for runaway adaptive radiation given enough evolutionary time.

According to the dated phylogeny of Kawahara and Barber (2015), the Gelechioidea diverged from its sister groups in the mid-Cretaceous and radiated further into its constituent families before the end-of-Cretaceous (K-T). event. Compared with other highly successful moth families, this is a relatively early origin and certainly ticks the box for the 'available time' part of the speciation hypothesis. At this time, the four southern continents were still connected, although somewhat tenuously towards the end of the period. The analyses of Kim *et al.* (2016) place the small subfamily, the Pleurotinae, as the basal group of the Oecophoridae and this group does not occur (or, perhaps, no longer occurs) in Australia.

Kim *et al.* (2016) have matched an ingenious analysis of ecological traits onto their molecular phylogeny which suggests strongly that the habit of concealed feeding, probably within silken tubes, and the use of dry (or, at least drier) larval resources are plesiomorphies within the oecophorids. These preadaptations for coping with drier environments subsequently served them well in Australia. But was the Australia they first encountered the dry continent we currently observe?

Here the history of the Myrtaceae is a key factor. Molecular evidence suggests that the tribe Eucalypteae and some related taxa originated and developed in Gondwana before the major split into the familiar southern continents. The earliest fossil evidence, from Patagonia, Antarctica and south-eastern Australia, however, is from the early to mid-Eocene (Macphail and Thornhill, 2016). In any case, Macphail and Thornhill (2016) suggest that the earliest eucalypt-like genera were fire associated. At this time 'Australia' was located in a more southerly position and supposedly had a wetter, cooler climate than developed subsequently. Earlier, stem Myrtaceae were quite possibly rainforest associated as, perhaps, were the earliest oecophorids. Several modern genera are associated with rainforest and wet sclerophyll leaf litter. As the continent drifted northwards and dried progressively, we may assume that the Eucalypteae gradually extended and diversified, driving a matching diversification in the oecophorids. I suggest, though, that this was not a species-to-species co-evolution but the result of an initial patchiness as drier habitats appeared piecemeal across the continent interacting with oecophorid populations that were of limited mobility.

The exceptional species diversity of the Australian oecophorids remains a challenge: the scenario I have developed is still a hypothesis.

4.4.2.1 The biogeographic scenario

The Oecophoridae originated in the Gondwana supercontinent, possibly in the later stages of that land mass's existence. The Australian clades of the family were early offshoots of the stem groups and were preadapted to using leaf litter as a larval resource. As a group, they evolved to cope with the plant defences presented by, in particular, the Myrtaceae. Subsequently they radiated dramatically in step with the radiations of the Eucalypteae in eastern Gondwana, subsequently in Australia/Antarctica and finally, and most prolifically, in the isolated continent of Australia. As the continent dried, so the dry-preadapted Myrtaceae thrived, driving the radiation and diversification of one of their primary insect associates, the Oecophoridae. The eons-long process of continental drying, the continuing role of fire at the landscape scale, the increasing dominance and ubiquity of the Eucalypteae and the sheer size of the continental landmass, interacting with a taxon composed of small, poorly vagile, moths produced the impressive and unprecedented diversity we observe currently. Overwhelmingly associated with sclerophyllic ecosystems, the taxon is strongly associated with eucalypt-dominated ecosystems and reaches its highest levels of speciosity in the southern half of the continent.

4.4.3 The Zygaenidae

The Zygaenidae are one of the two large families within the ancient superfamily Zygaenoidea. The superfamily, as currently conceived, contains 12, mostly small, families worldwide (Miric *et al.*, 2024). The nominate family, the Zygaenidae, contains globally about 1100 species classified into 184 genera (van Nieukerken *et al.*, 2011). The family has attracted the attention of naturalists over the years because of their day-flying habits, bright warning colouration and cyanogenic capacity.

The restricted Australian fauna are ideal subjects for a biogeographic analysis for many reasons. Two recent publications, in particular, facilitate such an analysis. First, Gerhard Tarmann (2004) has monographed the Australian fauna with distribution maps of all species as well as his own speculations on the geography of the fauna. Tarmann's fine work has been backed up more recently by a comprehensive, fossil-dated, family-level phylogeny by Mirela Miric and her colleagues (including Tarmann) (Miric *et al.*, 2024). This incorporated representatives across all the putative subfamilies of the Zygaenidae except the east Asian Inouelinae. Four well-established subfamilies were confirmed: the Zygaeninae, Procridinae, Chalcosiinae and Callizygaeninae. The first three of these were confirmed as monophyletic; the Callizygaeninae were represented in the analysis by a single species so monophyly could not be tested. Miric *et al.*'s (2024) phylogeny suggests that the Zygaenidae diverged from its sister taxa (Lacturidae and Phaudidae) in the late Cretaceous (*ca.* 86 mya). Of the well-established extant subfamilies, the Zygaeninae had diverged also in the late Cretaceous (*ca.* 79 mya) with the Procridinae diverging from the [Chalcosinae + Callizygaeninae] close to the Cretaceous/Paleogene boundary (*ca.* 67 mya). The remaining two subfamilies appear to have diverged in the early Palaeogene (*ca.* 58 mya). The family clearly weathered the end-of-Cretaceous mass extinction event without loss of any known higher clades. Miric *et al.* suggest that most modern genera appeared in the Eocene, diverging further in the Oligocene and Miocene.

The Australian fauna are entirely procridines. The Procridinae, commonly referred to as forester moths presumably because of their predominantly green colouration as adults, are generally organized into two large tribes, the Artonini and the Procridini. The Artonini are basal within the subfamily yet are paraphyletic as a taxon. Miric *et al.* included only four genera of Artonini in their analyses: *Thyrassia*, *Ischnusia*, *Pollanisus* and *Clelea*. In their phylogeny *Thyrassia* is sister to all other Procridinae. *Ischnusia* and *Pollanisus* are sister taxa, and *Clelea* is sister to the remaining species usually classified as Procridini.

Returning to the Australian fauna, Tarmann (2004) recognizes ten genera of which only two figured in the later Miric *et al.* (2024) work (*Thyrassia* and *Pollanisus*). All of these ten genera are placed within the Artonini which, Tarmann notes, is largely an Australian and Afrotropical taxon. Tarmann does present a phenetic phylogeny for the Australian genera based on 31 morphological and ecological characters. This remains useful for assessing within-fauna relationships but, in particular, misplaces *Thyrassia* in a complex polycteny with other genera. We await a more comprehensive molecular analysis of the Australian fauna.

Table 4.1 summarizes the richness, distribution and information on larval food plants for the ten Australian genera collated from Tarmann's (2004) monograph. Six of the ten genera are endemic to Australia. Two further genera stretch northwards to New Guinea and the Bismarcks. *Thyrassia* is widespread over southern Asia. Finally *Palmartona*, in Tarmann's opinion, is an introduced species found associated with palms throughout Asia and into the far-eastern Palaearctic. Most of Australia's native procridines occur in the southern half of the continent variously including the south-west and Tasmania.

Figure 4.2 illustrates the relative abundance of the Australian zygaenids across the continent. The node around the Cairns region of far northern Queensland should be treated with caution. It includes a group of poorly known, undescribed species which may represent separate species but remain of uncertain status due to a paucity of study material (Tarmann, 2004).

Examining the Australian fauna overall, two genus groups are of particular interest from a zoogeographical standpoint. The single genus *Pollanisus* has radiated widely in Australia with 20 described species and, potentially, further species awaiting description. This genus appears to have radiated together with Australian genera of the Dilleniaceae as larval food plants. The Dilleniaceae are a family of mostly Gondwanan plants, having arisen in the Palaeogene and radiated widely in the southern hemisphere subsequently. In Australia

Table 4.1. The Australian fauna of Zygaenidae (summarized from Tarmann, 2004).

	Number of species	Distribution		
Genus	Described	Within Australia	Extra-limital	Larval food plants
Pollanisus	20+7*	Eastern mainland 3 Northern Terr. 1 Tropical NE 11* SE mainland 5 SE plus Tasmania 4 South-west 3	-	Dilleniaceae (9 records) Urticaceae (1 record) Fabaceae (1 record)
Onceropyga	2	Coastal Qld & NSW	-	Vitaceae
Hestiochora	8	E mainland 1 SE mainland 3 NE mainland 1 SE, E and SW coast 1 SE plus Tasmania 1 South-west 1	-	Myrtaceae
Turneriprocris	1	SE plus Tasmania	-	Myrtaceae
Myrtartona	4	SE mainland 2 SE plus Tasmania 1 S Coast iincl. SW 1	-	Myrtaceae
Australartona	1	SE mainland 1	-	Poaceae
Homophylotis	3	Cape York	Bismarcks, New Guinea	Unknown
Pseudoamuria	2	Coastal Queensland	New Guinea	Unknown
Palmartona	1	SE Queensland (introduced)	Asia: Eastern Palaearctic	Arecaceae
Thyrassia	1	NE Qld	SE Asia; Sri Lanka; China	Vitaceae

*Includes several undescribed 'species' of uncertain status.

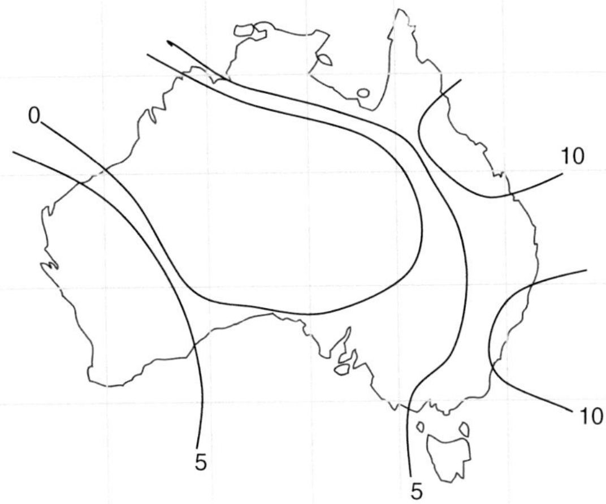

Fig. 4.2. Isopleths showing species richness of the Zygaenidae: Procridinae in Australia, collated from Tarmann (2004).

the family is represented by three genera. *Dillenia* (one species) and *Tetracera* (two species) are representatives of genera widespread in Asia and beyond. *Hibbertia* in contrast is an Australian endemic genus of more than 350 species. Food plant records for the zygaenid genus *Pollanisus*, in Australia, overwhelmingly identify species of *Hibbertia* with a single tropical *Pollanisus* species recorded from *Tetracera*.

The group of procridine genera, *Hestiochora*, *Turneriprocis* and *Myrtartona*, comprise 13 described species and appear as a single clade in Tarmann's phenetic anlysis and, insofar as is known, all exploit Myrtaceae as larval food plants. Species of *Hestiochora* exploit *Eucalyptus*. Larvae of the single described species of *Turneriprocis* feed on *Leptospermum*. *Myrtartona* larvae have been associated with *Leptospermum*, *Melaleuca* and *Eucalyptus* depending on the species examined.

The remaining endemic species, *Australartona mirabilis*, is recorded as feeding on grasses, a habit known in several related Asian Artonini. Of the remaining native genera, we have host plant records only for *Onceropyga* and *Thyrassia*. Larvae of both genera exploit Vitaceae.

4.4.3.1 The biogeographic scenario

At the putative period of origin of the Zygaenidae as a family, the modern continents had more or less formed. Naumann (1987) states quite clearly that the earliest extant subfamily, the Zygaeninae, are Afrotropical in origin but that the principal radiation of that group occurred in Eurasia. Current distributions bear out this scenario. Miric *et al.*'s (2024) phylogeny places the genus *Pryeria* as basal to that clade. The three extant *Pryeria* are from South-east Asia. Next in line in that phylogeny is the African/European genus *Neurosymploca*. However, in Miric *et al.*'s words, they sampled the southern genera only poorly. In Naumann's (1985) morphologically-based phylogeny (reproduced by Fernández-Rubio and Nel, 2000), a suite of African genera are basal to the zygaenine tree. The subsequent Eurasian radiation occurred in the Miocene and Oligocene with an early Oligocene fossil from France (Fernández-Rubio and Nel, 2000) and a largely Palaearctic current distribution of large genera.

In understanding the Australian fauna, however, it is the origin and spread of the Procridinae and, in particular, the Artonini which are critical. Tarmann (2004) erects credible hypotheses on the sequence of events based on comparative morphology and food plant associations. He suggests that the two procridine tribes, the Procridini and the Artonini, split somewhere in the south-east Palaearctic or the Oriental region. Miric *et al.*'s (2024) phylogeny would place this split some time in the early Palaeogene. Tarmann surmised that a very early product of the Artonini grouping headed south to Australia – so early indeed, that the characteristic apomorphic '*Artona*-finger' of the male genitalia is absent in what are presumed to be the earliest clades invading Australia. This encapsulates the genera *Pollanisus*, *Onceropyga*, *Hestiochora*, *Turneriprocis* and *Myrtatona*, supporting, according to Tarmann, the notion that this group of genera represent a radiation independent of other Artonini. The earliest split within this Australian clade separated *Turneriprocis* and *Myrtatona* which have a food plant association with the Australian Myrtaceae (but not so far recorded on *Eucalyptus*). Morphological features support this hypothesis. The largest radiation within Australia, that of *Pollanisus*, appears to have been driven by, or simply paralleled, co-radiation with the Australian Dilleniaceae. The remaining radiation in Australia, that of the genus *Hestiochora*, has exploited *Eucalyptus* as larval food plants.

Both *Pollanisus* and *Hestiochora* are described as 'uniform' (either in genitalia or habitus), suggesting a long period of isolation from any relatives. Tarmann suggest further invasions of Asian Artonini followed, including the grass-feeding *Australartona* and latest of all, the tropical, rainforest-associated genera *Pseudoamuria*, *Homophylotis* and *Thyrassia*. The Australian representatives of these genera all have congeners beyond Australia to the north.

4.4.4 The Heliothinae

The Heliothinae are a well-established monophyletic subfamily of the Noctuidae with a worldwide distribution (Fig. 4.3). The subfamily contains more than 350 species in total, of which 37 are recorded from Australia. The Heliothinae have attracted a disproportionate amount of attention by researchers in no small part because the subfamily includes some of the most serious crop pests across the world. In Australia, the most notorious of these are the earworms, *Helicoverpa armigera* and *H. punctigera* (see Chapter 32). Two sources make this group amenable to a biogeographic analysis in Australia. First, Marcus Matthews' (1999) monograph deals in great detail with the Australian heliothine fauna, including describing six new species. Matthews includes a morphologically-based phylogeny for the subfamily. This has now been expanded and revised based on the molecular work of Cho *et al.*

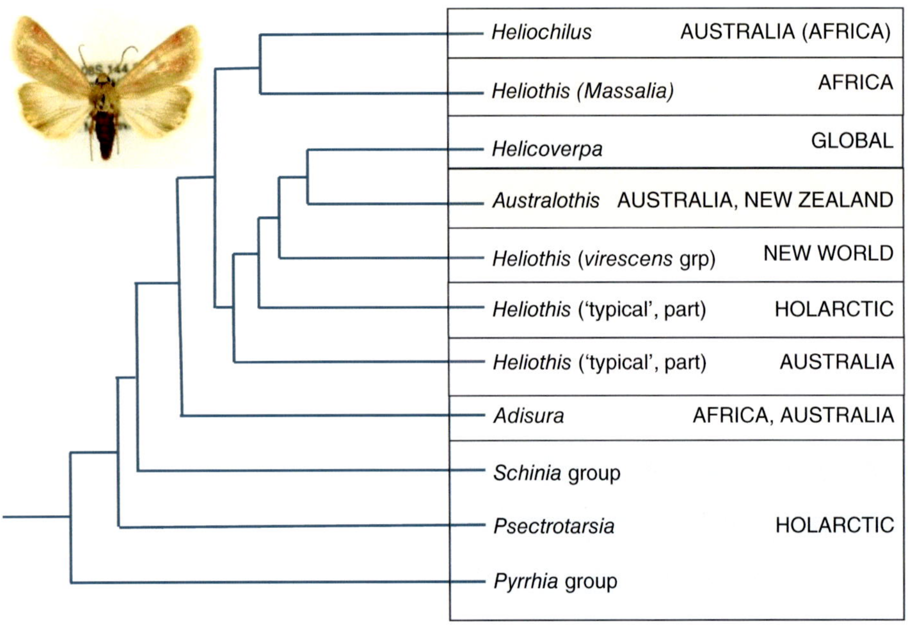

Fig. 4.3. Phylogenetic hypothesis for the Noctuidae: Heliothinae. Redrawn and simplified after Cho *et al.* (2008).

(2008). Figure 4.3 is a simplified version of Cho *et al.*'s Figure 2.

Kawahara *et al.*'s (2019) dated phylogeny of the Lepidoptera included two genera of heliothines and implies a separation from other related noctuids somewhere in the mid-Palaeogene (*ca.* 40 mya), making this a relatively recent radiation. Cho *et al.* sampled 71 species of heliothines for the state of two nuclear genes as well as the mitochondrial CO1 region. Their phylogeny identifies the *Pyrrhia* group of genera as the most basal, followed by two other groups (the *Schinia* group and the genus *Psectrotarsia* (= *Erythroecia*)). All of these three basal clades are restricted to the Holarctic region with no southern representatives.

The Australian fauna of 37 species includes 29 species that are not recorded beyond our shores (Matthews, 1999) although some of these may be found to occur in New Guinea and elsewhere once more intensive collecting is carried out. Our fauna comprises five genera (Matthews, 1999). Two species of *Adisura* share distributions with Asia. This genus is basal to the more derived genera (the '*Heliothis*' group). The genus *Heliothis* itself emerges from Cho *et al.*'s analysis as paraphyletic splitting into at least three sections across the remainder of the phylogeny. Three Australian species in the genus include the continent-wide endemic polyphagous species *H. punctifera.* Only one of the Australian species appears in the Cho *et al.* phylogeny (*punctifera*) and that is in the so-called 'typical' *Heliothis* section.

The Australasian genus *Australothis* includes only four species, of which three occur in Australia. The fourth, *volatilis*, is native to the South Island of New Zealand. The Australian species have somewhat complementary distributions: one is ubiquitous, one from Cape York, and one from the centre of the continent. Sister to *Australothis* is the genus *Helicoverpa.* With about 20 species globally, our fauna owns five species, three of which are endemic. Two species are very serious pests of a wide range of field crops and cereals in Australia: *Helicoverpa punctigera* is a widespread Old World species whereas *armigera* is a widespread endemic pest mostly in the eastern half of the continent. *H. assulta* is a further widespread Old World pest species (although, according to Matthews (1999), of little concern in Australia), occurring in the tropics and subtropics. The remaining two species are endemic, one to the Top End of the Northern Territory and the Kimberleys (*hardwicki*), the other very rare species (*prepodes*) is known from the temperate and subtropical East Coast and the extreme south west.

Finally, the genus *Heliocheilus* is the real Australian success story. Of the 37 species ascribed to the genus, 25 occur in Australia. Twenty-four of these species have not been recorded beyond Australia; one is also known from Sumba and Timor (that is, within the formal 'Australasian' region). The Australian species form a monophyletic clade within Cho *et al.*'s (2008) phylogeny. The extra-Australian 12 species occur in sub-Saharan Africa (five species), North America (five species), eastern Asia (one species) and the Middle East (one species). Most Australian *Heliocheilus* have widespread distributions either spread across the north and east of the continent (nine species) or across the tropical north (eight species). The remaining eight have localized distributions (or in one case an idiosyncratic spread across the western two-thirds of the continent).

Cho *et al.* (2008) suggest that grass-feeding and wide polyphagy (bases for their pest status) are derived features within the phylogeny of the Heliothinae. They match what is known about larval food plants against their phylogeny. The basal genera are overwhelmingly associated with Asteraceae (with Onagraceae and Scrophulariaceae as second and third best). For the *Heliothis*/*Helicoverpa* 'central' section of the phylogeny, some species are associated with specific plant families such as Asteraceae, Solanaceae, Malvaceae, Scrophulariaceae and Chenopodiaceae: a few, often the pest species, have become very widely polyphagous. Finally, what is known of *Heliocheilus* suggests it is a grass-feeding genus. Matthews (1999) records five Australian species each recorded from (or, in one case, suspected to feed on) a different genus of native grasses. Cho *et al.* (2008) suggest, in addition, that two of the African congeners are grass feeders.

4.4.4.1 The biogeographic scenario

It seems likely that the Heliothinae overall originated in the northern hemisphere. At the time of the subfamily's putative origin, the mid-Palaeogene, the modern continents were in place. All three extant basal lineages are restricted to the Holarctic. The modern *Schinia* lineage appears to have radiated subsequently within the Holarctic and, with about 135 named species, remains by far the most speciose of all heliothine lineages. The ancestors of *Adisura* appear to have invaded the southern continents, including Australia, presumably by active dispersal events. The earliest 'typical' extant *Heliothis* in Cho *et al.*'s (2008) phylogeny is currently Australian but other members of that genus are Holarctic. Where the genus actually originated, therefore, remains moot.

The genera *Helicoverpa* and *Australothis*, sisters to the typical *Heliothis*, complicate the picture. *Australothis* is fully Australasian whereas *Helicoverpa* is widespread across the Old World (including oceanic branches in Hawaii). The remaining, derived, lineages are overwhelmingly African or Australian with just the odd reinvasion (perhaps) of the Americas. The heliothines in general have clearly been (and remain) strong flyers and ancestral species have been able, on occasion, to move intercontinentally.

Finally, the Australian success story genus, *Heliocheilus*, nevertheless appears to have originated in Africa, but again with a small radiation in the Americas. In Australia, however, it has radiated impressively to be the most speciose of Australian heliothine genera. The modest number of host plant records imply a genus-level specialization within the Poaceae as a potential isolation mechanism especially for the co-occurring species. Much more life history information is required, however, before such a hypothesis can be confirmed.

4.4.5 The Sphingidae

The Sphingidae are perhaps the best known group of moths (see Chapter 19). The Australian fauna has been comprehensively revised, reviewed and described by Moulds *et al.* (2020). In addition, an available global checklist (Kitching and Cadiou, 2000) and recent comprehensive phylogenies (Kawahara *et al.*, 2009; Kawahara and Barber, 2015; Wang *et al.*, 2021) provide the background to a biogeographic analysis of the fauna. Almost unique among moth families, the Sphingidae have also been subject to previous biogeographic analyses, principally of the Malesian region (South-east Asia plus Wallacia and New Guinea) (Beck *et al.*, 2006a–c; Beck and Kitching, 2007).

The most recent phylogeny is the fossil-dated version of Kawahara and Barber (2015) to which can be added Wang *et al.*'s (2021) newly elevated subfamily, the Langiinae, basal to the [Smerinthinae + Sphinginae] clade. Figure 4.4 is a simplified version of this phylogeny. Kitching

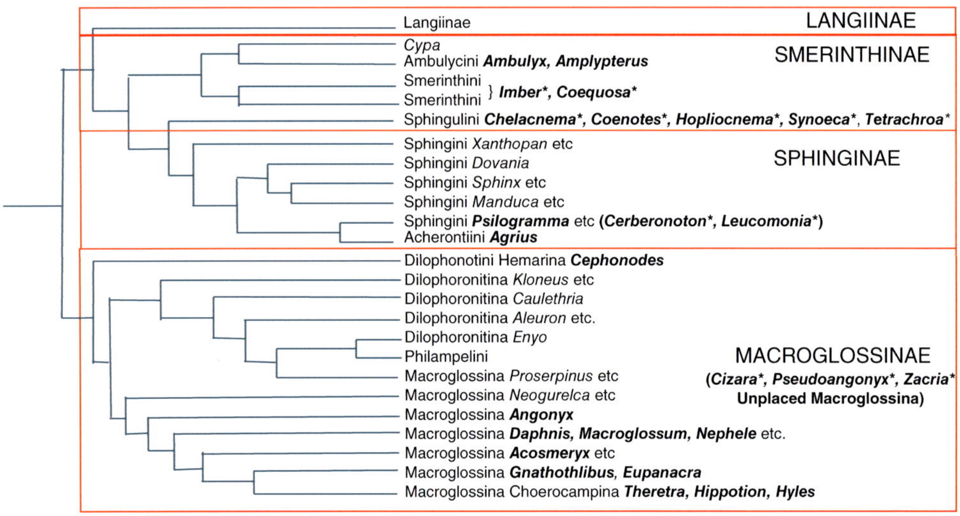

Fig. 4.4. Phylogenetic hypothesis for the Sphingidae. Australian genera are indicated in bold. Single asterisk indicates Australian genera not explicitly included in the underlying phylogenetic analyses. Redrawn and modified from Kawahara and Barber (2015) and Wang *et al.* (2021).

and Barber suggest the Sphingidae diverged from their sister group, the Saturnidae (*s.l.*), in the late Cretaceous (*ca* 60 mya). The clade divided into its two basic parts, the Macroglossinae and the stem [Sphinginae + Smerinthinae], in the early to mid-Palaeogene (*ca.* 47 mya) and then radiated extensively in the Eocene, Oligocene and Miocene, tracking that of the angiosperms of which they became common pollinators. Kawahara and Barber (2015) describe this radiation around their account of an 'arms race' with insectivorous bats. Within the Sphingidae, they identify six independent origins of sound production in response to touch and, within those clades, two origins of hearing and 'jamming' of bat sonar signals.

The Kawahara and Barber (2015) phylogeny confirms the monophyly of the three major sphingid subfamilies, the Macroglossinae, Sphinginae and Smerinthinae. The conventional tribes and what are called 'subtribes' are not uniformly supported across the phylogeny, particularly those conventionally ascribed to the Smerinthinae. Mapping the Australian fauna onto the family phylogeny is challenging and, conservatively, I have followed the checklist in Moulds *et al.* (2020). The Australian fauna is spread across almost the whole of the family classification with the exception of the basal Langiinae and the macroglossine tribes Dilophonotini and Philampolini.

The Australian sphingid fauna comprises 85 species excluding those found only on Christmas Island (part of the Asian region). For the analyses discussed here, I have also excluded species found exclusively on the Torres Strait Islands (mainly Dauan and Boigu) as well as *Marumba timora*, known only from a single specimen observed on the Kimberley Coast. This leaves a fauna of 74 species organized into 27 genera. Twenty-eight species (28%) are known only from Australia, as are 30% of the genera (8 of 27). This relatively low level of Australian endemicity strikes the theme for any consideration of the Australian fauna which, more than most, demands to be examined against the faunas to our north. In Table 4.2, I have compared species lists from Australia with those from New Guinea, Sumatra, Java, Borneo, Peninsular Malaysia and Thailand – that is, those readily available in databases or publications (Diehl, 1980; Holloway, 1987; Lödl, 1993; Inoue *et al.*, 1997; Lane, 2020; Moulds *et al.*, 2020; I.J. Kitching, pers. commun.).

What is clear in this set of numbers is the importance of Wallace's line separating true 'Asian' faunas from those of Australasia even for such a mobile set of species as the hawkmoths. High levels of similarity (that is, co-occurrence of species) occur between the faunas of Australia and New Guinea which share almost half their species. Similarly within the Greater Sunda region (the islands of Java, Sumatra, Borneo and Peninsular Malaysia), as well as Asian land masses further north, there are high degrees of similarity (sharing between 49% and 63% of their faunas at the species level). Both of these groupings are well explained by the fact that these land masses were in fact connected in the not too distant past at the height of the last Ice Age.

What is perhaps less obvious for such highly mobile organisms as the hawkmoths is that this similarity drops precipitously across these two

Table 4.2. Similarity indices among selected Indo-Australian faunas. Numbers in the table are values of the Sørensen Index and represent the proportion of shared species for each pair of faunas. Data from multiple sources.

	Australia	New Guinea	Sumatra	Java	Borneo	Pen. Mal.	Thailand
Australia	1	0.49	0.14	0.14	0.17	0.15	0.13
New Guinea	-	1	0.24	0.29	0.24	0.24	0.18
Sumatra	-	-	1	0.53	0.86	0.81	0.55
Java	-	-	-	1	0.49	0.54	0.50
Borneo	-	-	-	-	1	0.75	0.54
Pen. Malaysia	-	-	-	-	-	1	0.63
Thailand	-	-	-	-	-	-	1

regions (with levels of species sharing between 13% and 29%). Traditionally this contrast is defined by Wallace's line which runs between Borneo and Sulawesi on the one hand and Bali and Lombok on the other. This imaginary line marks the persistent ancient presence of deep-water barriers and was defined by Wallace based largely on known distributions of mammals and, to a lesser extent, birds. Why then should it mark differences among such highly mobile groups as the hawkmoths? This phenomenon, as applied to the sphingids, has been discussed at length by Beck *et al.* (2006b). They present a largely dispersalist explanation for these patterns which, on the one hand, maintains similarities amongst adjacent faunas, yet supports distance-driven differences based on differential flight capacity of subsets of the fauna. They suggest that the Smerinthinae in particular have a lesser dispersal capacity than other hawkmoths and this may drive the observed interregional contrasts.

Returning, then, to the specifics of the Australian fauna, an additional 32 species are found in New Guinea, Timor, the Moluccas and/or Sulawesi – all formally 'Australasian' inasmuch as they occur east of Wallace's line. Some of these also occur in the Pacific Islands. An additional 14 members of our fauna are variously widespread in the rest of Asia, as far west as Iran and Afghanistan in some cases. Finally, two species are widespread in the Old World, including Africa, parts of Europe and Asia.

From a biogeographical point of view, it is instructive to look at the endemic species at several levels of resolution. If we assume the degree of local differentiation is reflected in taxonomic distance then this is a rough measure of the 'penetration' of the fauna over millennia. Endemic species are, taxonomically, scattered through the Australian faunal list but only eight genera are designated by Moulds *et al.* (2020) as endemic. Further, six of those eight genera are monotypic and the remaining two contain but two species. Even among speciose non-endemic genera such as *Macroglossum*, *Psilogramma*, *Hippotion* and *Theretra*, levels of species-level endemism are low. *Psilogramma* tops the list with four endemic species out of seven.

Unlike other moth families examined, no one genus has exploited the Australian environment by radiating widely (*cf. Pollanisus* in the Zygaenidae, *Heliocheilus* in the Heliothinae). This phenomenon is further emphasized by a consideration of larval food plant preferences. Again thanks to the Moulds *et al.* (2020) *opus*, we have information on food plant preference for 70 of the 74 mainland species of hawkmoths – by far the best such dataset for any Australian moth family. At the level of the plant family, 26 species exploit members of the Rubiaceae, 11 the Vitaceae, eight the Araceae and six for each of the Apocynaceae, Bignoniaceae and Lamiaceae. Smaller numbers of species exploit a wide range of other plant families. Of particular note is that only single species are known to feed on the characteristically Australian families Myrtaceae and Proteaceae. Both of these moth species are in the endemic smerinthine genus *Coequosa*: *australasiae* larvae feed on *Angophora*, *Corymbia* and *Eucalyptus* (i.e. *Eucalyptus s.l.*); *triangularis* on *Banksia*, *Grevillea*, *Hakea*, *Macadamia*, *Persoonia* and *Stenocarpus*. Other families that, in contrast, have radiated widely in Australia have frequently done so by specializing in one or more of these plant families.

Figures 4.5 and 4.6 explore the distribution of hawkmoths within Australia. Figure 4.5 summarizes the number of species across the continent by plotting species-richness isopleths. These show clearly a trend of increasing species richness from the depauperate south west to the species-rich tropical north east – a pattern repeated in many taxa across the Australian insect fauna. Of course, as discussed in section 4.4.3 above, such isopleth maps reflect collecting and research activity as well as actual underlying distributions but, given the visibility and popularity of the sphingids, we can have confidence that the pattern is real even if the exact placing of the isopleths lines may be debated. At the species level a number of within-continent distribution patterns repeat themselves. A subset of eight species are highly mobile and probably migratory, being recorded across the entire continent, although only three reach Tasmania. Three additional species have very wide distributions across the northerly and easterly two-thirds of the continent. Figure 4.6 shows a range of other distributions exhibited by different species of sphingids. Some are commonplace, others are unusual 'special' distributions. These are discussed in the figure legend.

Fig. 4.5. Isopleths showing species richness of the Sphingidae collated from the distribution maps of Moulds *et al.* (2020).

4.4.5.1 The biogeographic scenario

I conclude that the key to understanding the Australian sphingid fauna as a whole is to look north, both within and beyond our continental bounds. A parsimonious explanation may well be that many elements of our fauna evolved to our north before and including the existence of the Sahul supercontinent with ancestral species seeding the continent in a series of invasions from that time on. This process of north to south invasion and colonization may be ongoing as evidenced by the southward spread of *Acherontia lachesis* into the Torres Strait and the record of *Marumba timora* in the Kimberleys. In general, the thesis presented by Beck *et al.* (2006b) that dispersalist mechanisms shape sphingid faunas is upheld when we extend their arguments to the Australian fauna. Of course, there are exceptions and the small coterie of species that are endemic to the centre and south west are cases in point. The failure (with just two exceptions) to exploit Myrtaceae and Proteaceae suggests two things. First, a physiological limitation may restrict the use of these chemically challenging plants as larval food plants (although I note other sphingids exploit the equally chemically challenging Apocynaceae). And second, the sheer mobility of these species may overwhelm isolationist tendencies and inhibit speciation in general.

4.5 Environmental Change and Assemblages of the Future

We have encountered and defined ecological, vicariance, dispersalist, historical and even synthetic biogeography; what remains elusive is something we might call *predictive biogeography*. There is considerable evidence that assemblages of moth species track spatial changes in the environment (e.g. Kitching *et al.*, 2023; see also Chapter 8). A smaller body of work shows how such assemblages change with time (Chen *et al.*, 2009, 2011; L.A. Ashton *et al.*, pers. commun.) even if those changes are systemic declines (Bell *et al.*, 2020; Wagner, 2020). Environments change through time and species assemblages

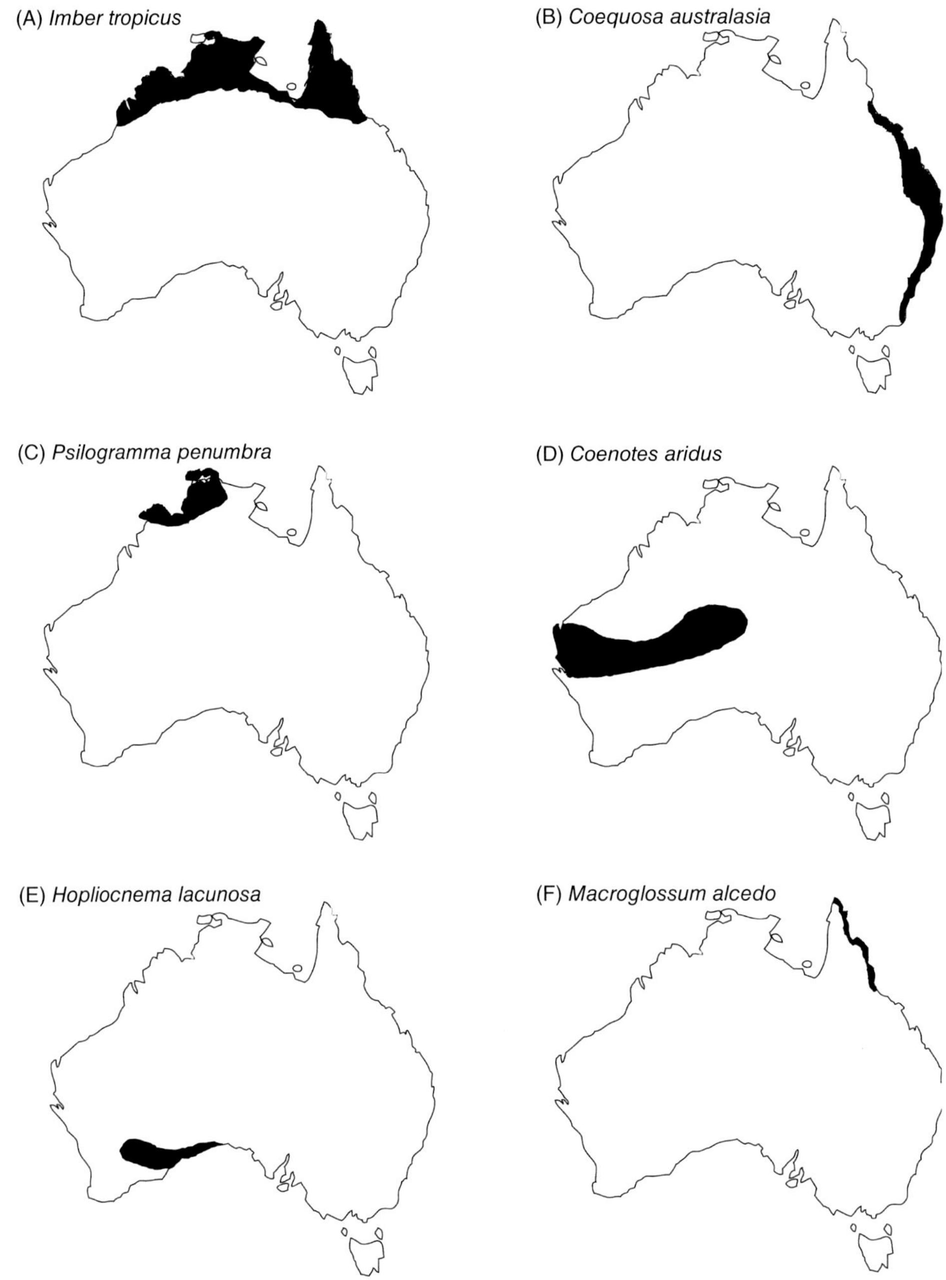

Fig. 4.6. Selected species' distributions of the Australian Sphingidae. (A) *Imber tropicus*, a 'general tropical' distribution. (B) *Coequosa australasia*, a coastal, temperate to tropical distribution. (C) *Psilogramma penumbra*, a 'Kimberleys' distribution. (D) *Coenotes aridus*, a rare 'arid zone only' distribution. (E) *Hopliocnema lacunosa*, a unique south coast, arid distribution. (F) *Macroglossum alcedo*, a 'Cape York' distribution. Simplified from the point records in Moulds *et al.* (2020).

mirror these changes. For the moths, these changes will often be cascading effects driven by changes in their larval food plants.

As Wagner (2020) points out, anthropogenic impacts have been changing the environments of moths (and everything else) since prehistoric times: clearing forests, burning grasslands, introducing (or removing) large herbivores, and so on. As *Homo sapiens* colonized the world, became more abundant and, in particular, engineered more and more intrusive and pervasive cultural and technological mechanisms, so environmental change accelerated. With currently observed anthropogenic climate change, these environmental changes and their biotic consequences are at a new high and continue to accelerate. We have already observed shifts in distribution of Lepidoptera tracking these climatic changes. From a biogeographic viewpoint this means that new mixes of species are appearing.

The key processes of ecosystem functioning are photosynthesis, herbivory, predation, mutualism and decomposition. These processes occur across all plant-based ecosystems regardless of their constituent species. It is functioning ecosystems that provide the much touted 'ecosystem services' which humans (and other organisms) rely upon as free goods provided by nature. Current conservation policy is based on the idea that we must make every effort to maintain the species mixes that we humans first encountered at some relatively recent time in the past. At the same time, much of the rhetoric as to *why* we should preserve biodiversity revolves around the ecosystem services idea.

I suggest (and predict) that future biogeographers will need to identify and monitor 'new' ecosystems made up of novel sets of species. Species characteristic of current ecological communities will disappear (or, optimistically, move elsewhere). Others commonly regarded as exotics will come to play vital ecological roles in maintaining new mixes of herbivores (such as moths) and the higher trophic levels these in turn support (Pearce, 2015).

Predicting which species will play these roles in the future is uncertain. Certainly, we may expect that currently highly specialized species, so dependent on other species for their wellbeing, will be most vulnerable. Robust generalists will dominate as must surely have happened after previous major and minor mass extinction events. Nevertheless, the inexorable processes of evolutionary change will roll on and new specialists will arise. I confidently assert that the moths will remain an important part of any new mix that arises. There will remain a role for biogeographers to monitor and explain these changes. Whether or not there will be biogeographers available to do so is moot.

Acknowledgement

I am grateful to Dr Ian Kitching for providing information about sphingid faunas used in my analyses.

References

Beck, J. and Kitching, I.J. (2007) Correlates of range size and dispesal ability: A comparative analysis of sphingid moths from the indo-Australian tropics. *Global Ecology & Biogeography* 16, 341–349.

Beck, J., Kitching, I.J. and Linsenmair, K.E. (2006a) Measuring range sizes of south-east Asian hawkmoths (Lepidoptera: Sphingidae): Effects of scale, size and phylogeny. *Global Ecology & Biogeography* 15, 339–348.

Beck, J., Kitching, I.J. and Linsenmair, K.E. (2006b) Determinants of regional species richness: an empirical analysis of the number of hawkmoth species (Lepidoptera: Sphingidae) on the Malesian archipelago. *Journal of Biogeography* 33, 694–706.

Beck, J., Kitching, I.J. and Linsenmair, K.E. (2006c) Extending the study of range-abundance relations to tropical insects: Sphingid moths in south-east Asia. *Evolutionary Ecology Research* 8, 1–14.

Belbin, L., Wallis, E., Hobern, D. and Zerger, A. (2021) The atlas of living Australia: History, current state and future directions. *Biodiversity Data Journal* 9, e65023.

Bell, J.R., Blumgart, D. and Shortall, C.R. (2020) Are insects declining and at what rate? An analysis of standardises, systematic catches of aphid and moth abundances across Great Britain. *Insects Conservation & Diversity* 13, 115–120.

Benton, M.J. (2003) *When Life Nearly Died: The Greatest Mass Extinction of All Time*. Thames & Hudson, London.

Blackburn, T.J., Olsen, P.E., Bowring, S.A., McLean, N.M., Kent, D.V. *et al.* (2013) Zircon U-Pb geochronology links the end Triassic extinction with the Central Atlantic magmatic province. *Science* 340, 941–945.

Boyle, M.J., Bonebrake, T.C., Dias da Silva, K., Dongmo, M., França, F.M. *et al.* (2025) The causes and consequences of insect decline in tropical forests. *Nature Reviews Biodiversity* 10.1038/s44358-025-00038-9.

Braga, M. and Janz, N. (2021) Host repertoires and changing insect–plant interactions. *Ecological Entomology* 46, 1241–1253.

Cardosa, A.W., Malhi, Y., Oliveras, I., Lehmann, D., Ndong, J.E. *et al.* (2019) The role of forest elephants in shaping forest-savannah coexistence. *Ecosystems* 23, 602–616.

Chen, I-C., Shiu, H-J., Benedick, S., Holloway, J.D., Chey, V.K. *et al.* (2009) Elevation increases in moth assemblages over 42 years on a tropical mountain. *Proceedings of the National Academy of Sciences* 106, 1479–1483.

Chen, I-C., Hill, J.K., Holloway, J.D., Benedick, S., Chey, V.K. *et al.* (2011) Asymmetric boundary shifts of tropical montane Lepidoptera over four decades of climate warming. *Global Ecology and Biogeography* 20, 34–45.

Cho, S., Mitchell, A., Mitter, C., Regier, J., Matthews, M. and Robertson, R. (2008) Molecular phylogenetics of heliothine moths (Lepidoptera: Noctuidae: Heliothinae), with comments on the evolution of host range and pest status. *Systematic Entomology* 33, 581–594.

Common, I.F.B. (1994) *Oecophorine Genera of Australia I. The Wingia group (Lepidoptera: Oecophoridae)*. Monographs on Australian Lepidoptera 3. CSIRO Publishing, Collingwood, Victoria.

Common, I.F.B. (1997) *Oecophorine Genera of Australia II. The* Chezala, Philobota *and* Eulechria *groups (Lepidoptera: Oecophoridae)*. Monographs on Australian Lepidoptera 5. CSIRO Publishing, Collingwood, Victoria.

Common, I.F.B. (2000) *Oecophorine Genera of Australia III. The* Barea *group and unplaced Genera (Lepidoptera: Oecophoridae*). Monographs on Australian Lepidoptera 8. CSIRO Publishing, Collingwood, Victoria.

Cranston, P.S. and Naumann, I.D. (1991) Biogeography. In: Naumann, I. (ed.) *Insects of Australia*, 2nd edn. Melbourne University Press, Carlton, Victoria, pp. 180–197.

Croizat, L. (1958) *Panbiogeography*. Privately Published, Caracas.

Darlington, P.J. (1965) *Biogeography of the Southern End of the World: Distribution and History of Far-southern Life and Land, with an Assessment of Continental Drift*. Harvard University Press, Cambridge, MA.

Diehl, E.E. (1980) *Heterocera Sumatrana 1. Sphingidae*. Classey, London.

Dirzo, R., Young, H.S., Ceballos, G., Isaac, N.J.B. and Collen, B. (2014) Defaunation in the anthropocene. *Science* 345, 401–406.

Dufour, P., Lees, A.C., Gilroy, J. and Crochet, P-A. (2024) The overlooked importance of vagrancy in ecology and evolution. *Trends in Ecology and Evolution* 39, 19–22.

Eeles, P. (2023) *British and Irish Butterfly Rarities: Migrants, Extinctions and Introductions*. Pisces Publications, Newbury, UK.

Fernández-Rubio, F. and Nel, A. (2000) *Neurosymploca? oligocenica* a new fossil species of Lepidoptera Zygaenoidea of the Oligocene of Cérestes (Luberon, France). *Boletín de la Sociedad Entomólogica Aragonesa* 27, 7–16.

Gordon, C.E., Greve, M., Henley, M., Allin, P. and Svenning, J.-C. (2023) Elephant rewilding affects landscape openness and fauna habitat across a 92-year period. *Ecological Applications* 33, e2810, 1–17.

Gibbs, G. (2006) *Ghosts of Gondwana. The History of Life in New Zealand*. Craig Potton Publishing, Nelson, New Zealand.

Gibbs, G.W. (2010) Micropterigidae (Lepidoptera) of the Southwestern Pacific: A revision with the establishment of five new genera from Australia, New Caledonia and New Zealand. *Zootaxa* 2520, 1–48.

Gibbs, G. (2016) *Ghosts of Gondwana. The History of Life in New Zealand*, 2nd edn. Potton & Burton, Nelson, New Zealand.

Gressitt, J.L. (1974) Insect biogeography. *Annual Review of Entomology* 19, 293–321.

Grimaldi, D. and Engel, M.S. (2005) *Evolution of the Insects*. Cambridge University Press, Cambridge, MA.

Hennig, W. (1950) *Grundzüge einer Theorie der Phylogenetischen Systematik*. Deutsche Zentralverlag, Berlin.

Holloway, J.D. (1977) *The Lepidoptera of Norfolk Island, Their Biogeography and Ecology*. Series Entomologica 13. W. Junk, The Hague.

Holloway, J.D. (1979) *A Survey of the Lepidoptera, Biogeography and Ecology of New Caledonia*. Series Entomologica 15. W. Junk, The Hague.

Holloway, J.D. (1986–2011) *The Moths of Borneo*, 18 vols. Southdene Sdn. Bhd., Kuala Lumpur.

Holloway, J.D. (1987) *The Moths of Borneo: Superfamily Bombycoidea: families Lasiocampidae, Eupterotidae, Bombycidae, Brahmaeidae, Saturniidae, Sphingidae*. Southdene Sdn. Bhd., Kuala Lumpur.

Holloway, J.D. (1993) Aspects of the biogeography and ecology of the Seram moth fauna. In: Macdonald, A.A., Edwards, I.D. and Proctor, J. (eds) *A Natural History of Seram*. Intercept, Andover, pp. 91–114.

Holloway, J.D. and Nielsen, E.S. (2003) Biogeography of the Lepidoptera. In: Kristensen, N.P. (ed.) *Lepidoptera, Butterflies and Moths,* 4(1) *Handbook of Zoology*. De Gruyter, Berlin & New York, pp. 423–462.

Inoue, H., Kennett, R.D. and Kitching, I.J. (1997) *Moths of Thailand 2. Sphingidae*. Chok Chai Press, Bangkok.

John, B. (ed.) (1979) *The Winters of the World: Earth under the Ice Ages*. Jacaranda, Milton, Queensland.

Jorge, L.R., Prado, P.I., Almeida-Neto, M. and Lewinsohn, T.M. (2014) An integrated framework to improve the concept of resource specialisation. *Ecology Letters* 17, 1342–1350.

Kawahara, A.Y. and Barber, J.R. (2015) Tempo and mode of antibat ultraspound production and sonar jamming in the diverse hawkmoth radiation. *Proceedings of the National Academy of Sciences* 112, 6407–6412.

Kawahara, A.Y., Mignault, A.A., Regier, J.C., Kitching, I.J. and Mitter, C. (2009) Phylogeny and biogeography of hawkmoths (Lepidoptera: Sphingidae): Evidence from five nuclear genes. *PLOS One* 4, e5719.

Kawahara, A.Y., Plotkin, D., Espeland, M., Meusemann, K., Toussaint, E.F.A. *et al.* (2019) Phylogenomics reveals the evolutionary timing and pattern of butterflies and moths. *Proceedings of the National Academy of Sciences* 116, 22657–22663.

Kim, S., Kaila, L. and Lee, S. (2016) Evoloution of life mode of Oecophoridae (Lepidoptera: Gelechioidea) inferred from molecular phylogeny. *Molecular Phylogenetics and Evolution* 101, 315–335.

Kolbert, E. (2014) *The Sixth Extinction*. Henry Holt, New York.

Kitching, I.J. and Cadiou, J.-M. (2000) *Hawkmoths of the World: An Annotated and Illustrated Revisionary Checklist*. Natural History Museum, London.

Kitching, R.L. (1981) The geography of the Australian Papilionoidea. In: Keast, A. (ed.) *Ecological Biogeography of Australia*. W. Junk, The Hague, pp. 977–1005.

Kitching, R.L. and Dunn, K.L. (1999) The biogeography of Australian butterflies. In: Kitching, R.L., Scheermeyer, E., Jones, R.E. and Pierce, N.E. (eds) *Biology of Australian Butterflies*. Monographs on Australian Lepidoptera 6. CSIRO Publishing, Collingwood, Victoria, pp. 53–74.

Leakey, R.E. and Lewin, R. (1995) *The Sixth Extinction: Patterns of Life and the Future of Humankind*. Doubleday, New York.

Levinton, J.S. (2013) Extinction in the fossil record. In: *Encyclopedia of Biodiversity*. 2nd edn. Elsevier, Amsterdam, pp. 413–425.

Lödl, M. (1993) The Malaysian hawkmoths: An annotated and illustrated checklist. *Tropical Lepidoptera* 4, 3–6.

Macarthur, R.H. and Wilson, E.O. (1967) *The Theory of Island Biogeography*. Princeton University Press, Princeton, NJ.

Macphail, M. and Thornhill, A.H. (2016) How old are the eucalypts? A review of the microfossil and phylogenetic evidence. *Australian Journal of Botany* 64, 579–599.

Matthews, M. (1999) *Heliothine Moths of Australia: A Guide to Pest Bollworms and Related Noctuid Groups*. Monographs on Australian Lepidoptera 7. CSIRO Publishing, Collingwood, Victoria.

Mey, W., Léger, T. and Lien V.V. (2021) New taxa of extant and fossil primitive moths from South-East Asia and their biogeographic significance (Lepidoptera, Micropterigidae, Agathiphagidae, Lophocoronidae). *Nota Lepidopterologia* 44, 29–56.

Miric, M., Efetov, K.A., Tarmann, G.M., Chiocchio, A., Heikkilä, M. *et al.* (2024) First comprehensive higher level phylogeny of Zygaenidae (Lepidoptera) including estimated ages of the major lineages and a review of known zygaenid fossils. *Systematic Entomology* 49, 610–623.

Moulds, M.S., Tuttle, J.P. and Lane, D.A. (2020) *Hawkmoths of Australia: Identification, Biology and Distribution*. Monographs on Australian Lepidoptera 13. CSIRO Publishing, Collingwood, Victoria.

Munroe, E. (1965) Zoogeography of insects and allied groups. *Annual Review of Entomology* 10, 325–344.

Naumann, C.M. (1985) Phylogenetiche Sistematik und klassic-typologische Systematic mit einigen Anmerkungen zu stammegeschichitlichen Fragen bei den Zygaenidae. *Mitteilungen Münchener Entomologische Gesellschaft* 74, 1–35.

Naumann, C.M. (1987) On the phylogenetic significance of two Miocene zygaenid moths (Insecta, Lepidoptera). *Paläontologische Zeitschrift* 61, 299–308.
Nielsen, E.S. and Common, I.F.B. (1991) Lepidoptera (moths and butterflies). In: Naumann, I. (ed.) *Insects of Australia*, 2nd edn. Melbourne University Press, Carlton, Victoria, pp. 817–915.
Pearce, F. (2015) *The New Wild: Why Invasive Species will be Nature's Salvation*. Icon, London.
Poisot, T., Bever, J.D., Nemri, A., Thrall, P.H. and Hochberg, M.E. (2011) A conceptual framework for the evolution of ecological specialisation. *Ecology Letters* 14, 841–851.
Rule, S., Brook, B.W., Haberle, S.G., Turney, C.S.M., Kersaw, A.P. and Johnson, C.N. (2012) The aftermath of megafaunal extinction: ecosystem transformation in Pleistocene Australia. *Science* 335, 1483–1486.
Schulte, P., Alegre, L., Arenillas, I., Arz, J.A., Barton, P.J. *et al.* (2010) The Chicxulub asteroid impact and mass extinction at the Cretaceous-Paleogene boundary. *Science* 327, 1214–1218.
Tarmann, G.M. (2004) *Zygaenid Moths of Australia. A Revision of the Australian Zygaenidae (Procridinae: Artonini)*. Monographs on Australian Lepidoptera 9. CSIRO Publishing, Collingwood, Victoria.
van Nieukerken, E.J., Kaila, L., Kitching, I., Kristensen, N.P., Lees, D.C. *et al.* (2011) Order Lepidoptera Linnaeus, 1758. In: Zhang, Z.Q. (ed.) Animal biodiversity: An outline of higher-level classification and survey of taxonomic richness. *Zootaxa*, 212–221. DOI: 10.11646/ZOOTAXA.3148.1.41
Wagner, D.L. (2020) Insect declines in the anthropocene. *Annual Review of Entomology* 65, 457–480.
Wallace, A.R. (1876) *The Geographical Distribution of Animals*. Macmillan, London.
Wang, X., Zhang, H., Kitching, I., Xu, Z.-B. and Huang, Y.-X. (2021) First mitogenome of subfamily Langiinae (Lepidoptera: Sphingidae) with its phylogenetic implications. *Gene* 789, 145667.
Waterhouse, D.F. (1991) Insects and humans in Australia. In: Naumann, I. (ed.) *Insects of Australia*, 2nd edn. Melbourne University Press, Carlton, Victoria, pp. 221–235.
Westaway, M.C., Olley, J.M. and Grün, R. (2016) At least 17000 years of coexistence: Modern humans and megafauna at the Willandra Lakes, south-eastern Australia. *Quaternary Science Reviews* 157, 206–211.
Xu, Y., Wang, Y., Li, L., Lu, N., Zhu, Y. *et al.* (2024) Plant–insect interactions across the Triassic–Jurassic boundary in the Sichuan basin, south China. *Frontiers in Ecology and Evolution* 11, 1338865.

5 Life Histories and Feeding Ecology

Tim R. New[1]* and Ying Y. Luo[2]
[1]*Department of Ecological, Plant and Animal Sciences, La Trobe University, Melbourne, Victoria, Australia;* [2]*yingyluo@outlook.com*

Abstract
The diversity of moth life histories and larval feeding habits enables them to exploit numerous environments and resources. Whilst moths are predominantly herbivores, particular moth taxa have broadened larval feeding habits to utilize foodstuffs such as keratin, animal faeces, fungi and others; a few have become predators or parasites. Larvae of many moths are polyphagous generalist feeders on a variety of different plants. Others are very specific and depend on particular food plants, to which they may be directed by female oviposition. Australia's moths include numerous endemic taxa that are linked intricately with native endemic vegetation, either as free-living external feeders or as miners or tunnellers inside the plants. This chapter is a summary of ecological and evolutionary variety and specializations among Australian moths as exemplified through their feeding ecology.

5.1 Introduction

Moths are most familiar as a predominant group of insect herbivores, with their larvae associated widely with plants across most terrestrial biomes. They are probably the most diverse radiation of phytophagous insects (Scoble, 1992) and, collectively, exploit almost all groups of gymnosperms and angiosperms, as well as bryophytes (Powell *et al.*, 1998). Larvae of the earliest Lepidoptera lineages are believed to have fed internally on non-vascular plants and their adults were mandibulate (see Chapter 3). As angiosperms proliferated, major changes and developments included externally feeding larvae and the evolution of the suctorial proboscis in adults – perhaps allowing them to disperse further and exploit new host plants (Kawahara *et al.*, 2019). These developments paved the way for increasing moth diversity as flowering plants themselves diversified, and led to intricate functional relationships between Lepidoptera and plants.

Perhaps 98% of all Lepidoptera are plant feeders, predominantly on angiosperms (Wagner and Hoyt, 2022). The innumerable variations on the 'insect–plant nexus', with a host of subtle (and often specialized) variations that encompass exploitation of almost any part of any accessible plant species, are complemented by evolutionary excursions leading particular taxa of moths to adopt larval feeding by predation, as parasites, and on foodstuffs such as fungi, lichens, general debris, keratin, faeces and stored products.

Whatever their diet, larvae of many moths have very specific feeding requirements and an

*Corresponding author: t.new@latrobe.edu.au

DOI: 10.1079/9781800627680.0005

associated gut microbiome (Shao *et al.*, 2024). The distributions of monophagous plant-feeding taxa (perhaps around three-quarters of all moth species) are restricted by those of their particular food plants; other moth species are more flexible and oligophagous (feeding on a limited array of, usually, related plants) and yet others are polyphagous, generalist feeders on numerous plant taxa. Many monophagous species are, by definition, 'specialists' with their fate dependent on that of their host and are vulnerable to its decline or loss; some show long-term chemical or behavioural associations with specific hosts. In some cases these species-to-species interrelationships are mutualisms by which both participants benefit and on which both depend. The specific pollination associations of Heliozelidae (Chapter11) are one such example.

This chapter is an overview of the evolutionary and ecological variety in the Australian moth fauna.

5.2 Setting the Scene

Diversification, including taxonomic radiations, of moths in association with plants characterizes the Australian fauna, some leading to rich endemic and poorly known complexes of difficult-to-distinguish taxa, for which biological knowledge is key to species recognition and diagnosis. Ecologically dominant and diverse plant genera support rich lepidopteran faunas (Wagner and Hoyt, 2022). Australia's two most prominent plant genera, *Eucalyptus s.l.* and *Acacia s.l.*, support large numbers of endemic moth species. Numerous other endemic flora also support host-restricted moths. While much of this diversification rests on the feeding habits of the larval stages, adult habits also vary considerably. As examples, (i) many moths are nectar feeders and can be relatively long-lived, some being important pollinators (many of them with specific flowering plants, such as obligate relationships of many Heliozelidae with species of Rutaceae) or their feeding is constrained by the length of their proboscis, while (ii) others do not feed and generally have only a short adult life. Some implications of these contrasting strategies were discussed by Janzen (1984) in a thoughtful comparison of Neotropical hawkmoths (Sphingidae) and silkmoths (Saturniidae). Other feeding habits occur among adult moths. Some species feed by piercing ripe fruit (see Chapter 30) and a few seek exudations from mammal tear-ducts or, rarely, take blood (Banziger, 2021) as less familiar liquid foods.

The variety of life history patterns and needs (especially food) emphasizes the importance of understanding ecology in interpreting Australia's moth evolution and diversity, and patterns of distribution. Essentially, many moths have evolved in conjunction with unique and localized endemic floristic resources across a variety of topographic and climatic regimes. This has produced an intricate mosaic of combinations and radiations associated with different Australian biomes that collectively characterize the country's moth fauna. Many species have very limited distributions. Dietary restriction may dictate geographical restriction and, more broadly, features of native flora both influence and determine moth distributions in Australia. Many moth habitats are defined broadly by vegetation type within environments ranging from tropical to cool temperate and arid to humid. It follows that gradients in elevation, latitude or aridity involve changes in vegetation and other environmental features. Most moths are terrestrial, but a group of Crambidae (the Acentropinae) feed on emergent aquatic plants and are thus aquatic (see Chapter 17; Hawking, 2014).

The main life history gradients of moths are thus associated with (i) dietary breadth, with the range from larval polyphagy to monophagy; and (ii) seasonality, with active development influenced by synchrony with food and other resources and suitable climatic conditions, notably temperature and moisture. In common with other insects, moths may show well-defined and predictable seasonal patterns of development. These are often familiar to hobbyist collectors who know that adults of 'species A' will appear around 'time X' each year, with well-defined generations in some cases interrupted by periods of diapause or related dormancy, or develop continuously with multiple generations. Development may be continuous and take only a few weeks, but can extend to 2–3 years of larval feeding (some Castniidae) or considerably longer (some large Hepialidae and Cossidae). Development may be interrupted by diapause that can sometimes become protracted and may be an adaptation to counter unpredictably adverse conditions: 12 years was reported for *Agathiphaga* (Upton, 1997). Weather may be influential – autumn rains are widely a cue for mass emergence of some moths, in some cases marking the imminent availability of fresh food plant growth suitable for oviposition and larval growth.

In addition to Australia's rich native flora, the array of plants reaching Australia since European settlement has provided further opportunities for host expansion, some involving switches to very different hosts. The term 'host switching' implies change of host plants, but more often is really host range expansion as opportunities arise from newly encountered, frequently introduced, plants. Some native moths, for example, have exploited Australia's predominant softwood crop, the imported Monterey pine (*Pinus radiata*), as larval food, even becoming significant pests in plantations. Other cases involve plants more closely related to native hosts; the apparent use of the declared noxious weed Chilean needle grass (*Nassella neesiana*) by the threatened golden sun moth (*Synemon plana*, Castniidae: see Chapter 35) poses a dilemma for weed control. Crop and ornamental plants can also be vulnerable to both native and exotic moth species (some becoming pests) and, conversely, Australian moths can become concerns elsewhere. In New Zealand, the polyphagous painted apple moth (*Teia anartoides*, Erebidae) was found in 1999 and prompted an eradication campaign for fears of its impacts on native vegetation (Suckling *et al.*, 2007). Only a few moth species in Australia are aliens, but some are well-established and significant pests. 'Clothes moths' and 'pantry moths' (Tineidae) are domestic pests. Codling moth (*Cydia pomonella*) and the noctuid *Spodoptera frugiperda* are economically important feeders on orchard crops and cereals, respectively. The latter, the fall armyworm, has rapidly become a major pest (Volp *et al.*, 2022). 'Wax moths' (Pyralidae) are pests in the apiary industry. The list of examples goes on (see Chapter 23). Continual biosecurity vigilance is needed to detect arrival of others, such as the gypsy moth (*Lymantria dispar*, Erebidae), that could become major pests.

As a further widespread trait, the amount of food available to a larva may be very limited and, accordingly, limiting. Living in a single leaf as a miner and being unable to move elsewhere clearly restricts the total food available as well as the time over which it occurs. This restricts the size the moth may attain. Many leaf miners are very small moths. Likewise, developing in a single koala scat on the forest floor (e.g. *Telanepsia*; Common and Horak, 1994) restricts the moth larva's feeding universe. In contrast, feeding on dead wood or inside living trees ensures a considerable surplus of potential food that persists for years, and larvae of some wood-feeding moths take correspondingly long periods to develop and become large. Some of the 'wood moths' (Cossidae) and 'swift moths' (Hepialidae) are among the heaviest Australian moths (see Chapters 10 and 15). *Endoxyla funereus* females are reported to be the world's heaviest moth, at around 30 g (see Chapter 26).

The dynamics of food selection, availability and use are key to understanding moth biology. Feeding externally on plants can also impose seasonality in life history development. Particular food needs such as flush-growth foliage, flowers or seeds may be available only over limited periods (so synchrony or longevity on the part of the moth is vital). In contrast, more generalized foliage feeders have a longer opportunity to encounter and use abundant food resources. The complex interactions between insect herbivores and plant chemistry have engendered a vast literature on so-called secondary plant chemicals as attractants and feeding deterrents to herbivores. Some such chemicals make the plant host distasteful or toxic to the herbivores, and lead to discussions of an 'evolutionary arms race' with continuing reciprocal adaptation and counteradaptation between consumer and food plant. Physical characters (such as toughness) also greatly influence food selection and use, and the seasonality and specificity of feeding.

For toxic plant chemicals, generalist feeders can evolve broad-spectrum mechanisms of detoxification that can apply across a variety of compounds, while specialist feeders on hosts containing highly toxic compounds may lose the ability to feed on many other hosts, exacerbating their functional specificity (Chialvo *et al.*, 2018). Food selection may be very subtle. Larval leaf miners, for example, may confine their feeding to particular tissue layers and so be very selective in their food choices (Body *et al.*, 2015). As an example from the Gracillariidae, larvae of the introduced oak leaf miner (*Phyllonorycter messaniella*) confine their feeding in oak leaves to the upper region, so avoiding the tannin-accumulating layer of palisade cells near the underside of the leaf.

In general, the main drivers of ecological restrictions imposed upon phytophagous moth larvae are the need to avoid harmful effects of plant defensive chemicals and to avoid attack by predators and parasitoids (see Chapter 6), so achieving what has been termed enemy-free space (Jeffries and Lawton, 1984). Choice of host plants (or other dietary needs) must accommodate these adversities and ensure that adequate nutrients are obtained within microclimate regimes suitable for development. The complexities of larval life reflect that caterpillars are both major consumers of plant material and major prey for a variety of predators – their ecological position of being 'in the middle' of tritrophic interactions (Marquis and Koptur, 2022) necessitates many very individualistic structural and behavioural developments.

The usual conventional understanding of moths based on larval and adult life styles should be augmented by attention to the intervening pupa stage. Pupation may last for several months and often is the longest phase of the life cycle. The pupal stage for many species is a refuge over inclement parts of the year. Pupae are generally on or close to larval feeding sites, but may be further afield. Again, the contrast between Saturniidae and Sphingidae is instructive (Janzen, 1984). Silkmoth larvae (Saturniidae) usually construct tough silken cocoons, many incorporating leaves, and pupate on the larval food plant. In contrast, hawkmoth caterpillars (Sphingidae) move to the ground and pupate in soil, tunnelling for at least several centimetres (sometimes up to 30–60 cm deep). Soil pupation is widespread among moths. Legal (2023) suggested that the advantages of this habit (in principle shared with pupation in other sheltered sites) include protection from predators (by camouflage and/or physical protection) and buffering against variations in temperature and humidity. Pupation in ground litter necessitates larvae descending or dropping from their food plants – either at the final larval instar or earlier and, in many taxa, using fallen foliage or other materials in their cocoon construction.

5.3 Major Habitats

Every vegetated terrestrial ecosystem supports moths, many of which are sufficiently restricted to be referred to typically as 'grassland moths', 'forest moths' or such other epithet that summarizes their key broad habitat limitations or preferences. Within each, monophagous and more generalist feeders intermingle. Many moths in Australia are poorly known, and their diets can only be inferred from related species for which we do have information. More broadly across free-living herbivorous insects, several problematical situations arise in understanding the food preferences and variety of particular species. First, for many species, nothing is known of larval feeding habits, and the species are known only from the adult stage. For some others, some larval feeding records exist but food plants are identified imprecisely ('grass', 'eucalypt', 'bipinnate wattle', etc.). Sometimes the known food plants of a given species are only part of the actual diet spread, with other foods unrecorded. Often, errors of plant identification can be perpetuated through repeated citation sometimes complicated by changes in plant taxonomy. On occasion, the association may be reported from a casual occurrence of a larva on a plant, with no evidence of feeding. Finally, the moth species itself may be imprecisely identified, for example as a member of a sibling group or taxonomic complex, and voucher specimens may not be available for examination. These issues are discussed further in Chapter 8.

These concerns span many taxa. By way of example, consider the Larentiinae (Geometridae) in Australia. Australian representatives of this subfamily are largely endemic and include more than 45 genera and about 270 described species, for which a comprehensive overview (Schmidt, 2016) yielded larval host plant records for only 51 species, with almost all those records isolated and opportunistic rather than accumulated systematically. They remain 'indicative' rather than comprehensive. The unique compendia of larval host plant records for South Australian geometrids by McFarland (1979, 1988) remain exemplary exceptions to the more general situation.

Many older host records for Australian moths cannot be confirmed. Indeed, the early stages of most species have not been associated clearly with the adult stage or described clearly. While broad habitat associations may be clear (many Oecophoridae have been reared from boxed samples of forest leaf litter, or light trap catches of moths reflect locally predominant vegetation), more precise needs remain unknown or at best obscure. Molecular barcodes and molecular identification of gut contents potentially could allow larval/adult associations to be determined without time-consuming captive breeding but this process remains expensive and technically demanding.

Australia's major biomes each support characteristic moths. Castniidae are restricted to grasslands, in which larvae feed underground on grass or sedge roots (see Chapters 15 and 35). The largely endemic Anthelidae split into two main ecological strategies – one group are grassland inhabitants and the larger is arboreal, mostly associated with eucalypts (Zwick, 2008), with the 'flexibility' of some species enabling them to feed on introduced conifers. Forests and savannah ecosystems, both very variable in vegetation composition, all support assemblages of characteristic moths, with their richness at any site reflecting the underlying diversity of plant life.

5.4 Modes of Life

The largest number of moth species are typified by free-living larvae that roam on vegetation and feed on living foliage as external 'chewers'. Many modifications to this pattern occur, and transitions in lifestyle as larvae develop are not uncommon. This can include change of larval form – heteromorphosis – such as change from cylindrical to flattened body shape and loss of prolegs with adoption of mining, or the converse if leaving mines to become free-living or change feeding habits. For example, the final instar of *Ogmograptis scribula* and its relatives – the iconic 'scribbly gum' moths (Bucculatricidae) - is transformed from an earlier legless borer of bark tissue to an active, legged larva feeding on callus within the earlier constructed tracks as an 'obligate behavioural switch' (Horak *et al.*, 2012; see Chapter 34).

We summarize here the variety of 'other' feeding strategies and demonstrate, *inter alia*, the difficulties of categorizing larval feeding habits reliably. The tabulations by Powell *et al.* (1998) summarize global trends across all moth taxa, and suggest some generalities relevant to the Australian fauna. Most notably, feeding on lower plants (bryophytes, gymnosperms) is confined largely to more primitive moths, whereas almost all larger superfamilies of 'higher moths' feed predominantly or solely on angiosperms, with no such group associated primarily with non-flowering plants. Fern feeding occurs in several families but is relatively uncommon, some caterpillars (Stathmopodidae: *Calicotis*, *Stathmopoda*) feeding on spores on the underside of fern fronds.

Larvae of most moths pass their developmental period on or near their food plants, and vary widely in appearance; some are well camouflaged to resemble vegetation, others brightly coloured (so-called 'warning colouration' indicative of supposed distastefulness) or are covered in spines or urticating hairs as protection against predators. Others characteristically construct some form of protective shelter, a habit referred to by Marquis *et al.* (2022) as 'construct building' and using the plant as 'a container, platform or scaffold'. Many Oecophoridae (mallee moths) live in cases of leaf fragments held together by silk (Chapter 12). Most larval Psychidae (bagworms or casebuilders) construct elaborate shelters that may include lengths of twig, and are relatively species specific (see Chapter 14). Many species of Tortricidae 'roll' leaves of their host, using silk to construct tunnel retreats from a single leaf. Others are leaf tiers, using silk to tie two or more leaves together as shelters. More extensive silk production can lead to webs covering individuals or groups of larvae, with 'webworms' (such as *Herpetogramma licarsisalis*, Crambidae) being occasional pests of lawns. Larvae of the notodontid *Epicoma protrahens* are initially communal under webs but later become solitary and finally communal again. Some web-producing olethreutines (Tortricidae) incorporate faeces and vegetable material into their coverings. Some others (representing many families) include urticating hairs that enhance protection from natural enemies and can cause urticarial 'itch' if handled by people. More broadly, Battisti *et al.* (2024) reported venomous Lepidoptera in 576 species, representing 14 families.

All the habits noted above are globally widespread.

5.4.1 Litter feeders

Larvae of most Lepidoptera feed on living foliage. Some, however, including many oecophorid larvae, are saprophagous and characteristically feed on dead leaf litter on the forest floor. They are among the few insect groups that have thrived on this dry nutrient-poor food. They are considered important comminutors of otherwise long-persistent fallen eucalypt foliage. Many oecophorids pass their earliest stages on trees and later drop to the ground, where their level of food species selection is unknown; they are classed mainly as 'litter feeders' without further qualification. The Oecophoridae, associated closely with eucalypt forest and woodland, is by far the most diverse family of Australian moths, encompassing as many as 5000 species (see Chapter 12; Common, 1994).

5.4.2 Enclosed environments

5.4.2.1 Leaf mining

Some form of 'leaf mining' occurs in about 35 moth families, in some of which both free-living and mining species occur. The two largest mining families (Nepticulidae, Gracillariidae) are both diverse in Australia with numerous undescribed species and only very sporadic host records. Leaf mining has been defined as 'consuming live foliage while simultaneously dwelling inside it' (Connor and Taverner, 1997). Mines are usually visible as linear tunnels or blotches on the leaf surface, and may be occupied throughout larval life or abandoned as caterpillars later become free-living.

Three major groups of hypotheses have been advanced to explain the adaptiveness of leaf-mining: escape or protection from natural enemies (parasitoids, predators, pathogens); protection from the physical environment (dislodgement from the food plant during rough weather, desiccation, ultraviolet radiation); and avoidance of plant defences (by selective feeding within leaves to use the most nutritious and least toxic tissues, avoiding those with noxious or deterrent properties). The form of the mine is often characteristic, reflecting the unique ecologies and morphologies of each species. *Phyllocnistis* mines, for example, are always shallow and visible on only one side of the leaf (generally the upper surface), as their larvae never develop into rotund 'tissue feeders', remaining flattened semi-liquid feeders just below the epidermis.

5.4.2.2 Gall inhabitants

Gall forming has arisen may times in insects, through induction of hyperplasia by feeding, in many cases by surface feeders or stem miners. Leaf miners may also induce galls (Guiget *et al.*, 2019). Gall-inducing moths are poorly known and the complexity of gall insect communities obscures the true role of many insect inhabitants that emerge from them; some, indeed, may simply be invaders of long-dead galls. Both in Australia (New, 1982) and South Africa (McGeoch, 1993), moths have been reared from large woody fungus galls (formed by *Uromycladium* and *Ravenelia*, respectively, both on *Acacia*), in which the larvae feed. The Gondwanan gall-inducing Cecidosidae, however, are not known from Australia, although present in New Zealand.

5.4.2.3 Wood borers and tunnellers

Larvae of many families of moths bore into wood, ranging from twigs to tree trunks, and many taxa have very specific habits. The large Cossidae (wood moths) and Hepialidae (swift moths) are both notable radiations in Australia. *Endoxyla cinereus* (Cossidae), with a development time of 3 years, can cause losses in eucalypt plantations, through directly weakening small trees and increasing their susceptibility to damage by cockatoos seeking the larvae as food (see Chapter 26; Thurman, 2022). Early instars of some *Aenetus* spp. (Hepialidae) feed in decaying leaf litter and fungi for several months before tunnelling into tree stems (see Chapter 10; Beaver *et al.*, 2020). Larvae of the fruit tree borer *Maroga melanostigma* (Xyloryctidae) feed in and on bark of *Acacia* spp. as well as on a range of fruit and ornamental trees, in some cases wholly ringbarking the plant (see Chapter 13). The iconic 'scribbles' of *Ogmograptis* spp. on eucalypts are the exposed feeding tunnels of the larvae, resulting from the unusual habit of larval feeding/tunnelling in the potential phellogen-forming tissue (see Chapter 34; Horak *et al.*, 2012).

5.4.3 Seed feeders/fruit feeders

As discussed in Chapter 3, the kauri moth, *Agathiphaga queenslandensis*, is an ancient taxon, its lineage extending over some 200 million years. Larvae feed in fallen seeds of the equally ancient kauri pine (*Agathis robusta*), itself much exploited for timber and now very localized. Larvae leave the hollowed-out seed before pupation (Dumbleton, 1952). The exotic codling moth, *Cydia pomonella*, is a serious pest of orchards. Larvae bore into apples to reach the seed on which they feed. Some noctuids also eat fruit; *Helicoverpa armigera*, for example, is sometimes a pest of tomatoes.

Complex obligate 'nursery-pollination mutualisms' are exemplified by some *Epicephala* spp. (Gracillariidae) in which females both pollinate the host plants and oviposit in the ovules of the plant, where the larvae subsequently feed (Finch *et al.*, 2021).

5.4.4 Keratin feeding

Many Tineidae ('clothes moths') can exploit fur, feathers, horns and a range of other foods only rarely eaten by animals. Several species of *Tinea* feed on wool or silk and are regarded as pests, and larvae of some species construct cases from their food.

5.4.5 Fungus and lichen feeding

Mycophagy is widespread among moths of many families, some having adopted fungal fruiting bodies or lichens as their primary larval food source. It has been suggested that the arctiine tribe Lithosiini (a predominantly lichen-feeding group) and others may use fungi as a source of defensive chemicals.

5.4.6 Feeding on faeces

Two genera of Oecophoridae are recorded feeding on marsupial scats in Australia. Five species of *Oxytheca* are found in scats of wombats, rock wallabies and wallabies. More than 30 species of *Telanepsia* (Oecophoridae) have been reared from koala and possum scats, both derived largely from eucalypt foliage, where the feeding habit could be a simple extension from the broader eucalypt litter-feeding habit in the family (see Chapter 12).

Parrot moths, *Trisyntopa* spp. (Oecophoridae), are associated with rare and localized species of parrots nesting in large termite mounds, where the moth larvae live in a mass of silken tunnels on the floor of the parrot nest, feeding on the excreta and faeces of the nestling chicks. The three species of *Trisyntopa* are all rare and each is specific to a particular parrot. Knowledge of their biology is incomplete. *Trisyntopa scatophaga* occurs with the golden-shouldered parrot, *Psephotus chrysopterygius* in Cape York Peninsula and *T. neossophila* is associated with the hooded parrot, *Psephotus dissimilis*, in the tropical savannahs of the Northern Territory. Adult moths do not feed, but the life cycle seems to be synchronized with that of the parrot host. Moths leave the parental nest to seek a newly formed nest in which to oviposit, so that larvae and parrot chicks occur at the same time (see Chapter 12; Cooney *et al.*, 2009).

5.4.7 Predation and ectoparasitism

A review of carnivorous Lepidoptera by Pierce (1995) summarized information then available and which has been augmented very little subsequently for Australian moths. Relatively unspecialized habits include sporadic cannibalism and necrophagy. Only two moth families are wholly carnivorous. Larvae of Cyclotornidae, an endemic Australian family with six described species (and others known), initially feed as parasites on Homoptera and later switch to feed on ant brood. Their biology, however, is poorly known, although most feed mainly on sessile prey. Epipyropidae, with some 32 described species (Meyer-Rochow *et al.*, 2023), is more widely distributed. Larvae are ectoparasitic on plant hoppers, leaf hoppers or cicadas, presumably feeding on the hosts' blood. Accordingly host availability is a key need and likely determinant of seasonal development. The full host range of any of the epipyropids remains unknown. Some species of *Stathmopoda* (Stathmopodidae) are predators of sedentary scale insects; one feeds on spider eggs. Pierce (1995) ventured that three traits found in close relatives of carnivorous Lepidoptera may have been important in the evolution of carnivory in Lepidoptera, namely scavenging on insect remains or spider eggs, mycophagy or lichen feeding (both widespread across several families) and associations with ants.

Australia appears not to harbour any moths that parallel some Hawaiian Geometridae (*Eupithecia* spp.) in becoming active ambush predators of small insect prey. Similarly, there are no Australian records of blood sucking on mammals as has been documented in the so-called 'vampire moths' (*Calyptra* spp., Erebidae) of south-east Asia and elsewhere (Banziger, 2021).

5.4.8 Nest parasitism

Moths of at least six families have been reported as parasites of the brood of social wasps, Vespidae (Shin *et al.*, 2023). One such pyralid, *Hypsopygia mauritialis*, occurs in northern Australia.

5.5 Finding Food

Two strategies for larval food discovery have rather different implications for development, and link with mode of oviposition. Females of many moth species actively select oviposition sites, and larvae eclose on or near suitable food, to which they may be committed by poor ability to move elsewhere. Larval food is thereby selected by the parental female. This strategy may be viewed broadly as linking with trends to specialization and low fecundity. Exceptions occur: first instar larvae of some species (perhaps most notably of some moths with flightless females as in some Erebidae: Lymantriinae) can disperse by 'ballooning' on silken threads, with their destination uncertain; if not polyphagous, mortality may be high. In contrast, oviposition may be more indiscriminate, and females do not select such precise oviposition sites. A perhaps extreme case is the low-level 'bombing runs' of some larger Australian Hepialidae (such as *Abantiades* spp.) whereby the non-feeding and short-lived females eject large numbers of eggs whilst flying over possibly suitable habitat areas. Larvae must then find food from the mixed vegetation available. In the rainy weather associated with the moths' emergence some eggs may be washed into crevices around trees, bringing them near to potential larval food within exposed tree roots. Nevertheless, mortality is likely to be high from failure to find food, and the very high fecundity of such moths (*Abantiades atripalpis* has the greatest reported fecundity of any moth, with an estimate of potentially laying up to about 44,000 eggs) presumably compensates for this. *Endoxyla* spp. (Cossidae) have a reported fecundity of around 20,000 eggs.

Larval food choice allows for greater generalization. Similarly, ectoparasites such as Epipyropidae, in which hatchling larvae must discover a host and mortality is likely to be high, may also produce many eggs: up to 1000 were noted by Meyer-Rochow *et al.* (2023) for the Japanese *Epipomponia nawai* (Epipyropidae).

5.6 Conclusion

As with many insect groups in Australia, knowledge of moth biology and life histories is substantially incomplete and any attempt at an embracing or definitive synthesis is fraught with uncertainty. Overgeneralization is inevitable when seeking to impose uniformity on such a biologically varied group, in which 'the next survey' is likely to yield previously unknown taxa or host records. Closely related species may have very different biologies and host plant ranges. Even for the best documented groups, the gaps in understanding counter any complacency flowing from extrapolating from information available. This is complicated further by the fact that many species are apparently both specialized and scarce (so, difficult to study or even find). Taxonomic uncertainty further confuses interpretation, thwarting attempts to compare and validate data across different studies or sites. In short, we can often not be certain that separate studies leading to cross-area comparisons of host range and other aspects of biology even involve the same moth species.

Many of the moth–plant associations, ranging from those entrenched on older floral elements to more recently developed interactions, demonstrate evolutionary subtleties that are still incompletely documented. The high levels of endemism in both moths and flora have provided many opportunities for development of intricate and obligate feeding relationshps, some based on long histories of co-evolution (or, at least, of association with continuing nuanced changes). This enhances specialization but also creates vulnerability if either partner becomes threatened or otherwise diminished.

Some intricate relationships between moths and plants are long-established in the geological record: feeding traces identified as recent genera of Nepticulidae and Gracillariidae, for example, occur in the mid-Cretaceous flora of North America, from 97 million years ago (Labandeira *et al.*, 1994). It follows that, even with very incomplete knowledge, conservation of remnant and ancient flora in Australia may benefit notable endemic moths that depend wholly on those plants. The association between ancient taxa is exemplified by the primitive enigma moth *Aenigmatinea glatzella* feeding on the conifer *Callitris gracilis* on Kangaroo Island. The ancient southern beech (*Nothofagus*) supports a number of endemic Geometridae and local endemics of several other families (see Chapter 18; McQuillan, 1993), and plant families with long fossil history (such as Proteaceae and grasses, Poaceae) host numerous endemic species (Austin *et al.*, 2004). Australia's vegetation has furnished series of unique ecological milieux for evolution of phytophagous insects and, in turn, of their predators and parasitoids (see Chapter 6). It is likely that very few species of native gymnosperm or angiosperm plants are truly free of moth consumers. Conversely, it is usually not clear what the full host plant range of any particular moth species may be, or what are the chemical or physical cues that influence plant selection.

The life histories, phenology, food plant range and distribution of most Australian moths are not known in detail; for many, early stages and adults have not been associated clearly, and larvae have not been described. Indeed, nothing is known of early stages and developmental needs of many moths and the species are known only from adults – many of them sufficiently intriguing or unusual to endorse the peculiarity of Australia's fauna and to stimulate searches for their early stages. However it is achieved, synchrony between consumers and food species is vital, and necessitates or drives patterns of seasonal incidence and development, many of which are only now becoming apparent. It is increasingly clear that taxonomic and ecological diversity are both very high among Australian moths, that much further study is needed to understand their variety and evolution, and that feeding ecology is a key determinant of their evolution and sustainability.

References

Austin, A.D., Yeates, D.K., Cassis, G., Fletcher, M.J., La Salle, J. *et al.* (2004) Insects 'down under' – Diversity, endemism and evolution of the Australian insect fauna: Examples from select orders. *Australian Journal of Entomology* 43, 216–234.

Banziger, H. (2021) *Vampire Moths. Behaviour, Ecology and Taxonomy of Blood-sucking* Calyptra. Natural History Publications, Kota Kinabalu, Malaysia.

Battisti, A., Walker, A.A., Uemura, M., Zalucki, M.P., Brinquin, A.-S. *et al.* (2024) Look but do not touch: The occurrence of venomous species across Lepidoptera. *Entomologia Generalis* 44, 29–39.

Beaver, E.P., Moore, M.D., Grehan, J.R., Velasco-Castrillon, A. and Stevens, M.I. (2020) Four new species of splendid ghost moths (Lepidoptera: Hepialidae: *Aenetus*) from Australia and Papua New Guinea. *Zootaxa* 4809, 449–474.

Body, M., Burlat, V. and Giron, D. (2015) Hypermetamorphosis in a leaf-miner allows insects to cope with a confined nutritional space. *Arthropod–Plant Interactions* 9, 75–84.

Chialvo, C.H.S., Chialvo, P., Holland, J.D., Anderson, T.J., Breinholt, J.W. *et al.* (2018) A phylogenomic analysis of lichen-feeding tiger moths uncovers evolutionary origins of host chemical sequestration. *Molecular Phylogenetics and Evolution* 121, 23–34.

Common, I.F.B. (1994) *Oecophorine Genera of Australia I: The Wingia group*. Monographs on Australian Lepidoptera 3. CSIRO Publishing, Collingwood, Victoria.

Common, I.F.B. and Horak, M. (1994) Four new species of *Telanepsia* Turner (Lepidoptera: Oecophoridae) with larvae feeding on koala and possum scats. *Invertebrate Taxonomy* 8, 809–828.

Connor, E.F. and Taverner, M.P. (1997) The evolution and adaptive significance of the leaf-mining habit. *Oikos* 79, 6–25.

Cooney, S.J.N., Olsen, P.D. and Garnett, S.T. (2009) Ecology of the coprophagous moth *Trisyntopa neossophila* Edwards (Lepidoptera: Oecophoridae). *Australian Journal of Entomology* 48, 97–101.

Dumbleton, L.J. (1952) A new genus of seed-infesting micropterigid moths. *Pacific Science* 6, 17–29.

Finch, J.T.D., Power, S.A., Welbergen, J.A. and Cook, J.M. (2021) Staying in touch: How highly specialised moth pollinators track host plant phenology in unpredictable climates. *BMC Ecology and Evolution* 21,161.

Guiget, A., Ohshima, I., Takeda, S., Laurans, F., Lopez-Vaamonde, C. and Giron, D. (2019) Origin of gall-inducing from leaf-mining in *Caloptilia* micromoths (Lepidoptera, Gracillariidae). *Scientific Reports* 9, 67694.

Hawking, J.H. (2014) Systematics and ecology of the Australian aquatic moths, Acentropinae (Insecta: Lepidoptera). PhD thesis, La Trobe University, Melbourne, Victoria.

Horak, M., Day, M., Barlow, C., Edwards, E.D., Su, Y.N. and Cameron, S.L. (2012) Systematics and biology of the iconic Australian scribbly gum moths *Ogmograptis* Meyrick (Lepidoptera: Bucculatricidae) and their unique insect-plant interaction. *Invertebrate Systematics* 2, 357–398.

Janzen, D.H. (1984) Two ways to be a tropical big moth. Santa Rosa saturniids and sphingids. *Oxford Surveys in Evolutionary Biology* 1, 85–140.

Jeffries, M.J. and Lawton, J.H. (1984) Enemy free space and the structure of ecological communities. *Biological Journal of the Linnean Society* 23, 269–286.

Kawahara, A.Y., Plotkin, D., Espeland, M., Meusemann, K., Toussaint, E.F.A. *et al.* (2019) Phylogenomics reveals the evolutionary timing and pattern of butterflies and moths. *Proceedings of the National Academy of Sciences* 116, 22657–22663.

Labandeira, C.C., Dilcher, D.L., Davis, D.R. and Wagner, D.L. (1994) Ninety-seven million years of angiosperm–insect association: Paleobiological insights into the meaning of coevolution. *Proceedings of the National Academy of Sciences* 91, 12278–12282.

Legal, L. (2023) "Lepidoptera flies", but not always ... interactions of caterpillars and chrysalis with soil. *Diversity* 2023, 15, 27.

Marquis, R.J. and Koptur, S. (eds) (2022) *Caterpillars in the Middle: Tritrophic Interactions in a Changing World*. Springer, Cham, Switzerland.

Marquis, R.J., Baer, C.S., Lill, J.T. and Wang, H.G. (2022) The impact of construct building by caterpillars on arthropod colonists in a world of climate change. In: Marquis, R.J. and Koptur, S. (eds) *Caterpillars in the Middle: Tritrophic Interactions in a Changing World*. Springer, Cham, Switzerland, pp. 509–537.

McFarland, N. (1979) Annotated list of larval foodplant records for 280 species of Australian moths. *Journal of the Lepidopterists' Society* 33(Supplement), 1–72.

McFarland, N. (1988) *Portraits of South Australian Geometrid Moths*. Allen Press, Lawrence, Kansas.

McGeoch, M.A. (1993) The microlepidoptera associated with a fungus gall on *Acacia karroo* Hayne in South Africa. *African Entomology* 1, 49–56.

McQuillan, P.B. (1993) *Nothofagus* (Fagaceae) and its invertebrate fauna – An overview and preliminary synthesis. *Biological Journal of the Linnean Society* 49, 317–354.

Meyer-Rochow, V.B., Namin, S.M. and Jung, C. (2023) Behavioural and phylogeographic observations on *Epipomponia nawai* (Dyer, 1904): An east Asian moth (Lepidoptera: Epipyropidae) whose larvae are ectoparasitic on cicadas (Hemiptera; Cicadidae; Sonatini). *Journal of Asia-Pacific Entomology* 26, 102007.

New, T.R. (1982) Lepidoptera from *Uromycladium* galls on *Acacia*. *Australian Journal of Zoology* 30, 357–364.

Pierce, N.E. (1995) Predatory and parasitic Lepidoptera: Carnivores living on plants. *Journal of the Lepidopterists' Society* 49, 412–453.

Powell, J.A., Mitter, C. and Farrell, B. (1998) Evolution of larval food preferences in Lepidoptera. In: Kristensen, N.P. (ed.) *Lepidoptera, Butterflies and Moths,* 4(1), *Handbook of Zoology.* De Gruyter, Berlin & New York, pp. 403–422.

Schmidt, O. (2016) Larval food plants of Australian Larentiinae (Lepidoptera: Geometridae) – A review of available data. *Biodiversity Data Journal* 4, e7938.

Scoble, M.J. (1992) *The Lepidoptera: Form, Function and Diversity*. Oxford University Press, Oxford.

Shao, Y., Mason, C.J. and Felton, G.W. (2024) Toward an integrated understanding of the Lepidoptera microbiome. *Annual Review of Entomology* 69, 117–137.

Shin, Y.-M., Lee, H.S., Kim, I.-K., Kim, C.-J. and Choi, M.B. (2023) Host range expansion of nest-parasitic moths *Pyralis regalis* and *Hypsopygia mauritialis* in social wasp nests: New findings and implications for biological control. *Diversity* 15, 789.

Suckling, D.M., Barrington, A.M., Chhagan, A., Stephens, A.E.A., Burnip, G.M. *et al.* (2007) Eradication of the Australian painted apple moth *Teia anartoides* in New Zealand: Trapping, inherited sterility, and male competitiveness. In: Vreysen, M.J.B., Robinson, A.S. and Hendrichs, J. (eds) *Area-Wide Control of Insect Pests: From Research to Field Implementation*. Springer Nature, Dordecht, The Netherlands, pp. 603–615.

Thurman, J.H. (2022) Beyond the pest: Life history, ecology and ethnoentomology of the giant wood moth (*Endoxyla cinereus*). *Austral Ecology* 47, 733–747.

Upton, M.S. (1997) A twelve-year diapause in the Queensland kauri moth, *Agathiphaga queenslandiae* Dumbleton (Lepidoptera: Agathiphagidae). *Entomologist* 116, 142–143.

Volp, T.M., Zalucki, M.P. and Furlong M.J. (2022) What defines a host? Oviposition behavior and larval performance of *Spodoptera frugiperda* (Lepidoptera: Noctuidae) on five putative host plants. *Journal of Economic Entomology* 115, 1744–1751.

Wagner, D.L. and Hoyt, A.C. (2022) On being a caterpillar: Structure, function, ecology, and behavior. In: Marquis, R.J. and Koptur, D. (eds) *Caterpillars in the Middle: Tritrophic Interactions in a Changing World*. Springer, Cham, Switzerland, pp. 11–65.

Zwick, A. (2008) Molecular phylogeny of Anthelidae and other bombycoid taxa (Lepidoptera: Bombycoidea). *Systematic Entomology* 33, 190–209.

6 Natural Enemies: Challenges and Responses

Tim R. New[1]*, Myron P. Zalucki[2] and Roger L. Kitching[3]
[1]*Department of Ecological, Plant and Animal Sciences, La Trobe University, Melbourne, Victoria, Australia;* [2]*School of the Environment, The University of Queensland, Brisbane, Queensland, Australia;* [3]*School of Environment and Science, Griffith University, Nathan, Queensland, Australia*

Abstract

All stages of moths are attacked by 'natural enemies', a variety of predators, parasitoids, parasites and pathogens that kill or debilitate them and affect the wellbeing of individuals and populations. The major categories are discussed in this chapter, with most information drawn from studies on predators and parasitoids, some related to pest management manipulations such as biological control. The limited information on the natural enemies of Australian moths reflects both the vast richness of small parasitoid wasps, in particular, and difficulties of identifying them and clarifying their biology, and almost complete ignorance of pathogen impacts in wild populations. Understanding the variety and impacts of all groups of natural enemies and the countermeasures developed by moths to evade or counter their effects helps to mould appreciation of moth variety. Many aspects of moth behaviour, appearance and ecology are related to interactions with natural enemies, especially parasitoids and predators.

6.1 Introduction

Moths live in a complicated, often inimicable, world. The wellbeing of individuals, populations and species, and the richness and distribution of assemblages, are influenced by (and depend on) two main suites of environmental features. The feeding stages (larvae, adults) depend intricately on the availability of food (Chapter 5) and all stages are constituents of food webs whereby they become the prey or hosts of 'natural enemies', or succumb to disease-causing pathogens and debilitating parasites. These two complexes combine in the idea of moths being 'in the middle' of trophic webs (Marquis and Koptur, 2022), with food supply considered as 'bottom-up' effects on growth, survival and reproduction, opposed by 'top-down' impacts of organisms causing mortality. In Lawton and McNeill's (1979) early expression, insect herbivores are 'between the devil and the deep blue sea'; that is, between the 'devil' of natural enemies and the 'deep blue sea' of utilizing their food supply. Understanding moth evolution and ecology involves both these pressures and the dynamic balance between them.

The term 'natural enemies' is generally restricted to free-living predators, parasitoid wasps and flies, true internal parasites and the wide range of pathogenic fungi, micro-organisms and viruses and thus excludes the numerous chemical and physical defences proffered by the food plants themselves.

*Corresponding author: t.new@latrobe.edu.au

DOI: 10.1079/9781800627680.0006

No life stage of any moth species is immune from attack by natural enemies. Even moth eggs and newly hatched larvae are susceptible (Zalucki *et al.*, 2002). Their impacts lead to varying reactions from people. Hobbyists frequently encounter small wasps or flies emerging from caterpillars or pupae that they are attempting to rear to adulthood – so regard these as undesirable – and, conversely such insects may be employed as biological control agents to counter pest moths, where they are considered beneficial. Indeed, investigations to assure the safety (minimal non-target impacts) of candidate agents proposed for introduction to a novel environment have been a major contributor to understanding the specificity and biology of numerous predatory and parasitoid arthropods (Howarth, 2000; van Driesche and Hoddle, 2016). Groups of natural enemies range from generalist species which include moths (either as adults, larvae or other stages) as part of a broader diet, and which may even feed on non-insects, to highly specialized narrow-diet consumers that may confine their focus to a single moth species or stages.

This chapter examines the challenges faced by moths in Australia from actions of their natural enemies, and the challenges to biologists of understanding and documenting the variety and impacts of those organisms. It comprises two main sections, the first dealing with the array of organisms (predators, parasitoids, parasites, pathogens) involved and the second appraising behavioural, ecological and evolutionary responses by moths to their presence and activities. These essentially represent the roles of moths in being attacked and in defending themselves against a host of natural enemies.

6.2 Perspective

Several distinct ecological groups are discussed briefly below to indicate their variety. Terminology and usage vary in the literature. We use the key terms as follows.

- *Predators* are free-living animals that eat and kill moths or their life stages. They may do this by complete consumption (as in the case of most vertebrate predators) or by feeding externally on part of the prey (as in blood-sucking bugs or liquid-feeding spiders)
- *Parasitoids* are mostly wasps or flies that attack early stages, and whose larvae feed in or on the host insect and eventually kill it. Parasitoids may in turn be attacked by '*hyperparasitoids*', adding further complexities to interpreting functional relationships. A few nematode groups also fit this definition although these are conventionally designated 'parasites'. These categories are referred to as parasite-like (parasit*oid*) rather than true parasites because they almost inevitably kill their hosts which true parasites seldom do.
- *Parasites* are organisms that feed on or in the host insect but rarely kill it, although they may severely affect growth and wellbeing. Parasites are conventionally divided into endo- and ectoparasites depending on whether they operate within the body of their hosts or by attaching to the exoskeleton, piercing into the body cavity and extracting sustenance from the host's haemolymph. External blood feeders vary from those that remain attached for whole stages of their life cycles (like some ticks and mites) through to those that visit, feed and move on (like midges and mosquitoes)
- *Pathogens* are disease-causing agents such as protozoa, nematodes, bacteria, fungi and viruses that lead to morbidity and mortality. Their impact at a population level can be significant.

Of these categories, the predators and parasitoids are the best documented and arguably the most important and tangible influences on moth lifestyles. A multi-authored volume has become a standard work on insect pathogens (Vega and Kaya, 2012) and we have used this work as a guide to our comments below.

We emphasize that these 'definitions' cannot be applied unquestioningly. The vast array of natural enemies which confront moths (and other insects) exhibit a wide variety of feeding strategies which do not always fit comfortably within a particular category.

For most Australian moths, the full spectrum of natural enemies has not been determined. Numerous isolated records, mostly of predators or parasitoids, have accumulated from field and rearing observations. The most complete information, however, has come from more intensive studies of pest insects as components of planning pest management programmes which include biological control. Even this rarely leads to comprehensive information, being based almost always on regionally restricted studies with limited knowledge of natural enemies across the wider distribution of the species involved. A 3-year survey for parasitoids of the light brown apple moth (*Epiphyas postvittana*, Tortricidae), a significant orchard pest, yielded 25 species (some of them hyperparasitoids), but with many gaps in biological knowledge (Paull and Austin, 2006). A survey of parasitoids of another important pest, the invasive fall armyworm (*Spodoptera frugiperda*, Noctuidae) yielded 12 species, of which five were new to science (Fagan-Jeffries *et al.*, 2024).

The labour and expertise involved in undertaking comprehensive surveys for natural enemies of moth species lacking any economic significance are rarely available. A notable exception to this generalization is the Sphingidae, the hawkmoths (see Chapter 19). The Australian Sphingidae were monographed by Moulds *et al.* (2020) based on many years of study involving the rearing of a large majority of the known Australian species. This painstaking approach inevitably led to the rearing and subsequent identification of many parasitoids. Moulds *et al.* (2020) present a list of 102 parasitoids reared from sphingid eggs, larvae and pupae with representatives of seven families of Hymenoptera plus the dipteran Tachinidae. These authors also present an elegant overview of 'natural enemies' which, although based on their knowledge of the hawkmoths, could well serve as a general introduction for the moths in general.

6.3 The Players

The limited historical accumulation of reliable prey or host records of the enemies of Australian moths is largely serendipitous. Likewise, identifying the various invertebrates and pathogens involved may be very difficult, and cross-study comparisons uncertain. Many of the 'pest surveys' above have yielded previously unknown species and host records of parasitoids, for example – and whereas most predatory or parasitoid insects can be identified to family (and often genus), species-level identification is commonly uncertain, and their distribution and fuller prey/host range can usually only be inferred. The literature reveals many such cases. The identification of pathogenic micro-organisms is even more challenging although increasingly sophisticated molecular techniques offer a clear way forward.

6.3.1 Vertebrate insectivores

Many kinds of animals eat moths, some as part of a wider insectivorous diet but others more closely associated with particular taxa, habitats or seasonal behaviour. Many insectivores are essentially opportunistic and feed on whatever prey is most accessible. Numerous terrestrial vertebrates (mostly reptiles, birds and mammals) have very broad invertebrate diets, and the significance of many specific prey records of moths remains uncertain; some are assuredly casual opportunism, whilst others can be more predictable.

6.3.1.1 Terrestrial mammals

Although Australia lacks members of the order Insectivora (other than the now extinct Christmas Island shrew), several clades of marsupials have adapted to fill that trophic gap either as specialists or opportunistically as omnivores.

The diets of all dasyurid marsupials are substantially insect based with larger prey featuring substantially only in the quolls and the Tasmanian devil (among extant species). All of the smaller species and especially the many species of *Antechinus* are all more or less obligate insectivores. Strahan's handbook (1983) described the diet of each species (where this was known). In general, dasyurid diets are described only at the level of the class or even phylum ('insects' or 'arthropods'). Moths are mentioned explicitly as part of the diet of various dunnarts, planigales and ningauis. The arthropod diet of these smaller dasyurids seems to reflect the fauna of the leaf litter in which they forage and, as such, inevitably, will have included many moth larvae and pupae. The only explicit species-level record presented is that of Tasmanian devils feeding on 'corbie grubs' (larvae of *Oncopera intricata*, Hepialidae). In a more focused account, Hall (1980) described the diet of two species of *Antechinus* based on analyses of faecal samples. Lepidopteran adults and larvae were eaten by both species but as minor diet constituents compared with amphipods and beetles. Consumption of moth larvae and adults by these mammalian species appears to be opportunistic, with intake perhaps reflecting the frequencies of encounters in general foraging activities.

Likewise, the predatory peramelomorphs, the bandicoots and bilbies, are generally omnivores feeding on 'insects, small vertebrates and plants' (Strahan, 1983). As ground-dwelling foragers like most dasyurids, they too will encounter larvae and pupae opportunistically.

A much-discussed exception to this 'opportunistic' generalization is the mountain pygmy-possum, *Burramys parvus* (Burramyidae). This endangered alpine marsupial has been supposed to have a strong dependence on aestivating populations of the bogong moth (*Agrotis infusa*, see Chapter 37). Broome (2001) showed that bogong moths made up between 30% and 50% of the mammals' diet, with this being of especial importance in their posthibernation life in the alpine spring. Recent declines in the moth numbers have been flagged as a major concern for possum conservation. More recently, Hawke *et al.* (2018) studied a newly discovered population of *Burramys* at lower alpine elevations, without seasonal access to bogong moths. These *Burramys* were far more opportunistic in their invertebrate diets, suggesting at least a level of adaptability for the species.

6.3.1.2 Bats

Bats are the predominant nocturnal or crepuscular vertebrate predators of adult moths. The extensive literature on the interactions between bats and moths reflects the intricate interplay between echolocation mechanisms of the bats and hearing mechanisms and, even, disruptive sounds on the part of moths.

Australia harbours more than 50 species of insectivorous microbats. Diets of many of them have been studied using stomach contents or faecal matter. Moths feature prominently but not universally in these results (Milne *et al.*, 2016). Bats whose sonar call frequencies most closely coincided with the hearing frequencies of 'eared' moths (i.e. those with tympani) have been shown to catch fewer of those groups of moths (at the expense of non-'eared' moth families). Bat species with call frequencies outside (usually above) this hearing range exploited mainly 'eared' moths (Pavey and Burwell, 1998).

The dynamic interaction between bats and moths and consequences for the distribution of the latter are further exemplified by the cave-roosting Australian noctuid *Speiredonia spectans* in eastern Queensland. The moths can detect most calls of two co-habiting bat species (Fullard *et al.*, 2008) but not most calls of another bat species. Indeed, the moth was excluded from day roosts occupied by this third species, *Hipposideros ater*.

Many examples of this kind led to the proposition that the evolution of hearing mechanisms and, in some cases, vocal abilities in moths was a direct response to the selective pressures imposed by insectivorous bats. The availability of well-supported and dated phylogenies for both taxa now show that, in all likelihood, hearing organs (tympani) in moths predated bats by many millennia. This raises the interesting question as to what drove the repeated evolution of hearing organs (across taxa and anatomical locations) in the moths if not predator pressure. Intraspecific communication has been suggested. Recent discoveries of 'bat-like' insectivorous pterosaurs, such as the Anurognathidae that hunted insects in Jurassic and Cretaceous skies, raise interesting speculations (Wei *et al.*, 2021) (discussed in Chapter 20). Nevertheless, during the long period in the Palaeogene when night-flying moths and insectivorous microbats did co-exist, a co-evolutionary 'arms race' appears to have occurred between the two groups building on pre-existing hearing organs in the case of the moths (Hofstede and Ratcliffe, 2016).

Concerns arise over the increased impacts of bats on moths and other phototactic insects as artificial night lighting proliferates and moths are attracted to light sources (Minnaar *et al.*, 2015; Stone *et al.*, 2015).

6.3.1.3 Reptiles and amphibians

Four of the five major families of Australian lizards are generalist insectivores, although agamids and some skinks may take some additional vegetable matter (Wilson and Swan, 2003). Manicum and Scharzkopf (2011) examined the stomach contents of three sympatric species of tropical skinks, *Carlia* spp. Lepidoptera adults and larvae featured in the diets of all three species but were not, apparently, equally attractive to all. For all three species, larvae were more prominent than adults in their lepidopteran intakes.

Atkins *et al.* (2018) examined the diet of the alpine skink, *Liopholis guthaga* across seasons. They analysed lizard droppings and scored arthropod prey items both numerically and volumetrically. Lepidoptera were present in most of their summer samples across sites and seasons but only as a minor part of the overall diet, in which ants and beetles were more prominent.

It is perhaps to be expected that the equally nocturnal geckoes (Geckonidae) might have the most impact on nocturnal moths – as implied by watching domestic geckoes capturing moths around house lights. Henle (1990) examined the diets of the arid-zone gecko, *Gehyra variegata* by comparing the content of stomach flushes against the profile of arthropods caught using sticky traps at three locations. No significant rank correlation was found between availability and diet at any of three sites examined. Moths amounted to between 3.1% and 10.8% of items eaten. These results need to be interpreted with care. The authors say that geckoes sought larger food items and, in general, moths would have been among the larger prey available (compared, say, with Diptera or psyllids).

Among the amphibians, the only likely predators of moths are adult anurans, usually regarded as generalist insectivores. Lima *et al.* (2000) compared the diets of three species of frogs with sympatric lizards in subtropical rainforests in Queensland. In general, the size of prey in frogs was smaller than for lizards and was a reflection of gape size. Apart from a few unspecified 'larvae', they found no Lepidoptera in their studies. In comparable studies Hodgkison and Hero (2003) found only trace amounts of Lepidoptera in faecal analyses of only one of three species of tropical frogs they examined in Queensland. Other studies have identified termites, ants and even mites as frog specialities. Without further evidence, we conclude that frogs probably have negligible impacts on moth populations.

6.3.2 Free-living invertebrate predators

A wide range of generalist arthropods are free-living predators that may include moths in their diet. Depending on hunting strategies, likely targets may be larvae or adult moths, seldom both. Possible targets are also likely to reflect behavioural features such as diurnal activity. Moth larvae are 'available' (physical access permitting) at all times of day. Adult moths are principally nocturnal and as such may avoid the many vision-oriented diurnal predatory insects. Tool-using predators such as web-building spiders or larval mycetophilids pose threats to adult moths at all times. There is little if any data on the impact insect predators might have on moth populations (unlike that of parasitoids, birds or bats) yet some have been proposed as biocontrol agents for outbreak pests, including forest and pasture moths.

We describe here examples of some of the more important and well-known taxa of invertebrate predators.

6.3.2.1 Odonata

Adult dragonflies and damselflies are free-flying predators, using their spiny legs to scoop prey insects either from perches or in the air. Although certainly frequent predators of adult Lepidoptera in general, their principally diurnal habits probably limit their impact to day-active moths.

6.3.2.2 Mantodea

Mantids are generalist sit-and-wait predators, using their raptorial forelimbs to strike any prey insect that approaches too closely. No doubt their prey include adult moths from time to time. Mantids detect prey visually and are probably most active during the day (although nocturnal species are known in the Asian tropics), limiting their likely impact on adult moths. They are attracted to light and may exploit the moth collectors' bounty from time to time.

6.3.2.3 Heteroptera

Most terrestrial true bugs, belonging to the suborder Heteroptera, are plant feeders. Several families or parts of families are in fact active predators piercing the epidermis of their prey, secreting enzymes via their proboscides and sucking the digestate back into their guts. Unlike some other 'suckers', their impact appears generally to be fatal. Slow-moving lepidopterous larvae are common prey items (Jones, 2023), adults presumably very infrequent ones. Gross (1975) identifies the asopine Pentatomidae and some Anthocoridae as well as the voracious Reduviidae (and related families), as insect predators targetting, *inter alia*, moth larvae.

6.3.2.4 Diptera

Several clades of Diptera are predatory as adults (Kitching *et al.*, 2005). Most notable among these are the Empididae, Dolichopodidae and Asilidae. All three are recorded as feeding on a range of smaller arthropods, no doubt including moths. Empidids in particular are active at night and are likely to capture smaller adult moths. Some large asilids ('robber flies') are no doubt capable of snatching moths from their resting places or in flight, with day-flying moths likely to be particularly vulnerable. They have been identified as a potential threat to adults of the endangered golden sun moth (*Synemon plana*), with which they may share similar flight periods (see Chapter 35).

A separate guild of flies act as temporary ectoparasites on moth larvae (and occasionally adults) as they suck blood from their host's haemolymph. There are many records of ceratopogonid adults feeding on lepidoptera larvae and adults (e.g. Kawahara *et al.*, 2006). The midge genus *Forcipomyia* has been recorded feeding on larvae and adults of at least eight families of Lepidoptera (Moulds *et al.*, 2020). A record from NSW of midges feeding on larvae of limacodids led to the death of the larvae and the suggestion that the *Forcipomyia* midges were actually vectors of pathogenic viruses (Moore, 1958). Similar fatal associations had been reported elsewhere (Mayer, 1955). Moulds *et al.* (2020), however, report several instances of midges feeding on hawkmoth larvae with no obvious negative consequences.

6.3.2.5 Hymenoptera

In addition to the great variety of parasitoid families (see sections below), the order Hymenoptera contains many free-ranging predatory families (Naumann, 1991) that may well have an impact on moth populations. Many species provision larval cells within their nests with paralysed caterpillars of Lepidoptera to act as food for their larvae. The principal taxa attacking caterpillars belong to two subfamilies of the Vespidae. The solitary Eumeniinae build larval cells in old nests of other insects or in mud nests they create from scratch. These are provisioned with lepidopteran caterpillars. The eusocial Polistinae build carton nests and collect caterpillars which are masticated and fed to the polistine larvae. Various species of the polistine genera *Polistia* and *Ropalidia* are very common in warmer regions and nests can reach substantial sizes; their impact on caterpillar populations must be significant but we are not aware of any quantifications of this impact.

Introduced *Vespula* are predatory as adults wherever they have become established.

Finally, many ants are ubiquitous generalist predators and must take a large toll of free-ranging caterpillars.

6.2.3.6 Spiders

Spiders (the arachnid order, Araneae) are undoubtedly the most common and ubiquitous group of arthropod-on-arthropod predators. Again, most are generalist feeders, although particular foraging strategies will mean that some are much more likely to target moths than others. Again there is little explicit data. Free-ranging spiders such as lycosids probably pose little threat to moths in contrast to either ambush predators or web builders of various kinds. Ambush predators, particularly so-called flower spiders (Thomisidae), have been reported feeding on adult moths (Morse, 1983; Sakagami *et al.*, 2021) and presumably are active at night as well as during the day.

The real impact of spiders on moths, however, must be from the web builders. Webs vary from the sometimes massive catch-all orb webs of the Araneae and Nephilidae to the messy, unstructured webs of theridiids such as the undeservedly notorious redbacks. The Australian fauna also includes variations such as the web-casting retarius spider (Deinopidae) and the fascinating bolas spiders (Araneidae: *Ordgarius* spp.). All these strategies will catch moths from time to time. The vast orb webs almost always have one or two moths entangled in them and, of course, operate just as effectively at night as during the day.

There are surprisingly few accounts, globally, of the range of prey captured by spiders (e.g. Morse, 1983; Sakagami *et al.*, 2021) and none that we are aware of for Australian species. This mirrors the general challenges associated with spider studies in Australia (New, 1999). The appearance of two guidebooks may help circumvent the many impediments (Framenau *et al.*, 2014; Whyte and Anderson, 2017). Projects recording prey type and frequency in the webs of some of the well-known species would be relatively straightforward, if time-consuming, and might be ideal projects for citizen scientists. Given the assumed importance of spiders in terrestrial food webs, such information would be useful and important – and would tell us more about the life challenges of the moths.

6.3.3 Mites

If the spiders present a methodological and taxonomic challenge to the biologists, this pales into insignificance compared with those who choose to study the Acari – the mites. The mites are a further group of arachnids many of which are predatory, ectoparasitic or, at least, phoretic upon the Lepidoptera. Asher Treat (1975) published an entire volume on *Mites of Butterflies and Moths*, engagingly described by Walter and Proctor (1999) as 'lyrical'. David Walter and Heather Proctor's volume *Mites: Ecology, Evolution, Behaviour* offers the best more general and popular introduction to the mites as an entire order.

The most obvious group of mites on Lepidoptera are the larvae of terrestrial Parasitengona, which scamper over substrates until they encounter an unlucky host, then attach, engorge and become bright red or orange blobs. These then drop off and their nymphs and adults become the large predatory mites best known as red velvet mites, although this common name applies to just one group (Trombidiidae), and others are variously bristly, coloured and patterned (Walter *et al.*, 2009). Several species of *Austrothrombium* (Trombidiidae) are widely distributed in Australia, and larvae of the austral genus *Clinotrombium* parasitize spiders and moths (Southcott, 1986; Walter *et al.*, 2009).

Other ectoparasites are hidden from view, finding the tympana of moths more to their liking. The best known of these are *Dicrocheles* (Laelapidae) which infest just one ear of noctuid moths – for infesting both would render the moth deaf and easy prey for bats (Treat, 1975). This extraordinary act is achieved simply: when a female mite climbs onto a moth, she senses if a mite has been there before. If not, she leaves a pheromonal trail for all subsequent mites to follow. Similarly, the 'ear-sparing mites' (Otopheidomenidae) occupy the tympana of noctuid moths but only lay eggs there, feeding elsewhere. Some infest sphingid moths and, as these lack tympanic organs, they reside in the thoraco-abdominal junction. Otopheidomenidae are yet to be discovered in Australia, but systematic sampling and assessments are lacking. Even the better known *Dicrocheles* are relatively rare, averaging 2% of a large collection of *Pseudaletia convecta* (Treat, 1975; Lindquist *et al.*, 2009).

Numerous groups of mites are phoretic; that is, they use other animals for transport among habitats. Many associations between mites and Lepidoptera seem accidental, but sometimes this transport is a critical part of the mite's life cycle. For example, the nectar and pollen-feeding mite *Proctolaelaps lobatus* (Melicharidae) rides between flowers on butterflies (Boggs and Gilbert, 1987). Similarly, *Afrocypholaelaps africana* (Ameroseiidae) most often moves between flowers on bees but has been found on moths in Australia as well (Owen Seeman, pers. commun.). More sinister are phoretic associations between mites such as *Blattisocius* (Blattisociidae) and *Pyemotes* (Pyemotidae). Once the phoretic females arrive in their preferred habitat, the hitchhiker turns upon their carrier – by eating their offspring. Certain species of *Blattisocius* are voracious egg predators, and *Pyemotes* are predators of eggs and larvae.

In Australia, Halliday *et al.* (1998) report *Blattisocius dentriticus* from macadamia nut borer, *Cryptophlebia omradelta*, and Treat (1975) records it from various noctuid moths on which they are likely phoretic. *Blattisocius tarsalis* occurs in insect cultures and has chelicerae reminiscent of a hand-held tree saw, which are thought to be useful for ripping into insect eggs. Halliday *et al.* (1998) reports this species from

Sitotroga cerealella and *Galleria mellonella*. These mites are phoretic on moths and then feed on the eggs, but the records from Australia are in laboratory cultures or bee hives, respectively, for each host. A North American species, *B. patagiorum*, appears to be parasitic on adult noctuid moths and can reproduce without other sources of food (Treat, 1967, 1975; Lindquist *et al.*, 2009).

Pyemotes species were recorded from 'heliothis' larvae in Australia but it is not clear if this was ex-field or in cultures. Other records are on *Phthorimaea* on tobacco in Tasmania, and possibly *Spodoptera exigua* – as '*Lyp. exigua*' (O. Seeman, pers. commun.). Acarologists often refer to *Pyemotes* as 'parasitoids' – a nice point to debate over a glass of wine. *Pyemotes* is the agent of the painful condition 'grain itch', caused when they attempt to feed on humans, although one species with more focused tastes in hosts shows promise as a biocontrol agent of the globally invasive *Spodoptera frugiperda* (Liu *et al.*, 2020).

6.3.4 Hymenopteran parasitoids

Early stages of moths are attacked by many kinds of small wasps, many of them highly host and stage specific. Particular wasps may attack eggs, larvae or pupae, either of particular host species or more broadly, and larvae develop in or on the host, which is maintained alive over the parasitoid's feeding period. Broadly, four life history patterns occur amongst larval parasitoids based on (i) whether the parasitoid is an external feeder (ectoparasitoid) or internal feeder (endoparasitoid) and (ii) whether the parasitoid causes the host to cease further development after attack (idiobionts) or whether the host continues to develop (koinobionts). Unlike many predators, parasitoids must occupy the same habitats as their hosts. Predators of moths tend to act individually, but numbers of parasitoids can increase vastly through polyembryony or multiple ovipositions, so that a single moth caterpillar or pupa may yield tens to hundreds of identical adult wasps. Parasitoids are sometimes claimed to be the most important causes of mortality in insect herbivores (Godfray, 1994).

Parasitoid wasps are extraordinarily diverse, both taxonomically and in their intricate biology. Naumann (1991) lists 71 families of Hymenoptera occurring in Australia. Of that huge diversity, some 50 families are parasitoidal to some degree and he records 17 families that utilize eggs, larvae or pupae of Lepidoptera (principally moths) as their hosts. They are dominated by two vast superfamilies, the Ichneumonoidea, with most species of the two constituent families (Braconidae, Ichneumonidae) being primary parasitoids of Lepidoptera (Quicke, 2015) and the Proctotrupomorpha, the latter embracing formerly distinct superfamilies including the Chalcidoidea and Platygastroidea, each with many egg parasitoids of Lepidoptera; relationships are discussed by Polaszek and Vilhelmsen (2023). Each of these groups contains numerous undiagnosed species in Australia, where levels of endemism (and implied host specialization) are high. Novel discoveries continue. The evolutionarily important enigma moth (*Aenigmatinea glatzella* – see Chapter 3) on Kangaroo Island hosts an apparently restricted braconid wasp described as a new genus by Glatz *et al.* (2022).

Most commonly, exploitation of the hosts is by direct endoparasites, in fewer cases as hyperparasites, and in a minority of cases as ectoparasites. Some species transition from one lifestyle to another as they mature or as the opportunity arises. Generally eggs are laid directly into the host although in a few cases, the eggs are laid on plants and then consumed by the lepidopteran larvae (e.g. species of Trigonalidae in *Anthela*). These hatch in the host's alimentary tract and become parasitoidal within the body cavity. Some species are gregarious, leading to multiple individuals within a single host ('superparasitism') (e.g. species of Braconidae: Agathidiinae). A similar end is achieved in a few species through polyembryony where somatic duplication occurs early in development within the parasitoid egg (e.g. some Encyrtidae). Parasitoid families may be more or less specialized in terms of both their preferred host stadia and the range of hosts exploited. Minute species target lepidopteran eggs (e.g. Aphelininae, Trichogrammatidae, Eupelmidae), other much more robust species target large larvae only (e.g. some Chalcididae, Ichneumonidae: Ichneumoninae). Other groups have focused on leaf-mining and leaf-rolling moth species (e.g. Ichneumonidae: Cremastinae, several subfamilies of Braconidae) or even on larvae in wood (e.g. Stephaniidae, Braconidae:

Braconinae). The list of variations on the parasitoidal theme is immense.

Parasitoids may be closely associated with a particular host species. In some cases different monophagous parasitoids may target different life-stages of the same host species. Other parasitoids, of course, are oligophagous or more or less broadly polyphagous. Complexities of identification of parasitoids (and, indeed, of their hosts that may also comprise species complexes) are confirmed continually through revisionary studies employing molecular methods that reveal numbers of undescribed species in complex relationships; some 'species' earlier regarded as 'generalists' through attacking many host moth species are now known to be a complex of specialist species each attacking, and restricted to, a single host. Many groups of parasitoid await such investigation. As an instructive example, the braconid genus *Glyptapanteles* was known from only three Australian species until a recent revision revealed a further 31 species, all undescribed (Fagan-Jeffries *et al.*, 2022).

As we have noted already, only a very few of our moths have been evaluated in terms of their parasitoidal 'loads' and it remains impossible to estimate the richness of the parasitoid fauna either locally or globally; suffice it to say, it must be immense.

6.3.5 Dipteran parasitoids

The 'bristle flies' (Tachinidae) are the most diverse and ecologically important group of parasitoids after Hymenoptera (Stireman *et al.*, 2021), and one of the richest families of the Diptera. Most species for which information is available attack larvae of holometabolous insects, especially Lepidoptera and Coleoptera. They also have a variety of strategies to obtain hosts, with proliferation on lepidopterous hosts perhaps linked with diversification among the flies. Moth larvae are sometimes reported with numerous small white spots – tachinid eggs – deposited on their body, whence the hatchling larvae burrow into the host and feed internally. Other tachinids oviposit on the plant substrate and caterpillars ingest the micro-eggs as they feed, and in others (e.g. the genus *Exorista*) the adults are 'ovolarviparous' (that is, adults lay a hatching larva) and the maggot infests its host on the plant.

As for hymenopterous parasitoids, many host records, levels of specificity and unambiguous identifications of tachinid parasitoids are imprecise and incomplete. Australian tachinids are largely endemic, with Rutiliini restricted to the continent and the Oriental region: known hosts of *Rutilia*, however, are scarab beetle larvae.

Parasitoid larvae of both Hymenoptera and Diptera may pupate within the host body, or leave it and pupate (most wasps in silken cocoons) externally.

6.3.6 Nematodes

These unsegmented worms are a ubiquitous phylum found in virtually all environments, playing almost all ecological roles available to multicelled animals.

Several nematode families have been described from moth hosts globally. In Australia, the families Steinernematidae and Heterorhabditidae were identified as pathogens of insects occurring naturally in soil. The infective stages are attracted to potential insect hosts and enter them via natural orifices and intersegmental membranes. Once within the host's haemocoel, they release symbiotic bacteria which cause septicaemia in the insect, thereby providing suitable nutrients that support nematode reproduction. They also inhibit other micro-organisms within the insect (Poinar, 1979; Akhurst and Bedding, 1986). The breakdown of the insect's tissues releases large numbers of infective juvenile nematodes into the soil, each potentially infecting further insect hosts. There is little doubt that they can cause significant mortality and, indeed, have been widely proposed as a potential control measure for pest species (Bedding *et al.*, 1993). Target Lepidoptera in Australia and beyond have included the currant clearwing moth (*Synanthedon tipuliformis*), the armyworm (*Spodoptera frugiperda*), wood-boring cossids (such as *Endoxyla* spp.), peach fruit moth (*Carposina niponensis*) and light-brown apple moth (*Epiphyas postvittana*). Simple cultures of a pyralid (*Galleria melonella*) were used by Bedding and Akhurst (1975) as bait before establishing cultures of entomopathogenic nematodes in Tasmania. How much mortality is caused in naturally-occurring non-pest species of moths is unknown but is likely to be significant.

Two species of Mermithidae have been recorded from aestivating conglomerations of bogong moths (*Agrotis infusa*), and were described by Welch (1963) – *Amphimermis bogongi* and *Hexamermis cavicola*. Like other mermithids, the free-living stage of these species is the adult stage rather than the juveniles (Nickle, 1972). The adult worm matures within the body of its host and then exits via an aperture it creates in the body wall. Once free, mermithids are supposed to release eggs which are ingested by herbivores, moth larvae in this case. The erstwhile host presumably perishes. Mermithids generally are supposed to have wide host ranges (Nickle, 1972).

There appear to be no records of the other parasitic worm phylum, the Nematomorpha, for Lepidoptera.

6.3.7 'Protista'

'Protista' is now an out-of-date term for what was regarded as a kingdom of unicellular (and loosely multicellular) organisms, including many pathogens. The organisms involved are now assigned to a growing number of high-level clades in their own right. We use the term here to cover a miscellaneous collection of unicellular eukaryotes that are recorded as pathogenic in the Lepidoptera. Lange and Lord (2012) identify five major clades of protistans recorded as pathogens within insects globally: the Amoebozoa, Apicomplexa, Ciliophora, Euglenozoa and Helicosporidia. Most records of these protistan pathogens in insects are from relatively well-known, economically important species such as locusts, mosquitoes and midges.

Only three of these groups, the Apicomplexa (subclades Neogregarinorida and Coccidia), Euglenozoa and Helicosporidia, have been recorded in Lepidoptera (according to Lange and Lord, 2012), with no records from Australia. We know of no general microbiologists who have surveyed the local biota: it seems inevitable given records elsewhere that species of moths in Australia will harbour a variety of protistan pathogens. Where these have been studied elsewhere, they in general seem to lead to morbidity rather than death. Several known species vary in their roles from mutualists to parasites to pathogens within the bodies of insects. This is a fertile field for future investigations.

6.3.8 Fungi

Representatives of the vast eukaryotic kingdom, Fungi, are ubiquitous across all environments. Fungal infestations are commonly associated with insects (and other arthropods) although just which are true pathogens and which are decomposers associated with dead or moribund individuals is often difficult to determine. Unlike most other pathogens, fungal spores germinate on the body surface of the insect. Emergent hyphae secrete enzymes such as chitinase, leading to penetration of the cuticle and subsequent invasion of the body mass of the host. Infections do not occur orally. Life cycles are completed after the death of the host when the infective fungi release either asexually produced conidiospores or various kinds of sexually produced spores (depending on the taxon of the fungus). Such spores are present ubiquitously in soil. Most confirmed pathogens of insects belong to the phyla Ascomycota, Entomophthoromycota or Microsporidia (now confirmed as fungi, see below) (Vega *et al*., 2012).

Most information on insect-pathogenic fungi has been derived from studies of major pest species in the interests, usually, of developing biopesticides. Two main groups have been targeted: the order Hypocreales within the Ascomycota (class Sordariomycetes) and the phylum Entomophthoromycota (a phylum of about 250 species divided into three classes and six families: Humber, 2012).

Within the Hypocreales, a cluster of families contain well-known insect-pathogenic genera such as *Beauvaria*, *Metarhizium*, *Isaria* and *Ophiocordiceps*. Approaching any understanding of these is complicated by the existence of anamorphs and teleomorphs (Vega *et al*., 2012). *Beauvaria*, *Metarhizium* and *Isaria* (with more than 22 other 'genera'), for example, are all 'anamorphs' – that is, asexually reproducing forms – of *Cordyceps sensu lato* (Sung *et al*., 2007). This last is a sexually reproducing 'teleomorph' much more restricted geographically. As *Beauvaria*, it is recorded from a wide range of moth families including tortricids, erebids, saturniids and sesiids. The related genus *Metarhizium* occurs in a wide array of moths, especially Noctuoidea and Bombycoidea. Most fungus-based biopesticides have been developed from *Beauvaria*, *Metarhyzium*, *Isaria* and related genera of the Hypocreales.

The Entomophthoromycota have proved less useful in this connection. Most are obligate entomopathogens and cause natural epizootics (Chandler, 2017).

The original genus *Cordyceps* (*s.l.*) is worthy of special mention. The global 400+ species assigned to the genus have now been divided into four genera. These fungi are well known as a key and highly lucrative part of traditional Chinese medicine where they have therapeutic claims over disorders ranging from fatigue to sexual dysfunction (Olatunji *et al.*, 2018). The fruiting bodies of many species emerge following death by mycosis of insect larvae and pupae. Sung *et al.* (2007) provide a global list of 61 species they used in their phylogenomic analysis. A third of these were associated with lepidopteran larvae and pupae. The Atlas of Living Australia (accessed March 2025) lists 17 species of *Cordyceps* (*s.s.*), plus about 14 species now in the genus *Ophiocordyceps*, and a single species of *Metacordyceps*. *Elaphocordyceps* does not appear to have been recorded in Australia. Most entries do not record the host of the fungus but there can be little doubt that many of them prey upon Lepidoptera, especially those with larvae or pupae in soil and leaf litter. *Cordyceps gunnii* is known as a pathogen of hepialid larvae in Tasmania.

The exact status of the fungoid unicellular organisms referred to as Microsporidia has been much debated. Recent molecular evidence suggests they are in fact simple fungi that sit somewhere close to the base of the fungal phylogenetic tree (Lee *et al.*, 2010). Microsporidia are obligate pathogens in all major taxa of animals (Pilarska *et al.*, 2017). There are few records from Australia although transmission to humans, possibly from insects, has been recorded (Sutrave *et al.*, 2018). Surveys of Lepidoptera elsewhere, however, have shown widespread occurrence across families such as Tortricidae, Erebidae, Noctuidae and others. Five genera of microsporidia have been recorded from Lepidoptera (up to 2012; Solter *et al.*, 2012). These genera and their relatives have been used principally to produce effective biopesticides (Chandler, 2017). The microsporidia are universally pathogenic and reduce reproductive rates and feeding efficiencies (Pilarska *et al.*, 2017).

Few workers have examined naturally-occurring assemblages of fungi associated with lepidopteran larvae in the field and we are unaware of any such work in Australia. Gielen *et al.* (2020) provide an exception to this global pattern and a model as to how such surveys might be approached. They examined five European species of moths, four geometrids and a noctuid, collected in the field and reared subsequently in the laboratory, and identified 24 species of pathogenic fungi from 15 genera and 11 families. Only anamorphs were detected, of which species of *Cordyceps* were the most abundant. It would be interesting and useful to carry out similar community-level surveys elsewhere and Gielen *et al.* (2020) lay out the cultural and molecular approaches they used for this purpose.

6.3.9 Bacteria

The ubiquitous, diverse kingdom of prokaryotic organisms referred to collectively as the Eubacteria challenges the terminology and methodology of those more familiar with the biology of eukaryotes (such as the arthropods). A wide range of 'species' across at least four phyla are known pathogens of insects and many occur in Lepidoptera (Jurat-Fuentes and Jackson, 2012).

The Gram-positive bacteria (that is, those which retain staining using Gram stains) with no outer cell membrane are the phylum Firmicules. Pathogens of insects belonging to at least five families sit within this phylum, spanning genera such as *Bacillus*, *Enterococcus* and *Clostridium*. *Bacillus thuringiensis*, from which both a class of insecticides and genetically modified pest-resistant crop plants were developed, belongs in this group (see below). Gram-negative bacteria with an outer cell membrane belong to the phylum Proteobacteria. At least four families within this clade contain genera pathogenic to insects, including *Photorhabdus*, *Pseudomonas*, *Chromobacterium* and the relatively well known rickettsias which include *Wolbachia* (see below). Finally, the bacterial phylum Tenericutes, recognized by its lack of a cell wall, contains the insect-pathogenic genus *Spiroplasma* (Spiroplasmataceae).

Insect-associated bacteria are gained from the environment orally. In Lepidoptera, the eggs of the moths have an associated coating of bacteria which are consumed by the newly emerged larva, beginning the processes of bacterial inoculation. The ingested bacteria develop within the larva into a bacteriome engaged in various non-pathogenic processes such as digestion and pheromone production. Bacterial cells which become either facultatively or obligately pathogenic break through the cells lining the midgut and proliferate in the haemolymph, producing a septicaemia which is fatal to the insect. Bacterial cells or cell fragments await further transmission following decomposition of the initial hosts. Some bacteria can sporulate and produce desiccation-resistant spores.

Turning explicitly to those bacteria associated with moths, Mereghetti *et al.* (2017) reviewed work on the types of bacteria associated with a wide range of more or lesss cosmopolitan pest species of moths. They added a fourth phylum to those described by Jurat-Fuentes and Jackson (2012), the Actinobacteria, and identified 19 families of bacteria across four well-studied species of forest and garden pests including species of Tortricidae, Noctuidae, Erebidae and Thaumatopoeidae. A similar analysis across eight species of agricultural pests showed that collectively they housed 63 families of bacteria. That set of moths included five species of Noctuidae, and single species of Crambidae, Yponomeutidae and Sphingidae. It must be stressed that this wide range of microbiota included commensals, mutualists and, presumably, pathogens. Many of the genera noted by Jurat-Fuentes and Jackson (2012) as confirmed pathogens were noted in Mereghetti *et al.*'s (2017) review.

As pointed out by Jurat-Fuentes and Jackson (2012), there seems to be considerable flexibility in the roles many of these bacteria play. Genetic borrowings across strains, dietary issues and other local environmental factors can cause otherwise benign bacterial species to become pathogenic (and, presumably, vice versa).

We conclude that the very wide diversity within the bacterial microbiota within moths includes many that will be either facultative or obligate pathogens. Few if any non-pest species of moths have been investigated and this will be a productive field for community ecologists interested in micro-organisms. We conclude this section by noting two taxa of considerable significance for moth wellbeing.

Bacillus thuringiensis (Bt) is a Gram-positive bacterium in the phylum Firmicules belonging to the family Bacilliaceae. It was first noted as causing heavy mortality in silkworm populations in Japan in 1901. Later a highly effective insecticide based on Bt toxins was developed and deployed against moth pests of crops in the USA (Bravo *et al.*, 2011). Many different strains of Bt exist, each with very specific targets among Lepidoptera hosts (Jurat-Fuentes and Jackson, 2012). Genetic engineering allowed key crops themselves to be manipulated so as to produce Bt toxins, effectively protecting them from attack by earworms, bollworms and related pests. This sparked significant debate on ethical, nutritional and environmental grounds, now largely muted.

Wolbachia is a genus of rickettsial Protobacteria generally supposed to be the most widespread 'endosymbiotic' bacterium on Earth (Hughes and Rasgon, 2012) occurring on six continents (Ahmed *et al.*, 2015). It is suggested to occur in 40% of all arthropod species (Zug and Hammerstein, 2012). It occurs across all insect orders that have been examined as well as Arachnida and Nematoda. Estimates of its occurrence in Lepidoptera rise as high as 80% but with a mean of about 24% (Ahmed *et al.*, 2015; Weinert *et al.*, 2025). The bacterium interferes with reproductive processes within its host, producing a variety of effects from male 'feminization' to increased susceptibility to other diseases (Graham *et al.*, 2012). Gene transfer with the host may also occur (Hotopp, 2011). Mass-released *Wolbachia*-infected male mosquitoes have been successful in control of vector mosquitoes in Australia (Hoffmann *et al.*, 2011) and elsewhere, and similar approaches hold considerable potential for the non-chemical control of other species including moth pests.

6.3.10 Viruses

Viruses are small segments of RNA or DNA enclosed in a 'capsid' with or without a further enclosing proteinaceous 'envelope'. They have no independent means of reproduction and must use their host's mechanisms to multiply within the host cells. Accordingly they are, universally, obligate endoparasites. Many are also pathogenic either within their primary host or within secondary hosts to which they are transmitted by the primary host. They persist outside the hosts' bodies only over such time as their outer coatings remain inviolate.

To quote Rothman and Myers (2000): 'Insects, mostly in the orders Diptera.., Hymenoptera.., Coleoptera.., and particularly Lepidoptera..are hosts to a variety of viruses...some cause spectacular epizootics and extensive mortality, while others are more benign...'. The emphasis on Lepidoptera reflects an early stocktaking in which 70% of all pathogenic viruses known from holometabolic insects were from Lepidoptera (Martignoni and Iwai, 1986). Insect virology has become a major field of applied entomology, with entire textbooks introducing the subject matter (e.g. Hurst, 2000; Asgari and Johnston, 2010; Smith, 2012). Here we can only note a few basics with examples from Lepidoptera. Most virus studies have been carried out on pest species of insects with a view to biocontrol. Many of the moth species involved are present in Australia although most of the comments below relate to studies across the globe.

Rothman and Myers (2000) provide an accessible introduction. Insect viruses may be classified in various ways. Fundamentally, they may be RNA or DNA based, in each case either single- or double-stranded. They may be further divided based on the presence or absence of so-called occlusion bodies. Eight different major groups of DNA viruses from insects and six of RNA forms are recorded.

6.3.10.1 DNA viruses

Of the eight major groups of DNA viruses known from insects, five have been associated with lepidopteran hosts principally in studies of economically important pest species of moths.

Ascoviruses are double-stranded DNA viruses which occur in noctuid moths such as *Spodoptera*, *Helicoverpa*, *Heliothis* and *Trichoplusia*. They are transmitted across moth larvae by parasitoidal wasps. They lead to cell death in their lepidopteran hosts and have potential for biological control (Prasad and Srivastava, 2016). Densoviruses have single strands of DNA and cause lethal infections in *Spodoptera* species. Polydnaviruses are double-stranded DNA viruses again vectored by parasitoidal wasps and particularly present in moths. The presence of the virus apparently suppresses the moths' immune system, enhancing survivorship of the parasitoid larvae (Drezen *et al.*, 2014). *Bracovirus* has been used for biocontrol of *Plutella xylostella*. Entomopoxviruses have double-stranded DNA. They have been used to assist control of armyworm in rice crops and entomopox genes have been incorporated in transgenic rice plants (Hukuhara *et al.*, 1999).

Baculoviruses are undoubtedly the best known and most widespread viruses infecting Lepidoptera. This group of double-stranded DNA viruses contains nucleopolyhedrosis (NPV) viruses and granuloviruses (GV). Prasad and Srivastava (2016) list over 50 alpha-baculoviruses plus 17 beta-baculoviruses from Lepidoptera across a wide range of families. As a group, these appear to be Lepidoptera specialists with just a very few known from Hymenoptera and Diptera. They are highly effective pathogens with an exceptional capacity to exist in a dormant state outside their insect hosts. They have been widely used as biopesticides (Moscardi, 1999).

6.3.10.2 RNA viruses

Five of the six major groups of RNA viruses associated with insects are recorded from Lepidoptera (Chen *et al.*, 2012). Flaviviruses in arthropods appear to be restricted to mosquitoes and ticks.

Dicistroviruses are single-stranded RNA viruses which occur across a range of insect orders, including the Lepidoptera. Infection causes leg paralysis and ultimately death. Iflaviruses are single-stranded RNA viruses recorded from a wide range of insect orders. Iflaviruses have been isolated from the geometrid moth *Ectropis obliqua*, a pest of tea plants in China, and the silkworm *Bombyx mori* (van Oers, 2010). Nodaviruses is another single-stranded group occurring in Lepidoptera. Investigations have generally used *Galleria* (Pyralidae) as a subject host species. The group appears to have a wide range of lepidopteran (and other) hosts (including Crustacea and mammals). Tetraviruses are single stranded and a range of types have been recorded from the moth *Nudaurelia cytherea*, an African saturniid. These viruses cause periodic epizootics (Hendry *et al.*, 1985). Cypoviruses are double-stranded RNA viruses infecting Hymenoptera, Diptera and, particularly, Lepidoptera. They are recorded from Noctuidae, Bombycidae and Lasiocampidae as well as pest species of Megalopygidae in the Neotropics.

We have used global references in summarizing the occurrence of viruses in moths. As noted, these viral hosts have almost always been species of economic importance. There can be no doubt that a range of viruses occur in Australian species and many agricultural and medically oriented laboratories have active virological programmes. Explicit studies of insect viruses focus, on the one hand, on viruses of medical or veterinary importance emphasizing vector species such as mosquitoes and midges and, on the other, on species of agricultural pests (as noted above). Indeed, to even contemplate the range of virus 'species' which must occur in wild populations of no particular economic importance is mind-boggling.

We are not aware of any Australian overview of insect viruses and their impacts on moths.

6.4 Overcoming Natural Enemy Impacts

Countering predators and parasitoids involves many 'antipredator' strategies and devices, spanning visual, chemical and acoustic 'defences', the last often less heeded but reflecting that 'caterpillars live in rather complex acoustic environments' (Yack, 2022). Many changes in behaviour, appearance, habitat, diet and seasonal appearance of moths can be postulated as some form of 'defence', wider 'resistance', 'retaliation' or 'avoidance' of attack by predators and/or parasitoids, with acknowledgement that few – if any – moth taxa are free from such mortality-causing agents. As with the dynamics between herbivorous insects and their food plants, continuing adaptations and counter-adaptations between natural enemies and their prey or hosts are likely to be universal and often subtle. The idea of 'enemy-free space' reflects the need for spatial or ecological environments in which natural enemy impacts may be avoided or lowered and has generated the general idea that caterpillar diet restriction may reflect response to predation, as well as responses to plant metabolites (Singer *et al.*, 2022). Defences by caterpillars may be pre-emptive, concurrent or subsequently during encounters with predators or parasitoids – essentially avoiding or preventing being noticed, repelling or eluding contacts and potential threats – and 'after the event' measures such as physiological elimination or incapacitating parasitoid offspring (Quicke *et al.*, 2023).

The methods by which moths may counter detection and attacks by predators and parasitoids, summarized below, demonstrate the many ways in which natural enemies may have been selective forces influencing moth evolution and ecology. The outcome, the great variety of appearances, habitats and behaviour shown among moths and their early stages, is powerful evidence of these impacts and supposedly of the means taken in moth evolution to thwart natural enemies in the continuing dynamic duel between attackers and attacked. Various adaptations may occur together – physical and chemical defences can coincide, for example. The autumn gum moth *Mnesampela privata* (Geometridae) may have reduced encounter rates with natural enemies by a combination of larval leaf shelters and autumnal activity, as a season in which enemies are less numerous (Steinbauer *et al.*, 2001).

Moths (embracing all life stages) may counter natural enemies by being inconspicuous so they are not found, being in some way physically or chemically deterrent, or becoming highly conspicuous so that they are avoided as advertising that they are harmful or potentially harmful through being distasteful or toxic. Many variations on these themes fall into a few broad categories, discussed briefly below. The vast relevant literature links closely with that on 'plant defences' against herbivore attack – and in many taxa plant chemicals sequestered by caterpillars are subsequently used as protection against predators or parasitoids in subtle and novel ways that continue to surprise and intrigue biologists – and strain credibility at first sight! Consider, for example, the hawkmoth caterpillar that feeds on native *Nicotiana* and has developed a method of passing ingested nicotine to its haemolymph and allowing it to be exhaled through the spiracles during spider attack, as a form of 'defensive halitosis' (Kumar *et al.*, 2014). In a rather different context, caterpillars of some bombycoid moths emit a variety of sounds (clicks, chirps, whistles and other vocalizations), evolved variously to startle predators or to warn of a chemical defence (Bura *et al.*, 2016). The 'whistles' (produced by forcing air through spiracles) of the North American sphingid *Amorpha juglandis* are believed to mimic the alarm calls of small predatory songbirds, causing the predators to cease searching (Bura *et al.*, 2011).

Many such initially unlikely associations have withstood detailed investigation at levels from species to community interactions. The core of much understanding is the immense variety of putative defences that caterpillars have developed against natural enemies (Greeney *et al.*, 2012). However, linking caterpillar features with specific functional contexts can be difficult and involve a 'trade-off' between one benefit or another. Some aposematic features (bright colours), for example, may increase defence but affect thermoregulatory success that depends on dark coloration (e.g. in the Arctiinae: Nielsen and Mappes, 2020).

Coloration and behaviour contribute to one or other of two main 'strategies' in either reducing detectability or issuing some form of 'warning' to natural enemies, predominantly predators. The three main categories of these are (i) characteristics that reduce encounters with enemies; (ii) morphological and behavioural features; and (iii) physiological defences. Their effectiveness is influenced strongly by way of life and the living environment occupied; contrast of colour with that of the host plant may affect detection, for example. Conspicuous caterpillar colours are associated with herbaceous host plants (many of them defended by toxins), while camouflage colours and patterns are often associated with grasses and woody plants (Robinson *et al.*, 2023).

Different defence strategies may be needed against large predators hunting by sight and small parasitoids relying on chemical or sonic detection.

6.4.1 Being cryptic: camouflage

Cryptic insects employ generalized or more specific coloration, form or behaviour as means of camouflage that reduce detectability and accessibility to natural enemies. Many adult moths resting on vegetation are variously brown, grey, striped or mottled in various ways that resemble their substrates, and caterpillars are often coloured to resemble their food plants. Both stages show innumerable variations in posture, shape, and structure that aid crypsis

'Hiding' also involves avoiding exposed environments, but with larval habits such as tunnelling and leaf mining (Chapter 5) often not now effective in avoiding parasitoid attack. Caterpillars feeding endophytically, such as inside tree branches, are attacked by Ichneumonoidea with ovipositors far longer than their body. Solitary or group living may also be linked to the balance between crypsis and aposematism (Wang *et al.*, 2021).

6.4.2 Being conspicuous: aposematism

Aposematic insects employ visual and chemical signals to advertise their unpalatability to predators. Many caterpillars and other herbivores sequester toxins from their food plants, with their defence process being that a predator catches/eats an individual, becomes ill and learns to avoid similar-looking or similar-smelling insects as future prey; the more conspicuous the potential food species, the more effective the protection for other individuals may be. Aggregations may increase the overall defensive effect. Chemical sequestration may compromise the caterpillar's immunity to parasitoids so that they become more susceptible – and suggestions that the combination of protection from predators and compromised immunity can make aposematic caterpillars ideal hosts for parasitoids have generated this as a 'safe haven hypothesis' (after Dyer and Gentry, 1999).

The 'warning coloration' of aposematic Lepidoptera can transcend growth stages. Patterns of adults and larvae of some moths are somewhat similar, and suggested to broaden the defence effect for the species as a whole.

6.4.3 Confusing the enemy: mimicry

Mimicry is a counter-predator strategy that benefits palatable species. Not all brightly coloured or conspicuous insects are actually toxic. Many are innocuous and suitable food for predators, but adopt the markings, colour or behaviour of toxic species, and gain protection when predators avoid them from previous learning experience as exhibiting a 'common pattern of potential harm'. Species from different taxa, even orders, may converge in appearance and benefit from co-existence with toxic species. Multiple species may benefit from adopting a common mimicry pattern, as long as they do not collectively outnumber the toxic model species. Likewise, several toxic species may converge in appearance to widen possible joint protection.

Mimicry has many forms, and much functional interpretation has come from studies on adult or larval Lepidoptera. Confusing colour and markings may be accompanied by behavioural changes that increase superficial resemblance between model and mimic. The topic has played extensively in discussions of Lepidoptera protection from predators.

6.4.4 Confusing the enemy: thwarting attack

The dense hairs and spines of many caterpillars aid both camouflage and defence. In many species, body hairs simply impede access to the body surface for parasitoid oviposition. In others the hairs are 'urticating', containing toxins that induce pain and inflammation to animals contacting them – some people can have serious anaphylactic reactions to such 'caterpillar rash' – and, in turn, some gregarious larvae incorporate these hairs into the webs that cover their communal aggregations, as is the case for *Ochrogaster* spp. (see Chapter 29).

Regurgitation or secretion of defensive or repellent chemicals from caterpillars or adult moths is not uncommon and can become both complex and costly. Male moths of the aposematic wood tiger (*Arctia plantaginis*) in Europe can produce two chemically different kinds of secretion when confronted by predators; the abdominal secretion is more effective against invertebrate predators such as ants, and that from the thoracic glands is directed toward bird predators. However, the moths do not feed and deploying the abdominal secretion may decrease reproductive capacity (Lindstedt *et al.*, 2020), so 'defence' incurs 'reproductive cost', together with that of earlier producing aposematic chemicals.

'Startle display' involves the sudden or unexpected appearance of features that alarm a potential predator. The well-defined 'eyespots' or other bright markings on hindwings of some moths are normally concealed by the forewings when the moth is resting – but when disturbed they are exposed, sometimes with the moth repeatedly moving forewings that gives the appearance of blinking eyes and causes the predator to curtail attack. Likewise, some caterpillars have 'eyespot' marks or more elaborate markings that can resemble the head of small vertebrate predators and may extend or rear up to increase exposure of these; the larva of *Phyllodes imperialis* (Chapter 36) superficially resembles a small aggressive predator, for example.

6.4.5 Visual and acoustic behaviour

Most predators of caterpillars and adult moths respond primarily to visual detection whereas many parasitoids rely little on sight but depend on vibrational/acoustic detection and chemical cues. This different emphasis translates in part to many predators being mobile and searching actively for prey, and many parasitoids searching more intensively over a smaller universe selected as likely to be rewarding.

6.4.6 Protection from within: the moth immune system

When pathogens, or even larger challenges such as parasitoid eggs, enter the haemocoel of an insect, they are faced with overcoming the physiological defence mechanisms innate to the insect body – its immune system.

Insect immune systems act at two levels (Sheehan *et al.*, 2020). The *cellular* level consists of haemocytes which may engulf infective bodies by phagocytosis or, for larger intrusions, encapsulate them. At the *humeral* level, haemocytes and fat-body cells produce antimicrobial peptides and melanins. The peptides interact with invasive bacteria by hydrolysing their cell walls. Others are directly lethal to both bacteria and fungi.

Insects do not possess a direct specific response immune system producing targeted antibodies as do vertebrates. There is, however, a system referred to as 'immune priming' whereby initial contact with low levels of potential pathogens can 'prime' the system so that a subsequent response to pathogen challenges is more efficient (Cooper and Eleftherianos, 2017).

Interestingly, much of the detail of these insect immune responses was worked out using *Galleria* (Pyralidae) and *Hyalophora* (cecropia moths, Saturniidae) and other Lepidoptera as model organisms.

6.5 Generalizations and Conclusions

The greatest contributions to understanding moth ecology and evolution have flowed from progressive documentation of caterpillar–parasitoid associations, and range from the species-level contexts emphasized above to having major implications for community structure. Vertical distribution of caterpillars was linked with their levels of parasitoid attack and defensive strategies in tropical forest in Cameroon (Finnie *et al.*, 2024), with defence strategies being important indicators of parasitoid attack rates at different vertical strata. Such patterns or putative correlations frequently invoke interactions between plants, caterpillars and natural enemies, with 'costs' in fitness influencing the balance; host toxicity is a widespread theme in attempts to 'explain' diversity and specificity of parasitoids. Consideration of all three levels in these tritrophic associations is necessary (Marquis and Koptur, 2022).

The 'safe haven hypothesis' (Lampert *et al.*, 2010) reflects a rather different scenario – that parasitoids which can tolerate higher levels of sequestered plant compounds in their host gain a protective refuge because the host is strongly avoided by predators and hyperparasitoids. The sequestering herbivore host becomes a 'safe haven' for the parasitoid. Lampert *et al.*'s study was of the North American catalpa sphinx (*Ceratomia catalpa*), whose larvae sequester the iridoid glycoside catalpol from the sole food plant, leading to refuge for the braconid parasitoid *Cotesia congregata* through its high tolerance of the compound.

Another widely discussed theme is the 'nasty host hypothesis', initially based largely on observations that (i) tropical plants commonly carry heavier loads of protective plant compounds that (ii) in turn may lead to higher levels of host specialization by insect herbivores that (iii) sequester these compounds from their defences against natural enemies and (iv) lead to higher specialization of parasitoids in the tropics than elsewhere.

However, as Quicke (2012) pointed out, interpreting geographical patterns of parasitoid diversity and distribution remains highly incomplete and reflects 'massive under-description of tropical parasitoid faunas'. As elsewhere, large numbers of parasitoid wasps are captured in insect surveys in Australia but are largely disregarded in favour of particular target species or groups of greater interest to the practitioners.

The influence of most predators reflects visual detection of prey. Most parasitoids lack such vision and searching for hosts is commonly linked more with chemical and acoustic cues within particular plant or microclimatically defined arenas. Cataloguing the subtleties of any such association has largely been opportunistic, and focused on individual taxa and interactions. Impacts on the parent communities are far less defined. Clearly, wider effects flow from changing relative abundance of co-occurring prey/host species and their natural enemies are key drivers of such changes, the extent of which may reflect local mortality rate as a specific outcome of the host community and state.

To many entomologists, the ubiquity and complexity of moth–parasitoid associations, in particular, provide a formidable hindrance to understanding – as well as unwelcome components of rearing exercises anticipated to provide perfect cabinet specimens of host moths. All recorded host-associated reared parasitoids can contribute to furthering knowledge of host ranges, distributions, seasonal incidence and other aspects of the intergrading nexus between natural history and science so necessary to understand biodiversity. Increased understanding and documentation of Australia's moths and their natural enemies is an important facet of that endeavour, and one to which hobbyists, citizen scientists and professional biologists can all contribute.

Acknowledgement

We thank Owen Seeman, Queensland Museum, for valuable input on mites.

References

Ahmed, M.Z., Araujo, E.V., Welch, J.J. and Kawahara, A.Y. (2015) *Wolbachia* in butterflies and moths: geographic structure in infection frequency. *Frontiers in Zoology* 12(16), 1–9.

Akhurst, R.J. and Bedding, R.A. (1986) Natural occurrence of insect pathogenic nematodes (Steinernematidae and Heterorhabditidae) in soil in Australia. *Journal of the Australian Entomological Society* 25, 241–2454.

Asgari, S. and Johnston, K.N. (eds) (2010) *Insect Virology*. Caister Academic Press, Norfolk.

Atkins, Z.A., Clemann, N., Schroder, M., Chapple, D.G., Davis, N.E. *et al.* (2018) Consistent temporal variation in the diet of an endangered alpine lizard across two south-eastern Australian sky-islands. *Austral Ecology* 43, 339–351.

Bedding, R.A. and Akhurst, R.J. (1975) A simple technique for the detection of insect pathogenic nematodes in soil. *Nematologica* 21, 109–110.

Bedding, R.A., Akhurst, R.J. and Kaya, H.K. (eds) (1993) *Nematodes and the Biological Control of Insect Pests*. CSIRO Publishing, Clayton South, Victoria.

Boggs, C.L. and Gilbert, L.E. (1987) Spatial and temporal distribution of *Lantana* mites phoretic on butterflies. *Biotropica* 19, 301–305.

Bravo, A., Likitvivatanavong, S., Gill, S.S. and Soberón, M. (2011) *Bacillus thuringiensis:* A story of a successful bioinsecticide. *Insect Biochemistry and Molecular* Biology 41, 423–431.

Broome, L.S. (2001) Density, home range, seasonal movement and habitat use of the mountain pygmy possum, *Burramys parva* (Marsupialia: Burramyidae) at Mount Blue Cow, Koskiuszko National Park. *Austral Ecology* 26, 275–292.

Bura, V.L., Rohwer, V.G., Martin, P. and Yack, J.E. (2011) Whistling in caterpillars (*Amorpha juglandis*, Bombycoidea): Sound-producing mechanism and function. *Journal of Experimental Biology* 214, 30–37.

Bura, V.L., Kawahara, A. and Yack, J.E. (2016) A comparative analysis of sonic defences in Bombycoidea caterpillars. *Scientific Reports* 6, 31469.

Chandler, D. (2017) Basic and applied research on entomopathogenic fungi. In: Lacey, L.A. (ed.) *Microbial Control of Insect and Mite Pests*. Academic Press, London, pp. 69–89.

Chen, Y.P., Becnel, J.J. and Valles, S.M. (2012) RNA viruses infecting pest insects. In: Vega, F.E. and Kaya, H.K. (eds) *Insect Pathology*, 2nd edn. Academic Press, London, pp. 133–170.

Cooper, D. and Eleftherianos, I. (2017) Memory and specificity in the insect immune system: Current perspectives and future challenges. *Frontiers in Immunology* 8, 539.

Drezen, J.M., Chevignon, G., Louis, F. and Huguet, E. (2014) Origin and evolution of symbiotic viruses associated with parasitoid wasps. *Current Opinion in Insect Science* 6, 35–43.

Dyer, L.A. and Gentry, G.L. (1999) Predicting natural-enemy responses to herbivores in natural and managed ecosystems. *Ecological Applications* 9, 402–408.

Fagan-Jeffries, E.P., McClelland, A.B., Bird, A.J., Giannotta, M.M., Bradford, T.M. and Austin, A.D. (2022) Systematic revision of the parasitoid wasp genus *Glyptapanteles* Ashmead (Hymenoptera: Braconidae: Microgastrinae) for Australia results in a ten-fold increase in species. *European Journal of Taxonomy* 792, 1–116.

Fagan-Jeffries, E.L., Miles, M.M. Atkin-Zalvidar, T.-C., Howe, A.G., Subramaniam, S. *et al.* (2024) Hymenopteran parasitoids of fall armyworm (*Spodoptera frugiperda* (J. E. Smith)(Lepidoptera: Noctuidae)) in Australia with the description of five new species in the families Braconidae and Eulophidae. *Austral Entomology* 63, 136–174.

Finnie, S., Butterill, P., Novoyny, V., Redmond, C., Jorg, L.R. *et al.* (2024) Vertical stratification and defensive traits of caterpillars against parasitoids in a lowland tropical forest in Cameroon. *Oecologia* 204, 915–930.

Framenau, V.W., Baehr, B.C. and Zborowski, P. (2014) *A Guide to the Spiders of Australia*. New Holland, London.

Fullard, J.H., Jackson, M.E., Jacobs, D.S., Pavey, C.R. and Burwell, C.J. (2008) Surviving cave bats: Auditory and behavioural defences in the Australian noctuid moth, *Speiredonia spectans*. *Journal of Experimental Biololgy* 211, 3808–3815.

Gielen, R., Meister, H., Tammaru, T. and Põldmaa, K. (2020) Fungi recorded on folivorous Lepidoptera: High diversity despite moderate prevalence. *Journal of Fungi* 7, 25–33.

Glatz, R.V., Fagan-Jefferies, E.P., Tetley, S.P. and Austin, A.D. (2022) Enigma moth parasitoid: A new Australian cyclostome genus and species, *Ovaustra aurantia* Tetley, Glatz & Fagan Jefferies, gen. et sp. nov. (Hymenoptera: Braconidae) from Kangaroo Island. *Austral Entomology* 61, 420–432.

Godfray, H.J.C. (1994) *Parasitoids. Behavioral and Evolutionary Ecology*. Princeton University Press, Princeton, NJ.

Graham, R.I., Grzywacz, D., Mushobozi, W.L. and Wilson, K. (2012) *Wolbachia* in a major African crop pest increases susceptibility to viral disease rather than protects. *Ecology Letters* 15, 993–1000.

Greeney, H.F., Dyer, L.A. and Smilanich, A.M. (2012) Feeding by lepidopteran larvae is dangerous: A review of caterpillars' chemical, physiological, morphological, and behavioral defenses against natural enemies. *Invertebrate Survival Journal* 9, 7–34.

Gross, G.F. (1975) *Plant-feeding and other Bugs (Hemiptera) of South Australia. Heteroptera – Part 1*. Handbook of the Flora and Fauna of South Australia. Government of South Australia, Adelaide, South Australia.

Hall, S. (1980) The diets of two coexisting species of *Antechinus* (Marsupialia: Dasyuridae). *Australian Wildlife Research* 7, 365–378.

Halliday, R.B., Walter, D.E. and Lindquist, E.E. (1998) Revision of the Australian Ascidae (Acarina: Mesostigmata). *Invertebrate Taxonomy* 12, 1–54.

Hawke, T., Bates, H., Hand, S., Archer, M. and Broome, L. (2018) Dietary analysis on an uncharacteristic population of the mountain pygmy possum (*Buramys parva*) in the Kosciuszko National Park, New South Wales, Australia. *PeerJ* 7, e6307.

Hendry, D., Hodgson, V., Clark, R. and Newman, J. (1985) Small RNA viruses co-infecting pine emperor moth (*Nudaurelia cytherea capensis*). *Journal of General Virology* 66, 627–632.

Henle, K. (1990) Population ecology and life-history of the arboreal gecko, *Gehyra variegata* in arid Australia. *Herpetological Monographs* 4, 30–60.

Hodgkison, S. and Hero, J.-M. (2003) Seasonal, sexual and ontogenetic variations in the diet of the 'declining' frogs *Litoria nannotis, Litoria rheocola* and *Nyctimystes dayi*. *Wildlife Research* 30, 345–354.

Hoffmann, A., Montgomery, B.L., Popovici, J., Iturbe-Ormaetxe, I., Johnsson, P.H. *et al.* (2011) Successful establishment of *Wolbachia* in *Aedes* populations to suppress dengue transmission. *Nature* 476, 454–457.

Hofstede, H.M. and Ratcliffe, J.M. (2016) Evolutionary escalation: The bat-moth arms race. *Journal of Experimental Biology* 219, 1589–1602.

Hotopp, J.C.D. (2011) Horizontal gene transfer between bacteria and animals. *Trends in Genetics* 27, 157–163.

Howarth, F.G. (2000) Non-target effects of biological control agents. In: Gurr, G. and Wratten, S. (eds) *Biological Control: Measures of Success*. Springer, Dordrecht, The Netherlands, pp. 369–403.

Hughes, G.L. and Rasgon, J.L. (2012) *Wolbachia* infections in arthropod hosts. In: Vega, F.E and Kaya, H.K. (eds) *Insect Pathology*, 2nd edn. Elsevier, Amsterdam, The Netherlands, pp. 351–366.

Hukuhara, T., Hayakawa, T. and Wijonarko, A. (1999) Increased baculovirus susceptibility of armyworm larvae feeding on transgenic rice plants expressing an entomopoxvirus gene. *Nature Biotechnology* 17, 1122–1124.

Humber, R.A. (2012) Entomophthoromycota: a new phylum and reclassification of entomophthoroid fungi. *Mycotaxon* 120, 477–492.

Hurst, C.J. (ed.) (2000) *Viral Ecology*. Academic Press, London.

Jones, R. (2023) *Shieldbugs*. HarperCollins, London.

Jurat-Fuentes, J.L. and Jackson, T.A. (2012) Bacterial entomopathogens. In: Vega, F.E. and Kaya, H.K. (eds) *Insect Pathology*, 2nd edn. Elsevier, Amsterdam, The Netherlands, pp. 265–349.

Kawahara, A.Y., Winkler, I.S. and Hsu, W. (2006) New host records of the ectoparasitic biting midge *Forcipomyia (Trichohelea) pectinunguis* (Diptera: Ceratopogonidae) on adult geometrid moths (Lepidoptera: Geometridae). *Journal of the Kansas Entomological Society* 79, 297–300.

Kitching, R.L., Bickel, D.J. and Boulter, S. (2005) Guild analysis of dipteran assemblages: A rationale and investigation of seasonality and stratification in selected rainforest faunas. In: Yeates, D.K. and Wiegmann, B.M. (eds) *The Evolutionary Biology of Flies*. Columbia University Press, New York, pp. 388–415.

Kumar, P., Pandit, S.S., Steppuhn, A. and Baldwin, I.T. (2014) Natural history-driven, plant-mediated RNAi-based study reveals CYP6B45's role in a nicotine-mediated antipredator herbivore defense. *Proceedings of the National Academy of Sciences* 111, 1245–1252.

Lampert, E.C., Dyer, L.A. and Bowers, M.D. (2010) Caterpillar chemical defense and parasitoid success: *Cortesia congregata* parasitism of *Ceratomia catalpa*. *Journal of Chemical Ecology* 36, 992–998.

Lange, C.E. and Lord, J.C. (2012) Protistan entomopathogens. In: Vega, F.E. and Kaya, H.K. (eds) *Insect Pathology*, 2nd edn. Elsevier, Amsterdam, The Netherlands, pp. 367–394.

Lawton, J.H. and McNeill, S. (1979) Between the devil and the deep blue sea; on the problem of being a herbivore. In: Anderson, R.M., Turner, B.D. and Taylor, L.R. (eds) *Population Dynamics*. Blackwell Scientific Publications, Oxford, UK, pp. 223–244.

Lee, S.C., Corradi, N., Doan, S., Dietrich, F.S., Keeling, P.J. and Heitman, J. (2010) Evolution of sex-related locus and genomic features shared in Microsporidia and Fungi. *PLOS One* 5, e10539.

Lima, A., Magnusson, W.E. and Williams, D.G. (2000) Differences in diet among frogs and lizards coexisting in subtropical forests of Australia. *Journal of Herpetology* 34, 40–46.

Lindquist, E.E., Krantz, G.W. and Walter, D.E. (2009) Order Mesostigmata. In: Krantz, G.W. and Walter, D.E. (eds) *A Manual of Acarology*, 3rd edn. Texas Tech University Press, Lubbock, TX, pp. 124–232.

Lindstedt, C., Suisto, K., Burdfield-Steel, E., Winters, A.E. and Mappes. J. (2020) Defense against predators increases high reproductive costs for the aposematic moth *Arctia plantaginis*. *Behavioral Ecology* 32, 844–850.

Liu, F.-F., Tian, T.-A., Li, X.-L., Chen, Y.-C., Yu, X.-F., Tan, X.-F. *et al.* (2020) Is *Pyemotes zhonghuajia* (Acari: Pyemotidae) a suitable biological control agent against the fall armyworm *Spodoptera frugiperda* (Lepidoptera: Noctuidae)? *Systematic and Applied Acarology* 25, 649–657.

Manicum, C. and Schwarzkopf, L. (2011) Diet and prey selection of sympatric tropical skinks. *Austral Ecology* 36, 485–496.

Marquis, R.J. and Koptur, S. (eds) (2022) *Caterpillars in the Middle: Tritrophic Interactions in a Changing World*. Springer, Cham, Switzerland.

Martignoni, M.E. and Iwai, P.J. (1986) *A Catalog of Viral Diseases of Insects, Mites, and Ticks*, 4th edn. US Department of Agriculture, Forest Service, Pacific Northwest Research Station, Oregon.

Mayer, K. (1955) Der Parasitismus de heleiden (Ceratopogoniden) Diptera. *Zeitschrift für Angewandte Zoologie* 1, 95–107.

Mereghetti, V., Chouaia, B. and Montagna, M. (2017) New insights into the microbiota of moth pests. *International Journal of Molecular Sciences* 18, 2450.

Milne, D.J., Burwell, C.J. and Pavey, C.R. (2016) Dietary composition of insectivorous bats of the Top End of Australia. *Australian Mammalogy* 38, 213–220.

Minnaar, C., Boyles, J.G., Minnaar, I.A., Sole, C.L. and McKechnie, A.E. (2015) Stacking the odds: Light pollution may shift the balance in an ancient predator–prey arms race. *Journal of Applied Ecology* 52, 522–531.

Moore, K.M. (1958) Observations on some Australian forest insects 2. A biting midge (Diptera: Ceratopogonidae) and mortalities among larvae of Lepidoptera (Limacodidae). *Proceedings of the Royal Zoological Society of New South Wales* 1956–1957, 77–78.

Morse, D.H. (1983) Foraging patterns and time budgets of the crab spiders *Xysticus emertoni* Keyserling and *Misumena vatla* (Clerck) (Araneae: Thomisidae) on flowers. *Journal of Arachnology* 11, 87–94.

Moscardi, F. (1999) Assessment of the application of baculoviruses for control of Lepidoptera. *Annual Review of Entomology* 44, 211–234.

Moulds, M.S., Tuttle, J.P. and Lane, D.A. (2020) *Hawkmoths of Australia: Identification, Biology and Distribution*. Monographs in Australian Lepidoptera 13. CSIRO Publishing, Clayton South, Victoria.

Naumann, I. (1991) Hymenoptera. In: Naumann, I. (ed.) *Insects of Australia*, vol. 2, 2nd edn. Melbourne University Press, Carlton, Victoria, pp. 916–1000.

New, T.R. (1999) Untangling the web: Spiders and the challenges of invertebrate conservation. *Journal of Insect Conservation* 3, 251–256.

Nickle, W.R. (1972) A contribution to our knowledge of the Mermithidae (Nematoda). *Journal of Nematology* 4, 113–146.

Nielsen, M.E. and Mappes, J. (2020) Out in the open: Behavior's effect on predation risk and thermoregulation by aposematic caterpillars. *Behavioral Ecology* 31, 1031–1039.

Olatunji, O.J., Tang, J., Tola, A., Auberon, F., Oluwaniyi, O. and Ouyang, Z. (2018) The genus *Cordyceps*: An extensive review of its traditional uses, phytochemistry and pharmacology. *Fitoterapia* 129, 293–316.

Paull, L. and Austin, A.D. (2006) The hymenopteran parasitoids of light brown apple moth, *Epiphyas postvittana* (Walker) (Lepidoptera: Tortricidae) in Australia. *Australian Journal of Entomology* 45, 142–156.

Pavey, C.R. and Burwell, C.J. (1998) Bat predation on eared moths: a test of the allotonic frequency hypothesis. *Oikos* 81, 143–151.

Pilarska, D., Takov, D., Hylis, M., Radek, R., Fiala, I. *et al.* (2017) Natural occurrence of microsporidia affecting Lepidoptera in Bulgaria. *Acta Parasitologia* 62, 858–869.

Poinar, G.O. (1979) *Nematodes for Biological Control of Insects*. CRC Press, Boca Raton, FL.

Polaszek, A. and Vilhelmsen, L. (2023) Biodiversity of hymenopteran parasitoids. *Current Opinion in Insect Science* 56, 101026.
Prasad, V. and Srivastava, S. (2016) Insect viruses. In: Omkar (ed.) *Ecofriendly Pest Management for Food Security*. Elsevier, Amsterdam, The Netherlands, pp. 411–442.
Quicke, D.J.L. (2012) We know too little about parasitoid wasp distributions to draw any conclusions about latitudinal trends in species richness, body size and biology. *PLOS One* 7, e32101.
Quicke, D.J.L. (2015) *The Braconid and Ichneumonid Parasitoid Wasps: Biology, Systematics, Evolution and Ecology*. John Wiley and Sons, Chichester.
Quicke, D.L.J., Moghaddan, M.G. and Butcher, B.A. (2023) Dietary challenges for parasitoid wasps (Hymenoptera: Ichneumonoidea); coping with toxic hosts, or not? *Toxins* 15, 424.
Robinson, M.L., Weber, M.G., Freedman, M.G., Jordan, E., Ashlock, S.R. *et al.* (2023) Macroevolution of protective coloration across caterpillars reflects relationships with host plants. *Proceedings of the Royal Society B* 290, 20222293.
Rothman, L.D. and Myers, J.H. (2000) The ecology of insect viruses. In: Hurst, C.J. (ed.) *Viral Ecology*. Academic Press, San Diego, pp. 385–412.
Sakagami, K., Funamoto, D. and Sugura, S. (2021) Nocturnal ambush predators and their potential impact on flower-visiting moths. *Ecology* 102, e03482.
Sheehan, G., Farrell, G. and Kavanagh, K. (2020) Immune priming: The secret weapon of the insect world. *Virulence* 11, 238–246.
Singer, M.S., Anderson, R.M., Henessy, A. B., Leggat, E., Prasad. A. *et al.* (2022) Predators and caterpillar diet breadth: Appraising the enemy-free space hypothesis. In: Marquis, R.J. and Koptur, S. (eds) (2022) *Caterpillars in the Middle: Tritrophic Interactions in a Changing World*. Springer, Cham, Switzerland, pp. 273–296.
Smith, K.M. (2012) *Insect Virology*. Academic Press, London.
Solter, L.F., Becnel, J.J. and Oi, D.H. (2012) Microsporidial entomopathogens. In: Vega, F.E. and Kaya, H.K. (eds) *Insect Pathology*, 2nd edn. Elsevier, Amsterdam, The Netherlands, pp. 221–263.
Southcott, R.V. (1986) Studies on the taxonomy and biology of the subfamily Trombidiinae (Acarina: Trombidiidae) with a critical revision of the genera. Australia. *Journal of Zoology* 123(Suppl. Ser), 1–116.
Steinbauer, M.J., McQuillan, P.B. and Young, C.J. (2001) Life history and behavioural traits of *Mnesampela privata* that exacerbate population responses to eucalypt plantations: Comparisons with Australian and outbreak species of forest geometrids from the northern hemisphere. *Austral Ecology* 26, 525–534.
Stireman III, J.O, Cerretti, O., O'Hara, J.E. and Moulton, J.K. (2021) Extraordinary diversification of the 'bristle flies' (Diptera: Tachinidae) and its underlying causes. *Biological Journal of the Linnean Society* 133, 216–236.
Stone, E.L., Harris, S. and Jones, G. (2015) Impacts of artificial lighting on bats: A review of challenges and solutions. *Mammalian Biology* 80, 213–219.
Strahan, R. (ed.) (1983) *The Australian Museum Complete Book of Australian Mammals*. Angus & Robertson, Sydney.
Sutrave, G., Maundrell, A., Keighley, C., Jennings, Z., Brammah, S. *et al.* (2018) *Anncaliia algera* microsporidial myositis, New South Wales, Australia. *Emerging Infectious Diseases* 24, 1528–1531.
Sung, G.-H., Hywel-Jones, N.L., Sung, J-M., Luangsa-ard, J.J., Shrestha, B. and Spatafora, J.W. (2007) Phylogenetic classification of *Cordyceps* and the clavicipitaceous fungi. *Studies in Mycology* 57, 5–59.
Treat, A.E. (1967) Mites from noctuid moths. *Journal of the Lepidopterists' Society* 21, 169–179.
Treat, A.E. (1975) *Mites of Moths and Butterflies*. Corbell University Press, New York.
van Driesche, R. and Hoddle, M. (2016) Non-target effects of insect biocontrol agents and trends in host specificity since 1985. *CAB Reviews* 11, 66.
Van Oers, M. (2010) Genomics and biology of iflaviruses. In: Asgari, S. and Johnston, K. (eds) *Insect Virology*. Caister Academic Press, Norfolk, pp. 231–250.
Vega, F.E. and Kaya, H.K. (eds) (2012) *Insect Pathology*, 2nd edn. Elsevier, Amsterdam.
Vega, F.E., Meyling, N.V., Luangsa-ard, J.J. and Blackwell, M. (2012) Fungal entomopathogens. In: Vega, F.A. and Kaya, H.K. (eds) (2012) *Insect Pathology*, 2nd edn. Elsevier, Amsterdam, The Netherlands, pp. 171–220.
Walter, D.E. and Proctor, H.C. (1999) *Mites: Ecology, Evolution and Behaviour.* University of New South Wales Press, Sydney, and CABI Publishing, Wallingford.

Walter, D.E., Lindquist, E.E., Smith, I.M., Cook, D.R. and Krantz, G.W. (2009) Order Trombidiformes. In: Krantz, G.W. and Walter, D.E. (eds) *A Manual of Acarology*. Texas Tech University Press, Lubbock, TX.

Wang, L., Cornell, S.J., Speed, M.P. and Arbuckle K. (2021) Coevolution of group-living and aposematism in caterpillars: Warning coloration may facilitate the evolution from group-living to solitary habits. *BMC Ecology and Evolution* 21, 25.

Wei, X., Pêgas, R.V., Shen, C., Guo, Y., Ma, W. *et al.* (2021) *Sinomacrops bondei*, a new anurognathid pterosaur from the Jurassic of China and comments on the group. *PeerJ* 9, e11161.

Weinert, L.A., Araujo, E.V., Ahmed, M.Z. and Welch, J.J. (2015) The incidence of bacterial endosymbionts in terrestrial arthropods. *Proceedings of the Royal Society B* 282, 20150249.

Welch, H.E. (1963) *Ephemeris bogongae* sp. nov. and *Hexameris cavicola* sp. nov. from the Australian bogong moth, *Agrotis infusa* (Boisd.), with a review of the genus *Amphimeris* Kaburaki & Imamura, 1932 (Nematoda: Mermithidae). *Parasitology* 53, 55–62.

Whyte, R. and Anderson, G. (2017) *A Field Guide to Spiders of Australia*. CSIRO Publishing, Clayton South, Victoria.

Wilson, S. and Swan, G. (2003) *A Complete Guide to Reptiles of Australia*. New Holland Publishers, Sydney, New South Wales.

Yack, J.E. (2022) Acoustic defence strategies in caterpillars. In: Marquis, R.J. and Koptur, S. (eds) *Caterpillars in the Middle: Tritrophic Interactions in a Changing World*. Springer, Cham, Switzerland, pp. 195–224.

Zalucki, M.P., Clarke, A.R. and Malcolm, S.B. (2002) Ecology and behaviour of first instar larval Lepidoptera. *Annual Review of Entomology* 47, 361–393.

Zug, R. and Hammerstein, P. (2012) Still a host of hosts for *Wolbachia*: analysis of recent data suggests that 40% of terrestrial arthropod species are infected. PLOS One 7: e38544.

7 Dispersal and Migration

V. Alistair Drake
School of Science, University of New South Wales, Canberra, ACT, Australia, and Institute for Applied Ecology, University of Canberra, Canberra, ACT, Australia

Abstract
Both dispersal and migration occur widely within the Australian moth fauna and have been studied with a wide range of technologies. Species of economic significance have received most attention, as movement has major implications for pest management, while the iconic bogong moth, *Agrotis infusa*, has been the subject of curiosity-driven research. Dispersal, over distances of a few hundred metres or a few kilometres, is well documented only for a few pests and biological control agents. Overnight migratory displacements of hundreds of kilometres, at heights above 100 m, have been reported for several moth species, mostly noctuids. The migrants show a degree of common orientation but the flights are primarily windborne. They can bring moths from the inland plains to cropping regions nearer the coasts and even across Bass Strait to Tasmania. Variability of wind direction leading to large-scale mixing may account for the continent-wide panmixia found in some populations.

7.1 Types of Movement

The ability to move, at least during one stage of an individual's life, is a fundamental requirement for an organism's capacity to obtain the resources needed to reproduce and ensure the viability of its offspring. Active, purposive movement is particularly developed in animals, where it serves to locate food, shelter and sites suitable for reproduction, to find a mate and to escape from danger. Movements to locate resources are sometimes classified as *foraging*, *ranging* or *migration*, the distinction depending on their spatial scale and frequency and on differences in locomotory behaviour (Dingle, 2014). Specifically, foraging individuals cease movement as soon as a resource they are seeking, a food item for example, is encountered, whereas migrants ignore such distractions and continue on their way. Foragers typically follow erratic paths as they search for the specific resource they are seeking; migratory movement, in contrast, is persistent and highly directed and the resulting trajectories typically straightened and rapid (Kennedy, 1961, 1985). Ranging is an intermediate form of movement, undertaken by animals seeking a new habitat because the existing one is no longer suitable or already occupied. Like migration, it probably involves a temporary suppression of responses to immediate cues such as individual food items, but it differs in being local in scale. Migration, in contrast, typically occurs over sufficient distances for the migrants to reach a climatically different region and is timed to coincide with resources becoming available there.

Corresponding author: a.drake@unsw.edu.au

DOI: 10.1079/9781800627680.0007

In addition to these individual-based definitions, movement types can be distinguished by the effects they have on ensembles of individuals, or even entire populations. When movements are in a variety of directions and on a limited scale, as is typical during foraging and ranging, the result is mixing within an unaltered range, a process termed *dispersal*. When all individuals move in similar directions, or expand outwards and later retreat inwards, the area occupied by the ensemble changes. Depending on the form and extent of the movement, the initial and final areas may or may not overlap. When movements of these latter types occur on a daily timescale, as animals move back and forth between communal roosts and feeding sites, they are referred to as *commuting*. On a large spatial scale and a seasonal timescale, and when populations are involved, these movements are forms of *migration*. Here the term is being used to describe the *outcome* for a population, a complementary sense to that of the previous paragraph where it referred to the *process* undertaken by the individuals comprising the population.

Because insects are short-lived, there are very few examples worldwide of individuals migrating back and forth between different climate zones, as typically occurs with vertebrates. Instead, insect migratory cycles are usually completed over two or more generations. Contemporary perspectives on insect migration are inclusive, incorporating all levels of biological organization – from the genetic to the evolutionary – and revealing a diverse range of adaptations and ecological interactions (Chapman *et al.*, 2015). They fall within the domain of integrative biology.

The dispersal versus migration distinction forms the basis for the organization of this chapter. These movement types are not always completely separable, and Sappington (2024) has recently proposed the term 'aseasonal undirected migration' to describe movements that involve the persistent flight characteristic of migrants but lead to mixing of the population and dispersal over a relatively large spatial scale. These do not carry a population to regions where the climate is becoming seasonally more favourable, but instead have the function of finding new areas of suitable habitat within the same climatic region; they would seem to represent an extended form of ranging. Movements of Australian moths in the inland, where habitat quality depends primarily on unpredictable rainfall and is typically patchy and ephemeral, may be partially of this type.

Both dispersal and migration have ecological impacts and these in turn can have economic consequences; most obviously, the larvae of many migratory moth species are important crop pests. Other species are now of conservation concern, and spectacular migrations such as that of the monarch butterfly *Danaus plexippus* of North America and Australia's bogong moth *Agrotis infusa* can be regarded as phenomena warranting conservation in their own right (Malcolm, 1993).

7.2 Modes of Locomotion

In pterygote insects, the winged adult has much greater mobility than the preceding stadia, and dispersal and migration are predominantly undertaken by flight. This often (though not always) occurs before reproduction begins (the 'oogenesis-flight syndrome'; Johnson, 1969). In some species, the female's flight apparatus is partially autolysed following migration and the resources converted to egg mass. Terrestrial movements by larvae, over relatively short distances, occur in some species, perhaps primarily as a response to overcrowding which leads them to 'march' *en masse* out of an exhausted habitat in search of fresher and more plentiful resources nearby. Such movements have been reported in Australia (e.g. for cotton bollworm *Helicoverpa armigera* and cotton web spinner *Achyra affinitalis*; Sloan, 1945) but appear to be less frequent, or at least less often noticed, today – perhaps due to changed agricultural practices. The common name 'armyworm' for some moth species arises from this behaviour of the larvae.

The remarkable marching behaviour of the processionary caterpillar *Ochrogaster lunifer* is described in Chapter 29 of this volume; these movements to food trees or pupation sites can be classified as ranging. First instar larvae of some lepidopteran species disperse in the air with the aid of silken threads. In an Australian study, first instars of the common armyworm, *Mythimna convecta*, were shown to move up to 2 m in this way, from a starting point 0.7 m above ground level (McDonald, 1991). Much longer distances might be achieved if the flights were made when the atmosphere was turbulent, as reported for spongy moth *Lymantria dispar* (an erebid) in North America (Taylor and Reling, 1986). The remainder of this chapter will consider only movement by flying adults.

7.3 How Moth Movement is Studied

Moths fly predominantly at night and, unlike some diurnal species (butterflies, dragonflies, hoverflies) that fly close to the ground and are highly visible, their movements are generally not noticed by the casual observer and, accordingly, are easily overlooked (Satterfield *et al.*, 2020). A range of methods is available for detecting insects flying at heights of only a few metres, both by day and by night, including many based on relatively novel technologies such as lidar or nets carried on drones (Drake, 2023). Large-aperture nets mounted on vehicles (Fig. 7.1B) are effective for this purpose and avoid biases that arise with attraction-type traps due to differing or changing responses to light or baits. To assess movement across crop boundaries, a bidirectional interception trap (Macfadyen *et al.*, 2015) might be effective, depending on the flight mode of the species of interest. Inferences about short-distance dispersal can be made from observations of these types, and also from laboratory measurements of flight duration

Fig. 7.1. (A) drogue net interception trap held aloft at a height of about 150 m by a kite, near Trangie, NSW, 1980. (B) Large-aperture truck trap, Narrabri, NSW, 1990. (C) CSIRO Entomological Radar (Mark 2), Trangie, 1980. Photos: V.A. Drake.

(e.g. Danthanarayana and Gu, 1992), but often firmer information connecting start- and end-points is needed.

For species newly introduced to an area, such as biological control agents on first release, dispersal distances can be inferred from the rate of spread, as assessed from periodic surveys for the agents on potential host plants. For established or native species, mark–recapture can be employed. Fluorescent powders allow marking with minimal contact and recapture is with a network of traps, usually pheromone baited, around the release point. Estimated characteristic dispersal distances may need to incorporate a correction if the distribution of traps is not uniform. Moths can be derived from a laboratory culture if necessary to obtain sufficient numbers to provide an adequate recapture sample, though this introduces uncertainty about whether the flight performance of culture specimens mirrors that of wild stock. Some Australian applications of these methods are described in the section on dispersal, below. In a study of two noctuid species, wild populations were built up by growing a 6 ha host crop specifically for them and then marked by spraying this crop with strontium chloride when the populations were at the larval stage; recaptures were confirmed by X-ray spectroscopic analysis, a time-consuming and expensive process (Fitt *et al.*, 1995). Use of natural markers such as pollen, chemical composition or DNA overcomes some problems but increases analysis complexity and introduces additional uncertainties.

For more extended movements, a range of methods is again available (Drake, 1991), with new technologies enhancing and supplementing those used in earlier studies. Occasionally, immigrants arrive in such large numbers that their sudden arrival is evident to all, especially when it occurs early in the season before local populations are active. A spectacular example was provided by *Agrotis infusa* and other species at the closing ceremony of the Olympic Games in Sydney in the spring of 2000 (Taouk, 2020). When moth numbers are monitored systematically, usually with an attraction trap (e.g. light, bait), sudden increases and first records for the season are readily detected. A light trap operated for many years at Stony Rise near Devonport in northern Tasmania has revealed likely migrations from the mainland for numerous species (Hill, 2011). Identification as immigrants is strengthened if, first, expected emergence dates for any local populations are known or can be estimated from temperature records; second, the increase follows a period with a warm wind blowing from the direction of a plausible source population; and, third, specimens of multiple putative migrant species arrive simultaneously. A study of the incidence of *Mythimna convecta* in Tasmania provides a nice example (Hill, 2013).

When it can be established that the captured specimens are immigrants, possible sources can be identified by back-tracking wind trajectories until regions where habitat is suitable, and emergence times are climatically plausible, are reached (e.g. Helm, 1975). In overseas studies, possible origin regions have been inferred from the chemical composition of the captured specimens, from isotope ratios of their elements or from pollen attached to their bodies (e.g. Suchan *et al.*, 2024, which uses all three). In Australia, resistance to pesticides has been used to recognize an immigrant population (Daly and Fitt, 1990). Genomic analyses can provide information about the source regions of trapped specimens (e.g. Rane *et al.*, 2023). When genomics indicate that populations over a wide area are genetically similar, it can be inferred that the population is well mixed ('panmictic') and that at least some individual movements are on a sufficient scale to traverse the area within a few generations (Chapuis *et al.*, 2011). Sophisticated multidisciplinary analyses of these various types can make heavy demands on time and resources and are unsuitable for routine or low-budget investigations.

Trapping can also be used to capture moths while they are in the process of migrating, either by intercepting them in flight or by attracting them down with a powerful upwards-directed light. In China, very powerful vertically pointing 'searchlight' (narrow beam) lamps are routinely used in migration research projects to draw down moths from heights of up to 500 m, with apparently only limited contamination of samples by low-flying local populations (Feng *et al.*, 2004; Huang *et al.*, 2024). This technology has not yet been widely employed in Australia, but upward-pointing lights mounted on towers were operated over five insect flight seasons (September–April) at two mountaintop sites in northern

New South Wales (NSW) from 1985 to 1990, and a range of ancillary observations supported the interpretation that the catches were almost entirely of migrants (Gregg *et al.*, 1993). These traps allowed data on moth migration to be obtained over an extended period, though only at specific sites away from the inland plains where many of the populations are believed to originate. In contrast, interception trapping, using a drogue-type net carried aloft on a kite, a tethered aerodynamic balloon or a light aircraft, is best undertaken over flat terrain but is too labour intensive to sustain for extended periods (Farrow and Dowse, 1984; Fig. 7.1A).

Moths flying at altitudes of hundreds of metres in darkness are visible only when they pass fleetingly across the face of the moon (Pruess and Pruess, 1971). However, they can be detected quite readily with a suitably designed radar, and this technology has been widely used to study insect migration (Drake and Reynolds, 2012), including in Australia. Initially (1970s and 1980s), observations were made with manually operated scanning-beam units which were deployed in the field during periods of expected flight activity by the species of interest, typically for durations of 10–20 days (Fig. 7.1C). Since the late 1990s, these have been largely superseded by fully automatic vertical-beam units which, in their latest versions, operate almost continuously and throughout the year and can detect individual larger insects to heights of 2.5 km (Drake *et al.*, 2020).

Modern weather radars also record echo from insects. In Australia, this has been demonstrated only for locusts (Orthoptera: Acrididae; Rennie, 2013); however, as new units are installed and expertise on using them develops, weather radars may become an important source of information on moth migration. Radar-derived migration data comprises numbers, directions and speeds of movement, and orientation directions, and how all of these vary with height and time.

Unfortunately, and in contrast to methods involving capture (where specimens can be passed to a taxonomist if necessary), radar can provide only rather broad indications of the identity of the insects it detects. The vertical-beam radar type is more effective than earlier scanning units and weather radars in this respect, and can classify the detected insects by their size, wing-beat frequency, and 'electromagnetic shape', although for all three parameters the resolution is coarse. In Australia, locusts and moths, which are believed to constitute the two most frequent categories of high-flying nocturnal migrants, can be reliably distinguished, and the moths can be classified as either 'large' or 'medium' in size (Hao *et al.*, 2020). Other categories emerge from the radar data from time to time, as clusters with a particular combination of the three classification parameters, but their taxonomic identities remain unknown. Ideally, radar observations would be complemented by sampling, but this would add greatly to the cost and in Australia has not been attempted since the 1980s.

A common way to track an animal is to attach a radio transmitter of some type, but most insects are too small to carry even the most lightweight devices available. In Europe, this method has been used to track individuals of a large sphingid, the death's-head hawkmoth *Acherontia atropos*, which were fitted with transmitters and followed (using a light aircraft) for up to 90 km as they migrated south in autumn (Menz *et al.*, 2022). The tagged moths had a mean mass of 2.65 g, ten times the 240 mg of the highly miniaturized transmitters. Most Australian moth migrants are much lighter – for example, *A. infusa* and native budworms *Helicoverpa punctigera* captured following migration weighed in at 170–250 mg and 100–160 mg respectively (Drake *et al.*, 2024a) – and would be incapable of carrying these devices.

An alternative that is much lighter than a transmitter, because it does not need to incorporate a power cell, is the harmonic tag which obtains its power from the incoming signal and transmits at double the received frequency (Drake and Reynolds, 2012). The tagged insect is tracked with an harmonic radar; as only reflections from the tag exhibit frequency doubling, the signal from the insect is not swamped by echoes from nearby terrain and vegetation. Scanning harmonic radars can follow insects in flight; the more affordable handheld direction-finders are only able to relocate them after their flight has terminated. In either case, the extent of movement that can be followed is no more than a few hundred metres, so the method is unsuitable for studying migration but potentially useful for short-range dispersal.

In Europe, responses of moths to pheromones were investigated with a scanning unit (Svensson *et al.*, 2001).

7.4 Dispersal

Studies of two biological control agents have shown slow range expansions from Australian release sites of the order of 100 m per year for the ragwort stem and crown boring moth, *Cochylis atricapitana* (a tortricid; McLaren *et al.*, 2000) and about five times this for the dock moth *Pyropteron doryliformis* (a sesiid; Morley *et al.*, 2004). A mark–recapture study showed that diamondback moths *Plutella xylostella* (a plutellid) moved distances of *ca* 100 m when released within growing brassica crops in South Australia (Mo *et al.*, 2003). Oriental fruit moths *Grapholita molesta* usually moved no further than 200 m, though a few individuals reached 1 km (Il'Ichev *et al.*, 1998). Light-brown apple moths *Epiphyas postvittana* moved a similar distance when released close to an orchard near Canberra, ACT (Geier and Briese, 1980), and up to 600 m in New Zealand where the species has established as a pest (Suckling and Brockerhoff, 2010).

Moths flying adjacent to crops in a cotton-growing area and sampled with an 8 m^2 vehicle-mounted net (Fig. 7.1B) driven over a 4 km loop transect included both *Helicoverpa* spp. and other similar-sized moth species, with the smaller noctuid *Athetis tenuis* particularly numerous (Drake and Fitt, 1990; V.A. Drake unpublished data). The study, using internal strontium marking described in the previous section, established movements of 13 km for *H. armigera* moths and 10 km for native budworms *H. punctigera* (both noctuids; Fitt *et al.*, 1995). The proportion of the moths produced that were recaptured was greater for *H. armigera* than for *H. punctigera*, and the *H. armigera* captures were on average at shorter distances. It was concluded that *H. armigera* disperses over distances of up to 10 km whereas *H. punctigera* is an obligate migrant and most individuals had immediately left the area covered by the trap network. These findings are consistent with inferences that had been drawn from patterns and timings of infestations of the two species and with subsequent research (see below). They have been incorporated into models for assessing the effectiveness of a strategy for preventing the development of resistance to insecticides (Baker *et al.*, 2016; Parry *et al.*, 2017), illustrating the value of firm information on dispersal for management decision making.

7.5 Migration

The aggregation of bogong moths, *Agrotis infusa*, in caves and crevices around the peaks of the Australian Alps was known to the Aboriginal inhabitants at the time of European contact around 1830 (Flood, 1980). Traditional practices for harvesting and processing the moths were documented and there is archaeological evidence of their use as food 2000 years ago (Stephenson *et al.*, 2020; Chapter 37 this volume). Whether the harvesters recognized that the moths had originated far away is uncertain, but their status as long-distance migrants was described to the scientific world by Common (1954) who connected the aestivating moths in the mountains to reports of their 'cutworm' larvae in the inland plains hundreds of kilometres to the north and west, where they were known as an occasional pest of winter or spring crops.

Individuals of this predominantly univoltine species make 'return' migrations, which as noted earlier are very unusual in the insect world – though whether each moth's destination, where it initiates the next generation, has any relation to where it developed as a larva is unknown. Wind trajectory calculations indicate that a single night of flight is potentially sufficient to cover the distance to the aestivation sites in spring and similarly to carry the moths to larval development areas in the plains in autumn (Ahrens *et al.*, 2025). Other evidence, however, indicates a more indirect, multi-step process, at least for the spring flights when immigrant moths are recorded taking nectar at lower-altitude locations far from the aestivation sites (Warrant *et al.*, 2016). Laboratory studies have shown that this species uses both visual and geomagnetic cues to maintain orientation in nocturnal flight (Dreyer *et al.*, 2018).

Winds during the spring and autumn migration periods blow in a wide variety of directions, potentially leading to large-scale mixing of the population and providing a mechanism for the panmixia shown by a genetic analysis using single-nucleotide polymorphisms (Ahrens *et al.*, 2025). Trap monitoring of *A. infusa* in the plains indicates that significant elements of the population pass the summer away from the Alps and summer activity and breeding have been recorded from several locations (McDonald *et al.*, 2024). Thus, aestivation, univoltinism and dependence on alpine caves and rock shelters to survive hot and dry summers are evidently not universal in this species. The bogong moth's remarkable adaptations and anthropological significance ensure continuing public and scientific attention and, following a precipitous decline in numbers at the aestivation sites (Green *et al.*, 2021), a degree of conservation concern.

Although the 1960s saw important developments in understanding of insect migration overseas (e.g. Kennedy, 1961; Southwood, 1962; Johnson, 1969), there was apparently little further interest in moth migration in Australia until the 1970s, when reports emerged of moths (along with butterflies and insects of other orders) flying over the sea and on a small island more than 60 km off the Queensland coast (Fletcher, 1973) and of moths from the Australian mainland reaching Norfolk Island (Holloway, 1982) and New Zealand (Fox, 1978). These last two occurrences required non-stop flights of over 1300 km and 2000 km respectively. Noctuids of a variety of species were involved in all three studies, with smaller numbers of crambids and sphingids. A close association of arrivals with westerly airflows was also recognized. Offshore flights like these must lead to very high mortality, as evidenced in 1980 by numerous moth bodies observed along the high-tide line along the southern New South Wales coast (Drake and Farrow, 1985). Return flights by the small minority that reach an island, or by their offspring if they find suitable hosts and are able to breed, will likely have an equally low prospect of success. Tasmania, however, is a large island with developed agriculture providing potential hosts for migratory pests. Catches, and occasional conspicuous flights, of the southern armyworm *Persectania ewingii* (a noctuid) there in early spring could not be accounted for by local emergence as temperatures were too low, and Helm (1975) boldly proposed that these moths were windborne migrants from a source area 900 km away in SA (Fig. 7.2).

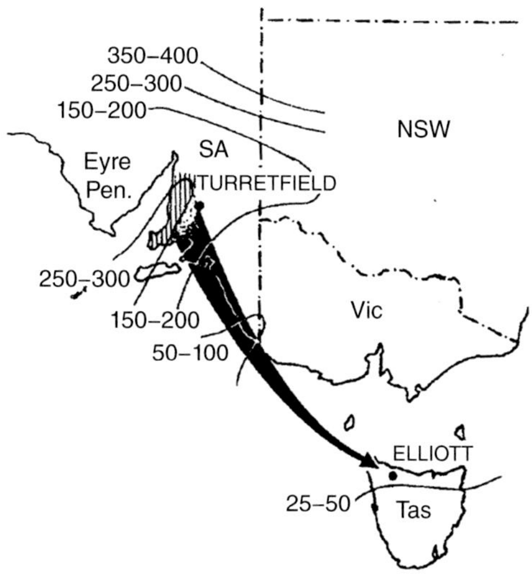

Fig. 7.2. Source area in the Yorke peninsula of South Australia (shaded) and migratory route of *Persectania ewingii* moths occurring in Tasmania in early spring, as proposed by Helm (1975). Contours show degree-days for June–September. Reproduced (with modifications) from Helm (1975).

These early observations have been confirmed by subsequent research. A light trap was operated on Willis Island, in the Coral Sea 450 km off the north Queensland coast, for 11 months in 1977–78, catching over 3000 specimens of 14 noctuid and three crambid species and small numbers of sphingids (seven species) and erebids (*Utetheisa lotrix* and the heliotrope moth *Utetheisa pulchelloides*); catches predominantly followed periods when winds were westerly, from the mainland (Farrow, 1984). Captures in the Stony Rise light trap and analyses of wind trajectories during the immediately preceding days have documented flights across Bass Strait of adults of the noctuid lesser armyworm *Spodoptera exigua* (Hill, 2014) and of the crambids cabbage-centre grub *Hellula hydralis* (Hill, 2012), beet webworm *Spoladea recurvalis* (Hill, 2016) and lucerne seed web moth *Etiella behrii* (Hill, 2018). These flights were often (though not exclusively) in early spring, before local populations, if any, would have reached adult stage. Usually, other probable migrant species – *P. xylostella*, *H. punctigera* and *U. pulchelloides* are specific examples – were captured at the same time, having evidently travelled on the same wind system (Hill, 2014, 2018). These species mostly do not have established populations in Tasmania and the captured specimens can be regarded as vagrants, their arrival in Tasmania an example of accidental movement; nevertheless, their long flights reveal a developed migratory capability that is presumably adaptive within each species' breeding range.

A more spectacular documented example of vagrancy is provided by records of two noctuid species, *P. ewingii* and the southern old lady moth *Dasypodia selenophora*, on Macquarie Island, which is 1200 km southeast of Tasmania and 2000 km from the Australian mainland, in the spring of 1996 (Greenslade *et al.*, 1999). Suitable wind transport was available on three occasions, with the most favourable conditions occurring on 15 October when winds of *ca.* 120 km h^{-1} (33 m s^{-1}) at heights of 600–1100 m (where temperatures were between 5°C and 6.5°C) would have carried the moths from Tasmania in 8–10 h or from the mainland in 14–16 h. The Macquarie Island records coincided with a period of unusually high *P. ewingii* catches in the Stony Rise trap.

Much more recently, wind trajectory analysis has identified plausible origins in Indonesia, in the Maluku Islands or the south of Sulawesi island, for specimens of the fall armyworm, *Spodoptera frugiperda*, detected in the Torres Strait and Cape York areas of far north Queensland in 2020 (Qi *et al.*, 2021). These were the first Australian records of this highly invasive species which was rapidly spreading through the Old World and had already become established in the two proposed source regions. Subsequent genetics-based research, however, indicates multiple introductions into northern Australia from diverse sources and provides no support for flights from the Maluku islands or Sulawesi (Rane *et al.*, 2023).

In 1971, the CSIRO's then Division of Entomology procured a purpose-built entomological radar and 2 years later it was deployed to a site on the north coast of Tasmania to test Helm's (1975) ideas about *P. ewingii* migrations from the mainland. During a 24-day period in September and early October 1973, the radar detected six separate windborne insect migrations across Bass Strait (Drake *et al.*, 1981). Light trapping at three Tasmanian sites showed that *P. ewingii*, *A. infusa*, *H. punctigera* and two other noctuid species, the brown cutworm moth *Agrotis munda* and the tobacco looper moth *Chrysodeixis argentifera*, were numerous on the island at this time. The radar's coastal location enabled direct observation of migrants arriving from over the sea, and wind back-tracks indicated origins in western Victoria and southern South Australia. A follow-up radar and trapping exercise on the Yorke peninsula of South Australia in September 1981 (Drake, 1984a) confirmed that windborne movements of moths towards the south-east were occurring at that time and location, just as Helm (1975) had proposed. The moths were observed flying over Spencer Gulf from source regions on the Eyre Peninsula and beyond, the migration becoming established about 1 h after local populations had been seen taking off at dusk.

Between 1978 and 1980, the CSIRO radar was operated at Trangie, in the central west of NSW, an area with flat terrain and a mix of pastoralism and cropping. Wind was measured once or twice each night by tracking a pilot balloon (or occasionally a radiosonde, which also measured temperature) with a theodolite and in the latter 2 years, the radar observations were sometimes complemented by aerial trapping using a drogue net carried aloft by a kite. The radar revealed large numbers of insects taking off at dusk and moving downwind for a few hours, and sometimes through the night, especially during warmer and settled weather (e.g. Drake, 1984b; Drake and Farrow, 1985). These movements were already strong in September, when the observation programme began, leading to the inference that moths, which were numerous at the radar site, were the likely source of these early-season radar echoes. (From November, adult Australian plague locusts *Chortoicetes terminifera* were also present in some years.)

In the winter of 1980, moth larvae were observed in plague numbers across an extensive area of south-western NSW and by September and early October large numbers of moths were being reported from NSW, Victoria and the Australian Capital Territory, with positive identifications of *A. infusa*, *H. punctigera* and *C. argentifera* (Drake and Farrow, 1985). Aerial trapping took place at Trangie on eight nights between 16 and 25 September, for a total of 20 h of sampling, with 36 moths being caught at heights of 100–200 m. As well as the three species just mentioned, there were six specimens of the lesser budworm *Heliothis punctifera*, a noctuid and two crambids, *A. affinitalis* and *Nomophila corticalis*. Nearby light and vehicle net trapping confirmed these species were common near the ground, with *P. ewingii* and *A. tenuis* also numerous. The radar consistently showed a take-off at dusk (with numbers aloft increasing 10–100 times) followed by downwind migration at speeds of 5–15 m s^{-1} for at least 3 h, to the south on some nights and to the east or north-east on others. Moths were detected flying at heights up to 1300 m, the upper limit of observations; densities, accumulated over all heights, were estimated at between 10 and 100 ha^{-1}. Later radar-and-trapping programmes undertaken further north and in summer (November–February), in cotton-growing areas of Queensland (Darling Downs, Emerald) and NSW (Narrabri), showed broadly similar but less intense patterns of moth flight activity; aerial trapping samples, however, were too small to establish species compositions (Drake and Fitt, 1990; V.A. Drake, unpublished data).

Much research in the 1980s was focused on *H. punctigera* and *H. armigera*, which have at times threatened the viability of the Australian cotton industry (Wilson *et al.*, 2018; Chapter 32 this volume). *Mythimna convecta*, an important pest of barley, especially in Victoria, also received considerable attention around that time. An essential early step in each case was establishing the breeding range through an extensive network of pheromone or bait traps for adults and wide-area surveys for larvae (Gregg *et al.*, 1995; McDonald, 1995; Walden, 1995). It was discovered that *H. punctigera* breeds, sometimes in extraordinary numbers, in native pastures of the far inland of eastern Australia. This region was soon identified as the source of the populations of *H. punctigera* moths that appeared in spring in cotton and legume crops in the agricultural zone to the east and south-east. Similar inland winter breeding leads to spring infestations in croplands in south-western Western Australian, with moths arriving on warm north-easterly winds. *H. armigera* was also sometimes found in the inland, less widely and in smaller numbers, but it is also known to overwinter in the northern parts of the agricultural belt (Gregg *et al.*, 1995). Direct evidence for migration was obtained from moths of both species that were caught in spring within the agricultural belt, as some of these carried phoretic pollen of plants that only grow, or would only have been in flower at the time, in the inland (Gregg, 1993). In 1996, an overnight migration in late September by *H. punctigera* and two other noctuids, *A. munda* and *H. punctifera*, was documented by larval surveys and trapping in both source and destination regions; the moths moved 400 km to the south-west, reaching the Wilcannia area of NSW, bringing them much closer to, but still beyond, the agricultural belt (Gregg *et al.*, 2001).

Genetic studies using a range of technologies and markers (electrophoreses of enzymes, insecticide resistance, microsatellites) indicated some short-term structuring of populations at the regional scale in *H. armigera*, but little systematic geographical variation, suggesting migration both within (distances *ca* 100 km) and between (*ca* 400 km) different cotton-growing regions in eastern Australia, with the continent-wide population effectively panmictic over longer periods (Daly and Gregg, 1985; Daly and Fitt, 1990; Scott *et al.*, 2005; Endersby *et al.*, 2007). This species is probably a facultative migrant (i.e. one that emigrates only if experiencing deteriorating conditions) and is a candidate Australian representative of Sappington's (2024) 'aseasonal undirected migration'. More limited genetic evidence for *H. punctigera* indicates panmixia and supports inferences of more extensive or frequent migratory movements by this species (Daly and Gregg, 1985).

Mythimna convecta, which unlike the two *Helicoverpa* species depends on gramineous hosts, has outbreaks that develop in the subtropics of eastern Australia, sometimes on native Mitchell grass (*Astrebla* spp.) pastures in the inland and sometimes on cereal crops in the agricultural zone or in irrigated cropping areas further inland, and moves southwards over a series of generations (McDonald *et al.*, 1990; McDonald, 1995). Their migrations seem rather cryptic, manifesting as unwelcome larval infestations in crops rather than obvious moth accumulations – in part because the species is relatively unresponsive to light (Smith and McDonald, 1986). In both this and the two *Helicoverpa* species, direct evidence for 'return migrations' taking populations back to the inland or the subtropics is generally absent. Plausible wind transport mechanisms for a return can often be identified, and high fecundity means that only a small proportion of a population needs to complete the cycle for the migratory adaptation to be maintained.

Farrow and McDonald (1987) provide a detailed account of earlier knowledge of the migrations of these and other Australian noctuid species, with a discussion of their 'strategies' – that is, how migration functions as an adaptation to the continent's specific geography and climate. The tower-mounted light traps operated in northern NSW in the late 1980s captured specimens of 47 noctuid species (though only in small numbers in many cases), suggesting migration may occur more widely in this family than generally recognized (Gregg *et al.*, 1994). These traps also captured large numbers of sphingids, adding to the rather limited firm evidence for migration in this family – known for its migratory species elsewhere – in Australia.

Information on the migrations of smaller moths is very limited, in most cases to occasional captures in circumstances indicative of immigration or migratory flight. Genetic studies of the diamondback moth, *Plutella xylostella*, show little variation across Australia (Perry *et al.*, 2020). This may be due primarily to a founder effect but the species was introduced more than a century ago and some long-distance movement appears necessary to maintain this degree of genetic uniformity. This cosmopolitan species is considered a migrant elsewhere and has been caught in an aerial net 200 m above ground level in the United Kingdom (Chapman *et al.*, 2002). Occasionally, peaks in *P. xylostella* moth numbers cannot be accounted for from previous local generations, indicating immigration from distances of tens or even hundreds of kilometres (Goodwin and Danthanarayana, 1984; Ridland and Endersby, 2008) and suggesting the species is a facultative migrant. A study of colonization of winter canola crops in southern Australia suggested a local origin, from wild-growing non-crop plants, at distances of not more than a few tens of kilometres (Perry *et al.*, 2019) – a type of movement that could equally be regarded as dispersal. As previously noted, mark–recapture studies have exposed even shorter movements, supporting the facultative assignment.

Observations from vertical-beam entomological radars operating at sites in inland eastern Australia indicate that flights of insects at heights of hundreds of metres and lasting several hours occur on most nights when the weather is fine, except in midwinter (June and July) (Drake *et al.*, 2001, 2024b). Targets of the 'large moth' type have estimated masses corresponding approximately to measured masses of *H. punctigera* and *A. infusa* moths (Hao *et al.*, 2020), and (taking account also of a range of ancillary information) these targets are identified as being predominantly moths of this general size. 'Medium moths' are smaller and may include the crambids *A. affinitalis* and *N. corticalis* which were confirmed as nocturnal migrants by capture in the aerial-netting operations at Trangie in 1980. The frequency with which both target categories are observed suggests that moth migration occurs almost continually from August to May, though of course with major variations in both intensity and direction.

A persistent theme in the preceding paragraphs is that moth migration is highly dependent on the wind. This is a direct consequence of the low airspeeds that moths can achieve: *ca* 4 m s^{-1} for larger species and probably only *ca* 1 m s^{-1} for smaller types (Dudley, 2000). In the fair-weather conditions in which moth migration predominantly occurs, a surface inversion develops after nightfall and confines the slowing effect of surface friction to a shallow layer. Winds above the inversion are *geostrophic* (Drake and Farrow, 1988), and speeds at 600 m of 10–20 m s^{-1} were recorded during moth flights across Bass Strait (Drake *et al.*, 1981). At the same height in inland NSW, winds at dusk (before the geostrophic flow would have been fully established) ranged between 2 m s^{-1} and 12 m s^{-1} over an entire (September–April) insect flight season (Drake *et al.*, 2001). Radar measurements of insect flight speeds exceeded these wind speeds by several metres per second at both locations, indicating that the moths were actively flying downwind. The dominance of the wind speed over the air speed means that, once in the air, the moths have only limited control over their direction of movement (Drake *et al.*, 2001).

Insects are poikilotherms, and migratory flight is possible only when temperatures are within the range that is compatible with sustained flight; as temperature generally decreases with height, this constraint also determines a flight ceiling. Active flight, however, generates heat, moths are well insulated by their body scales, and 'shivering' of wing muscles is used to raise thoracic temperatures before take-off (Coombs, 1993). Moths have been recorded flying at temperatures around 5°C in various circumstances (Drake and Reynolds, 2012). Catches in the tower traps in northern NSW occurred when temperatures at the tower base were as low as 3.2°C (Gregg *et al.*, 1994). Upper temperature limits for flight are little studied but seem unlikely to constrain nocturnal moth flight. Migratory activity is presumably suppressed or terminated by heavy rain.

Radar observations of moth flight at Trangie in November 1978 during a period of 'settled' weather under the influence of an anticyclone showed that moth numbers decreased steadily with height around dusk (when the moths were still making their initial ascent) but later developed a layered density profile with a single density maximum at around 200 m (Drake, 1984b; Fig. 7.3). The intensity of the layer, and the speed with which it formed, varied greatly among the three nights studied in detail. Radiosonde ascents confirmed that these layers were associated with surface temperature inversions – an increase in temperature with height immediately above the ground. In 1980, numerous moths were still flying an hour or so before first light on 26 September, when they had become concentrated into a layer at 250 m (Drake, 1985a). A pilot balloon ascent showed that the layer was in a 'low-level jet', a zone of faster moving air which in this case had its maximum at 150 m, where the speed was 14 m s^{-1}. These jets develop in association with a surface temperature inversion, and it is inferred that the moths had become concentrated where the air was warmest. Layers sometimes form well above the surface – at 1100 m and 1900 m, for example, during the 1980 Trangie observations (Drake and Farrow, 1985); these presumably either develop in response to some higher stratification feature of the night-time atmospheric environment or represent accumulation at a flight ceiling. Radar observations have also shown that steady nocturnal migrations can be disturbed, especially by outflows from distant thunderstorms and other small-scale wind systems (Drake and Reynolds, 2012 ch. 11), and moth migrations in inland Australia are frequently affected by these phenomena

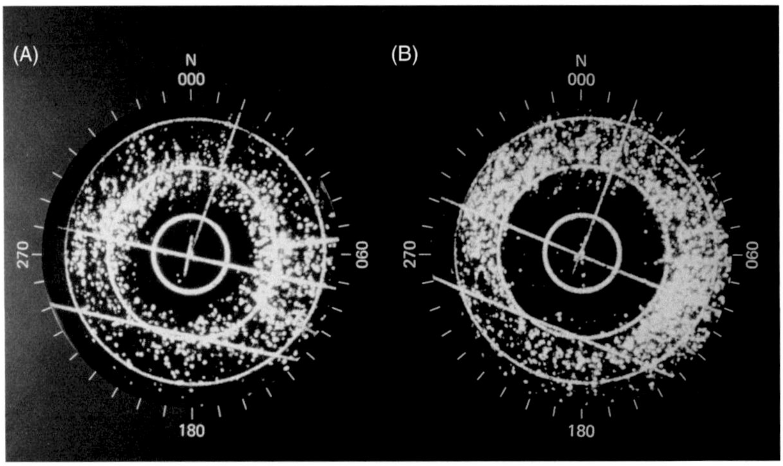

Fig. 7.3. Display of the CSIRO Entomological Radar operating at Trangie, NSW, on 26 November 1978 at 23.55 h, with the antenna at an elevation angle of 12° in (A) and 8° in (B). Range rings are at 0.25 n. mile (463 m) intervals. The moths had aggregated into a layer concentration with a maximum at about 200 m and a sharp lower boundary, visible at 0.35 n. mile in (A) and 0.5 n. mile in (B), corresponding to a height of ~130 m. Each dot-mark represents an echo from a single moth. Migration was towards the south. Noctuid and crambid moths of a variety of species recognized as migratory were abundant at Trangie at the time (Drake, 1984b). Reproduced with permission from Drake (1984b). n. mile: nautical mile

(Drake, 1984b, 1985b). They have also been observed in migrations over the sea (Drake, 1984a).

A well-known finding from radar observations is that larger insects very often exhibit a degree of collective orientation during nocturnal migratory flight (Drake and Reynolds, 2012), and this occurred frequently during the Bass Strait and Trangie moth migrations described above. Orientation is also evident in data from vertical-beam entomological radars. For both radar types, it is the alignment of the insects' body axes (range 180°) that is observed, not the heading direction (range 360°). When data are available for individual insects, however, as is the case with vertical-beam radars, the 180° ambiguity can be resolved using an ensemble approach (Drake *et al.*, 2021). This method exploits the spread of heading directions; for insects of the 'large moth' type observed at Bourke, NSW, over 15 nights in September and October 2007, this spread was typically ±20°. The corresponding spread for track directions was ±10°, illustrating how wind transport narrows the range of directions in which an insect can move (Drake *et al.*, 2001).

An example of orientation behaviour exhibited during another strong migration of 'large moths' is presented in Fig. 7.4. Wind direction varied smoothly, and only moderately, with both time and height through this night, and heading and track directions showed similarly smooth and minor variations from one hour and one height interval to the next. The general direction of the wind was towards the south-east while headings were towards the east – that is, to the left of both the wind and the resultant track. Behaviours of this type are presumably adaptive, and in the northern hemisphere's temperate zone they appear to increase a moth's chance of successfully moving to higher latitudes in spring or lower latitudes in autumn (Chapman *et al.*, 2008a, b). Differing geographical and climatic contexts make such analyses less straightforward for data from the Australian radars, and the function of the orientation behaviours exhibited by the several moth

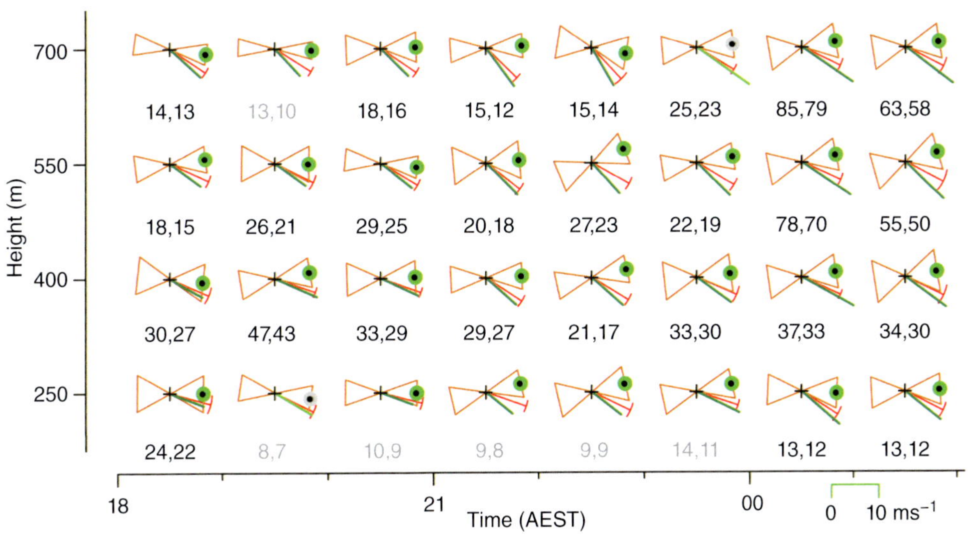

Fig. 7.4. Variation with height and time of the direction, speed, orientation and wind during a strong migration of large moths at Bourke, NSW on the night of 16–17 May 2008, derived from observations with a vertical-beam insect monitoring radar. For clarity, only the first 8 h and the bottom 750 m are shown. Orange hourglass symbols represent the distribution of body alignments (mean ±1 standard deviation), red pickaxe symbols the track directions (mean ±1 SD), green lines the speed of the inferred wind and the direction towards which it is blowing, and circles the orientation direction resolved by the method of Drake *et al.* (2021) at confidence levels of 0.05 (green) or 0.2 (grey). Numbers are sample sizes (total and with outliers removed) for each 1 h × 150 m analysis unit. Inferred winds were consistent with those from (reanalysed) operational meteorological observations. Unpublished data from V.A. Drake and co-workers.

species that migrate over the continent's inland plains remains open to speculation.

7.6 Discussion and Priorities for Future Research

Dispersal is a near-universal phenomenon among living organisms, necessary to locate the resources needed for survival and successful breeding, and migration is widely adopted as a strategy for exploiting resources that become available in different places at different times (Dingle, 2014). Moths in Australia disperse and migrate much like their counterparts on other continents. Differences arising from the specific forms of each continent's geography and climate could probably be identified, but they would not be fundamental in nature. Several larger moths appear adapted to life in the extensive drier plains of the inland, where favourable conditions arise rather erratically and unpredictably from predominantly winter, temperate-zone rainfall in the south and predominantly summer, subtropical rainfall in the north. Long-distance movements, combined with high fecundity, allow patchy and ephemeral habitats to be located and a new generation to be produced. Warm airflows associated with the passage of cold fronts sometimes sweep these migrants beyond the inland plains and even out to sea; most perish but a few are caught in traps or noticed by enthusiasts, briefly revealing the largely hidden phenomenon of nocturnal moth migration. *Agrotis infusa*, with its summer aestivation and two-way migration by individuals, presents one variation on this theme. There are surely others, perhaps within the tropical savannahs or the forests of the Great Dividing Range, perhaps over shorter distances or from lower to higher ground, that are yet to be discovered. For species other than noctuids, we know little more than that they turn up in traps in circumstances that strongly indicate a long-distance windborne flight.

Migration has been the primary focus of Australian studies of insect flight with entomological radars (studies of dispersal have been attempted but with limited success; Drake, 2023). Observations of moth migration comprise a large proportion of the accumulated radar dataset – in part because moths (of one species or another) occur so frequently in the night sky that they are observed almost whenever a radar is operating during the hours of darkness. Much has been learnt from these electronic detections and recorded echo signals, perhaps most significantly the frequency and ubiquity with which moth migration occurs, the height of flight and the dominant role of wind transport in determining migration distances and directions. A comprehensive statistical analysis of the full radar dataset is still to be completed. It should provide night-to-night, seasonal and year-to-year variations and variances for migration intensities, speeds, directions, heights of flight and (inferred) durations; with additional meteorological data, temperature thresholds and preferences may be revealed.

The utility of radar-based statistics will be limited, however, by the lack of specific identifications. Ancillary trapping operations, a feature of the short-term radar exercises prevalent in the 1980s, have long ceased, due to their high costs (including for sorting and identification) and logistical challenges. New detection and identification technologies are being developed to address these limitations (Drake, 2023) but seem unlikely to be deployed on the regional or larger scale needed for effective monitoring of migratory populations. A volunteer-operated light-trap network, a useful data source for interpreting entomological radar observations in the UK (Harrington and Woiwod, 2007; Chapman *et al.*, 2008a), has not developed in Australia – but perhaps may eventually do so through internet-mediated 'citizen science'. The lack of a central database and archive for reports from district agronomists and others of larval infestations in crops and pastures means that potentially valuable historical information on species, hosts and spatial population processes has effectively been lost.

Research on insect dispersal and migration in Australia has mainly been focused on pest species, with significant funding support from both government and industry bodies. Additionally, the unusual life history of *A. infusa*, and the intriguing flight orientation behaviours exhibited by many migratory species, have attracted curiosity-driven research supported by fundamental science funding. Entomological radar operations are primarily maintained for the information they provide about locusts; moths are studied in part to ensure they can be distinguished from the more sporadic orthopterans.

Much early radar research was concerned with how migratory flight was affected by features of the atmospheric environment, especially wind and temperature (and their variation with height) and disturbances of the lower atmosphere that disrupt the migrations. The surge of research in the 1980s that was driven by the recognition that many important pest species are migratory has largely dissipated, though new technologies (genomics, new-generation weather radars) provide opportunities for renewal. Perhaps the prospect of changes in migration patterns in response to global warming – as demonstrated for the rice leafroller, *Cnaphalocrocis medinalis*, a crambid, in China (Zhang *et al.*, 2025) – will restimulate interest. Recent endeavours overseas have focused on the collective orientation phenomenon and its possible mechanisms and functions, and with how migratory behaviours enhance survival and increase fitness (Chapman *et al.*, 2015, Hu *et al.*, 2016). Australia provides a fine natural laboratory for such research, especially in its inland plains – though perhaps one that will demand subtler or more complex hypotheses and analyses than are required for more predictable northern hemisphere temperate-zone environments.

Acknowledgements

Peter Gregg made valuable comments on a draft. This review showcases research from numerous scientists, students and technical staff, many of whom I have had the privilege and pleasure of collaborating with at some time or other during the past 47 years.

References

Ahrens, C.W., Slavenko, A., Kriesner, P., van Rooyen, A., Heinze, D. *et al.* (2025) Stochastic wind-driven migration likely maintains panmixia in the endangered bogong moth, *Agrotis infusa*. *Biological Conservation* 302, article 110993.

Baker, G., Parry, H. and Tann, C. (2016) Pigeon pea refuge crops are likely to provide patchy delivery of *Helicoverpa* (Lepidoptera: Noctuidae) within Bt cotton production regions in eastern Australia. *Austral Entomology* 55, 439–448.

Chapman, J.W., Reynolds, D.R., Smith, A.D., Riley, J.R., Pedgley, D.E. and Woiwod. I.P. (2002) High-altitude migration of the diamondback moth *Plutella xylostella* to the U.K.: A study using radar, aerial netting and ground trapping. *Ecological Entomology* 27, 641–650.

Chapman, J.W., Reynolds, D.R., Mouritsen, H., Hill, J.K., Riley, J.R. *et al.* (2008a) Wind selection and drift compensation optimize migratory pathways in a high-flying moth. *Current Biology* 18, 514–518.

Chapman, J.W., Reynolds, D.R., Hill, J.K, Sivell, D., Smith, A.D. and Woiwod, I.P. (2008b) A seasonal switch in compass orientation in a high-flying migrant moth. *Current Biology* 18, R908–R909.

Chapman, J.W., Reynolds, D.R. and Wilson, K. (2015) Long-range seasonal migration in insects: mechanisms, evolutionary drivers and ecological consequences. *Ecological Letters* 18, 287–302.

Chapuis, M., Popple, J.M., Berthier, K., Simpson, S.J., Deveson, E. *et al.* (2011) Challenges to assessing connectivity between massive populations of the Australian plague locust. *Proceedings of the Royal Society B Biological Sciences* 278, 3152–3160.

Common, I.F.B. (1954) A study of the ecology of the adult bogong moth, *Agrotis infusa* (Boisd) (Lepidoptera: Noctuidae), with special reference to its behaviour during migration and aestivation. *Australian Journal of Zoology* 2, 223–263.

Coombs, M. (1993) Endothermy and flight thresholds for *Helicoverpa punctigera* and *H. armigera* (Lepidoptera: Noctuidae). *Australian Journal of Zoology* 41, 577–587.

Daly, J.C. and Fitt, G.P. (1990) Resistance frequencies in overwintering pupae and the first spring generation of *Helicoverpa armigera* (Lepidoptera: Noctuidae): selective mortality and immigration. *Journal of Economic Entomology* 83, 1682–1688.

Daly, J.C. and Gregg, P. (1985) Genetic variation in *Heliothis* in Australia: species identification and gene flow in the two pest species *H. armigera* (Hübner) and *H. puntigera* Wallengren (Lepidoptera: Noctuidae). *Bulletin of Entomological Research* 75, 169–184.

Danthanarayana, W. and Gu, H. (1992) Influence of environmental conditions on flight duration of *Epiphyas postvittana* (Walker) (Lepidoptera: Tortricidae). *Australian Journal of Zoology* 40, 477–484.

Dingle H. (2014) *Migration: the Biology of Life on the Move*, 2nd edn. Oxford University Press, Oxford.

Drake, V.A. (1984a) A solitary wave disturbance of the marine boundary layer over Spencer Gulf revealed by radar observations of migrating insects. *Australian Meteorological Magazine* 32, 131–135.

Drake, V.A. (1984b) The vertical distribution of macro-insects migrating in the nocturnal boundary layer: A radar study. *Boundary-Layer Meteorology* 28, 353–374.

Drake, V.A. (1985a) Radar observations of moths migrating in a nocturnal low-level jet. *Ecological Entomology* 10, 259–265.

Drake, V.A. (1985b). Solitary wave disturbances of the nocturnal boundary layer revealed by radar observations of migrating insects. *Boundary-Layer Meteorology* 31, 269–286.

Drake, V.A. (1991) Methods for studying adult movement in *Heliothis*. In: Zalucki, M.P. (ed.) *Heliothis: Research Methods and Prospects*. Springer-Verlag, New York, pp.109–121.

Drake, V.A. (2023) Monitoring airborne movement of crop insect pests and beneficials. In: Fountain, M. and Pope, T. (eds) *Advances in Monitoring of Native and Invasive Insect Crop Pests*. Burleigh Dodds Science Publishing, Cambridge, pp. 147–194.

Drake, V.A. and Farrow, R.A. (1985) A radar and aerial-trapping study of an early spring migration of moths (Lepidoptera) in inland New South Wales. *Australian Journal of Ecology* 10, 223–235.

Drake, V.A. and Farrow, R.A. (1988) The influence of atmospheric structure and motions on insect migration. *Annual Review of Entomology* 33, 183–210.

Drake, V.A. and Fitt, G.P. (1990) Studies of *Heliothis* mobility at Narrabri, summer 1989/90. Fifth Australian Cotton Conference, Broadbeach, Queensland, 8–9 August 1990. Australian Cotton Growers Research Association, Narrabri, Australia, pp. 295–304.

Drake, V.A. and Reynolds, D.R. (2012) *Radar Entomology: Observing Insect Flight and Migration*. CABI, Wallingford, UK.

Drake, V.A., Helm, K.F., Readshaw, J.L. and Reid, D.G. (1981) Insect migration across Bass Strait during spring: A radar study. *Bulletin of Entomological Research* 71, 449–466.

Drake, V.A., Gregg, P.C., Harman, I.T., Wang, H.K., Deveson, E.D. *et al.* (2001) Characterizing insect migration systems in inland Australia with novel and traditional methodologies. In: Woiwod, I. and Reynolds, D.R. (eds) *Insect Movement: Mechanisms and Consequences*. CABI Publishing, Wallingford, UK, pp. 207–233.

Drake, V.A., Hatty, S., Symons, C. and Wang, H. (2020). Insect monitoring radar: Maximizing performance and utility. *Remote Sensing* 12, article 596.

Drake, V.A., Hao, Z., and Warrant, E. (2021) Heading variations resolve the heading-direction ambiguity in vertical-beam radar observations of insect migration. *International Journal of Remote Sensing* 42, 3873–3898.

Drake, V.A., Hao, Z., and Wang, H. (2024a) Monitoring insect numbers and biodiversity with a vertical-beam entomological radar. *Philosophical Transactions of the Royal Society B* 379, article 20230117.

Drake, V.A., Mirkovic, D. and Steinbauer, M. (2024b) Insects as radar targets: size, form, density and permittivity. *International Journal of Remote Sensing* 45, 2985–3002.

Dreyer, D., Frost, B., Mouritsen, H., Gunther, A., Green, K. *et al.* (2018) The Earth's magnetic field and visual landmarks steer migratory flight behavior in the nocturnal Australian Bogong moth. *Current Biology* 28, 2160–2166.e5.

Dudley, R. (2000) *The Biomechanics of Insect Flight: Form, Function, Evolution*. Princeton University Press, Princeton, NJ.

Endersby, N.M., Hoffmann, A.A., McKechnie, S.W. and Weeks, A.R. (2007) Is there a genetic structure in populations of *Helicoverpa armigera* from Australia. *Entomologia Experimentalis et Applicata* 122, 253–263.

Farrow, R.A. (1984) Detection of transoceanic migration of insects to a remote island in the Coral Sea, Willis Island. *Australian Journal of Ecology* 9, 253–272.

Farrow, R.A. and Dowse, J.E. (1984) Method of using kites to carry tow nets in the upper air for sampling migrating insects and its application to radar entomology. *Bulletin of Entomological Research* 74, 87–95.

Farrow, R.A. and McDonald, G. (1987) Migration strategies and outbreaks of noctuid pests in Australia. *Insect Science and its Applications* 8, 531–542.

Feng, H.-Q., Wu, K.-M., Cheng, D.-F. and Guo, Y.-Y. (2004) Northward migration of *Helicoverpa armigera* (Lepidoptera: Noctuidae) and other moths in early summer observed with radar in northern China. *Journal of Economic Entomology* 97, 1874–1883.

Fitt, G.P., Dillon, M.L. and Hamilton, J.G. (1995) Spatial dynamics of *Helicoverpa* populations in Australia: Simulation modelling and empirical studies of adult movement. *Computers and Electronics in Agriculture* 13, 177–192.

Fletcher, B.S. (1973) Observations on a movement of insects at Heron Island, Queensland. *Journal of the Australian Entomological Society* 12, 157–160.
Flood, J. (1980) *The Moth Hunters: Aboriginal Prehistory of the Australian Alps*. Australian Institute of Aboriginal Studies, Canberra, ACT.
Fox, K.J. (1978) The transoceanic migration of Lepidoptera to New Zealand – a history and a hypothesis on colonisation. *New Zealand Entomologist* 6, 368–380.
Geier, P. and Briese, D. (1980) The light-brown apple *moth, Epiphyas postvittana* (Walker): 4. Studies on population dynamics and injuriousness to apples in the Australian Capital Territory. *Australian Journal of Ecology* 5, 63–93.
Goodwin, S. and Danthanarayana, W. (1984) Flight activity of *Plutella xylostella* (L.) (Lepidoptera: Yponomeutidae). *Journal of the Australian Entomological Society* 23, 235–240.
Green, K., Caley, P., Baker, M., Dreyer, D., Wallace, J. and Warrant, E. (2021) Australian Bogong moth *Agrotis infusa* (Lepidoptera: Noctuidae), 1951–2020: decline and crash. *Austral Entomology* 60, 66–81.
Greenslade, P., Farrow, R.A. and Smith, J.M.B. (1999) Long distance migration of insects to a subantarctic island. *Journal of Biogeography* 26, 1161–1167.
Gregg, P. (1993) Pollen as a marker for migration of *Helicoverpa armigera* and *H. punctigera* (Lepidoptera: Noctuidae) from western Queensland. *Australian Journal of Ecology* 18, 209–219.
Gregg, P.C., Fitt, G.P., Coombs, M. and Henderson, G.S. (1993) Migrating moths (Lepidoptera) collected in tower-mounted light traps in northern New South Wales, Australia: species composition and seasonal abundance. *Bulletin of Entomological Research* 83, 563–578.
Gregg, P.C., Fitt, G.P., Coombs, M. and Henderson, G.S. (1994) Migrating moths collected in tower-mounted light traps in northern New South Wales, Australia: influence of local and synoptic weather. *Bulletin of Entomological Research* 84, 17–30.
Gregg, P.C., Fitt, G.P., Zalucki, M.P. and Murray, D.A.H. (1995) Insect migration in an arid continent. II. *Helicoverpa* spp. in eastern Australia. In: Drake, V.A. and Gatehouse, A.G. (eds) *Insect Migration: Tracking Resources through Space and Time*. Cambridge University Press, Cambridge, pp. 151–172.
Gregg, P.C., Del Socorro, A.P. and Rochester, R.A. (2001) Field test of a model of migration of moths (Lepidoptera: Noctuidae) in inland Australia. *Australian Journal of Entomology* 40, 249–256.
Hao, Z., Drake, V.A., Taylor, J.R. and Warrant, E. (2020) Insect target classes discerned from entomological radar data. *Remote Sensing* 12, article 673.
Harrington, R. and Woiwod, I. (2007) Foresight from hindsight: the Rothamsted Insect Survey. *Outlooks on Pest Management* 18, 9–14.
Helm, K.F. (1975) Migration of the armyworm *Persectania ewingii* moths in spring and the origin of outbreaks. *Journal of the Australian Entomological Society* 14, 229–236.
Hill, L. (2011) Continual insect migration across Bass Strait? *Victorian Entomologist* 41, 117–122.
Hill, L. (2012) Cabbage-centre grub, *Hellula hydralis* Guenée, not resident in Tasmania. *Plant Protection Quarterly* 27, 91–100.
Hill, L. (2013) The common armyworm, *Mythimna convecta* (Walker) (Lepidoptera: Noctuidae), a seasonal resident in Tasmania. *Plant Protection Quarterly* 28, 114–119.
Hill, L. (2014) Lesser armyworm, *Spodoptera exigua* (Hübner) (Lepidoptera: Noctuidae), a vagrant moth in Tasmania. *Plant Protection Quarterly* 29, 131–142.
Hill, L. (2016) An extreme weather event brings two rare vagrant moths to Tasmania. *Victorian Entomologist* 46, 88–89.
Hill, L. (2018) Vagrant status of lucerne seed web moth, *Etiella behrii* (Zeller, 1848) (Lepidoptera: Pyralidae) in Tasmania. *Crop Protection* 104, 65–71.
Holloway, J.D. (1982) Further notes on the Lepidoptera of Norfolk Island, with particular reference to migrant species. *Journal of Natural History* 16, 351–365.
Hu, G., Lim, K.S., Reynolds, D.R., Reynolds, A.M. and Chapman, J.W. (2016) Wind-related orientation patterns in diurnal, crepuscular, and nocturnal high-altitude migrants. *Frontiers in Behavioral Neuroscience* 10, article 32.
Huang, J., Feng, H., Drake, V.A., Reynolds, D.R., Gao, B. *et al.* (2024) Massive seasonal high-altitude migrations of nocturnal insects above the agricultural plains of East China. *Proceedings of the National Academy of Sciences* 121, article e2317646121.
Il'Ichev, A.L., Hossain, M.S. and Jerie, P.H. (1999) Migration of oriental fruit moth *Grapholita molesta* Busck. (Lepidoptera: Tortricidae) under wide area mating disruption in Victoria, Australia. *IOBC-WPRS Bulletin* 22, 53–62.

Johnson, C.G. (1969) *Migration and Dispersal of Insects by Flight*. Methuen, London.

Kennedy, J.S. (1961) A turning point in the study of insect migration. *Nature* 189, 785–791.

Kennedy, J.S. (1985) Migration: Behavioral and ecological. In: Rankin, M.A. (ed.) *Migration: Mechanisms and Adaptive Significance*. Contributions in Marine Science, Supplement 27. Marine Science Institute, University of Texas at Austin, Port Aransas, TX, pp. 5–26.

Macfadyen, S., Hopkinson, J., Parry, H., Neave, M.J., Bianchi, F.J.J.A. *et al.* (2015) Early-season crop colonization: Movement dynamics of pest and natural enemies across a native vegetation-crop ecotone. *Agriculture Ecosystems and Environment* 200, 110–118.

Malcolm, S.B. (1993) Conservation of monarch butterfly migration in North America: an endangered phenomenon. In: Malcolm, S.B. and Zalucki, M.P. (eds) *Biology and Conservation of the Monarch Butterfly*. Science Series 38, Natural History Museum of Los Angeles County, Los Angeles, CA, pp. 357–361.

McDonald, G. (1991) Oviposition and larval dispersal of the common armyworm *Mythimna convecta* (Walker) (Lepidoptera: Noctuidae). *Australian Journal of Ecology* 16, 385–393.

McDonald, G. (1995) Insect migration in an arid continent. I. The Common Armyworm *Mythimna convecta* in eastern Australia. In: Drake, V.A. and Gatehouse, A.G. (eds) *Insect Migration: Tracking Resources through Space and Time*. Cambridge University Press, Cambridge, pp. 131–150.

McDonald, G., Bryceson, K.P. and Farrow, R.A. (1990) The development of the 1983 outbreak of the common armyworm, *Mythimna convecta*, in eastern Australia. *Journal of Applied Ecology* 27, 1001–1019.

McDonald, G., Maino, J.L., Smith, A.M., Gregg· P.C. and Ridland, P.M. (2024) New insights from old data: the complex migration and breeding patterns of the bogong moth, *Agrotis infusa* (Boisduval) (Lepidoptera: Noctuidae), in eastern Australia. *Austral Entomology* 63, 466–484.

McLaren, D.A., Ireson, J.E. and Kwong, R.M. (2000) Biological control of ragwort (*Senecio jacobaea* L.) in Australia. In: Spencer, N.R. (ed.) Proceedings of the X International Symposium on Biological Control of Weeds, 4–14 July 1999, Bozeman, Montana, USA, pp. 67–79.

Menz, M.H.M., Scacco, M., Bürki-Spycher, H.-M., Williams, H.J., Reynolds, D.R. *et al.* (2022) Individual tracking reveals long-distance flight-path control in a nocturnally migrating moth. *Science* 377, 764–768.

Mo, J., Baker, G., Keller, M. and Roush, R. (2003) Local dispersal of the diamondback moth (*Plutella xylostella* (L.)) (Lepidoptera: Plutellidae). *Environmental Entomology* 32, 71–79.

Morley, T.B., Faulkner, S. and Faithfull, I.G. (2004) Establishment and dispersal of dock moth *Pyropteron doryliformis* (Ochsenheimer) (Lepidoptera: Sesiidae) in Victoria. In: Sindel, B.M. and Johnson, S.B. (eds) Fourteenth Australian Weeds Conference, Wagga Wagga, Australia, 6–9 September 2004. Weed Society of New South Wales, Sydney, pp. 381–384.

Parry, H.R., Paull, C.A., Zalucki, M.P., Ives, A.R., Hulthen, A. and Schellhorn, N.A. (2017) Estimating the landscape distribution of eggs by *Helicoverpa* spp., with implications for *Bt* resistance management. *Ecological Modelling* 365, 129–140.

Perry, K.D., Baxter, S.W. and Keller, M.A. (2019) Unveiling diamondback moth, *Plutella xylostella*, movement at landscape and regional scales. In: Srinivasan, R., Sotelo-Cardona, P., Zalucki, M.P. and Shelton, A.M. (eds) Proceedings, VIII International Conference on Management of the Diamondback Moth and Other Crucifer Insect Pests, Shanhua, Tainan, Taiwan, 4–8 March 2019. World Vegetable Centre, Shanhua, Tainan, Taiwan, pp. 33–41.

Perry, K.D., Keller, M.A. and Baxter, S.W. (2020) Genome-wide analysis of diamondback moth, *Plutella xylostella* L., from *Brassica* crops and wild host plants reveals no genetic structure in Australia. *Scientific Reports* 10, article 12047.

Pruess, K.P. and Pruess, N.C. (1971) Telescopic observation of the moon as a means for observing migration of the army cutworm, *Chorizagrotis auxiliaris* (Lepidoptera: Noctuidae). *Ecology* 52, 999–1007.

Qi, G.-J., Ma, J., Wan, J., Ren, Y.-L., McKirdy, S. *et al.* (2021) Source regions of the first immigration of fall armyworm, *Spodoptera frugiperda* (Lepidoptera: Noctuidae) invading Australia. *Insects* 12, article 1104.

Rane, R., Walsh, T.K., Lenancker, P., Gock, A., Dao, T.H. *et al.* (2023) Complex multiple introductions drive fall armyworm invasions into Asia and Australia. *Scientific Reports* 13, article 660.

Rennie, S.J. (2013) Common orientation and layering of migrating insects in southeastern Australia observed with a Doppler weather radar. *Meteorological Applications* 21, 218–229.

Ridland, P.M. and Endersby N.M. (2008) Seasonal phenology of diamondback moth populations in southern Australia. In: Shelton, A.M., Collins, H.L., Wu, Q. and Zhang, Y. (eds) Fifth International Workshop on the Management of Diamondback Moth and Other Crucifer Pests, 24–27 October 2006, Beijing, China. China Agricultural Science and Technology Press, Beijing, pp. 90–101.

Sappington, T.W. (2024) Aseasonal, undirected migration in insects: 'Invisible' but common. *iScience* 27, article 110040.

Satterfield, D.A., Sillett, T.S., Chapman, J.W., Altizer, S and Marra, P.P. (2020) Seasonal insect migrations: Massive, influential, and overlooked. *Frontiers in Ecology and the Environment* 18, 335–344.

Scott, K.D., Lawrence, N., Lange, C.L., Scott, L.J., Wilkinson, K.S. *et al.* (2005) Assessing moth migration and population structuring in *Helicoverpa armigera* (Lepidoptera: Noctuidae) at the regional scale: Example from the Darling Downs, Australia. *Journal of Economic Entomology* 98, 2210–2219.

Sloan, W.J.S. (1945) Migrations of the corn ear worm. *Queensland Agricultural Journal* 61, 272–274.

Smith, A.M. and McDonald, G. (1986) Interpreting ultraviolet-light and fermentation trap catches of *Mythimna convecta* (Walker) (Lepidoptera: Noctuidae) using phenological simulation. *Bulletin of Entomological Research* 76, 419–431.

Southwood, T.R.E. (1962) Migration of terrestrial arthropods in relation to habitat. *Biological Reviews* 37, 171–214.

Stephenson, B., David, B., Fresløv, J., Arnold, L.J., GunaiKurnai Land and Waters Aboriginal Corporation *et al.* (2020) 2000 year-old Bogong moth (*Agrotis infusa*) Aboriginal food remains, Australia. *Scientific Reports* 10, article 22151.

Suchan. T., Bataille, C.P., Reich, M.S., Toro-Delgado, E., Vila, R. *et al.* (2024) A trans-oceanic flight of over 4,200 km by painted lady butterflies. *Nature Communications* 15, article 5205.

Suckling, D.M. and Brockerhoff, E.G. (2010) Invasion biology, ecology, and management of the light brown apple moth (Tortricidae). *Annual Review of Entomology* 55, 285–306.

Svensson, G.P., Valeur, P.G., Reynolds, D.R., Smith, A.D., Riley, J.R. *et al.* (2001) Mating disruption in *Agrotis segetum* (Lepidoptera: Noctuidae) monitored by harmonic radar. *Entomologia Experimentalis et Applicata* 101, 111–121.

Taouk, M. (2020) Swarm of moths caused panic during the Sydney Olympics closing ceremony. ABC News. Available at: https://www.abc.net.au/news/2020-09-18/moths-panic-at-sydney-olympics-closing-ceremony/12674484 (accessed 27 August 2024).

Taylor, R.A.J. and Reling, D. (1986) Density/height profile and long-range dispersal of first-instar gypsy moth (Lepidoptera: Lymantriidae). *Environmental Entomology* 25, 431–435.

Walden, K.J. (1995) Insect migration in an arid continent. III. The Australian Plague Locust *Chortoicetes terminifera* and the Native Budworm *Helicoverpa punctigera* in Western Australia. In: Drake, V.A. and Gatehouse, A.G. (eds) *Insect Migration: Tracking Resources through Space and Time*. Cambridge University Press, Cambridge, pp. 173–190.

Warrant, E., Frost, B., Green, K., Mouritsen, H., Dreyer, D. *et al.* (2016) The Australian bogong moth *Agrotis infusa*: a long-distance nocturnal navigator. *Frontiers in Behavioral Neuroscience* 10, article 77.

Wilson, L.J., Whitehouse, M.E.A. and Herron, G.A. (2018) The management of insect pests in Australian cotton: an evolving story. *Annual Review of Entomology* 63, 215–237.

Zhang, S.-Y., Zhang, Y.-Y., Yang, F., Zhou, C., Shen, H.-M. *et al.* (2025) Climate change is leading to an ecological trap in a migratory insect. *Proceedings of the National Academy of Sciences* 122, article e2422595122.

8 The Community Ecology of Australian Moths

Louise A. Ashton[1]*, Erica H. Odell[2] and Roger L. Kitching[3]

[1]*School of Biological Sciences, The University of Hong Kong, Hong Kong SAR, China;* [2]*RMIT University, Bundoora, Victoria, Australia;* [3]*School of Environment and Science, Griffith University, Nathan, Queensland, Australia*

Abstract

Community ecology examines ecological patterns and processes that affect assemblages of species – such as moths – at a variety of spatial and temporal scales. The science examines diversity, trophic ecology, community assembly, landscape patterns and biogeography (this last dealt with in Chapter 4). Moths are particularly useful as targets for community-level study because of their familiarity, ease of sampling using lights, their abundance and species richness, and the history of previous taxonomic and ecological study. This chapter further discusses key Australian-based studies examining the impacts of climate, latitude, verticality and human impact. The open question of a general decline is discussed, as are the many knowledge gaps and current and future technological innovations which will open new opportunities in the study of moth assemblages.

8.1 Introduction

Strictly speaking, an ecological *community* is the living part of any particular ecosystem. Accordingly, it's strictly correct to speak of the 'lowland rainforest community' or the 'mangrove community' or 'the mallee community'. In practice, though, we seldom if ever study the whole community; we generally focus our studies either taxonomically or structurally within the ecosystem of choice. It then becomes commonplace to speak of the 'bird community', the 'bacterial community', the 'tree-hole community' or, indeed, the 'moth community'. Strictly these are *assemblages*. Semantics apart, there are many important fundamental and applied ecological questions best approached by studying co-occurring sets of species – the stuff of community ecology.

Insect ecology has traditionally been more focused on populations. This autoecological approach remains appropriate where management of individual species for control, exploitation or conservation is the goal. In parallel with these single-species approaches, other areas of ecology developed multispecies approaches mainly for classificatory purposes, reflecting in no small part the increasing availability of multivariate statistical methods. Although essentially descriptive and hypothesis generating, these statistical methods nevertheless allow large sets of species to be regarded as ecological objects able to be handled computationally. Initially these methods were applied extensively by plant ecologists, then by freshwater ecologists.

Of course, fundamental community concepts such as diversity, trophic structure, food chains and webs, succession and community assembly

*Corresponding author: lashton@hku.hk

DOI: 10.1079/9781800627680.0008

had been part of the ecological vocabulary from the earliest emergence of the discipline. The challenge was turning these concepts into measurable and useful statistics that could be used to test hypotheses about community structure and dynamics. When community concepts and statistical capacity did emerge, they coincided with a heightened interest in questions of conservation. Community-level measures were rapidly incorporated as questions arose concerning the impact of ecosystem modification, degradation and restoration on whole sets of organisms. In Australia, moths have featured prominently in these community-level approaches.

8.2 Components of Community Ecology

8.2.1 Diversity

Diversity in its simplest form is a measure of *richness*: simple counts of how many species occur at particular places and times. Underlying any measure of richness, however, are statistical distributions of relative abundance – from common to rare. Summarizing these complex distributions into simple indices has been an ongoing challenge to community ecologists. Useful measures include *evenness* (how much variation there is across the distribution of relative abundances for any set of species), *dominance* (the extent to which any one, or a few, species dominate the assemblage), the *proportions of rare species* ('singletons' or 'doubletons') in a sample, and so on. A wide range of diversity indices combine the concepts of richness and evenness. New analytical methods now allow statistically valid comparisons of some of these indices against null models based on random resampling of actual datasets.

It has become commonplace to speak of alpha, beta and gamma diversity (partially following Whittaker, 1960) where *alpha diversity* is a within-habitat measure (or sample or location), *beta diversity* is turnover across pairs of habitats, and *gamma diversity* summarizes diversity across the range of habitats in a landscape.

8.2.2 Trophic ecology

Trophic ecology extends simple assignments of the feeding habits of species into complex networks – *food webs*. Single unidirectional tracks through such food webs are *food chains*. Victor Summerhayes and Charles Elton summmarized the relatively simple food web of Spitzbergen island in 1923 and Alistair Hardy attempted the same task for the much more complex herring fishery of the North Sea in 1924. Fellow Oxford ecologist Owain Richards introduced insects into the equation with his study of Oxshott Common published in 1930. These descriptions were hard to evaluate statistically; only in the latter part of the 20th century did analytical tools and technology emerge that obviated this problem. Defining and quantifying feeding links remains a challenge. One of the most promising, recent approaches is based on the molecular analyses of gut contents allowing identification of feeding links directly rather than inferentially (Pompanon *et al.*, 2012). First developed in insects for predator–prey interactions (Paula *et al.*, 2016), recent advances tackle the more complex problem (molecularly speaking) of herbivore–plant interactions (Garcia-Robledo *et al.*, 2013; De Vere *et al.*, 2017; Hardwick, 2018; Hausmann *et al.*, 2020).

A variety of measures have been developed to summarize the structures of such food webs, including *connectivity* (the ratio of actual to potential trophic links in a web), *compartmentalization* (the extent to which one subset of a food web is independent of the rest), *nestedness* (where subsets of links and nodes nest within larger sets) and a variety of *ratios* summarizing the relative diversity, size or biomass of particular combinations of trophic levels.

8.2.3 Community assembly/succession

Succession, the sequence of changes in species composition, species interactions and ecological processes which occur over time within any particular ecosystem, has long been of interest to community ecologists, first from a botanical point of view (starting with Clements 1916). Traditionally ecologists have spoken of *primary* and *secondary* succession. 'Primary' succession envisages a 'bare earth' – post volcanic, post mining or the like – starting point developing until a so-called 'climax' community is reached – a closed canopy forest or a mature, self-sustaining community, such as mature deciduous woodland. In contrast, a 'secondary' succession such as begins after an ecosystem disturbance event – a cyclone or fire. The *intermediate disturbance hypothesis* (Connell and Slatyer, 1977) suggested various stages of disturbance and recovery following disturbance to create a mosaic of ecosystem states which, overall, maximizes diversity at a local landscape scale.

These ideas of succession and disturbance have been applied in restoration, even reconstruction, of damaged ecosystems. The related idea of creating functional and sustainable yet entirely novel ecosystems is also gaining traction.

8.2.4 Landscape ecology

When the spatial scale of study encompasses a range of habitat types, we are studying *landscapes*. These range from microhabitats such as a rotting log, exposed boulder or other resource 'unit' through to the patchwork of ecosystems within a river catchment, an elevational transition or a designated nature reserve. Habitat type to habitat type turnover is the key issue increasing gamma diversity across the landscape.

8.2.5 Biogeography

Inasmuch as it deals with assemblages of species across spatial scales, the subdiscipline of biogeography may legitimately be considered as part of community ecology, albeit with more focus on processes operating on longer time frames. This is discussed at length in Chapter 4.

8.3 Why Moths as Community Objects?

Approaches to addressing community ecology generally first identify one or more *focal taxa* for special attention. The Lepidoptera, especially the moths, have attracted a great deal of attention in this regard for a variety of biological, methodological and socio-scientific reasons. The moth focus has several underlying factors.

8.3.1 The 'butterfly effect' – amateur engagement, visibility, size range

The butterflies and moths have attracted popular attention from ancient times. The earliest surviving images date back more than 3500 years to Egyptian and Chinese paintings. The arrival of the printing press quickly led to Wooton's 1552 book which was one of the earliest even though, according to Dunbar (2010), he considered caterpillars to be plant products. Moffat's posthumously published 1634 book on insects was perhaps the first produced in England. From this point on, books in Latin, English and various European languages appeared (and have continued to appear) regularly.

The taxonomic 'revolution' following Linnaeus' (1758) tenth edition of *Systema Naturae* standardized binomial names for animals, stimulating a growing number of amateur and dilletante lepidopterists to become more 'scientific'. The biological renaissance a century later driven by Darwin, Wallace, Bates and many others again refocused attention on the familiar Lepidoptera now viewed in the light of evolution by natural selection.

A second major advantage of the Lepidoptera has been their 'collectability'. Properly prepared and curated specimens of butterflies and moths are virtually immortal, to be admired, gloated over, even studied scientifically by generations of enthusiasts. This is much less the case for immature stages (although dry preservation techniques do exist). Modern enthusiasts often use high-quality digital images as complements to physical collections.

Two additional features locked in the Lepidoptera as study objects. A great many species are relatively large and readily examined with no or just a little magnification. Second, members of the order are almost ubiquitous: if there are plants, there will be Lepidoptera.

It is no great surprise, then, that the first naturalists to visit Australia paid at least some attention to the butterflies and moths. Linnaeus' pupil Johannes Fabricius described the first Australian moths in 1775 from collections made on the voyage of the Endeavour, collected in Australia in 1770. The first locally produced colour illustration of a moth, *Nyctemera amica*, appeared in the journal of midshipman George Raper, painted in about 1788 (Braby and Olsen, 2011). A continuous trickle of illustrated books on Australian moths followed and continues to appear (Kitching and Edwards, 2020).

8.3.2 Light attractedness and sampling

A large majority of moths are both nocturnal and attracted to light, making it possible to sample them in large numbers using light traps. The reasons for this positive phototaxis have been the subject of much debate over many decades. Many different, generally untested, hypotheses have been proposed (Robinson and Robinson, 1950; Mazokhin-Porshnyakov, 1960; Callahan, 1965; Sotthibandhu and Baker, 1979). Finally in 2024, Samual Fabian and colleagues produced and tested an alternative model which has the virtue of being supported by experimental data. They assumed, based on earlier work, that insects use lighter areas in order to remain upright during uninterrupted flight. According to these authors, this is a 'highly conserved' response. In the presence of artificial light sources, they continue to orient their dorsal surfaces to the strongest light present, resulting in a gradual spiralling into the light source and their ultimate 'capture' by the light emitter.

A wide variety of moth traps have been developed and variously modified for use in different situations. Southwood and Henderson (2000) summarize a range of these. Notably, many lepidopterists have moved away from traditional mercury vapour and actinic bulbs to more durable and lower energy LED bulbs, such as the LepiLED model which can be run using small rechargeable batteries, allowing easier sampling in remote locations (Brehm, 2017).

Many variables affect the numbers and variety of moths that are attracted to light traps, including the inherent properties of particular taxa, the immediate surrounding environment, the ambient weather conditions and, finally, the structure and type of the traps themselves. Several well-designed experimental and/or statistical studies have explored these issues although no such studies have been made (that is, published) in Australia.

- In summary, light trap catches in general are negatively correlated with moon phase, with largest catches observed at the times of the new – that is, no – moon. The negative effects of bright moonlight may be mitigated by cloud and forest canopy cover.
- Catches are driven by temperature although not necessarily in a linear fashion.
- Exposure of traps to strong winds may reduce trap catches.
- Finally, in our experience, rainfall has little negative effect on catch sizes, all else being equal. Indeed, light rainfall under warm conditions will generally enhance catch size in tropical and subtropical sites.

So, simplistically, we expect catches to be largest on warm, still, humid, moonless nights. This does not mean that any other conditions should be avoided at all costs. In any real-world field study, constraints of time and cost obviate the possibility of waiting for such 'perfect' nights.

Monitoring changes in moth communities against a background of environmental change demands resampling over considerable time intervals. Different trap designs undoubtedly have different target efficiencies. When the identification of temporal change is a research goal, it is essential to use the same design of trap throughout. This alone is a compelling reason to keep older 'traditional' designs in working order. We use standard Pennsylvania 'upright' traps enclosed in a framework that both protects the catch from rain and holds the requisite 12v battery. A version of this trap is shown in Fig. 8.1 and an earlier version is described technically by Kitching *et al.* (2005).

8.3.3 Size of the taxa and statistical issues

A third reason why the moths make excellent subjects for community ecological analysis is the sheer size of the taxonomic group to which they belong.

Detecting community change based on mass sampling depends on various forms of mathematical analysis. Fundamentally, this involves examination of the statistical distributions exhibited by the sample obtained (that is, how many of each type of object – species of moths in this case – were encountered and how these conform to one or other of known mathematical distributions). From such distributions various summarizing statistics can be calculated and compared with values observed in other samples. We then compute probability values attached to any observed differences or similarities.

Fig. 8.1. A modified Pennsylvania light trap in action in rainforest. Photo: L.A. Ashton.

Two key issues are central to the probity of such analyses. First, we assume the two samples are independent of each other: that is, they are two quite different attempts to sample the same thing. In moth sampling, this means we do not place two traps adjacent to each other. As a rule of thumb, they should be out of line of sight of each other. This may circumvent the fact that different moth families have different radii of attraction to light sources (Merckx and Slade, 2014). Once we have confidence that we have enough independent samples of the same thing – that is, the moth assemblage in location A, for example – then we can compute our summary statistics, which are then available for comparison with other locations.

Second, we need to take account of randomness both in terms of how we place traps and how we interpret results. Within a vegetation plot, for example, we can select x and y co-ordinates using a random number generator as long as we observe the proximity 'rule' discussed above. Once we have a sample in hand, we sort to species (or 'morphospecies'). Inevitably there will be a few species represented by multiple individuals (the 'dominants') and, relatively speaking, many represented by just one ('singletons') or two ('doubletons') individuals. The question when we compare this sample with another is which if any of these species are characteristic of the sampling location and which just happen to have been captured randomly by a particular trap. If our sample is large and varied enough, however, we can make such comparisons. If, for example, a sample contains 100 singletons then it is unlikely that all will be random 'tourists' that just happened to arrive in our trap. Particularly useful tools for understanding hyperdiverse moth communities are rarefaction curves to estimate sampling sufficiency and extrapolation tools such as iNEXT which can interpolate total diversity based on abundance or incidence data (Hsieh *et al.*, 2016).

In any natural forested location in Australia, we can assume that there are many species of moths 'available' for sampling at any one time, sufficient to overcome the 'rareness' factor described, downplaying the statistical importance of any particular species' count while allowing clear identification of assemblage-to-assemblage differences. The same arguments can be extended to families within the order. Some researchers have focused on the Geometridae (Australian estimate, 2300+ species), the Pyraloidea (1600+ species) or the Noctuoidea (2600+ species), sometimes in combination with other well-known groups.

8.3.4 State of taxonomic knowledge

It is supposedly indisputable that the Lepidoptera are the best known of all insect orders and, if the goal is to place binomial tags on a set of species, then this may be a further reason for them to be a preferred target taxon. This assertion hides a much more complex situation.

For the Australian fauna, several key uncertainties emerge.

- Just how well known is our fauna overall?
- Which Lepidoptera in our fauna are taxonomically well known (or not)?
- How accessible and usable is our taxonomic base?

8.3.4.1 A well-known fauna

Assuming a depth of knowledge on moth diversity and evolution is a major oversimplification. This is especially the case when considering the Australian fauna.

First, we make a distinction between formal systematics and taxonomy. Although these two terms have on occasion been considered synonymous, recent advances in phylogenetics have drawn and broadened the gap between these two related fields. The primary goal of systematics has become the understanding of evolutionary relationships among taxa, increasingly using phylogenomic tools. Strictly, taxonomy is about classification, naming and description. Of course, these approaches are closely related, and professional (and quasi-professional) systematists are also generally taxonomists. The reverse is not necessarily the case.

With regard to systematics, then, it is probably true to say that the Lepidoptera, globally, are relatively well known. Indeed, the state of our knowledge is the subject of the greater part of Chapter 3. Where questions still remain, well-established phylogenomic techniques will likely resolve these in due course.

'Well known', however, may well be taken as what proportion of the Australian fauna is actually described and named. Yeates *et al.* (2003) published what they called 'new estimates' for the sizes of insect groups in Australia. A critical examination of their Table 8.2 suggests many of the order level estimates are conservative. Nevertheless, they give an idea of the level of knowledge of each order. Across all 26 insect orders (as then recognized) 22.4(±4.22)% of the estimated fauna remained undescribed.[1] If we restrict analyses to the 'big six' – that is, the six orders with estimated richnesses of greater than 2000, then the estimated average level of undescribed species is much higher at 48.2(±7.96)%. Their figures suggest that 30.7% of the Australian Lepidoptera remained to be described in 2003.

So, if at least 30% of the Australian Lepidoptera remain unnamed, where do we find the 10,000 or so anonymous species? Common's (1994, 1997, 2000) masterworks on the Oecophoridae: Oecophorinae lists over 1600 named species and a further 1400 or so known but unnamed. On this foundation, Zborowski and Edwards (2007) suggested there might be as many as 3400 unnamed species in this taxon alone. Robinson and Nielsen (1993) suggested as many as 1000 Tineidae were known and awaiting description without speculating on species not 'known'. For smaller groups, similar or even greater levels of novelty occur.

In one important respect, however, the Australian fauna is well known and potentially attractive to the community ecologist. In 2010 and 2011 as part of the Barcode of Life Project, an attempt was made to DNA barcode the entire Lepidoptera fauna held in the Australian National Insect Collection (ANIC). The barcode sequence used was the cytochrome c oxidase 1 (CO1) gene recognized as a universal tool for identifying insect species. A total of 41,650 specimens of 12,699 'species' (including undescribed morphospecies) was processed (Hebert *et al.*, 2013). These barcodes are publicly available and in principle allow the identification of unknown individuals in any subsequent sampling effort. The parallel development of metabarcoding approaches, in which mixed samples are barcoded *en masse* and subsequently identified to putative species without prior sorting (see, for example, Ji *et al.*, 2013), makes the existence of the barcoding library of the Australian fauna even more useful.

8.3.4.2 Using the taxonomic base

First, if simple community-level comparisons among samples is needed then, arguably, we have no need of taxonomy at all. Multivariate analyses of various kinds will tell us if two or more assemblages of moths sorted to putative 'species' are different, statistically speaking. We can obtain visual representation of such differences for descriptive purposes in the same fashion (Fig. 8.2). Similarly, if we simply wish to calculate the range of community ecological indices for alpha- or beta-diversity then there is no need to name the entities we have sorted to 'species'. Needless to say, this sort of comparison is of limited usefulness. Statistical analyses will tell us which 'species' contribute most to the overall assemblage level differences – so-called 'indicator species', together making a 'predictor set' (Kitching and Ashton, 2013). Calculating levels of dominance in a sample will single out the most abundant 'species'. In all of these cases, identifying the 'species' involved will then link us

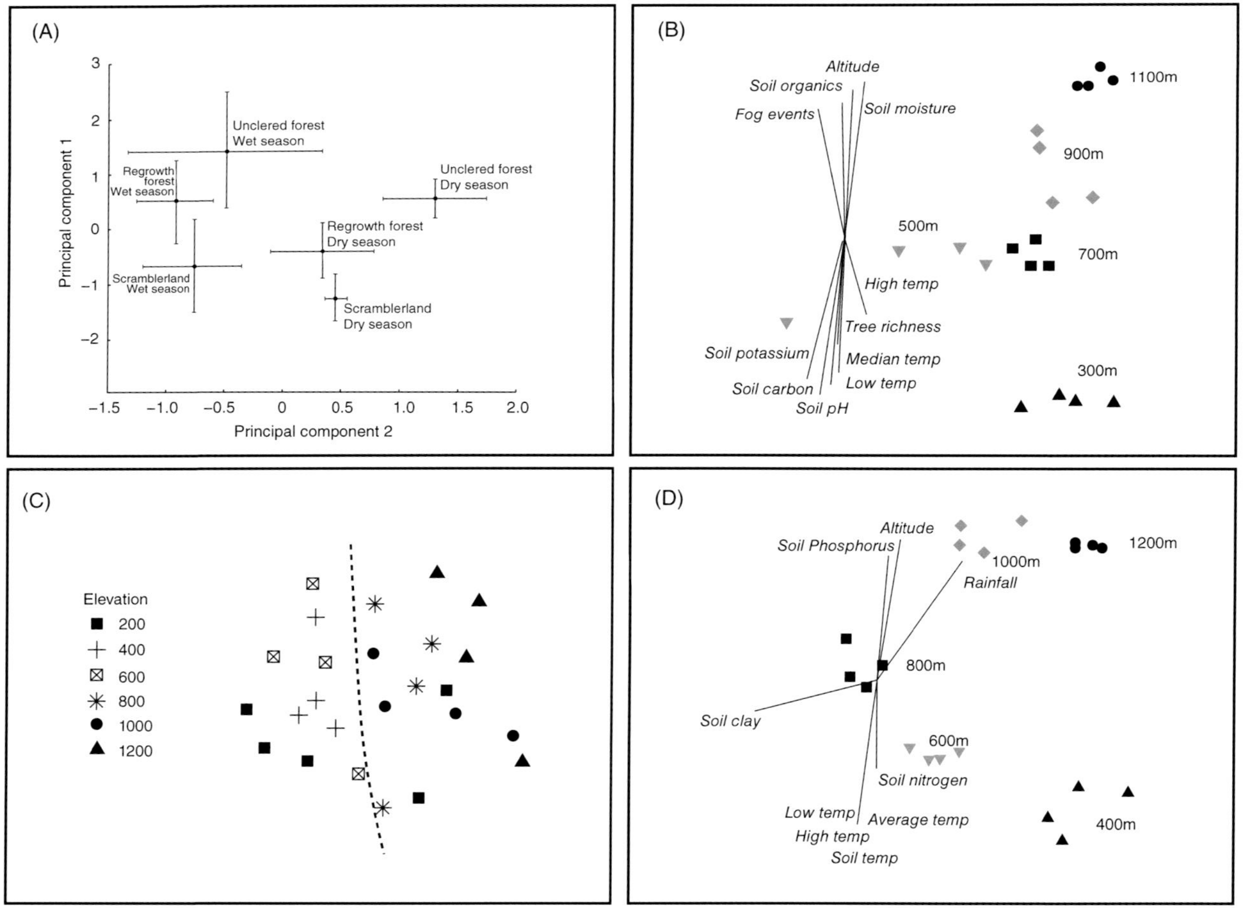

Fig. 8.2. Ordination plots from selected moth community studies. (A) Rainforest remnants on the Atherton Tablelands, Queensland. (B) Elevational change in Lamington National Park, Queensland. (C) Elevational change in assemblages of Pyraloidea at Eungella National Park, Queensland. (D) Elevational change at Mt Lewis, Queensland. A,C, from Kitching *et al.* (2000), with permission; B,D, from Ashton *et al.* (2016), with permission.

to available information on wider distribution or other aspects of biology on the species (or their close relatives). For all of these reasons, identifying to named (or known) species part or all of our mixed moth samples becomes very useful.

Two approaches are currently available. The first is collection based whereas the second requires extracting a DNA 'soup' from the mixed sample (but also, eventually, links to previously collected material).

The first approach involves making a local collection of the identified morphospecies generated by sampling. Photographing each morphospecies together with a scale bar is also useful from which further information can be drawn (e.g. various morphometric traits). Identification to known species can then be done by comparison with published images or directly with a museum collection (or by DNA barcoding).

There are hardcopy guides available for a few Australian taxa or regional faunas. General works such as Common's *Moths of Australia* (1990) and Zborowski and Edwards' *A Guide to Australian Moths* (2007) are well illustrated and useful for identification if your particular species happens to be included. More comprehensive regional guides include Cocking *et al.*'s *Moths in the A.C.T.* (2022), Richardson's *Tropical Queensland Wildlife from Dusk to Dawn* (2015) and the several booklets in the *Moths of Victoria* series (Marriott *et al.*, 2008–2020). The various volumes in the *Monographs on Australian Lepidoptera* are also useful for identification by comparison, especially the later volumes with colour illustrations (Matthews, 1999; Tarmann, 2004; Kaila, 2011; Simonsen, 2018; Moulds *et al.*, 2020).

We noted earlier that a database of the CO1 barcodes of almost all of the Lepidoptera in the Australian National Insect Collection was created in 2010–2011. The useful spin-off for the would-be identifier of moths is that there is, accordingly, at least one picture of almost all Australian moths held in the searchable BOLD database (boldsystems.org). It is a laborious process plodding through the images trying to match up a specimen but if some pointers have been gained from other sources then it can be useful.

Simpler 'compare the picture' approaches can be employed using two other web-based collections of images. The CSIRO resource *Australian Moths Online* is now available again via the Atlas of Living Australia (https://moths.csiro.au/). Finally, we draw attention to the set of images at http://lepidoptera.butterflyhouse.com.au. This site is particularly useful as it also lists known host plants for each species.

The second alternative is simply liquidizing an entire sample, extracting DNA and running this through one or other of the machines available for what is called metabarcoding. Fundamentally, this process examines the entire DNA profile of the mixed sample after labelling (or otherwise recognizing) the stop/start codons in the nucleotide sequences. This enables the machine used to generate sets of sequences, each a 'barcode' for a species, from within the mixed sample. With careful scrutiny (generally by appropriate software), each of the barcodes so obtained can be compared with existing data libraries and, accordingly, identified. This works well when the targeted fauna (or flora) has had barcodes extracted from well-identified reference collections and a regional barcode library so established.

The advantage of a metabarcoding approach is its semi-automated nature, requiring little entomological expertise on the part of the practitioner. The disadvantages are that no indication of relative abundance is generated (although this may change as the techniques mature) and the remaining reference material is a liquidized soup rather than a usable reference collection.

Both methods, the traditional collection-based approach and metabarcoding, are in many ways complementary and together are powerful tools which hold great promise for the community ecologist (Ji *et al.*, 2013).

8.3.5 The Lepidoptera–plant nexus and the state of ecological knowledge

All terrestrial ecosystems are driven by solar energy which is fixed through the photosynthetic activities of plants (and other green organisms). Whatever the ecosystem, however, there are only two ways in which that energy, locked into plant biomass, can be released to support higher trophic levels. Plant parts can die and decay through the activities of micro-organisms, fungi and other decomposer organisms, or plant parts may be eaten by herbivores. Just how much energy passes into the wider food web through herbivory varies widely. Warm-blooded vertebrate herbivores exhibit production efficiencies of 1.5 ± 0.12% to 3.1 ± 0.28%; non-insect invertebrate herbivores have an average production efficiency of 20.8 ± 1.38% whereas insect herbivores show an average of 38.8 ± 1.93% (Humphries, 1979). Insect herbivory clearly plays a vitally important role in maintaining ecosystems.

The Lepidoptera, as an order, are overwhelmingly herbivorous and their roles in liberating photosynthetically fixed energy into the dependent animal community are generally dominant and ubiquitous. In a few families, species have adapted to plant feeding but on moribund or dead leaves in the leaf litter (see Chapters 4 and 12). These exceptions apart, Lepidoptera as larvae exploit living plants, directly exploiting all above-ground plant parts from woody roots and stems, to buds, flowers, fruits and, of course, leaves. A few stimulate the plant to produce additional tissue in the form of galls. Lepidopterous larvae may feed on plant material externally or as leaf miners, internally. This last habit may be the ancestral relationship between Lepidoptera and angiosperms (see Chapters 3 and 5).

From a community ecological point of view, then, focusing on the Lepidoptera makes sense given their vital role within wider ecosystems. In principle, we should be able to connect the natural history of the set of moths we encounter directly with the plant assemblage which sits beneath them trophically speaking. So the question arises: which moths and which plants?

Most community analyses will try to relate 'significant' species to the vegetation among which they were collected. This means seeking larval host plant (or other feeding) records for the identified species. For some regions and taxa, there are complete or at least extensive databases on known larval host plants. So what about Australia? For an unbiased overview of the level of life history knowledge, we selected 43 species of moth at random from the Australian checklist (Nielsen *et al.*, 1996). Why 43? Simply because this fills a single page table (see Table 8.1). We then searched online and through a near complete library on Australian Lepidoptera seeking host plant records for these randomly selected species. For 12 (28%) species we found actual feeding records. For a further eight species (19%) some inference about likely host plants can be made from records for congeneric species or knowledge of the family as a whole. This, of course, left substantially more than half the sample as 'unknown'. Of course, a random sample of the entire moth fauna inevitably included many little-studied groups. Levels of knowledge for a few taxa – Sphingidae, Anthelidae, some noctuoids – are high and ever growing. A community study using moths can focus on just a few of the better known taxa. Nevertheless, we stress that carefully carried out, recorded and published life history studies of the majority of our moth fauna that are currently in the 'unknown' category remain one of the most desirable and potentially useful future contributions to our general understanding of moth biology.

8.4 Moth Communities in Australia

The availability of efficient sampling methods and powerful statistical techniques initiated an era of community-level studies of moths in Australia beginning in the early 2000s. The Kitching research group based at Griffith University led this effort with its focus on Queensland rainforests. This then expanded internationally with collaborative studies using essentially similar methodology in Panama, Malaysia, China, Papua New Guinea, Vietnam, Vanuatu, metropolitan France and Réunion island.

In all these studies the goal has been to determine to what extent assemblages of moth species change along ecological gradients of one sort or another and, if so, what subsets of species are statistically responsible for driving these changes. The ecological contrasts involved may be natural (based on elevation, latitude, inter-site distance, substrate type and so forth) or human induced (as in logging programmes or forest fragmentation).

8.4.1 Patches of rainforest

In the late 1990s the Wet Tropics Management Authority funded a pilot project focusing on nine patches of remnant rainforests on the Atherton

Table 8.1. Assessment of the level of life-history knowledge of a random selection of Australian moths (see text).

Family	Subfamily	Species	Food-plants
Psychidae		*Lepidoscia herbicola*	lichens
Bucculatricidae		*Bucculatrix acrogramma*	UNIKNOWN
Gracillariidae	Gracillariinae	*Caloptilia chlorella*	UNKNOWN
Yponomeutidae	Yponomeutinae	*Zelleria cryptica*	UNKNOWN
Glyphipterygidae		*Glyphipterix lamprosema*	UNKNOWN (family: grasses, sedges)
Oecophoridae	Oecophorinae	*Callithauma glycera*	UNKNOWN
Oecophoridae	Oecophorinae	*Chrysonoma habropis*	UNKNOWN (other *Chrysonoma* Myrtaceae)
Oecophoridae	unplaced	*Pholeutis analcis*	*Eucalyptus* leaf litter
Xyloryctidae		*Bassarodes syriaca*	UNKNOWN (wood-borer)
Xyloryctidae		*Lichenaula onychodes*	*Eucalyptus* (Myrtaceae)
Xylorictidae		*Xylorycta argentella*	UNKNOWN (wood-borer)
Ethmiidae		*Ethmia sphaerosticha*	*Ehretia* (Boraginaceae)
Momphidae		*Zapyrasta* sp A	UNKNOWN
Cosmopterygidae	Chrysopeleiinae	*Leptozestis cyclonica*	UNKNOWN
Gelechiidae	Dichomeridinae	*Dichomeris cirrhostola*	UNKNOWN
Gelechiidae	Gelechinae	*Scrobipalpa aptatella*	*Nicotiana, Solanum* (Solanacceae)
Gelechiidae	*Aristotelia* grp	*Aristotelia aphthoropa*	UNKNOWN
Tortricidae	Tortricinae	*"Phryctora"phaeosema*	UNKNOWN
Tortricidae	Tortricinae	*Procalyptis oncota*	UNKNOWN
Tortricidae	Olethreutinae	*Crocidosema polyphaea*	UNKNOWN (probably Malvaceae)
Limacodidae		*Comana resplendens*	UNKNOWN
immidae		*Birthana cleis*	*Decaisnina* (Loranthaceae)
Simaethistidae		*Metaprotus asuridia*	UNKNOWN
Pyralidae	Phycitinae	*Phycita trachystola*	UNKNOWN
Pyralidae	Phycitinae	*Assara holophragma*	UNKNOWN
Crambidae	Crambinae	*Hednota empheres*	UNKNOWN (other *Hednota* grasses (Poaceae))
Crambidae	Scopariinae	*Perimeceta niphotypa*	UNKNOWN
Geometridae	Ennominae	*Pholodes sinisraria*	*Macadamia* (Proteaceae) (etc.)
Geometridae	Geometrinae	*Oenospila flavifuscata*	UNKNOWN
Lasiocampidae	Lasocampinae	*Neurochyta edna*	UNKNOWN
Anthelidae	Anthelinae	*Anthela ochroptera*	UNKNOWN
Sphingidae	Sphinginae	*Meganoton rufescens*	Annonaceae, Bignoniaceae
Sphingidae	Macroglossinae	*Macroglossum vacillans*	*Strychnos* (Loganiaceae)
Sphingidae	Macroglossinae	*Hippotion rosetta*	Nepenthaceae, Rubiaceae
Erebidae	Arctiinae	*Macaduma toxophora*	UNKNOWN
Erebidae	Arctiinae	*Nyctemera secundiana*	*Crassocephalum, Senecio* (Asteraceae)
Erebidae	Herminiidae	*Adrapsa ablualis*	UNKNOWN
Erebidae	Hypeninae	*Artigisa lignicolaria*	UNKNOWN (other *Artigisa* fungivorous)
Erebidae	Scoliopteryginae	*Anomis definata*	UNKNOWN
Erebidae	Catocalinae	*Ischyja neocherina*	UNKNOWN
Euteliidae		*Gyrtona polionota*	UNKNOWN
Noctuidae	Acontiinae	*Autoba pectorora*	UNKNOWN (other *Autoba* polyphagous)
Noctuidae	Acontiinae	*Periscepta polysticha*	*Hibbertia* (Dilleniaceae)

Tablelands in tropical Queensland. These fell into three categories: three had never been cleared, three comprised 40–60 year regrowth and three were recently cleared 'scramblerlands'. Moths were sampled on each patch in both wet and dry seasons.

There was clear differentiation in both seasons across the three habitat types (Fig. 8.2A) and different groups of moths drove these differences (usually, and interestingly, at the subfamily level). In highly simplistic terms, the simple ratio between the diversity of Geometridae to Noctuoidea in a sample formed a rough and ready index of the degree of biodiversity disturbance on any remnant patch. More details plus a more or less complete species list are provided in Orr and Kitching (1999, 2003) and Kitching *et al.* (2000).

8.4.2 Elevation, climate and the legacy of IBISCA Queensland

In 2003, Yves Basset of the Smithsonian Institution's tropical research centre called on a large international team to make ground-canopy comparisons in the San Lorenzo rainforest on the Panamanian Caribbean shore. The approach proposed replacing many long years of research by smaller teams of scientists with short expeditions by large teams of different specialists. He christened the approach 'IBISCA' – **I**nvestigations of the **Bi**odiversity of **S**oil and **C**anopy **A**rthropods. The approach proved to be powerful and successful (Basset *et al.*, 2012). There followed a number of follow-up projects including 'IBISCA Queensland'. From 2006 to 2010, the Queensland project examined arthropod and plant diversity along an elevational transect of subtropical rainforest. Elevational gradients are useful 'natural laboratories' as a number of biotic and abiotic factors shift in a predictable way, allowing better understanding of how biodiversity responds to these factors, including temperature, humidity, wind speed and UV radiation as well as the associated changes in vegetation (Hodgkinson, 2005). This has potential uses in predicting how assemblages may respond to shifting climatic conditions.

The IBISCA Queensland study was carried out in Lamington National Park, south-east Queensland (28°S; Kitching *et al.*, 2011). This was later extended for some taxa, including moths, to include comparable transects on the southern aspect Border Ranges National Park, NSW (adjoining Lamington), in mid-tropical Queensland (Eungella, 21°S, 2012–2014) and far northern Queensland (Mt Lewis, 16°S, 2009–2011). For the moth data, Ashton *et al.* (2011) present detailed results for Lamington, Odell *et al.* (2016) for Eungella, and Ashton *et al.* (2016) describe a comparative study across these four sites. Kitching *et al.* (2020) examined the Pyraloidea from the Eungella survey in detail.

The descriptive analyses, including the entire set of moth taxa, showed similar outcomes across all transects, with sites at each elevation clustering and differing from those higher and lower. Examples of these outputs are shown in Figs 8.2B and 8.2D. A subsequent analysis of the Pyraloidea (alone) from the Eungella site showed a looser grouping into a set of 'higher elevation' sites and one of the lower elevation sites (Fig. 8.2C) (Kitching *et al.*, 2020): that is, there was a mid-elevation turnover rather than an elevation-to-elevation one.

Many of the studies outlined above were established with the specific goal of facilitating the resampling of moths through time. The permanent, 20 × 20 m plots established along each transect were designed to be useful to future researchers and the data for these sites is available online (https://osf.io/vsfr9/). The first major resampling effort on these plots was carried out along the Lamington transect in 2022–2023, 18 years after their first establishment and analysis of species redistributions is under way. The future resampling of these sites has the potential to provide important insights into our changing biodiversity.

8.4.3 The vertical dimension

Along with the large-scale environmental gradients of latitude and elevation, there are strong abiotic changes within forests from the understorey to the canopy (Nakamura *et al.*, 2017). Forest understoreys tend to be darker, cooler and moister than forest canopies, which have higher UV radiation and windspeeds. There is also a biotic gradient, with the communities of plants and animals changing from the forest floor to the canopy. These factors strongly shape moth communities, which almost universally show strong separation of understorey and canopy assemblages.

In the largest study of its kind, over 100,000 moths were sampled across elevational and vertical gradients from primary forest sites of Australia and Asia, ranging from Border Ranges National Park in NSW to the subalpine forests of Lijiang, China (Ashton *et al.*, 2015). This work demonstrated vertical turnover of moth communities across Australian forests, but the underlying mechanisms remain poorly understood. Future work should focus on how the vertical distributions of host plants and plant parts shape moth assemblage turnover, and the relative importance of biotic factors and abiotic factors. Better understanding of these patterns can help us to predict how moth communities will respond to further climate change and their capacity to shift vertically to track their physiological climatic envelopes.

8.4.4 The role of climate and seasonality

As ectotherms, moths are sensitive to temperature as well as the availability of their hosts plants. Activity patterns of moths can be highly influenced by climate and seasonality (Hickinbotham *et al.*, 2024). Moth abundance and community composition can change throughout the year, with greater moth abundances and species richness with higher temperatures found in Tasmania (McQuillan *et al.*, 1998), as well as during the previously described moth sampling at Lamington National Park, Queensland (Ashton *et al.*, 2011). Towards the tropics, seasonal differences in temperature are less pronounced and we can expect richness and abundance of moth communities to be more even throughout the year than further south. In Eungella National Park, Queensland, for example, moths were sampled at the beginning and end of the wet season. In November 2013, 13,861 moths belonging to 713 morpho-species were sampled, whereas in March 2014, 10,045 moths belonging to 607 morpho-species were collected (Odell *et al.*, 2016), showing relatively low seasonality across the beginning and end of the wet season.

8.4.5 Are moths declining?

Anthropogenic threats to insect biodiversity include climate change, land-use change, exotic species, pesticide use and shifting species interactions (Wagner *et al.*, 2021a). There has been increasing alarm in recent years about global insect declines. Much of the empirical evidence for insect responses to environmental change, however, come from a handful of taxonomic groups in temperate, northern hemisphere landscapes where long-term data are available. In the UK, for example, regular moth monitoring has been carried out as part of the Rothamsted Insect Survey since 1963, and these invaluable data have been key in highlighting the distribution shifts and declines of many moth species (Fox *et al.*, 2021). This type of long-term monitoring can demonstrate the complexity of insect dynamics, with the distributions of some species contracting while others expand (Fox *et al.*, 2021). There may be heterogeneous responses across taxonomic groups and locations (Wagner *et al.*, 2021b).

Recent data from a range of tropical sites indicate that stronger El Niños are driving declines in some insect groups, and this is correlated with declines in insect-mediated ecosystem processes such as herbivory (Sharp *et al.*, 2025). Adding complexity to understanding moth community dynamics are changes to interactions with other species; changes to plant nutrient quality can alter plant–herbivore interactions (Hamann *et al.*, 2021), as well as causing both spatial and temporal mismatches (Pelini *et al.* 2009; Forrest, 2016).

In Australia, long-term moth monitoring data are nearly non-existent (Chen *et al.*, 2020). We have generally been reduced to using space for time substitution in attempts to quantify community-level changes reflecting environmental differences. The elevational studies discussed above are examples. Elsewhere, similar approaches have examined the effects of logging activities and land use. These space-for-time substitutions are clearly useful but cannot substitute, in the long term, for carefully designed monitoring studies in which standard methods are applied over long time frames.

Our understanding of how Australian moths are responding to these threats is shaped by the availability of data. These issues are discussed at length in Chapter 9 and particular examples are the subjects of Chapters 35–37. Biodiversity monitoring programmes, especially of invertebrates, have been unpopular with both the scientific community and funding agencies. As the urgency of understanding broad-scale environmental change becomes more and more apparent, this neglect needs to be rectified.

8.5 Knowledge Gaps and Future Directions

Community ecology is a dynamic field and new directions and technologies arise continually. Many of these have been developed to the 'proof of concept' stage but have not yet been applied widely in Australia.

8.5.1 Life history information

Information on the ecology and life history of many Australian moth species is sparse, especially for non-pest species. Basic knowledge on host plant associations, larval and pupal development, habitat preferences, seasonality, ecological roles and interactions is lacking. Any or all of these are important in determining the conservation status of many species and their roles within dynamic ecosystems.

Amateur naturalists have played an important role in advancing our understanding of the ecology and life history of many Lepidoptera. Meticulous documentation of sightings, environmental conditions, interactions and behaviours, among other things, have yielded valuable datasets, specimen collections and opportunistic scientific insights (Miller-Rushing *et al.*, 2012). Citizen science-based platforms such as iNaturalist (see section 8.5.5) are expanding and proving to be both popular and useful.

As noted in several chapters in this volume, insects, including moths, feature in the mythology, song, place names, seasonal calendars and cultural practices of Australia's First Nations people. Insects were exploited for food, medicine, pigment and poisons, among other things (Si and Turpin, 2015), as well as serving as ecological indicators of resource availability, seasonality and other natural phenomena. There is undoubtedly further insight to be gained from this traditional knowledge of moths (Faast *et al.*, 2020).

8.5.2 Trait analysis

Functional traits are features such as morphology, phenology, ecology and behaviour that shape an organism's role in, and effects on, ecosystems and can determine how resilient species are to environmental change (Diaz *et al.*, 2013). The study of moth functional traits in Australia has so far been limited. Trait-based studies will lead to better understanding of the ecology of moth communities, and predictions on how species will respond to threats including climate change and habitat loss (Slade *et al.*, 2013).

Commonly measured morphological traits of moths include colour lightness, which is likely linked to the ability to thermoregulate, and resilience to ultra-violet and disease. In the northern hemisphere, moths generally are darker at higher elevations (Xing *et al.*, 2018) and latitudes (Heidrich *et al.*, 2021). Different families, however, can show contrasting patterns (Heidrich *et al.*, 2021). Size and wing aspect ratio are also important, related to flight capacity and dispersal ability. Generally, moths will be larger at lower elevations and latitudes, although again, contrasting patterns have been found across moth families (Heidrich *et al.*, 2021) determined by evolutionary history and morphological contingency. Further complexity is added through the interplay of latitude and elevation, which can result in hump-shaped patterns in body size (Alcantara *et al.*, 2024). Feeding traits such as host plant preferences and diet breadth are important determinants of resilience to environmental change. However, as there are limited data on host plant preferences for most Australian moth species, more evidence on the host plant associations of moths across regions and ecological gradients is needed.

8.5.3 Neglected ecosystems

When considering which Australian ecosystems are 'neglected' in our understanding of the community ecology of moth assemblages, it is perhaps easier to consider which ecosystems are *not* neglected.

Previous work has focused on moths of the eastern subtropical and tropical rainforests. Rainforests make up about 5% of Australia's forest estate, suggesting that the sclerophyllous ecosystems dominated by Myrtaceae, Proteaceae, Acacias, saltbush and so on, that make up most of Australia's vegetated landscape, have not been studied in detail. There are a few notable exceptions. McQuillan *et al.* (1998) examined seasonal variation in geometrid assemblages in lowland and highland eucalypt forests in Tasmania. The studies published by the Entomological Society of Victoria are particularly good examples. These have focused on the Gippsland Lakes region (Marriott, 2017), Morwell National Park (Harris and Marriott, 2018) and the Great Otway National Park (Marriott *et al.* undated, surveys ending in 2019). Comparable studies have been made elsewhere: Lake Broadwater and its surroundings on Queensland's Darling Downs (Scott, 1988); and, more recently, in the Cumberland Plains woodlands of the Sydney district (Lomov *et al.*, 2006). Other such local studies no doubt exist.

8.5.4 Neglected taxa

Many taxa have been neglected in community studies despite their probable ecological importance, reflecting challenges in their collection, curation and lack of taxonomic or systematic resolution. Limited subject matter experts, poor ecological knowledge or simply lack of aesthetic appeal also contribute to this bias.

Many of the neglected groups are within the so-called 'microlepidoptera' which are rarely considered in biodiversity surveys. Nevertheless, they occupy a wide range of ecological niches, probably wider than their 'macro' co-ordinals. They remain promising and versatile candidates for further work and possess some unique advantages over their larger counterparts. Significantly, micromoths hold great potential to investigate ecological processes over small spatial scales such as small forest fragments, even single trees (Uhl *et al.*, 2016). DNA barcoding and metabarcoding are rapidly changing the ability to include these taxa in studies.

8.5.5 Citizen science

'Citizen science' or 'community science' harnesses the efforts of amateur observers to record particular species in localities either in short, sharp 'bioblitzes' or by ongoing contributions to public data platforms such as iNaturalist. These efforts and data platforms have generally been designed by professional scientists to ensure potentially useful analytical outcomes. The widespread participation of volunteers allows the generation of data across wide spatial and temporal scales. It also encourages a shared responsibility for environmental preservation and improves scientific literacy.

There exists a continuum of such approaches (Arazy and Malkinson, 2021) that may be classified as unstructured, semi-structured and/or structured. Unstructured projects allow participants to make opportunistic observations with little or no training. This can lead to high rates of participation, perhaps at the expense of data quality and utility (Smith *et al.*, 2019; Callaghan *et al.*, 2021). Semi-structured projects attempt to reconcile data integrity with levels of participant engagement by providing a framework for observations but allowing flexibility. Structured projects attempt to control for variation in data quality by prescribing strict sampling protocols, creating opportunities to answer specific or more complex ecological questions. Such projects demand a higher level of commitment from participants.

Citizen science initiatives have potential to accelerate research and data capture on Australia's moths (Fox *et al.*, 2019; Tng *et al.*, 2021; Ware *et al.*, 2024). The internet, smartphones, photo recognition, specialized apps and open access data repositories provide volunteers with easy means to participate in global scientific efforts.

Per capita, Australia is one of the most prolific global contributors to iNaturalist, with moths making up nearly 40% of all uploaded insect observations (Mesaglio and Callaghan, 2021; Mesaglio, 2022).

Volunteer-driven 'bioblitzes' vary but, in general, focus on recording as many target taxa as possible from an area of interest over a concentrated period of time. A growing number of such activities have focused on moths. These may be integrated into wider initiatives such as National Moth Week (nationalmothweek.org) or the Great Southern BioBlitz. These moth blitzes improve local species inventories for biodiversity monitoring at a relatively low cost (Meeus *et al.*, 2023).

There has been reluctance to adopt citizen science approaches as a primary research tool (Burgess *et al.*, 2017) and few syntheses have focused on improving our understanding of community dynamics. Concerns about data quality (Mesaglio and Callaghan, 2021) are probably overstated. Verification of observations on the biodiversity monitoring platform iSpot found that citizen scientists accurately identified 92% of moths (Silvertown *et al.*, 2015). Statistical advances have also improved our ability to account for noise and bias in data (Bird *et al.*, 2014; Mesaglio and Callaghan, 2021).

8.5.6 Metabarcoding

We described the basics of metabarcoding earlier. There is no question that work done overseas has produced proof of concept. However, the approach remains to be tested, or used, in Australia. A targeted research project would introduce the approach nationally and would also be cutting-edge science. Extending the approach to diet analysis of gut contents of moths would also go some way to supplying some of the information demanded in section 8.5.1.

8.5.7 Remote sensing and other new technologies

Remote sensing is also being harnessed to better understand insect distributions and migrations. Vertical-beam entomological radar has been used to detect insects flying between 150 and 2250 m above sea level (Bauer *et al.*, 2024; Drake *et al.*, 2024; see also Chapter 7). Camera-based remote sensing has been used successfully for tracking the activity of bogong moths in Australia. In the UK, automated monitoring stations have been combined with deep learning tools (Burniston *et al.*, 2024), but require well-developed databases of moth species to function well. Combining technologies into multisensor apparatus such as the automated multisensor stations for monitoring species diversity (AMMODs) (Wägele *et al.*, 2022) allows the application of tools such as audio recordings, computer vision and eDNA methods to identify insects, along with plants, amphibians and mammals, to produce rapid, multitaxon monitoring data.

8.6 Conclusion

Further studies of multispecies assemblages – community ecology – of moths hold great promise as tools for understanding the variation of biodiversity across Australian landscapes, both to quantify spatial variation and to monitor change through time. In this fashion, we can define intrinsic heterogeneity in biodiversity across the Australian landscape and the ways in which this has been modified by human activities. Such approaches will interact positively with other dimensions of moth study such as increasing knowledge of natural history, more refined taxonomies and increasingly well-resolved phylogenies. New technologies will make such approaches more efficient. We conclude by suggesting that creative collaborations between naturalists, technologists and an interested public will be the optimal way forward in this enterprise.

Note

[1] Note that where Yeates *et al.* (2003) gave a range for the estimate of the total size of some orders, we have taken the midpoint of that range in calculating these averages.

References

Alcantara, M.J.M., Fontanilla, A.M., Ashton, L.A., Burwell, C.J., Cao, M. *et al.* (2024) Bugs and Bergmann's rule: A cross-taxon large-scale study reveals idiosyncratic altitudinal and latitudinal body size patterns for different insect taxa. *Entomologia Generalis* 44, 715–725.

Arazy, O. and Malkinson, D.A. (2021) Framework of observer-based biases in citizen science biodiversity monitoring: Semi-structuring unstructured biodiversity monitoring protocols. *Frontiers in Ecology and Evolution* 2021, 9.

Ashton, L.A., Kitching, R.L., Maunsell, S.C., Bito, D. and Putland, D.A. (2011) Macrolepidopteran assemblages along an altitudinal gradient in subtropical rainforest – exploring indicators of climate change. *Memoirs of the Queensland Museum. Nature* 55, 375–389.

Ashton, L.A., Nakamura, A., Basset, Y, Burwell, C.J., Cao, M. *et al.* (2015) Vertical stratification of moths across elevation and latitude. *Journal of Biogeography* 43, 59–69.

Ashton, L.A., Odell, E.H., Burwell, C.J., Nakamura, A. *et al.* (2016) Altitudinal patterns of moth diversity in tropical and subtropical Australian rainforest. *Austral Ecology* 41, 197–208.

Basset, Y., Cizek, L., Cuénoud, P., Didham, R.K., Guilhaumon, F. *et al.* (2012) How many arthropod species live in a tropical forest? *Science* 338, 1481–1484.

Bauer, S., Tielens, E.K. and Haest, B. (2024) Monitoring aerial insect biodiversity: A radar perspective. *Philosophiocal Transactions of the Royal Society* B 379, 20230113.

Bird, T.J., Bates, A.E., Lefcheck, J.S., Hill, N.A., Thomson, R.J. *et al.* (2014) Statistical solutions for error and bias in global citizen science datasets. *Biological Conservation* 173, 144–154.

Braby, M. and Olsen, P. (2011) *A Flutter of Butterflies*. National Library of Australia, Canberra, ACT.

Brehm, G. (2017) A new LED lamp for the collection of nocturnal Lepidoptera and a spectral comparison of light-trapping lamps. *Nota Lepidopterologica* 40, 87–108.

Burgess, H.K., De Bey, L.B., Froehlich, H.E., Schmidt, N., Theobald, E.J. *et al.* (2017) The science of citizen science: Exploring barriers to use as a primary research tool. *Biological Conservation* 208, 113–120.

Burniston, S., Faith, M., Kriukov, V., Laidlaw, R., Kalashami, M. *et al.* (2024) *Advancing Insect Biodiversity Monitoring through Automated Sensors, Deep Learning, and Citizen Science Data*. Data Study Group Final Report. UK Centre for Ecology and Hydrology, Wallingford, UK.

Callaghan, C.T., Poore, A.G.B., Hofmann, M., Roberts, C.J. and Pereira, H.M. (2021) Large-bodied birds are over-represented in unstructured citizen science data. *Scientific Reports* 11, 19073.

Callahan, P.S. (1965) Intermediate and far infrared sensing of nocturnal insects. Part I. Evidence for a far infrared (FIR) electromagnetic theory of communication and sensing in moths and its relationship to the limiting biosphere of the corn earworm. *Annals of the Entomological Society of America* 58, 727–745.

Chen, J., McQuillan, P.B., McDonald, E. and Hawkins, C. (2020) Citizen science reveals the Palaearctic poison hemlock moth *Agonopterix alstroemeriana* (Clerck) (Lepidoptera: Depressariidae) has established in Australia. *New Zealand Entomologist* 43, 86–92.

Clements, F.E. (1916) *Plant Succession. An Analysis of the Development of Vegetation*. Carnegie Institution, Washington, DC.

Cocking, G., Bond, S. and Edwards, T. (2022) *Moths in the A. C. T.* Glen Cocking, Canberrra, A.CT.

Common, I.F.B. (1990) *Moths of Australia*, Melbourne University Press, Carlton, Victoria.

Common, I. F. B. (1994) *Oecophorine Genera of Australia 1. The Wingia Group (Lepidoptera: Oecophoridae)*. Monographs on Australian Lepidoptera, 3. CSIRO Publishing, Collingwood, Victoria.

Common, I.F.B. (1997) *Oecophorine Genera of Australia II. The Chezala, Philobota and Eulechria Groups (Lepidoptera: Oecophoridae)*. Monographs on Australian Lepidoptera, 5. CSIRO Publishing, Collingwood, Victoria.

Common, I.F.B. (2000) *Oecophorine Genera of Australia III. The Barea Group and Unplaced Genera*. Monographs on Australian Lepidoptera, 8. CSIRO Publishing, Collingwood, Victoria.

Connell, J.H. and Slatyer, R.O. (1977) Mechanisms of succession in natural communities and their role in community stability and organization. *American Naturalist* 111, 1119–1144.

De Vere, N., Jones, L.E., Gilmore, T., Moscrop, J., Lowe, A. *et al.* (2017) Using DNA to investigate honey bee foraging reveals limited flower use despite high floral availability. *Scientific Reports* 7, 42838.

Díaz, S., Purvis, A., Cornelissen, J.H., Mace, G.M., Donoghue, M.J. *et al.* (2013) Functional traits, the phylogeny of function, and ecosystem service vulnerability. *Ecology and Evolution* 3, 2958–2975.

Drake, V.A., Hao, Z. and Wang, H. (2024) Monitoring insect numbers and biodiversity with a vertical-beam entomological radar. *Philosophical Transactions of the Royal Society B* 379, article 20230117.

Dunbar, D. (2010) *British Butterflies: A History in Books*. British Library, London.
Faast, R., Clarke, P.A., Taylor, G.S., Salagaras, R.L. and Weinstein, P. (2020) Indigenous use of lerps in Australia: so much more than a sweet treat. *Journal of Ethnobiology* 40, 328–347.
Fabian, S.T., Sondhi, Y., Allen, P.E., Theobals, J.C. and Lin, H.-T. (2024) Why flying insects gather at artificial light. *Nature Communications* 15, 689.
Fabricius, J.C. (1775) *Systema Entomologiae: Sistema insectorum classes, ordines, genera, species, adiectis synonymis, locis, descriptionibus, observationibus*. Korte and Flensberg, Leipzig.
Forrest, J.R.K. (2016) Complex responses of insect phenology to climate change. *Current Opinion in Insect Science* 17, 49–54.
Fox, R., Parsons, M.S. and Harrower, C.A. (2019) *A Review of the Status of the Macro-moths of Great Britain, Butterfly Conservation Report to Natural England*. Butterfly Conservation Report Number S19-17. Butterfly Conservation, England.
Fox, R., Bourn, N.A.D., Dennis, E.B., Heafield, R.T., Maclean, I.M.D. and Wilson, R.J. (2021) Opinions of citizen scientists on open access to UK butterfly and moth data. *Biodiversity and Conservation* 28, 3321–3341.
Garcia-Robledo, C., Erickson, D.L., Staines, C.L., Erwin, T.L. and Kress, J.W. (2013) Tropical plant-herbivore networks: reconstructing species interactions using DNA barcodes. *PLOS One* 8, e52967.
Hamann, E., Blevins, C., Franks, S.J., Jameel, M.I. and Anderson, J.T. (2021) Climate change alters plant–herbivore interactions. *New Phytologist* 229, 1894–1910.
Hardwick, J. (2018) The community responses of Orthoptera to tropical forest degradation with applications of DNA barcoding for food web analysis. PhD thesis, Griffith University, Brisbane.
Hardy, A.C. (1924) *The Herring in Relation to its Animate Environment. Part 1.The Food and Feeding Habits*. Ministry of Agriculture and Fisheries, Fisheries Investigations Series 7. Ministry of Agriculture and Fisheries, London.
Harris, K. and Marriot, P. (2018) *Morwell National Park Lepidoptera and Neuroptera*, Entomological Society of Victoria, Melbourne, Victoria.
Hausmann, A., Diller, J., Moriniere, J., Höcher, A, Floren, A. and Haszprunar, G. (2020) DNA barcoding of fogged caterpillara in Peru: a novel approach for unveiling host-plant relationships of tropical moths (Insecta: Lepidoptera). *PLOS One* 15, e0224188.
Hebert, P.D.N., deWaard, J.R., Zakharov, E.V., Prosser, S.W., Sones, J.E. *et al.* (2013) A DNA 'Barcode Blitz': rapid digitization and sequencing of a natural history collection. *PLOS One* 8, e68535.
Heidrich, L., Pinkert, S., Brandl, R., Bässler, C., Hacker, H. *et al.* (2021) Noctuid and geometrid moth assemblages show divergent elevational gradients in body size and color lightness. *Ecography* 44, 1169–1179.
Hickinbotham, E.J., Pattison, Z., Fox, R. and Rushton, A.P. (2024) Drivers of moth phenology in England and Wales. *Journal of Insect Conservation* 28, 969–979.
Hodgkinson, I.D. (2005) Terrestrial insects along elevation gradients: species and community responses to altitude. *Biological Reviews* 80, 489–513.
Hsieh, T.C., Ma, K.H.F. and Anne, A. (2016) iNEXT: an R package for rarefaction and extrapolation of species diversity (H ill numbers). *Methods in Ecology and Evolution* 7, 1451–1456.
Humphries, W.F. (1979) Production and respiration in animal populations. *Journal of Animal Ecology* 48, 427–453.
Ji, Y., Ashton, L.A., Pedley, S., Edwards, D.P., Tang, Y. *et al.* (2013) Reliable, verifiable and efficient monitoring of biodiversity via metabarcoding. *Ecology Letters* 16, 1245–1257.
Kaila, L. (2011) *Elachistine Moths of Australia (Lepidoptera: Gelechioidea: Elachistidae)*, Monographs on Australian Lepidoptera, 11. CSIRO Publishing, Collingwood, Victoria.
Kitching, R.L. and Ashton, L.A. (2013) Predictor sets and biodiversity assessment: The evolution and application of an idea. *Pacific Conservation Biology* 19, 418–426.
Kitching, R.L. and Edwards, E.D. (2020) Books about Australian moths 1805–2020: an annotated bibliography. *Australian Entomologist* 47, 119–154.
Kitching, R.L., Orr, A.G., Thalib, L., Mitchell, H., Hopkins, M.S. and Graham, A.W. (2000) Moth assemblages as indicators of environmental quality in remnants of upland Australian rain forest. *Journal of Applied Ecology* 37, 284–297.
Kitching, R.L., Boulter, S.L., Vickerman, G., Laidlaw, M., Hurley, K.L. and Grimbacher, P.L. (2005) *The Comparative Assessment of Arthropod and Tree Biodiversity in Old-World Rainforests: The Rainforest CRC/Earthwatch Protocol Manual*, 2nd edn. Rainforest CRC, Cairns, Queensland.
Kitching, R.L., Putland, D.A., Ashton, L.A., Laidlaw, M.L., Boulter, S.L. *et al.* (2011) Detecting biodiversity changes along climatic gradients: the IBISCA-Queensland Project. *Memoirs of the Queensland Museum. Nature* 55, 235–250.

Kitching, R.L., Ashton, L.A., Orr, A.G. and Odell, E.H. (2020) The Pyraloidea of Eungella: a moth fauna in its elevational and distributional context. *Proceedings of the Royal Society of Queensland* 125, 65–79.
Linnaeus, C. (1758) *Systema Natura*, 10th edn, vol. 1. Laurentius Salvius, Stockholm.
Lomov, B., Keith, D.A., Britton, D.R. and Hochuli, D.F. (2006) Are butterflies and moths useful indicators for restorartion monitoring? A pilot study in Sydney's Cumberland Plain woodland. *Ecological Management and Restoration* 7, 204–210.
Marriott, P. (2017) *Gippsland Lakes: Lepidoptera*, Entomological Society of Victoria, Melbourne, Victoria.
Marriott, P. *et al.* (2008–2020) *Moths of Victoria* (9 parts). Entomological Society of Victoria, Melbourne, Victoria.
Marriott, P., Hewish, M., Harris, K., Powers, C., Mules, D. *et al.* (2019) *Great Otway National Park Lepidoptera*. Entomological Society of Victoria, Melbourne, Victoria.
Matthews, M. (1999) *Heliothine Moths of Australia: a Guide to Pest Bollworms and related Noctuid Groups*. Monographs on Australian Lepidoptera, 7. CSIRO Publishing, Collingwood, Victoria.
Mazokhin-Porshnyakov, G.A. (1960) Why insects fly to light by night. *Entomologicheskoe Obozrenie* 39, 52–58.
McQuillan, P.B., Taylor, R.J., Brereton, R.N. and Cale, P.G. (1998) Seasonal patterns of activity in geometrid moths (Lepidoptera: Geometridae) fom a lowland and highland eucalypt forest in Tasmania. *Australian Journal of Entomology* 37, 228–237.
Meeus, S., Silva-Rocha, I., Adriaens, T., Brown, P.M.J., Chartosia, N. *et al.* (2023) More than a bit of fun: the multiple outcomes of a bioblitz. *BioScience* 73, 168–181.
Merckx, T. and Slade, E.M. (2014) Macro-moth families differ in their attraction to light: Implications for lighy-trap monitoring programmes. *Insect Conservation and Diversity* 7, 453–461.
Mesaglio, T. (2022) Moths are a big deal in Australia. Available at: www.greatsouthernbioblitz.org/post/moths-are-a-big-deal-in-australia (accessed 1 October 2025).
Mesaglio, T. and Callaghan, C.T. (2021) An overview of the history, current contributions and future outlook of iNaturalist in Australia. *Wildlife Research* 48, 289–303.
Miller-Rushing, A., Primack, R. and Bonney, R. (2012) The history of public participation in ecological research. *Frontiers in Ecology and the Environment* 10, 285–290.
Moulds, M.S., Tuttle, J.P. and Lane, D.A. (2020) *Hawkmoths of Australia: Identification, Biology and Distribution*, Monographs on Australian Lepidoptera, 13. CSIRO Publishing, Collingwood, Victoria.
Nakamura, A., Kitching, R.L., Cao, M., Creedy, T.J., Fayle, T.M. *et al.* (2017) Forests and their canopies: Achievements and horizons in canopy science. *Trends in Ecology and Evolution* 32, 438–451.
Nielsen, E.S., Edwards, E.D. and Rangsi, T.V. (1996) *Checklist of the Lepidoptera of Australia*. Monographs on Australian Lepidoptera, 4. CSIRO Publishing, Collingwood, Victoria.
Odell, E.H., Ashton. L.A. and Kitching, R.L. (2016) Elevation and moths in a central eastern Queensland rainforest. *Austral Ecology* 41, 133–144.
Orr, A.G. and Kitching, R.L. (1999) A checklist of macrolepidoptera collected from rain forest and former forest areas on basalt soils on the Atherton Tableland. *Australian Entomologist* 26, 15–27.
Orr, A.G. and Kitching, R.L. (2003). A faunistic analysis of Macrolepidoptera from complex notophyll vine forest, North Queensland, Australia. *Journal of Natural History* 37, 1537–1554.
Paula, D.P., Linard, B., Crampton-Platt, A., Srivathsan, A., Timmermans, M.J.T.N *et al.* (2016) Uncovering trophic interactions in arthropod predators through DNA shotgun-sequencing of gut contents. *PLOS One* 11, e0161841.
Pelini, S.L., Prior, K.M., Parker, D.J., Dzurisin, J.D.K., Lindroth, R.L. and Hellmann, J.J. (2009) Climate change and temporal and spatial mismatches in insect communities. In: Letcher, T.M. (ed.) *Climate Change: Observed Impacts on Planer Earth*. Elsevier, Amsterdam, pp. 215–231.
Pompanon, F., Deagle, B.E., Symondson, W.O., Brown, D.S., Jarman, S.N. and Taberlet, P. (2012) Who is eating what: Diet asssessment using next generation sequencing. *Molecular Ecology* 21, 1931–1950.
Richards, O.W. (1930) The animal community inhabiting rotten posts at Bagley Wood, near Oxford. *Journal of Ecology* 18, 131–138.
Richardson, B. (2015) *Tropical Queensland Wildlife from Dusk to Dawn: Science and Art*. LeapFrogOz, Kuranda, Queensland.
Robinson, G.S. and Nielsen, E.S. (1993) *Tineid Genera of Australia (Lepidoptera)*. Monographs on Australian Lepidoptera, 2. CSIRO Publishing, Collingwood, Victoria.
Robinson, H.S. and Robinson, P.J.M. (1950) Some notes on the observed behaviour of Lepidoptera in flight in the vicinity of light-sources, togetherwith a description of a light-trap designed to take entomological samples. *Entomologist's Gazette* 1, 3–20.

Scott, G. (ed.) (1988) *Lake Broadwater: The Natural History of an Inland Lake and its Environs*. Darling Downs Institute Press, Toowoomba, Queensland.
Sharp, A.C., Boyle, M.J., Bonebrake, T.C., Guo, Y., Kitching, R.L. *et al.* (2025) Elevated ENSO extremes are reducing tropical forest invertebrate diversity and function. *Nature* 645, 946–951.
Si, A. and Turpin, M. (2015) The importance of insects in Australian Aboriginal society: A dictionary survey. *Ethnobiology Letters* 6, 175–182.
Silvertown, J., Harvey, M., Greenwood, R., Dodd, M., Rosewell, J. *et al.* (2015) Crowdsourcing the identification of organisms: A case-study of iSpot. *ZooKeys* 480, 125–146.
Simonsen, T.J. (2018) *Splendid Ghost Moths and their Allies, A Revision of Australian Abantiades, Oncopera, Aenetus, Archaeoaenetus and Zelotypia (Hepialidae)*. Monographs on Australian Lepidoptera, 12. CSIRO Publishing, Collingwood, Victoria.
Slade, E.M., Merckx, T., Riutta, T., Bebber, D.P., Redhead, D. *et al.* (2013) Life-history traits and landscape characteristics predict macro-moth responses to forest fragmentation. *Ecology* 94, 1519–1530.
Smith, M.R., Parker, M.J. and Schaefer, J.A. (2019) Structured and unstructured citizen science: seven decades of expanding bird populations in central Ontario, *Canadian Journal for Nature Conservation* 50, 125717.
Sotthibandhu, S. and Baker, R.R. (1979) Celestial orientation by the large yellow underwing moth, *Noctua pronuba* L. *Animal Behaviour* 27, 786–800.
Southwood, T.R.E. and Henderson, P.A. (2000) *Ecological Methods*, 3rd edn. Blackwell Scientific, Oxford.
Summerhayes, V.S and Elton, C.S. (1923) Contributions to the ecology of Spitzbergen and Bear Island. *Journal of Ecology* 11, 214–286.
Tarmann, G.M. (2004) *Zygaenid Moths of Australia: a Revision of the Australian Zygaenidae (Procridinae: Artonini)*. Monographs on Australian Lepidoptera, 9. CSIRO Publishing, Collingwood, Victoria.
Tng, D.Y.P., Apgaua, D.M.G., Fisher, N.J. and Fazio, V.W. (2021) Moth species richness in an upland tropical rainforest: A citizen scientist assisted study. *bioRxiv* 2021.11.07.467659.
Uhl, B., Wölfling, M., Fiala, B. and Fiedler, K. (2016) Micro-moth communities mirror environmental stress gradients within a Mediterranean nature reserve. *Basic and Applied Ecology* 17, 273–281.
Wägele, J.W., Bodesheim, P., Bourlat, S.J., Denzler, J., Diepenbroek, M. *et al.* (2022) Towards a multisensor station for automated biodiversity monitoring. *Basic and Applied Ecology* 59, 105–138.
Wagner, D.L., Fox, R., Salcido, D.M. and Dyer, L.A. (2021a) A window to the world of global insect declines: Moth biodiversity trends are complex and heterogeneous. *Proceedings of the National Academy of Sciences* 118, e2002549117.
Wagner, D.L., Grames, E.M., Forister, M.L., May Berenbaum, M.R. and Stopak, D. (2021b) Insect decline in the Anthropocene: Death by a thousand cuts. *Proceedings of the National Academy of Sciences* 118, e2023989118.
Ware, C., Williams, K.J., Ferrier, S., Faith, D.P., Manion, G. *et al.* (2024) The value of targeted biological surveys: An assessment of Australia's Bush Blitz programme. *Diversity and Distributions* 30, e13806.
Whittaker, R.H. (1960) Vegetation of the Siskiyou Mountains, Oregon and California. *Ecological Monographs* 30, 279–338.
Wooton, E. (1552) *De Differentis Animalium*. Lutetiae Parisiorum, Paris.
Xing, S., Bonebrake, T.C., Ashton, L.A., Kitching, R.L., Cao, M. *et al.* (2018) Colours of the night: Climate-morphology relationships of geometrid moths along spatial gradients in southwestern China. *Oecologia* 188, 537–546.
Yeates, D.K., Harvey, M.S. and Austin, A.D. (2003) New estimates for terrestrial arthropod species-richness in Australia. *Records of the South Australian Museum Monograph Series* 7, 231–241.
Zborowski, P. and Edwards, E.D. (2007) *A Guide to Australian Moths*. CSIRO Publishing, Collingwood, Victoria.

9 Moth Conservation in Australia

Tim R. New
Department of Ecological, Plant and Animal Sciences, La Trobe University, Melbourne, Victoria, Australia

Abstract

Interest in moth conservation lags considerably behind that for butterflies. They are at least equally deserving, and many moth species have indeed been signalled across the world to be threatened as their habitats decline in extent and quality, invasive species impacts increase, pollution diversifies, land transformations for intensive agriculture and urban spread continue apace, and climates change. Conservation goals and efforts focus on both individual species considered threatened and on wider diversity, the latter offering the lure of greater benefits for 'biodiversity' through protection and management of key sites. Both approaches demand detection and amelioration of threats, long-term security of the focal habitats, sensible tailored and adaptive management, and continuing logistic support for undertaking and monitoring trends and outcomes against defined objectives. In general, moths evince little conservation sympathy, and garnering public and political support for such activities can be challenging. Much moth conservation devolves more passively on measures taken for other taxa or notable sites. This chapter discusses these issues for Australia's moths, many of which are vulnerable through being both ecologically specialized and having very localized distributions.

9.1. Introduction

Insect conservation has its historical foundations in concerns for butterflies, as the most 'popular' and iconic group of insects (e.g. Duffey, 1968; Pyle, 1976; Collins and Morris, 1985; New, 1997; Sands and New, 2013; Taylor *et al.*, 2018; Geyle *et al.*, 2021). Attention to the far larger component of the Lepidoptera, the moths, has lagged far behind – notwithstanding that some diurnal moths have indeed received considerable attention, a few dubbed as 'honorary butterflies'. Moths are still very much the 'poor cousins' of butterflies in conservation (New, 2023). They are at least equally deserving, and many species have indeed been signalled across the world to be threatened as their habitats diminish in extent and quality, invasive species impacts spread, pollution diversifies, land transformations for intensive agriculture and urban spread continue apace, and climates change. Urbanization includes the major component of artificial lighting at night ('ALAN'), of particular significance to phototactic insects and implicated strongly in moth declines elsewhere. Much popular opinion still regards moths as 'pests', so targets for suppression; overcoming that prejudice is a key aspect of promoting moth conservation, aided by emphasizing pragmatic themes such as the importance of moths as pollinators

Corresponding author: t.new@latrobe.edu.au

DOI: 10.1079/9781800627680.0009

and as food for insectivorous birds and mammals. These conditions are all evident in Australia, and apply to many different invertebrate taxa.

The twin levels of conservation attention are, on the one hand, single species, whereby individual species considered threatened become the targets of conservation effort, perhaps aided by formal 'listing' under some protective legislation and gaining priority for limited funding or other resources, and on the other, wider diversity, with emphasis on local faunas, with the hope of conserving the places and resources on which multiple co-existing species depend. These are often termed the 'fine filter' and 'coarse filter' approaches, respectively, with the latter emphasizing need for a broad 'landscape approach' to conservation. Either is likely to be initiated with very incomplete knowledge of either the biology, distribution and vulnerability of a focal species, or the richness and composition of an assemblage on a restricted site. The latter approach is receiving increased attention with the realization that the 'single-species approach' cannot accommodate all deserving species, that the methods of selecting priority species are debated (and sometimes contentious) and that each such campaign is both protracted and expensive.

Conservation of broadly defined moth diversity (largely through protection of key sites, accompanied by resource enhancement and wider restoration) offers the lure of greater benefits for biodiversity in general (more 'bang for the buck') in the context of the natural environment. Both approaches devolve on detection and amelioration of threats, long-term security of the focal sites, sensible tailored and adaptive management, and continuing logistic support for undertaking and monitoring trends in the species or assemblage.

Priorities for attention among numerous worthy candidates may be difficult to select, with the additional disadvantage that, in general, moths evince little conservation sympathy and gaining any public support for their wellbeing may be difficult. Without that sympathy and 'citizen science' interest, many necessary conservation measures may not be possible. Much moth conservation devolves more passively on measures taken for other taxa or notable sites.

These issues are discussed for Australia's moths in this chapter, in places drawing on an earlier synthesis (New, 2023). Much of our awareness of needs flows from extensive evidence of moth declines in Britain, western Europe and North America, where historical records of species incidence and distribution are far more extensive than in Australia.

9.2 Needs for Conservation: Threats to Moths

A major barrier to determining changes in Australia's moth fauna (and, so, possible needs for conservation) is simply the paucity of historical records of species and diversity, so that we have no well-informed baseline against which to appraise changes in abundance and distribution of most species, or the richness and composition of local assemblages (as discussed in Chapter 8). Anecdotal declines of emperor gum moths (*Opodiphthera* spp.) over recent decades express individual concerns and ecological surveys continue to reveal numerous 'singletons' of species – as a typical picture of insect surveys – without direct evidence of rarity or loss, or quantitative information on changes in abundance and distribution. Other species, some the focus of targeted surveys, appear to be genuinely rare despite suitable resources being more widespread: the Tasmanian pandani moth (*Proditrix nielseni*) is one such case (McQuillan, 2003). The limited knowledge of the natural history of most Australian moths, many known only from adults, single localities, single specimens and not seen since their description long ago, perhaps obscures the extent of their losses and reduced richness in assemblages over recent decades.

Very few species have been signalled directly as of conservation concern, but circumstances such as being in any sense 'rare', narrow range endemics, ecological specialists and/or having small populations can predispose species to increased vulnerability as their environments change. The few noted species may well be simply the tip of the iceberg of Australia's moth and other insect losses.

The major drivers of threatened status, of both species and assemblages, interact, often synergistically. Three of the most significant and widespread threats are noted here. Others, such as pollution and pesticide side-effects, are more sporadic. All are complex; impacts differ across environments, circumstances and species, and demand investigation in each conservation case. Detection and amelioration of 'threats' are central to conservation planning.

9.2.1 Habitat loss and fragmentation

The clearing of natural vegetation and the resulting losses and fragmentation of habitat are universal concerns in conservation. These range from forest to grassland and wetland plants, their loss of direct significance for the many moths that are specialized herbivores. Vast areas of Australia have long been converted to pasture and cropping areas, leaving many complex and sensitive floral communities occupying small and functionally isolated remnants of their former extent, and monoculture crops and grasslands dominated by introduced species supporting few of their former moth inhabitants. Land transformations for agriculture and urban and infrastructure development are the major causes of vegetation (habitat) loss. Both continue apace. For example, lowland native grasslands in the south-east have been designated as Australia's most threatened ecosystems (Kirkpatrick *et al.*, 1995), with more than 99% of their area no longer existing: that loss is a prime cause of concern for the golden sun moth (*Synemon plana*, Chapter 35) and other less well-known grassland specialist moths. In fact, some other species of *Synemon* are far more localized and vulnerable to losses of their narrow grassland ranges. The sun moth family, the Castniidae, are a significant flagship group for grassland conservation, many other inhabitants of which are less conspicuous. Highly modified grasslands in Tasmania support far fewer native Geometridae: Larentiinae than unmodified grasslands (McQuillan, 1999).

Broadly, habitat loss or despoilation, in conjunction with loss of connectivity among remaining patches, is the most universal and obvious conservation concern. It is likely that many localized forest moth species, in particular, may have suffered from either general environmental destruction or more selective exploitation of their natural habitats. The scarcity of the primitive *Agathiphaga* moths, with larvae feeding in *Agathis* seeds, is attributed in part to loss of the host trees to timber demands (Chapter 3). The southern pink underwing moth, *Phyllodes imperialis smithersi* (Chapter 36), occurs in undisturbed subtropical rainforest, with loss of this habitat recognized as a key threat to the localized moth (Sands, 1999). Native forests continue to be exploited, but with endemic marsupials being their primary 'flagships' for conservation advocacy.

Moth vulnerability is thus linked closely with anthropogenic changes, with the highly transformed landscapes of agroecosystems, plantation forestry and urban settlements commonly retaining few remnants of their former vegetational diversity.

9.2.2 Alien and invasive species

The processes of native vegetation loss have been accompanied by introductions of alien and invasive species, many of which have become pests of agriculture and forest production (see Chapters 23–32) as well as penetrating remaining natural vegetation. These invasions have further conservation consequences, including changing interactions with native species through competition or predation; increased pesticide use to suppress undesirable aliens, whether plants or animals; increased fertilizer applications for crop growth, as a pollution impact by changing nitrogen supply; and provision of novel acceptable resources for native species (New, 2016). There may well be other impacts marking broad ecological changes to which ecologically specialized species may be susceptible.

9.2.3 Climate change

All tangible threats to moths are overlain by the influences of climate change on the habitability of places, leading to movements of species (most typically to higher elevations or latitudes) to track changing conditions, and increased vulnerability from poor dispersal prowess and changed ecological interactions as community composition changes. Australia has very limited high elevation areas, essentially meaning that moths already restricted to the upper subalpine regions of the south-east have nowhere to go as warming occurs, so that some local alpine Geometridae and Arctiinae (and others) seem destined for extinction as warming continues and renders their current habitats untenable.

Larvae of the cushion plant moth (*Nemotyla oribates*, Oecophoridae) feed beneath the plants of low-growing Epacridaceae and Stylidiaceae, and occur only on the highest mountains of Tasmania where populations are very restricted (Nielsen *et al.*, 1992). Likewise, the isolated areas of the subtropical and tropical rainforests of Queensland, with their rich moth faunas, are regarded as vulnerable to warming. Lacking opportunities for elevational range shifts, distributional changes induced by climate must, rather, be latitudinal (Kitching *et al.*, 2013). This may be most practicable for moths living below the highest elevations or those with most flight mobility. Such resilience will depend on tracking critical resources (such as food plants) from which they might become decoupled if they respond at different rates to changed local conditions, or if landscape connectivity is low.

Other possible climate impacts include increased fires (of direct concern to litter-dependent Oecophoridae and other taxa in forests) and coastal inundations, to which narrow-range endemic moths are susceptible. Concerns of vulnerability to fire were expressed, for example, for two new species of *Abantiades* (Hepialidae) described from Kangaroo Island (Moore *et al.*, 2021).

9.3 Conservation of Single Species

Concerns for single species are the most familiar aspect of conservation, and arise from perceived declines, losses of restricted habitats and specific resources, and actual or anticipated threats such as spread of invasive species or urban or industrial development. Because 'species' (in popular parlance, anyway) are tangible entities, their conservation is also communicable and recognized as a worthwhile activity, paralleling better-known efforts to 'save the helmeted honeyeater' or 'save Leadbeater's possum' but lacking their public and political prominence. Many endemic Australian moths have equivalent scientific stature to such vertebrates. The methodologies, well established for mammal or bird species conservation, have major shortcomings and inadequacies for moths and other insects, but are attractive to non-entomologist managers. The lack of invertebrate expertise in most conservation agencies is difficult to redress, despite repeated calls to do so.

Very few moths have been listed formally under any Australian legislation, and this shortcoming could create a false impression that others are secure. Based on the IUCN Red Data List categories of threat, only three species are listed under the EPBC Act. *Phyllodes imperialis smithersi* (lower montane rainforest) and *Trisyntopa scatophaga* (associated with parrot nests in termite mounds) are listed as 'Endangered', *Synemon plana* (temperate grasslands) as 'Vulnerable' as at April 2024. This represents a very incomplete picture of the conservation status of Australia's moths – but still reflects considerable ecological variety. Both *Phyllodes* and *Synemon* were deemed eligible for listing on only one of the five criteria ('geographic distribution'), as no quantitative data on population size or trends were available. The formal criteria for evaluation of status and risk of extinction are difficult to apply. Adding taxa covered by regional legislations increases the total to 14 species (September 2022). Listing of the bogong moth (*Agrotis infusa*) (Chapter 37) as 'Endangered' by the IUCN in late 2021 attracted wide media attention, linked with massive declines in numbers of aestivating moths, and as one of the few moth species known to many people (at least by repute) as a culturally and ecologically significant species, because of its migrations in late spring, and as a predominant food of the endangered mountain pygmy possum, *Burramys parvus* (Green *et al.*, 2021). Declines of formerly abundant species are of particular concern because, as for *A. infusa*, they are likely to be ecologically significant and declines may well indicate or herald wider effects.

Many additional candidates could be suggested for listing, and lead to a more realistic appraisal of need; several species (of Anthelidae, Geometridae, Hepialidae) were noted by Hill and Michaelis (1988) but have not been pursued. But selecting the most deserving additions to enlarge this number is complex and can easily become subjective, and accumulating sound evidence to support a nomination is difficult, as implied above for even relatively well-studied species. Increasing the number of listed species may simply lead to even greater deficits in support from the limited funding, political interest, scientific capability and wider participation.

The information suggested for inclusion of an insect nomination for 'listing' in Australia (Taylor *et al.*, 2018) is rarely available other than in fragmentary form, but it is not unusual to find expressions of concern in taxonomic papers on Australian moths over possible conservation need due to few records or very narrow distributions known. Discussing new species of *Oxycanus* (Hepialidae), for example, Beaver *et al.* (2020) noted 'Concerns are raised about the conservation status of all three new species due to few or localised distribution records'. In short, many such moths are 'data deficient' with some subjectively regarded as 'near threatened'. The small whistling moth (*Hecatesia exultans*, Noctuidae) in Victoria exemplifies the latter case (New *et al.*, 2012), and almost all Australian moths have not been formally evaluated further. For many truly or apparently scarce moths, the difference between 'absence' and 'failure to collect or record' is blurred, and long-term targeted surveys are needed to clarify status. These are rarely possible, especially in the more remote parts of Australia. Collectively, at least several hundred Australian moth species could be nominated as needing conservation attention because they seem to be scarce and geographically and ecologically restricted – but on grounds inadequate to be formally persuasive and in many cases reflecting the precautionary principle that the species may indeed be threatened and should not be neglected.

Sound scientific evidence for conservation need is often lacking and the global predominance of butterfly species over most moths in conservation largely reflects the greater legacy of reliable biological and distributional information from interests of hobbyists. Increasing that information is arduous, long-term and expensive and, as for butterflies, cases from other parts of the world provide invaluable guides to what is needed and feasible in Australia. Every listed moth species becomes a 'flagship' or 'ambassador' (New, 2011) for wider diversity in its habitat, so that credibility in its worth is critical. Listing is simply a first step in promoting conservation need, as a prelude to further assessment and action, but is important also in that it may accord the species priority for limited funding and other resources – and so deprive other deserving (but non-listed) species of that attention. Listing may indeed be a condition for such support. The enigma moth (*Aenigmatinea glatzella*, perhaps the most phylogenetically remarkable endemic moth discovery in Australia in recent decades and the only member of the family Aenigmatineidae) was assessed as Critically Endangered under IUCN guidelines (Glatz and Young, 2021) but has no protection under Australian law, despite its vulnerability as a narrow-range endemic on Kangaroo Island. Likewise, the bogong moth (Chapter 37) lacks formal protection.

Taylor *et al.* (2018) suggested the dual categories of 'threatened species' and 'iconic species' for listing, as rather different groups that can both appeal to public interests. This approach helps to overcome uncertainty over threat status: spectacular species such as Ward's atlas moth (*Attacus wardi*) and others for which records and quantitative data are sporadic but which may capture public sympathy. *A. wardi* may indeed be 'Vulnerable' but also is a potential flagship or even umbrella species for the coastal tropical monsoon forest patches it inhabits in north-western Australia (Braby and Nielsen, 2011). The spectacular bent-wing swift moth (*Zelotypia stacyi*, the largest Australian hepialid) has historically been considered very rare, but also noted as a pest of plantation eucalypts through larval burrowing, leading to suggestions that its scarcity might have been exaggerated (DeBaar and Hockey, 2009). For each such species, some form of 'recovery plan' or 'action plan' is usually needed, pinpointing concerns and activities for conservation or promotion as accurately as possible, and increasing fundamental knowledge of biology and vulnerability, so setting out a clear perspective of conservation status and needs. Exceptions occur: for the endangered ant-bed parrot moth (*T. scatophaga*, see above), no separate plan was considered necessary because the moth's wellbeing depends entirely on that of the threatened host golden-shouldered parrot (*Psephotus chrysopterygius*) for which conservation measures envelop the moth's needs (Preece *et al.*, 2009).

A conservation action plan has several elements but must also be understandable by non-entomologists/managers whose support and activities may help to bring it into practice. The plan should include:

- a full summary of the moth's biology and distribution (identifying gaps and steps needed to improve this);
- the evidence for conservation need, specifying threats and local vulnerability (as on individual sites) where relevant;

- management steps (including site protection, site remediation, threat amelioration and prevention, any need for wider measures such as translocations, improved connectivity, and others); and
- allocation of responsibility for all actions needed (together with costing, provision for monitoring and review, and adaptive management as needed). At this early stage all stakeholders can participate usefully in harmonizing different sectoral priorities and defining goals.

Any such programme must incorporate such varying interests and priorities, together melding science and 'authority' with the need to gain long-term financial and community support for the enterprise. Collectively, these face several hindrances (or impediments) to acceptance:

- the need for clear and unambiguous recognition and identification of the target species by non-specialists;
- the wide lack of information on early stages and larval needs, and the basic life cycle of the species;
- lack of information on the species' environmental needs and conditions, and vulnerabilities, that enable honing general conservation measures for more focused effects;
- lack of information on the species' distribution, from local to regional scales; and
- the scarcity of funding, expertise and interest that can address any of these factors constructively and lead toward a sound conservation status assessment and management plan (New, 2023).

9.4 The Focus on Wider Diversity

The main alternative to the species-level approach, that of focusing on species richness, devolves largely on conservation of key habitats known to harbour high moth diversity and/or key focal species or unusual ecological milieux, 'protecting' those areas or ecosystems from anthropogenic disturbances and, where necessary and possible, restoring them. Many moths do indeed occur in existing protected areas such as National Parks, most of which are managed for a range of other taxa and for human uses perhaps inimicable with insect wellbeing, and in which insects are largely ignored as silent passengers in the overall conservation endeavour.

Growth of winter sports activities and summer tourism in alpine areas, for example, may damage sensitive areas of forb-dominated grassland inhabited by restricted Geometridae and others.

Surveys in Australia's National Parks and other 'reserves' continue to show rich moth faunas – but most records come from 'one-off' or 'snapshot' samples (mostly by light traps) rather than planned longer-term and year-round investigations, so fall far short of any definitive inventory. Such surveys are invaluable in providing incidence records, sometimes of moths far from their previously known range; one such is the capture of a specimen of the geometrid *Oenochroma barcodificata* in the Great Otway National Park, Victoria (Hewish, 2019). Previously believed to occur only in the alpine region, this incidence demonstrated the value of even short surveys in providing serendipitous records of notable species, and of defining their distributions fully. The highly seasonal flight pattern of many moths ensures that 'spot samples' contribute incomplete pictures of total richness. The task is also rendered difficult by the large numbers of undescribed or undifferentiated taxa likely to be encountered, so that uncertainties over species richness and delimitations will continue to thwart both characterizing local diversity and recognizing the most vulnerable taxa.

In conservation, and echoing many other aspects of studying moth biology in Australia, 'The real challenge facing any moth worker remains identifying specimens' (Kitching and Edwards, 2020). Usable quantitative information is largely from focus on the better-known groups of 'macromoths' that are attracted to light. Apart from their evident richness, the ecology and vulnerability of most 'micromoths' cannot be appraised realistically in assemblage samples. We simply know that there are a lot of them, and see their habitats disappearing.

Because many moths are very local, some as narrow-range endemics, conservation of individual sites may be needed and there is wide agreement that ongoing site management – for example, to control vegetation succession – may be necessary to sustain both heterogeneity and key resources. Site 'security' or tenure may become an issue – sound conservation depends on the focal site not being lost, for example as in the continuing tension between developers and conservation interests over remnant grasslands for the golden sun moth. Few 'reserves' have been designated, however, especially for moths, with a formal reserve in Victoria for two species of sun moths perhaps the first (Douglas, 2004; see Chapter 35). These reserves need not be large – the small remnants of natural vegetation found in many urban and agricultural areas are acknowledged widely as significant refuges for insect life and, if buffered adequately, even a few hectares can sustain insect populations. Promotion of connectivity between remnants, for example by vegetated (natural or restored) roadsides, is widely advocated for insect conservation (New *et al.*, 2021). Widening of field margins also provides refuge habitats and 'corridors' for insect movements, as a feature of numerous agrienvironment schemes.

Many designated protected ecological communities in Australia are founded on vegetation features, and so are likely to support insect herbivores that depend on either the individualities or representativeness of plants present, in either case constituting unique assemblages. Whether for single moth species or wider assemblages, plants are a key conservation consideration.

9.5 Bringing Ideas into Practice: Some Principles Involved

As a basic perspective shared with most other insects, moth conservation in Australia must usually proceed without detailed knowledge of either the biology of focal species or the richness and peculiarities of local assemblages, but also with awareness that many moths are indeed declining as their habitats succumb to human cupidity, and that we cannot await full scientific information before acting to ameliorate this. Progress must therefore follow the approach advocated by a number of recent commentators on insect conservation, that action is imperative and we know enough to make intelligent decisions now (Forister *et al.*, 2019). Allied to this is the need to redress both knowledge lack and public awareness as effectively as possible, and to communicate this need effectively.

'Citizen science' already has significant roles in this endeavour. Volunteer contributions to surveys, photographs submitted to iNaturalist, and protection and rehabilitation of key sites and native vegetation signalled for moth interest, and others, simply indicate the variety of activities for which benefits have already accrued. This emphasis reflects that most knowledge of moth distributions and biology resides among the wider community of naturalists, rather than with professional scientists or managers. A related need is to expand the information base from, largely, adults to include early stages, as has become progressively feasible for the better-documented faunas of Europe and North America (Di Cecco and Hulbert, 2022), in which many caterpillars are recognizable to species level.

The needs and principles for moth conservation apply equally to many other invertebrate groups. The two foremost requirements are, first, a secure 'place to live' with all critical resources assured and, second, freedom from threats – both deceptively simple in theory but difficult in practice! For example, some rare species for which the host plant range is not known fully but which are known from one or a few sites may simply be 'hanging on' in highly marginal situations rather than thriving. This lack of knowledge can lead to energetic (and perhaps costly) conservation programmes striving to emulate what are in fact suboptimal conditions. Further dilemmas arise when an invasive weed, targeted for eradication, is also an important food plant for a threatened moth on sites where the native food has diminished; Chilean needle grass (*Nassella neesiana*), for example, can be important for golden sun moth (*Synemon plana*) conservation.

Practical conservation incorporates social and economic decisions, harmonizing differing priorities of stakeholders and, for moths, reflects a number of widespread scientific and societal principles as a collective foundation for progress. All these factors apply to far wider taxonomic groups but are noted here to indicate the great variety of considerations that arise (Table 9.1); the list is indicative rather than comprehensive.

As for other insects, moth diversity in Australia is both a value and a hindrance in conservation! Richness, with accompanying problems of quantifying this, ensures that many species are localized, ecological specialists dependent on single or few plant species and susceptible to availability of those foods, found naturally in low numbers, with population structure and fluctuations unknown, and are poorly known or

Table 9.1. A series of general themes and principles listed to suggest the many considerations in developing practical moth conservation in Australia (slightly modified from New, 2023).

1. Many key moth habitats and vegetation types in Australia have undergone immense changes, and the impacts of those changes on the moth fauna cannot be wholly assessed; effects of direct losses of habitats are augmented by a variety of other anthropogenic stressors as 'threats'.
2. Many of those key habitats now exist only as small remnants, serving as refuges for taxa previously far more widespread, and their protection from further loss is critical for the survival of specialized taxa restricted to them; increasing the extent and representativeness of protected areas is a key component of moth diversity conservation.
3. Simply that a moth species or defined assemblage occurs in a defined 'protected area' does not guarantee its future security, for which proactive management may be needed.
4. Any moth habitat is embedded in a wider and more variable landscape matrix whose features may influence the habitat of interest and necessitate management – for example, to reduce edge effects and invasive species impacts.
5. The constitution and diversity of moth assemblages are continually dynamic, and management may be needed to sustain or restore features such as early successional stages of vegetation to conserve that variety.
6. Not all moth species in a survey will be identifiable easily; cross-site and cross-survey comparisons need very careful interpretation to avoid confusion that may invalidate inferences of impacts of change from failing to distinguish similar species with differing biology.
7. Different functional guilds (such as feeding groups) of moths respond differently to habitat changes, and this may not be evident from simple measures of species richness.
8. Use of changes in abundance or incidence of moth species or assemblages as indicators or predictors of environmental impacts is attractive, but may depend heavily on local conditions and the taxa selected.
9. The management of heavily transformed environments (such as agricultural, forestry and urban ecosystems) for moth conservation involves combination of (a) promoting and enhancing natural features and connectivity and (b) reducing and preventing further threats.
10. Heterogeneity and vegetational diversity benefit moth diversity, and much conservation management focuses on this, and the condition/provision of particular plant species.
11. Mosaic management that promotes variety across the landscape is generally preferable to large-scale uniform management.
12. Focus on single scarce or threatened species as 'conservation flagships' is an important component of moth conservation, but wider measures to benefit local moth faunas are also needed.
13. Public perceptions are not always supportive of moth conservation, and activities such as (a) education to improve appreciation of moth importance and diversity and (b) promoting citizen science through community awareness are integral to much conservation activity.
14. Increasing formal awareness of, and attention to, moth conservation needs (such as by increased representation on 'protected species lists') and increasing their profiles in conservation decisions and recommendations can have benefits such as (a) advocacy for conservation action and (b) complementing conservation efforts primarily targeting other taxa, while also demonstrating moth richness, ecological variety, functional importance and vulnerability.

documented. Selecting the most vulnerable species for conservation priority from a large array of deserving candidates becomes difficult.

Places and ecosystems that harbour many species, as an individual dynamic assemblage, are each unique, with significant moth species superimposed on a broader and more generalized assemblage founded on the predominant vegetation present. The extent of conservation need can only be inferred – albeit sometimes strongly as key vegetation communities are lost or increasingly fragmented. Perceived vulnerability of many individual species suggests that they deserve 'listing' for increased attention and protection – but while long lists of threatened moth species could indeed help to communicate the extent of concern, any such *ad hoc* increases are unlikely ever to receive adequate support and be rendered secure. Some clearer focus is necessary, building constructively on Lepidoptera being the most studied order of insects in conservation. Studies on a variety of moth species undertaken in Europe and North America, in particular, have demonstrated pathways to successful conservation management, some based on convincing evidence of moth declines and losses – and for some, extirpations (Fox, 2013). Each individual case adds to the corpus of knowledge and experience to increase confidence in future exercises undertaken across the range of moth taxa and environments.

Rather than simply increasing lengths of 'lists' of threatened species, the combination of threatened and iconic species suggested by Taylor *et al.* (2018) links with another aspect of their advocacy: that a small number of threatened or iconic insects selected as flagships to represent all main bioregions in Australia (that is, the 89 recognized IBRA [Interim Biogeographic Regionalisation for Australia] regions) could be promoted and perhaps listed through the scientific community. Moths are an ideal group to yield candidates for this, and any such co-ordinated programme of priorities could have far wider biodiversity benefits as well as endorsing the importance of moths in a range of contexts, establishing them as both targets and tools in conservation across Australia, and demonstrating the great variety of ecosystems they inhabit to managers and others unaware of the subtleties of insect life.

Such a transition from moths being 'passengers' in wider conservation to becoming 'drivers' as direct foci in a series of ecologically different contexts (and complementing butterflies through their far greater diversity) could establish them firmly on Australian conservation agendas, as a major need for their future. Nevertheless, and despite the welcome increasing attention moths now receive, the future of many taxa depends on wider measures to conserve Australia's biodiversity, including remnant ancient flora: *Nothofagus*, *Araucaria*, *Agathis*, *Callitris* and others each harbour moths found nowhere else. There are twin needs for moth conservation progress both representing critical, ongoing challenges. On the one hand, we need increased fundamental knowledge of Australian moth systematics, biology and distributions as a basis for assessing conservation need and, on the other, effective communication and education to increase public profiles of moths and overcome societal reluctance to support their conservation. These challenges must be met while ensuring that major ecosystems, and many individualistic sites key for survival of known notable species and wider moth diversity, are sufficiently protected to assure a future for Australia's unique insect life. Moths are a potent, irreplaceable component of the nation's insect fauna and are accessible to both constructive study and practical conservation.

References

Beaver, E.P., Moore, M.D., Velasco-Castrillon, A., Stevens, M.I. (2020) Three new ghost moths of the genus *Oxycanus*, Walker 1856 from Australia (Lepidoptera: Hepialidae). *Zootaxa* 4732, 351–374.

Braby, M.F. and Nielsen, J. (2011) Review of the conservation status of the Atlas moth, *Attacus wardi* _Rothschild, 1910 (Lepidoptera: Saturniidae) from Australia. *Journal of Insect Conservation* 15, 603–608.

Collins, N.M. and Morris, M.G. (1985) *Threatened Swallowtail Butterflies of the World. The IUCN Red Data Book*. IUCN, Gland, Switzerland.

De Baar, M. and Hockey, M. (2009) *Zelotypia stacyi* Scott (Lepidoptera: Hepialidae) – a conservation perspective. *Australian Entomologist* 36, 85–88.

Di Cecco, G.J. and Hulbert, A.H. (2022) Caterpillar patterns in space and time: insights from and contrasts between two citizen science datasets. In: Marquis, R.J. and Koptur, S. (eds) *Caterpillars in the Middle. Tritrophic Interactions in a Changing World*. Springer Nature, Cham, Switzerland, pp. 541–556.

Douglas, F. (2004) A dedicated reserve for conservation of two species of *Synemon* (Lepidoptera: Castniidae) in Australia. *Journal of Insect Conservation* 8, 221–228.

Duffey, E. (1968) Ecological studies on the large copper butterfly, *Lycaena dispar* Haw. *batavus* Obth. at Woodwalton Fen National Nature Reserve, Cambridgeshire, England. *Journal of Applied Ecology* 5, 69–96.

Forister, M.L., Pelton, E.M. and Black, S.H. (2019) Declines in insect abundance and diversity: we know enough to act now. *Conservation Science and Practice* 2019, e80.

Fox, R. (2013) The decline of moths in Great Britain: A review of possible causes. *Insect Conservation and Diversity* 6, 5–19.

Geyle, H.M., Braby, M.F., Andren, M., Beaver, E.P., Bell, P. *et al.* (2021) Butterflies on the brink: identifying the Australian butterflies (Lepidoptera) most at risk of extinction. *Austral Entomology* 60, 98–110.

Glatz, R.V. and Young, D.A. (2021) *Aenigmatinea glatzella. The IUCN Red Data Book*. IUCN, Gland, Switzerland.

Green, K., Caley, P., Baker, M., Dreyer, D., Wallace, J. and Warrant, E. (2021) Australian Bogong moths, *Agrotis infusa* (Lepidoptera, Noctuidae), 1951–2020: decline and crash. *Austral Entomology* 60, 66–81.

Hewish, M. (2019) Moths. In: Walker-Smith, G. and Phillips, K. (eds) *Otway Bioscan. Great Otway National Park, July 2018–June 2019*. Museums Victoria, Melbourne, Victoria, pp. 55–68.

Hill, L. and Michaelis, F.B. (1988) *Conservation of Insects and Related Wildlife*. Occasional Paper no. 13. Australian National Parks and Wildlife Service, Canberra, ACT.

Kirkpatrick, J., McDougall, K. and Hyde, M. (1995) *Australia's Most Threatened Ecosystem: the Southeastern Lowland Native Grasslands*. Surrey Beatty and Sons, Chipping Norton, NSW.

Kitching, R.L. and Edwards, E.D. (2020) Books about Australian moths 1805–2020: an annotated bibliography. *Australian Entomologist* 47, 119–154.

Kitching, R.L., Ashton, L.A., Burwell, C.J., Boulter, S.L., Greenslade, P. *et al.* (2013) Sensitivity and threat in high elevation rainforests: outcomes and consequences of the IBISCA-Queensland project. In: Lowman, M., Devy, S. and Soubadra, T. (eds) *Treetops at Risk*. Springer, Berlin, Germany, pp. 131–139.

McQuillan, P.B. (1999) The effect of changes in Tasmanian grasslands on the geometrid moth tribe Xanthorhoini (Geometridae: Larentiinae). In: Ponder, W. and Lunney, D. (eds) *The Other 99%. The Conservation and Biodiversity of Invertebrates*. Royal Zoological Society of New South Wales, Mosman, NSW, pp. 21–128.

McQuillan, P.B. (2003) The giant Tasmanian 'pandani' moth *Proditrix nielseni*, sp. nov (Lepidoptera: Yponomeutoidea: Plutellidae *s.l.*). *Invertebrate Systematics* 17, 59–66.

Moore, M.D., Beaver, E.P., Velasco-Castrillon, A. and Stevens, M.I. (2021) Two new endemic species of *Abantiades* Herrich-Schaffer (Lepidoptera: Hepialidae) from Kangaroo Island, Australia. *Zootaxa* 4951(3), 571–597.

New, T.R. (1997) *Butterfly Conservation*, 2nd edn. Oxford University Press, Oxford, UK.

New, T.R. (2011) Launching and steering flagship Lepidoptera for conservation benefit. *Journal of Threatened Taxa* 3, 1805–1817.

New, T.R. (2016) *Alien Species and Insect Conservation*. Springer, Cham, Switzerland.

New, T.R. (2023) *The Other Lepidoptera: Moth Conservation in Australia*. Springer, Cham, Switzerland.

New, T.R., Van Praagh, B.D. and Yen, A.L. (2012) Invertebrate conservation status and the limits of reliable information: examples from Victoria, Australia. *Victorian Naturalist* 129, 68–76.

New, T.R., Sands, D.P.A. and Taylor, G.S. (2021) Roles of roadside vegetation in insect conservation in Australia. *Austral Entomology* 60, 128–137.

Nielsen, E.S., McQuillan, P.B. and Common, I.F.B. (1992) The Tasmanian cushion plant moth *Nemotyla oribates* gen. n., sp. n.: Systematics and biology (Lepidoptera: Oecophoridae: Xyloryctinae). *Journal of the Australian Entomological Society* 31, 47–56.

Preece, N., Shephard, S., Shephard, T. and Garnett, S. (2009) *Re-assessment of the Status of the Golden-shouldered Parrot* Psephotus chrysopterygius. Queensland Department of Environment and Resource Management, Brisbane, Queensland.

Pyle, R.M. (1976) Conservation of Lepidoptera in the United States. *Biological Conservation* 9, 55–75.

Sands, D.P.A. (1999) Conservation status of Lepidoptera: Assessment, threatening processes and recovery actions. In: Ponder, W. and Lunney, D. (eds) *The Other 99%. The Conservation and Biodiversity of Invertebrates*. Royal Zoological Society of New South Wales, Mosman, NSW, pp. 382–387.

Sands, D.R.P. and New, T.R. (2013) *Conservation of the Richmond Birdwing Butterfly in Australia*. Springer, Dordrecht, The Netherlands.

Taylor, G.S., Braby, M.F., Moir, M.L., Harvey, M.S., Sands, D.P.A. *et al.* (2018) Strategic national approach for improving the conservation management of insects and allied invertebrates in Australia. *Austral Entomology* 57, 124–149.

10 The Hepialidae: The Ghost Moths

Michael Moore
South Australian Museum, Adelaide, South Australia, Australia

Abstract

The ghost moths, the Hepialidae, are a globally distributed family of moths but reach their highest diversity in the southern hemisphere. Long-lived larval stages result in very short-lived, non-feeding adults. Larval resources include fungi and lichens as well as plant roots and stems, into which some larvae bore. Taxonomically, the Australian hepialids divide into the *Fraus* group of 26 smaller species. The remainder of the family form the Hepialidae *sensu stricto*. The Hepialinae comprise five genera totalling 91 species overall. By far the largest of these genera is *Abantiades* with 47 species. The Oxycaninae contains three genera comprising 50 species, dominated by *Oxycanus* with 47 species. This chapter looks at the importance of the Australian hepialid fauna in a global context and discusses some general geological events that might have driven its development historically.

10.1 Introduction

The Hepialidae is a worldwide family of moths at its most diverse in the southern hemisphere, particularly in the Australasian region (including New Guinea, New Caledonia and New Zealand) and South America.

Having similar wing venation on the fore- and hindwings, hepialids are considered primitive moths. Evidence suggests that they appeared on Earth before flowering plants. Their unusual 'external' fertilization of the egg, the lack of a strong wing-linking mechanism, random scattering of eggs in flight, and their long, subterranean existence as a larva all add to this notion. Simonsen (2018) looked at the fossil evidence available and decided that there is no reliable fossil material available for the Hepialidae so, at present, conclusions about times of origin have been based on morphology of extant taxa and their geographic distribution, and, more recently, from genetic work.

The family demonstrates a close relationship to rain and rainfall. They probably evolved in a wet environment – one that persisted in Australia until this continent's final break with Antarctica about 45 million years ago, as indicated by biogeographic studies showing extant species' distributions corresponding to Mesozoic geology (Grehan and Mielke, 2018). Since then, Australia's movement north has taken it closer to drier latitudes in the southern hemisphere, and opened up a cold Southern Ocean. This changed Australia from a wet continent to one with large-scale desertification as evidenced by the large number of arid or semi-arid hepialid

Corresponding author: michael.moore@samuseum.sa.gov.au

DOI: 10.1079/9781800627680.0010

species and their differing forms. This continental drying seems to have driven a more rapid genetic development of the dry-adapted hepialid forms and seemingly only slow or non-existent development in the original wet-adapted species found today in the Australian coastal margins.

As the taxonomy currently stands, Australia is home to the largest hepialid in the world and, probably, the smallest too, as well as the heaviest and the lightest. In terms of sheer numbers of species, however, Australia probably ranks second to South America.

10.2 Life Cycles

Adult hepialids have non-functioning mouthparts (either absent or reduced), meaning that the adults cannot feed and have to survive on the fat stored in their bodies during the larval stage. Consequently, as emerged adults they only have short lives. Indeed, available evidence would indicate that after emergence, the vast majority of adult moths do not survive into a second evening because they fall prey to all manner of predators. Their existence as a larva, however, is much longer, a year at minimum, with larger species requiring a few years to develop to an adequate size for pupation.

Their emergence as adults is linked to unsettled, rain-producing weather, where the pupae react to pressure variation or to the precipitation itself. In the southern half of Australia this is usually in autumn, while some (but not all) northerly species instead react to spring or early summer rainfall.

We believe that females are mated very soon after emergence, maybe even before wing drying, and then disperse and simply release eggs when flying or resting. The eggs fall into the leaf litter to develop further.

Those eggs that survive develop to maturity within a couple of weeks and their caterpillars enter the leaf litter to feed on lichen and/or fungi that have developed since the triggering rainfall. Norman Tindale once estimated that a female *Abantiades argentata* (Tindale, 1932) had at least 40,000 eggs within her abdomen, these huge numbers reflecting a potentially very large mortality rate amongst the young larvae.

After a period of above-ground growth, some caterpillars dig into the soil, either to avoid predation and/or to feed on the roots of particular host plants, whereas others bore into shrub or tree stems. Some subterranean species have been found living several metres under the surface. Those species that just hide underground emerge from the ground at night to feed upon the leaves or stems of their host plants, while stem borers remain in their cells feeding on the more nutritious stem tissues. More permanent residents of the subterranean world feed on the external or internal portions (dependent on the species) of their host plant's roots.

Little is known about the host plant species for many species of Australian hepialids, nor how these are utilized. This lack of knowledge mostly originates from the difficulties in studying their long, subterranean larval lifespans, and a lack of direct observations during the larval feeding stage. Nearing maturity, the caterpillars will construct a vertical tunnel to the surface and then block the opening. Mature caterpillars pupate in these predug tunnels and await suitable conditions for emergence. It is recognized that the pupae are surprisingly agile within the tunnels, and it is thought that they actively monitor the environment to determine the most suitable time for emergence. They may even be able to stall their emergence for some period of time.

Most species exhibit mass emergence behaviour, resulting in thousands of individuals flying and mating all at once, after which the ground can be littered with pupal cases sticking out from their tunnel or scattered over the ground after the moth has dragged it with them a short distance from the pupal tunnel. Under these conditions, as adults, competition for mates must be intense and it is believed that mating is coincidental with emergence in most genera. After mating, females fly off into the woodland, releasing eggs whilst the males continue to seek other females.

In most genera, it is the female that calls chemically to attract mates. In this situation females release pheromones that males detect and follow the scent trail back to the female. In contrast and definitely in one, but possibly two, Australian genera, *Aenetus* and *Oncopora*, it is the male calling while females fly and seek out the males. In these genera, males have substantial tibial hair tufts at the bases of the hind leg which are expanded and contain fluid that might function as pheromone-producing glands. Some collectors have commented that they detect odours when collecting individuals of these species. The non-calling gender are the individuals most collected at light traps.

Hepialids are collected at night mostly during rain or, with some species, pre-rain events. It is vital that the moon is not visible; indeed, a night's collecting (with a lamp to attract the moths) can come to a rapid halt once the moon appears. The moths themselves are very much attracted to bright lights, including street lights, and in suburban or rural areas, moths will fly into windows in an effort to get to a light source. Moths in the field are attracted to set-up lights, and when settled will stay so, quite peacefully on the sheet or net. Indeed, trying to release surplus specimens at the end of the evening can be difficult as they return to the light after being placed back into darkened woodland. The moths sit with their relatively large forewings tent-like against the body, with their long forelegs extended, making them easily distinguished from other moth families, even on a crowded light sheet.

10.3 The Australian Fauna

The taxonomy of the family in Australia has been and continues to be intensively studied and what is discussed below will almost certainly change as newly acquired knowledge becomes incorporated into our understanding of relationships within the family.

Difficulty with the family taxonomy in Australia arises immediately with the family being divided into two major subgroups – the hepialids (*sensu stricto* or 'true' hepialids) and the 'Fraus group' (Nielsen and Kristensen, 1989). The *Fraus* group of species is exclusive to Australia and is one of a few so-called primitive hepialid groups within the family, the status of which is continually discussed by researchers. The 'true' hepialids can be divided into two groups, the hepialine group (five genera) and the oxycanine group (three genera), a split based on wing venation and DNA sequencing. In the hepialine group, wing vein Rs3 arises from Rs4 whereas in the oxycanine group, Rs3 arises from the common R/Sc vein (Kristensen, 1998). It should be noted that the terms 'hepialine' and 'oxycanine' are at this stage only grouping terminology, rather than holding any defined scientific status.

10.3.1 *Fraus* group

Originally named by Francis Walker (1856), *Fraus* was the focal group in a significant generic revision by Nielsen and Kristensen (1989) (Fig. 10.1). These researchers reviewed all that was then known, carried out morphological dissections, described many new species, and grouped species according to morphological features. Unfortunately, molecular methods were not available to them at that date and could not be used to support the morphological taxonomy. Since their work, only one new species has been added (Beaver and Moore, 2020) – *Fraus extremepoda* – bringing the species list to 26. Much recent genetic work has been undertaken demonstrating that concerns about taxonomic grouping, commented on by the original authors, need further investigation.

Fraus spp. are amongst the smallest moths in the Hepialidae, and definitely the smallest in Australasia, with some species having a forewing length of only 14 mm. They are sombre coloured, with brown to grey as the dominant colour, with either small dark spots on the wings or, if brown winged, a white or black and white line. In some species the males have the lines and the females the spots, others vice versa. It is thought that the caterpillars of these species emerge at night to feed on native grasses and some species, when present in large numbers, can retard lawn or pasture growth.

Fig. 10.1. *Fraus simulans*, male. Photo: Paul Kay, with permission.

Nielsen and Kristensen characterized some *Fraus* species as more primitive: those having vestiges of a proboscis and an epiphysis (a gripping structure on the foreleg tibia).

It is possible that a well-provisioned collector or researcher might be able to raise species from this genus in a laboratory or home situation. One would need a source of healthy host plants in a suitable substrate, and a source of caterpillars or eggs to achieve this.

Fig. 10.3. *Abantiades argentangulum*, female uppermost, male below. Photo: Paul Kay, with permission.

10.3.2 The Hepialids *sensu stricto*

As mentioned above, the 'true' hepialids are split into two groups based on wing venation – the hepialine group and the oxycanine group.

10.3.2.1 Hepialine group

The hepialine group consists of five genera: *Abantiades*, *Aenetus*, *Archaeoaenetus*, *Zelotypia* and *Oncopora*.

ABANTIADES. This is the largest genus in the hepialine group and was originally described by Herrich-Schäffer (1855) (Figs 10.2–10.6). The genus now consists of 47 described species with six more in pre-publication and at least seven more known to exist. The moths in this uniquely Australian genus range from medium to very large in size, are robust and large bodied, and the males have long, elaborate, pectinate antennae. An *Abantiades atripalpis* female collected in 2021 had a wingspan of almost 200 mm and, with a load of 40,000+ eggs, could be the heaviest

Fig. 10.4. *Abantiades auriregulus*, male. Photo: Paul Kay, with permission.

Fig. 10.2. *Abantiades albofasciatus*, male. Photo: Paul Kay, with permission.

Fig. 10.5. *Abantiades hydrographus*, male. Photo: Paul Kay, with permission.

Fig. 10.6. *Abantiades paradoxa*, female. Photo: Paul Kay, with permission.

moth on Earth. *Abantiades* can fly strongly and, on cold nights, they can warm themselves up by rapid wing fluttering. They seem to use all their efforts not in flying long distances but remaining local, searching for mates or releasing eggs. Their lives are, however, short.

Simonsen (2018)'s revision of the genus established a much stronger scientific basis for their taxonomy. He realized that in this genus, 'all' species had monopectinate antennae (that is, with a single stem on each segment), which then forked or not, producing unforked, bi-forked or tri-forked species. This realization led him to collapse the genera *Trictena* and *Bordaia* into the first named *Abantiades*, which also had the advantage of keeping the taxonomy simple and monophyletic. Understandably, the phylogeny of *Abantiades* does seem to demonstrate a deeper evolutionary internal structure, with the species which have tri-forked rami being of particular interest. This tri-forked group is monophyletic, seems to be the best adapted to drier environments and is widespread in that environment. It is also the group which we believe to be changing, genetically, the most rapidly. Most of the species in this tri-forked clade are brown winged with two large white marks on the forewing. Unfortunately, with the high degree of variation observed within the species, and the similar patterning among species, they are particularly difficult to identify. Place of capture will aid in identification, as will mitochondrial COI gene sequencing, but genitalia dissection will be needed to define some clades. As this tri-forked group inhabits the driest habitats, it will be interesting to follow this group from an evolutionary perspective as Australia continues to dry out.

The unforked *Abantiades* species tend to inhabit the wetter areas of the continent and are boldly patterned, some beautifully so, and can be identified, with practice, using their wing patterning and antennal structure, though mitochondrial COI sequencing can also be used. Their preference for wet conditions means that most species are near the coast and/or in elevated locations.

There are several bi-forked species and this feature is known to have arisen separately in at least two different branches of the *Abantiades* evolutionary tree. All of the known bi-forked species live in Western Australia, with only one of those species having a range extending into South Australia and Victoria.

The *Abantiades* genus inhabits *Eucalyptus* woodland and it is thought that most have a *Eucalyptus* or *Corymbia* species as their host plant. It is known, however, that at least three species utilize *Allocasuarina* spp. Western Australia has the largest number of eucalypt species and consequently may be home to the largest number of *Abantiades* species. The genus was originally thought to occur only as far north as Rockhampton on the East Coast, but recent collecting indicates that it ranges as far north as the Atherton Tablelands in the East and Millstream in the West.

Due to the unpredictability of rainfall, collection of dryland species can be extremely difficult. In some places rain might fall only once or not at all during the hatching period, so being present to collect these 'gems' can be a low chance outcome, resulting in some species being represented in collections by only few specimens. The larval lifestyle and their requirement for large host trees make laboratory rearing extremely difficult. Unfortunately, too, colours fade over time, frustrating collectors of this genus, particularly for the rather brilliant and attractive purple, pink or magenta, fat-soluble shading present in many species.

AENETUS. The genus *Aenetus* was erected by Herrich-Schäffer (1855). Moths in the genus are found predominantly in Australia (22 species) and Papua New Guinea, (10 species), with a further three species on Indonesian islands, one species in New Caledonia and one in New Zealand (Figs 10.7, 10.8). Species in this genus are found over the wetter parts of Australia where they vary in size from medium to very large. In contrast to the large elaborate antennae of *Abantiades*, *Aenetus* spp. have relatively short (few segmented) and unadorned (with no pectinations) antennae.

All species could be described as handsome and their beautiful wing patterning has attracted collectors from the Victorian era to modern times. The colours on these moths are chemically stable and older specimens still enthral with their beauty. The females of most species are pink and green, while most of the males are mixtures of green, white and aqua.

The larvae of *Aenetus* are stem and branch borers, with the larvae utilizing the bore hole they produce in the wood for protection as well as a food resource. They come to the bore entrance at night to consume the nutritionally richer, self-renewing cambium layers in the plant stem. They cover the hole's entrance with a mixture of wood detritus, sawdust, spider-silk and their own silk, allowing an observant collector to locate them on the host plants. These bores are termed vestibules and collecting from them can be difficult; raising larvae in captivity is even more so as one can be facing a lengthy process requiring ongoing dedication. Those that do this, though, are usually richly rewarded with perfect specimens. *Aenetus djernaesae* has been raised from egg to adult in captivity (Kay *et al.*, 2020).

This genus utilizes a wide variety of host plants. Some species have many possible host plant options including introduced species, leading to questions as to why this is so. Intriguingly, though, collectors in Western Australia think that four of the five south-western species might utilize a single species of *Taxandria* (Myrtaceae). The fifth species, *Aenetus djernaesae*, feeds on *Myoporum insulariae* (Scrophulariaceae), a common seaside shrub, and *Dodonaea hackettiana* (Sapindaceae).

It is possible that members of this genus might emerge in response to changes in air pressure ahead of rain-bearing fronts rather than the rain itself. This correlation could possibly result in some emerging at unusual times during the year. This genus seems to be male-calling, meaning the females fly in search of the males, resulting in many more females being attracted to light traps. The males have large hair tufts on their rear legs, and it is thought these release pheromones that attract the female moths.

Fig. 10.7. *Aenetus dulcis*, male left, female right. Photo: Paul Kay, with permission.

Fig. 10.8. *Aenetus cyanochlora*, male. Photo: Paul Kay, with permission.

ARCHAEOAENETUS. Clearly closely related to *Aenetus*, this genus was described by Simonsen (2018) to contain a single described species, although preliminary molecular work hints that this might not be case. The green/grey species, *A. nielseni*, has a restricted range, occupying hill-top locations in central New South Wales where it flies in the somewhat unusual time frame of November to early January. Few specimens of this little known genus have been collected.

ZELOTYPIA. This is another single species genus and was described by Scott (1869). Females of this species are generally accepted as the largest hepialids in the world. *Zelotypia staceyi* is a trunk borer, usually high up in its *Eucalyptus* host. As these moths also choose the tallest of the *Eucalyptus* species, the caterpillars can be well out of reach of collectors. Historically, it was found along most of the NSW coast and just over the border into Queensland, but habitat clearance has severely reduced its range and abundance.

This is a spectacular reddish brown and white moth, with a large eyespot on its forewing. Adults can be collected in suitable locations using light traps but have also been attracted to car headlights travelling slowly through woodland, usually when raining.

ONCOPORA. The genus *Oncopora* was erected in 1856 by Francis Walker but it was only in 1933 that Tindale studied the genus with any purpose, naming eight new species (Tindale, 1933). Common (1966) named one new species and produced some fine genitalia drawings while Simonsen (2018) produced the most complete and organized analysis of the genus, naming one new species, making 13 in all.

These are small brown moths with short, unremarkable antennae. Some species have a leopard spot-like forewing pattern, but this is variable and not particularly helpful in identifying the species in hand. Other species have plain forewings, again unhelpful in identification. All the species can be identified using genitalia dissection but an incomplete study utilizing mt COI genetics suggests that that method is suitable for identification purposes. The study also suggests several unrecognized species still awaiting description.

Some *Oncopora* species are known to feed on the roots of various grass species and it is likely that they all do. All species live in areas where rainfall is above 600 mm per annum, restricting the genus to coastal or upland habitats where the larvae of some species can be pasture pests.

As in *Aenetus*, the males have hindleg tibial scale tufts, indicating the possibility of male calling. However, there seems not to be the disproportionate gender capture as present in that genus.

10.3.2.2 Oxycanine group

The oxycanine group of Hepialidae in Australia comprises currently three recognized genera. The genus *Elhamma* has only one Australian species; the genus *Jeana* has only two described species; and, the genus *Oxycanus* has 47 Australian species and 32 New Guinea species. However, unpublished research carried out by a variety of parties over the past 15 years is challenging the above taxonomy. This work has shown a range of morphological variability particularly in the antenna and genitalia structures and, supported by mitochondrial DNA, challenges the monophyly of the group. When this is added to a greater knowledge about host plant species, it is expected that a major revision will result.

ELHAMMA. Erected by Francis Walker (1856), Simonsen (2015) reviewed the genus and, after erecting two 'new' species from New Guinea, left the genus with six species in total, one of those being unique to Australia.

The single Australian species, *Elhamma australasiae*, is widespread along the east coast of Australia, around Victoria and extending just into South Australia. The female moth is considerably larger, but plainer, than the male. Both genders have simple filiform antennae which are highly variable in size, colour and patterning. Common on most males is a small, inverted, white, V-shaped mark on the forewing.

JEANA. Tindale created this genus in 1935 and assigned a single species, *Jeana delicatula* (Tindale, 1935). A second, *Jeana robiginosa*, was added by Turner (1939). Turner's decision is now under question, as is the status of the genus. These moths are considerably smaller than *Oxycanus* but have the same wing venation. This group is at its most diverse in south-eastern Australia, particularly in Tasmania where undescribed species are likely to be found in the future.

OXYCANUS. This is currently by far the largest genus on this side of the family. It was described by Francis Walker (1856). The moths in the genus range in size from the medium small, Tasmanian endemic *Oxycanus sphragidias*, to the large Queensland endemic *Oxycanus bulwandgi*. Many species are strikingly marked and highly coloured, though as with other hepialids, within any one species they are extremely variable in this regard (Fig. 10.9).

Fig. 10.9. *Oxycanus determinata*, male. Photo: Paul Kay, with permission.

In 1935 Norman Tindale accumulated as much material as he could across Australia and developed the first useful taxonomy for the genus (Tindale, 1935). His work focused on the structure of the genitalia, particularly the shape of the pseudotegumen, which varies widely throughout the genus. These pseudotegumenal shapes hint at possible evolutionary links, but Tindale did not elaborate on that in his work. Unfortunately, the number of specimens available for some of his determinations was extremely limited and this did not allow him to investigate the full range of variation within many of the species. Tindale's work resulted in his describing many new species and organizing what had up to that point been a very confusing array of material. Tindale's work is still the most complete for this group, though new genetic and life cycle information is inviting a review.

Oxycanid-related species exist in New Guinea and those species are currently retained in the *Oxycanus* genus. There is some evidence to suggest that a number of Australian species started their ancestral evolution in New Guinea but subsequently moved to Australia where they have continued their development. New Zealand has three closely related genera comprising 13 species. Unsurprisingly, these genera were originally considered as *Oxycanus* until the work of Jack Dumbleton and John Dugdale showed that they were different enough to warrant their separate classification (Dumbleton, 1966; Dugdale, 1994).

The antennae of *Oxycanus* are unique in their structure, being either bipectinate or quadripectinate with the expanded pectinate ends being expanded and covered with setae.

The members of this genus seem to have retained their close association with wetter regions, being most speciose along the east coast and in Tasmania, and are somewhat uncommon west of Melbourne but reappear in numbers in the south-west of Western Australia. With the drying of central Australia and the creation of the Nullarbor Plain as a potential and long-term major isolating factor between populations, it will be interesting to see how these eastern and western species are related.

Some *Oxycanus* groups seem to have an association with *Acacia* species whereas others seem to rely on grasses or sedges. One grass-eating species, *Oxycanus antipoda*, has been raised in a laboratory situation.

10.4 Conclusion

The mass emergences of hepialids are regularly commented on in the media. Unsurprisingly, people are fascinated by their behaviour and are eager to learn more about them. The group is also very popular with amateur collectors, not just because of their emergence but also their variety and colouration. Unfortunately, their flight times can be very short so material is hard to come by. Many specimens in institutional collections come from amateur collectors and the public. This collecting is to be encouraged as our understanding of the family's taxonomy and life cycle relies heavily on such specimens.

References

Beaver, E.P. and Moore, M.D. (2020) A new species of *Fraus* Walker 1856 (Lepidoptera: Hepialidae) from the south-west of Western Australia. *Australian Entomologist* 47, 39–50.

Common, I.F.B. (1966) A new species of *Oncopera* (Lepidoptera: Hepialidae) from New South Wales, with taxonomic notes on the genus. *Journal of the entomological Society of Queensland* 5, 18–28.

Dugdale, J.S. (1994) Hepialidae (Insecta: Lepidoptera). *Fauna of New Zealand, Canterbury, New Zealand* 30, 1–164.

Dumbleton, L.J. (1966) Genitalia, classification, and zoogeography of New Zealand Hepialidae (Lepidoptera). *New Zealand Journal of Science* 9, 920–981.

Grehan, J.R. and Mielke, C.G.C. (2018) Evolutionary biogeography and tectonic history of the ghost moth families Hepialidae, Mnesarchaeidae, and Palaeosetidae in the Southwest Pacific (Lepidoptera: Exoporia). *Zootaxa* 415, 243–275.

Herrich-Schäffer, G.A.W. (1855) *Sammlung neuer oder wenig bekannter aussereuropäischer Schmetterlinge*. G.J. Manz, Regensburg, Germany.

Kay, P., Hutchinson, P. and Grehan, J. (2020) New rearing method, life cycle, tunneling behavior and ecological notes on the splendid ghost moth *Aenetus djernaesae* Simonsen, 2018 from Western Australia (Lepidoptera: Hepialidae). *Journal of Insect Biodiversity* 15, 26–39.

Kristensen, N.P. (1998) The homoneurous Glossata. In: Kristensen, N.P. (ed.) *Lepidoptera: Butterflies and Moths 1. Handbook of Zoology/Handbuch der Zoologie IV/35*. De Gruyter, Berlin and New York, pp. 51–63.

Nielsen, E.S. and Kristensen, N.P. (1989) *Primitive Ghost Moths: Morphology and Taxonomy of the Australian Genus Fraus Walker(Lepidoptera: Hepialidae s. lat.)*. Monographs on Australian Lepidoptera, 1. CSIRO Publishing, Collingwood, Victoria.

Scott, A.W. (1869) Description of a new genus belonging to the family Hepialidae of Stephens. *Transactions of the Entomological Society of New South Wales* 2, 36–39.

Simonsen, T.J. (2015) *Elhamma* Walker (Lepidoptera: Hepialidae) revisited: Assessment of recently proposed synonyms and descriptions of two species. *Zootaxa* 3955, 301–328.

Simonsen, T.J. (2018) *Splendid Ghost Moths and their Allies, A Revision of Australian Abantiades, Oncopera, Aenetus, Archaeoaenetus and Zelotypia (Hepialidae)*. Monographs on Australian Lepidoptera, 12. CSIRO Publishing, Collingwood, Victoria.

Tindale, N.B. (1932) Revision of the Australian Ghost Moths (Lepidoptera, Homoneura, Family Hepialidae) Part I. *Records of the South Australian Museum* 4, 500–505.

Tindale, N.B. (1933) Revision of the Australian Ghost Moths (Lepidoptera, Homoneura, Family Hepialidae) Part II. *Records of the South Australian Museum* 5, 13–43.

Tindale, N.B. (1935) Revision of the Australian Ghost Moths (Lepidoptera, Homoneura, Family Hepialidae) Part III. *Records of the South Australian Museum*, 5, 275–332.

Turner, A.J. (1939) A second revision of the Lepidoptera of Tasmania. *Papers and Proceedings of the Royal Society of Tasmania* 1938, 57–115.

Walker, F. (1856) *List of the Specimens of Lepidopterous Insects in the Collection of the British Museum. Part VII.—Lepidoptera Heterocera*. British Museum, London.

11 The Heliozelidae: The Shield-bearers

Liz Milla[1]* and David A. Young[2]
[1]*Centre for Australian National Biodiversity Research, CSIRO, Canberra, ACT., Australia;* [2]*D'Estrees Entomology & Science Services, MacGillivray, South Australia, Australia*

Abstract
The Heliozelidae are mostly day-flying microlepidoptera with a significant yet underappreciated presence in the Australian landscape. Commonly known as shield-bearers, they are part of the superfamily Adeloidea. Australia is a key region for their diversity, offering important insights into their evolution and ecology. Some species of Heliozelidae have formed unique pollination relationships with their host plants. Recent studies highlight the diversity of both genera and species in the southern hemisphere, suggesting a likely evolutionary origin in the Australian region. This chapter discusses key Australian genera, providing an overview of the diversity and ecological roles of the heliozelid fauna in Australia.

11.1 Introduction

The Heliozelidae, commonly known as shield-bearer moths, are a widespread and cosmopolitan family of small, day-flying moths with a global distribution. The family was first recognized and described by Heinemann and Wocke (1876). Heliozelidae are part of the superfamily Adeloidea, together with the Adelidae, Incurvariidae, Prodoxidae, Cecidosidae and the recently described family Tridentaformidae (Regier *et al.*, 2015). As members of a monotrysian superfamily, Heliozelidae are an evolutionarily early diverging group of moths. Monotrysian moths represent a basal lineage within Lepidoptera, yet account for only 2% of all described species (Mitter *et al.*, 2017).

Although Heliozelidae are found across the world, Australia stands out as a major centre of heliozelid diversity, making it a crucial region for understanding the evolution and ecology of this often overlooked family of moths. This chapter examines the distinctive characteristics of Heliozelidae, emphasizing the diversity, evolutionary history and ecological roles of the Australian species.

11.2 Morphology

While a comprehensive morphological treatment of the entire Heliozelidae family is lacking, several studies have provided insights into their morphology. Davis (1998) published an introductory description, Common (1990) offered an overview of heliozelid morphological characters, while Nielsen (1980) and Karsholt and Kristensen (2003) proposed four apomorphies

*Corresponding author: liz.milla@csiro.au

DOI: 10.1079/9781800627680.0011

for the family. In a recent publication to describe the new genus *Aspilanta*, van Nieukerken and Eiseman (2020) reviewed key characteristics based on the work by Nielsen (1980) and other studies dealing with individual genera. These definitions were primarily based on northern hemisphere taxa, with limited inclusions of Australian genera (van Nieukerken and Eiseman, 2020), stemming from a scarcity of available material at the time.

When identifying apomorphies for Heliozelidae, Nielsen (1980) and Karsholt and Kristensen (2003) considered *Plesiozela*, an endemic South American genus, part of the family due to shared characteristics. Based on this hypothesis, they proposed the following four heliozelid apomorphies: dorsoventrally curved anterior tentorial arms (parallel or converging in other Adeloidea), the absence of the M-CuA cross-vein in the hindwing (usually present in other adeloid genera), the absence of pilifers on the labrum (present in other Adeloidea) and reduced mandibles, appearing as small sclerotized protrusions in *Plesiozela* and *Hoplophanes* (absent in other genera). The study by Milla *et al.* (2020), however, showed that *Plesiozela* was invariably placed within Incurvariidae, while the rest of Heliozelidae were consistently recovered as a fully supported monophyletic clade (see section 11.3). This result suggested that further assessment of the family apomorphies was required (Milla *et al.*, 2020; van Nieukerken and Eiseman, 2020).

The four aforementioned characters broadly define Heliozelidae within an Australian context but are not diagnostic for the global fauna as some of those traits are present in at least one genus of Incurvariidae. The following additional characters were proposed by van Nieukerken and Eiseman (2020) as potential Heliozelidae apomorphies, or additional synapomorphic traits: the 'smooth' head covered by flattened lamellar scales (as opposed to the 'rough' head covered by raised piliform scales found in other Adeloidea), the absence of an epicranial suture (which is part of the Adeloidea ground plan), maxillary palps with four or fewer segments (the plesiomorphic state is five-segmented) and veins 1A+2A in the forewing joined at the base. Nielsen (1980) also highlighted the length of the proboscis as an important characteristic for Heliozelidae. In most heliozelids, the proboscis is much longer than the labial palps, whereas in most other adeloids, it is of similar length or shorter. Notably, a longer proboscis is present in *Nemophora* (Adelinae), the sister group to Heliozelidae, although some genera of Heliozelidae have short proboscides.

Overall, the proposed morphological characters that define Heliozelidae require further testing and refinement, with future studies requiring the inclusion of key elements of endemic Australian taxa, such as the large group containing *Pseliastis* and *Prophylactis* (Fig. 11.1) species. The development of a firm, character-based framework for the family through formal taxonomic revision, with a broader focus on the Australian fauna, is an ongoing effort.

11.3 Systematics and Phylogeny

Several molecular and morphological studies have supported the monophyly of the Heliozelidae (Nielsen, 1980; Friedlander *et al.*, 2000; Wahlberg *et al.*, 2013; Regier *et al.*, 2015; Milla *et al.*, 2018, 2020). Early morphological studies, such as a comprehensive Hennigian cladistic

Fig. 11.1. Close-up of female *Prophylactis tetrandrallax* moth on a flower of its *Boronia tetrandra* (Rutaceae) host, which it also pollinates. Near Hopetoun, WA, July 2017. Photo by Liz Milla.

analysis by Nielsen (1980), were crucial in establishing the monophyly of the Heliozelidae and its placement within the Lepidoptera.

In recent years, the incorporation of molecular methods has significantly advanced our understanding of many Lepidopteran families (Mitter *et al.*, 2017), including Heliozelidae. Phylogenetic studies of Heliozelidae using COI sequences began to emerge from 2012 onwards. Initial COI phylogenies were produced by van Nieukerken *et al.* (2012b), Bernardo *et al.* (2015) and van Nieukerken and Geertsema (2015). These studies suggested that the large genus *Antispila* was likely not monophyletic. However, these analyses focused on northern hemisphere species, leaving a gap in the understanding of the southern hemisphere lineages. Later, Milla *et al.* (2018) sequenced two mitochondrial (COI and COII) genes and two nuclear genes (28S and H3) to generate the first multigene phylogeny of Heliozelidae. This study included eight of the 12 known genera at the time and provided strong support for five major clades: a widespread clade containing *Antispilina*, *Coptodisca* and *Holocacista*, a clade containing most *Antispila* species, a clade with *Heliozela* and *Tyriozela*, and two Australian clades (one with *Pseliastis* and one with *Hoplophanes*). The study confirmed the monophyly of *Coptodisca*, *Holocacista*, *Antispilina*, *Pseliastis* and *Hoplophanes*, while confirming that *Antispila* was polyphyletic and that *Heliozela* was paraphyletic with respect to *Tyriozela*. The study revealed the need for a taxonomic revision of the family and a more complete molecular phylogeny.

Using hundreds of genes from transcriptome-derived sequences, Milla *et al.* (2020) produced a comprehensive phylogeny of Heliozelidae that included most known genera. The study resolved the evolutionary relationships within the Heliozelidae genera, providing strong support for two major Australian groups containing *Hoplophanes* and *Pseliastis* respectively. It also recovered a large clade of cosmopolitan taxa containing four groups: *Heliozela* + *Antispila sensu stricto*, *Holocacista* + *Antispilina*, *Antispila* 'group II' and *Coptodisca*. Subsequently, van Nieukerken and Eiseman (2020) erected the genus *Aspilanta* and placed within it six Nearctic species previously categorized as *Antispila* 'group II'. Milla *et al.* (2020) also estimated the origin of the Heliozelidae to the Late Cretaceous and proposed that the family originated in the Australian region. These results are further discussed in section 11.8.

11.4 Global Diversity and Distribution

Despite being a widespread family, Heliozelidae are underrepresented in the literature, with few species described in the past 50 years, including several from Japan and the Americas (Inoue *et al.*, 1982; Karsholt and Kristensen, 2003; Lee *et al.*, 2006a, b; van Nieukerken *et al.*, 2012b; Lee and Hirowatari, 2013) and, more recently, from Australia (Hilton *et al.*, 2025). Globally, Heliozelidae genera comprise approximately 128 described species (E.J. van Nieukerken and D.J. Hilton, unpublished checklist), with many more yet to be described. Currently, the family contains 13 described genera, including the recently described Nearctic genus *Aspilanta* (van Nieukerken and Eiseman, 2020) and the endemic Australian genus *Prophylactis*, recently reinstated from synonymy and containing the type and 15 newly described species (Hilton *et al.*, 2025).

Some heliozelid genera including *Antispila*, *Heliozela* and *Holocacista* are widespread, whereas others like *Coptodisca*, *Pseliastis* and *Hoplophanes* are restricted to specific regions, such as North and Central America and Australia respectively. Four genera – *Antispila*, *Heliozela*, *Coptodisca* and *Hoplophanes* – contain over 90% of the described species (van Nieukerken *et al.*, 2012b). Recent studies indicate that heliozelids are significantly more diverse in the southern hemisphere than previously documented, though absent from New Zealand. For instance, species in the genus *Heliozela* have been found in Asia and Australia (Milla *et al.*, 2018). Species of the genus *Antispilina*, once known only from Central Europe, have been discovered in Vietnam and Japan (Milla *et al.*, 2018). Much of the recently uncovered diversity in the genus *Holocacista* has been identified from South Africa, South-east Asia and Australia (van Nieukerken and Geertsema, 2015). *Pseliastis* is more widespread and diverse in the southern parts of Australia than previously thought, while numerous new putative species of *Hoplophanes* have been found in south-west Western Australia.

Additional undescribed and species-rich genera are known from Australia.

As the taxonomy of these groups is explored, they will undoubtedly contribute many new species to the total Australian fauna. Several other large regions, including India, South America and Africa, still remain poorly surveyed for their Heliozelidae diversity.

11.5 The Australian Fauna

Australia is considered a major hotspot of heliozelid diversity (Milla *et al.*, 2018, 2020; Hilton *et al.*, 2025). With three described endemic genera (*Hoplophanes*, *Pseliastis* and *Prophylactis*) and several others suspected, Australia showcases remarkable diversity. There are estimates of over 200 undescribed species of Heliozelidae in south-west Australia, with the total number of Australian species likely to exceed 300, representing a significant portion of the global heliozelid fauna.

The Australian taxa exhibit a unique array of characteristics. Although many heliozelid larvae are leaf miners, we have observed numerous Australian species that have specialized to feed on seeds, flowers or galls. Predominant host plant families in Australia include Myrtaceae, Rutaceae (Fig. 11.2) and Ericaceae. Notably, a group of Western Australian *Prophylactis* species have formed a pollination relationship with a group of species in the plant genus *Boronia* (Rutaceae) (Hilton *et al.*, 2025). This association is discussed in section 11.7.1.

11.6 Biology and Life Histories

Based on our observations, the early life histories of the Australian Heliozelidae are characterized by a high degree of host specificity. In many Australian species where the host is either known or strongly suspected, it appears that a single host species is utilized as the larval food source (Hilton *et al.*, 2025). Several species have been observed to lay eggs that are transparent spindle or ovoid forms, and relatively large in size when compared with the adult moth (Fig. 11.3). They are often inserted into the host plant tissue, proximate to the larval feeding site, although some species of *Prophylactis* lay their eggs externally, hidden within *Boronia* flowers (Hilton *et al.*, 2025).

Upon emergence, Heliozelidae larvae feed internally. The orthodoxy has been that most Heliozelidae create a tortuous internal mine between the dermal layers of a host leaf, then cut an ovoid case or 'shield' from the terminal portion of the mine, hence the common name shield-bearers. They then lower themselves to the ground carrying the case to a substrate where they silk the case and eventually pupate. Common (1990), citing McGiffen and Neunzig (1985), mentioned a North American species that forms leaf galls, and noted that other species

Fig. 11.2. Adult moth of the Rutaceae-feeding group of Heliozelidae on its *Geleznowia verrucosa* (Rutaceae) larval host. Meanarra Track, Kalbarri National Park, WA, August 2014. Photo by Andy Young.

Fig. 11.3. Egg of an undescribed Heliozelidae, laid into a *Hibbertia subvaginata* (Dilleniaceae) bud. Larvae emerged 12 hours later. Badgingarra, WA, August 2012. Photo by Andy Young.

may mine leaf petioles to access the host leaf in which they then live.

The predominance of a leaf-mining strategy across the Australian fauna has been challenged by recent work, with various other larval feeding strategies adopted by endemic genera. These range from internal seed feeding to the use of stem and flower galls within which the larvae have been found to develop. While the specific cause of gall formation has yet to be identified, we suspect mite initiation may be involved. Our observations of larval morphology of many Australian heliozelid fauna follow the plan described in Common (1990). Legs and prolegs are often reduced to the point of being undetectable or only present as tiny, crescent-shaped scars. In some members of the *Hoplophanes* group, however, we have observed that larval morphology is more typical of other Lepidoptera, with a complement of well-developed legs and prolegs. We have observed gall-inhabiting species pupate within the galls and emerge as adults from a weakened section or hole created by the larvae. Based on our limited observations of species that feed on seeds or flower parts, these prepupal larvae bury themselves partially below the soil and create a silk puparium.

Fig. 11.4. Female moth of an undescribed *Hoplophanes* species on its *Leucopogon* (Ericaceae) larval host. The larvae live within galls formed on the plant, possibly initiated by mites. Fitzgerald River NP, WA, 2015. Photo by Andy Young.

11.6.1 Key Australian genera

11.6.1.1 Hoplophanes

This Australian endemic genus is characterized by one of the largest wingspans in the Heliozelidae of up to 18 mm, and a long, pointed ovipositor in females (Fig. 11.4). Nielsen (1980) proposed the distinct ovipositor shape and larger size as possible synapomorphies for this group, although this generalization may be challenged as more Australian species are described (Hilton *et al.*, 2025). Antennae in *Hoplophanes* have around 28–30 flagellomeres, more than other examined heliozelid genera (van Nieukerken and Eiseman, 2020). The larvae of some *Hoplophanes* are gall feeders on Ericaceae plants. *Hoplophanes* forms a strongly supported monophyletic group containing one described (*H. niphochalca*) and several putative species (Milla *et al.*, 2018, 2020). Notably, molecular analyses place *Hoplophanes* and related taxa as a sister group to all other Heliozelidae (see section 11.3). This basal position is supported by the presence of remnant mandibles, which are absent from all the other Heliozelidae groups (Karsholt and Kristensen, 2003).

11.6.1.2 Pseliastis

Another endemic genus in Australia, *Pseliastis* was once thought to be restricted to Tasmania (Common, 1990). However, our recent field collections show that *Pseliastis* species are more widespread and diverse, with their range spreading from the alpine regions of Tasmania and Victoria to subtropical Queensland, offshore on Kangaroo Island in South Australia and the dry inland regions of Western Australia. Most, if not all, species of *Pseliastis* are associated with plants in the family Rutaceae, and several species within this genus are known to be seed feeders. Observations from preliminary morphological examinations suggest that the *Pseliastis* group represents more than one genus. For example, the type species *P. trizona*, *P. spectropa* and other putative *Pseliastis* species share characteristics such as forewings with white fasciae; however, *P. xanthodisca* lacks these typical fasciae, and groups most closely with a species within the recently resurrected sister genus *Prophylactis* (Hilton *et al.*, 2025).

11.6.1.3 *Heliozela*

This genus has a widespread global distribution, including multiple species in Australia. *Heliozela* are mostly leaf miners, although the larvae of a few species mine the petiole or midrib of the leaf, or feed in a gall (Davis, 1998). Major host families are Myrtaceae, Fagaceae and Betulaceae, with *Eucalyptus* an important genus for *Heliozela* species in Australia, while Myrtaceae, in general, provide an important host family for tropical species. Nielsen (1980) placed *Tyriozela* as sister to *Heliozela* based on shared synapomorphies, such as the spear-shaped ovipositor of the females, and suggested a strongly developed scent organ in the male abdomen as a possible autapomorphy. The strongly supported *Heliozela* + *Tyriozela* group contains all described and putative *Heliozela* species, as well as the only known *Tyriozela* species, *T. porphyrogona* (van Nieukerken and Geertsema, 2015; Milla *et al.*, 2018). Taxonomic revision of this group will be necessary to determine whether it contains multiple genera or whether *Heliozela* and *Tyriozela* should be collapsed into a single genus.

11.6.1.4 *Holocacista*

This widespread genus of leaf-mining heliozelids has a diverse presence in Australia, Africa, East and South-east Asia (Milla *et al.*, 2018). *Holocacista* moths are very small (3–5 mm wingspan) and often display a pattern of metallic-silvery spots. Distinguishing features include strongly reduced venation compared with *Antispila*, and a typically long appendix on the phallus along with a small epiphysis in males, while females have a flattened, deeply dented ovipositor (van Nieukerken and Geertsema, 2015; van Nieukerken and Eiseman, 2020). Larvae are typically yellow or whitish, lacking legs and prolegs, and construct a case for pupation. *Holocacista* are mostly associated with the Rubiaceae and Dilleniaceae plant families, and Australian species were collected from *Psychotria* and *Coprosma* (Rubiaceae) and *Logania* (Loganiaceae) (Milla *et al.*, 2018). *Holocacista* appears to have expanded into Australia through the Oriental region from its putative Palaearctic origins (Milla *et al.*, 2020).

Molecular evidence suggests that *Holocacista* is split into two groups, one including all the seven described species and three putative additional species, and another containing six unnamed species from Australia (Milla *et al.*, 2018). Detailed morphological analyses are required to determine whether this clade of Australian species belongs to *Holocacista* or requires the erection of a separate genus.

11.6.1.5 *Prophylactis*

Originally described by Meyrick (1897), this genus (see Fig. 11.1) contained three species, the type *P. argochalca* and two other species (*P. aglaodora* and *P. chalcopetala*). A fourth species, *P. memoranda*, was added later by Meyrick (1922). Nielsen (1996) later synonymized *Prophylactis* with *Hoplophanes*. However, a recent study by Hilton *et al.* (2025) which re-examined the *P. argochalca* syntypes and thousands of Australian specimens reinstated the genus *Prophylactis* and named 15 new species, with many other species awaiting description. Species of *Prophylactis* can be distinguished from *Pseliastis* and other heliozelid genera by their red eyes and plain forewings lacking fasciae or spots (Hilton *et al.*, 2025). *Prophylactis* includes a clade of species that exhibit a unique abdominal cleft in females that has been linked to pollination (Hilton *et al.*, 2025). This association is discussed in detail in section 11.7.1 below.

11.6.1.6 *Undescribed genera*

Based on the examination of thousands of specimens collected from around Australia in the past decade, several putative new genera exist within the Australian Heliozelidae, particularly those associated with the plant families Rutaceae, Myrtaceae, Dillenaceae, Rhamnaceae and Ericaceae.

11.7 Ecological Roles and Significance

Heliozelidae exhibit a high degree of niche specialization and play a role in the functional ecology of ecosystems. Australian species in particular show remarkable diversity and unique interactions with their host plants.

As herbivores, Heliozelidae are part of the food web and serve as food sources for insectivorous animals. Their leaf-mining activities can reduce the photosynthetic capacity of plants, potentially affecting plant growth with knock-on effects for other dependent organisms. Flower- and seed-mining heliozelid larvae likely affect the reproductive capacity of their host plants, although this impact has yet to be quantified.

The global Heliozelidae genera feed on a wide variety of plant families. For example, *Antispila sensu stricto* primarily feeds on Cornaceae and Vitaceae, whereas *Heliozela* species feed on Myrtaceae, Fagaceae, and Betulaceae. *Holocacista* species are leaf miners feeding mainly on Vitaceae and Rubiaceae. Some heliozelids are more host specific, with several genera feeding on only one or two plant families, one genus or even a single species (Milla *et al.*, 2020; Hilton *et al.*, 2025). This specificity can lead to intricate relationships with their hosts. The genus *Antispilina*, for example, feeds exclusively on herbaceous Polygonaceae, whereas *Hoplophanes* is predominantly associated with Ericaceae, and *Pseliastis* and *Prophylactis* with Rutaceae (Milla *et al.*, 2018). Due to their host specificity, some Heliozelidae could act as indicator species, reflecting the health and diversity of their host plant communities, with any changes to moth populations indicating changes to the environment and habitat disturbance.

11.7.1 Pollination mutualism with *Boronia*

Perhaps the most striking of the ecological roles of Heliozelidae is the pollination mutualism that has arisen between species of *Prophylactis* and the plant genus *Boronia* (Rutaceae) (Hilton *et al.*, 2025). These pollination associations are examples of highly specialized plant–insect interactions. They comprise a brood pollination mutualism, where the adult moth pollinates the host plant while its larvae feed on the host, consuming developing seeds (Sakai, 2002). Because of the highly co-evolved relationships required to maintain them, brood pollination mutualisms are important model systems for the study of co-evolutionary biology, with the best-known and best-studied examples involving yuccas and yucca moths (Riley, 1892; Pellmyr, 1997) and figs and their fig wasps (Wiebes, 1979; Cook *et al.*, 2004).

Recent work has shown that a group of 15 *Prophylactis* species endemic to south-west Western Australia are pollinators of *Boronia* (Hilton *et al.*, 2025). The 13 *Boronia* hosts of the pollinator species are also endemic to the region and form a single clade, but not all *Boronia* species within the clade host a *Prophylactis* pollinator. The females of the pollinator species have evolved a pollen-carrying cleft on the dorsal side of their last two abdominal tergites (Hilton *et al.*, 2025). This feature is not known from any other Lepidoptera – indeed, it may be unique within the insects – and it is used to collect and deposit pollen, representing a novel morphological adaptation. In some cases the pollination mechanism involves active pollen deposition (Milla, 2018), where the moths actively collect or deposit pollen using this unique cleft. This is distinct from passive pollination, where insects indirectly collect pollen as a by-product of foraging in flowers.

Prophylactis pollinators are invariably collected in the field from their *Boronia* host species or as they perch on nearby plants. DNA barcode sequences of both pollinating and non-pollinating *Prophylactis* collected from the same *Boronia* species across different geographical areas form well-supported groups (Hilton *et al.*, 2025). This suggests that all known *Prophylactis* species, including the 15 identified pollinators, are specific to a single species of host. Conversely, each *Boronia* species hosts a single *Prophylactis* pollinator, except for two *Boronia* species that host two pollinator species each (Hilton *et al.*, 2025). In some cases, *Prophylactis* pollinators share *Boronia* hosts with one or more non-pollinating *Prophylactis*. This high level of host specificity observed in *Prophylactis* has likely contributed to the evolution of the pollination specialization.

Another remarkable feature of this mutualism is that some *Boronia* species and their *Prophylactis* pollinators are in a reciprocally obligate relationship, where both the plant and insect are entirely dependent on each other for reproduction (Milla, 2018). This means that some *Boronia* species rely entirely on their specific moth for pollination, while the moth depends on its *Boronia* host as its larval food source. There are very few other brood pollination mutualisms that are known to have evolved such reciprocally obligate relationships (Kato and Kawakita, 2017). The reciprocally obligate relationship between *Boronia megastigma* and its newly described pollinator moth, *P. megastigmallax*, had been suspected for some time (Mactavish and Menary 1997; Plummer *et al.*, 1999), but was not confirmed until recently. Reciprocally obligate relationships are thought to exist between *Prophylactis* pollinators and at least three other *Boronia* hosts (Hilton *et al.*, 2025).

The *Boronia–Prophylactis* mutualism represents a relatively recent evolutionary development. The group of pollinators is estimated to have diverged from the rest of *Prophylactis* around twelve million years ago (Milla *et al.*, 2020) with the distinctive abdominal cleft evolving only once in *Prophylactis* (Hilton *et al.*, 2025). This unique system offers valuable insights and exceptional opportunities to study the evolution of mutualistic plant–insect interactions, particularly in understanding how these relationships have generated some of the rich biodiversity within the Australian region.

11.8 Biogeographical Origins

Ancestral range reconstruction analyses have proposed an Australian origin of Heliozelidae, with some models indicating an additional region of origin, either in the Palaearctic or Oriental regions (Milla *et al.*, 2020). The family is estimated to have appeared at least 95 million years ago during the Late Cretaceous, coinciding with Australia's separation from East Antarctica (McLoughlin, 2001). The first expansion of the Heliozelidae outside Australia is thought to be the lineage leading to *Antispila sensu stricto*, estimated to have emerged in the Palaearctic region between the Late Cretaceous and Middle Paleogene (Milla *et al.*, 2020). *Antispila sensu stricto* has a wide range in the Nearctic region and utilizes multiple plant hosts (van Nieukerken *et al.*, 2012b), suggesting better dispersal and colonization abilities than *Hoplophanes* and *Pseliastis*. *Heliozela* species likely spread from Australia to the Neotropics (Eocene), the Oriental region (Oligocene) and the Palaearctic region (early Miocene). *Holocacista* is the only genus present in Australia not thought to have originated there; rather, it is thought to have expanded from the Palaearctic through the Oriental region.

The two earliest diverging clades, *Hoplophanes* and *Pseliastis*, are both restricted to Australia. The split between the *Hoplophanes* group and the rest of the family is estimated to have occurred more than 84 million years ago, and the most recent common ancestor of the *Pseliastis* group dates to the late Eocene, around 78 million years ago. Five of the currently recognized genera (*Hoplophanes*, *Heliozela*, *Pseliastis*, *Prophylactis* and *Holocacista*) are found in Australia, with commensurate high species diversity, particularly in the southern part of the continent. In addition, two other heliozelid genera, *Phanerozela* and *Ishnocabana*, are known only from the southern hemisphere, supporting a southern hemisphere origin.

However, possible inclusion of species from other small and monotypic genera from large, undersampled regions may influence future ancestral range reconstructions. Due to the scarcity of confidently assigned Heliozelidae fossils within the relatively poor Lepidoptera record (Doorenweerd *et al.*, 2015; Sohn *et al.*, 2015), Milla *et al.* (2020) relied on fossils external to the family for dating calculations. Future studies of fossil plant leaves may reveal the characteristic mines and oval cases typical of this group. This material, more resilient in the fossil record than the adult moths and their scales, may allow greater resolution around these questions, but such studies have not yet been attempted.

11.9 Economic Importance

Although the Heliozelidae are a relatively small family of moths, they have global economic importance due to some species being considered pests of commercial crops (Maier, 1988; van Nieukerken *et al.*, 2012a; Bernardo *et al.*, 2015; van Nieukerken and Geertsema, 2015). Leaf-mining Heliozelidae are predominantly pests on trees and vines, causing notable damage to grapevines, cranberry and walnut crops. *Aspilanta oinophylla* and *Coptodisca lucifluella*, both originating in North America, are known pests of grapevines and walnut trees in Europe. *Antispila uenoi*, native to Japan, has also been reported as a pest of commercial vineyards (Wang *et al.*, 2018). In South Africa, *Holocacista capensis* is a pest of commercial grapevines (Steyn *et al.*, 2020). To date, however, heliozelids are not reported as a major problem in commercial crops in Australia.

Most described heliozelid species are leaf miners but many recently discovered species also mine petioles, midribs, twigs, flowers and seeds and initiate galls. The impact of these non-leaf-mining behaviours on economic crops has not been reported in the literature. In addition, their small size and the morphological similarities among heliozelid species often make their identification challenging. Combined with a lack of comprehensive taxonomic, molecular and ecological data, this has complicated species identification, as seen in the case of several invasive species overseas (Bernardo *et al.*, 2015; van Nieukerken and Geertsema, 2015). More information is needed to appreciate fully the economic impact of heliozelids in commercial crops.

References

Bernardo, U., Nieukerken, E. van, Sasso, R., Gebiola, M., Gualtieri, L. and Viggiani, G. (2015) Characterization, distribution, biology and impact on Italian walnut orchards of the invasive North-American leafminer *Coptodisca lucifluella* (Lepidoptera: Heliozelidae). *Bulletin of Entomological Research* 105, 210–224.

Common, I.F.B. (1990) *Moths of Australia*. Melbourne University Press, Carlton, Victoria.

Cook, J.M., Bean, D., Power, S.A. and Dixon, D.J. (2004) Evolution of a complex coevolved trait: Active pollination in a genus of fig wasps. *Journal of Evolutionary Biology* 17, 238–246.

Davis, D.R. (1998) The Monotrysian Heteroneura. In: Kristensen, N.P. (ed.) *Lepidoptera, Butterflies and Moths. Handbook of Zoology* 4(1). De Gruyter, Berlin and New York, pp. 65–90.

Doorenweerd, C., van Nieukerken, E.J., Sohn, J.-C. and Labandeira, C.C. (2015) A revised checklist of Nepticulidae fossils (Lepidoptera) indicates an Early Cretaceous origin. *Zootaxa* 3963, 295–334.

Friedlander, T.P., Regier, J.C., Mitter, C., Wagner, D.L. and Fang, Q. (2000) Evolution of heteroneuran Lepidoptera (Insecta) and the utility of dopa decarboxylase for Cretaceous-age phylogenetics. *Zoological Journal of the Linnean Society* 130, 213–234.

Heinemann, H.V. and Wocke, M. (1876) *Die Schmetterlinge Deutschlands und der Schweiz. Zweite Abteilung. Kleinschmetterlinge. Band 2. Die Motten und Federmotten*. Heft 2. CA Schwetschke und Sohn, Braunschweig.

Hilton, D.J., Young, D.A., Milla, L., Jin, M., Wilcox, S. *et al.* (2025) Phylogeny and taxonomy of a new clade of Australian Heliozelidae in the genus *Prophylactis* Meyrick, 1897 (Lepidoptera, Adeloidea) pollinating *Boronia* (Rutaceae: Sapindales). *Arthropod Systematics & Phylogeny* 83, 175–210.

Inoue, H., Sugi, S., Kuroko, H., Moriuti, S., Kawabe, A. and Owada, M. (1982) *Moths of Japan*, vol. 1. Text Kodansha, Tokyo, Japan.

Karsholt, O. and Kristensen, N.P. (2003) *Plesiozela*, gen. nov. from temperate South America: Apparent sister-group of the previously known Heliozelidae (Lepidoptera: Incurvarioidea: Heliozelidae). *Invertebrate Systematics* 17, 39–46.

Kato, M. and Kawakita, A. (eds) (2017) *Obligate Pollination Mutualism*. Springer Japan, Tokyo, Japan.

Lee, B.-W. and Hirowatari, T. (2013) Heliozelidae. In: Hirowatari, T., Nasu, Y., Sakamaki, Y. and Kishida, Y. (eds) *The Standard of Moths in Japan*. Gakken Kyoiku Shuppan, Tokyo, Japan.

Lee, B.-W., Hirowatari, T. and Kuroko, H. (2006a) Five new species of the genus *Heliozela* Herrich-Schaffer (Lepidoptera, Heliozelidae) from Japan. *Transactions of the Lepidopterological Society of Japan* 57, 81–91.

Lee, B.-W., Hirowatari, T. and Kuroko, H. (2006b) A new species of *Antispila* (Lepidoptera: Heliozelidae) from the Ryukyus, Japan, with descriptions of immature stages. *Annals of the Entomological Society of America* 99, 1019–1027.

Mactavish, H.S. and Menary, R.C. (1997) The effect of flower maturity and harvest timing on floral extract from *Boronia megastigma* (Nees). *Annals of Botany* 80, 299–303.

Maier, C.T. (1988) Life cycle of *Coptodisca negligens* (Lepidoptera: Heliozelidae) on cranberry. *Journal of Economic Entomology* 81, 497–500.

McGiffen, K.C. and Neunzig, H.H. (1985) *A Guide to the Identification and Biology of Insects feeding on Muscadine and Bunch Grapes in North Carolina*. North Carolina Agricultural Research Service, North Carolina State University, Raleigh, NC.

McLoughlin, S. (2001) The breakup history of Gondwana and its impact on pre-Cenozoic floristic provincialism. *Australian Journal of Botany* 49, 271–300.

Meyrick, E. (1897) Descriptions of Australian Microlepidoptera. XVII. Elachistidae. *Proceedings of the Linnean Society of New South Wales* 22, 297–435.

Meyrick, E. (1922) *Exotic Microlepidoptera* 2, 513–544. Available at: www.biodiversitylibrary.org/item/37719#page/9/mode/1up (accessed 1 October 2025).

Milla, E. (2018) Evolution and ecology of the Australian Heliozelidae (Adeloidea, Lepidoptera). PhD thesis, University of Melbourne, Parkville, Victoria.

Milla, L., van Nieukerken, E.J., Vijverberg, R., Doorenweerd, C., Wilcox, S.A. *et al.* (2018) A preliminary molecular phylogeny of shield-bearer moths (Lepidoptera: Adeloidea: Heliozelidae) highlights rich undescribed diversity. *Molecular Phylogenetics and Evolution* 120, 129–143.

Milla, L., Moussalli, A., Wilcox, S.A., van Nieukerken, E.J., Young, D.A. *et al.* (2020) Phylotranscriptomics resolves phylogeny of the Heliozelidae (Adeloidea: Lepidoptera) and suggests a Late Cretaceous origin in Australia. *Systematic Entomology* 45, 128–143.

Mitter, C., Davis, D.R. and Cummings, M.P. (2017) Phylogeny and evolution of Lepidoptera. *Annual Review of Entomology* 62, 265–283.

Nielsen, E.S. (1980) The Incurvarioid Genera of the World (Lepidoptera: Incurvarioidea). Unpublished thesis, University of Copenhagen.

Nielsen, E.S. (1996) Heliozelidae. In: Nielsen, E.S., Edwards, E.D. and Rangsi, T.V. (eds) *Checklist of the Lepidoptera of Australia*. Monographs on Australian Lepidoptera 4, CSIRO Publishing, Collingwood, Victoria.

Pellmyr, O. (1997) Pollinating seed eaters: Why is active pollination so rare? *Ecology* 78, 1655–1660.

Plummer, J., Wann, J. and Spadek, Z. (1999) Intraspecific variation in oil components of*Boronia megastigma* Nees (Rutaceae) flowers. *Annals of Botany* 83, 253–262.

Regier, J.C., Mitter, C., Kristensen, N.P., Davis, D.R., van Nieukerken, E. *et al.* (2015) A molecular phylogeny for the oldest (nonditrysian) lineages of extant Lepidoptera, with implications for classification, comparative morphlogy and life-history evolution. *Systematic Entomology* 40, 671–704.

Riley, C.V. (1892) *The Yucca Moth and Yucca Pollination*. Missouri Botanical Garden, 3rd Annual Report, St Louis, MO, pp. 99–158.

Sakai, S. (2002) A review of brood-site pollination mutualism: Plants providing breeding sites for their pollinators. *Journal of Plant Research* 115, 161–168.

Sohn, J.-C., Labandeira, C.C. and Davis, D.R. (2015) The fossil record and taphonomy of butterflies and moths (Insecta, Lepidoptera): Implications for evolutionary diversity and divergence-time estimates. *BMC Evolutionary Biology* 15, 12.

Steyn, L., Geertsema, H., Malan, A. and Addison, P. (2020) A review of leaf-mining insects and control options for their management, with special reference to *Holocacista capensis* (Lepidoptera: Heliozelidae) in vineyards in South Africa. *South African Journal of Enology and Viticulture* 41, 218–232.

Van Nieukerken, E.J. and Eiseman, C.S. (2020) Splitting the leafmining shield-bearer moth genus *Antispila* Hübner (Lepidoptera, Heliozelidae): North American species with reduced venation placed in *Aspilanta* new genus, with a review of heliozelid morphology. *ZooKeys* 957, 105–161.

Van Nieukerken, E.J. and Geertsema, H. (2015) A new leafminer on grapevine and *Rhoicissus* (Vitaceae) in South Africa within an expanded generic concept of *Holocacista* (Insecta, Lepidoptera, Heliozelidae). *ZooKeys* 507, 41–97.

Van Nieukerken, E.J., Doorenweerd, C., Stokvis, F. and Groenenberg, D. (2012a) DNA barcoding of the leaf-mining moth subgenus *Ectoedemia* s. str.(Lepidoptera: Nepticulidae) with COI and EF1-α: Two are better than one in recognising cryptic species. *Contributions to Zoology* 81, 1–24.

Van Nieukerken, E.J., Wagner, D., Baldessari, M., Mazzon, L., Angeli, G. *et al.* (2012b) *Antispila oinophylla* new species (Lepidoptera, Heliozelidae), a new North American grapevine leafminer invading Italian vineyards: Taxonomy, DNA barcodes and life cycle. *ZooKeys* 170, 29–77.

Wahlberg, N., Wheat, C.W. and Peña, C. (2013) Timing and patterns in the taxonomic diversification of Lepidoptera (butterflies and moths). *PLOS One* 8, e80875.

Wang, N., Liu, T., Xu, J. and Jiang, B. (2018) The leaf-mining genus *Antispila* Hübner, 1825 feeding on Vitaceae in Shandong Peninsula, China with one new species (Lepidoptera, Heliozelidae). *ZooKeys* 744, 49–65.

Wiebes, J. (1979) Co-evolution of figs and their insect pollinators. *Annual Review of Ecology and Systematics* 10, 1–12.

12 The Oecophoridae: The Mallee or Concealer Moths

Robert J.B. Hoare
New Zealand Arthropod Collection (NZAC), Manaaki Whenua Landcare Research, Auckland, New Zealand

Abstract
This chapter reviews the enormously diverse Oecophoridae, the largest family of Lepidoptera in Australia. Nearly 3300 Australian species are recognized, almost all of which are endemic. After brief accounts of the family circumscription and diversity, I cover the history of its study, with emphasis on the massive contributions by Edward Meyrick and the late Ian Common. A brief section deals with our incomplete understanding of oecophorid phylogeny, before a more substantial summary of known life histories: the larvae of most species feed on myrtaceous leaf litter, but a few feed on living foliage, some are saproxylic and others specialize on grasses or the nests of birds. I then consider adult coloration, crypsis and mimicry, before discussing the few species of economic importance and the handful of Australian Oecophoridae that have become established overseas, especially in New Zealand.

12.1 Introduction

Historically, few lepidopterists have specialized on Oecophoridae. It is a family with only a few species of economic importance, and forms a relatively small and generally inconspicuous element of the fauna of most countries worldwide; for example, there are about 150 species of Oecophoridae (as currently defined) in Europe, out of a total Lepidoptera diversity of over 8500 species (Lepiforum website: www.lepiforum.de). The larvae, which are mostly concealed in dead wood or in leaf litter, also do not attract much attention.

Australia is the exception to this rule. The dominance of the Oecophoridae in this country, where it forms the most diverse group of Lepidoptera, means that the family can scarcely be ignored here. There are probably more oecophorid species in Australia than in the rest of the world combined. Further incentives to their study are the diversity of (often very beautiful) wing patterns and the importance of their larvae as recyclers of nutrients in the sclerophyllous vegetation that dominates so much of the landscape (e.g. Common, 1994). Nonetheless, a truly comprehensive overview of the vast Australian oecophorid fauna was a long time in coming, although Edward Meyrick and Alfred Jefferis Turner provided descriptions of hundreds of species before 1950. It was Ian Common's great genus-level revisions (Common, 1994, 1997, 2000), heavily drawn on below, that brought the family into the limelight and

Corresponding author: hoarer@landcareresearch.co.nz

DOI: 10.1079/9781800627680.0012

gave us new insights into its true diversity at genus and species level. Common also investigated life histories, and although our knowledge of oecophorid biology in Australia is still extremely incomplete, we now have an excellent insight into the various ecological niches inhabited by the larvae. Some unusual and fascinating larval behaviours have been described, including a variety of case-making strategies, storage of eucalypt oils in a gut diverticulum as a defensive adaption (genus *Myrascia*), specialization on marsupial scats as a pabulum (e.g. *Telanepsia*) and potential symbioses with parrots in their nests in termite mounds (genus *Trisyntopa*). All of these are briefly discussed in this chapter. Doubtless other remarkable life histories will be revealed in future in such an enormous family.

Our understanding of how the Australian oecophorid fauna diversified into its current 'hyperspeciose' state, and how exactly it relates to overseas faunas, is still in its infancy. As with the myrtaceous hosts, adaptation and an 'evolutionary explosion' almost certainly followed the aridification of the continent from the Miocene onwards. The few available phylogenetic studies on the family do not contain very many Australasian taxa (Kim *et al.*, 2016; Sohn *et al.*, 2016). Common recognized a number of genus groups, based on shared morphological characters; the monophyly of these groups and their interrelationships have not been further tested by molecular phylogenetic methods. Here is another fertile area for future investigation.

A superb, richly illustrated introduction to the Australian Oecophoridae can be found in Cocking *et al.* (2022), which covers a wide selection of genera and species from the ACT and contains excellent colour photographs of live adult moths together with notes on identification and life histories.

12.2 Definition of the Family

The family Oecophoridae belongs to the highly diverse worldwide superfamily Gelechioidea. In common with other gelechioid families, it has suffered from considerable taxonomic instability, and its circumscription has varied widely over the last few decades. A number of groups formerly recognized as subfamilies of Oecophoridae (e.g. by Toll, 1964; Common, 1990, 1996) are now given full family status. The Autostichidae, Depressariidae, Stathmopodidae and Xyloryctidae are cases in point. Others are now included as subfamilies but in other families: Stenomatinae in Depressariidae, Chimabachinae (unknown from Australia) in Lypusidae (Heikkilä *et al.*, 2014). The chequered history of the definition of Oecophoridae and inclusion or exclusion of various subfamilies were well covered by Kim *et al.* (2016) and are not repeated here. Heikkilä *et al.* (2014) restricted the Oecophoridae to two subfamilies: Oecophorinae and Pleurotinae, with *Deuterogonia*, formerly assigned subfamily status (Toll, 1964), nested tentatively within the Oecophorinae. Only the Oecophorinae are currently recognized from Australia.

Morphological characters defining the Oecophoridae are few and not necessarily unique to the group, or constant throughout it. Common (1994) mentions the parallel condition of veins Rs and M1 in the hindwing and the presence of dorsal scobinations on the gnathos in the male genitalia as recognition features. The gnathos lacks the close-set rows of spinules that characterize some other gelechioid groups such as Depressariidae and Elachistidae. Heikkilä *et al.* (2014) refer to the fusion of the base of the gnathos with the tegumen (a feature shared with Stathmopodidae) as an almost constant character throughout the family. The basally scaled proboscis and acute, recurved labial palpi, characters traditionally used to define the superfamily Gelechioidea, are almost universal throughout the Oecophoridae.

12.3 Diversity and Endemism of the Australian Fauna

Little has been published since Ian Common completed his monumental genus-level revision of the entire subfamily Oecophorinae for Australia (Common, 1994, 1997, 2000). Common (2000) ultimately recognized 271 genera of Australian Oecophorinae, containing 1768 named and 1372 unnamed species, a total of 3140 species. In addition, he listed 132 species originally named as Oecophoridae but unassigned to genus for various reasons, noting that some of these may well prove not to be oecophorids.

Common's 'grand total' of recognized species was therefore 3272.

I am aware of only four minor changes that have been made to the taxonomy in more recent years, as follows: (i) a new species, *Trisyntopa neossophila*, was described by Edwards *et al.* (2007); (ii) *Leistarcha*, placed in Xyloryctidae (as Xyloryctinae) by Common (1996), was rejected from Xyloryctidae by McMillan (2013) and shown to belong in the Oecophoridae by Sohn *et al.* (2016); (iii) *Casmara* (two Australian species) was removed to the Ashinagidae by Wang and Li (2020); and (iv) a new species, *Tachystola mulliganae*, was described by Sterling *et al.* (2023). The two new species are discussed later in this chapter. With regard to *Leistarcha*, four species are listed as belonging to this genus by Common (1996), but one of these, *Bathydoxa tenuistria*, is in fact a xyloryctid (I. McMillan, pers. commun.). I also argue below that *Mermeristis* should be removed from the Australian list (see Postscript).

In consequence of these works, Common's species total is slightly adjusted to 3274; an exact figure is, of course, premature, since many further taxa will surely come to light with further detailed study, while some of the 'unassigned' genera and species may turn out not to be oecophorines. But to put even this preliminary count in context, there are fewer than 2000 species of the whole order Lepidoptera recognized in New Zealand (Dugdale, 1988; Hoare, 2010a), and only *ca* 3300 named species of Oecophoridae worldwide (van Nieukerken *et al.*, 2011). With the exception of the two cosmopolitan pest species *Hofmannophila pseudospretella* and *Endrosis sarcitrella*, probably almost the entirety of this vast Australian fauna is endemic at the species level. The oecophorines of Australia therefore represent possibly the single greatest endemic radiation of a family of Lepidoptera in one country anywhere in the world.

From Common's revisionary work, it became clear that endemism at the genus level was much higher than previously supposed. Edward Meyrick had frequently referred taxa from New Zealand and/or South-east Asia to Australian genera such as *Eulechria*, *Euthictis*, *Leptocroca*, *Locheutis* and *Philobota*, but it is now apparent that these genera are most likely Australian endemics and their supposed overseas representatives have been incorrectly placed, with the possible exception of a few species from New Guinea (*cf.* Common 1997 on *Eulechria*). New species have been described in the genus *Locheutis* from China in the last 25 years (Wang, 2002, 2004; Wang and Tao, 2021); regrettably, Common's (2000) revised circumscription of the genus was ignored in all cases, and it is obvious from the illustrations that the Chinese species are misplaced.

Conversely, many Australian species were wrongly assigned to European genera, as was the case with species of *Thema*, referred to *Pleurota* by Meyrick and Turner because of their superficially similar porrect labial palpi (Common, 1997). The New Zealand fauna almost entirely belongs to the *Barea* group of Common (2000), and is apparently quite closely related to the Australian, but my assessment is that only one genus is naturally shared between the two countries: that is, *Prepalla*, which has an endemic New Zealand species (*P. austrina*) in addition to nine Australian endemics (Hoare, 2005). This genus is placed in the *Chezala* group (Common, 1997). The other supposed example, *Mermeristis*, accepted by Common (2000) as naturally shared between Australian and New Zealand, is almost certainly endemic to New Zealand and, I believe, was recorded in Australia only as a result of mislabelling (see Postscript, below).

Apart from *Endrosis* and *Hofmannophila*, non-endemic genera are very few. A fascinating example is *Aristeis*, which has species in India and Sri Lanka as well as Australia; the Indian species, such as *A. dorothea*, have a similar case-making larval biology to the Australian (see below), but feed on the soft-fruited myrtaceous genus *Syzygium* (as *Eugenia* in Common, 1997).

12.4 History of Study of the Oecophoridae

Common (1994) treated the history of Oecophorinae taxonomy in Australia: I give a briefer summary of the early years here but also an overview of Common's own work. As with most antipodean insects, for many years almost all taxonomic work was done in Europe, and the earliest type specimens remain in overseas institutions. The first species to be named was

Phalaena fascialis by Fabricius in 1775 (now *Chrysonoma fascialis*); this was based on a specimen collected by Joseph Banks on Captain Cook's first voyage (1768–1771). English lepidopterist Edward Newman named 11 new species from a collection made by T.R. Oxley in the Mt Alexander Range, Victoria (Newman, 1856). Until Meyrick's first publication on the group (Meyrick, 1883), only two other authors described more than two species: these were Philipp Christoph Zeller (1855, 1877), who named nine, and the prolific Francis Walker (1863, 1864a, b, 1869), who named 97. Walker's work was notoriously fast and often inaccurate, and 15 of the 97 names have proved to be synonyms (Common, 1994). However, he did name some important genera such as *Chezala*, *Barea* and *Garrha*.

Edward Meyrick's (1854–1938) contribution was vast; he had the advantage of living in Australia (and New Zealand) from 1877 to 1886 as a schoolteacher, and collected avidly during this time. As a taxonomist, he was highly industrious but far more careful and insightful than Walker, and made an especial study of wing venation as an aid to classification at the genus level and above. In a series of papers published in the 1880s, he began to document the extraordinary and unsuspected diversity of Australasian Oecophoridae (see references in Common, 1994); the New Zealand fauna, which is diverse but far less so than that of Australia, was also included in this series. Of the 66 new Australian genera described in these papers, 62 are now recognized as belonging to Oecophorinae. Meyrick also described no less than 626 new Australian species. His productivity continued after his return to England in 1886, many new species now being based on specimens sent to him by correspondents such as George Lyell and Oswald Lower. These further taxa were chiefly named in a paper published in the *Transactions of the Royal Society of South Australia* (Meyrick, 1902) and in his own journal *Exotic Microlepidoptera* (Meyrick, 1912–1923). His revision of the world fauna of Oecophoridae was published in *Genera Insectorum* (Meyrick, 1922). Following revision of the fauna by Common, Meyrick can be credited with naming 88 valid genera of Australian Oecophorinae and around 850 species (Common, 1994; *Leistarcha* added and *Mermeristis* omitted, see above and Postscript below).

Alfred J. Turner (1861–1947), the Queensland paediatrician, ranks second only to Meyrick in the number of new species of Oecophorinae he described. Turner collected and studied most groups of Lepidoptera, but later took a particular interest in Oecophoridae and, between 1932 and his death in 1947, published a revision of the Australian fauna of this family. According to Common (1994), he can be credited with naming 49 currently valid genera and 835 valid species.

Revision of the Australian Oecophorinae at the genus level was Ian Common's (1917–2006) major retirement project and one of his crowning achievements as a lepidopterist, completed when he was over 80 years old (Common, 1994, 1997, 2000). It was in full progress when I came to Canberra as a PhD student in 1995, and undoubtedly inspired me to work later on the New Zealand Oecophoridae (though ironically, the two 'oecophorine' genera I ended up revising, *Hierodoris* Meyrick and *Izatha* Walker (Hoare, 2005, 2010a), have since been transferred to Xyloryctidae!). The monumental nature of Ian's project is difficult to conceive. Given that most genera are hard to recognize without examination of the male genitalia, one assumes from the precise species counts given for each genus, including undescribed species, that specimens of almost every species (over 3000) were dissected. Much of this demanding work was undertaken by Leanne Regan, who provided expert technical assistance at every stage of the revision. The revolution in understanding the diversity and endemicity of the Australian oecophorines has been alluded to above. No less than 120 new genera were described in these three volumes. Common also made a special effort to rear as many species as possible by making collections of potential larval substrates (mainly leaf litter) wherever he went. He was thus able to document the biology of many genera and species. His first volume contains a superb, illustrated summary of the morphology of the early stages, including scanning electron micrographs of eggs and pupae, and chaetotaxy of larvae; the larval account was based on examination of species belonging to 32 different genera (Common, 1994).

12.5 Phylogeny

As pointed out by Kim *et al.* (2016), most recent phylogenetic work involving Oecophoridae has been focused on resolving the complex issues of the family and subfamily classification of the Gelechioidea (Hodges, 1998; Kaila, 2004; Bucheli and Wenzel, 2005; Heikkilä *et al.*, 2014; Sohn *et al.*, 2016; Wang and Li, 2020). Phylogeny of the family itself has been comparatively neglected. Common (1994, 1997, 2000) divided the Australian fauna into six genus groups: the *Wingia* group, *Chezala* group, *Philobota* group, *Eulechria* group, *Barea* group and *Tisobarica* group. These were mainly defined based on putative synapomorphies of the male and female abdomen base and male genitalia. Common (1997) did not find convincing synapomorphies for the *Chezala* group, so recognized that this group might be paraphyletic. He left a number of genera unplaced to group (Common, 2000). He also constructed morphological data matrices for each of the genus groups he recognized amongst the Australian fauna and produced phylogenies based on these (Common, 1994, 1997, 2000). In the case of the *Wingia* group at least, he regarded the phylogenetic analysis as 'very tentative' (Common, 1994).

From a more recent perspective, these analyses are potentially compromised by several aspects of methodology: the exclusion of overseas taxa, the inclusion of a hypothetical ancestor as an outgroup and the treatment of all characters as ordered. The assumption of monophyly for all genus groups and their treatment in separate analyses are further problems. It is beyond the scope of this chapter to discuss details of these analyses; they must have cost a great deal of work and were probably undertaken only reluctantly by their author, who was aware of their limitations. Without further work, including molecular analyses covering a range of Australian and overseas taxa, we cannot know how useful their tentative results really are.

The study of Kim *et al.* (2016) aimed more explicitly to resolve the composition of Oecophoridae and relationships within the family, with a special emphasis on evolution of larval feeding modes. Sequences from eight genes were used to reconstruct the phylogeny of 82 species of Gelechioidea. However, only 28 of these were Oecophoridae (*s.s.*), and half of those were species of the genus *Promalactis*. Only three Australian taxa of Oecophoridae (*s.s.*) were included in this analysis: *Wingia aurata* (Fig. 12.1) (*Wingia* group of genera), *Philobota* sp. and *Palimmeces* sp. (both belonging to the *Philobota* group). All of these genera were recovered as belonging to a single clade, together with the Holarctic *Borkhausenia*, the cosmopolitan pests *Hofmannophila* and *Endrosis*, and the New Zealand *Tingena* (a member of the *Barea* group according to Hoare, 2005). This clade was sister to a clade that included the northern hemisphere (Palaearctic) genera *Deuterogonia*, *Harpella*, *Oecophora* and *Callimodes* and the mostly Oriental *Promalactis*. This result is consistent with that of Sohn *et al.* (2016) who recovered the Australian genera *Wingia*, *Mimobrachyoma* and *Leistarcha* as sister to *Deuterogonia* + *Promalactis*. Clearly, from an Australian perspective, there is room for much more work on phylogenetic relationships within the Oecophorinae.

Fig. 12.1. *Wingia aurata*. Photo used with permission from D. Hobern.

12.6 Biology and Life Histories

A superb summary of the biology of the subfamily was provided by Common (1994), and this section draws on that account, as well as on the more detailed information given by Common (1994, 1997, 2000) in his accounts of each genus.

The life history of the vast majority of Australian Oecophorinae is unknown but from the numerous species that have been reared, it is clear that the strongly favoured larval pabulum is leaf litter, more specifically the dead sclerophyllous leaves of dry-fruited Myrtaceae, especially *Eucalyptus* species (Common, 1994). A number of species (especially in the *Wingia* group) are known to feed on living leaves of Myrtaceae, but as pointed out by the late

J.S. Dugdale (pers. commun.), the distinction between a mature live leaf and a dead leaf in these sclerophyllous plants may be rather minor from a digestive point of view. Indeed, different species of the same genus may exploit living or dead myrtaceous leaves, as is the case with *Heteroteucha* (Fig. 12.2) (Common, 1994). In order to exploit this resource, the larvae must have evolved special adaptations to deal with the low nitrogen content and high concentration of phenolic compounds (e.g., tannins) in these myrtaceous leaves. The most likely scenario involves adoption of a specialized gut flora of microbes (Common, 1994). Having achieved this evolutionary breakthrough, they had access to an extremely abundant food source following the aridification of Australia from the Miocene onwards and the increasing dominance of sclerophyllous vegetation (see further discussion in Chapter 4).

The default larval behaviour, as in so many Gelechioidea, appears to be spinning leaves together to form a shelter and feeding from within this. Larvae may occasionally be gregarious, as in *Ageletha* on *Eucalyptus* and *Tanyzancla argutella* on *Kunzea* (Myrtaceae) or *Monotoca* (Epacridaceae), with several sharing a communally spun webbing.

Larvae of a number of genera have adopted a case-making habit. This is of interest, since larval case making can be almost universal within certain families of Lepidoptera – Incurvariidae, Psychidae and Coleophoridae, for example – whereas in others it occurs much more sporadically – Tineidae, Oecophoridae, Xyloryctidae, Gelechiidae and Tortricidae are examples. Larvae of *Garrha* construct their flattened, oval cases from two pieces of dead *Eucalyptus* leaf fixed one above the other. By climbing clear of the ground and orienting one edge of the case towards the sun, the larva is able to minimize its exposure to the extreme heat in the arid areas that this genus often inhabits (Common, 1990). Larvae of *Chrysonoma fascialis* have a similar mode of case construction but they feed on green (mainly juvenile) leaves of *Eucalyptus* or *Lophostemon* (Common, 1990, 1997); other members of this genus form shelters but not portable cases (Common, 1997). The known larvae of *Hemibela* (nine of the 26 species) have the remarkable habit of constructing a portable case not from the live *Eucalyptus* leaves on which they feed but by cutting off and hollowing out a section of twig. Progressively larger twigs are utilized as the larva grows. The case is attached at right angles to a leaf and the larva then eats a hole through the leaf to graze on the other surface. The head and anal plate are heavily sclerotized and allow the larva to block both ends of the case when it is disturbed. Pupation takes place in the case, which is attached to a branch or trunk (Common, 1994). In *Pycnocera hypoxantha*, the larval case is constructed from a rolled piece of *Eucalyptus* bark and the larva feeds on dead leaves (Common, 1990, 1997); *Stictochila* larvae have very similar habits (Common, 1994). Remarkable cases more or less resembling snail shells are made by larvae of the genus *Aristeis*; these are formed from small, overlapping pieces of their live myrtaceous host leaves bound to an inner helical silk tube containing the larva. As in *Hemibela*, the head capsule of *Aristeis* larvae is strongly sclerotized and can be used to block the open end of the case. Host plants in Australia include *Eucalyptus*, *Leptospermum*, *Lophostemon*, *Melaleuca* and *Lithomyrtus* (as *Fenzlia* in Common, 1997). As noted above, Indian members of *Aristeis* feed on *Syzygium* (Common, 1997).

Fig. 12.2. *Heteroteucha dichroella*. Photo used with permission from D. Hobern.

Some species, such as *Eochrois callianassa*, only construct a case for pupation, cutting out part of its larval shelter between two leaves of the host *Banksia* (Proteaceae) and attaching this with silk to the host plant (Common, 1994). The larva of *Syringoseca mimica* (Meyrick) makes its case for pupation from a fresh *Eucalyptus* leaf (not from the larval spinning) and drops to the ground in this (Common, 1994). Other examples of larvae making cases only for pupation are those of *Ironopolia* on dead leaves of *Eucalyptus* and *Parocystola* on live phyllodes of *Acacia* (Common, 1994).

Larvae of *Myrascia* are remarkable in possessing a diverticulum of the foregut in which they store the essential oils from their myrtaceous food plants (Common, 1994; Common and Bellas, 1997). These are then regurgitated as a defensive measure when the larva feels threatened. The diverticulum is possibly unique amongst Lepidoptera as a whole, and is shed, complete with contents, on pupation (Common, 1994, fig. 680). *Myrascia bracteatella* feeds on living leaves of *Leptospermum* spp. from a portable case that is enlarged as the larva grows (Common, 1994), while *M. megalocentra* initially spins together leaves of *Melaleuca* spp. with silk, and only makes a case in the final instar.

Species of the genus *Telanepsia* have taken the habit of feeding on dead myrtaceous leaves one step further by selecting those leaves that have already passed through the gut of a herbivorous mammal; that is, they specialize on marsupial droppings. Four species were described by Common and Horak (1994); three of these (*T. stockeri*, *T. scatophila* and *T. tidbinbilla*) were reared from scats of koalas and larvae of these completed their development within a single scat, pupating therein. Larvae of the fourth species, *T. coprobora*, feeding in scats most likely from brushtail possums in western Victoria, were soil dwelling and seemed to come up at night to feed on the scats when moisture conditions were suitable. They pupated in the sandy soil. *Scatochresis* also has at least two species with larvae feeding in marsupial scats, and *Oxythecta* (a genus of 20 known species) has at least four species with this habit (Common, 1997). The genus *Ioptera* has been reared from wallaby as well as tree kangaroo scats (Cocking *et al.*, 2022).

A few genera have saproxylic larvae; that is, they feed in dead wood. The most notable example is the diverse genus *Barea*, with 103 named and 182 unnamed species (Common, 2000). Larvae generally feed under the bark in silk tunnels and, like other saproxylic larvae, are probably more or less dependent on the fungal hyphae in the wood for their nutrition (Common, 2000; Hoare, 2010b). Saproxylic moths are rarely host specific to a particular species of dead wood, so, for example, *Barea exarcha* has been reared in New Zealand from dead stems of the exotic *Ulex europaeus* (gorse). As might be expected in such a large genus, not all species have a similar life history, and one unnamed *Barea* with a brachypterous female has been reared in New South Wales from soil-dwelling larvae feeding on native pasture plants, presumably including grasses (Common, 2000).

Some species of *Philobota* (Fig. 12.3) also have soil-dwelling larvae that feed on live foliage of grasses and herbs, although, as for other oecophorines, myrtaceous leaf litter is a more typical pabulum for this huge genus (464 named and unnamed species recognized by Common, 1997). The pasture-feeding species include *P. productella*, *P. agnesella*, *P. pilipes*, *P. chionoptera*, *P. diaereta* and *P. physaula*. Most of these belong to the *P. hydara* species group of Common (1997), except for *P. productella* (*P. laxeuta* group) and *P. agnesella* (*P. arabella* group). Typical larval behaviour is as described by Swan (1941) for a species identified by Turner as *P. productella* (Common, 1997): larvae live in vertical tunnels about 8 cm long in the soil, cutting grass from around the tunnel entrance at night and dragging it into the tunnel in a manner similar to that of pasture-feeding Hepialidae. Clover (*Trifolium* spp.) may also be included in the diet. The tunnel entrance is raised above ground level by a small 'chimney' of grass particles bound with silk. Larvae seem to be particularly

Fig. 12.3. *Philobota* sp. Photo used with permission from D. Hobern.

damaging in areas of sown pasture such as lawns and golf fairways, and less prevalent in natural grasslands (Swan, 1941); areas of wetter climate and heavier soils seem also to be favoured. Pupation is in the tunnel. Two of these grass-feeding *Philobota* species (*P. chionoptera* and an unnamed species of the *P. hydara* group) have become established in the North Island of New Zealand in recent years; the unnamed species has been reported feeding on ryegrass (*Lolium perenne*) and tall fescue (*Festuca arundinacea*) (Gerard, 2013; Hoare and Hudson, 2018). There have been few records of them since they arrived and they do not seem to have spread far or become noticeable pests in the country.

The two species of the genus *Conobrosis* exhibit a particularly unusual life history, with larvae feeding in the dead male cones of cycads (*Macrozamia* spp.), but only of those species that retain the cones after the pollen has been shed (Common, 1997). Larvae are gregarious and numerous adults usually emerge from a single cone.

Lichen feeding is a specialized form of fungivory but has been reported for a few Australian Oecophorinae. Larvae of *Psaroxantha calligenes* were reported by Meyrick (1886) to feed on lichens on wooden fences from a portable case. He also stated that *Zonopetala decisana* larvae made oval cases from fragments of dead wood and fed on lichen particles as well as the wood (Meyrick, 1889). Other species of *Zonopetala* (Fig. 12.4) have the more standard habit of feeding on dead *Eucalyptus* leaves (Common, 1994).

Of special interest are those Oecophorinae that have adapted to complete their life cycles in birds' nests. Both of the cosmopolitan pest species *Hofmannophila pseudospretella* and *Endrosis sarcitrella* make use of this microhabitat but are by no means confined to it. The endemic *Boroscena phaulopis*, the only known member of its genus, has been reared on a few occasions from nests of finches, especially the red-browed finch (*Neochmia temporalis*), which are constructed from grass and lined with feathers (Common, 1997). The most famous Australian examples are the three species of *Trisyntopa*, whose larvae live in the nests of parrots (Psittacidae). The bird collector William McLennan provided the first details of the life history of *T. scatophaga* (White, 1922), which inhabits the nest hollows of the (now endangered) golden-shouldered parrot

Fig. 12.4. *Zonopetala clerota*. Photo used with permission from D. Hobern.

(*Psephotellus chrysopterygius*) in termite mounds. Larvae emerge from silk galleries to feed on the droppings of nestlings and, unusually for a coprophagous moth, are reported to begin tucking into these as soon as they are voided (White, 1922; Turner, 1923). Pupation is in silk cocoons spun gregariously on the side of the nest chamber nearest the entrance. E.D. Edwards found that cocoons of *T. scatophaga* can produce moths up to 5 years after collection, indicating the potential of the moth to diapause as larva or pupa and await the host's reuse of the nest chamber in successive seasons (Common, 2000).

More recently, a new species, *T. neossophila*, was described from nests of the hooded parrot (*Psephotellus dissimilis*), another termite mound nester (Edwards *et al.*, 2007). Its biology seems to be largely similar to that of *T. scatophaga* and has been reported on in considerable detail by Cooney *et al.* (2009). The adults are also very similar but may be distinguished on male genitalia characters (Edwards *et al.*, 2007). As noted by Edwards *et al.* (2007), there used to be a third termite mound-nesting *Psephotellus*, the now-extinct paradise parrot (*Psephotellus pulcherrimus*) of south-east Queensland, last seen in 1927. Following the discovery of *T. neossophila*, it is plausible that the paradise parrot had its own unique *Trisyntopa*, now lost forever along with its host. The larvae of the third *Trisyntopa* species, *T. euryspoda*, live in the nests of the much commoner eastern rosella (*Platycercus eximius*) in tree hollows, but have also been reported from a nest of the mulga parrot (*Psephotellus varius*); their habits are similar to those of the other species. All *Trisyntopa* species are reported to keep the nests of their hosts clean and it has been thought this may be favourable for the parrots. This probably applies chiefly to *T. euryspoda* and *T. neossophila*, as they seem to be nearly constant nest companions of their host species. Most nests of the hosts of *T. scatophaga* lack the moth (Edwards *et al.*, 2007), and the eastern rosella is well established and appears to be doing very well in New Zealand even in its absence!

Fig. 12.5. *Arachnographa micrastrella*. Photo used with permission from D. Hobern.

12.7 Adult Coloration: Crypsis, Mimicry and Aposematism

Adults of Australian oecophorines exhibit a huge range of colour patterns, wing shapes and resting postures, as would be expected for such a diverse group. There is an unusual preponderance of yellow, orange and pink coloration; in many cases, this is probably related to camouflage amongst myrtaceous leaf litter. The sample shown on Plate 4 (figs 7–24) of *Moths of Australia* (Common, 1990) gives a good impression of the range. A leaf-like effect may be enhanced by having a falcate wing shape (i.e. with the forewing apex slightly produced and hooked): the strongest examples of this belong to the *Wingia* group and include *Wingia aurata* (Fig. 12.1), *Ancistromorpha phryganophanes*, *Deigmoesta* spp. and *Arachnographa* spp. (Fig. 12.5) (Common, 1994). In some *Tachystola* species (*Chezala* group), such as *T. anthera* and *T. acroxantha*, the forewing has a falcate appearance due to a pallid semi-circle of fringe that does not extend to the tornus.

Bright yellow or orange hindwings are relatively frequent in the Australian oecophorines; in other moths, these have often been attributed to 'flash coloration'. A bird may easily be able to keep the bright colour in sight while chasing a moth on the wing, but as soon as the moth settles and conceals its hindwings under the more cryptic forewings, the 'search image' is lost (e.g. Walsingham, 1890). It is remarkable that amongst the endemic New Zealand Oecophoridae, which seem to be closely related to their Australian counterparts, not a single species is known with yellow hindwings, although many species are at least partially diurnal and insectivorous birds would have been abundant in New Zealand forests before human colonization. Intriguingly, hindwing colour may be variable within a single species, as is the case for *Sphyrelata amotella* in which the hindwing may have a bright orange base or be entirely dull greyish (Common, 2000); both forms occur in New Zealand, where the moth has been accidentally introduced.

A form of concealment that has evolved repeatedly amongst Lepidoptera worldwide is bird-dropping mimicry. There are excellent examples amongst Australian Oecophorinae in the genera *Callimima* and *Piloprepes* (Figs 12.6 and 12.7). In both cases, the realism is enhanced by the presence of irregularly scattered pure white scaling running across one of the dark forewing bands, suggesting a slightly more liquid and reflective area of the dropping. Interestingly, although *Callimima* exhibits a flat resting posture, with the wings parallel to the substrate,

Fig. 12.6. *Piloprepes anassa*. Photo used with permission from D. Hobern.

Fig. 12.7. *Piloprepes antidoxa*. Photo used with permission from D. Hobern.

Piloprepes species rest with the wings slightly domed (Fig. 12.6) (Common, 1994).

Clear examples of bark mimicry are found, for example, in the genera *Anomozancla*, *Phloeograptis* and *Phloioletes*. Here the forewings are dull greyish with numerous black longitudinal streaks; the moths are adapted to rest concealed on the trunks of rough-barked *Eucalyptus* species (Common, 1997, 2000). At the other end of the scale, species of the genus *Lophopepla* exhibit an extraordinary forewing pattern of bright red on a pale background, often with a few large silvery white spots (Common, 1994). They also have small tufts of raised scales. It is not clear whether this pattern is aposematic or whether it may be cryptic against certain backgrounds.

Diurnal species are often characterized by the small compound eyes, and by brightly patterned, often partly metallic and reflective wings. Rather few genera consistently exhibit these traits, and these are often rare in collections. Examples are *Trichomoeris*, *Petalanthes*, *Aspasiodes* and *Metaphrastis* (Common, 1997, 2000). Colours in some of these species may be aposematic, and Common (1997) speculated that *Trichomoeris*, with its dark coloration and whorls of expanded scales on the midlegs and hindlegs, might be an effective mimic of certain wasps; this genus occurs in the Northern Territory and north-east Queensland. *Metaphrastis* species have been found visiting native flowers by day; this genus is the only one currently referred to Oecophorinae in Australia that possesses simple eyes (ocelli) above the compound eyes. Common (2000) mentions two endemic New Zealand genera with ocelli; these have since been synonymized under the senior name *Hierodoris* (Hoare, 2005), and this genus is now accepted as belonging to Xyloryctidae (Kaila *et al.*, 2011; Kaila, 2013). Accordingly, some doubt must exist over the continued assignment of *Metaphrastis* to Oecophorinae.

Male scent scales (androconia) are frequent in many groups of Australian moths (e.g. Nepticulidae; Hoare and van Nieukerken, 2013), but seem to be rare amongst the Oecophorinae. An exception is *Guestia chaetophora*, which has an expansible hair pencil of long, slender, white scales at the base of the male hindwing (Common, 2000). Brachyptery (wing reduction) is also very rare and known only in females. Examples reported by Common (1994, 1997, 2000) are a few *Philobota* species (see Common, 1997, fig. 544), *Micropeteina dryinodes* and an unnamed *Barea* species (referred to above in section 12.6). In common with many other species (worldwide) which have brachypterous, flightless females, *M. dryinodes* emerges in the cooler months of the year from late autumn into winter. The female forewings are elliptical and apically acute, and the hindwings minute (see Common, 2000, fig. 564).

12.8 Economic Importance

Very few Oecophoridae are recognized as pests but the family does contain two of the most common and widespread household moths in the world: the brown house moth, *Hofmannophila pseudospretella*, and the white-shouldered house moth, *Endrosis sarcitrella*. Both of these are present in Australia but surprisingly, their full distribution has not been well documented (Common, 1997). Records on the online iNaturalist platform suggest that both species are concentrated along the eastern rim of the country, and are relatively sparse in South Australia and Western Australia, and very sparse or absent from the arid interior and most of the tropical north. These moths are notorious for the catholic tastes of their larvae. Both species have been reported feeding on dried plant material including stored cereals and seeds, on fur of dead animals, on museum insect specimens, in birds' nests and on wool and carpets (Clarke, 1941; Common, 1997). They rarely seem to build up very large populations and are therefore generally less damaging than, for example, clothes moths of the genus *Tinea* (Tineidae) or than phycitine stored product pests such as *Plodia interpunctella* (Pyralidae).

The wild origins of these two oecophorid species have never been definitively traced; they have been transported around the world by humans and are now almost cosmopolitan. A second supposed *Endrosis* species, *E. braziliensis*, has been shown by Becker (2024) not to belong in this genus. A third species, *E. psammodora*, has been described from South Africa but since examination of the genitalia is crucial for classification of Oecophoridae and Meyrick never made dissections, its generic placement needs checking (Common, 1997). The recent description of an almost certainly endemic Chilean species of *Endrosis* (*E. apablazai*) in Urra *et al.* (2023) suggests that *E. sarcitrella* may in fact originate from southern South America. Larval characters quoted by Clarke (1941) suggest *Hofmannophila* and *Endrosis* may be related, so it seems plausible that *Hofmannophila* could also be South American; Meyrick (1895) had already suggested an American origin. It remains a monotypic genus.

There are not many reports of economic damage caused by endemic Oecophorinae. Pasture damage by *Philobota* species was alluded to above. Moore (1964) recorded larvae of *Tortricopsis semijunctella* (Fig. 12.8) boring into and killing shoots of young *Pinus radiata* and also feeding on foliage of this tree and of *Cupressus*; the species normally feeds on dead *Eucalyptus* leaves (Common, 1994). He also found larvae of *Arachnographa micrastrella* (Fig. 12.5) feeding on needles of several species of *Pinus* as well as on *Juniperus* and *Hakea*. Damage to structural timber, posts and poles caused by *Barea* spp. was discussed by Moore (1959); the chief culprit was *B. consignatella* and it was mainly wood in damp situations that sustained damage. Such damage, however, may have been exceptional, as it occurred during especially high rainfall in central New South Wales between 1953 and 1956 (Moore, 1959; Common, 1994). Of course, climate change is beginning to alter our perspective of what is 'exceptional', and extreme weather conditions are becoming more frequent.

12.9 Australian Oecophorinae Overseas

Moths with detritivorous larvae are relatively easily transported accidentally with plant material. Although there may be some degree of host plant specialization (e.g. restriction to dead leaves of Myrtaceae), this is likely to be less

Fig. 12.8. *Tortricopsis semijunctella*. Photo used with permission from D. Hobern.

stringent than for their oligophagous counterparts feeding on living plant material. After arrival in a new locality outside their native range, they thus have a potential advantage over specialist phytophages in their ability to find suitable substrates on which to lay their eggs and establish a population. It is therefore no surprise that, for example, many of the adventive Lepidoptera recently established as breeding species in New Zealand belong to the detritivore guild, with Oecophoridae and Tineidae being prominent among them (Hoare, 2001; Hoare and Hudson, 2018).

New Zealand has accumulated no less than 14 Australian species of Oecophoridae, all of which appear to be well established (Dugdale, 1988; Hoare, 2001; Hoare and Hudson, 2018), apart from *Heteroteucha dichroella*, of which there is only a single, old record (Hoare, 2001, 2005). There are no fewer than five species of the dead wood-feeding genus *Barea*: *B. codrella*, *B. confusella*, *B. consignatella*, *B. exarcha* and an unnamed species related to *confusella*. All of these are fairly common in their adopted country, except *B. codrella* which is rather scarce. It is interesting that a different, Tasmanian species, *B. asbolaea*, has recently become established in the south-west of England (Cornwall and the Isle of Wight), where its life history has been reported on in impressive detail by Heckford and Beavan (2020). *Philobota* has two recently established species (see section 12.6 above). The third genus with more than one Australian adventive in New Zealand is *Tachystola*, discussed below. The remaining species are an unnamed *Atalopsis*, *Leptocroca sanguinolenta*, *Opsitycha squalidella* and *Sphyrelata amotella*.

There is one more possibly Australian species in the country: '*Trachypepla*' *indolescens*. As discussed by Hoare (2005), this is the only representative of the *Wingia* group of genera established in New Zealand, and it is certainly wrongly placed in *Trachypepla* (a *Barea* group genus endemic to New Zealand). Rather, it seems to belong to an unnamed genus closely related to the endemic Australian *Limothnes*, *Basiplecta* and *Idiozancla*. Because of this, and because it has shown a pattern of increasing abundance after being discovered in New Zealand relatively late, it seems plausible that it is an overlooked Australian species, in spite of having apparently never been collected in Australia (Hoare, 2005).

Tachystola is unique in that the two species established in New Zealand, *T. acroxantha* and *T. hemisema* (*s.l.*), have also turned up elsewhere in the world. The former species has been known for a long time from the south-west of England (Clutterbuck, 1910) but it has since spread east and north (Agassiz, 2002). The latter, a recent arrival in New Zealand (Hoare and Hudson, 2018), is also recorded from California and Hawaii (Sterling *et al.*, 2023). A moth discovered by amateur lepidopterist Barbara Mulligan in west London in 2021 was at first thought to be *T. hemisema*, but further study showed that specimens under this name represent a species complex, and differences in DNA barcode and morphology indicated that the London moth was an undescribed species. This was named as *Tachystola mulliganae* by Sterling *et al.* (2023). These authors managed to match the London material with a specimen from Western Australia in the Meyrick collection in the Natural History Museum, London, and this was designated as holotype. It has not yet been confirmed whether the New Zealand and North American specimens represent the true *T. hemisema*.

Postscript: A Note on the Genus *Mermeristis* Meyrick, 1915

Meyrick (1915) described *Mermeristis* (type species *M. spodiaea*) on the basis of a single specimen supposedly from Georges Bay, Tasmania. Ever since then, this confusing genus of small brown oecophorines has been the subject of understandable misinterpretation. Hudson (1939) recorded the type species, *M. spodiaea*, from Wellington, New Zealand (Wilton's Bush and Wainuiomata), and he quoted Meyrick as saying that its occurrence in this country was probably natural. Later, Dugdale (1988) pointed out the great similarity between the genitalia of *M. spodiaea* and those of *Trachypepla ocneropis* (type locality Nelson), listing both species as occurring in New Zealand. In his revision of Australian Oecophorinae, Common (2000) redescribed the genus in detail, and transferred *ocneropis* to *Mermeristis*. He implied that all New Zealand records of *spodiaea* (including Hudson's from Wellington) referred in fact to *ocneropis*. He also considered that the occurrence of the genus on both sides of the Tasman Sea was probably natural. Not a single specimen of *Mermeristis*, however, has ever been taken in Australia since the original capture.

I suggest here that the holotype was mislabelled and that *M. spodiaea* is endemic to New Zealand. First, at least three species of *Mermeristis* are now recognized in the New Zealand fauna (including one or two unnamed species). It is most unusual in Lepidoptera genera shared between Australia and New Zealand for New Zealand to have the greater number of taxa. If *Mermeristis* is discounted, the only genus of Oecophorinae naturally shared between the two countries is *Prepalla*, and this has only one New Zealand species but nine in Australia (Common, 1997). Second, *M. spodiaea* can now be confirmed as present in New Zealand, based on numerous specimens matching the holotype in genitalia and external appearance (specimens and slides in NZAC, Auckland). It occurs almost throughout the North Island but may be most common around Wellington.

This is particularly interesting as, while Meyrick was indeed at Georges Bay, Tasmania, on 3 January 1885 (the label date on the holotype of *spodiaea*), on 3 January 1886 he was in Wellington. His diary of captures makes no mention of the species on either date (K. Sattler, pers. commun.), so this does not help. It seems likely that Meyrick labelled the specimen at first with a date only, relying on his diary to supply the locality details at a later stage. Often, in the first few days of a new year, one persists in writing the previous year's date through force of habit, and I suggest that on this occasion, Meyrick wrote '1885' instead of '1886'. When he came to write out the full label (perhaps years later; it was, after all, 30 years before he described the species), he did not recall the circumstances of capture so he did not notice the dating error and wrote the wrong locality on the label. Even a worker as meticulous as Meyrick could easily have made such a mistake.

Although the evidence is circumstantial, I suggest that *Mermeristis* be removed from the list of Australian Lepidoptera, until such time as further proof of its presence is forthcoming.

References

Agassiz, D.J.L. (2002) The invasion and spread of Lepidoptera in Britain. *Bulletin de l'Institut Royale des Sciences Naturelles de Belgique* 72 (supplement), 159–164.

Becker, V.O. (2024) The identity of *Endrosis braziliensis* Moore, 1883 (Lepidoptera: Oecophoridae, Oecophorinae). *SHILAP Revista de Lepidopterologia* 52, 143–147.

Bucheli, S.R. and Wenzel, J.W. (2005) Gelechioidea (Insecta: Lepidoptera) systematics: A reexamination using combined morphology and mitochondrial DNA data. *Molecular Phylogenetics and Evolution* 35, 380–394.

Clarke, J.F.G. (1941) Revision of the North American moths of the family Oecophoridae, with descriptions of new genera and species. *Proceedings of the United States National Museum* 90, 33–286.

Clutterbuck, C.G. (1910) *Ocystola aethopis* Meyrick in Devonshire. *Entomologist* 43, 96–97.

Cocking, G., Bond, S.J. and Edwards, E.D. (2022) *Moths in the A.C.T.* Glenn Cocking, Canberra, ACT.

Common, I.F.B. (1990) *Moths of Australia*. Melbourne University Press, Melbourne, Victoria.

Common, I.F.B. (1994) *Oecophorine Genera of Australia I. The* Wingia *Group (Lepidoptera: Oecophoridae)*. Monographs on Australian Lepidoptera 3. CSIRO Publishing, Collingwood, Victoria.

Common, I.F.B. (1996) Oecophoridae. In: Nielsen, E.S., Edwards, E.D. and Rangsi, T.V. (eds) *Checklist of the Lepidoptera of Australia*. Monographs on Australian Lepidoptera 4. CSIRO Publishing, Collingwood, Victoria, pp. 59–89.

Common, I.F.B. (1997) *Oecophorine Genera of Australia II. The* Chezala, Philobota *and* Eulechria *groups (Lepidoptera: Oecophoridae)*. Monographs on Australian Lepidoptera 5. CSIRO Publishing, Collingwood, Victoria.

Common, I.F.B. (2000) *Oecophorine Genera of Australia III. The* Barea *Group and Unplaced Genera*. Monographs on Australian Lepidoptera 8. CSIRO Publishing, Collingwood, Victoria.

Common, I.F.B. and Bellas, T.E. (1997) Regurgitation of host-plant oil from a foregut diverticulum in the larvae of *Myrascia megalocentra* and *M. bracteatella* (Lepidoptera: Oecophoridae). *Australian Journal of Entomology* 16, 141–147.

Common, I.F.B. and Horak, M. (1994) Four new species of *Telanepsia* Turner (Lepidoptera: Oecophoridae) with larvae feeding on koala and possum scats. *Invertebrate Taxonomy* 8, 809–828.

Cooney, S.J.N., Olsen, P.D. and Garnett, S.T. (2009) Ecology of the coprophagous moth *Trisyntopa neossophila* Edwards (Lepidoptera: Oecophoridae). *Australian Journal of Entomology* 48, 97–101.
Dugdale, J.S. (1988) Lepidoptera – Annotated catalogue, and keys to family-group taxa. *Fauna of New Zealand* 14. Manaaki Whenua Press, Lincoln, New Zealand.
Edwards, E.D., Cooney, S.J.N., Olsen, P.D. and Garnett, S.T. (2007) A new species of *Trisyntopa* Lower (Lepidoptera: Oecophoridae) associated with the nests of the hooded parrot (*Psephotus dissimilis*, Psittacidae) in the Northern Territory. *Australian Journal of Entomology* 46, 276–280.
Gerard, P.J. (2013) Observations on *Philobota* sp. (Lepidoptera: Oecophoridae) near Hamilton. *New Zealand Plant Protection* 66, 148–152.
Heckford, R.J. and Beavan, S.D. (2020) Discovery in Cornwall, England, of the larva of the Tasmanian species *Barea asbolaea* (Meyrick, 1883) (Lepidoptera: Oecophoridae), together with an account of all the early stages. *Entomologist's Gazette* 71, 75–92.
Heikkilä, M., Mutanen, M., Kekkonen, M. and Kaila, L. (2014) Morphology reinforces proposed molecular phylogenetic affinities: a revised classification for Gelechioidea (Lepidoptera). *Cladistics* 30, 563–589.
Hoare, R.J.B. (2001) Adventive species of Lepidoptera recorded for the first time in New Zealand since 1988. *New Zealand Entomologist* 24, 23–47.
Hoare, R.J.B. (2005) *Hierodoris* (Insecta: Lepidoptera: Gelechioidea: Oecophoridae), and overview of Oecophoridae. *Fauna of New Zealand* 54. Manaaki Whenua Press, Lincoln, New Zealand.
Hoare, R.J.B. (2010a) Lepidoptera. In: Gordon, E.P. (ed.) *New Zealand Inventory of Biodiversity. Kingdom Animalia: Chaetognatha, Ecdysozoa, Ichnofossils*. Canterbury University Press, Christchurch, New Zealand.
Hoare, R.J.B. (2010b) *Izatha* (Insecta: Lepidoptera: Gelechioidea: Oecophoridae). *Fauna of New Zealand 65*. Manaaki Whenua Press, Lincoln, New Zealand.
Hoare, R.J.B. and Hudson, N. (2018) Adventive moths (Lepidoptera) established in mainland New Zealand: Additions and new identifications since 2001. *Australian Entomologist* 45, 273–324.
Hoare, R.J.B. and van Nieukerken, E.J. (2013) Phylogeny and host plant relationships of the Australian Myrtaceae leafmining moth genus *Pectinivalva* (Lepidoptera, Nepticulidae), with new subgenera and species. *ZooKeys* 278, 1–64.
Hodges, R.W. (1998) *The Gelechioidea*. In: Kristensen, N.P. (ed.) *Handbook of Zoology, Vol. IV, Arthropoda: Insecta, part 35, Lepidoptera, Moths and Butterflies*. Springer, Berlin, pp. 131–158.
Hudson, G.V. (1939) *A Supplement to the Butterflies and Moths of New Zealand*. Ferguson & Osborn Ltd, Wellington, New Zealand, pp. 387–481.
Kaila, L. (2004) Phylogeny of the superfamily Gelechioidea (Lepidoptera: Ditrysia): An exemplar approach. *Cladistics* 20, 303–340.
Kaila, L. (2013) Identity of *Eumenodora encrypta* Meyrick, a cryptic Australian moth (Lepidoptera: Gelechioidea). *Zootaxa* 3616, 165–172.
Kaila, L., Mutanen, M. and Nyman, T. (2011) Phylogeny of the mega-diverse Gelechioidea (Lepidoptera): Adaptations and determinants of success. *Molecular Phylogenetics and Evolution* 61, 801–809.
Kim, S., Kaila, L. and Lee, S. (2016) Evolution of larval life mode of Oecophoridae (Lepidoptera: Gelechioidea) inferred from molecular phylogeny. *Molecular Phylogenetics and Evolution* 101, 314–335.
McMillan, I. (2013) Xyloryctine Moths of Australia. Genera and Species. Available at: https://xyloryctinemothsofaustralia.blogspot.com/p/xyloryctinae.html (accessed 17 February 2025).
Meyrick, E. (1883) Descriptions of Australian Micro-lepidoptera. VIII. Oecophoridae. *Proceedings of the Linnean Society of New South Wales* 7, 415–547.
Meyrick, E. (1886) Descriptions of Australian Micro-lepidoptera. XII (continued). Oecophoridae. *Proceedings of the Linnean Society of New South Wales* 10, 765–832.
Meyrick, E. (1889) Descriptions of Australian Micro-lepidoptera. XV. Oecophoridae. *Proceedings of the Linnean Society of New South Wales* 3, 1565–1703.
Meyrick, E. (1895) *A Handbook of British Lepidoptera*. Macmillan & Co., London.
Meyrick, E. (1902) Descriptions of new species of Lepidoptera (Oecophoridae). *Transactions of the Royal Society of South Australia* 26, 133–174.
Meyrick E. (1912–1930) *Exotic Microlepidoptera* v. 1–3.
Meyrick, E. (1915) *Exotic Microlepidoptera* 1, 289–320.
Meyrick, E. (1922) Lepidoptera Heterocera. Fam. Oecophoridae. *Genera Insectorum* 180, 1–224.
Moore, K.M. (1959) Observations on some Australian forest insects. 3. The biology and larval taxonomy of some Lepidoptera attacking trees and timber. *Australian Zoologist* 12, 337–350.

Moore, K.M. (1964) Observations on some Australian forest insects. 16. Lepidoptera attacking *Pinus* spp. in New South Wales. *Proceedings of the Linnean Society of New South Wales* 88, 340–360.
Newman, E. (1856) Characters of a few Australian Lepidoptera, collected by Mr Thomas R. Oxley. *Transactions of the Entomological Society of London* 3, 281–300.
Sohn, J.C., Regier, J.C., Mitter, C., Adamski, D., Landry, J.-F. *et al.* (2016) Phylogeny and feeding trait evolution of the mega-diverse Gelechioidea (Lepidoptera: Obtectomera): New insight from 19 nuclear genes. *Systematic Entomology* 41, 112–132.
Sterling, M.J., Plant, C.W., Price, B.W., Cuber, P., Hall, A. *et al.* (2023) *Tachystola mulliganae* sp. nov. (Lep.: Oecophoridae), a Western Australian species adventive in West London. *Entomologist's Record and Journal of Variation* 135, 225–241.
Swan, D.C. (1941) Insects and mites of economic importance in South Australia during the period July, 1936, to June, 1940. *Journal of Agriculture, South Australia* 44, 647–761.
Toll, S. (1964) *Lepidoptera—Oecophoridae*. Klucze do oznaczania owadów Polski. Ser. 43 (Part 27), no. 35, pp. 1–174.
Turner, A.J. (1923) A lepidopterous scavenger living in parrots' nests. *Transactions of the Entomological Society of London* 1923, 170–175.
Urra, F., Araneda, S. and Moreira, G.R.P. (2023) New species of Oecophoridae (Lepidoptera: Gelechioidea) from Chile. *Zootaxa* 5254, 209–230.
Van Nieukerken, E.J., Kaila, L., Kitching, I., Kristensen, N.P., Lees, D.C. *et al.* (2011) Order Lepidoptera Linnaeus, 1758. In: Zhang, Z.Q. (ed.) Animal biodiversity: An outline of higher-level classification and survey of taxonomic richness. *Zootaxa*, 212–221.
Walker, F. (1863) *List of the specimens of Lepidopterous Insects in the Collection of the British Museum* 28, 287–561.
Walker, F. (1864a) *List of the specimens of Lepidopterous Insects in the Collection of the British Museum* 29, 562–835.
Walker, F. (1864b) *List of the specimens of Lepidopterous Insects in the Collection of the British Museum* 30, 837–1096.
Walker, F. (1869) *Characters of Undescribed Lepidoptera Heterocera*. E.W. Janson, London.
Walsingham, L. (1890) The President's address. *Proceedings of the Entomological Society of London* 52, 48–60.
Wang, Q.-Y. and Li, H.-H. (2020) Phylogeny of the superfamily Gelechioidea (Lepidoptera: Obtectomera), with an exploratory application on geometric morphometrics. *Zoologica Scripta* 49, 307–328.
Wang, S.X. (2002) First record of the genus *Locheutis* Meyrick from China, with description of one new species (Lepidoptera: Oecophoridae). *Acta Entomologica Sinica* 45 (Supplement), 61–63.
Wang, S.X. (2004) A new species of *Locheutis* Meyrick (Lepidoptera: Oecophoridae) from China. *Entomotaxonomia* 26, 32–34.
Wang, S.X. and Tao, Z. (2021) The genus *Locheutis* Meyrick (Lepidoptera: Oecophoridae) from the Palaearctic and Oriental regions, with descriptions of four new species. *Zootaxa* 4952, 561–570.
White, H.L. (1922) Description of the eggs of the golden-shouldered parrot (*Psephotus chrysopterygius*). *The Emu* 22, 98–99.
Zeller, P.C. (1855) Genus *Cryptolechia*. *Linnaea Entomologica* 10, 145–168.
Zeller, P.C. (1877) Exotische Microlepidoptera. *Horae Societatis Entomologicae Rossicae* 13, 3–493.

13 The Xyloryctidae: The Timber Moths

Ian McMillan
Imbil, Queensland, Australia

Abstract
The Xyloryctidae are a moderately sized family of gelechioid moths with a particularly rich Australian fauna. This chapter describes the complex and convoluted taxonomic history of the Xyloryctidae from the description in 1805 of the first species now placed in the family to the modern basis for the family given by molecular genetics and integrative taxonomy in 2020. Additional notes on biology and morphology conclude the account.

13.1 Introduction

Although the insects we now call xyloryctid moths were known for millennia by the Indigenous people of Australia, who regarded them as a delicious food source, we owe the first written observations and illustrations of these moths to John Lewin (1770–1819). Lewin was an English natural history artist who came to Australia in 1800, intending to work as a natural historian and collector for patrons back in England. Lewin was granted a farm at Parramatta but soon after is recorded living in Sydney, where he became Australia's first professional artist. He was active in colonial life, being appointed coroner by Governor Macquarie, and travelled frequently, making trips to Tahiti and across the Blue Mountains in 1815 into the newly discovered territory there.

Lewin began producing not only the first works on the subject of Australian natural history but some of the earliest colonial landscape painting. The copper plates for his engravings are the earliest made in Australia. They were sent to London to be printed and coloured by his brother Thomas. In 1805 his *Prodromus Entomology, Natural History of Lepidopterous Insects of New South Wales* was published in London with a text which may have been written or assisted by the Scottish entomologist Alexander Macleay (later to become the builder of Elizabeth Bay House in Sydney) as Lewin, apparently, lacked confidence with descriptive or scientific writing. In 1808 he published *Birds of New Holland with their Natural History* (Lewin, 1808). The two books were supposed to fund his passage home but although they were scientifically significant, they were not a commercial success. Lewin remained in Australia until his death in 1819.

In his 1805 publication, Lewin recorded an extraordinary caterpillar which, like a hepialid or ghost moth, or a cossid or wood moth, lived in a burrow it made in a tree branch (Lewin, 1805). This caterpillar, however, did not feed on the tree as it made its tunnel, but emerged at night to forage for leaves. These it collected and brought

Corresponding author: icmcmi@gmail.com

DOI: 10.1079/9781800627680.0013

back, tying them to the burrow entrance with silk and feeding on them while safely back in the burrow. He records this caterpillar's status as a sought-after food among the Indigenous people, who took considerable risks climbing to the tops of trees to find them, and goes on to describe four species with similar burrowing and nocturnal larval habits.

So what distinguished this peculiar animal, apart from its residential tunnels and strange nocturnal feeding habits as a larva? The adults Lewin described were not particularly unusual or distinguished creatures. They held their wings roofwise and were easily recognizable because of their long recurved 'horns', the labial palpi, which are a set of mouthparts which are recurved back over the top of the head. Apart from that, there seemed to be nothing in particular worth noting about their appearance. It was the larval behaviour which was exceptional.

Although additional species were later found that we now assign to the Xyloryctidae, described by Newman (1856), Wallengren (1861), Scott (1864), Walker (1864), Butler (1876) and Meyrick (1885, 1886a, 1887), 85 years passed between Lewin's identification of this group of moths and the formal description of the taxon by the British entomologist and microlepidoptera specialist Edward Meyrick, in 1890, as a separate family of the Lepidoptera, the Xyloryctidae. Following this, many new species were described by Meyrick and the Australian entomologists Turner, Lower and Lucas. Moths of this family are found all over the Indo-Pacific region, from India to Hawaii, from China to New Zealand. There are about 109 genera in total, consisting of 603 species, with 46 genera and 271 species in Australia. The Australian Xyloryctidae vary from tiny moths a few millimetres long to the largest with a wingspan of over 70 mm. These are the largest gelechioids known.

The Xyloryctidae are at their most numerous in Australia, where they form a unique part of the Australian fauna. As the entomologist Diakonoff said, they are 'as Australian as Kangourou' and yet, he laments, very little is known about them. They are not easy creatures to study because of their unique life habits and much remains to be discovered about the life histories of most of them.

13.2 Taxonomic History of the Xyloryctidae

13.2.1 Establishing the family

In 1805, four moths of what would later become known as the family Xyloryctidae were first described by Lewin, in his *Prodromus Entomology*. He described three of them (*Cryptophasa irrorata*, *C. albacosta* and *C. rubescens*) according to Alexander Macleay's eccentric adaptation of the Linnaean taxonomy of the time, under the 'genus' heading *Phalaena* (night-flying moths), in the 'species' group *bombyx* (silk spinners), into which Lewin introduced two new categories, 'family' Phalaenoides and 'section' Cryptophasa. A fourth species, *C.* (now *Xylorycta*) *strigata*, was described under a different heading: Gen. Phalena. Fam. Noctua, Sect. Cryptophasa. While it may be mystifying to have the same genus in two separate families, it may be an attempt to account for morphological differences found in *strigata*, which does not have the pectinated male antennae found in *Cryptophasa*.

Swainson (Swainson and Shuckard, 1840) elaborated on Lewin's imprecise terminology, creating five divisions for the Bombycides: Hepialidae, Bombycidae, Arctiadae, Lithosiadae and Cryptophasidae. For the last of these, he introduced a new family term which he called 'hermit moths', reflecting their burrowing habits. He compared their larvae with those of the Hepialidae, noting the absence of a pupal case, as the pupa was protected by the tunnel excavated by the larva. He noted especially their long recurved palpi, which were seen as being unique to this group. Long antennae and glossy wings were also considered significant. Cryptophasidae Swainson and Shuckard, 1840, therefore, has precedence over Xyloryctidae Meyrick, 1980. Later authors were either unaware of Swainson's 1840 coining of the term (relying entirely on Meyrick's authority) or saw Kirby's 1897 use of Cryptophasidae as the primary source. The term Cryptophasidae is still current in India (see, for example, Jayanth and Nagarkatti, 1981), where it was introduced by Fletcher in 1929. In 1841, Newman reviewed Swainson and Shuckard's (1840) work, making a single reference to the family in the course of a merciless criticism of Swainson's systematics, where he mocked his association with the apparently discredited 'Macleayian or quinarian' system of classification (Newman, 1841).

Forty years later, in 1883, when Meyrick began approaching the notion that there might be a family equivalent to the Xyloryctidae, he suggested a diagnostic trait (the parallel veins 6 and 7 of the hindwings) for separating the Oecophoridae from a new family, the Cryptolechiadae (Meyrick, 1883).

In 1886 Meyrick described *Scieropepla* in the Gelechiadae, described *Gonioma* in the Gelechiadae and, in 1897, placed two new species, *Cryptophasa leucadelpha* and *C. eccleciastis*, in the family Cryptolechiadae. Before 1890, however, he was informed by Lord Walsingham that the type of *Cryptolechia* belonged in the Oecophoridae, and realized he could not use the name Cryptolechiadae. This prompted him to rename this already recognized group as the Xyloryctidae (Meyrick, 1886b, 1890, 1897).

Meyrick sought to distinguish the Xyloryctidae from the closely related Oecophoridae. The smooth head, recurved and pointed labial palpi, absence of a basal pecten on the antennae and differences in wing venation were all noted, as was a 'connecting bar between vein 8 and the cell' which some subsequent observers claimed was often absent although Meyrick insisted it was always there. These morphological characteristics may not seem to be much on which to base the separation of a family but, in addition, the unique biology of the tunnelling species and their predominantly Australian origin may have been decisive factors. Walsingham confirmed Meyrick's new family in 1891, agreeing with him that the Cryptoplechiadae was an inappropriate name. Meyrick's description of the Xyloryctidae included the Stenomatidae and the Autostichinae; Meyrick did not establish the Stenomidae until 1906 (Walsingham, 1907) and the Autostichinae were established by Le Marchand (1947). A single species of Depressariinae is included in Meyrick's concept of the Xyloryctidae. He had already established the Depressariidae in 1883.

Kirby, in 1897, made a single reference to the name Cryptophasidae, in connection with a description of the genus *Cryptophasa*, but without elaboration; this bald reference has been taken as the origin of the term by subsequent authors.

Insecurities concerning this family group persisted for some time. Somewhere between 1890 and 1897, Meyrick repudiated 'the group formerly maintained as a distinct family under the name Xyloryctidae' and returned them to the Gelechiadae. He was, however, still using the term in 1898. Blandford (1898) recounts that Walsingham maintains the use of Meyrick's term in the form of 'the Xyloryctinae, a subfamily of the *Gelechiidae*'. Turner (1898), in his review of Queensland species, returned to the use of Xyloryctidae, 'an important section of the Tineina'. He repeated Meyrick's description verbatim and noted that most differences are in the neuration of the hindwing. His contribution, along with that of Lower and Lucas, consists of the identification of many new species.

In 1900 Turner noted that Meyrick had merged the Xyloryctidae with the Gelechiadae but, since the Australian Gelechiadae were not well described, decided to continue to refer to this group as the Xyloryctidae. He published *The Xyloryctidae of Queensland*, increasing available knowledge of the genera and introducing some new species (Turner, 1900).

Writing in the United States, Dyar (1903) included the Xylorictidae (*sic*) in his classification and included within it the two North American stenomatine genera *Stenoma* and *Ide*. It was this idea of the family that formed the basis for Sir George Hampson's (1918) creation of the Uzuchidae.

Meyrick, meanwhile, was still having doubts in 1904, expressing much concern over various details of wing venation which to him separated the Gelechiadae from the Oeccophoridae. His insecurity about these details was evident but he decided to continue to make use of Xyloryctidae as a matter of convenience.

13.2.2 Defining the family

In 1906, Meyrick proposed the Stenomidae, containing *Agriophara* and *Hypeuryntis*; this decision was based on the separation of veins 7 and 8 of the forewings as a means of distinguishing them from the Xyloryctidae (Meyrick, 1906). This was followed in 1907 by Walsingham's Stenomatidae, another name for the same group. In 1914, Meyrick once more conflates the Xyloryctidae with the Stenomidae; the separation of veins 7 and 8 in the forewings was forgotten in favour of 7 and 8 being either stalked or separate (Meyrick, 1914). Indecision reigned.

In an attempt to rationalize the formation of family group names by relating them to the first described genus in each family, Hampson created the Uzuchidae in 1918 from *Uzucha* Walker, 1864 (although the earliest generic name in the Xyloryctidae after *Cryptophasa* is in fact *Boydia* Newman, 1856). He adopted Dyar's 1903 definition, including the Stenomidae. Flying in the face of the rules of priority, this system was never widely adopted, but the family Uzuchidae survived for some time in some American publications. Hampson gave no reason for abandoning *Cryptophasa* as the nominative genus of the Uzuchidae, other than suggesting that Lewin's claim to be the author was insecure, or perhaps that Meyrick's 1890 emendation had long been abandoned by 1918.

The first advance in techniques of classification that would enable further progress to be made came with the use of dissected male or female genitalia as structural indicators of phylogenetic and taxonomic traits. These structures are involved in an 'evolutionary arms race' that has resulted in rapid rates of differentiation, and as such are appropriate targets for taxonomic study. Alexei Diakonoff reviewed the family in 1954, based on Papuan species. He noted that some of the genera, based on wing venation, are most ill defined, a perception that remains with us today. Diakonoff made the first published dissections and descriptions of genitalia, noting that the male genitalia are 'of a simple and uniform type' and that lack of difference between male genitalia of different species made identification difficult. Female genitalia are described as 'oecophorid in type'. Despite the addition of genital dissection to his arsenal, Diakonoff contributed little to the redefinition of the family, and his review is primarily a critique of earlier authors, opening up possibilities for further work.

Duckworth (1964, 1973) in a comprehensive review of the Stenomidae narrowed the definition of the Xyloryctidae, both morphologically and geographically, with the Stenomidae being associated with the Americas and the Xyloryctidae with the Indo-Pacific. Duckworth noted that Meyrick (1915) merged the two families under the older name Xyloryctidae, and took this as a starting point for establishing their separate existence.

Gates Clark (1955) made the first serious attempt to understand the Xyloryctidae based on their genital structure. He dissected 30 genera in Meyrick's collection in the Natural History Museum, London; his remarks on these investigations were concerned with finding groups of related species, and finding significant characteristics upon which to base a definition of the family. His criticism of Meyrick's reliance on wing venation was severe, ridiculing it as fanatical. In a similar vein, Meyrick was dismissive of all attempts to use genital structure as a taxonomic tool and ridiculed the idea that it might be worthwhile. Gates Clark took Meyrick's assertion that there is 'a short cross-vein between the cell and Vein 8 of the hindwing' as a starting point, and managed to find some species that have this feature but found it is absent in many others. Comparable inconsistencies were noted for other supposedly characteristic patterns of wing venation. When he turned his attention to aspects of male genitalia, he found some utility when it came to the definition of genera, but concluded that the family as a whole was extremely insecure, moving some species to the Oecophoridae but leaving much in doubt. In Gates Clark's view, Meyrick had unnecessarily expanded the definition of the family to include many genera and species, both his own and those of other authors, and as a result he concluded the Xyloryctidae 'comprise a large assortment of types and representatives of several families' with 'little to recommend the close association of these genera and much less to support their relationship to the Xyloryctidae'.

Unfortunately, Gates Clark did not examine any Australian genera, with the exception of the (Indian) type of *Thymiatris*; however, because of a taxonomic controversy, he did publish examples of *Phthonerodes scotarcha*. Diakonoff (1954) records that although Fletcher (1929) had sunk a number of genera (*Lichenaula*, *Tymbophora*, *Clerarcha*, *Xylorycta* and *Chalarotona* of Meyrick, *Illidgea* and *Neopdrepta* of Turner) into *Phthonerodes* Meyrick, Gates Clark had been able to verify that *Xylorycta* and *Phthonerodes* were different genera by examining the male genitalia of the type species of both genera. It should also be noted that although he dealt extensively with the Stenomidae, Gates Clark, following Meyrick, included *Agriophara* in the Xyloryctidae. It was left to Duckworth (1964) to clarify the position of *Agriophara*.

Ian Common had been using genital morphology to address taxonomic problems within the group since at least 1964, when he dissected two species of *Phthonerodes*. In his 1970 chapter on the Xyloryctidae in *The Insects of Australia*, however, he redescribed the Xyloryctidae very much in terms of wing venation, making no reference to genital morphology. He noted that relationships among the genera are little understood. Remarks on the biology and behaviour of individual species are attached to his description of the family.

Duckworth (1964) noted that Meyrick included more and more Stenomidae into the Xyloryctidae after 1915 and, by so doing, enormously enlarged the size of the family. He maintains that Busck (1934) had kept the two-family system, but excluded all Old World moths from his Stenomidae. This contributed to the Old World Stenomidae and, in consequence, the largely Australian Xyloryctidae being left uncatalogued in his scheme of things and largely unstudied, although Duckworth notes the work of Diakonoff and Common as exceptions to his generalization.

13.2.3 Placing the family

For very tenuous reasons, the American entomologist Elwood Zimmerman (1978) decided, in his review of the Hawaiian species, to redefine the Xyloryctidae as a subfamily of the Gelechiidae. In this, he was following Meyrick (1904) who, at that time, was having doubts about his own taxonomic creation, being unable to define its characters with sufficient clarity. Ultimately this was a retrograde decision. In fact, the Xyloryctidae had hardly been studied in Zimmerman's time. Genital dissections, particularly of males, enabled him to reduce all of Walsingham's Hawaiian genera to a single genus, *Thyrocopa* of Meyrick. He found the male genitalia of all species to be remarkably similar.

Following Zimmerman, Common (1990) reduced the status of the Xyloryctidae to a subfamily, but in the Oecophoridae, not the Gelechiidae. He gave a very concise description of the subfamily, compared it with the Oecophorinae and found many similarities that enabled him to conclude that they share a common ancestry. Otherwise very little evidence was given.

Minet (1990) placed the Xyloryctidae as a family within the Gelechioidea, using 32 characters from larval, pupal and adult morphology. Kaila (2004) later found Minet's characters insufficient. Scoble (1992), Robinson *et al.* (1994) and Common (1996) all followed Common's 1990 determination and retain the Xyloryctidae as a subfamily of the Oecophoridae. Sinev (1993) used so-called morphofunctional complexes to elevate the Gelechioidea to a new infraorder, the Coleophoromorpha, containing the superfamilies Oecophoroidea, Coleophoroidea, Elachistoidea, Gelechioidea, Chrysopeleioidea, and Cosmopterigoidea. His Oecophoroidea contained the families Oecophoridae (*s.s.*), Xyloryctidae, Symmocidae, Chimabachidae and Autostichidae. Xyloryctidae was now proposed as the sister group of the Oecophoridae.

Hodges (1998) was the first to make use of the principles of phylogenetic systematics, employing a variety of morphological characters in a parsimony analysis. He saw the Xyloryctidae as a family within the Gelechioidea with two contained subfamilies, Xyloryctinae and Scythridinae. Kaila (2004) questioned his methodology.

Holloway (2001) followed Hodges in recognizing the Xyloryctidae as a separate family, while noting the subordinate position given to it by previous authors.

Kaila (2004) conducted a review of the entire Gelechioidea, based on a thorough cladistic analysis using a large number of characters. He found that the Gelechioidea were situated within the clade Obtectomera. The Gelechioidea divided into two lineages, the 'gelechiid' and the 'oecophorid'. The Xyloryctidae (*s.l.*) fall within the 'oecophorid' lineage, but were seen as a paraphyletic group incorporating some oecophorids, deuterogoniines and blastobasines. The Xyloryctidae, along with other families, received quite low support values on each of their branches, which raised the question of their monophyly. Kaila's analysis removed the Scythrididae from the Xyloryctidae.

Hoare (2005, 2010), following the approach of Common (1994, 1997, 2000), took a cautious and conservative approach to Kaila's (2004) proposed classification of the Xyloryctidae.

He acknowledged that there is evidence to suggest that the New Zealand '*Hierodoris* group' represents the basal genera of the Xyloryctidae rather than belonging in the Oecophoridae, but ultimately chose to leave them, as a group, within the Oecophoridae, regarding the data, at that time, as not comprehensive or reliable enough, proposing further phylogenetic analysis as necessary before any stable conclusions can be drawn.

The first work that combined morphological *and* genetic data to examine any of the Gelechioidea was that of Bucheli and Wenzel (2005), although this only examined North American Scythridinae. Hodges (1998) placed the Xyloryctidae as a family of the Gelechioidea containing two subfamilies, the Xyloryctinae and Scythridinae.

Zborowski and Edwards (2007), in their popular *A Guide to Australian Moths*, placed the Xyloryctidae as a separate family, without subfamilies being mentioned. In *Zootaxa*'s influential review of animal biodiversity, van Nieukerken *et al.* (2011) placed the Xyloryctidae as a family of the Gelechioidea without any subfamilies being mentioned.

Lvovsky (2012) placed the Xyloryctidae as a family within the superfamily Oecophoroidea.

Heikkilä *et al.* (2014) carried out an important and extensive study based on parsimony analyses of both morphological and genetic data, recognizing the Xyloryctidae as a separate monophyletic family (without any of the previously assigned subfamilies) and a sister group to the (Lecithoceridae + Austostichinae). They confirmed the position of the *Hierodoris* group within the Xyloryctidae and established its relationship with *Eumenodora encrypta*, a basal member of the Xyloryctidae that has no distinctive xylorictid morphological traits: *Eumenodora* has only been placed in the Xyloryctidae using genetic characters. This study, however, did not confirm the close relationship between *Scieropepla* and *Hierodoris* as was previously suggested by Hoare (2010). It also placed *Nemotyla* in the Xyloryctidae (previously placed in the Oecophoridae: Xyloryctinae or Symmocidae). The effect of Heikkilä *et al.*'s (2014) work was, first, to consolidate the Xyloryctidae's rank as a family and, second, to suggest basal relationships within the Gelechioidea. Despite the support values for a monophyletic Xyloryctidae in this study being low and only a handful of xyloryctid exemplars having been used in the analysis, this nevertheless brought an air of finality to the problem of placing the Xyloryctidae.

The results of these earlier cladistic analyses were largely confirmed by the phylogenetic analyses of Sohn *et al.* (2016) who examined five nuclear genes. Within the Gelechioidea, a 'Depressariid assemblage' was identified accomodating the Autostichidae, Xyloryctidae, Lecithoceridae, Oecophoridae, Depressariidae and Lypusidae. This grouping was one of three major divisions within the Gelechioidea. Although the branches of families Autostichidae, Xyloryctidae and Elachistidae showed better supported monophyly than those of Heikkilä *et al.* (2014), they sampled fewer terminal taxa than the latter, leaving the monophyly and inner relationships within these families unresolved. Saprophagy was concentrated within the 'AXLO clade' (Autostichidae, Xyloryctidae, Lecithoceridae, Oecophoridae). Xyloryctidae does contain saprophages but is the only family in this clade that consists mainly of live plant feeders. This is perhaps a debatable statement given the fact that most *Cryptophasa*, for instance, feed on plant material that they harvest and bring back to their burrows, where it is attached to the opening of the burrow with silk and consumed as it dries out – a kind of half-way detritivory.

In a more recent study of the phylogeny of the Gelechioidea, Wang and Li (2020) used molecular genetic and morphological data to examine representatives of gelechioid families, including the Xyloryctidae. They found an 'ALX' clade consisting of Autostichidae (*sensu nov.*), Lecithoceridae and Xyloryctidae as having high support values, and although the relationship between the Autostichidae and the Xyloryctidae was found to be strong, they concluded that without further investigation questions of monophyly could not be settled. They did find, however, a core of Australian genera around which a robust monophyletic family might be constructed. Once again, small sample sizes contributed to many ongoing problems.

Despite this small xyloryctid dataset, genetic data clearly suggested the separation of the Australian + Asian genera (*Tymbophora*, *Xylorycta*, *Metathrinca*, *Cryptophasa*, *Scieropepla*) from the New Zealand genera (*Izatha*, *Gymnobathra* (the *Hierodoris* Group of Hoare, 2005) together with the basal Australian genus *Eumenodora*), both groups sharing a common ancestor. In this paper, a new technique, geometric morphometric analysis, used data obtained from fore- and hindwing; comparison shows its phylogenetic results are in conformity with genetic analysis. Only two species, *Metathrinca tsugensis* and *Xylorycta* sp., were used in this study, which shows little more than an affinity with each other. It is to be hoped, therefore, that in future this tool and others will be used to further assist the development our understanding of the Xyloryctidae.

13.3 Morphology and Biology

13.3.1 Morphology

The Xyloryctidae are sometimes difficult to identify from morphological features. The most prominent features in most species are, of course, the long, pointed labial palpi, which are recurved back over the head (Figs 13.1 and 13.2). In some genera such as *Metathrinca* or *Thysiarcha*, however, the labial palpi are very short. Another typical feature is a band of spiniform setae ('tergal spines') on the posterior areas of terga 2–6. I am informed by Mark Sterling (pers. commun.) that almost all the Asian species have spiniform setae on Tergum 7 as well. The adults of some *Cryptophasa* species often have an area of orange spines and scales dorsally on segment 2. Similar spines are found, however, in other families including the Oecophoridae, Gelechiidae and Cosmopterigidae. Many species are sexually dimorphic, in which case the female is often much larger than the male.

Turning to features of the genitalia, Common (1970) suggests 'the male genitalia of most species are characteristic, with a short, thick,

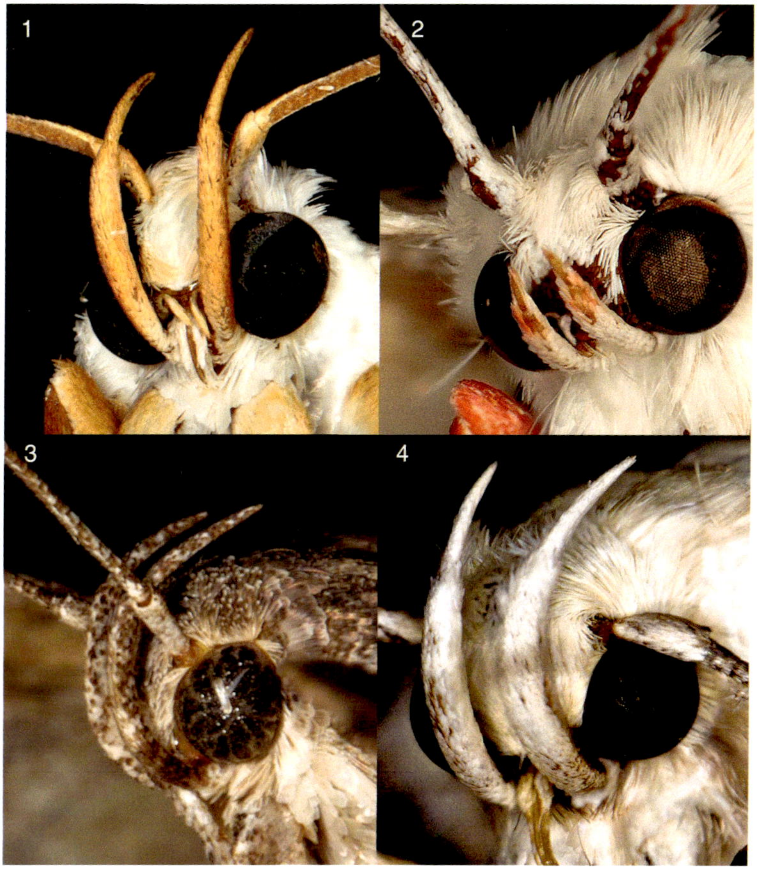

Fig. 13.1. Xyloryctidae from Queensland. Labial palpi: 1. *Xylorycta* sp.ANIC 45; 2. *Brachybelistis neomorpha*; 3. *Illidgea* sp. ANIC21; 4. *Cryptophasa porphyritis*.

Fig. 13.2. Xylorictidae: adults, near Imbil, Queensland. 1. *Xylorycta argentella*; 2. *Philarista* sp.; 3. *Lichenaula undulatella*; 4. *Uzucha borealis*.

heavily sclerotized uncus, joined laterally to a short, thick, upturned, heavily sclerotized gnathos beneath it'. I have used the term 'psittaciform' to describe the parrot's beak-like structure of the distal termination of the uncus and gnathos; this is a feature most typical of *Cryptophasa* (McMillan, 2019). Common (1970) also states that 'the valva is long and narrow with a curved apical process of the sacculus.' I have found that there is a difference between the termination of the sacculus and a separate postsaccular process developing from the the outer margin of the valva adjacent to the tip of the sacculus, which is distinctive especially in species of *Cryptophasa* (McMillan, 2019) (Fig. 13.3).

Various authors have proposed peculiarities of wing venation, most of which have been found to be unreliable. The male antennae are pectinated, ciliated or simple; the scape of the antenna lacks a pecten, a comb-like structure found in the Oecophoridae. A rule of thumb current among collectors is that the hindwings of some species, especially the smaller ones, are trapezoidal in shape and have a defined anal angle between the termen and the hindmargin.

Two Hawaiian species, *Thyrocopa apatela* and *T. kikaelekea*, have evolved to become wingless 'grasshopper moths'. This is presumed to be an adaptation to the extremely windy conditions under which they live (Medeiros, 2008, 2009).

13.3.2 Larval biology

A far more reliable way of identifying species within this family is their unique and strange larval biology. The most commonly cited mode of living for the larva is in a residential tunnel, from which the caterpillar emerges by night to collect leaves for food, eating them as they dry

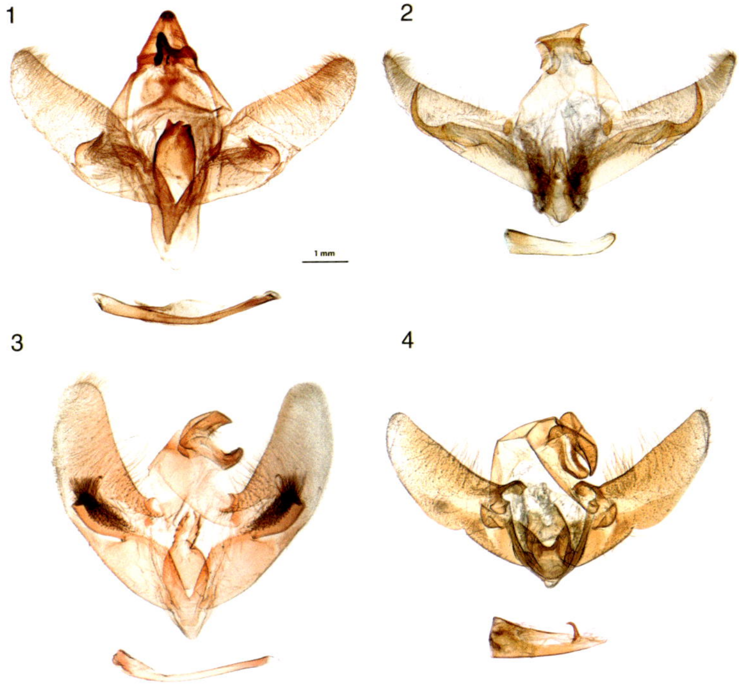

Fig. 13.3. Xylorictidae, male genitalia. 1. *Uzucha borealis*; 2. *Cryptophasa xylomima*; 3.*Tymbophora peltastis*; 4. *Lichenaula undulatella*. Various magnifications, not to scale.

out. This saprophagy, in this case consuming dead plant biomass, is said to be less general in the Xyloryctidae than it is in other gelechioid families where saprophagy is present.

Most xyloryctids are associated with trees. Larvae of the Xylorictidae feed on hosts in 21 plant families, including lichens, but almost half of the known hosts are species in the typically Australian plant families, Proteaceae and Myrtaceae (Hodges, 1998).

One small species, *Plectophila electella*, makes a spiral case out of silk, camouflaged with grass and bits of leaf, which enables it to travel over a shrub in search of food while retaining the xyloryctid habit of bringing leaves back to the entrance of the (in this case, mobile) burrow. The larva of *Xylorycta luteotactella* lives in either a webbing shelter among twigs and leaves or a short tunnel in a twig or the woody fruits of Proteaceae, including *Banksia* and *Hakea*, and is a pest of *Macadamia* (Common, 1970). Other species feed on inner or regrowing bark. Larvae of *Telecrates laetiorella* form a gallery out of silk under *Eucalyptus* bark, from where they emerge to feed on the inner bark. One genus, *Uzucha*, makes a shallow residential tunnel under the bark of *Eucalyptus* and emerges from it to gnaw off the outer layer of bark; it then emerges to feed on the cambium as it regrows (Fig. 13.4). *Scieropepla typhicola* larvae burrow into the seedheads of *Typha* (bulrushes), uncharacteristically abandoning trees. Others bore into flower heads of various Proteaceae: other species of *Scieropepla* have larvae which tunnel within flower spikes, usually of *Banksia*.

Some species feed on lichens growing on fences and rocks, sheltering in a gallery of silk and refuse particles (Common, 1970). *Perixestis rhizophaga* lives in an underground burrow (a mode of living also encountered in *Procometis*, Austostichidae). According to Turner (1902), it feeds on *Persoonia falcata*, making a funnel-shaped entrance at ground level, tunnelling into the stem for 6–10 inches, sometimes into a root.

13.4 Economic Importance

Xyloryctidae are sometimes economic pests of commercial tree crops, including tea, coffee,

Fig. 13.4. *Uzucha borealis* life history, Julatten, Queensland. 1.Pupal case *in situ*; 2. larva in uncovered retreat; 3, 4. mobile tunnel of silk and frass allowing larva to graze on cambium. Photos: Dick Whitford, Julatten, Qld.

cocoa, coconut, cloves and macadamia. One Australian species, whose natural hosts are species of *Acacia*, *Maroga melanostigma*, has crossed over to introduced species and become a pest of stone fruits (cherries in particular), pecans, citrus, pome fruits, figs and ornamentals. *Maroga* species ringbark their host by tunnelling right around the stem, often resulting in death of the tree. *Cryptophasa albacosta* has also crossed over to become a pest of fruit trees.

13.5 Parasitoids

Despite all their adaptations to remain hidden, Xyloryctidae are still subject to parasitism. *Xylorycta strigata* is parasitized by the chloropid fly, *Conioscinella xyloryctae*; *X. luteotactella* has been recorded being parasitized (among many other Lepidoptera) by *Echthromorpha intricatoria* (Ichneumonidae) and, most oddly, by the moth *Prosthenia xyloryctella* (Pyralidae, Gallerinae) which has been suggested as a parasite, as it was found in a 'xylo twig', although moths of this kind are detritivores (another species in this genus feeds on faeces and refuse in parrots' nests). It is perhaps more likely to feed on the frass and refuse left behind in xyloryctid burrows. Outside Australia, *Opisina arenosella*, the coconut black-headed moth, is parasitized by a diverse complex that includes members of the Braconidae, Eulophidae, Chalcididae, Bethylidae, Ichneumonidae, Tachinidae and Trichogrammatidae (Hymenoptera); egg, larval, pupal and multiparasitic wasps attack this very destructive pest species, which is rapidly spreading from India and Sri Lanka through South-east Asia and is invasive as far as Hainan and Guangxi in China (Lu *et al.*, 2023).

References

Blandford, W.H.F. (1898) Report on an exhibition of Xyloryctinae presented by Lord Walsingham, Entomological Society of London, 3 March 1898. *Entomologists' Monthly Magazine* 34, 91–92.

Bucheli, S.R. and Wenzel, J. (2005) Gelechioidea (Insecta: Lepidoptera) systematics: A reexamination using combined morphology and mitochondrial DNA data. *Molecular Phylogenetics and Evolution* 35, 380–394.

Busck, A. (1934) Stenomidae. *Lepidopterorum Catalogus* 67, 1–73.

Butler, A.G. (1876) Descriptions of new species of Lepidoptera from New Guinea, with notice of a new genus. *Proceedings of the Zoological Society of London* 4, 765–768.

Common, I.F.B. (1970) Lepidoptera (Moths and Butterflies). In: *The Insects of Australia*. Melbourne University Press, Collingwood, Victoria, pp. 765–866.

Common, I.F.B. (1990) *Moths of Australia*. Melbourne University Press, Collingwood, Victoria.

Common, I.F.B. (1994) *Oecophorine Genera of Australia. I. The* Wingia *group (Lepidoptera: Oecophoridae)*, Monographs on Australian Lepidoptera 3. CSIRO Publishing, Collingwood, Victoria.

Common, I.F.B. (1996) Oecophoridae. In: Nielsen, E.S., Edwards, E.D. and Rangsi, T.V. (eds) *Checklist of the Lepidoptera of Australia*. Monographs on Australian Lepidoptera 4. CSIRO Publishing, Collingwood, Victoria, pp. 59–89.

Common, I.F.B. (1997) *Oecophorine Genera of Australia II. The* Chezala, Philobota *and* Eulechria *groups(Lepidoptera: Oecophoridae*. Monographs on Australian Lepidoptera 5. CSIRO Publishing, Collingwood, Victoria.

Common, I.F.B. (2000) *Oecophorine Genera of Australia III The* Barea *group and unplaced Genera (Lepidoptera: Oecophoridae)*. Monographs on Australian Lepidoptera 8. CSIRO Publishing, Collingwood, Victoria.

Diakonoff, A. (1954) Microlepidoptera of New Guinea. Results of the third Archbold Expedition (American-Netherlands Indian Expedition 1938–1939). Part 4. *Verhandelingen der Koninklijke Akademie van Wetenschappen* 2 ser. 50, 1–191.

Duckworth, W.D. (1964) North American Stenomidae (Lepidoptera: Gelechioidea). *Proceedings of the United States National Museum* 116, 23–72.

Duckworth, W.D. (1973) The old world Stenomidae: A preliminary survey of the fauna, notes on relationships, and revision of the genus *Eriogenes* (Lepidoptera: Gelechioidea). *Smithsonian Contributions to Zoology* 147, 1–21.

Dyar, H.G. (1903) A list of the North American Lepidoptera and key to the literature of this order of insects. *Bulletin of the United States National Museum* 52, 1–723.

Fletcher, T.B. (1929) A list of generic names used for Microlepidoptera. *Memoirs of the Department of Agriculture of India, Entomology* 11, 1–244.

Gates Clark, J.F. (1955) *Catalogue of the Type Specimens of Microlepidoptera in the British Museum (Natural History) described by Edward Meyrick*, vol. 1. British Museum (Natural History), London.

Hampson, G.F. (1918) Some small families of the Lepidoptera which are not included in the key to the families in the Catalogue of Lepidoptera Phalaenae, a list of the families and subfamilies of the Lepidoptera with their types, and a key to the families. *Novitates Zoologicae* 25, 366–394.

Heikkilä, M., Mutanen, M., Kekkonen, M. and Kaila, L. (2014) Morphology reinforces proposed molecular phylogenetic affinities: A revised classification for Gelechioidea (Lepidoptera). *Cladistics* 30, 563–589.

Hoare, R.J.B. (2005) *Hierodoris* (Insecta: Lepidoptera: Gelechioidea: Oecophoridae), and overview of Oecophoridae). *Fauna of New Zealand* 54, 13–25.

Hoare, R.J.B. (2010) *Izatha* (Insecta: Lepidoptera: Gelechioidea: Oecophoridae). *Fauna of New Zealand, 65*. Manaaki Whenua Press, Lincoln, New Zealand, pp. 1–201.

Holloway, J.D., Kibby, G. and Peggie, D. (2001) *The Families of Malesian Moths and Butterflies*. Fauna Malesiana Handbook 3. Brill, Leiden, The Netherlands.

Hodges, R.W. (1998) The Gelechioidea. In: Kristensen, N.P. (ed.) *Lepidoptera, Butterflies and Moths, Handbook of Zoology* 4(1). De Gruyter, Berlin and New York, pp. 131–158.

Jayanth, K.P. and Nagarkatti, S. (1981) Feasibility of rearing *Nephantis serinopa* Meyr. (Lepidoptera: Cryptophasidae) on an artificial diet. *Journal of Entomology Research* 5, 43–74.

Kaila, L. (2004) Phylogeny of the superfamily Gelechioidea (Lepidoptera: Ditrysia): An exemplar approach. *Cladistics* 20, 303–340.

Kirby, W.F. (1897) *A Handbook to the Order Lepidoptera*. Lloyd's Natural History, Vol. V, Part III. Edward Lloyd, London.

Le Marchand, S. (1947) Les Tineina. *Revue Française de Lépidoptérologie* 11, 145–163.

Lewin, J, (1805) *Prodromus Entomology. Natural History of Lepidopterous Insects of New South Wales*. Thomas Lewin, London.

Lewin, J. (1808) *Birds of New Holland with Their Natural Histoy*. J. White and S. Bagster, London.

Lu, H., Lyu, B., Tang, J., Wu, Q., Wyckhuys, K.A.G. *et al.* (2023) Ecology, invasion history and biodiversity-driven management of the coconut black-headed caterpillar *Opisina arenosella* in Asia. *Frontiers in Plant Science*, 14, 1116221.

Lvovsky, A.L. (2012) Comments on the classification and phylogeny of broad-winged moths (Lepidoptera, Oecophoridae *sensu lato*). *Entomological Review* 92, 188–205.

McMillan, I.C. (2019) Redescription of *Cryptophasa irrorata* Lewin, 1805 (Lepidoptera: Gelechioidea: Xyloryctidae). *Zootaxa* 4615, 192–200.

Medeiros, M.J. (2008) A new species of flightless, jumping, alpine moth of the genus *Thyrocopa* from Hawaii (Lepidoptera: Xyloryctidae: Xyloryctinae). *Zootaxa* 1830, 57–62.

Medeiros, M.J. (2009) A revision of the endemic Hawaiian genus *Thyrocopa* (Lepidoptera: Xyloryctidae: Xyloryctinae). *Zootaxa* 2202, 1–47.

Meyrick, E. (1883) On the classification of some families of the Tineina. *Transactions of the Royal Entomological Society of London* 1883, 119–131.

Meyrick, E. (1885) Descriptions of New Zealand Microlepidoptera. Gelechiadae VIII. Tineina (part). *New Zealand Journal of Science* 2, 589–592.

Meyrick, E. (1886a) Descriptions of Lepidoptera from the South Pacific. *Transactions of the Entomological Society of London* 1886, 283.

Meyrick, E. (1886b) Descriptions of New Zealand Microlepidoptera. Gelechiadae VIII. Tineina (part). *New Zealand Journal of Science* 2, 589–592.

Meyrick, E. (1887) Descriptions of new Lepidoptera. *Proceedings of the Linnaean Society of New South Wales* 1, 1037–1048.

Meyrick, E. (1890) Descriptions of Australian Lepidoptera. Part I. Xyloryctidae. *Transactions of the Royal Society of South Australia* 13, 23–81.

Meyrick, E. (1897) Descriptions of new Lepidoptera from Australia and New Zealand. *Transactions of the Entomological Society of London* 1897, 367–390.

Meyrick, E. (1904) Descriptions of Australian Micro-lepidoptera. XVIII. Gelechiadae. *Proceedings of the Linnaean Society of New South Wales* 29, 255–441.

Meyrick, E. (1906) Descriptions of Australian Tineina. *Transactions of the Royal Society of South Australia* 30, 33–66.

Meyrick, E. (1914) Descriptions of New Zealand Lepidoptera. *Transactions and Proceedings of the New Zealand Institute* 47, 205–244.

Meyrick, E. (1915) *Exotic Microlepidoptera* 1, 289–320, 321–352, 353–384, 385–416, 417–448, 449–480, Taylor & Francis, London.

Minet, J. (1990) Remaniement partiel de la classification des Gelechioidea, essentiellement en fonction de caractères pré-imaginaux (Lepidoptera, Ditrysia). *Alexanor* 16, 239–255.

Newman, E. (1841) Analytical Notice of the 129th Volume of Lardner's Cabinet Cyclopaedia, entitled 'On the History and Natural Arrangement of Insects' by William Swainson, A.C.G., F.R.S. & L.S., Hon. F.C.P.S., and of several Foreign Societies; and W. E. Shuckard, Lib. R.S., &c. *The Entomologist* 3, 33–48.

Newman, E. (1856) Characters of a few Australian Lepidoptera, collected by Mr. Thomas R. Oxley. *Transactions of the Entomological Society of London* 1856, 281–300.

Robinson, G.S., Tuck, K.R. and Shaffer, M. (1994) *A Field Guide to the Smaller Moths of South-East Asia*. Natural History Museum, London & Malaysian Nature Society, Kuala Lumpur.

Scoble, M.J. (1992) *The Lepidoptera: Form, Function and Diversity*. Natural History Museum, London.

Scott, A. W. (1864) *Australian Lepidoptera and their Transformations drawn from the Life by Harriet and Helena Scott: with Descriptions General and Systematic*, vol. 1. Jan van Voorst, London.

Sinev, S.Y. (1993) On the systematics and phylogeny of the Gelechioidea *s.l.* (Lepidoptera). *Entomological Review* 72, 10–25.

Sohn, J.C., Regier, J.C., Mitter, C., Adamski, D., Landry, J.F. *et al.* (2016) Phylogeny and feeding trait evolution of the megadiverse Gelechioidea (Lepidoptera: Obtectomera): New insight from 19 nuclear genes. *Systematic Entomology* 141, 112–132.

Swainson, W. and Shuckard, W.E. (1840) On the history and natural arrangement of insects. In: *Lardner's Cabinet Cyclopedia*. Longmans, London, pp. 106–107.

Turner, A.J. (1898) The Xyloryctidae of Queensland. *Annals of the Queensland Museum* 4, 1–32.

Turner, A.J. (1900) New Micro-lepidoptera – Mostly from Queensland. *Transactions of the Royal Society of South Australia* 24, 6–23.

Turner, A.J. (1902) New Australian Lepidoptera. *Transactions of the Royal Society of South Australia* 26, 175–207.

van Nieukerken, E.J., Kaila, L., Kitching, I., Kristensen, N.P., Lees, D.C. *et al.* (2011) Order Lepidoptera Linnaeus, 1758. In: Zhang, Z.Q. (ed.) Animal biodiversity: An outline of higher-level classification and survey of taxonomic richness. *Zootaxa*, 3148, 212–221.

Walker, F. (1864) *List of the Specimens of Lepidopterous Insects in the Collection of the British Museum. Tineites* 29, 562–1096, British Museum (Natural History), London.

Wallengren, H.D.J. (1861) Lepidoptera. Species novas descripsit. *Svenska vetenskap Akademien* 2, 351–390.
Walsingham de Grey, T. (1891) African Micro-Lepidoptera. *Transactions of the Entomological Society of London* 39, 63–133.
Walsingham de Grey, T. (1907) Descriptions of new North American tineid moths, with a generic table of the family Blastobasidae. *Proceedings of the United States National Museum* 33, 197–228.
Wang, Q.-Y. and Li, H.-H. (2020) Phylogeny of the superfamily Gelechioidea (Lepidoptera: Obtectomera), with an exploratory application on geometric morphometrics. *Zoologica Scripta* 49, 307–328.
Zborowski, P. and Edwards, E.D. (2007) *A Guide to Australian Moths*. CSIRO Publishing, Collingwood, Victoria.
Zimmerman, E.C. (1978) Microlepidoptera, Gelechioidea. In: *Insects of Hawaii*, 9. University Press of Hawaii, Honolulu, pp. 928–931.

14 The Psychidae: The Bagworm Moths

Ethan P. Beaver

Research School of Biology, Australian National University, Acton, ACT, Australia, and Australian National Insect Collection, CSIRO, Black Mountain, ACT, Australia

Abstract

The Australian Psychidae fauna is introduced and discussed in the context of the known global assemblage. Several key taxa of relevance to Australia are noted and figured. This family of moths is poorly studied, particularly in the southern hemisphere, and many knowledge gaps are identified here for future exploration. Psychidae remain a fascinating group of moths where further research across any aspect of their biology or systematics will undoubtedly reveal many unexpected and interesting results.

14.1 Introduction

The bagworm or case moths, the family Psychidae, are a diverse group of globally distributed moths characterized by several distinctive attributes. The adult females exhibit a wide range of wing reduction or winglessness. Males often have transparent or discreetly patterned wings. Psychid larvae are generally polyphagous and always case bearing. The adults of many species are ephemeral and rarely seen, while the larvae are more readily observed, relatively long-lived, and frequently encountered on garden plants or affixed to both natural and man-made structures within the environment. The propensity for early instar larvae to disperse aerially via ballooning, coupled with the polyphagy of many species upon higher plants, has resulted in some species being considered as agricultural or plantation pests in tropical regions. This can be exacerbated by their eruptive population dynamics and parthenogenesis in some taxa. Nonetheless, most Psychidae are inoffensive with no capacity as pests, and the biology and the complex systematics of many taxa particularly in Australia remain unresolved.

14.2 Taxonomy and Phylogeny

The Psychidae are a monophyletic family of small to medium-sized moths within the superfamily Tineoidea, a paraphyletic assemblage of clades that represent the earliest diverging lineages of the Dytrisia (Regier *et al.*, 2015). The Psychidae comprise ten subfamilies, with over 1300 species spread amongst approximately 240 genera (Sobczyk, 2011). A deep split within psychid phylogeny that divides the fauna into two broad clades can be somewhat simplistically understood to comprise the 'tineiform' psychids, those smaller

Corresponding author: ethan.beaver@anu.edu.au

DOI: 10.1079/9781800627680.0014

species in which females may be either winged or wingless, and the bombyciform, 'psychine' lineage, usually larger species in which the females are always wingless and usually remain concealed within their larval bag. Winglessness has evolved multiple times independently in the Psychidae (Regier *et al.*, 2015; Niitsu *et al.*, 2017). The Psychidae are the only family of Lepidoptera with a significant cohort of species in which the females are primarily wingless (Sattler, 1991).

Monophyly of the Psychidae is supported morphologically by several diagnostic features of the larvae including abdominal crochets on segments 3–6 being arranged in a uniordinal lateral penellipse; the pronotum is fused with the lateral pinnaculum so as to include the spiracle and three prespiracular setae; and there are four pairs of epipharyngeal setae on the larval labrum (Davis, 1987). Fewer diagnostic apomorphies exist in the adult Psychidae, although it would appear that fused metathoracic furcal bridges would be one such feature given the recent treatment of the Arrhenophaninae as a subfamily of the Psychidae (see Robinson, 1988; Mutanen *et al.*, 2010).

The ten subfamilies are variously defined on the basis of the morphology of both sexes and aspects of their larval life histories (Sauter and Hättenschwiler, 1999; Sobczyk, 2011; Regier *et al.*, 2015). The Oiketicinae are the most diverse of these, comprising 67 genera with approximately 270 accepted species (Sobczyk, 2011). The monophyly of the Oiketicinae is strongly supported by molecular phylogenetics (Regier *et al.*, 2015) and by the morphology of the adult male, in which the eighth abdominal sternite has two pronounced and fork-like basal processes (Sauter and Hättenschwiler, 1999). This morphological character appears to be robust, though the similar Metisinae, which lack this feature, have not been included in any molecular phylogenetic study to date. There are four tribes within the Oiketicinae, the largest being the Acanthopsychini with over 90% of named genera, with the remainder shared among the Phalacropterigini, Oreopsychini and Apteronini, tribes that are presently known only from the Palaearctic region. Despite the high diversity and endemism in the region, the faunal assemblage and systematics of the Australasian Psychidae remain poorly known.

14.3 The Australian Fauna

The last major census of names attributed to this family from Australia (Nielsen and Edwards, 1996) recorded 188 species within 25 genera. Lack of recent study on the Australian members of this family presented two clear problems for those authors: the assignment of genera to tribes and subfamilies, and the placement of 25 species within appropriate genera. Since the publication of their checklist, several changes have occurred. The distinctive *Notiophanes fuscata* was described from northern Queensland by Davis and Edwards (2003) and placed in the Arrhenophanidae, which has been treated subsequently as a subfamily of the Psychidae (Robinson, 1988; Mutanen *et al.*, 2010; Regier *et al.*, 2015), and is the only known member of this lineage in Australia. Mutanen *et al.* (2010) further found molecular evidence that *Piestoceros* has an affinity to the *Heliocosma* group and is allocated to the Cossoidea-Sesioidea assemblage. Sobczyk (2011) recognized the subfamily status of several Australian genera and allocated them to Oiketicinae, although 14 genera of the mainly 'tineiform' assemblage remain without a satisfactory subfamilial placement. The unusual Australian genus *Trigonocyttara* (Fig. 14.22) was unexpectedly recovered as part of the psychine lineage (Regier *et al.*, 2015), and not within the tineiform lineages. Further systematic work is required to determine if this genus represents an undescribed subfamily.

The exact numbers of undescribed Psychidae in Australia are difficult to confirm as many are known in research collections from only single specimens. The diurnally active and crepuscular fauna remain very poorly sampled by comparison with the nocturnal light-attracted species. From the specimens available in museums and major scientific institutions, there remain at least another 120–150 undescribed species of Psychidae, and further, at least four undescribed genera.

Although the composition of the psychine fauna of Australia and New Guinea is similar to that of South-east Asia, and to some extent that of tropical Africa particularly among the oiketicine genera, the Australian non-psychine genera are mostly endemic and diverse, with presently untested biogeographic relationships with non-Australian genera. The Arrhenophaninae are

Fig. 14.1–14.23. Psychidae. 1, *Metura aristocosma* bag, 3, male; 2, *M. elongatus* bag, 4, male, 5, female; 6, *Lomera zophopepla* bag, 7, male; 8, *Hyalarcta huebneri* male; 9, *L. pantosemna* male; 10, *L. xanthochrysa* bag; 11, *Clania* sp., bag; 12, *L. crocobathra* male; 13, *L. boisduvalii* male; 14, *Iphierga chrysopa* male; 15, *I. stasiodes* male; 16, *Clania* sp., female; 17, *Bathromelas hyaloscopa* male; 18, *Ardiosteres sporocosma* female; 19, *A. moretonella* male, 20, *Clania lewinii* male; 21, *Clania* sp., larva; 22, *Trigonocyttara clandestina* male; 23, *Cebysa leucotelus* male. Scale bar 10 mm. All photos by the author.

known from tropical areas of Central America, Asia and north-eastern Australia, although it has been stressed that a biogeographic analysis of these known distributions would be premature without further sampling (Davis, 2003). In terms of species richness by land area, Australasia is distinctly more diverse than the entire Americas (see Davis, 1964) and, when revised, may eventually be comparable with the fauna of Europe or Asia in size (see Sauter and Hättenschwiler, 1991; Arnscheid and Weidlich, 2017). The Australian arid zone remains understudied and appears to retain many endemic undescribed species.

14.3.1 Taxonomic history

Systematic work on Psychidae in Australia began with the brief description of *Thyridopteryx nigrescens* by Edward Doubleday (1846) (now in the genus *Hyalarcta*), from a specimen collected in Adelaide during the Eyre expedition through Central and Western Australia in the 1840s. Over the following 80 years most of the Psychidae fauna was described in a succession of works by Francis Walker, Edward Meyrick and later by Alfred J. Turner and Oswald B. Lower. Most described specimens hailed from the temperate areas of eastern and south-western Australia, and tropical northern Queensland. At that time, many taxa which we now understand to belong to the Psychidae were variously placed within other microlepidopteran families such as the Tortricidae and Tineidae, many examples of which were resolved by Robinson and Nielsen (1993). Works that claim to revise the Australian Psychidae, although still major and important references, refer specifically to what are now considered as the subfamily Oiketicinae, such as work by Meyrick and Lower (1907) and Turner (1947). The global list by Dalla Torre and Strand (1929) included all Psychidae then known. No major revision of Australian fauna has been attempted since these publications.

More recently, Walker's oiketicine genus *Metura* has been revised, adding two new species to the Australian fauna and resolving the placement of a further two names (Beaver, 2020). Aspects of the life histories of several species of Australian Psychidae were first published and described in detail by Westwood (1854), Scott (1864) and Froggatt (1907), works that were beautifully and accurately illustrated. A modern summary of the known life histories of various Australian taxa was compiled by Common (1990).

Nothing is apparent in the western scientific literature of the significance or use of bagworms by Aboriginal Australians in precolonial Australia, to the best of my knowledge. The conspicuous nature of many species, and the continued connection between Aboriginal Australians and their lands, suggests there must be associated knowledge and connections. A modern, equitable anthropological study of Lepidoptera and Aboriginal culture, from an Aboriginal perspective, would generate significant interdisciplinary interest. Elsewhere, such as in Central Africa, bagworms are consumed as food and associated with Indigenous folklore (Scalercio and Malaisse, 2009) and, in Papua, the bags of larger, intricately adorned species are used as decoration in armbands, necklaces, belts and nose-pin covers (Peigler and Oberprieler, 2017) and in net-carrying bags ('bilums') (Beaver, 2020).

14.3.2 Systematics

Recent advancements in molecular phylogenetics have revolutionized systematics. The phylogenetics of Lepidoptera have undergone significant recent changes with a wide range of high-profile studies, including those utilizing transcriptomes (Kawahara *et al.*, 2019) and target or anchored hybrid enrichment and Illumina sequencing (Hamilton *et al.*, 2019; Mayer *et al.*, 2021; Kawahara *et al.*, 2023) to sequence deeply across large datasets of species. There appears to be concordance among phylogenetic studies that utilize nuclear and mitochondrial genome data (Kim *et al.*, 2020). Emerging research has highlighted the utility of data on whole mitochondrial genomes in even deep-level phylogenetics within Lepidoptera (Cameron, 2023; Timmermans *et al.*, 2023). Some of these studies have included one or up to a handful of Psychidae species, with the most comprehensive phylogenetic analysis of the paraphyletic Tineoidea being that of Regier *et al.* (2015).

Early systematic research into Australian Psychidae mainly placed genera into 'tineiform' or 'bombyciform' informal groupings, before the popularization of the formal subfamilial classification systems within the global Psychidae. Subfamilial systems for the Psychidae were incrementally proposed by various authors based on morphology (Tutt, 1900; Sauter and Hättenschwiler, 1991). The history of this classification is nuanced and complex, with many names that are today considered subfamilies being proposed originally as families (Tutt, 1900). Detailed classifications of genera and species in monographic treatments exist for specific geographic areas, such as the New World (Davis, 1964, 1975) and Europe (Kozhanchikov, 1956; Sauter and Hättenschwiler, 1991, 2004; Arnscheid and Weidlich, 2017). The remainder of the world fauna is discussed in many works of more limited scope addressing smaller numbers of genera or species. The larger morphologically based assessments of Davis (1964) and Sauter and Hättenschwiler (1991, 2004) have been largely accepted, are thorough and are supported in part by recent molecular analysis (Regier *et al.*, 2015). The work of Regier *et al.* (2015) found clear support for these informal groupings and identified monophyletic bases for many subfamilies.

There has been no phylogenetic study concerning the Australian Psychidae, with the recent work on Tineoidea including mainly Palaearctic and New World genera. Taxonomic and phylogenetic work is now required to determine the subfamilial placement of Australian genera, to resolve the unplaced Australian species and to describe the remaining new species and genera. Australian entomologists have assembled large collections of specimens of bagworm moths, the result being a rich fauna of undescribed species represented by specimens in museums and scientific research institutions. The Oiketicinae are the subject of ongoing research (Beaver, 2020, 2021, 2023) and are undergoing systematic revision by the present author.

14.4 Biogeography and Ecology

The age of the current extant radiation of subfamilies and genera within the Psychidae is unstudied. The family is diverse and globally distributed, with high diversity in tropical regions. Adding to this problem, there are few fossilized examples of Psychidae in general, but due to their phylogenetic position the group is expected to be old. Some estimates place this to approximately at least the late Jurassic, *ca.* 170 to 135 million years ago (Kawahara *et al.*, 2019).

Psychidae occur in all terrestrial ecosystems except true polar regions (Sobczyk, 2011). At the local scale, first instar larvae that are dispersed via ballooning may spread widely and can appear scattered in the landscape. Larvae that hatch on windless days may shroud the tree; otherwise open landscapes can accumulate more larvae than dense forested areas (Moore and Hanks, 2004). Ground-associated taxa such as some species of *Lomera* that specialize on widespread forbs, herbs and grasses may have different, unstudied dispersal dynamics and appear to be further capable of wandering widely across herbfields and grasslands as older larvae. Although some attention has been given to the effects of Oiketicinae dispersal mechanisms upon their population dynamics (Cox and Potter, 1986; Rhainds *et al.*, 2009), this has not untangled the wider biogeographic implications of this dispersal capacity on the processes of speciation.

Psychidae experience high predation pressures and are commonly parasitized by tachinid flies and groups of wasps such as Braconidae, Ichneumonidae and Chalcidoidea, often to an extreme degree (Rhainds *et al.*, 2009), all of which have little trouble negotiating the defences of the protective larval bag. The bag is more effective for camouflage and mechanical protection from larger non-parasitoid predators (Sugiura, 2016), but very large predators such as some birds and mammals are occasionally able to overcome this barrier with considerable effort. Bagworms expend significant energy in maintaining and repairing their bag from any damage. Their empty bags can remain in the environment for several years, attached to tree trunks or other substrate. Vacated bags provide a shelter for a wide variety of in-fauna, generally a range of terrestrial arthropods, and can provide food for Tineidae and dermestid beetles. This aspect of their ecology is complex and unstudied.

14.5 General Biology

The life histories of Psychidae are varied and complex. For most members of the psychine lineage, eggs are laid within the larval bag, concealed entirely within the female pupal exuvium and buried in protective, deciduous body scales from the ovipositing female. In some other species, females may attach eggs to the outside of the bag or in nearby concealed areas. The number of eggs laid varies taxonomically and can be between 40 and 5700 eggs (Rhainds *et al.*, 2009). Parthenogenesis is recorded in several genera of Psychidae. Upon hatching, larvae are capable of ballooning aerially upon silk threads, this being the primary form of dispersal for Oiketicinae (Cox and Potter, 1986). In this subfamily, secondary dispersal occurs due to the developed flight capacity of adult males, though this is greatly restricted by their short adult lifespan and influenced by temperature (Morden and Waldbauer, 1971) and photoperiod. Females of many taxa from the micropsychine lineage of Psychidae are winged and capable of flight. In these taxa it is presumed that dispersal occurs in the adult rather than larval stage.

The otherwise defenceless larvae shelter within the portable larval bag which they build immediately upon alighting on a suitable host or substrate. This bag is often conical in shape, constructed from silk and foliage or substrate fragments and has two openings, one each at the anterior and posterior of the bag. This structure is carried by the larva via the anterior aperture and can be temporarily sewn shut from within when the larva is threatened or is undergoing diapause, ecdysis or pupation. Although the bag is a distinctive characteristic of the Psychidae, there are at least ten other families of Lepidoptera in which some or many species construct broadly similar bags (Rhainds *et al.*, 2009) for much the same purpose. The bag is usually constructed from the inside and can be heavily adorned with foliage or substrate fragments. In most species, adornments are first threaded through the anterior aperture, passed through a temporary slit made within the bag wall and subsequently sewn back together internally with the adornment in place. Phylogenetic signal in bag design or bag construction behaviour in Psychidae is likely present to some extent but is not well studied in modern literature. The larva does not leave the bag under normal conditions and is almost always unable to build a new bag if removed entirely, but they can repair extreme damage surprisingly quickly.

Psychidae present a very wide breadth of feeding strategies and embrace many ecological niches: detritivores, fungivores, lichenophages, both polyphagous and monophagous herbivores, and, in some rare examples, true predators. Myrmecophiles, such as the Australian *Ardiosteres* and *Iphierga*, live with arboreal and terrestrial ants, respectively (Dodd, 1906; Dodd and Meyrick, 1911). The great diversity of feeding strategies occurs within micropsychine lineages, whereas the psychine lineage has specialized mainly upon the foliage of plants. The vast majority of Oiketicinae are highly polyphagous or express some degree of broad polyphagous behaviour within related families of plants, with up to 50 different families recorded for the North American *Thyridopterix* (Rhainds *et al.*, 2009). There is, however, preference in host species which may be induced by larval experience (Davis, 1964) in that mature larvae reared upon a single species may refuse to feed when transferred between otherwise viable host plants. This may incur a fitness cost associated with host transfer, though this is not unambiguously supported by the work of Ward *et al.* (1990). Despite polyphagous ability, there are clear host preferences and survival costs associated with the use of certain plant species (Ali *et al.*, 2023). A notable exception to otherwise broadly polyphagous dietary performance is found in the endemic Australian *Bathromelas hyaloscopa* (Fig. 14.17) which feeds exclusively on Casuarinaceae (Beaver, 2023). Taxa with more restricted diets may result in niche partitioning associated with specific foliage types – that is, some species appear to prefer broad-leaved, flat foliage, others narrow or rounded, sclerotized foliage, still others appear to consume only the flowers of their hosts. This aspect of Psychidae feeding biology is little studied.

The number of larval instars varies from six to 12 and can be difficult to record due to the concealed nature of the larva. Diapause usually occurs in the larval stage, often at the prepupal stage where the larva can rest for several months during suboptimal conditions. Pupation itself is of a shorter duration of usually a few weeks. Egg diapause is recorded for at least some species (Morden and Waldbauer, 1980). Larval development time is usually 1–2 years but can be significantly shorter in some tropical species, and significantly longer in some arid zone or temperate species that undergo a larval diapause. Before pupation, the larva reverses position so as to face the posterior aperture of its bag for easy emergence. There is a significant male-biased sex ratio in bagworms, and sex segregation of pupation sites is not uncommon in some species (Cox and Potter, 1988).

Adult females usually live for several weeks, whereas non-feeding males usually live no longer than 2 days. Mate attraction is facilitated by production of volatile pheromones, with females of concealed species passing corethrogyne scales over a pore in the ventral thorax before disseminating pheromones outside the bag via its posterior aperture. Some pheromone components have been identified and synthesized (Leonhardt *et al.*, 1983; Gries *et al.*, 2006; Rahmani *et al.*, 2020) with clear applications for conservation or pest monitoring and management. Within the larval bag, females of concealed species rest with the head facing the posterior aperture. In these taxa, the males have a highly developed, telescopic abdomen which they insert into the bag to mate with the female. This structure can extend several times the length of the resting abdomen and is partially devoid of scales on the medial parts of the sternites and on parts of the abdomen membrane. Due to these modifications, the male is able to mate without entirely entering the larval bag. This complex mating process leads to a high degree of mating failure, exacerbated by the short lifespan of males, and by the presence of both already mated females in the environment and of pheromone-impregnated empty bags of vacated females (Rhainds, 2018).

Many gaps exist in the understanding of Oiketicinae biology and life history. Although little is recorded in the Australian literature, the adult males have a wide though taxon-specific breadth of diel range, with species that fly only at specific times of the day – morning, evening or night. Ecological partitioning by environment structure exists: the taxa that occur on trees or shrubs are different from those that occur at the ground layer (associated with herbs, forbs, grasses or ground covers) and different again from those that occur on rocks or cave openings (unpublished results). More broadly, Oiketicinae are clearly underutilized model organisms for the study of population dynamics (Sheppard, 1975). Their evolution of mating systems and male-female fitness pressures, the variation inherent in reproductive success of individuals or populations, and how this differs in tropical and temperate species, as well as the top-down and bottom-up effects that regulate their population structures all invite further study (Cronin, 1989; Rhainds *et al.*, 2009).

14.6 Selected Life Histories

The following is an account of selected genera that are either those most commonly encountered or well studied of the Australian Psychidae, or are otherwise relevant or diverse in the Australian region.

14.6.1 *Metura* (Oiketicinae: Acanthopsychini): giant bagworms

The largest and most spectacular of the Psychidae are the Australian members of the genus *Metura*, the giant bagworms (Figs 14.1–5). In the largest and rarest species, the north Queensland *M. aristocosma* (Figs 14.1, 14.3), the larval bag can be in excess of 30 cm in length (Davis and Robinson, 1999; Beaver, 2020) and is constructed from intricately layered fine wood fibres extracted from stems of the host plants. The more common Saunder's case moth, *M. elongatus* (Figs 14.2, 14.4–14.5), constructs a bag approximately 10–18 cm in length (Fig. 14.4), adorned irregularly with twigs set between patches of bare silk, and is a widespread species in eastern Australia where they are frequently encountered in parks and gardens in major urban centres. A third species, *M. phyllosacca*, occurs uncommonly in eastern Australia in higher rainfall areas and exhibits a more variable bag structure, usually incorporating leaves or fine twigs in the bag design. Both the adult males and larvae of all species are brightly coloured in yellow and black scales.

14.6.2 *Hyalarcta* (Oiketicinae): leaf and ribbed bagworms

Hyalarcta is represented by two species endemic to Australia, each with distinctive bag designs. *Hyalarcta nigrescens* constructs a naked, cream-coloured pod-like bag which bears longitudinal ridges in mature specimens, whereas *H. huebneri* (Fig. 14.8) adorns the bag entirely in leaf or fine twig fragments. Both species are distinctive and have completely transparent wings, with a great degree of vein reduction in the hindwings, and the body black. The adult males of the two species can be quickly differentiated by the colour of their antennae, black in *H. nigrescens* and red-orange in *H. huebneri*. Both species can have sudden population explosions, where they can defoliate host plants. Both species are polyphagous although *H. nigrescens* is more often found on *Eucalyptus* and other related Myrtaceae.

14.6.3 *Clania* (Oiketicinae: Acanthopsychini): log cabin bagworms

The genus *Clania* (Fig. 14.11, 14.16, 14.20–21) is common and widespread across Africa, Asia and Australasia from tropical to temperate regions. There is a great degree of conformity of the larval morphology and larval bag design among species, all of which utilize a simple design with parallel-arranged twigs. The adults of *Clania* are more often distinctive in appearance with a wide range of striated wing patterns and degree of wing transparency. The recent literature in Australia pertains to two temperate zone species, *C. ignobilis* and *C. lewinii*, but a further 17 undescribed species found across Australia are known (Beaver, in preparation), as well as a further three described but unplaced species that require reclassification to recognize their position as members of the genus *Clania*. Most *Clania* are polyphagous where known, but seem to prefer Myrtaceae and Fabaceae in Australia, with several (undescribed) species specializing on specific genera within these families,

14.6.4 *Lomera* (Oiketicinae): fluffy bagworms

This genus (Figs 14.6, 14.7, 14.9, 14.10, 14.12, 14.13) is no doubt the most species rich of the Oiketicinae in Australia, presenting a wide range of ecological specializations with larvae diversely associated with grasses, ground covers, herbs and shrubs in almost every part of Australia. The adult males are distinctive in appearance with short, rounded wings, plumose bipectinate antennae and bodies that are densely scaled, giving a distinctly fluffy appearance. Most species are concolorous in various shades of brown or grey; others exhibit rich orange or yellow scales on the body; yet others are partially hyaline over the wings. The larval bags often differ among species and, in terms of structure, can appear anywhere from highly disorganized in plant adornment, to species with very intricately arranged whorled placement of twigs of incrementally increasing size. At present, there are 18 described species and more than 40 undescribed species are known, with several generic synonyms to be resolved.

14.6.5 *Iphierga* (unplaced to subfamily): meat ant moths

Iphierga (Figs 14.14, 14.15) have speckled or alternately barred forewings, sometimes with brightly coloured yellow hindwings, and a distinctive tuft of scales at the frons. Males and females, where both are known, are similar with the immediate difference being in the antennae of the males which are longer and bipectinate compared with those of females. *Iphierga* are known from across the continent where 12 described species are recognized, and many remain undescribed. This fascinating genus is associated with the meat ant, *Iridomyrmex* (Dolichoderinae), with which the moth larvae are ant-nest inquilines and are suspected to be scavengers (Dodd and Meyrick, 1911; Common, 1990) with little else known of the specifics of this ecological strategy. A similar biology is shared with the probably related genus *Ardiosteres* (Figs 14.18, 14.19).

14.6.6 *Cebysa* (unplaced to subfamily): metallic bagworms

Cebysa leucotelus (Fig. 14.23) is highly distinctive with a brachypterous, aposematic yellow and iridescent blue female, and a fully winged male of a more muted colour. Both sexes are diurnal and usually associated with lichens and detritus on rock ledges, tree bark or exposed dead wood in natural settings, though realistically are more often encountered in man-made partially enclosed environments such as carparks or garages near to wooded areas. The larval bag is brown, dorsoventrally compressed and composed of soil and wood fragments. As with *Iphierga* and many of the other micropsychine taxa, the phylogenetic placement of this highly distinctive genus is unknown with respect to other Psychidae. Several undescribed *Cebysa* are known.

14.7 Economic and Conservation Significance

The economic impact of Psychidae as pests in terms of global cost in damage per annum has not been calculated. Extensive literature, particularly from Asia (Wahid *et al.*, 1988; Agrawal *et al.*, 1997; Firake *et al.*, 2018; Lelana *et al.*, 2022; Manurung and Anwar, 2023), the Americas (Stephens, 1962; Gara *et al.*, 1990; Rhainds and Cabrera-La Rosa, 2010), Africa (Ripley and Webb, 1954; Okore *et al.*, 2016) and Australia (Heather, 1975; Common, 1990), documents crop and plantation damage by Psychidae, reduction in the health of crops, and potential control methods (Sulaiman and Talip, 2021). We are yet to see confirmed incursions in Australia of non-native species, but the risk of this should be a concern.

Several Australian native species are known to be minor pests locally, mainly of urban gardens or plantation crops, where they can sometimes build up to large numbers (Common, 1990). In temperate Australia, these urban outbreaks are generally uncommon. The effects in tropical Australia are unrecorded. In Indonesia, bagworms regularly defoliate plantations of various crops (Lelana *et al.*, 2022) with severe consequences. Although not yet established as such, Psychidae should be of concern as emerging pests within northern Australia, presenting some clear biosecurity risk particularly with ongoing trade and continued land use change in this region. Several Asian genera with recognized pest species are present in northern Australia, though this is unrecorded in the literature, and the species-level identification is ongoing.

Many Psychidae are not recognized as pests, and most species in Australia are endemic. Some may be of conservation concern due to their apparent restricted distributions within threatened ecosystems. Damage to fragile alpine ecosystems such as in the Southern Highlands and at Barrington Tops by feral animals may impact host plants in these places (personal observation) and this in turn may have flow-on effects for the associated insect communities. The alpine bagworm *Lomera caespitosae* is known only from alpine or subalpine ecosystems (Oke, 1948). Similarly, *Phasmyalea pelloceros* is known from a handful of specimens only from eastern Australian highland ecosystems and may be restricted in a similar way. Both species are clearly data deficient and require survey to understand their current distribution, both at the patch level and within the landscape.

Bagworms present model organisms for survey as the bags can be easily collected, identified as morphospecies, and barcoded relatively cost-effectively with reference to identified sequences. Alternatively, the approach used by Rahmani *et al.* (2020) to synthesize pheromone compounds could be applied to management programmes for these species. Synthesis and use of pheromones may be a further way to collect specimens or population data representing other rarely seen psychid species.

A great deal of the ecology, biology, systematics and taxonomy of the Psychidae remains unknown in Australia, and there is much to be uncovered by anyone interested in studying this enigmatic group of moths.

References

Agrawal, A., Pati, A.K. and Karkun, D. (1997) Seasonal variation in infestation characteristics of bagworm (Lepidoptera: Psychidae) in avenue plantation of *Acacia* and *Peltophorum* in the Chaattishgarh region. *Current Science* 72, 211–214.

Ali, S.M.F.S., Zaludin, M.S.M., Zabidi, M.S., Nasir, J., Naning, M.S. *et al.* (2023) Feeding preference of bagworm, *Metisa plana* Walker (Lepidoptera: Psychidae) on four different palm species. *The Planter, Kuala Lumpur* 99, 91–103.

Arnscheid, W.R. and Weidlich, M. (2017) *Psychidae.* Microlepidoptera of Europe, 8. Brill, Leiden, The Netherlands.

Beaver, E.P. (2020) Revision of the genus *Metura* (Lepidoptera: Psychidae) with description of two new species. *Zootaxa* 4861, 188–210.

Beaver, E.P. (2021) First record of *Pseudoclania* Bethune-Baker from Australia (Lepidoptera: Psychidae: Oiketicinae). *Australian Entomologist* 48, 207–214.

Beaver, E.P. (2023) Rediscovery and life history of *Bathromelas hyaloscopa* (Meyrick & Lower, 1907) Lepidoptera: Psychidae: Oiketicinae. *Memoirs of the Queensland Museum, Nature* 64, 260–272.

Cameron, S.L. (2023) Mitochondrial phylogenomics of the Australian scribbly gum moth *Ogmograptis* (Lepidoptera: Bucculatricidae) and an examination of deep-level relationships within Lepidoptera. *Austral Entomology* 62, 449–463.

Common, I.F.B. (1990) *Moths of Australia*. Melbourne University Press, Carlton, Victoria.

Cox, D.L. and Potter, D.A. (1986) Aerial dispersal behavior of larval bagworms, *Thyridopteryx ephemeraeformis*. *Canadian Entomologist* 118, 525–536.

Cox, D.L. and Potter, D.A. (1988) Within-crown distributions of male and female bagworm (Lepidoptera: Psychidae) pupae on juniper as affected by host defoliation. *Canadian Entomologist* 120, 559–567.

Cronin, J.T. (1989) Inverse density-dependent parasitism of the bagworm, *Thyridopteryx ephemeraeformis* (Lepidoptera: Psychidae). *Environmental Entomology* 18, 403–407.

Dalla Torre, K.W. and Strand, E. (1929) Psychidae. *Lepidopterorum Catalogus*, 34, 1–215, W. Junk, Berlin, Germany.

Davis, D.R. (1964) Bagworm moths of the Western Hemisphere (Lepidoptera: Psychidae). *Bulletin of the United States National Museum* 244, 1–233.

Davis, D.R. (1975) A review of the West Indian moths of the family Psychidae with descriptions of new taxa and immature stages. *Smithsonian Contributions to Zoology* 188, 1–66.

Davis, D.R. (1987) Micropterigidae, Eriocraniidae, Acanthopteroctetidae, Nepticulidae, Opostegidae, Tischeriidae, Heliozelidae, Adelidae, Incurvariidae, Prodoxidae, Tineidae, Psychidae, Ochsenheimeriidae, Lyonetiidae, Gracillariidae. In: Stehr, F.W. (ed.) *Immature Insects*. Kendall/Hunt, Dubuque, IA, pp. 341–378, 456, 459–460.

Davis, D.R. (2003) A Monograph of the Family Arrhenophanidae (Lepidoptera, Tineoidea). *Smithsonian Contributions to Zoology* 620, 1–80.

Davis, D.R. and Edwards, E.D. (2003) *Notiophanes* Davis and Edwards, new genus. *Notiophanes fuscata* Davis and Edwards, new species. In: Davis, D.R. (ed.) A Monograph of the Family Arrhenophanidae (Lepidoptera, Tineoidea). *Smithsonian Contributions to Zoology* 620, 21–22.

Davis, D.R. and Robinson, G.S. (1999) The Tineoidea and Gracillarioidea. In: Kristensen, N.P. (ed.) *Lepidoptera, Moths and Butterflies, Volume 1: Evolution, Systematics and Biogeography*. Handbuch der Zoologie Band IV Arthropoda: Insecta. 35. Walter de Gruyter, Berlin, Germany, pp. 91–118.

Dodd, F.P. (1906) Notes on some remarkable parasitic insects from north Queensland. *Transactions of the Entomological Society of London* 1906, 119–123.

Dodd, F.P. and Meyrick, B.A. (1911) Some remarkable ant-friend Lepidoptera of Queensland. *Transactions of the Entomological Society of London* 1911, 577–590.

Doubleday, E. (1846) Descriptions of some new or imperfectly characterized Lepidoptera from Australia. In: Stokes, J.L. (ed.) *Discoveries in Australia: With an Account of the Coasts and Rivers Explored and Surveyed during the Voyage of H. M. S. Beagle, in the Years 1837–1843*, 1 (Appendix). T. & W. Boone, London, pp. 513–520.

Firake, D.M., Behere, G.T., Arnscheid, W.R., Kumar, R. and Ngachan, S.V. (2018) First report of coconut case caterpillar, *Mahasena corbetti* Tams (Lepidoptera: Psychidae) from India. *Phytoparasitica* 46, 355–364.

Froggatt, W.W. (1907) *Australian Insects*. William Brooks & Co, Sydney, NSW.

Gara, R.I., Sarango, A. and Cannon, P.G. (1990) Defoliation of an Ecuadorian mangrove forest by the bagworm *Oiketicus kirbyi* Guilding (Lepidoptera: Psychidae). *Journal of Tropical Forest Science* 3, 181–186.

Gries, R., Khaskin, G., Tan, Z.X., Zhao, B.G., King, G.G.S. *et al.* (2006) (1*S*)-1-ethyl-2-methylpropyl dimethylpentadecanoate: Major sex pheromone component of *Paulownia* bagworm, *Clania variegata*. *Journal of Chemical Ecology* 32, 1673–1685.

Hamilton, C.A., St Laurent, R.A., Dexter, K., Kitching, I.J., Breinholt, J.W. *et al.* (2019) Phylogenomics resolves major relationships and reveals significant diversification rate shifts in the evolution of silk moths and relatives. *BMC Evolutionary Biology* 19, 182.

Heather, N.W. (1975) Life history and biology of the bagworm moth, *Hyalarcta huebneri* (Westwood) (Lepidoptera: Psychidae). *Journal of the Australian Entomological Society* 14, 353–361.

Kawahara, A.Y., Plotkin, D., Espeland, M., Meusemann, K., Toussaint, E.F.A. *et al.* (2019) Phylogenomics reveals the evolutinary timing and pattern of butterflies and moths. *Proceedings of the National Academy of Sciences* 116, 22657–22663.

Kawahara, A.Y., Storer, C., Carvalho, A.P.S., Plotkin, D.M., Condamine, F.L. *et al.* (2023) A global phylogeny of butterflies reveals their evolutionary history, ancestral hosts and biogeographic origins. *Nature Ecology and Evolution* 7, 903–913.

Kim, M.J., Kim, I. and Cameron, S.L. (2020) How well do multispecies coalescent methods perform with mitochondrial genomic data? A case study of butterflies and moths (Insecta: Lepidoptera). *Systematic Entomology* 45, 857–873.

Kozhanchikov, I.V. (1956) Nasekomye tscheschuelrylye, Fauna of the USSR, Lepidoptera, Psychidae. *Zoological Institute of the Academy of Sciences of the USSR* 3, 1–516. [1969 English Translation].

Lelana, N.E., Utami, S., Darmawan, U.W., Nuroniah, H.S., Darwo, A. *et al.* (2022) Bagworms in Indonesian plantation forests: Species composition, pest status, and factors that contribute to outbreaks. *Diversity* 14(471), 1–20.

Leonhardt, B.A., Neal, J.W., Klun, J.A., Schwarz, M. and Plimmer, J.R. (1983) An unusual lepidopteran sex pheromone system in the bagworm moth. *Science* 219, 314–316.

Manurung, C.F.B. and Anwar, R. (2023) Population dynamics of the bagworm, *Clania tertia* Templeton (Lepidoptera: Psychidae) on immature oil palm at Pelalawan Regency, Riau. *IOP Conference Series: Earth and Environmental Science* 1208(012019), 1–8.

Mayer, C., Dietz, L., Call, R., Kukowka, S., Martin, S. and Espeland, M. (2021) Adding leaves to the Lepidoptera tree: Capturing hundreds of nuclear genes from old museum specimens. *Systematic Entomology* 46, 649–671.

Meyrick, E. and Lower, O.B. (1907) Revision of the Australian Psychidae. *Transactions of the Royal Society of South Australia* 31, 192–208.

Moore, R.G. and Hanks, L.M. (2004) Aerial dispersal and host plant selection by neonate *Thyridopteryx ephemeraeformis* (Lepidoptera: Psychidae). *Ecological Entomology* 29, 327–335.

Morden, R.D. and Waldbauer, G.P. (1971) The developmental rates of *Thyridopteryx ephemeraeformis* from four latitudes and notes on its biology (Lepidoptera: Psychidae). *Entomological News* 82, 151–156.

Morden, R.D. and Waldbauer, G.P. (1980) Diapause and its termination in the psychid moth, *Thyridopteryx ephemeraeformis. Entomologia Experimentalis et Applicata* 28, 322–333.

Mutanen, M., Wahlberg, K. and Kaila, L. (2010) Comprehensive gene and taxon coverage elucidates radiation patterns in moths and butterflies. *Proceedings of the Royal Society B* 277, 2839–2849.

Nielsen, E.S. and Edwards, E.D. (1996) *Psychidae*. In: Nielsen, E.S., Edwards, E.D. and Rangsi, T.V. (eds) *Checklist of the Lepidoptera of Australia.* Monographs on Australian Lepidoptera, 4, CSIRO Publishing, Collingwood, Victoria, pp. 33–35.

Niitsu, S., Sugawara, H. and Hayashi, F. (2017) Evolution of female-specific wingless forms in bagworm moths. *Evolution & Development* 19, 9–16.

Oke, C.G. (1948) Description of a new species of casemoth (Lepidoptera, Psychidae). *Memoirs of the National Museum of Victoria* 15, 178–179.

Okore, O.O., Ekedo, C.M., Ibediugha, B.N. and Ozua, G.O. (2016) Pest status of *Thyridopteryx ephemeraeformis* (Lepidoptera: Psychidae) (Bagworms) on *Delonix regia* (Fabales: Fabaceae), an ornamental tree. *International Journal of Agriculture and Earth Science* 2, 43–51.

Peigler, R.S. and Oberprieler, R. (2017) Ethnographic description of cocoons and silk of the moth families Saturniidae, Lasiocampidae and Psychidae. *Nachrichten des Entomologischen Vereins Apollo* 38, 113–120.

Rahmani, R., Carrasco, D., Svensson, G.P., Roweck, H., Ryrholm, N. *et al.* (2020) Identification and synthesis of putative pheromone components of the threatened salt marsh bagworm moth, *Whittleia retiella* (Lepidoptera: Psychidae). *Journal of Chemical Ecology* 46, 115–127.

Regier, J.C., Mitter, C., Davis, D.R., Harrison, T.L., Sohn, J.-C. *et al.* (2015) A molecular phylogeny and revised classification for the oldest ditrysian moth lineages (Lepidoptera: Tineoidea), with implications for ancestral feeding habits of the mega-diverse Ditrysia. *Systematic Entomology* 40, 409–432.

Rhainds, M. (2018) Natural mating disruption in a protogynous bagworm (Lepidoptera: Psychidae). *Ecological Entomology* 43, 543–546.

Rhainds, M. and Cabrera-La Rosa, J.C. (2010) *Oiketicus kirbyi* (Lepidoptera, Psychidae), a key pest in Peruvian orchards of avocado. *International Journal of Pest Management* 56(2), 103–107.

Rhainds, M., Davis, D.R. and Price, P.W. (2009) Bionomics of bagworms (Lepidoptera: Psychidae). *Annual Review of Entomology* 54, 209–226.

Ripley, L.B. and Webb, D.V.M. (1954) Aerial oil-spraying on wattle plantations against bagworms. *Farming in South Africa* 29, 343, 495–499.
Robinson, G.S. (1988) A phylogeny for the Tineoidea. *Fauna Entomologica Scandinavica* 19, 117–129.
Robinson, G.S. and Nielsen, E.S. (1993) *Tineid Moths of Australia (Lepidoptera)*. Monographs on Australian Lepidoptera, 2. CSIRO Publications, Collingwood, Victoria.
Sattler, K. (1991) A review of wing reduction in Lepidoptera. *Bulletin of the British Museum of Natural History* 60, 243–288.
Sauter, W. and Hättenschwiler, P. (1991) Zum System der palaearktischen Psychiden (Lep. Psychidae) 1. Teil: Liste der palaearktischen Arten. *Nota Lepidopterologica* 14, 69–89.
Sauter, W. and Hättenschwiler, P. (1999) Zum System der palaearktischen Psychiden (Psychidae) 2. Teil: Bestimmungsschlüssel für die Gattungen. *Nota Lepidopterologica* 14, 69–89.
Sauter, W. and Hättenschwiler, P. (2004) Zum System der palaearktischen Psychidae. 3. Teil: Bestimmungsschlüssel für die Säcke. *Nota Lepidopterologica* 27, 59–69.
Scalercio, S. and Malaisse, F. (2009) Between species and ethnospecies: Edible Psychidae in tropical Africa. *Entomologie Faunistique* 62, 17–24.
Scott, A.W. (1864) *Australian Lepidoptera and Their Transformations, Drawn from the Life*, vol. 1. John van Voorst, London.
Sheppard, R.F. (1975) The bagworm, *Thyridopteryx ephemeraeformis*: A model system for studying the principles of insect population dynamics. *Bulletin of the Entomological Society of America* 21, 153–156.
Sobczyk, T. (2011) *Psychidae (Lepidoptera). World Catalogue of Insects*, 10. Apollo Books, Stenstrup, Denmark.
Stephens, C.S. (1962) *Oiketicus kirbyi* (Lepidoptera: Psychidae): A pest of bananas in Costa Rica. *Journal of Economic Entomology* 55, 381–386.
Sugiura, S. (2016) Bagworm bags as portable armour against invertebrate predators. *PeerJ* 15(4), e1686.
Sulaiman, N.M. and Talip, M.S.A. (2021) Sustainable control of bagworm (Lepidoptera: Psychidae) in oil palm plantation: A review paper. *International Journal of Agriculture, Forestry and Plantation* 11, 47–55.
Timmermans, M.J.T.N., Lees, D.C. and Simonsen, T.J. (2023) Towards a mitogenomic phylogeny of Lepidoptera. *Molecular Phylogenetics and Evolution* 79, 169–178.
Turner, A.J. (1947) Revision of the Australian Psychidae (Lepidoptera). *Proceedings of the Royal Society of Queensland* 57, 57–64.
Tutt, J.W. (1900) *A Natural History of the British Lepidoptera 2*. Swan Sonnenschein, London.
Wahid, M.B., Hassan, H.A.H. and Masijan, Z. (1988) Bagworms (Lepidoptera: Psychidae) of oil palms in Malaysia. *PORIM Occasional Paper*, Palm Oil Research Institute, Malaysia 23, 37.
Ward, K.E., Ramaswamy, S.B. and Nebeker, T.E. (1990) Influence of host type and host switching on nutritional performance of the bagworm *Thyridopteryx ephemeraeformis* Haworth (Lepidoptera: Psychidae) *Mississippi Agricultural and Forestry Experiment Station Technical Bulletin* 170, 13–20.
Westwood, J.O. (1854) Description of some species of lepidopterous insects belonging to the genus *Oiketicus*. *Proceedings of the Zoological Society of London* 22, 219–243.

15 The Cossoidea: Wood Borers, Sun Moths, Clearwing Moths and their Relatives

Axel Kallies
University of Melbourne, Parkville, Victoria, Australia

Abstract
Cossoidea are a morphologically diverse superfamily, ranging from huge, cryptically patterned Cossidae to butterfly-like colourful Castniidae or clearwinged Sesiidae. Members of the Cossoidea also show a wide range of behaviours, including nocturnal activity, as seen in most families, and flight strictly limited to the brightest time of the day, as seen in Castniidae, Sesiidae and most Brachodidae. All cossoid families, however, share an endophagous larval lifestyle and have mobile pupae with specialized features that allow them to move freely within their tunnels from which they protrude before eclosion. This chapter summarizes our current knowledge of the Cossoidea and places the rich Australian fauna in a global context

15.1 Introduction

Cossoidea are a morphologically exceptionally diverse superfamily, which includes some of the largest and heaviest moths in the world with powerful bodies and narrow wings (in the family Cossidae) as well as butterfly-like moths with slender bodies (e.g. in Castniidae) or small Hymenoptera-mimicking moths (in the family Sesiidae). Although the precise composition of the superfamily is still uncertain, it is generally accepted that Cossidae, Metarbelidae, Ratardidae, Castniidae, Sesiidae, Brachodidae and Dudgeoneidae make up the 'cossoid assemblage' of families (Edwards *et al.*, 1999; Regier *et al.*, 2009; Mutanen *et al.*, 2010; van Nieukerken *et al.*, 2011). The Cossoidea can be found in all zoogeographic regions of the world apart from New Zealand. Some families, however, are restricted to particular parts of the world. Metarbelidae, for example, are limited to Africa and southern Asia, Ratardidae to south-east Asia and Dugeonidae to the Old-World tropics and subtropics including Australia. Brachodidae are absent from North America, whereas Castniidae may have a Gondwanan origin, being species rich in Australia and South America, with just a few species known from South-east Asia.

The larvae of almost all members of the superfamily are endophagous and feed in the roots, stems, petioles or fruits of their host plants where they also pupate. The adults of most cossoid families are active at night; however, Castniidae, Sesiidae and Brachodidae are almost exclusively active during the day and prefer to fly in bright sunshine.

Corresponding author: axel.kallies@unimelb.edu.au

DOI: 10.1079/9781800627680.0015

15.2 Taxonomic History of the Superfamily

Over the past two decades, considerable progress has been made in resolving the phylogenetic relationships among the Lepidoptera, with robust superfamily-level frameworks now available for most major clades (Regier *et al.*, 2013; Heikkila *et al.*, 2015; Regier *et al.*, 2015a, b). The cossoid assemblage, however, remains one of the more challenging groups to delimit both morphologically and molecularly.

Early classifications proposed several different superfamilies to accommodate these lineages, including Cossoidea, Sesioidea and Castnioidea (Brock, 1971; Scott, 1986). Cossoidea initially included Cossidae and Metarbelidae, which were associated in different ways, either as subfamilies in Cossidae or distinct families in a Cossoidea superfamily concept. The superfamily Sesioidea was established by Brock (1971) who also included Choreutidae, Dudgeoneidae and Hyblaeidae in this concept. Heppner and Duckworth (1981) excluded the latter two and instead included Brachodidae. Subsequent studies, however, failed to find any support for a closer relationship between Choreutidae and other sesioid or cossoid families (Regier *et al.*, 2009; Mutanen *et al.*, 2010; Cho *et al.*, 2011). Castniidae were initially considered as constituting their own superfamily (Brock, 1971), but were later incorporated into Sesioidea based on several adult thoracic and larval synapomorphies (Minet, 1986, 1991). Edwards *et al.* (1999) proposed a cossoid assemblage of families that also included Dudgeoneidae and Ratardidae, which previously had been associated with various families and superfamilies (see below). Molecular studies failed to recover Sesioidea and Cossoidea as monophyletic units, often intermixing their constituent taxa (Regier *et al.*, 2009; Mutanen *et al.*, 2010; Cho *et al.*, 2011). Consequently, van Nieukerken *et al.* (2011) proposed collapsing Sesioidea into an expanded Cossoidea, a phylogenetic concept that is followed here.

Although this broader concept of Cossoidea is now widely recognized, its composition and family-level relationships remain unresolved, largely due to limited taxon sampling and potential rapid diversification events that obscure deep splits.

15.3 Morphological Definition

Although several morphological traits in combination generally enable the identification of Cossoidea, the superfamily currently lacks robustly supported autapomorphies. The most distinguishing feature of cossoid moths is their endophagous lifestyle, with larvae tunnelling in live plant tissue. This trait may represent an ecological synapomorphy for the superfamily. A notable exception are the Phycodinae (a subfamily of Brachodidae), the larvae of which feed between spun leaves.

Another consistent feature among most cossoids is the morphology of their highly mobile pupae, which possess two rows of tooth-like dorsal projections on each abdominal segment and a prominent projection on the head. Thus, Cossoidea fall within the non-obtectomeran grade of Apoditrysia, a clade defined by the presence of such mobile pupae with spinose abdomens that are extruded from their larval tunnels before adult emergence. Two additional larval characters proposed by Minet (1991) further support the monophyly of Cossoidea: a melanized oblique stripe across the prothoracic shield linking setae D1 and D2, and dorsally or anterodorsally concentrated, irregularly distributed spinules on most larval segments.

Many cossoid taxa retain plesiomorphic wing venation, such as a forked M-stem in both wings, a trait suggesting that Cossoidea may represent one of the earliest-diverging lineages within the Apoditrysia (Edwards *et al.*, 1999).

Although molecular data support the coherence of the cossoid group, further phylogenetic resolution will require more extensive genomic datasets and broader taxon sampling.

15.4 Cossoid Families

15.4.1 Cossidae

The Cossidae are a well-established family, members of which usually can be readily recognized by a combination of characters. They typically have long and narrow forewings that are cryptically patterned often greyish to brown with fine dark speckles, striae or reticulation, bodies that are densely covered in roughened scales and they lack a proboscis. Males have bipectinate antennae, in the subfamily Zeuzerinae only to about two-thirds of their length. Despite the general acceptance of the family concept, well-defined synapomorphies are harder to pinpoint. Minet (1986) and Schoorl (1990) suggested that the family is defined by an autapomorphy of the prothorax, namely the medially adjacent, well-projecting parapatagia. Notably, comprehensive modern studies that support the monophyly of Cossidae based on molecular analyses are lacking.

Cossidae are typically robust, medium to very large moths. Indeed, the females of some Australian species, including *Endoxyla cinereus*, are considered to be among the largest and heaviest moths in the world. Although Cossidae are mostly nocturnal, some species are strictly day flying, including the Palaearctic Stygiinae (Yakovlev, 2011a; A. Kallies, personal observations).

Cossidae are distributed worldwide, with the exception of New Zealand. Worldwide, there are more than 1000 species in about 170 genera known (van Nieukerken *et al.*, 2011); however, many new species and genera have been described in the last decade, mainly from Asia and Africa (e.g. Yakovlev, 2011b). Traditionally, Cossidae were split into two main subfamilies, the Cossinae and Zeuzerinae. More recently, numerous subfamilies have been erected that have yet to be tested in molecular studies (e.g. Yakovlev, 2011a, b).

The larvae of most Cossidae bore into woody plant tissue, sometimes taking several years to develop. The larvae of some European and Asian species develop in the roots of herbaceous plants. Almost all feed in dicotyledonous plants; exceptions include the Palaearctic genus *Phragmataecia*, the larvae of which bore in large reeds (Gramineae) (Kalshoven, 1981), *Eogystia* which tunnel in *Asparagus officinalis* (Liliaceae) (Tian *et al.*, 2010), *Dyspessa* which infest *Allium* bulbs (Amaryllidaceae) (e.g. Özgen *et al.*, 2024) and the Australian *Cossodes lyoneti* and *Ptilomacra senex*, which are thought to develop in grass trees (*Xanthorrhoea*, Asphodelaceae).

Australia has a rich fauna of Cossidae. The most detailed studies of the Australian Cossidae were conducted by the Australian entomologist Edward Turner, who described the majority of the currently recognized 91 species and some genera. He summarized the known data relating to Australian cossid moths in his monograph (Turner, 1945). Schoorl (1990) completed a revision of the world Cossidae at the generic level, resulting in the majority of Australian Cossidae being placed in the species-rich genus *Endoxyla*. Common (1990) gave an overview of the Australian Cossidae, figured a small number of species and described some life histories. Edwards (1996a) provided a comprehensive checklist, in which many synonyms were identified. Kallies and Hilton (2012) revised the subfamily Cossinae and the genus *Sympycnodes* (Zeuzerinae). *Endoxyla* is Australia's most diverse genus with 56 species currently recognized. *Sympycnodes* and *Culama* are the second and third most speciose genera with ten and seven species respectively (Edwards, 1996a; Kallies and Hilton, 2012). Although most cossids are conspicuous and large moths, many Australian species remain undescribed. Cossidae can be found in all parts of Australia, from the rainforests in the north-east to the dry inland regions in the centre. Notably, at least two species have brachypterous females, including the common *Endoxyla amphiplecta* (Figs 15.1 and 15.2) which occurs in inland Victoria, South Australia and New South Wales. Currently only two subfamilies are known from Australia – Zeuzerinae and Cossinae. The origin of the diverse Australian cossid fauna is unknown but parts of the Australian Zeuzerinae show strong affinities to the Oriental fauna (Edwards, 1996a). This may also be the case for the Australian Cossinae, most of which share the general appearance and basic genitalic features of typical Palaearctic and Asian Cossinae. Some genera, however, including *Cossodes* (Fig. 15.3), *Ptilomacra*, *Brevomorpha*, *Idioses*, *Charmoses*, *Eusthenica* and *Brevicyttara*, show few if any similarities to taxa outside Australia, and their subfamily status is uncertain. The genus *Archaeoses*, traditionally included in Cossidae, was recently moved to Dudgeoneidae (Mutanen *et al.*, 2010; Bazinet *et al.*, 2013; Regier *et al.*, 2013; Kawahara *et al.*, 2019).

15.4.2 Dudgeoneidae

The palaeotropical moth family Dudgeoneidae has long presented taxonomic challenges. Initially described as part of the Cossidae, *Dudgeonea* was later elevated to familial status (Berger, 1957). Subsequently, different authors proposed varying superfamily placements. Brock (1971) provisionally transferred Dudgeoneidae to the Sesioidea based on unspecified morphological similarities, whereas Minet (1983) placed it within the Pyraloidea, citing the presence of abdominal tympanal organs as a key character. Common (1990), however, emphasizing morphological coherence and biological similarity with cossoid taxa, reassigned Dudgeoneidae to Cossoidea. This placement was reinforced by Minet (1991), who identified two shared synapomorphies uniting Dudgeoneidae with Cossidae: the reduction or complete absence of the adult proboscis and the presence of a pair of small, distinct pits on the lower frontoclypeus.

Dudgeoneidae were long considered monobasic, including only a single genus, *Dudgeonea* (Holloway, 1986; Common, 1990). Adults of the genus *Dudgeonea* (Fig. 15.4) are medium-sized, robust moths with reddish-brown coloration and distinctive forewings adorned with large, circular, silvery-white spots. Their wing venation resembles that of more plesiomorphic members of Cossinae (Common, 1990). *Dudgeonea* is known from Africa, Madagascar, the Indian subcontinent,

Fig. 15.1. *Endoxyla amphiplecta* (Cossidae, Zeuzerinae), Big Desert, Victoria, male. Photo: John Lenegan.

Fig. 15.2. *Endoxyla amphiplecta* (Cossidae, Zeuzerinae), Big Desert, Victoria, female. Photo: Fabian Douglas.

Fig. 15.3. *Cossodes lyoneti* (Cossidae), Stratton, Western Australia. Photo: Jean and Fred Hort.

Fig. 15.4. *Dudgeonea* species (Dudgeoneidae), Fraser Hill, Malaysia. Photo: David Fisher.

South-east Asia, New Guinea and northern Australia (Holloway, 1986; Edwards, 1996b).

Schoorl (1990) proposed that the 'cossid' genus *Acritocera* from Fiji should be included in Dudgeoneidae, a view supported by Edwards *et al.* (1999) who reported tympanal organs to be present in this genus. *Dudgeonea* and *Acritocera* exhibit several morphological characters not observed in typical Cossidae (Schoorl, 1990), further supporting their combination at the family level. Van Nieukerken *et al.* (2011), following Mutanen *et al.* (2010), also included Cossulinae in the Dudgeoneidae, a relationship already suggested by Edwards *et al.* (1999) but not universally accepted. The Australian genus *Archaeoses* was recently assigned to Dudgeoneidae based on molecular and morphological analyses (see below). Accepting this broader concept of the family, there are about 60 species known worldwide.

The biology of Dudgeoneidae is poorly known. Larvae of the Australian species *Dudgeonea actinias* tunnel in trunks of *Psydrax attenuata* (Rubiaceae) (Turner, 1902). The larvae of the Australian *Archaeoses polygrapha* tunnel in *Myoporum insulare* (Scrophulariaceae) (A. Kallies, personal observation). *Acritocera* negligens feeds in cocos palms (Simmonds, 1921), a trait otherwise only found in Brachodidae. The morphology of the eggs, larvae and pupae remains unknown. Adults are taken at light at night (Common, 1990).

In Australia, three species of the genus *Dudgeonea* are known, but additional undescribed species have been recognized, with records from northern New South Wales, throughout Queensland and from central Western Australia. *Archaeoses* (Fig. 15.5) is an Australian endemic genus. It currently contains three named species, but a larger number of species remain to be described. The genus is mainly found in the southern half of Australia and is relatively species rich in Western Australia. It was excluded from Cossidae by Schoorl (1990), a view supported by recent molecular evidence that places it within the Dudgeoneidae (Mutanen *et al.*, 2010; Bazinet *et al.*, 2013; Regier *et al.*, 2013). This is also supported by morphological studies, which identified the presence of pilifers in both genera, that are otherwise absent in the Cossidae (Edwards *et al.*, 1999). Furthermore, Davis (2006) described small, paired tuberculate evaginations on abdominal tergites 2–8 shared by *Archaeoses* and *Dudgeonea*. *Archaeoses*, however, lacks the unique tympanic organs that are a characteristic of *Dudgeonea*.

15.4.3 Metarbelidae

Metarbelidae include medium-sized strong-bodied cossoid moths, often characterized by intricately patterned forewings (Fig. 15.6). The group is extraordinarily species rich in Africa, with many genera and more than 250 described species. The family is also present in tropical and subtropical Asia where several genera with more than 50 species are known. It is, however, absent from Australia. Metarbelidae have been variably included as a subfamily in the Cossidae or considered a distinct family. Minet (1986) and Edwards

Fig. 15.5. *Archaeoses polygrapha* (Dudgeoneidae), Mornington Peninsula, Victoria. Photo: Roger Standen.

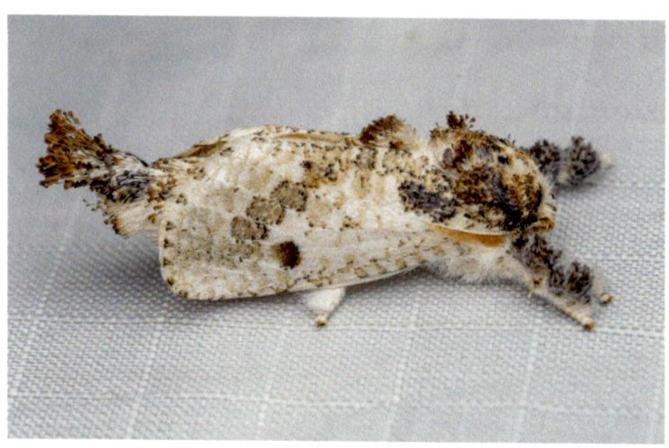

Fig. 15.6. *Indarbela obliquifasciata* (Metarbelidae), Hongkong, China. Photo: Artur Tomaszek.

et al. (1999) included them as a subfamily in the Cossidae, primarily based on the presence of highly developed parapatagia. In contrast, Roepke (1957) (as Squamuridae), Brock (1971), Holloway (1986) and Schoorl (1990) recognized Metarbelidae as a distinct family within the superfamily Cossoidea. Schoorl (1990) argued that although the presence of parapatagia may represent a synapomorphy shared between Metarbelidae and Cossidae, Metarbelidae are more appropriately treated as the sister group to Cossidae. This arrangement was followed by van Nieukerken (2011). Metarbelidae larvae are bark feeders, sheltering in short tunnels during the day (Gardner, 1945; Roepke, 1957; Holloway, 1986). The figure 8-shaped ovipositor lobes are apomorphic and provide an obvious link with the Ratardidae (Edwards *et al.*, 1999).

15.4.4 Ratardidae

Ratardidae are an elusive and poorly known family. They currently comprise three genera encompassing approximately ten species distributed from India through South-east Asia to Taiwan. The slender, butterfly-like morphology and rounded wings of the Ratardidae (Fig. 15.7) are considered highly apomorphic within the Cossoidea.

Fig. 15.7. *Ratarda excellens* (Ratardidae), Taiwan. Photo: Licheng Shih.

Throughout the 20th century, the systematic position of this group has been subject to extensive debate, having variously been placed within the Cossoidea, Bombycoidea and Calliduloidea (see Minet, 1987). The genus *Ratarda* was originally described within the Zygaenidae and subsequently elevated to family status by Hampson (1898). Subsequently, it has been included as a subfamily in the Cossidae (Edwards *et al.*, 1999) or considered a distinct family (Hering, 1926; Holloway, 1986; Heppner and Wang, 1987; Kobes and Ronkay, 1990; Owada, 1993). The biology and immature stages of Ratardidae remain unknown. Adults are presumed to be diurnal based on observational data, although individuals have also been collected at light, suggesting potential crepuscular or facultative nocturnal activity. The family is absent from Australia.

15.4.5 Brachodidae

Brachodidae are a small family of small to medium-sized moths that are characterized by strong bodies and narrow wings. The forewings are usually cryptically coloured, whereas the hindwings are often brightly coloured yellow, orange or white with a black border. Species of the subfamily Phycodinae often have intricate, sometimes metallic, patterns on their forewings.

Brachodidae occur worldwide with the exception of North America. Brachodinae are the largest subfamily, occurring in the Palaearctic, South America, South Africa, Asia and Australia (Heppner and Duckworth, 1981). Phycodinae contain fewer species and are mainly found in South America, throughout Africa, tropical and subtropical Asia and northern Australia (Heppner and Duckworth, 1981). The Pseudocossinae are currently only described from Madagascar. Recently, however, a species was discovered in far north Queensland, which may prove to belong to this group (A. Kallies, unpublished). The Brachodidae of the Palaearctic, Afrotropical and Oriental regions and adjacent areas including Papua New Guinea were revised relatively recently (e.g. Kallies, 1998a, b, 2001a, 2002, 2004, 2013, 2016).

Historically, members of Brachodidae were usually included in various families such as

Glyphipterigidae (Meyrick, 1914) or Tineidae (Rebel, 1901). Later, Agenjo (1966) erected Brachodinae as a subfamily in Glyphipterigidae. The current concept of Brachodidae was proposed by Heppner and Duckworth (1981) and included Brachodinae and Phycodinae. The Pseudocossinae with the single genus *Pseudocossus* were transferred to Brachodidae by Minet (1991) but are still included as a subfamily of Cossidae in some studies (Yakovlev, 2011b). Currently there is no strong evidence that would support the monophyly of the Brachodidae. Morphological and molecular data, however, provide good evidence that Brachodidae are part of the Cossoidea (Heppner and Duckworth, 1981; Minet, 1991; Edwards *et al.*, 1999; Mutanen *et al.*, 2010; van Nieukerken *et al.*, 2011). Possible autapomorphies were discussed by Minet (1991), Kozlov *et al.* (1998), Edwards *et al.* (1999) and Kallies (2004). The inclusion of Phycodinae into the Brachodidae appears poorly supported. Indeed, based on morphology, life history and preliminary molecular data, Phycodinae may not be part of the Cossoidea. Accordingly, Brachodidae may represent one or several lineages within a broader concept of Cossoidea.

Within the superfamily Cossoidea, Brachodinae and Pseudocossinae are unusual due to their exclusive association with monocotyledonous host plants, including grasses, sedges and palms, a characteristic only shared with Castniidae and some unusual cossoid taxa from Australia and the Palaearctic region (Common, 1990; Edwards *et al.*, 1999; Kallies and Hilton 2012, and unpublished observations of the author). In contrast, Phycodinae feed on a variety of dicotyledons, including and predominantly *Ficus*. Furthermore, Phycodinae differ from typical cossoid moths in their biology, with larvae feeding externally among spun leaves (Fletcher, 1920; Diakonoff, 1986; Arita, 1989; Gielis, 2008; Kumar and Ramamurthy, 2010; Kallies, 2013). Brachodidae are usually active during the day, tend to fly rapidly in sunshine and are rarely attracted to light. A few species are crepuscular. Currently, there are more than 160 species known worldwide. Australia has a relatively rich fauna of Brachodidae, with 22 species in four genera currently known and a similar number of species that await description. The family is mainly represented by the subfamily Brachodinae. The largest genus is *Miscera* (Figs 15.8 and 15.9) with about 19 described species. They are mainly distributed in the southern half of the continent, with south-western Australia being a

Fig. 15.8. *Miscera homotona* (Brachodidae), Talegalla Weir, Queensland. Photo: Scott Gavins.

Fig. 15.9. *Miscera* species (Brachodidae), East of Perth, Western Australia. Photo: Jean and Fred Hort.

clear centre of diversity. *Miscera* represents an endemic Australian element. Another group of genera with close affinities to the Oriental fauna is known from tropical Queensland. This includes *Synechodes coniophora* and an unnamed species of the genus *Saccocera* that was recently discovered (A. Kallies, personal observation). Phycodinae are represented by two undescribed species in the genus *Nigilgia* (one often misidentified *as Nigilgia adjectella*) (Fig. 15.10) and one species of *Paranigilgia* (Common, 1990; A. Kallies unpublished).

The biology of Brachodidae is poorly known. *Miscera* species are associated with sedges (Cyperaceae) and *Lomandra* (Asparagaceae). Larvae have been found in the roots of different Cyperaceae, including *Lepidosperma* (A. Kallies, unpublished). Two species from far-northern Queensland that belong to an unnamed genus are known to feed in the fruits of *Calamus* palms. *Synechodes coniophora* feeds in the petioles of different palms, including the non-Australian oil palm (A. Kallies, personal observations). The biologies of Australian Phycodinae have not been documented but based on related species from Asia, it is likely that their larvae develop between the spun leaves of Moraceae.

15.4.6 Castniidae

Castniidae, commonly referred to as sun moths, are a distinctive family of diurnal cossoid moths that are characterized by their slender bodies, clubbed antennae and morphological convergence with butterflies. The cryptic forewings contrast with brightly coloured and patterned hindwings,

Fig. 15.10. *Nigilgia* species (Brachodidae, Phycodinae), Airlie Beach, Queensland. Photo: Alison Pearson.

contributing to their mimicry and aposematic signalling strategies. Adults are obligate day flyers, and larvae are internal stem or root borers specialized on monocotyledonous host plants.

The family is currently divided into two subfamilies. Castniinae is the more species-rich subfamily, comprising approximately 100 species in 18 genera in Central and South America and 42 species in Australia (Miller, 1986; Lamas, 1995; Edwards, 1996c; Moraes and Duarte, 2014; Kallies and Edwards, 2025). Tascininae consists solely of the genus *Tascina*, which includes four described species distributed across South-east Asia (Fukuda, 2000).

Diagnostic morphological traits of Castniidae include broad wings with primitive venation (including forked M-stems), abruptly clubbed antennae, a bifurcate saccus in the male genitalia, an enlarged phallobase and recurved phallus, and a highly extensible ovipositor in females. Although these characters are not entirely exclusive to Castniidae, their combination provides a strong suite of diagnostic features within the Lepidoptera (Edwards *et al.*, 1999).

Larvae across Castniidae are endophagous within monocotyledonous plants. In South America, documented host families include Arecaceae (palms), Bromeliaceae, Orchidaceae, Musaceae (bananas) and Poaceae (sugarcane) (Miller, 1986; González and Cock, 2004; Díaz, 2022). Australian taxa are associated with Poaceae (grasses), Cyperaceae (sedges) and Asparagaceae (e.g. *Lomandra*), occupying dry sclerophyll habitats such as sedge-heathlands, savannahs and grasslands (Edwards *et al.*, 1999; A. Kallies, personal observation). The biology and larval host associations of *Tascina* remain unknown. Notably, monocot specialization is a recurrent trait across Castniidae, Brachodinae and Pseudocossinae, the latter two encompassing the majority of brachodid Cossoidea. This shared ecological preference may represent a convergent or potentially synapomorphic trait relevant to higher-level phylogenetic relationships within the Cossoidea. In Australia, the family Castniidae is represented exclusively by the endemic genus *Synemon*. Until relatively recently, this genus included 24 recognized species (Edwards, 1996c). Systematic revisions and descriptions of new taxa (e.g. Grund, 2011; Grund and Stolarski, 2012; Williams *et al.*, 2016, Williams and Williams, 2016; Kallies *et al.*, 2016, 2020; Kallies and Edwards, 2018, 2025; Braby *et al.*, 2020) have raised the total to 42 formally described species. Several distinct species groups have been identified within *Synemon* (Figs 15.11, 15.12, 15.13), some of which may warrant recognition at the generic level. The *Synemon magnifica* group is characterized by grey forewings and vivid red and black hindwings. Larvae feed on the roots of sedges (*Lepidosperma*, *Gahnia*), *Lomandra* and *Chamaexeros* (Common and Edwards, 1981; Kallies *et al.*, 2016, 2020; Williams and Williams, 2016; Williams *et al.*, 2016). The *Synemon sophia* group, comprises species mainly with yellow and black hindwings with larvae primarily associated with sedges (Williams *et al.*, 2016; Kallies *et al.*, 2020). The *Synemon collecta* group is distinguished by striated forewings and larvae feeding on grass roots. With 15 described species and at least three additional undescribed taxa, this group is currently the most species rich within *Synemon* (Kallies and Edwards, 2025). The *Synemon directa* group is a smaller lineage containing only two known species, both of which feed on sedges and exhibit a simplified wing pattern: brown forewings with a singular black-and-white transverse band (Williams *et al.*, 2016; Kallies *et al.*, 2020).

Fig. 15.11. *Synemon magnifica* (Castniidae), Royal National Park, New South Wales. Photo: David Fisher.

The high species diversity and ecological specialization within *Synemon*, particularly the grass-feeding *collecta* group, hint at adaptive diversification in response to climatic shifts and biome transitions in Australia and could be linked to the aridification and expansion of Australian grasslands during the Pleistocene.

15.4.7 Sesiidae

Clearwing moths (Sesiidae) are a globally distributed family of Cossoidea known for their striking mimicry of stinging Hymenoptera, such as wasps and bees, a classic case of Batesian mimicry. These moths exhibit a wasp-like appearance with narrow, partially or entirely transparent wings and cingulated bodies. Although they are found across a wide range of habitats, from the tundra of the northern Palaearctic to the tropics, they are notably absent from New Zealand and most Pacific islands.

Fig. 15.12. *Synemon laeta* (Castniidae), Wrattens National Park, Queensland. Photo: Braden McDonald.

Fig. 15.13. *Synemon sophia* (Castniidae), D'Entreacasteaux National Park, Western Australia. Photo: Liza Mazzella.

Their greatest diversity is observed in South America, Africa and South-east Asia (Pühringer and Kallies, 2025), with relatively few species recorded from the Australian region, including New Guinea (Kallies, 2020).

Due to their unique appearance, Sesiidae have been recognized as a distinct family for a long time. One of the first systematic studies focusing on their phylogeny was by Naumann (1971). Morphologically, Sesiidae are distinguished by a strong autapomorphy involving a unique wing-coupling mechanism, composed of a ventrally bent posterior margin of the forewing that engages with a dorsally bent anterior margin of the hindwing, both of which are lined with rows of stiff interlocking scales. This mechanism allows for the tight coupling of the fore- and hindwings along their entire length, enabling the moths to achieve a wasp-like mode of flight. As a result, Sesiidae are swift fliers that are predominantly active during the day (Naumann, 1971; Heppner and Duckworth, 1981; Edwards *et al.*, 1999; Špatenka *et al.*, 1999). Although Sesiidae have been included in most molecular phylogenetic studies, taxon sampling has usually been limited. Individual groups, however, have been analysed in detail (e.g. Bartsch *et al.*, 2021).

Larvae of Sesiidae are endophagous, feeding internally within dicotyledonous plants. Their feeding sites range from the woody tissues of trees and shrubs to the roots and stems of herbaceous plants. This feeding strategy is comparable to that observed in members of the Cossidae, another family within the cossoid assemblage

Historically, Sesiidae have been rarely encountered and are poorly represented in museum collections due to their secretive habits and limited availability through traditional sampling methods. The introduction of synthetic pheromone lures, however, has significantly improved detection rates, leading to the discovery and formal description of many new species and genera over the past 30 years including in Australia. At present, the family comprises approximately 1600 species distributed among more than 170 genera (Pühringer and Kallies, 2025).

Australia has a relatively poor sesiid fauna, with only 21 native species known to occur, in comparison with other regions of the world such as Europe with more than 100 species (Figs 15.14 and 15.15). Duckworth and Eichlin (1974) were the first to review the Australian members of the family before Kallies (2020) added a series of species and provided another comprehensive revision. Within Australia, Sesiidae are largely restricted to the tropical and subtropical north and north-eastern regions, with only a single species, *Paranthrenella chrysophanes*, extending into

Fig. 15.14. *Sphecodoptera cupreifascia* (Sesiidae), Cairns, Queensland. Photo: Dominic Funnel.

Fig. 15.15. *Melittia doddi* (Sesiidae), Cairns, Queensland. Photo: Bridgette Gower.

temperate south-eastern areas. It is noteworthy that Sesiidae appear to be absent from the semi-arid and arid interior of Australia, which is in sharp contrast to other dry parts of the world, including the Nearctic, the Palaearctic and Africa, all of which support a rich fauna of Sesiidae (Pühringer and Kallies 2025). The restricted distribution, low genus-level endemism and limited diversification into arid environments suggest a relatively recent dispersal of Sesiidae into Australia. Nevertheless, several genera, including *Micrecia*, *Lophocnema*, *Diapyra* (Tinthiinae) and *Archaeosesia* (Sesiinae), are either endemic or near-endemic to the region.

Knowledge of the biology of Australian Sesiidae remains limited, with confirmed host plant associations for only a few species. In the Tinthiinae, the larvae of *Diapyra igniflua* are known to tunnel under the bark of *Elaeocarpus grandis* (Elaeocarpaceae) (Lucas, 1894; Turner, 1917). Within Sesiinae, *Pseudosesia oberthuri* larvae induce large galls in *Ampelocissus* vines (Vitaceae), whereas *Nokona carulifera* was reared from a vine in the genus *Cissus* (Vitaceae) (Kallies, 2001b, 2020). Larvae of *Paranthrenella chrysophanes* have been found feeding beneath the bark of various tree species, most commonly *Exocarpos cupressiformis* (Santalaceae) (Duckworth and Eichlin, 1974; Common, 1990; Kallies, 2020).

15.5 Conservation of Australian Cossoidea

Some Cossoidea species are of exceptionally high conservational value. Australian sun moths, for example, have experienced significant habitat loss and, as a result, several species must now be considered endangered and are thus of significant conservation interest. Many species, including several species of the *S. collecta* group or the recently described *Synemon edwardsi*, inhabit Australia's most arable regions, including what is now the Australian Wheatbelt, comprising parts of Western Australia, South Australia, Victoria, New South Wales and south-eastern Queensland, dominated by broad-scale agriculture (Douglas, 1993; Williams and Williams, 2016; Williams *et al.*, 2016; Kallies and Edwards, 2025). These regions have been almost completely cleared of natural vegetation. In consequence, the grass, heath, sedgelands and open woodlands favoured by many sun moth species have been severely diminished and have become some of the most endangered habitats in Australia.

The golden sun moth, *Synemon plana*, the subject of Chapter 35, is one of the best-known sun moth species in Australia, occurring in open grasslands in the western and northern parts of Victoria, southern New South Wales and the Australian Capital Territory, where it is still one of the more frequently encountered sun moth species. It is clear, however, that *S. plana* has suffered substantial habitat loss, and it is therefore legally protected (O'Dwyer and Attiwill, 1999; Braby and Dunford, 2006; New *et al.*, 2007; Richter *et al.*, 2013).

One of the most acutely endangered species inhabiting temperate grasslands is *Synemon selene*, a large and distinctive species known from less than ten small locations in South Australia and Victoria (Douglas, 2008; Grund *et al.*, 2018). This species is particularly vulnerable not only due to its small and fragmented range but also because all but one of the extant colonies are parthenogenetic (Douglas, 2008), resulting in limited genetic diversity and resilience to environmental changes. Despite its precarious state, *S. selene* remains without official protection. Another species that inhabits temperate grass and open woodlands and is in urgent need of protection is *S. theresa*, currently only known from a few locations in the Adelaide Hills and the Clare valley area. Historically, it was also known from Victoria but the last records date back more than 100 years.

In addition to the grassland species, some Castniidae, Brachodidae and Cossidae that are associated tightly with winter wet swamps in coastal areas of southern Australia should be considered of high conservational value. This includes species such as *Synemon sophia*, *S. notha*, *Cossodes lyoneti* and *Miscera desmotoma* as well as several undescribed species, all of which inhabit small ranges and are known from only a few high-quality ephemeral swamp areas.

15.6 Economic and Cultural Importance

Members of the superfamily Cossoidea are not generally considered of great economic importance. There are, however, some accounts of serious damage to agriculture or forestry by Cossidae. The larvae of some species are occasionally reported to cause damage to fruit and forest trees (e.g. Kutinkova *et al.*, 2006; Gumus *et al.*, 2015; Blanco *et al.*, 2019; Scaccini *et al.*, 2021) or vegetables (e.g. Tian *et al.*, 2010; Özgen *et al.*, 2024). In Australia, the giant wood moth *Endoxyla cinereus* has been recorded as a pest of *Eucalyptus* plantations (Carne *et al.*, 1974; McInnes and Carne, 1978; Wylie and Peters, 1993; Chapter 26).

There are several species of Sesiidae that cause some damage to fruit and forestry trees. In Australia, the introduced *Synanthedon tipuliformis* damages currant bushes (*Ribes* spp., Grossulariaceae), and the native *Paranthrenella chrysophanes* and *P. xanthogyna* have adopted persimmon (*Diospyros kaki*, Ebenaceae) and avocado (*Persea americana*, Lauraceae), respectively, as host plants (Kallies, 2020; J. Watson, pers. commun.). On the other hand, several Sesiidae species were introduced into Australia as control agents for weeds: *Carmenta mimosa* for the control of *Mimosa pigra* (Fabaceae); *Pyropteron doryliformis* for the control of *Rumex* (Polygonaceae); *Chamaesphecia mysiniformis* to limit the spread of white horehound (*Marrubium vulgare*, Lamiaceae); and *Carmenta ithacae* (or a species close to it) to combat the spread of parthenium weed (*Parthenium hysterophorus*, Asteraceae). Notably, all of these species appear to have established stable populations in Australia, although their impact on their host plants has not been well documented (summarized in Kallies, 2020).

In countries other than Australia, Brachodidae have been observed to cause damage to coconut palms (*Cocos nucifera*) in Papua New Guinea (Kakul *et al.*, 2006) or found boring in oil palms in South-east Asia (*Elaeis guianeensis*) (Kallies, 1998a). Similarly, *Acritocera* (Dudgeoneidae) has been reported to cause damage to cocos palms in Fiji (Simmonds, 1921). Some Castniidae species in South America are considered serious pests of bananas (*Musa* spp., Musaceae), sugar cane, pineapple (*Ananas comosus*) and palms (including oil palm and coconut) (Arecaceae) (Myers, 1931; Cock, 1985; Miller, 1986; González and Cock, 2004). *Paysandisia archon*, introduced accidentally into Europe, is now a serious pest of palm trees (Monteys and Aguilar, 2005).

Historically, Cossidae larvae were a staple in the diets of some of the First Nations people of Australia, particularly in the dry inland. *Maku*, an aboriginal term for edible grubs, includes a range of different taxa, including Cossidae. This includes most famously 'witchetty grubs' (also spelled witchety or witjuti grub) derived from the Arabana language in the western Lake Eyre region in northern South Australia. It is a term used for the large, white insect larvae found in the roots of the witchetty bush (*Acacia kempeana*) that is widespread throughout the southern half of the Northern Territory and parts of Western and Southern Australia. 'Witchetty grub' was generally thought to apply to the larvae of *Endoxyla leucomochla*; however, *E. leucomochla* feeds on the roots of *Acacia ligulate* (Tindale, 1953) and does not occur in the range of *A. kempeana*. Indeed, barcode analysis of cossid larvae found in *A. kempeana* indicated a species different from *E. leucomochla* (Yen *et al.*, 2018). Thus, 'witchetty grub' refers to a possibly unnamed species of *Endoxyla*. The term, however, may also apply to larvae of other cossid moths, ghost moths (Hepialidae) and longhorn beetles (Cerambycidae) and is used mainly when the larvae are being considered as food. *Endoxyla cinereus* and related species were also an important Indigenous food resource (Dodd, 1916; Turpin and Si, 2017; Thurman, 2022).

References

Agenjo, R. (1966) Catalogo ordenador de los lepidopteras de Espana [part]. *Graellsia* 22, 147–155.

Arita, Y. (1989) Records of two species of *Phycodes* Guenee (Lepidoptera: Brachodidae) from Thailand. *Microlepidoptera of Thailand* 2, 15–21.

Bartsch, D., Pühringer, F., Milla, L., Lingenhöle, L. and Kallies, A. (2021) A molecular phylogeny and revision of the genus *Pyropteron* Newman, 1832 (Lepidoptera, Sesiidae) reveals unexpected diversity and frequent hostplant switch as a driver of speciation. *Zootaxa* 4972, 1–075.

Bazinet, A.L., Cummings, M.P., Mitter, K.T. and Mitter, C.W. (2013) Can RNA-Seq Resolve the rapid radiation of advanced moths and butterflies (Hexapoda: Lepidoptera: Apoditrysia)? An exploratory study. *PLOS One* 8(12), e82615.

Berger, L.A. (1957) Clé pour la détermination des familles de macrolépidoptères et des groupes supérieures de microlépidoptères (Faune éthiopienne). *Lambillionea* 57, 72–84.

Blanco, V.C.P., Burbano, O.I.I., Sarmiento-Naizaque, Z.X. and Duran, J.R. (2019) Guava borer worm (Lepidoptera: Cossidae), a limiting pest in guava: biology, lifecycle and management alternatives. *Heliyon* 5, e01252.

Braby, M.F. and Dunford, M. (2006) Field observations on the ecology of the golden sun moth *Synemon plana* Walker (Lepidoptera: Castniidae). *Austral Entomology* 33, 103–110.

Braby, M.F., Edwards, E.D. and Williams, A.A.E. (2020) A new species of sun-moth, *Synemon kimberleyensis* sp. nov. (Lepidoptera: Castniidae), from the Kimberley in Western Australia. *Journal of the Lepidopterist's Society* 74, 31–42.

Brock, J.P. (1971) A contribution towards an understanding of the morphology and phylogeny of the ditrysian Lepidoptera. *Journal of Natural History* 5, 29–102.

Carne, P.B., Greaves, R.T.G. and McInnes, R.S. (1974) Insect damage to plantation-grown eucalypts in north coastal New South Wales, with particular reference to Christmas beetles (Coleoptera: Scarabaeidae). *Journal of the Australian Entomological Society* 13, 189–206.

Cho, S., Zwick, A., Regier, J.C., Mitter, C., Cummings, M.P. *et al.* (2011) Can deliberately incomplete gene sample augmentation improve a phylogeny estimate for the advanced moths and butterflies (Hexapoda: Lepidoptera)? *Systematic Biology* 60, 782–796.

Cock, M.J.W. (ed.) (1985) *A Review of Biological Control of Pests in the Commonwealth Caribbean and Bermuda up to 1982*. Commonwealth Institute of Biological Control, Technical Communication 9. Commonwealth Agricultural Bureau, Farnham Royal, UK.

Common, I.F.B. (1990) *Moths of Australia*. Melbourne University Press, Carlton, Victoria.

Common, I.F.B. and Edwards, E.D. (1981) The life history and early stages of *Synemon magnifica* Strand (Lepidoptera: Castniidae). *Journal of the Australian Entomological Society* 20, 295–302.

Davis, S.R. (2006) Description of a new lepidopteran structure, the abdominal tubercles. *Journal of the Lepidopterists' Society* 60, 194–202.

Diakonoff, A.N. (1986) Glyphipterigidae *auctorum sensu lato* (Glyphipterigidae *sensu* Meyrick, 1913). In: Amsel, H.G., Gregor, F., Reisser, H. and Roesler, T.U. (eds) *Microlepidoptera Palaearctica* 7. Braun, Karlsruhe, Germany, pp. 1–436.

Díaz, J.J.G. (2022) Synopsis of *Athis thysanete* (Dyar, 1912) (Castniidae: Castniinae) populations, courtship behavior and other observations on its biology. *Tropical Lepidoptera Research* 32, 91–99.

Dodd, F.P. (1916) IV. Notes on the great Australian Cossidae. In: Oberthür, C., Houlbert, C. and Dodd, F.P. (eds) *Contributions à l'Etude des Grands Lépidoptères d'Australie*. Imprimerie Oberthür, Rennes, France, pp. 33–37.

Douglas, F. (1993) *The Conservation Status, Distribution and Habitat Requirements of Diurnal Lepidoptera in Central and Western Victoria. Part 1: Family Castniidae*. Department of Conservation and Natural Resources, Melbourne, Victoria.

Douglas, F. (2008) The Sun-moths (Lepidoptera: Castniidae) of Victoria, with a Detailed Study of the Pale Sun-moth (Synemon selene Klug, 1850). Master of Applied Science thesis, University of Ballarat, Ballarat, Victoria.

Duckworth, W.D. and Eichlin, T.D. (1974) Clearwing moths of Australia and New Zealand (Lepidoptera: Sesiidae). *Smithsonian Contributions to Zoology* 180, 1–45.

Edwards, E.D. (1996a) Cossidae. In: Nielsen, E.S., Edwards, E.D. and Rangsi, T.V. (eds) *Checklist of the Australian Lepidoptera*. Monographs on Australian Lepidoptera 4. CSIRO Publishing, Collingwood, Victoria, pp. 119–121.

Edwards, E.D. (1996b) Dudgeoneidae. In: Nielsen, E.S., Edwards, E.D. and Rangsi, T.V. (eds) *Checklist of the Australian Lepidoptera*. Monographs on Australian Lepidoptera 4. CSIRO Publishing, Collingwood, Victoria, p. 122.

Edwards, E.D. (1996c) Castniidae. In: Nielsen, E.S., Edwards, E.D. and Rangsi, T.V. (eds) *Checklist of the Australian Lepidoptera*. Monographs on Australian Lepidoptera 4. CSIRO Publishing, Collingwood, Victoria, p. 138.

Edwards, E.D., Gentili, P., Horak, M., Kristensen, N.P. and Nielsen, E.S. (1999) The cossoid/sesioid assemblage. In: Kristensen, N.P. (ed.) *Lepidoptera, Moths and Butterflies. 1. Evolution, Systematics and Biogeography. Handbook of Zoology,* 4 (1). *Lepidoptera.* Walter de Gruyter, Berlin, Germany, pp. 181–197.

Fletcher, T.B. (1920) Life-histories of Indian Insects. Microlepidoptera. V. Heliozelidae, Heliodinidae, Glyphipterigidae, Blastobasidae and Hyponomeutidae. *Memoirs of the Department of Agriculture in India, Entomological Series* 6, 117–167.

Fukuda, H. (2000) A new species of the genus *Tascina* Westwood (Castniidae, Tascininae) from Vietnam. *Transactions of the Lepidopterological Society of Japan* 52, 45–48.

Gardner, J.C.M. (1945) Immature stages of Indian Lepidoptera (Cossidae, Indarbelidae). *Journal of the Bombay Natural History Society*, 45, 390–396.

Gielis, C. (2008) Order Lepidoptera, family Brachodidae. *Arthropod Fauna of the UAE* 1, 444–446.
González, J.M. and Cock, M.J.W. (2004) A synopsis of the Castniidae (Lepidoptera) of Trinidad and Tobago. *Zootaxa* 762, 1–19.
Grund, R. (2011) A new species of *Synemon* Doubleday (Lepidoptera: Castniidae) from the Colona Plains, South Australia. *Australian Entomologist* 38, 167–178.
Grund, R. and Stolarski, A. (2012) Taxonomy and biology of *Synemon discalis* Strand and *S. parthenoides* R. Felder (Lepidoptera: Castniidae) in South Australia. *Australian Entomologist* 39, 121–160.
Grund, R., Stolarski, A. and Stevens, M. (2018) Rediscovery of *Synemon selene* (Lepidoptera: Castniidae) in the mid-north of South Australia. *Transactions of the Royal Society of South Australia* 143, 86–91.
Gumus, A., Karagoz, M., Shapiro-Ilan, D. and Hazir, S. (2015) A novel approach to biocontrol: Release of live insect hosts pre-infected with entomopathogenic nematodes. *Journal of Invertebrate Pathology* 130, 56–60.
Hampson, G.F. (1898) *Catalogue of the Lepidoptera Phalaenae in the British Museum*, 1. Trustees of the British Museum (Natural History), London.
Heikkila, M., Mutanen, M., Wahlberg, N., Sihvonen, P. and Kaila, L. (2015) Elusive ditrysian phylogeny: An account of combining systematized morphology with molecular data (Lepidoptera). *BMC Evolutionary Biology* 15, 260.
Heppner, J.B. and Duckworth, W.D. (1981) Classification of the Superfamily Sesioidea (Lepidoptera: Ditrysia). *Smithsonian Contributions to Zoology* 314, 1–144.
Heppner, J.B. and Wang, H.Y. (1987) A rare moth, *Ratarda tertia* Strand (Lepidoptera: Ratardidae), from Palin, Taiwan. *Journal of the Taiwan Museum* 40, 91–94.
Hering, M. (1926) Die Familie der Ratardidae (Lep.) mit Beschreibung 5 neuer Arten. *Mitteilungen aus dem Museum in Berlin* 12, 151–158.
Holloway, J.D. (1986) The moths of Borneo. Part 1: Key to families; families Cossidae, Metarbelidae, Ratardidae, Dudgeoneidae, Epipyropidae and Limacodidae. *Malayan Nature Journal* 40, 1–166.
Kakul, T., Aloysius, M. and Samai, K. (2006) Coconut inflorescence borer, *Synneschodes* [sic] *papuana* (Lepidoptera: Brachodidae), an important new pest of coconut in Papua New Guinea. In: Price, T.V. (ed.) *Pest and Disease Incursions: Risks, Threats and Management in Papua New Guinea*. Papers presented at the 2nd Papua New Guinea Plant Protection Conference, Kokopo, East New Britain Province, 8–10 November 2004. ACIAR Technical Reports, Canberra, ACT, pp. 146–150.
Kallies, A. (1998a) A contribution to the knowledge of the oriental and Australian Brachodidae (Lepidoptera, Sesioidea). *Tinea* 15, 312–337.
Kallies, A. (1998b) First contribution to the knowledge of the Palaearctic Brachodidae: Revision of *Brachodes fallax* with descriptions of new species from Central Asia. *Nota Lepidopterologia* 21, 170–193.
Kallies, A. (2001a) Revision of the *Brachodes pumila* (Ochsenheimer, 1808) species group (Sesioidea: Brachodidae). *Nota Lepidopterologica* 24, 7–19.
Kallies, A. (2001b) New records and a revised checklist of the Australian clearwing moths (Lepidoptera: Sesiidae). *Australian Journal of Entomology* 40, 342–348.
Kallies, A. (2002) *Brachodes flagellatus* sp. n. from Tibet (Lepidoptera: Sesioidea, Brachodidae). *Nachrichten des Entomologische Vereins Apollo* 23, 159–160.
Kallies, A. (2004) The Brachodidae of the Oriental region and adjacent territories (Lepidoptera: Sesioidea). *Tijdschrift voor Entomologie* 147, 1–19.
Kallies, A. (2013) New and little known Brachodidae from tropical Asia and Papua New Guinea (Lepidoptera, Cossoidea). *Zootaxa* 3641(3), 241–259.
Kallies, A. (2016) The Brachodidae of Sub-Saharan Africa (Lepidoptera, Cossoidea), with implications for the origin of the family. *Zootaxa* 4083(1), 1–39.
Kallies, A. (2020) The clearwing moths (Lepidoptera, Sesiidae) of Australia, New Guinea and the Pacific Islands. *Zootaxa* 4833, 1–64.
Kallies, A. and Edwards, E.D. (2018) A new sun moth species from the Flinders Ranges in South Australia (Lepidoptera, Castniidae). *Zootaxa* 4369, 292–300.
Kallies, A. and Edwards, E.D. (2025) New and critically endangered Sun Moth species associated with Australian grasslands (Lepidoptera, Castniidae) (Lepidoptera, Castniidae). *Zootaxa* 5689, 505–537.
Kallies, A. and Hilton, D. (2012) Revision of the Cossinae and small Zeuzerinae from Australia (Lepidoptera: Cossidae). *Zootaxa* 3454, 1–62.
Kallies, A., Edwards, E.D., Young, A. and Douglas, F. (2016). *Synemon ignita* sp. nov., a new sun moth species from southern Australia (Lepidoptera, Castniidae). *Zootaxa* 4092, 436–444.
Kallies, A., Edwards E.D. and Williams A.A.E. (2020) New and little-known sun-moth species from Australia (Lepidoptera, Castniidae). *Zootaxa* 4895, 151–195.

Kalshoven, L.G.E. (1981) *The Pests of Crops in Indonesia*. Ichtiar Baru-Van Hoese, Jakarta, Indonesia.

Kawahara, A.Y., Plotkin, D., Espeland, M., Meusemann, K., Toussaint, E.F.A. *et al.* (2019) Phylogenomics reveals the evolutionary timing and pattern of butterflies and moths. *Proceedings of the National Academy of Sciences* 116, 22657–22663.

Kobes, L.W.R. and Ronkay, L. (1990) The Ratardidae of Sumatra. *Heterocera Sumatrana* 6, 79–100.

Kozlov, M.V., Kuznetzov, V.I. and Stekolnikov, A.A. (1998) Skeleto-muscular morphology of the pterothorax and the male genitalia of *Synemon plana* Walker (Castniidae) and *Brachodes appendiculata* (Esper) (Brachodidae), with notes on phylogenetic relationships of tortricoid-grade families (Lepidoptera). *Invertebrate Taxonomy* 12, 245–256.

Kumar, R. and Ramamurthy, V.V. (2010) Morphology and bionomics of *Phycodes radiata* Ochsenheimer (Lepidoptera: Brachodidae) from New Delhi, India. *Tijdschrift voor Entomologie* 153, 15–24.

Kutinkova, H., Andreev, R. and Arnaoudov, V. (2006) The leopard moth borer, *Zeuzera pyrina* L. (Lepidoptera: Cossidae) – an important pest in Bulgaria. *Journal of Plant Protection Research* 46, 111–115.

Lamas, G. (1995) A critical review of J.Y. Miller's checklist of the Neotropical Castniidae (Lepidoptera). *Revista Peruana de Entomología* 37, 73–87.

Lucas, T.P. (1894) Descriptions of new Australian Lepidoptera, with additional localities for known species. *Proceedings of the Linnean Society of New South Wales, Series* 2, 8(1893), 133–166.

McInnes, R.S. and Carne, P.B. (1978) Predation of cossid moth larvae by yellow-tailed black cockatoos causing losses in plantations of *Eucalyptus grandis* in north coastal New South Wales. *Australian Wildlife Research* 5, 101–121.

Meyrick, E. (1914) Lepidoptera Heterocera, Fam. Glyphipterigidae. In: Wytsman, M.P. (ed.) *Genera Insectorum, 164*. Verteneuil and Desmet, Brussels, Belgium.

Miller, J.Y. (1986) The Taxonomy, Phylogeny, and Zoogeography of the Neotropical Moth Subfamily Castniinae (Lepidoptera: Castnioidea: Castniidae). PhD thesis. UMI Dissertation Information Service, Ann Arbor, MI.

Minet, J. (1983) Etude morphologique et phylogénétique des organes typaniques des Pyraloidea. 1. Généralités et homologies (Lep. Glossata). *Annales de Sociéte Entomologique de France* 19, 175–207.

Minet, J. (1986) Ebauche d'une classification moderne de l'ordre des Lépidoptères. *Alexanor* 14, 291–313.

Minet, J. (1987) Les Ratardidae proposés comme groupe-frère des Callidulidae [Lep. Ditrysia]. *Bulletin de la Société Entomologique de France* 92(1–2), 39–44.

Minet, J. (1991) Tentative reconstruction of the ditrysian phylogeny (Lepidoptera: Glossata). *Entomologica Scandinavica* 22, 69–95.

Monteys, V.S. and Aguilar, L. (2005) The Castniid palm borer, *Paysandisia archon* (Burmeister, 1880), in Europe: Comparative biology, pest status and possible control methods (Lepidoptera: Castniidae). *Nachrichten des Entomologischen Vereins Apollo* 26, 61–94.

Moraes, S. and Duarte, M. (2014) Phylogeny of Neotropical Castniidae (Lepidoptera: Cossoidea: Castniidae): Testing the hypothesis of the mimics as a monophyletic group and implications for the arrangement of the genera. *Zoological Journal of the Linnean Society* 170, 362–399.

Mutanen, M., Wahlberg, N. and Kaila, L. (2010) Comprehensive gene and taxon coverage elucidates radiation patterns in moths and butterflies. *Proceedings of the Royal Society B: Biological Sciences* 277, 2839–2848.

Myers, J.G. (1931) *A Preliminary Report on an Investigation into the Biological Control of West Indian Insect Pests*. Empire Marketing Board 42, London.

Naumann, C. (1971). Untersuchungen zur Systematik und Phylogenese der holarktischen Sesiiden (Insecta, Lepidoptera). *Bonner Zoologische Monographien* 1, 1–190.

New, T.R., Gibson, L. and Van Praagh, B.D. (2007) The Golden Sun-moth *Synemon plana* (Castniidae) on Victoria's remnant southern native grasslands. *Victorian Naturalist* 124, 192, 154–157.

O'Dwyer, C. and Attiwill, P.M. (1999) Restoration of a native grassland as habitat for the Golden Sun Moth *Synemon plana* Walker (Lepidoptera; Castniidae) at Mount Piper, Australia. *Biological Conservation* 89, 131–141.

Owada, M. (1993) The systematic position of *Shisa excellens* (Lepidoptera, Ratardidae). *Japanese Journal of Entomology* 61, 251–260.

Özgen, I., İçgil, Y.D. and Karan, H. (2024) The first record of economic pest on Tunceli mountain garlic (*Allium tuncelianum*): garlic moth *Dyspessa ulula* (Borkhausen, 1790) (Lepidoptera: Cossidae). *Munis Entomology & Zoology* 19, 425–430.

Pühringer, F. and Kallies, A. (2025) Checklist of the Sesiidae of the world (Lepidoptera: Ditrysia). Available at: www.sesiidae.net/Checklst.htm (accessed 6 March 2025).

Rebel, H. (1901) Famil. Pyralidae – Micropterigidae. In: Staudinger, O. and Rebel, H. (eds) *Catalog der Lepidopteren des palaearctischen Faunengebietes*, 1. R. Friedländer, Berlin, Germany.

Regier, J.C., Zwick, A., Cummings, M.P., Kawahara, A.Y., Cho, S. *et al.* (2009) Toward reconstructing the evolution of advanced moths and butterflies (Lepidoptera: Ditrysia): An initial molecular study. *BMC Evolutionary Biology* 9, 280.

Regier, J.C., Mitter, C., Zwick, A., Bazinet, A.L., Cummings, M.P. *et al.* (2013) A large-scale, higher-level, molecular phylogenetic study of the insect order Lepidoptera (moths and butterflies). *PLOS One* 8, e58568.

Regier, J.C., Mitter, C., Davis, D.R., Harrison, T.L., Sohn, J.-C. *et al.* (2015a) A molecular phylogeny and revised classification for the oldest ditrysian moth lineages (Lepidoptera: Tineoidea), with implications for ancestral feeding habits of the mega-diverse Ditrysia. *Systematic Entomology* 40, 409–432.

Regier, J.C., Mitter, C., Kristensen, N.P., Davis, D.R., van Nieukerken, E. *et al.* (2015b) A molecular phylogeny for the oldest (nonditrysian) lineages of extant Lepidoptera, with implications for classification, comparative morphology and life-history evolution. *Systematic Entomology* 40, 671–704.

Richter, A., Weinhold, D., Robertson, G., Young, M., Edwards, T. *et al.* (2013) More than an empty case: A non-invasive technique for monitoring the Australian critically endangered golden sun moth, *Synemon plana* (Lepidoptera: Castniidae). *Journal of Insect Conservation* 17, 529–536.

Roepke, W. (1957) The cossids of the Malay region (Lepidoptera: Heterocera). *Verhandelingen der Koninklijke Akademie van Wetenschappen te Amsterdam* 52(2), 1–60.

Scaccini, D., Ruzzier, E. and Daane, K.M. (2021) *Givira ethela* (Neumoegen and Dyar, 1893) (Lepidoptera: Cossidae): a previously unidentified pest on *Vitis vinifera* (L.). *Insects* 12, 239.

Schoorl, J.W. (1990) A phylogenetic study on Cossidae (Lepidoptera: Ditrysia) based on external adult morphology. *Zoologische Verhandelingen* 263, 1–295.

Scott, J.A. (1986) On the monophyly of the Macrolepidoptera, Including a reassessment of their relationship to Cossoidea and Castnioidea, and a reassignment of Mimallonidae to Pyraloidea. *Journal of Research on the Lepidoptera* 25, 30–38.

Simmonds, H.W. (1921) The big spathe boring moth of coconuts. *Fiji Department of Agriculture Bulletin* 2, 101.

Špatenka, K., Gorbunov, O., Laštůvka, Z., Toševski, I. and Arita, Y. (1999) Sesiidae, Clearwing Moths. In: Naumann, C.M. (ed.) *Handbook of Palaearctic Macrolepidoptera* 1. Gem Publishing Company, Wallingford, UK, pp. 1–569.

Tian, J., Hua, B.-Z. and Zhang, H.-J. (2010) Morphology of *Eogystia sibirica* (Alphéraky) (Lepidoptera: Cossidae) attacking *Asparagus officinalis* in Northern China with descriptions of its immature stages. *Journal of Natural History* 44, 2581–2595.

Tindale, N.B. (1953) On some Australian Cossidae including the moth of the witjuti (witchetty) grub. *Transactions of the Royal Society of South Australia* 76, 56–65.

Thurman, J.H. (2022) Beyond the pest: Life history, ecology and ethnoentomology of the giant wood moth (*Endoxyla cinereus*). *Austral Ecology* 47, 733–747.

Turner, A.J. (1902) New Australian Lepidoptera. *Transactions of the Royal Society of South Australia* 26, 175–207.

Turner, A.J. (1917) Lepidopterological gleanings. *Proceedings of the Royal Society of Queensland* 29, 70–106.

Turner, A.J. (1945) A revision of the Australian Cossidae (Lepidoptera). *Proceedings of the Royal Society of Queensland* 56, 47–70.

Turpin, M. and Si, A. (2017) Edible insect larvae in Kaytetye: Their nomenclature and significance. *Journal of Ethnobiology* 37, 120–140.

Van Nieukerken, E.J., Kaila, L., Kitching, I.J., Kristensen, N.P., Lees, D.C. *et al.* (2011) Order Lepidoptera Linnaeus, 1758. In: Zhang, Z.-Q. (ed.) Animal biodiversity: An outline of higher-level classification and survey of taxonomic richness. *Zootaxa* 3148, 212–221.

Williams, A.A.E. and Williams, M.R. (2016) A new species of *Synemon* Doubleday (Lepidoptera: Castniidae) from Western Australia. *Australian Entomologist* 43, 119–134.

Williams, A.A.E., Williams, M.R., Edwards, E.D. and Coppin, R.A.M. (2016) The sun-moths (Lepidoptera: Castniidae) of Western Australia: an inventory of distribution, larval food-plants, habitat, behaviour, seasonality and conservation status. *Records of the Western Australian Museum* 31, 90–162.

Wylie, F.R. and Peters, B.C. (1993) Insect pest problems of eucalypt plantations in Australia. *Australian Forestry* 56, 358–362.

Yakovlev, R.V. (2011a) Stygiinae Yakovlev, subfam. n., a new subfamily of Palaearctic carpenter-moths (Lepidoptera, Cossidae). *Entomological Review* 91, 508–512.

Yakovlev, R.V. (2011b) Catalogue of the family Cossidae of the old world. *Neue Entomologische Nachrichten* 66, 1–129.

Yen, A., Bilney, C., Shackleton, M. and Lawler, S. (2018) Current issues involved with the identification and nutritional value of wood grubs consumed by Australian Aborigines. *Insect Science* 25, 199–210.

16 The Limacodidae: The Australian Cup Moths

Marc E. Epstein[1]* and Yu-Chi Lin[2]

[1]*California Department of Food and Agriculture, Sacramento, CA, USA;* [2]*Department of Life Science, National Taiwan Normal University, Taipei, Taiwan*

Abstract

The Limacodidae are a temperate and tropical family in the Zygaenoidea containing primarily medium to small-sized nocturnal moths with buzzy erratic flights at light after sunset. They are best known for their slug-like larvae that have a modified undersurface of suckers as well as their hard and oval cocoons. These cocoons resemble cups because they lose their lids post adult emergence. While several authors have generated a great deal of information on the mostly endemic Australian fauna of 70 species in 25 genera of Limacodidae, we present here new knowledge of the fauna along with a global perspective on their phylogeny and the ontogeny of their larvae. This includes more global comparison of unusual morphologies and life histories that can be quite similar or very different. Similarities include sexually dimorphic species with clear-winged males and larger opaque-winged females, gregarious larvae and cocoons. A difference is that the lineage of the tribe Eucleini, the larvae of which possess urticating hairs, do not feed as first instars. In comparison with other temperate and tropical regions, the Eucleini is much more species rich in the Australian fauna than is the Limacodini, the larvae of which have no urticating hairs. We review local food plant information and life histories, and make some progress on resolving some taxonomic problems.

16.1 Introduction

Limacodidae have fascinated humanity in different parts of the globe for centuries. The earliest evidence is from reports of oval white cocoons with black stripes in Japan from the 14th century (Shimizu, 1989). These stripes have since been found to be caused by the blocking of calcium oxalate spray during cocoon construction (Epstein, 1996a). In Australia, the common name 'cup moth' is based on the appearance of the post-emergence cocoon. These may be attached to twigs, branches or tree trunks, in leaf litter and between leaves. Although Common (1990) did not use the term 'cup moths', he did note the 'characteristically cup-shaped faecal pellets' produced by the larvae.

Even though limacodids form a relatively small family in Australia of 70 species, their cocoons and characteristic and unusual-looking larvae often seen on myrtaceous trees and shrubs are a part of cultural awareness. Common names abound for these larvae due to painful encounters and their dazzling colours: 'rainbow battleship', 'warship' and 'wrathful militia' among them. Locally, they have even been referred to as 'Bondi trams' (Cocking *et al.*, 2022). Elsewhere their resemblance to ships has led them to be called Chinese junks and, in the United States, the skiff moth (*Prolimacodes*). Packard (1864)

*Corresponding author: marc.epstein@cdfa.ca.gov

DOI: 10.1079/9781800627680.0016

named *Euclea monitor* (= *Euclea delphinii*) for the heavily fortified 'monitor' battleships of US Civil War times.

More generally, they are referred to as slug caterpillar moths or simply slug moths (Scoble, 1992; Wagner, 2025). This can be interpreted in two ways. The Latin *limax* was first applied to the larvae and translated directly to 'slug' in English, even though less than half of known limacodid larvae are smooth and gastropod-like on their upper surfaces. The name could be based on the small European fauna consisting mainly of the smooth versions. A second interpretation is that the ventral surface of both the 'smooth' larvae (referred to elsewhere, but not in Australia, as 'gelatines') and the spiny larvae (known, again elsewhere, as 'nettles') have prolegs modified to sucker discs so, for lack of close examination, could be compared with that of a slug's foot.

16.2 Definition and Placement of the Limacodidae

Limacodidae is the largest family in the superfamily Zygaenoidea of over 1800 sp. (A. Giusti *et al.*, unpublished; see also van Nieukerken *et al.*, 2011). Limacodidae has recently become more broadly defined to include two subfamilies that were previously families: Dalcerinae from the New World (78 spp.) and Chrysopolominae from Africa (59 spp.), and one previous subfamily was revived in status, Crothaeminae from Africa (19 spp.). Currently, most limacodids, including all from Australia, are in the global subfamily Limacodinae at approximately 1600 species. Among the Limacodinae are two tribes, the Eucleini and the Limacodini. The Eucleini are known for their spiny, stinging larvae whereas the Limacodini, rare in Australia and common elsewhere, have smooth, more truly slug-like larvae (Epstein *et al.*, 2025; Giusti *et al.*, unpublished).

The Limacodidae are a lineage of Zygaenoidea along with others referred to as the Limacodid group families, which possess additional prolegs or crochets on the the second to seventh abdominal segments that can be modified into sucker discs (Epstein, 1996a; Epstein *et al.*, 1999). Sister to the Limacodidae is the 'core' Megalopygidae, consisting of the Megalopygidae (*s.s.*) with the Aididae and Himantopteridae. The African family Somabrachyidae may also belong among this core group (Epstein *et al.*, 2025).

Other families in the Zygaenoidea found in Australia include the Lacturidae (with 28 Australian spp. in a global fauna of *ca* 120 spp.) and the Zygaenidae (51 spp. of *ca* 1036 spp. globally) (with only Procridinae in Australia). The Epipyropidae (with *ca* 4 of 32 spp. globally) and the wholly endemic Cyclotornidae (5 spp.) also occur in Australia and have parasitic larvae considered to be zygaenoid by some, including Common (1990), and not by others depending on morphology and/or molecular data. The Epipyropidae were long considered close to the Limacodidae or Dalceridae (now Dalcerinae) (van Nieukerken *et al.*, 2011).

Zygaenoidea have slug-like larvae usually with retractile heads. Their pupal characters, including their eyes and spiracles, do not define them as once thought. For example, Epstein *et al.* (2025) show that the sculpted eyepiece also occurs in Cossidae. Molecular data support the superfamily, although the membership of parasitic families Epipyropidae and Cyclotornidae within the Zygaenoidea remains in flux (Regier *et al.*, 2013; Heikkilä *et al.*, 2015).

Common (1990) was steadfast in his view that the Limacodidae belonged in the same superfamily as the Zygaenidae, the Zygaenoidea, along with other families including the Lacturidae, Epipyropidae and Cyclotornidae largely based on characteristics of the juvenile stages. However, Brock (1971) and others including Heppner (1984, 1995) placed the Limacodidae in the Cossoidea based on adult morphology. In recent decades, evidence supporting the Zygaenoidea placement based on morphology (Epstein, 1996a) and molecular data (Regier *et al.*, 2009, 2013) has accrued even though affiliation with the Cossoidea has persisted in places like the Australian Museum website (https://australian.museum/learn/animals/insects/cup-moths/).

16.3 How to Recognize Limacodidae

Even though Limacodidae are best known for their larvae and cocoons in Australia, the less conspicuous adults are often recognized by their 'buzzy' flight at lights, appearing early, shortly after sundown. Forewings often appear relatively short compared with the body, giving them a stubby appearance, usually more pronounced in males than in females, which often have rounder wings. Females may appear an hour or two after the males and tend to have a less buzzy flight.

Once moths are settled on a sheet or other surface, they often hold their wings below their bodies, propped up on their legs, which have a furry appearance from pilose scales often splayed at the front or on the sides (Figs 16.1A–S). Other types of poses, however, occur in non-Australian species (e.g. males and females of the neotropical *Perola* complex raise their front legs above the head) (Epstein *et al.*, 2025). In addition to the wings being held below the body, some species exhibit a raised abdomen (Figs 16.1J, 16.1O), including some where the abdomen is curled back towards the head. In contrast, many males simply have abdomens that extend horizontally past the wing extremities, whereas females, usually gravid, have shorter and rounder abdomens.

Sexual dimorphism in Limacodidae occurs across many genera in all regions, with the most extreme version being narrow-winged males with a range from completely clear wings to those with smaller clear patches on either or both wings, contrasting with larger opaque-winged females with more rounded wings. Examples include *Doratifera oxleyi* (Figs 16.1E–F) in Australia and the hagmoth *Phobetron pithecium* in North America. Along related lines, clear-winged males appear to be trending towards diurnal flight, as suggested for *D. oxleyi* by McFarland (1979) and Common (1990). Data on *D. oxleyi* from iNaturalist suggests that males occur less frequently at lights than females (Fig. 16.6B).

The antennae of females of all Australian limacodids are filiform (thread-like) (Figs 16.1H–I), whereas those of males vary from broadly to narrowly bipectinate, the individual ramus long throughout, gradually decreasing in length (Fig. 16.1E), or broad at its base and abruptly shortening (Figs 16.1Q–S). In Australia no males are known to have filiform antennae.

Some adult characters may be region specific. Some, such as the number of spurs on the hind tibia, may indicate ancestral states, others are merely convergent. Nevertheless, these remain useful identification markers within isolated local faunas, including Australia. Such features may include contrasting white scale patches at the base of the antennae and the coxae of the front legs but also can be on individual tarsi (Fig. 16.1O). Dark-winged and -bodied *Doratifera stenora* (Fig. 16.1R), for example, can be separated from *Pygmaeomorpha plagioscia* (Fig. 16.1S) by the presence of white scales at the base of the antennae, absent in the latter species. *D. pinguis* (Fig. 16.1C) has white scales on the front legs not found in rather similar *Calcarifera ordinata* (Fig. 16.1B).

Limacodids have a variety of larval types. Larvae of *Doratifera* species are among Australia's best known not only because of their bright colours and aggregating behaviour, but also for their defensive behaviour using venom-carrying spines on their dorsal surfaces (Figs 16.3J–O). At rest, these spines are appressed to the body (Fig. 16.3K) and spring into action as rosettes on pedicels when touched, much like sea anemones (Figs 16.3L–M). *Doratifera* spp. also have tiny, non-venomous 'caltrop' spines in a circular depression at the base of their subdorsal scoli (Fig. 16.3I) (see below).

Genera which always have erectile urticating spines on scoli are frequent in the Australian fauna. Brightly coloured species with such spines are found in *Anaxidia*, *Comana*, *Hedraea*, *Lamprolepida* and *Praesusica*. Other Eucleini may be either covered with spines not on scoli (e.g. *Eloasa*, Figs 16.3F, 16.4P), the spines may be reduced to spinules (e.g. *Mecytha fasciata*, Fig. 16.4K) or they may be smooth like the Limacodini (e.g. unknown sp., Fig. 16.3G, *Pseudanapaea* sp., Fig. 16.3H). Less common in Australia among the Eucleini are the 'greater flat nettles', a cosmopolitan group that are often flat in cross-section (Epstein *et al.*, 2025). In Australia, these include *Thosea* and *Anepopsia* (Fig. 16.7). The final larval type and the least common, represented by only two species of *Chalcocelis*, are the Limacodini (Fig. 16.3E), which are separated by ancestry and behaviour.

Even though the larvae of the Eucleini can be quite beautiful in appearance, it is wise to be cautious with them as they can give nasty stings. Many are brilliantly patterned, such as the rainbow battleship caterpillar *Comana albibasis* (Fig. 16.3C). The venoms of Limacodidae from Australia and other regions are currently under investigation. In *Doratifera vulnerans* the venoms are complex and made up of many proteinaceous toxins (venom peptides) (Walker *et al.*, 2021).

Fig. 16.1. Adults of Australian Limacodidae (male unless indicated otherwise). (A) *Comana albibasis*. Photo: Ian McMillan. (B) *Calcarifera ordinata*. Photo: Ellura Sanct and Brett Smith. (C) *Doratifera pinguis*. Photo: Doug Evans. (D) *Doratifera vulnerans*. Photo: Ian McMillan. (E) *Doratifera oxleyi*. Photo: John Lenagan. (F) *Doratifera oxleyi* (female). Photo: Marianne Broug. (G) *Comana collaris*. Photo: Greg Neill. (H) *Pseudanapaea denotata* (female). Photo: Christopher and Sarra Peacock. (I) *Eloasa infrequens* (female). Photo: Diane Clark. (J) *Anaxidia lozogramma*. Photo: Larney Grenfell. (K) *Hedraea quadridens*. Photo: Todd Burrows. (L) *Squamosa barymorpha.* Photo: David Tng. (M) *Pygmaeomorpha ocularis*. Photo: Nick Lambert. (N) *Mecytha fasciata*. Photo: Larney Grenfell. (O) *Praesusica placerodes*. Photo: David Tng. (P) *Eloasa callidesma*. Photo: Ian McMillan. (Q) *Chalcocelis albiguttatus*. Photo: Dominic Funnell. (R) *Doratifera stenora*. Photo: Larney Grenfell. (S) *Pygmaeomorpha plagioscia*. Photo: Lily Kumpe.

In limacodid larvae, largely hidden from view is the unique ventral surface, which consists of tiny thoracic legs and abdominal prolegs modified into sucker discs (Fig. 16.3S). Viewed laterally or from beneath, these larvae are extraordinary. They have been said to 'glide' during

Fig. 16.2. Key larval features for differentiating Australian species using *Doratifera quadriguttata* as an exemplar. Scoli (warts, protuberances) are in two rows, dorsal (D) and subdorsal (SD). This species has only rudiments of spines except for those on the D scoli on the mesothorax (T3) and the first abdominal segment (A1). The venomous spines on these two segments are held to the body in these two larvae, only to spring out in defence when disturbed. Some *Doratifera* larvae have venomous spines on these same two segments, including *D. casta*, whereas others have additional venomous spines on D scoli on segments A6 and A7. These are *D. vulnerans*, *D. corallina*, *D. oxleyi* and *D. pinguis*. All key features of the thorax and abdomen are visible in these two figures except for the spiracles on abdominal segments A2–A8, which are hidden beneath the SD scoli. Oval patches that contain tiny caltrop spines are found at the bases of the SD scoli on A2–A9: there is no scolus on A1, as is typical for nettle larvae in tribe Eucleini. Note that the scoli are shorter in Fig. 16.2A than in Fig. 16.2B. This could be due to either differences in instar, one perhaps earlier or later, or differences in populations. *Doratifera stenora* is atypical of the genus in not having appressed spines on D scoli and have typical spiny D and SD scoli. (Photos: James Peake and Stephen Murray.)

locomotion as their highly flexible 'foot' contracts and expands in waves in either direction; their specialized spinneret brushes debris from the foot while applying semi-fluid silk ribbons to the substrate (Epstein, 1996a). Ventral views of postecdysis larvae expose the ventral nerve chord, the tracheal system, the silk gland, the gut and the Malpighian tubules, the last often in motion in the open circulatory system.

Like close relatives within the Zygaenoidea, limacodid larvae exhibit retractile feeding: feeding while the head is hidden by the prothorax (Fig. 16.3Q) (Epstein, 1996a). In early instars they scrape one surface of the leaf (Fig. 16.3P). Often by the third or fourth instar, they feed on leaf edges either on the periphery or on the sides of holes after breaking through the leaf's lamina. Those that feed in groups arise from eggs

Fig. 16.3. Larvae of Australian Limacodidae (late instar unless indicated). (A) *Calcarifera ordinata*. Photo: Ellura Sanct. (B) *Calcarifera* sp. Photo: Gunter Maywald. (C) *Comana albibasis*. Photo: Julien Rouard. (D) *Hedraea quadridens*. Photo: Todd Burrows. (E) *Chalcocelis albiguttatus*. Photo: Sarah Renneberg. (F) *Eloasa infrequens* with moulted skin (r). Photo: Dianne Clarke. (G) Eucleini sp. Photo: Dianne Clarke. (H) *Pseudanapaea transvestita*. Photo: Gunter Maywald. (I) *Doratifera quadriguttata* (note: caltrop patches and spiracles on T1 and A1). Photo: Stephen Murray. (J) *Doratifera stenora* (no appressed spines). Photo: Nick Lambert. (K, L) *Doratifera oxleyi*: K. Spines on D scoli in repose. Photo: Simon Grove, L. Spines on D scoli partly splayed out on T3. Photo: Ian D.B. Moodie and Ron I. Greer. (M) *Doratifera vulnerans* spines splayed out. Photo: Nicole Harman. (N) *Doratifera corallina* spines in repose. Photo: Gunter Maywald. (O) *D. pinguis* spines in repose on short scoli. Photo: Greg Bellion. (P–S) *Anaxidia lozogramma*: P. Early instar feeding on leaf surface. Photo: Donna Tomkinson. Q. Late instar retractile feeding on leaf edge. Photo: Dianne Clarke. R. Larva feeding on edge on right as seen from above. Photo: Dianne Clarke. S. Larva viewed from beneath through smooth translucent surface (note: the oval sucker discs and cloudy internal portion and beads of liquid suggest that the larva is about to construct a cocoon). Photo: Dianne Clarke.

Fig. 16.4. Eggs, fasting first-instar larvae, moulting, gregarious nettle larvae and others (Eucleini). (A–C) *Doratifera oxleyi*: A. Ovipositing female viewed from under a clear surface (note: entire abdomen coating second group of eggs). Photo: George Vaughan. B. Close-up of flat, shingled eggs from beneath, unobscured by hairy scale covering. Photo: Alan Melville. C. Eggs hidden by dense hairy scales from above. Photo: Alan Melville. (D) Darker scale covering egg mass – perhaps another species of *Doratifera*. Photo: Paul George. (E–H) *Doratifera stenora*: E. Eggs covered with scales about to hatch. F. Recently hatched first instar larvae (note: lightly coloured simple tubercles above). Photo: Larney Grenfell. G. Second instars soon after moulting (note: dark-coloured clumps of spines). Photo: Larney Grenfell. H. Anterior ends of gregarious late instars feeding on the edge of a leaf (note: black is dorsal sclerite on prothorax). Photo: Chris Pease. (I) *Doratifera quadriguttata* second instar with multispined scoli. Photo: Marc Epstein. (J) *Pseudanapaea transvestita* first instar. Photo: Marc Epstein. (K) *Mecytha fasciata* late instar. Photo: Gunter Maywald. (L) *Praesusica placerodes* late instar. Photo: Christina Suttner. (M–P) *Eloasa callidesma* induced to moult to second instar with water droplet at posterior end. Fasting first instar with droplet (note: unbranched tubercles on D and SD rows with second instar spines beneath the cuticle), N. Droplet absorbed from opposite side of droplet on anterior end, O. Second instar begins to push out of first instar pelt (note: multiple spines now visible on left as the larva crawls out of skin), P. After moulting the second instar larva turns around before it begins to feed for the first time. Photos: Marc Epstein.

that are laid in clusters. In *Doratifera* spp. the eggs are covered by thin, hair-like scales from the mother (Figs 16.4A–E). Unlike *Monoleuca* in the New World (Epstein, 1996b), which coat the eggs with scales from the terminal abdomen, we see in *Doratifera oxleyi* that the scales appear to come from a more extended portion of the ventral abdomen (Fig. 16.4A). On tough, fibrous leaves like palms, larvae may also feed much like the early instars, scraping one surface rather than feeding at leaf edges. Before cocoon construction, the prepupa secretes clear droplets and its ventral surface becomes cloudy (Fig. 16.3S); the voiding of fluid helps make it more compact for building an oval cocoon. They remain as prepupa inside the cocoon until pupating near the time of adult emergence.

The cup-like cocoons are well known. Cocoons of *Doratifera* are pyriform rather than oval or spheroid, as is more typical of limacodids. On the narrower end, *Doratifera* also have a noticeable 'hatch' or line of weakness or are flattened.

Fig. 16.5. Cocoons of Australian Limacodidae. (A–C) *Doratifera vulnerans*: A. Uneclosed cocoon (note: preformed hatch line on left, which is narrower than opposite side). Photo: Kim McCloughan. B. Eclosed cocoon (= cup). Photo: Kay Taranto. C. Eclosed cocoon with pupal skin remaining (right), uneclosed (upper left), cocoon with parasitoid exit hole (bottom left). Photo: Brendan Watts. (D) *Chalcocelis albiguttatus*. Photo: Sarah Renneberg. (E) *Praesusica placerodes*. Photo: Christina Suttner.

Doratifera cocoons can be found in groups on branches, tree trunks or under bark (Figs 16.5A–C). Such groups of cocoons also occur in New World *Acharia*, which also feed in aggregations. Many, if not all, limacodid cocoons are impregnated with a liquid calcium oxalate (CO) that is sprayed from the prepupal rectum during the final stages of construction (Epstein, 1996a), giving them a chalky appearance when dry (Fig. 16.5D). However, some genera or species can have more CO in general. The CO protects the prepupa or pupa while also ensuring the cocoons endure over several years.

The most common type of limacodid egg is flat, scale-like and oval shaped. They have a very thin chorion and are translucent, allowing embryonic development to be followed clearly. Eggs, laid on smooth leaf surfaces, are often in groups (Fig. 16.4B) but can also be laid singly. Groups remain together to form larval aggregations after hatching and while feeding as larvae. This habit is common in defoliating species (e.g. *Doratifera*, Figs 16.4G, 16.4H). Similar scale-covered eggs occur in *Monoleuca semifascia* in the New World (Epstein, 1996b; Epstein *et al.*, 2025) and probably also in *Ctenolita* in Africa, which have females with piliform scales on the ventral abdomen (M. Epstein, personal observation).

16.4 A History of Limacodid Study in Australia

Colonial encounters with limacodids in Australia began with a painful caterpillar encounter in 1770, when Joseph Banks referred to *Doratifera stenora* and its gregarious larvae as a 'wrathful militia'. He had experienced their venom in Bustard Bay's mangroves after landing in Queensland (Herbison-Evans and Crossley, 2025). As noted by Edwards (1996): 'The first Australian Limacodidae was collected on Cook's first voyage in 1770 at the present site of Cooktown ... described as *Bombyx* (= *Comana*) *corones* by Fabricius in 1775 and has been consistently misidentified until the present'. Walker (1855) noted two specimens of *Panisa circumdata* (= *Doratifera oxleyi*) from New Holland from 'Capt. Parry's collection'. Sir William Edward Parry (1790–1855) arrived in NSW in 1829 and was there until 1834 (https://adb.anu.edu.au/biography/parry-sir-william-edward-2539).

Hering (1931) provided the most comprehensive treatment of the Australian limacodid fauna, although he named only a few species (Edwards, 1996). Again, Swinhoe (1892) described few taxa but catalogued Australian limacodid specimens in the Oxford University Museum, including Walker's types. He did name the genus *Anaxidia* and its type species *A. lactea*, accompanied by an illustration of the wing venation and male antenna.

Turner described by far the most Australian Limacodidae (31 species), followed by Lower and Walker (seven each), Bethune-Baker and Hering (five each), Swinhoe (three) and several with two including Lucas, Felder, Scott and Meyrick. Turner (1942) described three species of limacodids including *Susica* (= *Squamosa*) *barymorpha*, a large stout moth, which perhaps gets its name from the Greek word *barys* (heavy, impressive, etc.) or the type locality of Lake Barrine in the Atherton Tableland where it was collected by E.J. Dumigan. The other two were *Comana* (then

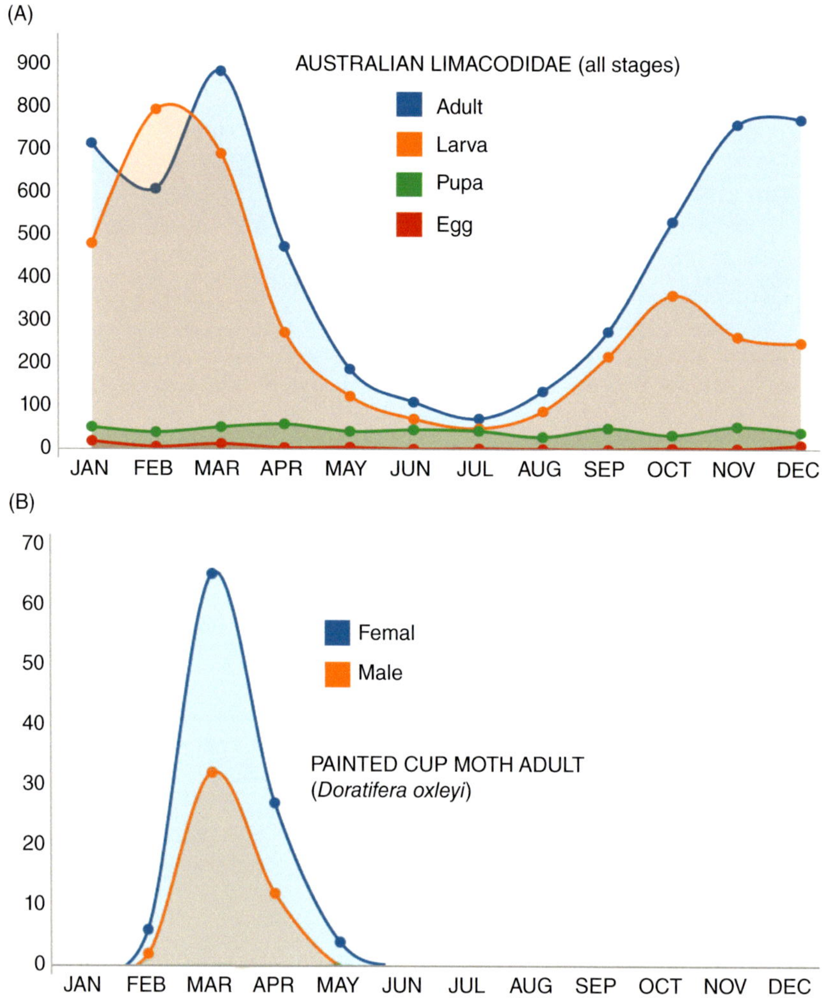

Fig. 16.6. Phenology of Australian Limacodidae from iNaturalist.org. (A) Life history stages of Australian Limacodidae (n = 1572 observations) (note: pupa = cocoons with living occupants or eclosed; eggs = scale-covered *Doratifera* ova that persist post hatch). (B) Greater numbers of females than males attracted to lights of sexual dimorphic painted cup moth (*Doratifera oxleyi*) (n = 148 adults, 69% females).

Susica) *idiomorpha*, also collected by Dumigan from the same location as *S. barymorpha*, and *Pseudanapaea transvestita* (then *Anabaena trigona* Turner) from throughout Australia.

Among the earliest descriptions and illustrations of an Australian limacodid caterpillar (= *Bombyx* (= *Doratifera*) *vulnerans*) was that by Macleay in 1805; illustrations and description of the species were reproduced in Lewin (1822). This included a caterpillar feeding on *Eucalyptus* while in defensive posture, with spines in repose on a stem, the cocoon, an adult male and female. He mentioned the caterpillar's 'singular power of darting forth wounding stings as a property unknown in any larva of Lepidopterous Insects before observed...' and '... darting out [of] eight rays or bunches of little stings from as many small knobs or protuberances on the back' (six of these in *D. vulnerans* are shown in Fig. 16.3M and eight in *D. oxleyi* in Fig. 16.3K–L).

Dyar (1898) reviewed the known Australian limacodid larvae at a time when Kirby (1892) listed merely 17 species in eight genera. Dyar, who had never visited Australia (Epstein,

2016), paraphrased Lewin's published description of *Doratifera vulnerans*, writing 'the power of everting eight little tufts of stinging spines... concealed when the larva is not irritated' to be 'most curious and apparently inexplicable'. He 'was at a loss to imagine the origin of mechanism of this structure'. He acknowledged receipt of a specimen from Mr. E.A.C. Olive of Queensland.

More recently, McKeown (1942) gave accounts of the life histories of *Doratifera vulnerans*, *D. oxleyi* and *Mecytha fasciata* (under the name *Susica semicana*). He describes the flattened eggs, the erectile spines of the larvae and the cup-like cocoons of *D. vulnerans*. He also describes outbreaks of *D. oxleyi* in south-eastern NSW leading to defoliation of red river gums (*Eucalyptus camaldulensis*) along the Murrumbidgee River. He identifies *Telopea* (Proteaceae) as the food plant of *M. fasciata* and notes the lack of stinging spines in that species' larvae. Perhaps the most recent account of the family is that of Cocking *et al.* (2022) referring to the fauna of the Australian Capital Territory. Cocking *et al.* (2022) illustrate adults and larvae of *Pseudanapaea transvestita*, *D. quadriguttata*, *D. oxleyi*, *D, pinguis* and *Mecytha fasciata*.

16.5 The Limacodid Fauna of Australia

All Australian limacodids are in the subfamily Limacodinae and all, except *Chalcocelis*, are in the tribe Eucleini. The adjacent Asian fauna is, similarly, entirely Limacodinae but does contain more Limacodini (Epstein *et al.*, 2025). The small percentage of Limacodini in Australia is unusual compared with both temperate and tropical Asia, Africa and the New World.

Seventy species of Limacodidae are known to occur in Australia, few in comparison with tropical faunas of Asia, the New World and sub-Saharan Africa. Based on distributions in iNaturalist, the totals for Australian states and territories are Queensland (50 spp.), NSW (21 spp.), Western Australia (22 spp.), Northern Territory (18 spp.), South Australia (11 spp.), Victoria (10 spp.) and Tasmania (3 spp.). The greater species richness in the more tropical regions mirrors the greater affinity of limacodids for tropical rainforests globally.

The phenology of Australian limacodids (Fig. 16.6A) correlates with peaks in precipitation in adult flight from September to April. Larvae show an appreciable peak in February. There is a noticeable dip in both adults and larvae from March to July after which adults increase from that month until November to March. Records of cocoons and eggs, while at much lower numbers, are more consistent throughout the year. Cocoon data includes overwintering prepupal and pupae inside cocoons as well as persistent spent cocoons. Egg numbers mostly reflect those of *Doratifera*, which occur as recognizable scale-covered masses (Figs 16.4B–E).

Edwards (1996) suggested there were 'a few undescribed species... in Australia but fewer than for most families'. We suspect some are cryptic, with similar-looking adults 'hiding' very different larvae, as we shall see with *Calcarifera*. Edwards also noted that Australia shares some genera with Borneo and South-east Asia, including those known to feed on palms. These include *Chalcocelis*, *Thosea*, *Scopelodes*, *Squamosa* and *Praesusica*.

Nearly the entire Australian limacodid fauna is endemic to Australia plus New Guinea. Several species are shared by northern Queensland and New Guinea. These include species of *Scopelodes*, *Hydroclada*, *Thosea*, *Anepopsia*, *Praesusica*, *Anaxidia* and *Pygmaeomorpha*. Among the northern Australian fauna, at least several species, including one Limacodini, *Chalcocelis albiguttatus* and *Scopelodes nitens* occur in New Guinea and in tropical Asia.

In Australia, *Doratifera* spp. are known defoliators of *Eucalyptus* forests but there have been no reported outbreaks of other limacodid genera on palms, as occur in South-east Asia (Edwards, 1996). A few limacodid larvae are known to be pests on nuts in Australia or PNG, including *Mecytha* on *Macadamia* and *Scopelodes* on cashews in PNG (Common, 1990). There are potential invasive limacodids of concern, particularly to northern Australia (Byrne *et al.*, 2021). These include *Parasa lepida*, *Setora nitens* and *Darna trima* from PNG or Asia, which can be defoliators and/or cause medical issues with larval venom. That this concern is warranted is shown by the recent establishment of *Oxyplax pallivitta* – closely related to *Darna* – in Hawaii (Nagamine and Epstein, 2007).

16.6 Phylogeny

16.6.1 A multigene maximum likelihood analysis

Figure 16.7 illustrates the results of a new phylogenetic analysis of Limacodidae using maximum likelihood (ML). This expands on Epstein *et al.* (2025) by adding further Australian taxa to others from Africa, Asia and the New World. Australian taxa included by Epstein *et al.* (2025) and repeated here are: *Anaxidia lozogramma, Doratifera quadriguttata, Comana albibasis, C. dochmosema, Eloasa callidesma, Mecytha fasciata, Praesusica placerodes, Pseudanapaea transvestita, Pygmaeomorpha modesta,* and *Squamosa barymorpha*. New additions are: *Anepopsia eugyra, Apodecta monodisca, Calcarifera ordinata, Chalcocelis albiguttatus, Comanula uniformis, Doratifera stenora, D. vulnerans, Ecnomoctena brachyopa, Hydroclada antigona, Hypselolopha hypodrosa, Lamprolepida chrysochroa, Parasoidea paroa, Pseudanapaea denotata, Pygmaeomorpha plagioscia, P. delocrossa* and *Thosea penthima*.

This seven-gene analysis includes the same six nuclear sequences and one mitochondrial gene as in Epstein *et al.* (2025). The nuclear genes were elongation factor 1 alpha (EF-1a); wingless (wg); carbamoylphosphate synthetase domain protein (CAD); cytosolic malate dehydrogenase (MDH); ribosomal protein subunit 2 (RpS2); arginine kinase (Arg K); and the mitochondrial cytochrome oxidase I (COI).

The in-group taxa comprise 92 Eucleini, 25 of which occur in Australia, and a large clade of 36 Limacodini, all used in Epstein *et al.* (2025). Only one species of the latter group occurs in Australia – *Chalcocelis albiguttatus*. The outgroup used is Chrysopolominae, the sister group of Limacodinae according to Epstein *et al.* (2025), which includes the same five species: two *Ectropa* (*E. ancilis* and *E. alberici*), *Chrysopoloma isabellina* and two *Strigivenifera* (*S. venata* and *S. eborea*). No Crothaeminae (Africa) or Dalcerinae (New World) were included.

The results show four clades of Eucleini consisting mostly of genera from Australia or New Guinea (Fig. 16.7). The relationships and high support levels of the five Chrysopolominae matched that of Epstein *et al.* (2025), as did the two early eucleine clades, which do not share common ancestry – the 'lesser flat nettles' (LFN) and the 'greater flat nettles' (GFN). The former is only known from Asia and Africa, whereas in Australia both *Thosea* and *Anepopsia* are with the GFN. All the other Eucleini in Australia are in the clade referred to as the 'raised caltrop nettles' (RCN) (Epstein *et al.*, 2025). Caltrops are a type of tiny deciduous spines associated with the subdorsal (SD) scoli; these tiny spines occur in a sunken oval at the base of these scoli in *D. stenora*, as noted by Dyar (1898). The *Parasa* complex, the most species-rich subgroup in the RCN, is found pantropically but does not occur in Australia.

The four clades of Eucleini principally of Australian genera are: Clade 1: *Pseudanapaea* + *Parasoidea*; Clade 2: [*Calcarifera* + *Comana* + *Hydroclada* + *Comanula*] + [*Mecytha* + *Hypselolopha* + *Ecnomoctena*]; Clade 3: *Anaxidia* + [*Doratifera* + *Lamprolepida*] + [*Preasusica* + *Squamosa* + *Eloasa*], and Clade 4: *Mambarona* + [*Pygmaeomorpha* + *Apodecta*].

The two other eucleine clades are primarily Asian: Clade 5: *Anepopsia eugyra* + *Thosea* – part of the globally distributed greater flat nettle group, and Clade 6: *Scopelodes*, a member of the long palp group found in other Asian genera and those in the Neotropics (for description of these two groups, see Epstein *et al.*, 2025).

Some of the genera in these clades are paraphyletic. Given the small taxon sampling for the most taxon-rich genera such as *Comana, Doratifera* and *Eloasa*, this finding is to be expected. *Comana albibasis*, for example, forms a clade with *Hydroclada antigona*, although *Comana dochmosema* + *Comanula uniformis* is sister to it. *Lamprolepida chrysochroa* is sister to *Doratifera quadriguttata* + *D. vulnerans* and *D. stenora* is sister to all three. Although the larvae of *L. chrysochroa* appear to lack the springing mechanism found in *Doratifera*, it is interesting to note that the cocoons figured by Herbison-Evans and Crossley resemble a common pyriform cocoon found in *D. vulnerans*.

Regarding Clade 4, Holloway (1987) moved both *Birthama plagioscia* and *B. delocrossa* to *Pygmaeomorpha* and in this analysis they form a clade with *P. modesta* and *Apodecta monodisca* and with *Mambarona congrua* as sister to all. This, of course, suggests more taxon sampling for future revisionary work.

Hypselolopha and *Elassoptila* are unknown in their larval stages but have adults that look

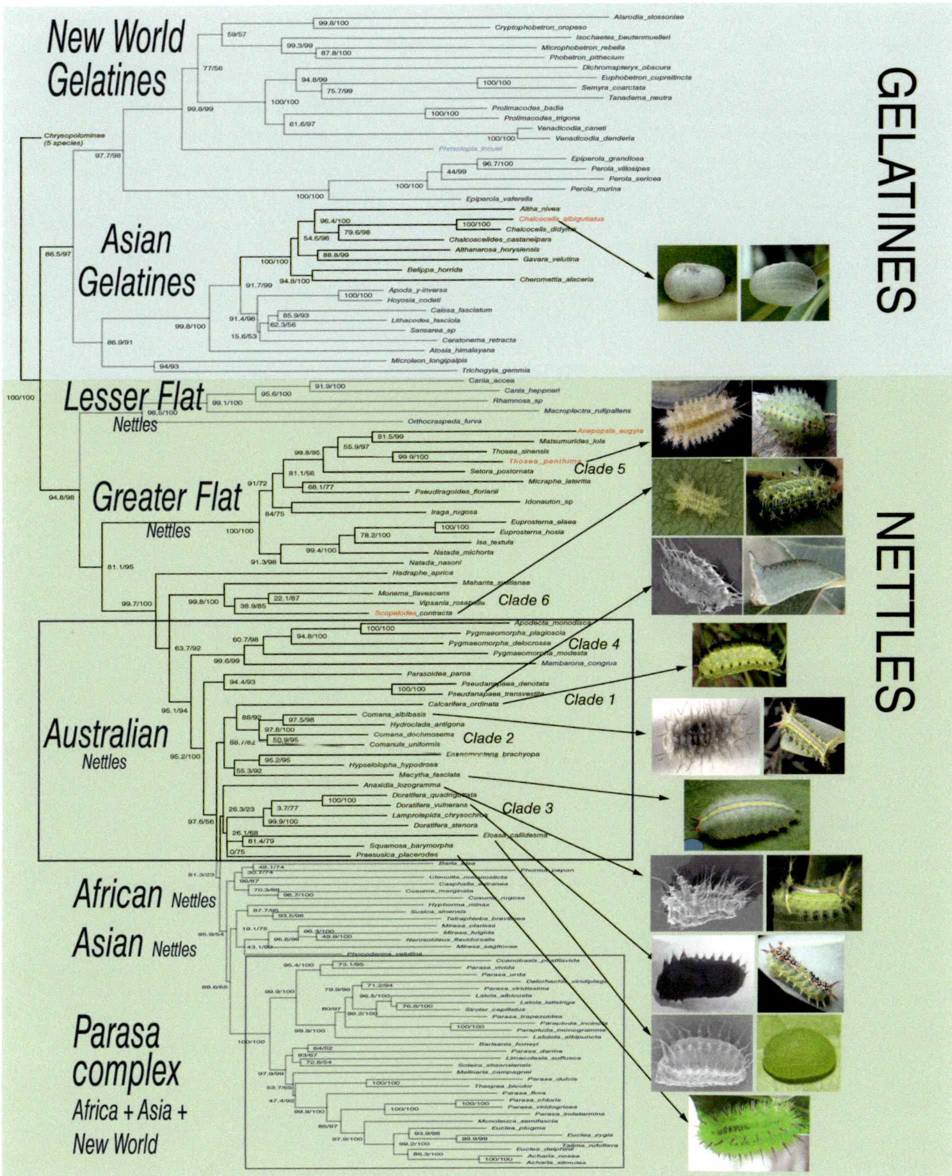

Fig. 16.7. Phylogeny of world Limacodinae with larval stages of Australian taxa primarily in the nettle tribe Eucleini. This is modified from Epstein *et al.* (2025) to include additional taxa from Australia: *Calcarifera*, *Hypselolopha*, *Lamprolepida*, *Doratifera stenora*, *Hydroclada*, *Comanula*, *Pseudanapaea denotata*, *Pygmaeomorpha delocrossa*, *P. plagioscia*, *Apodecta monodisca* and *Doratifera vulnerans*. The outgroup taxa were reduced from the previous phylogeny to include only the gelatine tribe Limacodini. Maximum likelihood analysis was used with six nuclear genes and one mitochondrial gene. Clades from Australia or that contain Australian taxa are bold, black or red. For example, *Chalcocelis*, *Thosea*, *Scopelodes* and *Anepopsia* – genera that occur in Australia – are in red type and occur in clades with primarily Asian taxa. Figures of first and late instar larvae are primarily from Epstein *et al.* (2025) or from the authors Epstein and Lin. Note that images of *Thosea* first instar and *Scopelodes* first and late instar are from Asia.

suspiciously like those of Asian Limacodini. However, in our molecular analyses both appear to be Eucleini. *Hypselolopha* is in a eucleine clade along with *Mecytha fasciata* and *Ecnomoctena brachyopa*. *Elassoptila* emerged as sister to all three eucleine clades in our COI analysis (below) (Fig. 16.8).

16.6.2 A COI analysis

Given the small representation of Australian taxa in the seven-gene analysis, we now present a larger one that includes related Asian fauna but is based solely on the COI mitochondrial gene. This analysis includes an in-group of Eucleini with 47 taxa primarily from Australia or congeneric taxa from Asia or Papua New Guinea (e.g. *Squamosa brevisunca brevisunca* (= *svetlanae*), *S. ferruginea*, *Scopelodes contracta* and *Thosea sinensis*). The outgroup consists of five taxa in the 'lesser flat nettles' from Asia and Africa.

Even though COI has limitations in tracking deeper phylogenetic relationships, the results (Fig. 16.8) suggest that many relatively large Australian genera, including *Comana*, *Eloasa*, and *Doratifera* and others, are paraphyletic. Other examples of paraphyletic genera include *Anaxidia*, *Anepopsia* and *Squamosa*.

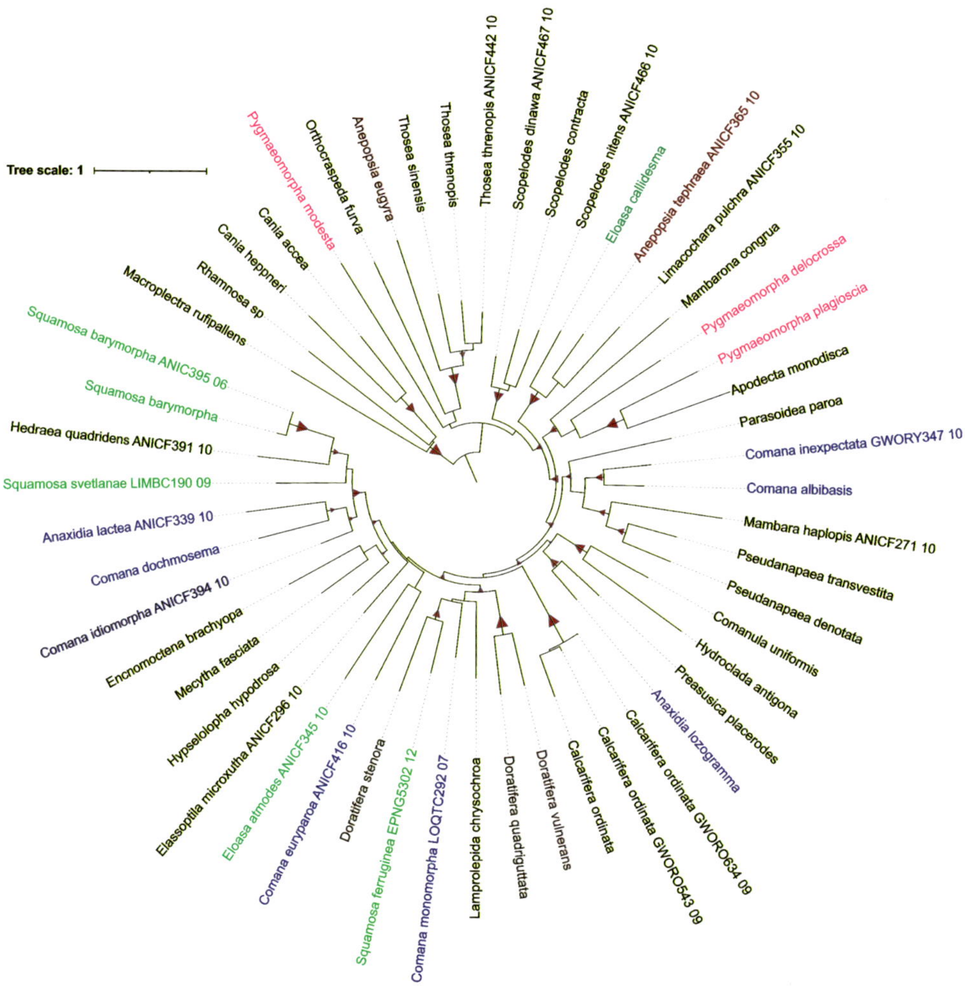

Fig. 16.8. Phylogeny using maximum likelihood analysis with mitochondrial COI. The outgroup is an early branch of nettle limacodids (Limacodinae: Eucleini) known as the lesser flat nettles (taxa only known from Asia and Africa). Genera found to occur in Australia that are suggested to be paraphyletic are colour coded.

Cook *et al.* (2025), using the COI sequence, also found considerable paraphyly among ten species of *Comana*, including *Comana idiomorpha* + *Anaxidia lactea*, and 'considerable genetic divergence' within many species of Australian limacodids. Like our findings, they found *Anaxidia lactea* separate from *A. lozogramma*.

16.7 The Hungry Caterpillar Strategy

Common (1990) noted that 'first-instar larvae of *Pseudanapaea trigona* [now *P. transvestita*] and *Parasa* [= Eloasa] *atmodes*, and probably other genera, do not feed but, after a period of about two days, undergo their first ecdysis before they begin to feed...'. Other examples of this phenomenon occur in North America. It occurs in all Eucleini in the eastern United States (Dyar, 1899) as well as the introduced *Monema flavescens* (Dyar, 1909).

This was referred to as 'the hungry caterpillar strategy' by Zaspel *et al.* (2015), who suggested that this was a uniquely derived behaviour shared by all spiny Limacodidae. This strategy has been further corroborated in a much larger group of genera (Epstein *et al.*, 2025) and in the phylogeny of the Australian Eucleini presented herein (Fig. 16.7) (see also Figs 16.4M–P). It has been suggested that this strategy protects the chorion from rupture, given that the first instar is less spiny than the second. Larvae of Limacodini that do feed shortly after hatching tend to be from eggs that are much more scattered than those of the Eucleini and are less likely to eat the unhatched embryos (Nagamine and Epstein, 2007).

16.8 Australian Limacodidae: Food Plants and Life Histories

Of the 40% of Australian limacodids that are known from larvae, there is only a fraction in which the first instars are known. Common (1990) discussed and figured the newly hatched larvae of *Pseudanapaea trigona* [=*transvestita*], showing a series of single-spined tubercles on the dorsum and the fasting first instar of *Eloasa atmodes*. As we have discussed, other Australian genera possess this unusual feature in the first instar yet follow very different pathways in subsequent instars.

16.8.1 Food plants

Limacodid larvae prefer smooth tough leaves and tend to be polyphagous. A continual update of the food plants of the Australian limacodids is given by Herbison-Evans and Crossley (2025). Among records from their site, the literature and iNaturalist, we found 57 food plants. Twenty-two of these records were from just two plant families. Fourteen species fed on species of Myrtaceae and eight on Proteaceae. Among the Myrtaceae, *Eucalyptus* was a common choice, with the genus *Doratifera*, reported from six eucalypt species, perhaps more specialized on these gum trees than hitherto suspected. The other 13 myrtaceous genera had only one limacodid record for each. Among the Proteaceae, seven of eight food plants were favoured by *Mecytha fasciata*, including *Macadamia integrifolia*. *Mecytha dnophera* was recorded feeding on *Grevillea* spp. and *Anaxidia lozogramma* on *Macadamia*. Food plant records from other plant families mirror those known to be utilized by limacodids in other geographical regions, including (number of records in parentheses) Arecaceae (3), Anacardiaceae (1), Celastraceae (1), Euphorbiaceae (2), Rosaceae (2), Rhizophoraceae (1), Rubiaceae (3), Rutaceae (3), Sapindaceae (3), Salicaceae (2), Sterculiaceae (1), Theaceae (1) and Verbenaceae (1).

Comana albibasis had the most food plant records at seven across seven plant families, although interestingly, none in the Myrtaceae or Proteaceae. *Comana monomorpha* has seven records in six families, including two in Myrtaceae. *Chalcocelis albiguttatus* (Limacodini) has six records in five families, with two records favouring Rubiaceae (*Caffea* and *Gardenia*). *Calcarifera ordinata* and an undescribed congeneric were reported only from Fabaceae: *Cassia* for *C. ordinata* and *Acacia* for the western species. *Eloasa infrequens* has four records in four different plant families, including the only records from Podocarpaceae and Celastraceae. *Hedraea quadridens* (Fig. 16.3D) is only known to feed on bracken fern (Lucas, 1901).

16.8.2 Life histories and morphological notes

Many Australian limacodid larvae remain unknown (Cook *et al.*, 2025), including some genera and many species. The best-known genus is undoubtedly *Doratifera* for which larvae are known from seven species. Less well known are *Comana* and *Eloasa* with larvae known for nine of 12 species in each case. There are no larval records for *Anepopsia* (2 sp.), *Apodecta* (1 sp.), *Comanula* (1 sp.), *Ecnomoctena* (3 sp.), *Elassoptila* (1 sp.), *Hydroclada* (1 sp.), *Hypselolopha* (1 sp.), *Limacochara* (1 sp.), *Mambara* (1 sp.), *Pygmaeomorpha* (5 sp.), *Squamosa barymorpha* (1 sp.) or *Parasoidea* (3 sp.). Larvae are unknown or only poorly known for about 60% of the Australian fauna.

16.8.2.1 *Doratifera*

Doratifera spp. have striking, gregarious larvae which, together with their numerous, clustered cocoons, make them Australia's best known limacodids. The nature of the first instars of any of its nine species globally has only recently been documented: for *D. pinguis* (Yu-Chi Lin, personal observation; Fig. 16.7) and *D. stenora* by Larney Grenfell (Fig. 16.4F). First instars of each have a single spined tubercle like those found in *Pseudanapaea transvestita* (Fig. 16.4J) and *Eloasa callidesma* (Figs 16.4M–N). *D. quadriguttata* larvae in the ANIC collection are mislabelled as first instars, having typically the appearance of other second instar Eucleini by virtue of their heavily spined dorsum (see Fig. 16.4I). Six of seven species of *Doratifera* found in Australia feed on *Eucalyptus*. The genus is also reported on other Myrtaceae, including *D. vulnerans* on *Angophora* and *Melaleuca* (Zborowski and Edwards, 2007).

Common (1990) noted that *Doratifera* has two combinations of segments with spiny D scoli held to the body in repose (= appressed) and which spring into action when disturbed (Figs 16.3K–L). These special scoli occur on two or four segments: T3 and A1 only, or on A6 and A7 in addition to the first two. In most late instar *Doratifera* the subdorsal (SD) row consists of scoli of roughly the same length except for the longer terminal pair on A9. It is surprising to note that in late instars, all *Doratifera* with appressed spines on two or four segments have non-functional remnants on all remaining D and SD scoli. Although lacking spines, we still refer to these as scoli given spines are present in earlier instars.

Doratifera, as typical of the tribe Eucleini, has no SD scolus on A1 and there is a spiracle in its place, more dorsally located than the others on segments A2–A8, which are hidden by SD scoli above them (Figs 16.2, 16.3I). Also common among many Eucleini are caltrop patches at the base of SD scoli. In *Doratifera* the caltrops occur in sunken ovals, apparently unique to the genus (Figs 16.2, 16.3I).

The larva of *Doratifera stenora* are clearly distinct from others in the genus by having typical D and SD scoli rather than the appressed or remnant spines (Fig. 16.4H).

Here is a summary of the three main types of scoli in the genus *Doratifera*, as presently defined, and acknowledging the help from many of images on iNaturalist.

- *Doratifera quadriguttata* and *D. casta* (Figs 16.2, 16.3I, 16.3J). Appressed spines on D scoli on T3 and A1 only. Spines are absent on all remaining D and SD scoli including the elongate D scoli on T2 and A8 in *D. quadriguttata*. SD scoli on A2 to A9 has caltrops in sunken ovals and narrow, elongate SD scoli. Caltrop ovals largest on A8 and A9. *D. casta* (Fig. 16.3J) are mostly black and yellow with short scoli mostly of equal length.
- *Doratifera oxleyi*, *D. vulnerans*, *D. corallina*, and *D. pinguis* (Figs 16.3K–O). Appressed spines are found on D scoli on T3 and A1, and on A6 and A7. Remnant or absent spines are found on all remaining D and SD scoli. Relatively long SD scoli in *D. oxleyi*, *D. vulnerans* and *D. corallina* give the larvae a flatter appearance compared with *D. pinguis* (Fig. 16.3O), which has short D and SD scoli.
- *Doratifera stenora* (Fig. 16.4H). These larvae have spines on all D and SD scoli, and none are appressed. It remains undetermined from images whether caltrop spine patches are present or absent in *D. stenora*. One feature they share with other *Doratifera* larvae is that they are gregarious, and their eggs, like others in the genus, are covered by female scales (Fig. 16.4E). There is also little overlap with the other *Doratifera* in

food plant records, feeding on mangroves and not *Eucalyptus*.

The closest known examples to the specialized defence and spines found in *Doratifera* were summarized by Dyar (1898). He noted the similarity of *Parasa chloris* with spines appressed toward the median and the differences from *Natada* (as *Sisyrosea*) *nasoni* with appressed spines held toward the lateral margins. Aididae, a small Neotropical family related to the Limacodidae among the Core Megalopygidae, also has species with appressed spines like many New World *Natada* (Epstein *et al.*, 2025). Today we know of many more examples of the same behaviour and morphology in other species of New World *Parasa* and *Natada*. To what extent these conditions are synapomorphic for lineages awaits further investigation.

16.8.2.2 *Pseudanapaea (Fig. 16.3H)*

Pseudanapaea spp. have atypical Eucleini larvae that become, by their late larval instars, more like smooth Limacodini. First instars are like those of *Doratifera* and *Eloasa callidesma* in having rows of single, elongate D and SD tubercles (Fig. 16.4J). By late instars, they have an anterior horn, perhaps from fusion of scoli on T3 in earlier instars. *Pseudanapaea* is also like many Limacodini in being solitary feeders and in having variable patches of grey on a green background. This combination of behaviour and pattern occurs in the *Prolimacodes* complex and in Palaearctic and Nearctic Limacodini (Epstein *et al.*, 2025).

16.8.2.3 *Eloasa (Fig.16.3F)*

Thus far, we know that three of the 12 *Eloasa* species are green, without pattern, and are covered by spines not borne on scoli scattered on the dorsum. First instar *Eloasa callidesma* has single-spined tubercles, only to become spines without scoli (Figs 16.4M–P). This is probably the same developmental path found in *E. infrequens* and *E. symphonistis*. Although we know the early instars of *E. callidesma*, including the first instar, their food plants remain unknown. The larva shown in Fig. 16.3G could be *Eloasa* without spines or a member of another Australian genus not presently known from early stages (e.g., *Parasoidea* or *Ecnomoctena*).

16.8.2.4 *Mecytha fasciata (Fig. 16.4K)*

Another of the seemingly 'unicorn' limacodids that include *Pseudanapaea*. The horns at each end occur from the fusion of two. First instars are unknown.

16.8.2.5 *Calcarifera ordinata (Fig. 16.3A)*

The confusion caused by the wrong association between the adult of *Calcarifera ordinata* and larva of *Comana albibasis* has been resolved. For many years the iconic turquoise, red and yellow 'rainbow battleship' larvae were referred to as *Calcarifera ordinata* in numerous publications and websites whereas they are in fact *Comana albibasis*. Common (1990) described the larvae of *Calcarifera ordinata*, on food plants in the Fabaceae, as 'brightly coloured blue-green and yellow... with numerous tubercles...'. Although these caterpillars are quite distinct, this general larval description can fit *Calcarifera ordinata* or *Comana albibasis*, which may have contributed to this misassociation.

The confusion began to abate when a green larva covered with blue warts, termed the 'blue warts caterpillar' by Brett Smith and M. Epstein on iNaturalist, was reared to an adult that matched *Calcarifera ordinata* at Ellura Sanctuary in South Australia (www.ellura.info/Lepidoptera.html#Calcarifera). Concurrently, the iconic turquoise, red and yellow caterpillar was reared to adults of *Comana albibasis*. Another step in resolution was Epstein's insistence that the uncoloured larval voucher in fluid in the ANIC labelled *Calcarifera ordinata* from Western Australia was a closer match to the blue warts caterpillar than to those of the rainbow battleship, yet differed in having more elongate scoli at each end (compare Figs 16.3A–D and see section 16.9.2). Both *C. ordinata* and the other *Calcarifera* sp. feed on Fabaceae, *Cassia* and *Acacia* respectively. This does not appear to be a common food plant family for Australian Limacodidae, although we know of at least one record of *Comana albibasis* feeding on *Acacia flavescens* in northern Queensland (C. Suttner, iNaturalist). More recently there has been molecular corroboration of the resolved associations by Cook *et al.* (2025).

16.8.2.6 *Comana albibasis (Fig. 16.3C)*

The famous 'rainbow battleship' larva, as mentioned above, was misassociated with *Calcarifera ordinata*. The first instar, obtained from female eggs by the senior author, is presently the only known example of a first instar Eucleini with two spines per tubercle (Fig. 16.7). Typical of many Eucleini, the scoli add multiple spines in later instars, although at which instar the exotic colour pattern becomes visible is for future study.

16.8.2.7 *Anaxidia lozogramma* (Figs 16.3P–S)

Late instar larvae of this species resemble *A. lactea* in having a bold white longitudinal band down the centre of the dorsum but, unlike *lactea*, have pink or purple along with white. The first instar has 3–5 spines per tubercle and more elongate tubercles at each end (Fig. 16.7). This species is reported from many non-native plants including citrus, blueberries and *Camelia*.

16.8.2.8 *Hedraea quadridens* (Fig. 16.3D)

The presence of caltrop patches on the apex of SD scoli on A2 to at least A8 (A9 obscured) in this species is presently the only known example of caltrops in Australian limacodids, outside of *Doratifera*. Until recent posts by Eugene Burrows and Matt Nimbs on iNaturalist, there may have been no reference to the larva or its food plant since it was reported on *Pteris esculentum* (bracken fern) by Lucas (1901).

16.8.2.9 *Praesusica placerodes* (Fig 16.4L)

The larva of this species was recently discovered by Christina Suttner, who reared the late instar to adult in a eucalypt/rainforest ecotone in far north Queensland. The larvae are green and covered with spiny verrucae, each with a longer central spine that goes from reddish to bluish purple at the sharp tip; no dorsal band is present between the dense verrucae. It fed on '*Symplocos cochinchinensis* (Symplocaceae) in wet sclerophyll forest at 1000 m elevation' (C. Suttner, pers. commun.). The cocoon is a typical ovoid shape (Fig. 16.5E).

16.9 Taxonomic Issues

16.9.1 *Squamosa barymorpha* and *Hedraea quadridens*

Two large and similar-looking limacodids with hyaline forewing spots found along the East Coast of Australia are currently in different genera: *S. barymorpha* (Fig. 16.1L) and *H. quadridens* (Fig. 16.1K). Common (1990) suggested that these are probably congeneric. Geographically, there appears to be a separation, with *S. barymorpha* found in northern Queensland and *H. quadridens* with a more southern distribution, centred around Brisbane from 20 km NE of Gympie to Urliup Wildlife Sanctuary south of the Gold Coast (iNaturalist).

The type species of *Squamosa*, *ferruginea*, is from Papua New Guinea, sharing hyaline forewing spots, a small, dark discal spot on the ventral hindwing (HW) and similar male genitalia with both *S. barymorpha* and *H. quadridens* hyaline. This probably places *Hedraea* as a junior synonym of *Squamosa*. The Asian members of *Squamosa*, in turn, lack the dark HW discal spot and have male genitalia with elongate juxtal spines below the phallus or saccular processes (Wu *et al*., 2022), which are not found in *S. ferruginea*, *S. barymorpha* or *H. quadridens*.

The two *Squamosa barymorpha* in our COI ML tree have a 98/100 and a 91/99 match with *Hedraea quadridens* (Fig. 16.8) in agreement with the morphology. This analysis, however, places *S. ferruginea* with clearly unrelated taxa including *Doratifera* and *Comana*. Likewise, and contrary to morphology, the COI analysis places *H. quadridens* and two *S. barymorpha* in a clade with an Asian *Squamosa*, *S. brevisunca brevisunca*. Given the parameters of this present publication, a more formal review of *Squamosa* is needed and will occur elsewhere.

16.9.2 *Calcarifera*

As mentioned above, there appear to be two *Calcarifera* species in Australia, each with similar adults to the one shown here (Fig. 16.2), but with very distinct larvae (compare Figs 16.3A–B and see section 16.8.2 above). The larval voucher in ethanol at ANIC is labelled as *C. ordinata* and #2836, which matches a logbook entry at the same collection that states it was found on *Acacia* sp. in April 1971, 67 Km SSE Carnavon (coastal Western Australia). Three years later N. McFarland reared larvae to adults of both sexes of apparently the same species, also from coastal Western Australia but 500 km to the south from Geralton. In our COI analysis (Fig. 16.8), two adults in BOLD from Western Australia

(GWORO634 09 and GWORO543 09) are grouped on the same clade as one collected in eastern Australia by Yu-Chi Lin, although their separation is not supported.

16.9.3 *Anaxidia*

As shown above in our COI analysis (Fig. 16.8) and in Cook *et al.* (2025), *Anaxidia lactea* and *A. lozogramma*, the only taxa in this genus, are in different clades. Although the larvae of each appear to be quite similar, this general larval habitus is found in other genera in Australia, including *Comana*. Likewise, note that in our COI analysis, *A. lactea* is in a clade with *Comana dochmosema* and *C. idiomorpha*, although the larvae of these *Comana* species are unknown. *Anaxidia lozogramma* is in a clade with *Praesusica placerodes* but they are quite dissimilar. The adult male resting postures of the two *Anaxidia* are very different: *A. lozogramma* raises the abdomen nearly perpendicularly (Fig. 16.1J) whereas *A. lactea* does not.

16.10 Conclusion

Our state of knowledge on the Australian Limacodidae, known locally as cup moths, is a work in progress given that we know less than half of the life histories of the 70 Australian species. Clearly, revisionary work especially on the larger genera awaits future efforts given the current molecular evidence. Given the growing interest in cup moths by scientist and non-scientist alike, we envision continued growth in our knowledge of this captivating group of moths over the next 30 years, as has occurred since the insights of Common (1990) and Edwards (1996).

Acknowledgements

It would not have been possible to complete this chapter without the people we've met through iNaturalist.org and their efforts in rearing larvae and providing images. Among these are the late Vic Fazio III, Ian McMillan, Sandra Clark (Urliup reserve), Christina Suttner, Dianne Clarke, Larney Grenfill, Dominic Funnell and Vireya Jacquard. It was a pleasure working with Brett Smith (Ellura Reserve), particularly on sleuthing the identity of the 'blue warts' caterpillar. We also thank Andy Walker (University of Queensland) for information on limacodid venoms and for providing specimens used for the chapter. The senior author would also like to acknowledge Todd and Mike Gilligan for a memorable and important collecting trip to Queensland in 2009. Michael Braby has provided key support for Yu-Chi Lin both in Australia and beyond. Finally, we acknowledge Marianne Horak, who was a key inspiration for our efforts on Limacodidae, which began during a visit to ANIC with Todd Gilligan. The loans she facilitated and literature she provided have been indispensable for this and projects of broader scope.

References

Brock, J.P. (1971) A contribution towards an understanding of the morphology and phylogeny of the Ditrysian Lepidoptera. *Journal of Natural History* 5, 29–102.

Byrne, C.J., Epstein, M.E. and Moyle, D.I. (2021) The Caterpillar Key Fact Sheet: Limacodidae. Available at: https://keys.lucidcentral.org/keys/v3/the-caterpillar-key/key/caterpillar_key/Media/Html/entities/limacodidae.pdf (accessed 4 October 2025).

Cocking, G., Bond, S. and Edwards, E.D. (2022) *Moths in the A.C.T.* Privately Published, Glenn Cocking, Canberra, ACT.

Common, I.F.B. (1990) *Moths of Australia*. CSIRO Publishing, Collingwood, Victoria.

Cook, L.G., Thurman, J.H. and Walker, A.A. (2025) DNA barcoding solves the mystery of the rainbow battleship caterpillar (Lepidoptera: Limacodidae) and reveals cryptic diversity in Australian slug moths. *Austral Entomology* 64, e12727.

Dyar, H.G. (1898) Larvae of the Australian Eucleidae. *Psyche* 8(262), 171–174.

Dyar, H.G. (1899) The life-histories of the New York Slug Larvae. (Conclusion). *Journal of the New York Entomological Society* 7(12), 234–253.

Dyar, H.G. (1909) The life history of an oriental species of Cochlidiidae introduced into Massachusetts (*Cnidocampa flavescens* Walk.). *Proceedings of the Entomological Society of Washington* 11, 162–170.

Edwards, E.D. (1996) Limacodidae. In: Nielsen, E.S., Edwards, E.D. and Rangsi, T.V. (eds) *Checklist of the Lepidoptera of Australia*, Monographs on Australian Lepidoptera, 4. CSIRO, Collingwood, Victoria, pp. 145–147.

Epstein, M.E. (1996a) Revision and phylogeny of the limacodid-group families, with evolutionary studies on slug larvae (Lepidoptera: Zygaenoidea). *Smithsonian Contributions to Zoology* 582, 1–102.

Epstein, M.E. (1996b) A new name and generic placement for the misidentified type species of *Epiclea* Dyar, 1905 (Lepidoptera, Limacodidae). *Proceedings of the Entomological Society of Washington* 98, 812–817.

Epstein, M.E. (2016) *Moths, Myths, and Mosquitoes: the Eccentric Life of Harrison G. Dyar, Jr.* Oxford University Press, Oxford.

Epstein, M., Geertsema, H., Naumann, C. M. and Tarmann, G. H. (1999) The Zygaenoidea. In: Kristensen, N.P. (ed.) *Lepidoptera, Butterflies and Moths*, 4(1), *Handbook of Zoology*. De Gruyter, Berlin and New York, pp. 159–180.

Epstein, M.E, Dupont, S., Garzon, I., Solovyev, A.V., Pierce, N. and Espeland, M. (2025) Phylogeny and reclassification of Limacodidae worldwide with emphasis on the ontogeny of larval defense (Lepidoptera: Zygaenoidea). *Insect Systematics and Diversity* 9(4), 2.

Heikkilä, M., Mutanen, M., Wahlberg, N., Sihvonen, P. and Kaila, L. (2015) Elusive ditrysian phylogeny: an account of combining systematized morphology with molecular data (Lepidoptera). *BMC Evolutionary Biology* 15, 1–27.

Heppner, J.B. (1984) Outline of classification of the Lepidoptera. In: Heppner, J.B. (ed.) *Atlas of Neotropical Lepidoptera. Checklist: Part I. Micropterigoidea – Immoidea*. W. Junk, The Hague, The Netherlands, pp. 19–23.

Heppner, J.B. (1995) Lacturidae, new family (Lepidoptera: Zygaenoidea). *Tropical Lepidoptera* 6, 146–148.

Herbison-Evans, D. and Crossley, S. (2025) Larvae of Australian Limacodidae. Available from: https://lepidoptera.butterflyhouse.com.au/lima/lima-cats.html (accessed 4 October 2025).

Hering, M. (1931) Limacodidae (Cochliopodidae). In: Seitz, A. (ed.) *The Macrolepidoptera of the World. 10. Bombyces and Sphinges of the Indo-Australian Region*. Alfred Kernen Verlag, Stuttgart, Germany, pp. 667–728.

Holloway, J.D. (1987) Limacodid pests of tropical Australasia. In: Cock, M.J.W., Godfray, H.J.C. and Holloway, J.D. (eds) *Slug and Nettle Larvae: The Biology, Taxonomy and Control of the Limacodidae of Economic Importance on Palms in South-east Asia*. CAB International, Wallingford, UK, pp. 119–121.

Kirby, W.F. (1892) *A Synonymic Catalogue of Lepidoptera Heterocera (moths). 1. Sphinges and Bombyces*. Gurney and Jackson, London.

Lewin, J.W. (1822) *Natural History of Lepidopterous Insects of New South Wales*. J.H. Bohte, London.

Lucas, T.P. (1901) Queensland Lepidoptera. *Proceedings of the Royal Society of Queensland* 16, 73–95.

McFarland, N. (1979) Annotated list of larval foodplant records for 280 species of Australian moths. *Journal of the Lepidopterists' Society*, Suppl., 1–72.

McKeown, K.C. (1942) *Australian Insects: An Introductory Handbook*. Royal Zoological Society of New South Wales, Sydney, NSW.

Nagamine, W. and Epstein, M.E. (2007) The chronicles of *Darna pallivitta* (Lepidoptera: Limacodidae): Biology and larval morphology of a new pest in Hawaii. *Pan-Pacific Entomologist* 83, 120–135.

Packard, A.S. (1864) Synopsis of the Bombycidae of the United States. *Proceedings of the Entomological Society of Philadelphia* 3, 337.

Regier, J.C., Zwick, A., Cummings, M.P., Kawahara, A.Y., Cho, S. *et al.* (2009) Toward reconstructing the evolution of advanced moths and butterflies (Lepidoptera: Ditrysia): an initial molecular study. *BMC Evolutionary Biology* 9, 280–301.

Regier, J.C., Mitter, C., Zwick, A., Bazinet, A.L., Cummings, M.P. *et al.* (2013) A large-scale, higher-level, molecular phylogenetic study of the insect order Lepidoptera (Moths and Butterflies). *PLOS One* 8(3), e58568.

Scoble, M.J. (1992) *The Lepidoptera: Form, Function and Diversity*. Oxford University Press, Oxford.

Shimizu, Y. (1989) *Japan: The Shaping of Daimyo Culture, 1185–1868*. George Braziller, New York.

Swinhoe, C. (1892) *Catalogue of Eastern and Australian Lepidoptera Heterocera in the collection of the Oxford University Museum. Part 1. Sphinges and Bombyces*. Clarendon Press, Oxford.

Turner, A.J. (1942) Fragmenta Lepidopterologica. *Proceedings of the Royal Society of Queensland* 53, 78–80.

Van Nieukerken, E.J., Kaila, L., Kitching, I.J., Kristensen, N.P., Lees, D.C. *et al.* (2011) Order Lepidoptera Linnaeus, 1758. In: Zhang, Z.-Q. (ed.) Animal biodiversity: An outline of higher-level classification and survey of taxonomic richness. *Zootaxa* 3148, 212–221.

Wagner, D.L. (2025) *Moths of the World: A Natural History*. Princeton University Press, Princeton, NJ.

Walker, A.A., Robinson, S.D., Paluzzi, J.-P.V., Merritt, D.J., Nixon, S.A. *et al.* (2021) Production, composition, and mode of action of the painful defensive venom produced by a limacodid caterpillar, *Doratifera vulnerans*. *Proceedings of the National Academy of Sciences* 118(18), e2023815118.

Walker, F. (1855) *List of the Specimens of Lepidopterous Insects in the Collection of the British Museum* 4, 969.

Wu, J., Solovyev, A.V. and Han, H.-L. (2022) Two new species of *Squamosa* Bethune-Baker, 1908 (Lepidoptera, Limacodidae) and first female record of *S. chalcites* Orhant, 2000 from southern Asia. *ZooKeys* 1090, 149–159.

Zaspel, J., Weller, S.J. and Epstein, M.E. (2015) Origin of the hungry caterpillar: Evolution of fasting in slug moths (Insecta: Lepidoptera: Limacodidae). *Molecular Phylogenetics and Evolution* 94, 827–832.

Zborowski, P. and Edwards, T. (2007) *A Guide to Australian Moths*. CSIRO Publishing, Collingwood, Victoria.

17 The Pyraloidea: The Snout and Grass Moths

Roger L. Kitching
School of Environment and Science, Griffith University, Nathan, Queensland, Australia

Abstract
This chapter describes recent work on the Pyraloidea, one of the most speciose of moth superfamilies. The chapter presents the taxonomic history of the superfamily and its current morphological definition. An overview of recent molecular and morphological phylogenies precedes an account of the Australian fauna. Summaries of the species richness, distributions and food plant relationships of the Pyralidae and Crambidae are tabulated and discussed. Particular attention is paid to the crambid subfamilies Spilomelinae and Pyraustinae, the largest clade within the family Crambidae. Finally, species of economic importance are discussed: first the biocontrol agents and then pest species.

17.1 Introduction

The Pyraloidea, commonly referred to as snout or grass moths (depending on the family), are a very large, monophyletic clade of 'micro'-moths. Ditrysians, they are part of the massive clade Obtectomera and sister to the so-called Macroheterocera (see Chapter 3). According to Kawahara *et al.*'s (2019) phylogeny, they emerged as a clade in the mid-Cretaceous, splitting into their constituent families at the end of the Cretaceous and subsequently radiating widely through the Palaeogene. The fossil record of pyraloids is poor. In an assessment of 11 putative fossil pyraloids, Heikkilä *et al.* (2018) were able to confirm only three as unequivocally pyraloid. *Eopyralis morsae* was the oldest of these from the earliest Eocene (33.9 mya) of Denmark. These authors suggest that this species is possibly representative of a stem group of the [Phycitinae + Pyralinae + Epipaschiinae] clade of Pyralidae, or a stem group phycitine. The two other confirmed pyraloid fossils are younger: the larva *Baltianania yantarnia* (probably Crambidae: Pyraustinae) from the mid-Eocene of Russia and *Glandotricha olgae* (probably Pyralinae: Endotrichini) from the mid-Eocene Baltic region. None of these fossils informs the origin of the superfamily.

Globally, the Pyraloidea are the third most speciose superfamily of moths, exceeded only by the Noctuoidea and Geometroidea (Solis, 2007) in terms of the number of described species. Indeed, this league table may change should the less well-worked Pyraloidea ever be fully described and catalogued. There are about 16,000 species named with, perhaps, as many as three times that number awaiting attention (Munroe and Solis, 1998). All recent phylogenies confirm pyraloid monophyly and that of the two constituent families, the Pyralidae and the Crambidae

Corresponding author: r.kitching@griffith.edu.au

DOI: 10.1079/9781800627680.0017

(Regier *et al.*, 2012, 2013; Mitter *et al.*, 2017; Kawahara *et al.*, 2019). Each of these families contains a wealth of, generally, well-established subfamilies: five in the case of the Pyralidae and as many as 17 for the Crambidae (Regier *et al.*, 2012; Léger *et al.*, 2021).

Biologically, members of the superfamily exhibit a very wide range of habits, from free-living feeders on a wide range of plants, including non-vascular species, to predators on scales and ant larvae. Others are detritivores or fungivores, part of the 'brown' food web of leaf litter and decaying wood (Common, 1990; Munroe and Solis, 1998). Grimaldi and Engel (2005) designate them as the most ecologically diverse of any group of Lepidoptera. Many species are significant pests of human agriculture and its products, often having been transported around the world with produce of one kind and another (see below). Others have been of immense benefit to humankind. Perhaps the most celebrated of these is *Cactoblastis cactorum* (Pyralidae: Phycitinae) which was instrumental in controlling *Opuntia* cacti on the arable lands of subtropical Australia (Dodd, 1940).

The Pyraloidea occur naturally and prolifically on all continents except Antarctica. They are most speciose and diverse in the tropics, particularly in wet forests, but do occur and have speciated on remote islands of the world's oceans (see, for example, Schaffer, 1990; Schaffer and Munroe, 2003; Guillermet, 2009).

17.2 Taxonomic History of the Superfamily

The taxonomy of the Pyraloidea has had a complex history since the name was first proposed (as 'Pyralites') by Pierre André Latreille in his 14-volume *magnum opus* on the arthropods (1806–1809). The type genus *Pyralis* is a Linnaean name from 1758. Latreille also erected the family Crambidae (as Crambites) based on the European genus *Crambus*. The earliest use of a superfamily designation is harder to pin down. Edward Meyrick used the term Pyralidina in 1884 to encompass a host of 'families' many of which we now recognize as subfamilies of either Pyralidae or Crambidae (Meyrick,1884). He did also include taxa which have now been removed to other superfamilies – the Oxychirotidae (now within the Tineodidae), Tineodidae, Siculodidae (now part of the Thyrididae), Pterophoridae and Alucitidae. Since that time various other families have been included in or excluded from the Pyraloidea depending on the authors' opinions. Fletcher and Nye (1984) tabulate these uses up to and including their 1984 work.

The Australian checklist (Shaffer *et al.*, 1996) excludes all but the nominate family the Pyralidae (*s.l.*), which is divided in that work into two informal groupings, the Pyraliformes and the Crambiformes, following earlier authors from Börner (1925) on (e.g. Munroe, 1972; Minet, 1982). Subsequently, these two groups have become regarded as families, the Pyralidae and Crambidae (Shaffer, 1990; Munroe and Solis, 1998; Solis, 2007). Detailed phylogenies of the family (Regier *et al.*, 2012) confirm this position and both families have been shown to be monophyletic with high degrees of confidence (see below).

17.3 Morphological Definition

The most authoritative and comprehensive account of the morphology of the Pyraloidea is that of Munroe and Solis (1998) in their chapter on the superfamily in the *Handbook of Zoology, Lepidoptera 1*. Their chapter predates definitive molecular/morphological phylogenies but nevertheless concludes that the superfamily should comprise the Pyralidae and Crambidae only. These two families exhibit a suite of shared derived characters of considerable antiquity and functional importance. They identify four key synapomorphies.

- Members of the superfamily possess paired tympanal organs on segment 2 of their abdomens.
- Pyraloids in general possess a maxillary palp which is porrect – beak-like or ascending.
- Pyraloid proboscides are scaled at their bases.
- The forewing radial veins R3 and R4 are stalked or fused; veins Sc+R1 and Rs of the hindwings either join (anastomose) or run closely together for some distance beyond the discal cell (terminology as used by Munroe and Solis, 1998). These authors

note that tympanal organs or proboscides may be reduced or absent in a few cases.

These synapomorphies exclude other families which have from time to time been included in the superfamily. The diagnostic characters are those used in other recent definitive works, notably Nielsen and Common (1991).

Members of both the Pyralidae and Crambidae share the abdominal tympanal organ but there are significant differences in the structure of these organs across the two families. Further family-to-family differences occur in the forewing venation, male genitalia and characteristics of some larval setae (Munroe and Solis, 1998).

17.4 Phylogeny

The Australian checklist (Schaffer *et al.*, 1996) recognizes six pyraliform subfamilies –Galleriinae, Epipaschiinae, Endotrichinae, Chrysauginae, Pyralinae and Phycitinae. Munroe and Solis (1998) recognized only five subfamilies within their Pyralidae, subsuming the Endotrichini as a tribe into the Pyralinae. The subfamily divisions within the Crambidae (crambiforms of some writers) are more complex. The Australian checklist recognizes 13 crambiform subfamilies. Munroe and Solis (1998) recognized 15 crambid subfamilies, subsuming the Heliothelinae into the Scopariinae and the Dichogaminae and Alatuncusiinae into the Glaphyriinae. They retained a broad view of the Pyraustinae and do not recognize the Spilomelinae as a separate subfamily (in line with the Checklist viewpoint). Two of their subfamilies, the Linostinae and Cathariinae, simply do not occur in Australia, so up to that point in time these authors were in agreement. The 13 Checklist subfamilies they recognized within what we now call the Crambidae were the Crambinae, Midilinae, Nymphulinae (subsequently to be known as the Acentropinae), Musotiminae, Scopariinae, Schoenobiinae, Wurthiinae, Cymbalomiinae, Evergestinae, Odontiinae, Noordinae, Glaphyriinae and Pyraustinae (*s.l.*).

These designations were based on both formal (cladistic) and informal morphological analyses. As systematic hypotheses, they were put to the test as molecular and molecular/morphological analyses appeared. For the Pyraloidea, the most important of these was that of Regier *et al.* (2012). Jerome Regier and his co-workers sampled 42 species of pyraloids spanning most putative subfamilies across both Pyralidae and Crambidae. In each case they sequenced up to five nuclear genes, with additional genes targeted in about half of the species. As well as confirming the deep time monophyly of both families, they confirmed the identities and monophyly of all but one of the 'traditional' subfamilies. For the Pyralidae they obtained strong support for all within-family relationships and all five 'traditional' subfamilies were recovered in their analyses. Their final phylogenetic hypothesis paired the Galleriinae and Chrysauginae as basal, sister to the three other subfamilies. They paired the Pyralinae and the Epipaschiinae which, together, were sister to the Phycitinae. In general, their outcomes for this family mirrored earlier morphological analyses. They suggested that when some Australian Chrysauginae were better sampled (especially the genus *Polyterpnes*), they may prove to be basal either within the Galleriinae+Chrysauginae pairing or, even, to all of the Pyralidae.

For the Crambidae, their analyses recovered 13 subfamilies but the within-clade relationships included a number of new groupings which, they suggest, need further morphological investigation. They clearly separated the Spilomelinae and the Pyraustinae (*s.s.*) in one clade (the 'PS clade'). They placed the single species of Wurthiinae within the Spilomelinae. This PS clade is sister to a second clade containing all other subfamilies (not surprisingly, the 'non-PS clade'). Within this second very large clade, two groupings emerged – the 'CAMMS clade' and the 'OG clade'. Within the CAMMS grouping, they hypothesize that the Musotiminae occupy a basal position. The five families that are the sister group to the Musotiminae form two groups: [Crambinae + Scopariinae] and [Schoenobiinae + Midilinae + Acentropinae]. The Crambinae and Scopariinae share some larval features as well as some larval habits (see below). The other three subfamilies within their CAMMS clade, as far as is known, are all associated with wet habitats and are referred to by Regier and colleagues as the 'wet habitat clade'. Of course, this includes the acentropines, many of which have fully aquatic larvae.

The final OG clade comprised the Odontiinae, Glaphyriinae, Evergestinae and Noordinae. Within this grouping, the Odontiinae may be basal. The

Glaphyriinae, Evergestinae and Noordinae share an affinity with the Brassicales (with some exceptions) in consequence of which they were designated the 'Mustard Oil clade'. Affinities within this clade were not clear and the authors urged more work resolving their relationships and biologies (see comments on Léger *et al.*, 2021, below).

Various follow-up studies attempted to confirm (or not) Regier *et al.*'s conclusions. Both Ye *et al.* (2013) and Liu *et al.* (2021) examined selected mitochondrial genomes of pyraloids and reassessed phylogenies accordingly. Ye *et al.*'s results confirmed previous studies except for the placement of the Schoenobiinae, which they postulated as sister to the Crambinae, in contrast to Regier *et al.*'s hypothesis. Liu *et al.* (2021) similarly recovered results that, in general, mirrored earlier outcomes. They did show that, based on their analyses of 90 mitogenomes, the south-east Asian genus *Orybina*, generally classified as a pyraline, emerged as basal to all of the Pyralidae. Meanwhile, Ghanavi *et al.* (2022) raised questions about the accuracy of phylogenies based on mitogenomes for family and subfamily recovery, albeit based on analyses of the Erebidae rather than the Pyraloidea. The debate continues!

Of more general interest, perhaps, is the work of Léger *et al.* (2021). These authors noted that the ground-breaking analysis of Regier and colleagues did not sample four of the morphologically designated subfamilies which, nevertheless, they recognized. Léger and his colleagues accordingly worked to complete the story by a reanalysis which included examples of the Cymbalomiinae, Heliothelinae, Linostinae and Canthariinae. They also included examples of the Lathrotelinae, earlier resurrected from synonymy with the Acentropinae (Minet, 2015). They sampled up to eight genes from 39 taxa and combined these data with those available from previous studies. The final outcome of the Léger *et al.* (2021) analysis represents the most complete set of phylogenetic hypotheses for the Pyraloidea to date and, as such, is reproduced here as Fig. 17.1.

In summary, all of the 'new' subfamilies Léger *et al.* analysed fall within Regier *et al.*'s 'non-PS clade'. Within the 'OG clade', they now locate the Linostinae although with varying degrees of confidence depending on the analysis used. Also within this larger clade, they propose that the Cymbalomiinae and Canthariinae along with the Evergestinae and Noordinae be included within a broadened concept of the Glaphyriinae ('Glaphyriinae *s.l.*'). Cymbalomiines in general are *Brassica* feeders: cathariines are specialized high-altitude species feeding on species of *Saxifraga* and *Heliosperma*. The remaining

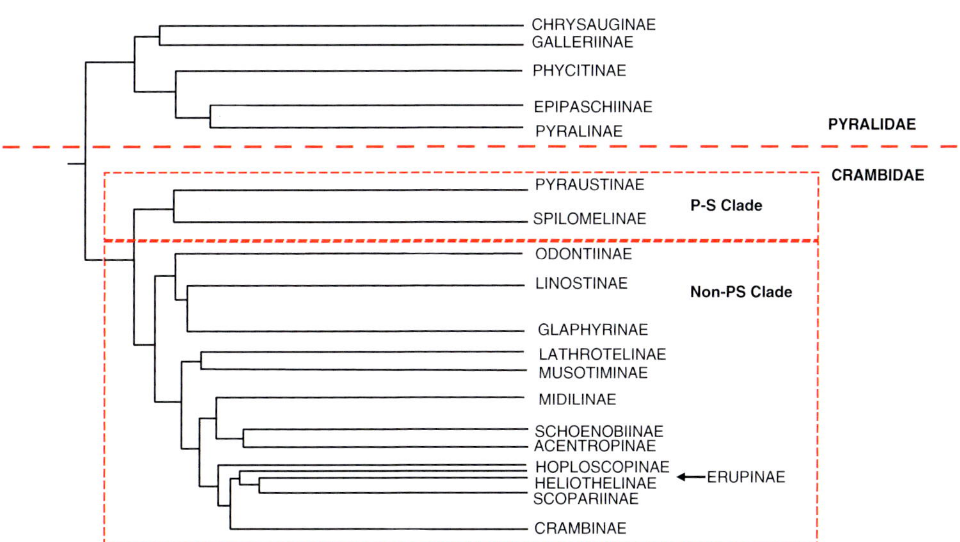

Fig. 17.1. The phylogeny of the Pyraloidea (redrawn and modified after Leger *et al.* 2021 and others). Branch length not to scale.

subfamilies, the Heliothelinae and Lathrotelinae, together with two newly proposed groupings, the Neotropical Erupinae and the Asian/Australasian Hoploscopinae, fall within Regier *et al.*'s 'CAMMS clade'. Léger *et al.* conclude with a summary of the 'current classification of the Pyraloidea', recognizing the five familiar subfamilies of Pyralidae and 15 subfamilies of Crambidae. All of the pyralid and 12 of the crambid subfamilies have Australian representatives (see Tables 17.1 and 17.2).

17.5 The Australian Fauna and Its Global Context

In this penultimate but extended, section of the chapter, I place each of the subfamilies which occur in Australia into a global context in terms of numbers of species and distribution. I also summarize what is published about their life histories and distributions in Australia. For all but the Spilomelinae and Pyraustinae (Crambidae), I summarize this information as two tables (Tables 17.1 and 17.2) with minimal additional notes. For the Spilomelnae and Pyraustinae, Regier *et al.*'s SP clade, I go into further detail stimulated by Mally *et al.*'s (2019) in-depth phylogeny of the clade, as well as recent research by myself and others. Again knowledge is tabulated (Table 17.3). Finally, I review the many pest species introduced or native to Australia. Again, this list is summarized in Table 17.4. All four tables are constructed in a similar manner.

I have extracted larval resource records from Don Herbison-Evans and Stella Crossley's website (https://lepidoptera.butterflyhouse.com.au/superfamilies.html#pyra). For species with more than one larval feeding habit, more than one entry into the tables has been made. Distribution summaries are extracted from the Australian-filtered BOLD database (https://bold-au.hobern.net) and the distribution notes attached to species' accounts in the broader BOLD database. This has the advantage of including unnamed but expert-designated species from the Australian National Insect Collection, but also means that the sums across the distribution columns will not match those in Column 3, which refers only to named species. In a few instances, I have ignored obvious distributional outliers in making decisions as to which subregion a particular species should be assigned. The 'Cosmopolitan' designation captures those several pest species which have spread globally as human commensals as well as those few species which, naturally or otherwise, occur in both the Old and New Worlds. The 'Australasia' column sums species which occur in Australia (usually the north) as well as elsewhere in the Australasian region, usually but not always New Guinea. The 'Old World' category is for those species which include Africa as well as Australia together, sometimes, with parts of Asia, in their distributions. For the within-Australia categories, 'Widespread' indicates species found across the continent from east to west. It does not necessarily indicate commonness. I have separated out species from the rich far northern regions (Cape York, the Top End of the Northern Territory and the Kimberleys) as well as the somewhat more uncommon far south (the southern coastal regions and Tasmania). Otherwise species are simply assigned to the 'east', 'west' or, rarely, the 'centre' of the continent. No species is counted more than once in the distribution columns. Cells in the tables for regions with no records for particular taxa remain unfilled rather than designated as zeroes to indicate the state of our current knowledge rather than, necessarily, a real absence. I have entered raw numbers in both the larval resources and distribution columns as a better summary of the published information available rather than obscure this by using percentages – which in any case can be calculated from the tabulated information. These distribution assessments are, inevitably, very conservative assessments but do identify emerging patterns of relative species richness across taxa and regions.

17.5.1 The Pyralidae

Table 17.1 summarizes knowledge on the Pyralidae. A few features are potentially of special interest.

- In terms of numbers of species, both the global and Australian faunas are dominated by the Phycitinae (making up 58% and 55% respectively).
- The relative proportions of the other four subfamilies in the Australian fauna are different

Table 17.1. Summary of the natural history of the Pyralidae. For methodological details, see text.

PYRALIDAE				DISTRIBUTION OF AUSTRALIAN SPECIES										LIFE HISTORY INFORMATION. Larval Resources of Australian Species.													
SUB-FAMILY	GLOBAL RICHNESS. Species (Genera)[1]	AUSTRALIAN RICHNESS Named Species. (Genera)[2]	Endemism	Cosmopolitan: exotic in Australia	Australasia	Asia/Australia	Old World	Endemic: widespread	Endemic: eastern	Endemic: western	Endemic: Far North	Endemic: Far South	Endemic: Central	Stored Products	Living Fruits & Nuts	Bee-hive Inquilines	Arable Crops[3]	Other Herbaceous Plants	Eucalypts & other Myrtaceae	Other Trees/Palms	Cacti	Monocotyledones	Dry plant Material	Fungi/Galls	Dung	Predatory: including oophagy	Parasitoid
Galleriinae	259(65)	60(19)	81%	3	1	1	0	3	33	5	30	3	1	1	–	3	1	–	1	3	–	2	–	–	–	–	1
Chrysauginae	391(133)	6(4)	100%	0	0	0	0	1	4	0	3	0	0	no information													
Pyralinae (incl. Endotrichinae)	>1100(129)	64(24)	85%	3	7	2	2	8	20	1	46	3	1	2	–	–	1	–	1	1	–	–	1	2	1	2	–
Epipaschiinae	705(94)	80(20)	95%	0	8	0	0	18	68	19	48	4	0	–	–	–	–	1	5	4	–	–	–	–	–	1	–
Phycitinae	*ca*3450(635)	*ca*262(*ca*102)	91%	8	14	4	2	66	78	24	104	6	7	6	1	–	3	2	1	6	3	4	–	–	–	1	–

[1]Based on Nuss *et al,* 2003-2011 as quoted by Regier *et al.* (2012).
[2]Based on Shaffer, Nielsen & Horak (1996).
[3]Mainly legumes.

Table 17.2. Summary of the natural history of the Crambidae. Subfamily assignments follow Léger *et al.* (2020). For other methodological details, see text.

CRAMBIDAE SUB-FAMILY	GLOBAL RICHNESS. Species (Genera)[1]	AUSTRALIAN RICHNESS Named Species. (Genera)[2]	Endemism	DISTRIBUTION OF AUSTRALIAN SPECIES: Cosmopolitan:	Australasia	Asia/Australia	Old World	Endemic: widespread	Endemic: eastern	Endemic: western	Endemic: Far North	Endemic: Far South	Endemic: Central	LARVAL RESOURCES: Living Fruits & Nuts	Arable Crops[3]	Other Herbaceous Plants	Eucalypts & other Myrtaceae	Other Trees & Shrubs/ Palms	Mistletoes	vines	Cacti	Monocotyledones	Ferns	Mosses, Liverworts	Aquatic Plants	Dry plant Material	Predatory: including oophagy
THE NON-SP CLADE																											
Acentropinae	700(78)	48(17)	90%	2	2	1	1	9	13	1	27	2	–	–	–	–	–	–	–	–	–	1	–	1	8	–	–
Crambinae	1987(179)	122(35)	96%	–	1	6	1	24	43	30	66	13	4	–	–	–	–	1	–	–	–	11	–	–	1	–	–
Glaphyriinae (s.l.)[4]	*ca* 390(70)	19(10)	86%	–	1	1	1	1	4	1	11	–	1	–	3	–	–	1	–	–	–	–	–	–	–	–	–
Heliothelinae	49(5)	6(1)	100%	–	–	–	–	6	3	–	–	–	1							no information							
Hoploscopinae	46(2)	2(1)	100%	–	–	–	–	–	2	–	–	–	–							no information[5]							
Midilinae	56(9)	7(2)	100%	–	–	–	–	1	6	1	–	–	–	–	–	–	–	4	–	–	–	–	–	–	–	–	–
Musotiminae[6]	162(24)	11(4)	100%	–	–	–	–	–	6	–	9	1	–	–	–	–	–	–	–	–	–	–	3	–	–	–	–
Odontiinae	360(91)	25(13)	91%	–	2	1	–	4	9	2	15	–	–	–	–	–	–	2	1	–	–	–	–	–	–	–	–
Scopariinae	555(24)	78(5)	100%	–	–	–	–	9	36	4	0	20	3	–	–	–	–	–	–	–	–	–	–	1	–	–	–
Schoenobiinae	201(29)	21(7)	96%	–	–	1	–	6	11	–	10	–	–	–	–	–	–	–	–	–	–	7	–	–	–	–	–
THE SP CLADE																											
Spilomelinae	4097(338)	*ca* 360(125)	56%	12	45	100	28	21	70	9	115	12	7	1	15	11	3	37	–	3	–	7	1	–	2	1	1
Pyraustinae	1413(190)	39(17)	37%	1	7	18	4	2	6	–	11	–	–	–	2	1	–	5	–	2	1	2	–	–	–	–	–

[1]Based on Nuss *et al*, 2003-2011 as quoted by Regier *et al.* (2012), modified after Léger *et al.* (2020).
[2]Based on Shaffer, Nielsen & Horak (1996), modified after Léger *et al.* (2020).
[3]Mainly legumes.
[4]Subsumes Cymbalomiinae, Evergestinae and Noordinae (see Léger *et al.* 2020).
[5]Likely to be ferns (Mally *et al.*, 2017)
[6]See Yen, Solis & Goolsby, 2004.

Table 17.3. Tribes of the Spilomelinae and Pyraustinae as defined by Mally *et al*. (2019). Summaries of global host plant records from Mally *et al*. (2019) are compared with those known for Australian genera included in their analyses. Australian genera that can be assigned to tribes are listed.

SUB-FAMILY & MAJOR CLADES	TRIBE	GLOBAL LARVAL RESOURCE PATTERNS	AUSTRALIAN GENERA	KNOWN AUSTRALIAN LARVAL RESOURCE USE
SPILOMELINAE				
NON-EUSPILOMELIINE	HYDRIRINI	Sapindaceae & a few single records on Fababeae, Ancardiaceae, Lamiaceae	*Hydriris*	Polygonaceae, Rosaceae, Convolvulaceae, Mimosaceae
	LINEODINI	Solanaceae (almost exclusively)	*Leucinodes*	Solanaceae
	UDEINI	Wide range - Polygonaceae, Urticaceae, Asteraceae etc etc	*Udea*	No records
	WURTHIINI	Brood parasites of *Oecophylla* (Formicidae)	*Niphopyralis*	Brood parasites
EUSPILOMELINE	AGROTERINI	Usually leaf tiers - Urticaceae, Convolvulaceae, Amaranbthaceae, Rosaceae etc	*Aetholix, Agrotera, Haritalodes, Notarcha, Patania, Pycnarmon, Tetracona*	Malvaceae, Lamiaceae, Clusaceae, Myrtaceae, Melastomataceae, Sterculiaceae
	MARGARONIINI	Many clades, many plant families. Genus level specialisations on Moraceae, Apocynaceae etc	*Agathodes, Agrioglypta, Analyta, Antigastra, Glyphodes, Nausinoe, Aphytoceros, Arthroschista, Botyodes, Caprinia, Cirrhochrista, Conogethes, Cydalima, Diaphania, Didymostoma, Dysallacta, Filodes, Maruca, Meroctena, Omiodes, Palpita, Parotis, Pygospila, Rhimphalea, Stemorrhages, Synclera, Talanga, Terastia, Tyspanodes*	Convolvulaceae, Cucurbitaceae, Fabaceae, Linaceae, Malvaceae, Moraceae, Oleaceae, Pedaliaceae, Phyllanthaceae, Plantaginaceae, Poaceae, Rubiaceae, Salicaceae, Solanaceae, Theaceae, Verbenaceae, Zingiberaceae
	SPILOMELINI	Heterogeneous including major radiation on grasses	*Aetholessa, Cnaphalocrocis, Orphanostigma*	Poaceae
	HERPETOGRAMMATINI	Ferns to angiosperms - wide range of families	*Eurrhyparodes, Herpetogramma,*	Poaceae, Menispermaceae, Convolovulceae, ferns
	HYMENIINI	Polyphagous - many vegetable crops	*Hymenia, Spoladea*	Poaceae, Fabaceae, Chenopodiaceae, Aizoaceae
	ASCIODINI	Caryophyllales	NONE	–
	TRiCHAEINI	Rubiaceae, Urticaceae, Celastraceae	NONE	–

Continued

Table 17.3. Continued.

SUB-FAMILY & MAJOR CLADES	TRIBE	GLOBAL LARVAL RESOURCE PATTERNS	AUSTRALIAN GENERA	KNOWN AUSTRALIAN LARVAL RESOURCE USE
	STENIINI	Few records - decaying plant material, even mangrove litter	*Bradina, Hymenoptychis, Metasia, Symmoracma, Tatobotys*	Poaceae, Pedaliaceae, mangrove litter
	NOMOPHILINI	Leaf miners principally on Fabaceae, Onagraceae, Rubiaceae, Vitaceae. Many other single records	*Ategumia, Bocchoris, Desmia, Diasemia, Diasemiopsis, Diathrausta, Niphograpta, Pardomima, Samea, Sameodes*	Araceae, Pontederiaceae, Ferns
PYRAUSTINAE				
	EUCLASTINI	Apocynaceae (only)	*Euclasta*	Apocynaceae
	PORTENTOMORPHINI	Phyllanthaceae (Malpighiales) also Solanaceae, Fabaceae, Asteraceae	*Hyalobathra, Isocentris*	Malvaceae
	PYRAUSTINI	Polyphagous; Lamiaceae, Verbenaceae, Amarantahceae, Asteraceae, Scrophulariaceae, Fabaceae (etc)	*Archyra, Circobotys, Crypsiptya, Ostrinia, Pagyda, Prooedema, Pyrausta, Tetridia*	Asteraceae, Fabaceae, Linaceae, Malvaceae, Poaceae, Chenopodiaceae, Amaranthacea, Polygonaceae, Salicaceae, Clusiaceae, Lamiaceae

from the global mix. Our fauna is richer in Galleriinae (13% vs 4%) and Epipaschiinae (17% vs 12%) than the global fauna and poorer in Chrysauginae (1% vs 7%) and Pyralinae (14% vs 19%). Some of these contrasts may be real – the Chrysauginae, for example, are known to have radiated extensively in the Neotropics and the Pyralinae especially in Africa (Munroe and Solis, 1998). In other cases this may simply reflect current levels of knowledge.

- Distribution information in the table highlights the expected high levels of endemism in the Australian fauna. Indeed, if we discount introduced cosmopolitan pest species, the figures are even higher. Within Australia, distributions are more or less consistent across subfamilies, with about two-thirds of species found either in temperate and subtropical eastern Australia or along the tropical northern coastline. A much smaller set of species (*ca* 8%) are restricted to the temperate west.
- Life history information is available for a mere 10% of species.
- The 61 records of larval resources show the diversity of the family ranging from leaf litter feeders to insect egg eaters. Most, though, exploit living plants. A majority of the Epipaschiinae for which information is available appear to have adapted to feed on eucalypts and other Myrtaceae.
- Species of economic importance are discussed in section 17.6 below.

17.5.2 The Crambidae

Since increasingly accurate phylogenies of the Pyraloidea have appeared with their confident definitions of subfamilies within the Pyraloidea (Regier *et al.*, 2012; Léger *et al.*, 2021), there has been an increased focus on within-family relationships. Two key papers focus on four large crambid subfamilies. Léger *et al.* (2019) used analyses of five nuclear and one mitochondrial gene from 56 genera of crambines and 12 genera of scopariines to revise suprageneric structures within those subfamilies. For the Crambinae, this led to a reassessment and revision of earlier tribal structures. Ten genera (of about 35) with Australian representatives were included within this study, allowing their confident tribal placements subsequently. In a considerably larger study, Mally *et al.* (2019) tackled Regier *et al.*'s SP clade using six molecular markers and 114 morphological characters in a phylogenetic analysis of genera of the Spilomelinae and Pyraustinae. I examine this study in detail in section 17.5.2.2 below.

17.5.2.1 The 'non-SP clade'

Table 17.2 summarizes global and Australian richness by subfamily together with distributions and what is known about larval food resources. A few features in the table are especially noteworthy.

- Both globally and locally, the non-SP clade is dominated by the Crambinae. This dominance is somewhat weaker in the Australian fauna (35%) than globally (44%).
- The relative dominance of all the other subfamilies matches global figures except for the Scopariinae which are more dominant in Australia (23% vs 12%) than they are globally.
- Unique among pyraloid subfamilies in Australia, the Scopariinae show a rich cool temperate fauna ('Far South' endemics in Table 17.2) with 27% of the fauna restricted to these regions. Further, unlike other groups, there are none that are restricted to the northern tropical fringes of the continent.
- Information on larval resources is scarce across the board. A few data points reflect global generalizations. Crambinae commonly exploit grasses, sedges and rushes (*cf.* Léger *et al.*, 2019). Grasses and similar monocotyledons are also the food plants of the Schoenobiinae, many of which are pests of cereal crops (see below). Crambinae and Schoenobiinae are widely separated within crambid phylogenies (Fig. 17.1) and their monocotyledonous specializations are probably independent evolutionary developments. Scopariinae are globally associated with mosses, with a few recorded as liverwort feeders in New Zealand. A few Crambinae are also moss feeders, suggesting bryophagy as a possible stem habit for the Crambidae in general (Munroe and Solis, 1998; Léger *et al.*, 2019). Fern feeding is relatively rare in

the Lepidoptera in general. Within the Crambidae, what little is known of the larval habits of musotomines identifies ferns as food plants (Yen and Solis, 2004). This habit is also ascribed to the Hoploscopinae although there are no explicit Australian records (Mally *et al.*, 2017). These pteridophyte-feeding groups are also widely separated phylogenetically, leading Léger *et al.* (2021) to suggest that the habit has arisen within the Crambidae at least twice independently.

- The unquestionable evolutionary success of the non-SP clade of the Crambidae, I suggest, lies in the wide range of larval resources that have been exploited. I note, however, that this is an extrapolation of global information confirmed by what little local information exists. The Spilomelinae and Pyraustinae (the 'SP clade') are in many ways even more successful, in no small part because they are largely angiosperm feeders. The 'explosive' radiation of the angiosperms since the Cretaceous has allowed the co-radiation of groups like these two most derived of the crambid subfamilies.
- Species of economic importance are discussed in section 17.6 below.

17.5.2.2 The Spilomelinae and Pyraustinae

The so-called SP clade of crambids contains an immense range of species often clearly recognizable by their characteristic triangular appearance at rest. Many are also boldly patterned, making casual identification (somewhat) simpler than in many other pyraloid groups. The two currently recognized subfamilies were long combined in a single 'Pyraustinae' subfamily and, as such, appear in the Australian Checklist and earlier accounts of Ian Common and others. The Spilomelini and Pyraustini were recognized as tribes within this single subfamily. Minet (1982) elevated them to subfamily level, arguing that the few autapomorphies are parallel developments. Morphological cladistic analyses (Solis and Mae, 2003) also suggested that the two taxa were not closely related within the Crambidae. Molecular and then molecular plus morphological analyses subsequently reunited the two taxa as sister clades (Regier *et al.*, 2012; Mally *et al.*, 2019). Their monophyly and subtle yet consistent set of autapomorphies have lead to general acceptance of their separate subfamilial status. Mally *et al.* (2019) provide a more detailed account of this taxonomic history.

The fine paper by Mally *et al.* (2019) presented a molecular and morphological analysis of 95 genera from within Regier *et al.*'s (2012) 'SP clade', showing the affinities among these taxa and allowing the authors to erect a well-based tribal classification. Their analysis included six molecular markers (five nuclear genes and one mitochondrial sequence) plus 114 coded morphological characters. This enables more than half of the described genera within the clade to be positioned within their classification. Their phylogeny clearly confirms both subfamilies as monophyletic entities (Fig. 17.2). As a result of their molecular analyses, Mally and his colleagues also transferred several genera from the Spilomelinae to the Pyraustinae, including the taxa *Prooedema* and *Tetridia* which have Australian representatives. The authors also discussed known food plant relationships tribe by tribe. Table 17.3 summarizes their findings and places Australian genera and their food plants into this context wherever possible.

THE SPILOMELINAE. The Spilomelinae are by far the most species-rich subfamily within the Crambidae, with well over 3500 species named globally. As always, many species await description. Mally *et al.* (2019) defined 13 monophyletic tribes within the Spilomelinae. Nine of these formed a superclade within which intertribal relationships were generally well defined. They christened this large clade the 'Euspilomelinae'. Four other more basal clades (the 'non-Euspilomelinae'), although clearly defined individuallly, were not well structured across tribes. The Australian spilomeline fauna comprises about 400 described species organized into more than 120 genera. As currently known, this includes genera assigned to 11 of the tribes defined by Mally *et al.* Their analyses allow about 50% of Australian spilomelines to be assigned to one or other of these tribes.

There have been several 'modern' revisionary papers of relevance to the Australian fauna in relatively recent times. Inoue (1996) revised the genus *Palpita*. Hari Sutrisno *et al.* (2006) worked with the Australian species of *Glyphodes*. Haines and Rubinoff (2012) examined the

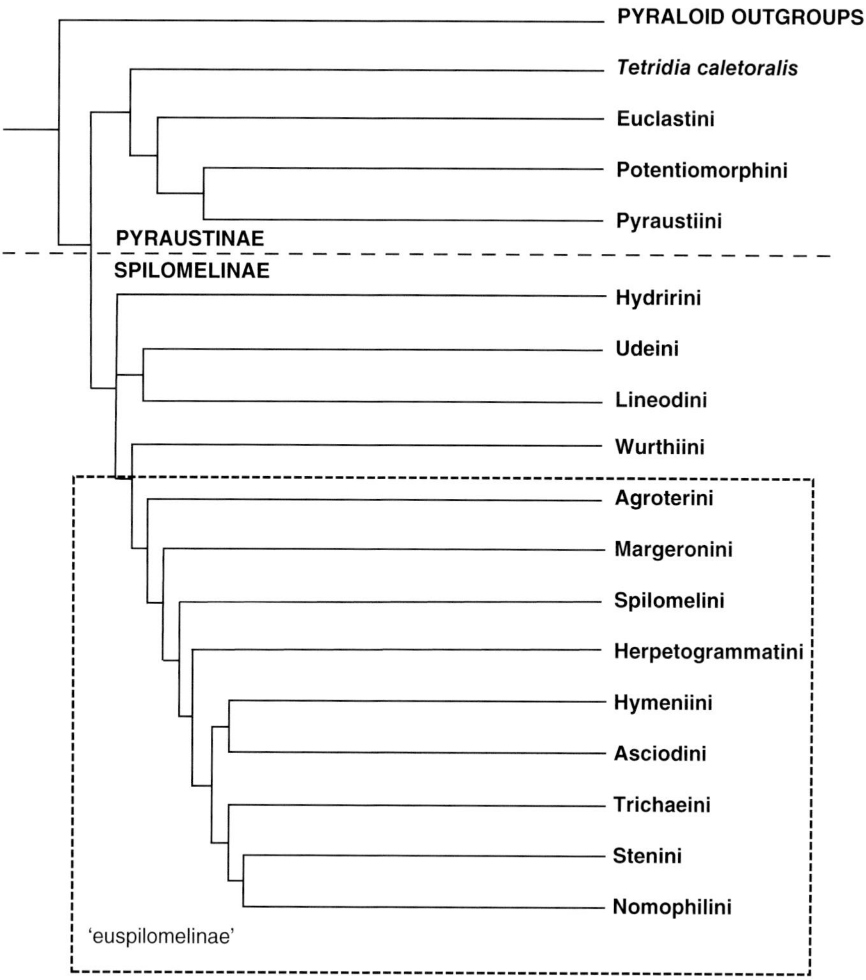

Fig. 17.2. Phylogeny of the Spilomelinae and Pyraustinae (redrawn and simplified from Mally *et al.* 2019). Branch length not to scale.

genus *Omiodes*, commenting on the Australian fauna within a Pacific context. Finally, Chen *et al.* (2017) revised the genus *Agrotera*. The greater part of the subfamily, nevertheless, awaits much needed re-examination.

Information on the geographical distribution of the Australian Spilomelinae is given in Table 17.2. Perhaps the most striking thing about these data is the low level of endemicity (56%) compared with most other pyraloid subfamilies in Australia (but not Pyraustinae – see below). Many Australian species have widespread distributions over Asia. Others have more restricted northerly distributions but within the bounds of Wallace's line; that is, they occur in Australasia. A surprising number of species are more widespread still, including Africa as well as Asia in their known distributions. Many species that have become widespread have done so due to human agency. Many are crop or pasture pests, others have been deliberately introduced as more or less successful biocontrol agents. For spilomeline species currently known only from Australia, by far the greatest proportion (almost 50%) are known from the northern coastal areas of Queensland, the Northern Territory and the tropical coast of Western Australia often associated with rainforest or rainforest-like patches.

For the Australian Spilomelinae as a whole, we have information on larval resources on a mere 71 species – somewhat less than 20% of the fauna. This means that any generalizations must be qualified against the need for more data. Some patterns, however, are clear, others warrant informed speculation. Overall, the Australian Spilomelinae exploit a wide range of angiosperm families as larval hosts with a few outliers feeding on ferns, leaf litter or even ant larvae. A total of 48 angiosperm families are recorded as used by one or more species, six of the Monocotyledones and 41 of the Dicotyledones. The phylogenetic range exploited within the dicotyledons is immense, including basal eucotyledous families such as the Proteaceae and Ranunculaceae, and spread across both the super-rosid and super-asterid mega-clades.

Of course, simply tallying the range of plant families used at the subfamily level hides much detail. A few tribes have affinities with particular plant taxa. The Lineodini, for example, are exclusively associated with the Solanaceae and the Australian genus *Lineodes* conforms to this generalization. As far as we know, the Wurthiinae, including the Australian *Niphopyralis*, are all brood parasites of ants. For the other nine tribes that have Australian representatives, the range of families of host plants is wide and heterogeneous. Some plant families have defences which seem to generate specialized herbivores. Within the Australian Spilomelinae, the genera *Agrioglypta*, *Aphytoceros*, *Cirrhocrista* and *Talanga* are recorded only from the latex-rich Moraceae (the figs). The species-rich genus *Glyphodes* appears to have specialised on the Apocynaceae (now including Asclepiadaceae) and Moraceae – both latex-producing families. It is possible that the striking appearance of adults of one or more of these five genera could be warning colouration reflecting sequestered plant semiochemicals – a hypothesis awaiting testing!

Monocot feeding is relatively rare among the Australian spilomelines but grass feeding seems to have evolved more than once – in some *Herpetogramma* (Herpetogrammatini) and in *Cnaphalocrocis* (Spilomelini), for example. Finally, I note two Australian records of fern feeding, for *Herpetogramma polycapna* (Herpetogrammatini) and the introduced *Samea multiplicalis* (Nomophilini). In the case of *S. multiplicalis*, this feeding habit underpinned its introduction as a potential control agent for the aquatic fern *Salvinia rotundifolia*. The species also feeds on several genera of aquatic angiosperms (Knopf and Habeck, 1976). As noted earlier, Léger *et al.* (2021) discuss fern feeding in Crambidae in general, suggesting that these are independent origins across the phylogeny of the group. The fern-feeding *Herpetogramma* and *Samea* noted here fall into different tribes within Mally *et al.*'s (2019) phylogeny, once again supporting the idea that this trait has arisen independently.

Finally, the widespread pneumatophore moth, *Hymenoptychis sordida*, is recorded feeding on mangrove detritus. Feeding on leaf litter is, according to Mally *et al.* (2019), a common feature of the Stenini. Other Australian Stenini are associated with grass tussocks (including rice plants) but exactly what they are feeding on remains to be confirmed.

THE PYRAUSTINAE. The Pyraustinae as currently defined comprise about 1400 species globally, organized into 190 genera. The Mally *et al.* (2019) analysis examined 20 species of 18 genera. *Tetridia caletoralis*, which they transferred from the Spilomelinae, appeared basal to all other Pyraustinae. Doubts about the exact identity of the genus and confusion about the identity of material used by others meant that *Tetridia* remained unplaced tribally in their analyses. The remainder of the subfamily they placed into three monophyletic clades which they nominated as tribes (except for *Uresiphita*). The basal Euclastini contains the single genus *Euclasta*. They placed three genera in the Portentiomorphini including *Hyalobathra*. A further genus, *Uresiphita*, consistently appeared in their analyses as sister to the Potentiomorphini but remained unplaced tribally because it shares no morphological synapomorphies with the other three genera. Finally, a large assemblage of genera made up the third tribe, the Pyraustini. Forty-two species in 19 genera make up the named Australian pyraustine fauna. Eleven of these 19 genera can be assigned to one or other of Mally *et al.*'s three named tribes. Both *Tetridia caletoralis* and two species of *Uresiphita* occur in Australia and their placement awaits further clarification.

Very little revisionary work has been carried out on the Australian Pyraustinae in recent years. Only *Hyalobathra* has received attention. The Australian species were reviewed and revised by Sutrisno and Horak (2003) who described the new species *H. crenulata*. The next most recent species description was that of *Emphylica xanthocrossa* described by Turner in 1913.

One other species which occurs in Australia has a further call to fame. *Ostrinia furnacalis*, the Asian corn borer, as well as being a major agricultural pest, has also been used as a laboratory model to demonstrate sound production as an adjunct to mate finding and courtship (Nakano *et al.*, 2006, 2010).

Information on the geographical distribution of the Australian Pyraustinae is given in Table 17.2. As with the Spilomelinae, levels of endemism are unusually low (37%) compared with most pyraloid subfamilies. This confirms that the SP clade – or, at least, the sections that are relevant to Australia – are more an Asian/Australasian development than Australian *per se*. Of the endemic species, the tropical far north again dominates. Just how many of these 'endemic' species will also have New Guinea connections awaits further study of that fauna.

Larval feeding records for Australian Pyraustinae are scarce, especially for non-pest species. The native *Euclasta maceratalis* like other Euclastini utilizes Apocynaceae. Its congener *E. gigantalis* was introduced to Australia as part of a biocontrol campaign against rubber vine (*Cryptostegia grandiflora*). In a similar fashion, *Loxomorpha flavidissimilis* was also introduced as a biocontrol agent of prickly pear (Cactaceae). Two species of *Hyalobathra* are known to exploit species of *Glochidion* (Euphorbiaceae) (Sutrisno and Horak 2003) and *Uresiphita ornithopteralis* has larvae that feed on various Fabaceae. *Hodebertia testalis* larvae are recorded from Malvaceae and Apocynaceae. Apart from these records, all others relate to pest species ranging from the extremely polyphagous cotton web-spinner, *Achyra affinitalis*, to the much more specialized teak skeletonizer, *Paliga damastealis*. Pest species are discussed further below.

17.6 Economic Importance

As indicated earlier in this chapter, several species of Pyraloidea have been introduced into Australia as biocontrol agents targetting a variety of environmental weeds from water hyacinth to prickly pear. None has been more succesful than the phycitine *Cactoblastis cactorum* which effectively put up to 24 million hectares (exact figures differ but are all extremely large) of grazing land back into use in southern Queensland and northern New South Wales in the 1920s and 1930s. A recent case history study of the prickly pear control programme by the Invasive Species Council (ISC) suggested that the cost of the programme to the federal government was repaid by a factor of 42 in terms of land returned to productive use (ISC, undated; Dodd, 1940). In terms of current costs, this suggests the government investment of A$900 million (£240,000 at the time) reaped a benefit of more than A$37 billion. So spectacular was this outcome that a memorial stone to the moth was erected in 1965 on the Queensland Darling Downs (Fig. 17.3). Ironically, the self-same species of moth is now invading the southern USA and Mexico, posing an existential threat to the highly profitable cactus-growing industry there.

The economic importance of many pyraloids, however, falls on the other side of the ledger. About 40 species of pyraloids (Table 17.4) are actual or potential pests of crops, pastures and plantation industries in Australia. Some of these are chronic, posing persistent threats to agricultural industries of one sort or another. Honey producers are continually threatened by wax moths such as *Galleria mellonella*. A range of species are ongoing threats to stored products of one kind or another (see Chapter 28). Sets of species are ongoing threats to rice and legume growers. Others are serious pests in their 'break-out' phases reflecting occasional favourable environmental conditions. The pasture pest, the spilomeline *Herpetogramma licarsisalis*, is a case in point. Finally, a set of species pose occasional or specialized threats to particular industries: aubergines, bananas, millet, figs and others all have associated pests or potential pests. A few species have effectively prevented the development of particular industries at any economic scale in Australia. Here perhaps the phycitine cedar tip moth, *Hypsipyla robusta*, is the outstanding example, having either prevented or severely restricted the development of plantation forestry for some tree species such as red cedar

Fig. 17.3. Monument to *Cactoblastis*, Dalby, Queensland. Photo: John McPherson, WikiMedia Commons, used under Creative Commons licence CC-BY-4.0

Table 17.4. Pyraloidea: pest species in Australia

SPECIES	COMMON NAME	DISTRIBUTION	IMPACT
PYRALIDAE			
Galleriinae			
Achroia grisella	Lesser Wax Moth	Cosmopolitan	Predator in beehives
Corcyra cephalonica	Rice Moth	Cosmopolitan	Stored products pest of cereals, flour, dried fruit
Galleria melonella	Greater Wax Moth	Cosmopolitan	Kleptoparasite in beehives. Destroys combs
Pyralinae			
Pyralis farinalis	Meal moth	Cosmopolitan	Stored products pest of flours
Phycitinae			
Assara seminivale	Macadamia kernel moth	South Asia; Australia	Borer in maturing macadamia nuts
Cadra cautella	Tropical warehouse moth	Tropics & Subtropics	Stored products pest of cereals and cereal products
Cadra figulilella	Raisin moth	Cosmopolitan	Stored products pest of dried fruits
Citripestis eutraferea	Mango fruit borer	Asia; Australia (N.T.)	Mango and cashew fruits
Cryptoblabes adoceta	Sorghum head moth	Australia	Feeding inside sorghum heads
Cryptoblabes seminivale	Macadamia kernel moth	Australia	Macadamia inflorescences and early fruit sets
Ectomyelois ceratoniae	Locust bean moth	Cosmopolitan	Many species of developing fruit
Ephestia kuehniella	Mediterranean flour moth	Cosmopolitan	Stored products pest of cereals and cereal products
Ephestia elutela	Tobacco moth (etc.)	Cosmopolitan	Stored products pest of dried plant produce

Continued

Table 17.4. Continued.

SPECIES	COMMON NAME	DISTRIBUTION	IMPACT
Etiella behrii	Lucerne seed web moth	South-east Asia; Australia	Legumes especially lucerne and lentils
Etiella zinckenella	Pulse pod borer	Cosmopolitan	Legumes especially beans
Hypsipyla robusta	Cedar tip moth	Africa; Asia; Australia	Meliaceae including *Toona* spp.
Plodia interpunctella	Indian meal moth	Cosmopolitan	Stored products - cereals, flour, dried fruit, etc.
CRAMBIDAE			
Crambinae			
Chilo suppressalis	Striped rice stem borer	Asia; introduced to Northern Territory	Rice
Culladia cuneiferellus	Grass moth	Australia, New Zealand, New Caledonia	Occasional pest of lawns and pastures
Schoenobiinae			
Scirpophaga innotata	Rice white stemborer	Asia; tropical Australia	Rice
Acentropinae			
Parapoynx stagnalis	Rice case-bearer	Cosmopolitan	Rice
Glaphyrinae			
Hellula hydralis	Cabbage centre grub	Australia	Brassicas, turnips, canola
Hellula undalis	Cabbage centre grub	Australia	Cabbages and other brassicas
Spilomelinae			
Antigastra catalaunalis	Sesame leaf-roller	Tropical Asia, Africa, Australia; migratory	Sesame
Cnaphalocoris medinalis	Rice leaf-roller	Subtropical & tropical Asia; Australia	Rice
Crocidolomia pavonana	–	Africa; Asia; Australia; Pacific islands	Cabbages and other brassicas
Conogethes punctiferalis	Yellow peach moth	Asia, Australia	Fruit trees
Diaphania indica	Cucumber moth	Asia; now pantropical	Occasional pest of cucurbits, cotton
Herpetogramma licarsisalis – pest fo pasture	Grass webworm	Cosmopolitan	Turf and pastures
Hymenia recurvalis	Beet webworm	Cosmopolitan	Spinach, beet, cotton, maize, soybean
Maruca vitrata	Bean pod borer	Pan-tropical	Leguminous crops
Nacoleia octasema	Banana scab moth	Australasia, Pacific	Bananas
Hedylepta diemenalis	Soybean leaf-folder	Asia, Australasia, Pacific	A leaf-roller on legumes
Paliga damastesalis	Teak skeletoniser	Asia, Australia (Northern Territory)	Plantation timber
Scleiodes cordalis	Eggfruit caterpillar	Australia, New Zealand, Sulawesi	Solanaceae
Talanga tolumnialis	Figleaf moth	Australia, PNG	Cultivated figs
Pyraustinae			
Achyra affinitalis	Cotton web-spinner	Australia, New Zealand	lucerne, cotton, maize, sorghum, linseed, sunflowers
Achyra massalis	Ombava	Old World Tropics	millet
Ostrinia furnacalis	Asian corn borer	Tropical Asia, Australia, Pacific	Principally maize
Pyrausta panopealis	Perilla leaf moth	Pan-tropical	Perilla, mint, rosemary, ornamentals. All Lamiaceae

(*Toona ciliata*, Meliaceae) in Australia (Griffiths, 2001; Cunningham *et al.*, 2005).

Most pest species of Pyraloidea recorded from Australia have originated elsewhere and represent but a small fraction of their global distributions. A few, however, may be Australian natives. The occasional serious pasture and lawn pest, the crambine *Culladia cuneiferellus*, appears to have spread (or, more likely, been carried) to New Zealand, New Caledonia, Vanuatu and other Pacific destinations from Australia. The two glaphyrines, *Hellula hydralis* and *H. undalis*, pests of the Brassicaceae and other vegetable crops, appear restricted to Australia. The macadamia kernel moth, the phycitine *Cryptoblabes hemigypsa*, and the macadamia flower caterpillar, the phycitine *Assara seminivale*, are also native to Australia, reflecting the origin of their host (Horak, 1994).

References

Börner, C. (1925) Lepidoptera, Schmetterlinge. In: Brohmer, P. (ed.) *Fauna von Deutschland. vol. 3. Verbesserte Auflage*. Quelle and Meyer, Leipzig, Germany, pp. 358–387.

Chen, K., Horak, M., Du, X. and Zhang, D. (2017) Revision of the Australian species of *Agrotera* Schrank (Lepidoptera: Pyraloidea: Crambidae: Spilomelinae). *Zootaxa* 4362, 213.

Common, I.F.B. (1990) *Moths of Australia*. Melbourne University Press, Carlton, Victoria.

Cunningham, S.A., Floyd, R.B., Griffiths, M.W. and Wylie, F.R. (2005) Patterns of host use by the shoot-borer *Hypsipyla robusta* (Pyralidae: Lepidoptera) comparing five Meliaceae tree species in Asia and Australia. *Forest Ecology and Management* 205, 351–357.

Dodd, A.P. (1940) *The Biological Campaign against Prickly Pear*. Commonwealth Prickly Pear Board, Brisbane, Queensland.

Fletcher, D.S. and Nye, I.W.B. (1984) *The Generic Names of Moths of the World. 5. Pyraloidea*. British Museum (Natural History), London.

Ghanavi, H.R., Twort, V., Hartman, T.J., Zahiri, R. and Wahlberg, N. (2022) The (non) accuracy of mitochondrial genomes for family level phylogenetics: the case of erebid moths (Lepidoptera: Erebidae). *Zoologica Scripta* 51, 695–707.

Griffith, M. (2001) The biology and ecology of *Hypsipyla* shoot borers. In: Floyd, R.B. and Hauxwell, C. (eds) Hypispyla *Shoot Borers in Meliaceae*. Proceedings of a Workshop. Australian Centre for International Agricultural Research, Canberra, ACT, pp. 74–80.

Grimaldi, D. and Engel, M.S. (2005) *Evolution of the Insects*. Cambridge University Press, Cambridge.

Guillermet, C. (2009) *Les Hétérocères, ou Papillons de Nuit de l'île de La Réunion. 3. Familles des Pyralidae et Crambidae*. Association Nature, Découverte et Partage, Parc National de la Réunion, Saint-Denis.

Haines, W.P. and Rubinoff, D. (2012) Molecular phylogenetics of the moth genus *Omiodes* Guenée (Crambidae: Spilomelinae) and the orgins of the Hawaiian fauna. *Molecular Phylogenetics and Evolution* 65, 305–316.

Heikkilä, M., Simonsen, T.J. and Solis, M.A. (2018) Reassessment of known fossil Pyraloidea (Lepidoptera) with descriptions of the oldest fossil pyraloid and a crambid larva in Baltic amber. *Zootaxa* 4483, 101–127.

Horak, M. (1994) Identity of two phycitine pests on macadamia (Lepidoptera: Pyralidae: Phycitinae). *Journal of the Australian Entomological Society* 33, 235–244.

Inoue, H. (1996) Revision of the genus *Palpita* Hubner (Crambidae: Pyraustinae) from the eastern Palaearctic, Oriental and Australian regions. Part 1: Groups A (*annulifer* group). *Tinea* 15, 12–46.

Invasive Species Council (ISC) (n.d.) Threats to Nature: Taming a Cactus. Available at: https://invasives.org.au/wp-content/uploads/2021/03/Case-study-Prickly-pear.pdf (accessed 4 October 2025).

Kawahara, A.Y., Plotkin, D., Espeland, M., Meusemann, K., Toussaint, E.F.A. *et al.* (2019) Phylogenomics reveals the evolutionary timing and pattern of butterflies and moths. *Proceedings of the National Academy of Sciences* 116, 22657–22663.

Knopf, K.W. and Habeck, D.H. (1976) Life history and entomology of *Samea multiplicalis*. *Environmental Entomology* 5 539–542.

Latreille, P.A. (1806–1809) *Histoire Naturelle Générale et particulière des Crustacés et des Insectes*. F. Dufart, Paris.

Léger, T., Landry, B. and Nuss, M. (2019) Phylogeny, character evolution and tribal classification in Crambinae and Scopariinae (Lepidoptera, Crambidae). *Systematic Entomology* 44, 757–776.

Léger, T., Mally, R., Neinhuis, C. and Nuss, M. (2021) Refining the phylogeny of the Crambidae with complete sampling of subfamilies (Lepidoptera, Pyraloidea). *Zoologica Scripta* 50, 84–99.

Linnaeus, C. (1758) *Systema Natura*, 10th edn, vol. 1. Laurentius Salvius, Stockholm.

Liu, X., Qi, M., Xu, H., Wu, L., Yang, M. and Li, H. (2021) Nine mitochondrial genomes of the Pyraloidea and their phylogenetic implications. *Insects* 12, 1039.

Mally, R., Léger, T., Vairappan, C.S., Sutton, S. and Nuss, M. (2017) Discovery of another fern-feeding group of moths: the larvae of Hoploscopini Iinsects: Lepidoptera: Pyraloidea from Borneo. *Raffles Bulletin of Zoology* 65, 100–108.

Mally, R., Hayden, J.E., Nienhuis, C., Jordal, B.H. and Nuss, M. (2019) The phylogenetic systematics of Spilomelinae and Pyraustinae (Lepidoptera: Pyraloidea: Crambidae) inferred from DNA and morphology. *Arthropod Systematics and Phylogeny* 77, 141–204.

Meyrick, E. (1884) On the classification of the Australian Pyralidina. *Transactions of the Entomological Society of London* 1884, 277–350.

Minet, J. (1982) Les Pyraloidea et leurs principales divisions systématiques (Lep. Ditrysia). *Bulletin de la Société Entomologique de France* 86, 262–280.

Minet, J. (2015) Lathrotelidae Clarke 1971: A rehabilitated name deserving subfamily rank (Lepidoptera, Crambidae). *Bulletin de la Société Entomologique de France* 120, 109–112.

Mitter, C., Davis, D.R. and Cummings, M.P. (2017) Phylogeny and evolution of Lepidoptera. *Annual Review of Entomology* 2017, 265–283.

Munroe, M. (1972) Pyraloidea: Pyralidae comprising subfamilies Scopariinae, Nymphulinae. In: *The Moths of America North of Mexico*, Fascicle 13.1. Classey, Faringdon, pp. 1–134.

Munroe, M. and Solis, M.A. (1998) The Pyraloidea. In: Kristensen, N.P. (ed.) *Lepidoptera, Butterflies and Moths*, 4(1), *Handbook of Zoology*. De Gruyter, Berlin and New York, pp. 233–256.

Nakano, R., Ishikawa, Y., Tatsuki, S., Surlykke, A., Skals, N. and Takahashi, T.(2006) Ultrasonic courtship song in the Asian corn borer moth, *Ostrinia furnacalis*. *Naturwissenschaften* 93, 292–296.

Nakano, R., Takanashi, T., Skals, N., Surlykke, A. and Ishikawa, Y. (2010) Ultrasonic courtship songs of male Asian corn borer moths assist copulation attempts by making the females motionless. *Physiological Entomology* 35, 76–81.

Nielsen, E.S. and Common, I.F.B. (1991) Lepidoptera (moths and butterflies). In: Naumann, I.D. (ed.) *Insects of Australia*, 2nd edn. Melbourne University Press, Carlton, Victoria, pp. 817–915.

Regier, J.C., Mitter, C., Solis, M.A., Hayden, J., Landry, B. *et al.* (2012) A molecular phylogeny for the pyraloid moths (Lepidoptera: Pyraloidea) and its implications for higher-level classification. *Systematic Entomology* 37, 635–656.

Regier, J.C., Mitter, C., Zwick, A., Bazinet, A.L., Cummings, M.P. *et al.* (2013) A large-scale, higher-level, molecular phylogenetic study of the insect order Lepidoptera (moths and butterflies). *PLOS One* 8, e58568.

Schaffer, J.C. (1990) Crambidae of Aldabra Atoll. I. Peoriinae. *Tropical Lepidoptera* 1, 21–24.

Schaffer, J.C. and Munroe, E. (2003) Crambidae of Aldabra Atoll (Lepidoptera: Pyraloidea). *Tropical Lepidoptera Research* 14, 1–110.

Shaffer, M., Nielsen, E.S. and Horak, M. (1996) Pyralidae. In: Nielsen, E.S., Edwards, E.D. and Rangsi, T.V. (eds) *Checklist of the Lepidoptera of Australia, Monographs on Australian Lepidoptera*, 4. CSIRO Publishing, Collingwood, Victoria, pp. 164–199.

Solis, A. (2007) Phylogenetic studies and modern classification of the Pyraloidea (Lepidoptera). *Revista Columbiana de Entomologia* 33, 1–9.

Solis, M.A. and Maes, K.V.N. (2003) Preliminary phylogenetic analysis of subfamilies of Crambidae (Pyraloidea Lepidoptera). *Belgian Journal of Entomology* 4, 53–95.

Sutrisno, H. and Horak, M. (2003) Revision of the Australian species of *Hyalobathra* Meyrick (Lepidoptera: Pyraloidea: Crambidae: Pyraustinae) based on adult morphology and with a description of a new species. *Australian Journal of Entomology* 42, 233–248.

Sutrisno, H., Azuma, N. and Higashi, S. (2006) Molecular phylogeny of the Indo-Australian *Glyphodes* and its allied genera (Insecta: Lepidoptera: Crambidae: Spilomelinae) inferred from mitochondrial CO1 and COII and nuclear EF-1a gene sequences. *Species Diversity* 11, 57–69.

Turner, A.J. (1913) Studies in Australian Lepidoptera, Pyralidae. *Proceedings of the Royal Society of Queensland* 24, 111–163.

Ye, F., Shi, Y., Yu, H. and You, P. (2013) The complete mitochondrial genome of *Paracymoriza prodigalis* (Leech, 1889), with a preliiminary phylogenetic analysis of Pyraloidea. *Aquatic Insects* 35, 71–88.

Yen, S.-H. and Solis, M.A. (2004) *Austromusotima*, a new musotomine genus (Lepidoptera: Crambidae) feeding on Old World climbing fern, *Lygodium microphyllum* (Schizaeaceae). *Annals of the Entomological Society of America* 97, 397–410.

18 The Geometridae: Waves, Emeralds, Carpets and their Relatives

Peter B. McQuillan[1]* and Catherine J. Byrne[2]

[1]*School of Geography, Planning and Spatial Sciences, University of Tasmania, Hobart, Tasmania, Australia;* [2]*Tasmanian Museum and Art Gallery, Department of State Growth, Hobart, Tasmania, Australia*

Abstract

The Geometridae, with 1200+ species, are the second largest macromoth family in Australia but significant numbers remain undescribed. A strongly endemic profile is dominated by the Ennominae (especially Diptychini and Boarmiini). Oenochrominae and Epidesmiinae are largely Australian clades with only minor presence in Asia. Geometrinae features a peculiar set of Pseudoterpnini while Larentiinae increase in importance with latitude and altitude.

The mesic to perhumid coastal fringe of Australia is richest in Geometridae. The fauna of the north-east Wet Tropics is enriched with Malesian forms and outliers of palaeotropical tribes. The woodlands and shrublands of the semi-arid inland support both widespread and endemic taxa mostly related to forest forms. Adaptations of geometrids to stressful habitats include rain-hatched eggs, extended pupal aestivation underground and frontal horns for penetrating soil. The south-eastern mountains feature various diurnal species while cool temperate Tasmania supports some unusual highland taxa. Disjunctions in distributions reflect habitat gaps and point to the long-term retreat of mesic and perhumid biomes or inundation, e.g. the Nullarbor, mid-coast Queensland and Torres and Bass Straits. Uplands, including the northern Tablelands and Mt Lofty Ranges, serve as important refugia.

There is scant evidence so far of Gondwanan connections among the geometrids. Open habitat taxa may have arrived in the Tertiary, with highland forms appearing later. Most relationships to New Zealand and New Caledonia appear to be via transoceanic dispersal. None are strictly migratory and passive movement within Australia involves just a few small species.

Food plant relationships show strong links to Myrtaceae and Fabaceae, but Proteaceae supports most Oenochrominae (*s.s.*). Loopers are a dominant group of free-living caterpillars in eucalypt canopies but are also common in the extensive *Acacia* woodlands. Almost none eat graminoids and few loopers take chenopods. Many associations remain unknown but polyphagy is rare.

Geometridae in Australia are pressured from ongoing habitat loss, declining environmental moisture and aggressive fire regimes.

18.1 Introduction

The loopers or geometer moths are the second largest family of macromoths in Australia with at least 1200 described species – about 5% of the global total (Scoble *et al.*, 1999). The Geometridae is by far the most species rich and widespread of the five families which make up the superfamily Geometroidea (Bazinet *et al.*, 2013; Regier *et al.*, 2013; Rajaei *et al.*, 2015,

*Corresponding author: p.b.mcquillan@utas.edu.au

DOI: 10.1079/9781800627680.0018

2022; Wang *et al.*, 2019a, b; Zhang *et al.*, 2020). Within this clade, geometrid moths are recognized by a pair of tympanal organs of distinctive form at the base of the abdomen, present in all but a few females where they have been lost. Its sister family has been thought to be the Uraniidae (Murillo-Ramos *et al.*, 2019; Õunap *et al.*, 2024), a small and largely tropical group which feature abdominal tympana of a different type and are mostly reliant on Euphorbiaceae as food plants. This relationship, however, has been challenged by Ding *et al.* (2023) whose analysis returned a relatively closer association between Uraniidae and the Oriental Epicopeiidae plus the Neotropical Sematuridae. The crown age of Geometridae is estimated at *ca* 54 mya (Yamamoto and Sota, 2007; Zhang *et al.*, 2020) near the early Eocene climatic optimum, and the Neotropics has been proposed as the most likely ancestral area for the family (Ghanavi *et al.*, 2025). The family has a thin fossil record, however, with less than 20 reliable fossils reported so far (Sarto i Monteys *et al.*, 2022).

Globally, eight or nine subfamilies of Geometridae are recognized based on multiple lines of evidence, including multilocus studies (Sihvonen *et al.*, 2011; Murillo-Ramos *et al.*, 2019; Ghanavi *et al.*, 2025). Each geometrid subfamily so far, however, lacks a single unique morphological character which defines them. Although a consensus topology is still emerging, high-level relationships consistently recovered include Sterrhinae as the sister group to the rest of the family, and a pairing of Geometrinae + Oenochrominae.

Of generally gracile form, geometrid moths are highly diverse in size, colour and markings. Australian species span an order of magnitude in wingspan, ranging from 8 mm to 100 mm, and include some of the most colourful and attractive of all Lepidoptera. Australia seems unusually rich in robust-bodied forms, in contrast to the more slender-bodied and broad-winged habitus of most groups elsewhere.

The larvae of Geometridae are widely known as loopers, geometers or inchworms. Their familiar body plan features a smooth and slender form, almost always lacking secondary setae, an extra lateral seta (L4) on the abdominal segments and the loss of prolegs on abdominal segments A3, A4 and A5. The majority progress with a slow looping gait due to a reduction in the number of functioning prolegs and those on A6 and A10 (claspers) are often strongly developed in compensation. A more complete set of prolegs, however, has been recovered in some genera, notably among certain Ennominae: Diptychini and some Oenochrominae (McFarland, 1988; Young, 2008a). Larvae pass through five or six instars and mature caterpillars usually abandon the canopy via silken threads, although some descend to the ground via plant stems. Pupation most commonly occurs in a weak cocoon under or amongst litter on the ground, or in a cell formed in the soil. A few taxa such as Nemoriini and the sterrhine *Traminda* pupate in a loose cocoon spun among leaves. The geometrid pupa is rarely exposed but if so, it may be aposematically coloured.

18.2 Taxonomic History

Geometrids were formally classified as a distinct lineage of moths by Carl Linnaeus in 1758 (as *Geometra*, a subgenus of *Phalaena*). Early authors of our fauna, such as Achille Guenée in Paris and Francis Walker in London, intuitively assigned new material to familiar European genera since the broad facies of many groups, such as the Boarmiini and Macariini, are globally conserved; more divergent taxa were allocated to new genera or even new families (e.g. the Oenochromidae of Guenée [1858]). Indeed, some species clearly belong to cosmopolitan genera, but the novelty of the austral fauna sometimes resulted in early misallocations even at family level. The ennomine *Corula* was described as a noctuid (Walker, 1856) and the geometrine '*Cerura*' *habrostola* as a notodontid (Lower, 1905) while the bombycoid *Carthaea* was erroneously allocated to the Oenochrominae by Prout (1910).

The earliest collectors of the fauna were mainly visiting naturalists or surgeons attached to expeditions from the British, French and Austrian empires, exemplified by Honoré Jacquinot and Elie Le Guillou on the Dumont d'Urville expedition (French) and Rudolf Felder and Alois Rogenhofer on the Novara expedition

(Austrian). Later, resident naturalists were urban professionals, school teachers and rural squatters, some connected to local learned societies which emerged in the 19th century. George F. Angas (1847) illustrated macromoths from the vicinity of Adelaide and later donated specimens for description by Felder and Rogenhofer (1875). Also in South Australia, Edward Guest, Oswald Lower (McQuillan *et al.*, 2023) and J.G. Otto Tepper collected and described geometrids now largely preserved in the state museum. Morton Allport, a Tasmanian lawyer and naturalist, submitted material to the British Museum in the 1850s where it was described by Francis Walker. Similarly, the businessman George Lyell in Victoria was an important member of informal networks of collectors, dealers and describers which flourished at the time. Pioneering residents and collectors in the remote north included Frederick Dodd in north Queensland, the grazier George Barnard in central Queensland and the itinerant collector Albert Meek at Cedar Bay (Parker, 1966) and British New Guinea. The Prussian ethnographer Emile Clement made important collections in north-western Australia in the 1890s.

In the late 19th century, the English lepidopterist and schoolmaster Edward Meyrick collected widely in southern Australia and New Zealand. A talented and influential taxonomist, he was able to combine local knowledge with a broader Old World perspective, evident in a series of foundational papers he published on the geometrid fauna (Meyrick, 1888, 1891, 1892). His work was extended by the Brisbane paediatrician Alfred J. Turner who revised most groups in the decades following Meyrick's return to England (Turner, 1904, 1922, 1929, 1930). The volume by Louis B. Prout (1933) in the illustrated *Macrolepidoptera of the World* series added new genera and usefully summarized the Indo-Australian fauna in a biogeographical context.

The utility of genitalic features in diagnosis and classification emerged in the early 20th century (Pierce, 1914) and gradually prevailed as a signifier of relationships until supplemented by molecular data which became widely available from the 1990s. Taxon sampling strategies, however, are consequential when exploring molecular-based relationships and recent reworkings of the Geometridae tend to poorly sample Australian taxa, with the recent exception of Murillo-Ramos *et al.* (2021a, b) for Epidesmiinae.

An important modern synthesis of the Australian geometrids was that of Ian Common (1990) while a more complete understanding of the tropical fauna emerged from Jeremy Holloway's *Moths of Borneo* series (Holloway, 1994, 1996, 1997). A remarkable contribution to our knowledge of the immature stages of the temperate zone geometrids was made by Noel McFarland (1973, 1988). A phylogeny of the Australian Ennominae using molecular and morphological data was proposed by Young (2006a, 2008a). She also described the eggs of 85 local ennomines but was unable to propose unifying characters from this evidence (Young, 2006b). Using a suite of 43 larval characters, 67 Australian Ennominae and related taxa were documented, helping to contextualize these within the global fauna (Young, 2008b).

18.3 Checklists

The global geometrid checklist of Scoble (1999) is now an online facility (Rajaei *et al.*, 2022). The most recent annotated Australian list is McQuillan and Edwards (1996). State-level checklists include Victoria (Marriott, 2011, 2012; Hewish *et al.*, 2014, 2016) and Tasmania (Turner, 1926a, b, 1928, 1939; McQuillan, 2004), and regional lists serve the Mt Lofty Ranges (Guest, 1887), the Broken Hill area (Lower, 1918) and Norfolk Island (Smithers, 1998).

Early revisions of various subfamilies were issued over three decades by Meyrick (1888, 1890, 1891, 1892) and Turner (1904, 1907, 1910, 1922). Other contributors included Thomas P. Lucas (1888) and Gilbert M. Goldfinch (1929, 1944). One of the most conspicuous Australian groups, the Diptychini, were placed on a modern footing by Young (2006a,b; 2008a,b, as 'Nacophorini'). Some genus-level reviews are also recent and are listed in the tribal overviews which follow. Autecological studies of pest and non-pest species include Madden and Bashford (1977a,b), Elliott and Bashford (1978), Elliott *et al.* (1980) and Young and McQuillan (2001, 2003).

18.4 Biogeography

While the Australian looper moth fauna is large and distinctive, it has complex but poorly investigated origins. Given the strong trophic connections to the flora, many geometrid distributions map to recognized biomes and phytogeographical regions. Regional climates defined by the annual total and seasonality of rainfall together with temperature range affect the distribution of many groups. Strongly apparent is the existence of a temperate Bassian fauna partly defined by the winter dominant rainfall zone and distinct from the Torresian fauna in the north where summer rainfall prevails.

Assuming a degree of conservatism in food plant associations among many Lepidoptera clades, insights derived from better studied plant groups may offer an outline of geometrid historical biogeography. Before the evolution of the Geometridae, the Australian flora was already rich in Proteaceae, of which three subfamilies had reached Australia by 104 mya (Lamont *et al.*, 2024). These underpinned the later diversification of the largely Australian subfamily Oenochrominae and some '*Psilosticha*' (Boarmiini) but few others. Similarly, Cretaceous Australia supported widespread Gondwanan conifers, mostly extinct but for some taxa now confined to perhumid refugia such as in Tasmania and Queensland (Carpenter *et al.*, 2011).

In contrast to these food plant families, there seems to be scant evidence of unequivocally Gondwanan taxa in the present-day geometrid fauna, although land masses derived from the supercontinent support some diverse but isolated groups. Notably, an endemic New Zealand ennomine clade centred on the genus *Declana* (Dugdale *et al.*, 2023) was recovered as a geographically isolated sister to the Diptychini which is rich in Australia (Murillo-Ramos *et al.*, 2019). In addition, a distinctive ennomine tribe in southern Africa, the Drepanogynini (sister to a group including Hypochrosini), also appears to be absent from Australia.

The northwards drift of the Australian plate towards Asia since the end of the Cretaceous has been a key driver of habitat change on the continent. During an extended climatic optimum in the early Eocene, Australia was confined south of *ca* 30°S and connected to Antarctica and South America by a corridor of mesic Gondwanan forest (Lamont *et al.*, 2024). At this time, Tasmania was near the polar circle yet supported mesothermal rainforest and mangroves. Notably, important food plant groups such as austral conifers, Lauraceae, Proteaceae and Rhizophoraceae were present (Pole, 2007), with *Eucalyptus* (Myrtaceae) evolving in the early Eocene (Macphail and Tornhill, 2016). Other austral tree taxa included Monimiaceae and Atherospermaceae (also in Eocene Patagonia; Knight and Wilf, 2013), both of which are used by modern geometrids in Australia. At this time, a zonal band with conditions favourable for generating tropical cyclones at high latitudes extended from *ca* 30°E to 120°W across the southern hemisphere. The south Indian Ocean was the basin most conducive to storm formation during this period (rather than the western north Pacific) (Yan *et al.*, 2019) and therefore supported a mechanism for helping disperse Lepidoptera over water.

The low level of genus endemism in the basal subfamily Sterrhinae argues against Australia as an arena of early diversification in the Geometridae. The contemporary sterrhine fauna largely comprises either dispersive cosmopolitan groups or outliers of probable Eurasian origin. Assuming geometrids first colonized a mesic Australia in the Eocene, their descendants were subjected to the gradual aridification of the continent beginning in the Miocene. This process drove waves of extinction, migration and adaptation with the remnants of the original biota contracting to habitat refugia along the coasts, in the uplands and at lower latitudes. The legacy of these events is still apparent in the profile and distribution of the fauna today.

Concurrent impacts on the flora had consequences for phytophagous insects, most notably an expansion in woodlands and open habitats at the expense of botanically richer rainforest canopies. *Eucalyptus*, a major food plant for many geometrids, is Gondwanan with early Eocene fossils known from Patagonia (Hermsen *et al.*, 2012; Zamaloa *et al.*, 2020). It is known from fossil pollen in south-east Australia dated at 45 mya and was favoured by a drying climate. A rapid diversification in eucalypt species in the recent past (*ca* 2 mya) drove their spread across southern Australia (Thornhill *et al.*, 2019).

The low-nutrient soils which typify much of low-relief Australia have promoted sclerophylly in the flora. Nitrogen-fixing *Acacia* (Fabaceae), a genus with almost 1000 species, is locally dominant or co-dominant in woodlands and shrublands and is also widely targeted by geometrids. Estimated crown divergence dates for the *Acacia* (*s.s.*) lineage span 21.0–23.9 mya and suggest that global Miocene aridification was a driver for its radiation (Miller *et al.*, 2013). The clade of Australian *Acacia* now occupies most climatic zones including the arid interior, monsoonal tropics and temperate south of the continent. The much older family Proteaceae also spans the continent. Evolving about 60 mya, *Banksia* (170 extant spp.) is mainly coastal and has had a long association with relatively open, sclerophyllous vegetation (Carpenter *et al.*, 2016) but unlike *Hakea* and *Grevillea*, is absent from the arid interior. It is noteworthy that the east coast *Banksia* separated about 26 mya from the more diverse south-western flora (Cardillo and Pratt, 2013) but support fewer species of geometrids despite a much greater latitudinal range.

The Pliocene (5.3–2.6 mya) saw global cooling after the warmer Miocene which contributed to the expansion of grasslands and savannahs, and hence fire, across large areas of the continent (Bryceson *et al.*, 2023). In contrast to noctuids, however, no geometrid clades diversified to take advantage of the rapid expansion of tussock and spinifex grasslands. The Plio-Pleistocene transition drove a rapid climatic shift to more arid conditions in southern Australia in the early Pleistocene. Extensive chenopod shrublands also emerged during the late Plio-Pleistocene and now support important genera such as *Chlenias* and its allies.

In many groups of Australian plants and animals, the Pleistocene climatic oscillations left genetic signatures at multiple levels of divergence and with complex geographical structuring, even in areas of low relief and few obvious barriers (Byrne, 2008; Byrne *et al.*, 2011). Barcode of Life (https://boldsystems.org/) BIN data for widespread eucalypt-feeding genera such as *Rhuma*, *Hypobapta*, *Arhodia* and *Gastrinodes* yield totals which are multiples of the number of currently described species, suggesting a similar but undocumented outcome for geometrids.

The contemporary moth fauna shows considerable regional endemism reflecting historic and present-day climates and other environmental gradients. South-western Australia, a biodiversity hotspot (Rix *et al.*, 2015), has a rich looper fauna which includes isolated endemics (e.g. *Arcina*, *Cathaemacta*, *Hypographa*, *Symphylistis*) and local diversifications of widespread austral genera (e.g. in *Crypsiphona*, *Dinophalus*, *Dichromodes* and *Oenochroma*). In addition, the temperate south-west features multiple vicariant populations or sister species of formerly transcontinental taxa of mesic biomes (e.g. *Heliomystis electrica*) now separated by the arid and treeless Nullarbor Plain, an important biogeographic barrier over the last 2.5 million years (Byrne and Murphy, 2020; Dröllner *et al.*, 2023).

The continental island of Tasmania is relatively mountainous and was the last part of Australia to depart a forested Antarctica. Northward drift has always kept it above the polar front and, unlike in Patagonia, a high-latitude cold-adapted fauna did not emerge. The South Tasman Rise finally separated from Antarctica at the time of the Eocene/Oligocene boundary (*ca* 33.5 mya). The deepening and widening of the Tasmanian Gateway disrupted warm-water circulation at high latitudes and led to one of the major climatic shifts of the Coenozoic which affected the local vegetation (Amoo *et al.*, 2022) and, by implication, its phytophagous fauna.

Climate cycles over the last 4 million years have seen Tasmania intermittently separated from Victoria by the 200 km-wide Bass Strait. The last iteration of the Bassian Plain emerged 43,000 BP and was colonized by humans soon afterwards; the land bridge was broken again by 14,000 BP (Adeleye *et al.*, 2024). The island's moth fauna below 800 m elevation is largely an attenuation of that of the eastern Australian forests and heathlands. However, it offers multiple examples of sister species endemism, based on habitats above and below this datum (e.g. in *Dolabrossa*, *Plesanemma*) although species turnover is generally subdued along these gradients (McQuillan *et al.*, 1998).

Although the Australian uplands have been an important domain for species evolution, the age and origin of the uplands themselves are not resolved. Multiple uplifts may have occurred in eastern Australia and age estimates of the current mountains remain contentious (Engel *et al.*, 2020). The eastern ranges may date from the Late Cretaceous (*ca.* 90 mya) when the Tasman Sea between Australia and New Zealand opened. The Australian and Pacific plates share a common margin but forces at this boundary, which deform New Zealand, may propagate into the Australian plate as tectonic stress released as uplift of the mountains in south-eastern Australia (Sandiford *et al.*, 2004). In response to the break-up of Gondwana in the Jurassic, magma intruded as sills, now exposed by erosion, and form the distinctive doleritic high plateaux of Tasmania. These cool moist habitats have long supported ancient conifers *in situ* (Carpenter *et al.*, 2011; Jordan *et al.*, 2015) the origins of which predate the evolution of the Geometridae. Notable among the island's endemics is a clade of small alpine ennomines of 'archiearine' facies and behaviour (*Acalyphes*, *Dirce*) with their larvae dependent on endemic conifers (Cupressaceae, Podocarpaceae) or *Epacris* (Epacridaceae).

As the interior of Australia became gradually more arid, many lepidopteran lineages retreated to the mesic fringes and uplands, becoming extinct in the centre. Onshore movement of moisture and orographic rainfall ameliorated the effect of the Miocene drying and created refugia, allowing some taxa to persist and adapt *in situ*. Although refugia are mainly northern (e.g. the Queensland Wet Tropics) and coastal, scattered outliers include the Central Ranges, Carnarvon Range, Clarke Range, Kroombit Tops, Grampians, Mt Lofty Ranges and Stirling Ranges (Byrne, 2008; Das *et al.*, 2019).

Seasonally cold mountain environments in the south-east now cover less than 1% of Australia, but some clades of geometrids have either diversified there (mostly larentiines) or have spawned local endemic taxa from lowland relatives; *Oenochroma alpina*, *Amelora oritropha* and various species of *Melitulias* and *Chrysolarentia* exemplify this pattern (McQuillan, 1986, 1996). The disjunct Australian Alps and alpine Tasmania share a subset of montane species, including *Cyneoterpna alpina*, but not others. For example, *O. alpina* is absent from Tasmania where its widespread congener *O. vetustaria* occurs to the treeline. Relict periglacial landscape features outline the former extent of cold moist environments at the Last Glacial Maximum (Slee and Shulmeister, 2015) which largely contain the present 'sky island' distribution of this upland fauna.

As a destination and source for geometrids, Papua New Guinea, the most tropical part of Sahul, is pertinent to understanding the profile of the present-day Australian fauna due to its proximity. In its perhumid highlands especially, Australian elements such as the geometrine genera *Aeolochroma*, *Hypodoxa*, *Eucyclodes* and *Oenochlora* mix with a richer endemic fauna probably sourced from Malesia to the west (Holloway, 1996). Interchange among the southern lowland savannah and coastal moth communities was possible via the Sahul land bridge as recently as 8000 years ago.

Along the uplifted margin of eastern Australia, disjunctions are apparent in many montane geometrids, suggesting their ranges were more continuous in the past when tree lines were lower and historical connectivity of populations is assumed. The high tablelands north of the Hunter Valley (e.g. Barrington Tops, New England National Park) reach up to 1500 m and have outlier populations of taxa more common at higher latitudes or in Tasmania. Similarly, uplands in the Wet Tropics of Queensland such as the Atherton Tableland support local endemics separated from their sister taxa 1000 km to the south (e.g. *Mnesampela athertonensis*).

Compared with vagile noctuoid groups in Australia, most geometrids have more limited powers of dispersal. Few seem to move the large, wind-assisted distances which enable some heliothines, noctuines and others to track and exploit ephemeral food plant biomass across the low-rainfall zones. Two small polyphagous larentiines show elements of this behaviour. *Phrissogonus laticostatus* can be irruptive inland and *Epyaxa sodaliata* reaches Tasmania episodically but does not breed there. New Zealand shares several temperate-zone genera with Australia (e.g. *Scopula*, *Chrysolarentia*, *Epyaxa*), which suggests they established across the 1000 km wide Tasman Sea with westerly wind assistance. A few tropical species of *Cleora*, *Thalassodes*, *Agathia* and others have dispersed eastwards to remote Pacific islands in the contemporary cyclone belt and most of these are associated with adventitious food plants (Holloway, 1977, 1979). New Caledonia, 1200 km to the east, supports species of Australian origin such as *Urolitha bipunctifera* and an endemic singleton each of *Oenochroma* and *Chlorocoma* with probable dispersalist ancestry.

18.5 Major Innovations

A suite of innovations has probably favoured the diversification of geometrid moths but limited knowledge of the natural history of many taxa restricts our capacity to identify the most important drivers.

Reduction in the number of prolegs has facilitated the angled-stick or petiole mimicry prevalent in many caterpillars (and usually anchored by a silk-line from the spinneret to the substrate). The flexible, elongated form of looper larvae enables them to reach adjacent leaves more easily, thereby conserving energy as they forage. Similar selection may have driven the emergence of a semi-looper phenotype in some noctuoids such as arboreal Erebinae.

The larvae of many Australian Diptychini and some Oenochrominae spin copious amounts of coarse, tough silk. *Plesanemma* and *Fisera* use their extra prolegs to engage silken pads which they spin along leaf midribs – a likely adaptation to the windswept open canopies of eucalypt woodlands. The claspers of some eucalypt-feeding geometrine larvae, including *Crypsiphona*, *Cyneoterpna* and *Hypobapta*, are notably enlarged to offer extra purchase when resting along leaf margins.

Australian canopies are rich in foraging ants such as *Iridomyrmex* spp. which must be avoided, engaged or appeased by larvae but little is known of looper strategies in this regard. Behaviour after provocation was documented for some species by McFarland (1988). Dropping on a silk line ('belaying') and hanging motionless is a common escape response in arboreal loopers, a facility also developed in uraniines and epiplemines (Lees and Smith, 1991). Contact stimuli can invoke regurgitation and may involve compounds repellent to predators and parasitoids (Steinbauer *et al.*, 2001). Reflex bleeding as a defence strategy, common in arctiine moths, has been recently reported in an Indonesian geometrid (Tautel and Stüning, 2023).

Abdominal tympanal organs are almost universal in geometrid moths (Cook and Scoble, 1992) and although their function is controversial, a capacity for directional hearing is likely to be useful for mate recognition and predator avoidance, since it has evolved independently in the biodiverse Noctuoidea and some other successful groups.

Physiological tolerance or adaptation to foliar antifeedants such as essential oils, glycosides and alkaloids has been a likely gateway to diversification in some clades via host capture. Such mechanisms may underlie the prevalence of eucalypts in the diet of many local Ennominae and 'Pseudoterpnini'. The acquisition and expression of detoxifying gut enzymes underlie polyphagy and insecticide resistance in some geometrids but are poorly known beyond a handful of pests such as *Hyposidra talaca*.

18.6 Food Plant Relationships

Most geometrid larvae are family-level monophagous (i.e. feeding on a single plant family), some are oligophagous, while polyphagy (feeding on >3 families) across plant families is relatively rare. Australia is remarkable for the dominance of the family Myrtaceae in its woodland and forest canopies where its high apparency and long-lived leaves may help explain its wide use as a food plant. Even in diverse canopies, however, some plant genera are disproportionately important for supporting interactions among plants and moths (Narango *et al.*, 2020). Conservation of food plant choices within some clades suggests novel host capture in the past might have been an enabler of diversification. Examples include *Idiodes* and its relatives on ferns, *Milionia* on austral conifers, *Austrocidaria* on *Coprosma*, *Corula* on *Callitris*, *Rhynchopsota* on Casuarinaceae and *Agathia* on Apocynaceae. The Oenochrominae associated with Proteaceae use hosts in rainforest as well as others in sclerophyllous, fire-prone woodlands. An important clade of robust ennomines in New Zealand, the 'declanoids', has diversified on a remarkably wide range of plant groups (Dugdale *et al.*, 2023).

On shared food plants, partitioning among caterpillars may help avoid competition. On eucalypts, *Didymoctenia exsuperata* is multibrooded, the larger *Gastrinodes* is bivoltine and the very large *Gastrophora henricaria* has a single generation annually. Small eupitheciine larvae on the flowers of ephemeral daisies in the inland probably encounter voracious *Helicoverpa* spp. and *Vanessa kershawi* larvae but the large size differences may mute competition. In southern woodlands, more than ten species of eucalypt feeders may exploit a pool of only two or three local hosts (McQuillan *et al.*, 1998). Food plant shifts probably assist lineages to both colonize new frontiers and persist in locations as the flora changes in response to environmental change over time.

Adaptation to the formidable chemical defences of *Eucalyptus* (Wallis *et al.*, 2010;

Jensen *et al.*, 2015) may underlie the diversification of disparate groups, notably austral Ennominae, Oenochrominae (*s.s.*) and Australian 'Pseudoterpnini'. Conversely, very few Sterrhinae, Larentiinae or Geometrinae: Hemitheini exploit eucalypts. There is opportunity for, and evidence of, fine tuning to host plant species within genera as well as contrasting consumption strategies as larvae grow (Steinbauer and Matsuki, 2004). Continent-wide genera such as *Hypobapta* merit further study in this regard. Non-eucalypt Myrtaceae such as *Melaleuca* support important genera including *Dichromodes*, *Syneora* and *Aeleochroma*. Wattles (*Acacia* spp.), often co-extensive with eucalypts, are another major food plant resource and the fauna of each genus shows very little overlap at species level. Although Diptychini and Boarmiini occur on both these hosts, Macariini seem to avoid eucalypts.

Polyphagy on unrelated woody hosts is a common feature of the tribe Boarmiini worldwide. In Australia, examples occur in both tropical (*Hyposidra talaca*) and temperate regions (*Ectropis* spp., *Pholodes sinistraria*, *Zermizinga sinuata*). Another member, *Cleora injectaria*, includes mangrove trees of at least four families in its host range. Among Diptychini, some *Capusa*, *Chlenias* (Fig. 18.2.14) and *Nisista* larvae are polyphagous, with corresponding non-cryptic colouration, but this capacity is otherwise uncommon in the tribe. *Mochlotona phasmatias* includes *Acacia*, *Olearia* and *Tasmannia* among its hosts, while some larentiines such as *Epyaxa subidaria* feed across multiple families of herbs.

A minority of geometrids reliant on annual food plants have adopted weeds as hosts. Livestock pastures, now an extensive open habitat rich in herbs of European origin, support *Scopula rubraria*, various xanthorhoines (*Epyaxa* spp.) and the ennomines *Zermizinga sinuata* and *Ciampa arietaria*.

The environmental extremes of Australia are a significant challenge to phytophagous insects, with few parts of the continent offering benign conditions across all seasons. Leaf flushing in local foodplants is often ephemeral, with the 'caterpillar season' correspondingly short. In a contrasting strategy, some robust larvae such as *Mnesampela*, *Niceteria* and *Paralaea* feed on mature leaves and are examples of slow-growing senescence feeders, a trophic subguild of insect herbivores (White, 2015).

18.7 Morphological Features

18.7.1 Wing phenotypes

Wing phenotypes are largely a response to the local environment. Cuticle melanism occurs in montane *Dirce* and some *Chrysolarentia* and may help manage heat flux. Other phenotypes represent solutions to avoid or alarm predators and can be recurrent across unrelated groups. Ensembles of shape, colour and pattern signal to other individuals and to potential predators, sometimes reinforced by rhythmic movement or other display behaviours. Although the wing margins in most geometrids are rounded, angled margins occur in some Macariini, *Epicompsa* and *Tessarotis* and scalloped margins feature in *Amphiclasta lygaea*, *Nycticleptes lechriodesma* and genera such as *Scioglyptis*. While green leaf crypsis is prevalent in Geometrinae-Hemitheini, it is also seen in *Oenochroma cerasiplaga* and the tropical ennomine *Aplochlora pisochroa*. Bark mimicry is prevalent in drab Boarmiini although a small diurnal tropical clade is colourful. Well developed in the local fauna is crypsis on fire-blackened ground or trunks, as seen in *Dinophalus incongrua*, *Melanodes anthracitaria*, *Nycterephes anthracica*, *Hypomecis pissinopa* and *Xanthorhoe anthracitaria*. Reddish-brown tones are seen across unrelated species from the arid and semi-arid zones where soils are tinted by iron oxide, as displayed by *Dichromodes rufula* and some *Loweria*.

Pink *Notiosterrha* mimic the anthocyanin-rich tones of *Tecticornia* stems in saltmarshes and chenopod shrublands; green moss and macrolichens are models for *Austrocidaria erasta*, *Sauris* spp., *Dirce aesiodora*, *Archephanes zalosema* and various *Aleochroma* spp. Bold disruptive patterns feature in *Thalaina* spp., *Lychnographa heroica* (Fig. 18.1.14), *Epidesmia tricolor*, *Phrataria* spp., *Dichromodes confluaria* and *D. personalis*. Startle defence has evolved independently many times and involves colourful hindwings in numerous species including *Dirce* spp., *Larophylla amimeta*, *Niceteria macrocosma*, *Mnesampela heliochrysa*, *Gastrophora henricaria*, *Heliomystis electrica*, *Dichromodes ainaria* or undersides in *Cassephyra plenimargo*, *Rhuma aurata* and *Parepisparis* spp. Many Oenochrominae (*s.s.*) possess a large purple blotch beneath the forewing which

Fig. 18.1. Geometridae: adults. 1. Sterrhinae: Scopulini – *Scopula rubraria*. Photo: James Peake. 2. Sterrhinae: Sterrhini – *Notiosterrha pulcherrima*. Photo: Daniel Meier. 3. Geometrinae: Eumeliini – *Eumelea rosalia*. Photo: Larney Grenfell. 4. Geometrinae: Agathiini – *Agathia lycaenaria*. Photo: Bridgette Gower. 5. Geometrinae: Hemitheini – *Mixochroa gratiosata*. Photo: John Walter. 6. Geometrinae: Nemoriini – *Eucyclodes pieroides*. Photo: Bridgette Gower. 7. Desmobathrinae: *Nearcha pseudophaes*. Photo: Jason van Weenen. 8. Epidesmiinae: *Phrataria bijugata*. Photo: Reiner Richter. 9. Epidesmiinae: *Dichromodes euprepes*. Photo: Charles Porter. 10. Oenochrominae – *Oenochroma quardrigramma*. Photo: David H. Fischer. 11. Larentiinae: Xanthorhoini – *Chrysolarentia perornata*. Photo: Martin Lagerway. 12. Ennominae: Diptychini – *Dirce solaris*. Photo: Larney Grenfell. 13. Ennominae: Diptychini – *Thalaina angulosa*. Photo: John Walter. 14. Ennominae: Diptychini – *Lychnographa heroica*. Photo: Samuel Frankel. 15. Ennominae: unplaced – *Xenomusa metallica*. Photo: David H. Fischer.

can be displayed when provoked. Aposematism may explain the showy facies of diurnal *Milionia* and *Dysphania*.

Accumulations of refractive leaf litter are ubiquitous in the sclerophyll forests and xeromorphic shrublands of Australia. Among geometrids which rest on or near the ground, dead leaf mimicry is very common in smaller, slender species (e.g. *Nearcha*, *Taxeotis*, many *Dichromodes*). Wing and body twisting can reinforce the dead leaf illusion and is seen in more robust species such as *Circopetes obtusata*, *Oenochroma turneri*, *Parepisparis* spp., *Gonodontis orthotoma* and *G. stramenticea*. In wetter forests, mouldy leaves and stems are models for intricately patterned species such as *Oenochroma polyspila*, *Parepisparis* spp., *Larophylla amimeta* and *Xylodryas*. *Polyacme dissimilis* also resembles rainforest litter.

Resting posture and roosting behaviour can be informative. Alighting behaviour varies among taxa which hold the wings flat. Xanthorhoini often roost head downwards in shady situations beneath tree-fern canopies or in hollowed tree trunks and a brief period of raising and lowering the wings often occurs before settling. A planiform resting posture is prevalent in Scopulini, Epidesmiinae and Boarmiini, whereas the uncommon veliform stance is typical of *Chrysolarentia tristis* and most *Hypographa*. A rare planiform type, with the forewings dislocated to point forward, occurs in some *Pingasa* and wing rolling is a feature of *Neogyne elongata* and *Thaumatographe singularis*, the latter resembling *Phazaca*, an epiplemid. A tectiform posture is typical of many Diptychini. *Mochlotona* typically rests with its woolly forelegs projecting, most *Rhuma* species expose the fore and mid legs, while *Sauris* splay their elongated forelegs.

18.7.2 Larval structure and feeding behaviours

The smooth-bodied geometrid ground plan has evolved into a wide array of forms and behaviours, especially in relation to exploiting crypsis or masquerade as a defence against predators (Higginson *et al.*, 2012). Dorsal processes on the body are mostly absent or subdued but conspicuously adorn many Nemoriini larvae whereas the ennomine *Mochlotona* (Fig. 18.2.8) is unusual for possessing dorsal scoli. Pointed or bifid head capsules are widely present in the larvae of Geometrinae but are not unique to this group.

Crypsis on food plants is reinforced by impressive colour and pattern matching (McFarland, 1988). For example, the larvae of *Rhynchopsota* spp. and *Boarmia suasaria* are almost invisible against *Allocasuarina* foliage and the green and white larva of *Thalaina clara* closely resembles the leaves of bipinnate *Acacia*. *Plesanemma* resembles *Eucalyptus* leaves, including the petiole. The larva of *Cycloprorodes* rests flat against the bark of its *Dodonaea* food plant while broken twig mimics are very common among the Boarmiini.

Aposematic colouration is uncommon in Australian geometrids but occurs in both the larvae and adults of *Bracca* and *Dysphania*. In contrast, the colourful larvae of *Chlenias* are adorned with fine stripes but give rise to very plain adults. Local crowding may induce aposematic colour forms in the larvae of some defoliating populations of *Arhodia*, *Antictenia* and *Paralaea*. Colour polymorphism in *Capusa cuculloides* is evident in bright green or orange caterpillars with only rare intermediates, while *Dysphania numana* has yellow and green forms. Counter-shading, a particular form of crypsis, is rare in loopers but apparent in the larvae of *Fisera* and some other Diptychini.

Feeding modes vary among groups but most larger larvae are edge feeders on leaves. Small larvae or early instars of *Casbia* may scrape the underside of leaves, leaving transparent 'windows' of epidermal tissue. Some nemoriines erode or channel the upper surface of leaves, especially when young. The softer tissues of flower buds and petals are targeted by some eupitheciines. Gregarious feeding is rare in geometrids but is typical of younger *Mnesampela privata* larvae which erode leaf surfaces before graduating to chewing on leaf margins (Steinbauer *et al.*, 2001).

18.7.3 Pupal features

Geometrid cocoons are typically flimsy and may incorporate soil or detritus in the silk matrix. The pupae of some *Eois* are attached to the undersides of leaves by the hooked spines of the cremaster. Most pupae are smooth and plain but patterned integument occurs in many geometrine genera. The well-sclerotized pupae of robust ennomines and oenochromines in the semi-arid zone can be located up to 15 cm deep in sandy soils where they are thermally buffered from extremes of heat and cold. The synchronous appearance of some semi-arid species suggests the pharate adults may be dormant for some time, awaiting suitable conditions to emerge.

18.7.4 Adult biology

Adult geometrids reinforce their camouflage through background matching and disruptive coloration. Behavioural strategies such as the choice of substrate and body orientation improve efficacy against predators (Kang *et al.*, 2015). Phenotypic variability among geometrids is likely an antipredator strategy. Camouflaged moths are more variable in wing patterning than aposematic species (Nokelainen *et al.*, 2024); *Cleora* species illustrate this principle well.

Male secondary characters are diverse across the family and may involve modified hindwings and hind tibiae, androconia, hair pencils and their associated structures. Sexual dimorphism is widespread in the antennae with filiform or subdentate antennae, thicker in males. Long rami are present in some species. Bipectinate antennae are common but unipectinate male antennae feature in epidesmiines and many oenochromines.

Pheromones are only known for a few species such as *Mnesampela privata* (Steinbauer *et al.*, 2004). Female calling behaviour is documented for some *Chlenias* (Madden and Bashford, 1977a).

Generation timespans in Australian geometrids range from biennial to multivoltine with continuous breeding apparent in some small species such as *Chloroclystis filata* (Marriott, 2011, 2012; Hewish *et al.*, 2014, 2016). *Chlenias inkata* can spend 2 years in the pupal stage (Tindale, 1961) and multivoltinism is less common at higher latitudes and elevations. Outbreak and boom–bust population cycles can occur in some ennomines (*Ciampa*, *Chlenias*, *Mnesampela* and *Paralaea*) (Madden and Bashford, 1977a,b; Steinbauer *et al.*, 2001) and a few oenochromines (*Antictenia punctunculus* and *Arhodia* sp.).

Behaviours mediated by climate such as aestivation and hibernation help tune life cycles to local conditions, especially in the extensive semi-arid zone where rainfall is highly variable. Facultative dormancy in eggs occurs in some groups (McFarland, 1973, 1988), probably serving to align caterpillar activity with peaks in leaf palatability or nutrition. Rain-hatched eggs help synchronize neonate larvae to unpredictable leaf flushes. Dispersal by first instar larvae occurs if females are sedentary or wingless and ballooning behaviour by newly hatched loopers is documented in *Chlenias* (Madden and Bashford, 1977a). The very active first instar larvae of *Arhodia* may wander for several days before settling to feed (McFarland, 1988).

Aestivation in the buried pupal stage is common in this zone and can exceed 9 months in duration. Eclosion is often locally synchronous and influenced by heavy rain or low temperatures. Structural adaptations which help newly eclosed adults reach the surface include a thickened frons (e.g. *Fisera*), horny facial prongs (*Ciampa*, some *Thalaina*) and fore-tibial spines (*Dinophalus*, *Thalaina*, *Loweria*). Some high-latitude moths are notably cold adapted, with adult activity peaking in winter (e.g. *Amphiclasta lygaea*, *Capusa cuculloides*) when bats are dormant. *Chlenias* emerges from its pupae into air temperatures as low as 4°C in Tasmania. *Ciampa arietaria* males fly in the predawn of late autumn (Guest, 1887).

Most Australian geometrids have a functional proboscis, but exceptions include *Aglossophanes* and *Anomogenes*. Visitation to flowers for nectar has been observed in many diurnal species but is largely unreported for nocturnal taxa, with the exception of *Oenochroma vinaria* carrying pollen of *Stackhousia* (Celastraceae) (Blackall *et al.*, 2023). The role of looper moths as pollinators is potentially considerable and warrants further study.

18.8 The Australian Fauna

Two centuries of study have yielded a first draft of the relationships among geometrid species and their higher groups. Historically, this emphasized superficial features such as body vestiture, wing shape and colour, venation and male antenna forms. On this basis, Guenée (1857) recognized, as Oenochromidae, a distinctive Australian clade of robust moths. Monophyly for this clade, the Oenochrominae (*s.s.*), is now reinforced by molecular analysis (Murillo-Ramos *et al.*, 2019). As newer methods allow homoplasy to be progressively identified, consensus is building around more robust classifications. The novel Australian 'Archiearinae' have been shown to be alpine ennomines with archiearine facies (Young, 2008a) and Pseudoterpnini (Geometrinae) is now believed to exclude the Australian taxa formerly included in that tribe; instead, the latter may form a new taxon arguably sister to the Hemitheini (Murillo-Ramos *et al.*, 2019).

Recent phylogenies integrate molecular and morphological evidence together with improved analysis and hypothesis-testing frameworks. A major challenge is to include more complete coverage in taxon sampling.

In the past, numerous tribes have been defined based on shared features of the genitalia, especially of males. Molecular evidence has been useful in testing the membership and arrangement of some groups. Improved resolution of groups at the level of tribes seems a useful framework for addressing biogeographical questions in particular.

The following overview of the fauna is largely resolved to the level of tribes.

18.8.1 Subfamily Sterrhinae

Commonly known as 'waves' from a wing pattern widely repeated across the group, most sterrhines are relatively small moths; indeed, some *Idaea* species are among the smallest of all Geometridae. Commonly, the forewing venation features one or two areoles and the hindwing veins Sc + R_1 and Rs are fused for a short distance before diverging.

Young (2006a, 2008a), using a wide range of Australian Geometridae, was one of the first authors using both molecular and morphological data to demonstrate that the basal lineages of the looper family were most likely the Sterrhinae and Larentiinae. Later work showed that the Sterrhinae may be the sister lineage to the rest of the Geometridae (Sihvonen *et al.*, 2020; Murillo-Ramos *et al.*, 2023) or sister to Larentiinae (Zheng *et al.*, 2022). Sterrhines comprise about 3000 species globally and are most diverse in the Neotropics followed by the Palaearctic and Afrotropics.

Only four of the nine recognized tribes occur in Australia, but both informal groupings are represented – the 'Timandrini lineage' including Cosymbiini and Rhodometrini and the 'Scopulini lineage' encompassing Scopulini and Sterrhini (Sihvonen and Kaila, 2004).

18.8.1.1 Cosymbiini

This is a cosmopolitan group featuring naked, square-shouldered pupae which are supported by a silken girdle around abdominal segment A3. Australian examples include *Chrysocraspeda*, *Cyclophora* and *Dizuga*; *Perixera* extends eastwards into the Pacific islands. Some feed on woody shrubs including *Syzygium* (Myrtaceae).

18.8.1.2 Rhodometrini

The Rhodometrini are an Old World tribe and richest in the Afrotropics. *Traminda* (seven local spp.) extends to tropical and subtropical Australia. Its larvae are polyphagous on flowers and new foliage, notably *Acacia*, and the flanged pupae are suspended in a netting of silk threads.

18.8.1.3 Scopulini

Scopulini is the largest cosmopolitan clade of sterrhines with *ca* 1000 species globally but is only modestly represented in Australia. The multivoltine *Scopula rubraria* (Fig. 18.1.1) is ubiquitous in the temperate zone and *S. lydia* largely replaces it in the semi-arid interior along with *S. episcia*. *Scopula achroa* is a habitat specialist in saltmarshes. The very slender larvae of *Scopula* feed mainly on native herbs but some have adopted adventive and naturalized weeds in their diet. *Problepsis* is monophagous on Oleaceae in the Old World tropics and subtropics.

18.8.1.4 Sterrhini

Sterrhines are the sister tribe to the Scopulini. They are distributed worldwide but most diverse in the Palaearctic and Neotropics. The endemic *Notiosterrha* (Fig. 18.1.2) found in halophytic habitats may belong here. The cosmopolitan genus *Idaea* includes a group of small, rather similar species, some of which can be very abundant. Some *Idaea* larvae have a feeding preference for withered or decaying leaves.

18.8.2 Subfamily Geometrinae

The emerald moth subfamily contains about 2700 species globally resolved into 18 tribes (Ban *et al.*, 2018) and is moderately well represented in Australia. Geometrinae and Oenochrominae (*s.s.*) may be sister groups, although no geometrines are known from Proteaceae and only a few feed on *Eucalyptus*. There is little evidence so far that Australia preserves basal taxa of geometrines, a subfamily absent from New Zealand.

The green colouration of emerald moths is derived from geoverdin, a pigment of unknown structure and origin, present in the scales (Cook *et al.*, 1994). The male abdomen often bears a pair of setal patches on sternite A3. The hindwing venation generally has M_2 arising closer to M_1 than M_3 while the frenulum is typically rather weak or even absent. Females commonly possess oblique, papillate ovipositor lobes (Holloway, 1996) and saccular pheromone glands (Bendib, 2001). Most adults rest with the wings splayed and flat but others, including *Mixochroa* and some *Euloxia* spp., hold their wings roofwise.

Geometrine eggs are often oval with planar sides and a sculpted chorion but in the Australian 'Pseudoterpnini', the sides are rounded and the eggs relatively featureless (Young, 2006b). The larval head capsule is commonly bifid or conical and the integument can be finely granular. The thoracic legs are often crowded in Hemitheini, the larvae of which can be rather turgid. Various larvae mimic leaves, petioles and stems. Pupae are typically pale grey or tan, speckled and flecked with darker spots and streaks and contained in a flimsy network cocoon.

The South-east Asian tropics are especially rich in geometrines and home to most of the known tribes. The subfamily includes a blend of food plant specialists, some exploiting toxic hosts and rather fewer generalists.

18.8.2.1 Eumeleini

This tribe contains a dozen species confined to the Indo-Australian tropics. The single genus *Eumelea* is of problematic placement (Sihvonen *et al.*, 2011) but is possibly sister to the Oenochrominae, Geometrinae or is sister to Oenochrominae + Geometrinae (Murillo-Ramos *et al.*, 2019, 2023; Ghanavi *et al.*, 2025). Three species inhabit coastal Queensland, with *E. rosalia* (Fig. 18.1.3) extending into the subtropics. Larvae are likely to be on understorey Euphorbiaceae such as *Macaranga*.

18.8.2.2 Agathiini

Agathia (6 local spp.; Fig. 18.1.4) and *Alloeopage* (1 sp.) are lurid green moths oligophagous on toxic Apocynaceae and Asclepiadaceae. *Agathia* extends from tropical Africa to east Asia and subtropical Australia, reaching Norfolk Island, New Caledonia and Fiji.

18.8.2.3 Hemitheini

In its broad concept, this tribe includes speciose Australian genera such as *Chlorocoma*, *Euloxia* and *Mixochroa* (Fig. 18.1.5) typical of the temperate zone. The green or rosy larvae tend to be elongated and the head capsule is bifid (Fig. 18.2.2). Collectively, the tribe exploits a wide range of woody host plants including *Acacia*, *Leptomeria* (Santalaceae), *Pimelea* (Thymelaeaceae), *Eucalyptus* and *Leptospermum* (Myrtaceae). Two species of *Prasinocyma*, *P. ocyptera* and *P. rhodocosma*, have an enigmatic distribution across the continent that features large disjunctions.

18.8.2.4 Comibaenini

A distinctive feature of this tribe is the bifid vinculum in the male genitalia. Four genera cover the Australian fauna. *Argyrocosma* (3 spp.), *Comibaena* (2 spp.) and *Protuliocnemis* (2 spp.) are tropical to subtropical in contrast to the cool-adapted *Iulops argocrana* which is confined to south-eastern Australia including Tasmania. The flanged larvae of some taxa conceal themselves with fragments of attached detritus and may exhibit swaying and trembling behaviours. Most pupate in a flimsy cocoon spun among leaves. *Protuliocnemis* feeds on *Acacia* and some *Comibaena* elsewhere are polyphagous.

18.8.2.5 Nemoriini

Male Nemoriini have a rod-like uncus and a central longitudinal midrib on the eighth sternite distinguishing them from the Comibaenini; the vinculum is rarely bilobed. Possibly a sister clade, *Eucyclodes* accounts for about 100 species in the Indo-Australian region, especially in Papua New Guinea. Sexual dimorphism can be marked in this genus. The genotype, *E. buprestaria*, feeds exclusively on *Cassytha* (Lauraceae) whereas *E. pieroides* (Fig. 18.1.6) is polyphagous. The monotypic *Chlorodes* (Fig. 18.2.3) is on *Leptospermum* with a relative on *Cassinia* (Asteraceae). The reddish-purple *Xenochlaena porphyropa* is an unusual inland species from the *Acacia* woodlands. The flanged larvae of some nemoriines also decorate themselves with detritus and show trembling behaviour.

18.8.2.6 Dysphaniini

The sole genus of this Indo-Australian tribe, *Dysphania*, is somewhat aberrant but clearly allied to Geometrinae (e.g. geoverdin is present) (Sihvonen *et al.*, 2011). The tropical *D. numana* (Fig. 18.2.4) is the only Australian representative. A fovea, a raised blister largely devoid of scales situated at the base of the forewing, is present in both sexes. Gigantism and aposematic colouration of both the larva and adult suggest they are repulsive to predators. In Asia, the genus is mimicked by species of the boarmiine genus *Praesos* and a chalcosiine zygaenid genus, *Psaphis*. *Dysphania* larvae feed mainly on tropical *Carallia* (Rhizophoraceae) foliage which contains hygroline, a distasteful alkaloid.

18.8.2.7 Australian 'Pseudoterpnini'

This is a group of relatively robust, predominantly grey moths with well-marked and sometimes colourful undersides. The male genitalia of most genera have bifid socii. Australia is home to some 15 genera, about half the global total for this Old World tribe (Pitkin *et al.*, 2007). Eight genera extend to Tasmania, including two species of the unusual genus *Rhuma*. In contrast to other geometrines, most of this tribe are stoutly built, including their larvae (e.g. *Pingasa*). Most Australian habitats support several pseudoterpnines but the mesic south-east is particularly rich in genera (Goldfinch, 1929), a few of which span the continent (e.g. *Crypsiphona, Hypobapta, Rhuma*). Indeed, *Crypsiphona ocultaria* may be the most frequently recorded geometrid in Australia. *Apodasmia, Cyneoterpna, Heliomystis* and *Hypobapta* are largely Bassian in distribution, including Tasmania, whereas *Austroterpna, Lophothorax* and *Paraterpna* do not cross Bass Strait.

Camouflaged on epiphyte-rich surfaces, *Hypodoxa* ranges from South-east Asia, northern and eastern Australia to Fiji. Its larvae are usually on Myrtaceae. *H. muscosaria* extends to Tasmania. The similar-looking *Aeolochroma* is mainly coastal in Australia but is common in Papua New Guinea: its food plants include *Beyeria* (Euphorbiaceae). Other elements of the rich East Asian pseudoterpnine (*s.s.*) fauna, such as *Epipristis* and *Pingasa*, reach Australia. *Epipristis* is tropical and subtropical while the latter penetrates eastern Australia as far south as Tasmania (*P. cinerea*).

Many of the Bassian genera are associated with eucalyptoid Myrtaceae as food plants, notably *Angophora, Corymbia* and *Eucalyptus*, but also leptospermoid shrubs such as *Melaleuca* and *Leptospermum*. The modest semi-arid fauna depends mostly on *Eucalyptus* (*Rhuma*) or *Acacia* (*Lophothorax*), although the unusual *Anomogenes* takes native *Citrus* (Rutaceae). Across a wide range in the Old World some *Pingasa* species are polyphagous. *P. chlora* in Australia takes *Euroschinus* (Anacardiaceae) and *Flindersia* (Rutaceae), food plants neglected by most geometrids.

Some geometrine taxa remain unassigned to tribes and more clarity is needed around the relationships of some local 'pseudoterpnines'. The unusual *Oenochlora* is disjunct with a restricted eastern subtropical and tropical upland distribution and an extension to the highlands of Papua New Guinea. Its life history is unknown.

18.8.3 Subfamily Desmobathrinae

These are typically slender-bodied moths with long and narrow appendages (Holloway, 1996), commonly with the apex of the forewing sharply pointed. Less than ten species of the rich south-east Asian fauna reach tropical north-eastern Queensland. Four genera are involved: *Celerena* (1 sp.), *Derambila* (4 spp.), *Noreia* (1 sp.) and *Ozola* (2 spp.). *Nearcha* (13 spp.; Fig. 18.1.7) is the largest local genus and prefers open woodland habitats across the continent; its larvae feed on woody subshrubs including *Dillwynia* (Fabaceae). Male secondary characters are diverse in this genus and diagnostic for species.

18.8.4 Subfamily Epidesmiinae

Epidesmiinae is a largely Australian subfamily as recently formalized (Murillo-Ramos *et al.*, 2021a, b) for a lineage of 'slender-bodied Oenochrominae' (*sensu* Turner, 1929, 1930) with unipectinate antennae. It includes eight local genera, of which *Dichromodes* (Fig. 18.1.9) is the largest genus. It is most diverse in heathlands and the semi-arid zone. The few *Dichromodes* that have been reared were associated with Myrtaceae shrubs. *Adeixis* also occurs in New Zealand and the maquis heathlands of southern New Caledonia. *Epidesmia* (9 spp.) and *Phrataria* (4 spp.; Fig. 18.1.8) include some very striking species. Various *Phrixocomes* (6 spp.) are typical of southern heathlands, while *Arcina*, *Ecphyas* and *Systatica* are monotypic genera.

18.8.5 Subfamily Oenochrominae

18.8.5.1 Oenochrominae (s.s.)

The core Oenochrominae (Prout, 1910; Scoble and Edwards, 1990) are a particularly attractive group of Australian macromoths. Two clades are apparent (Murillo-Ramos *et al.*, 2019) with diversification centred on two host plant families, the Proteaceae and Myrtaceae respectively. Unipectinate antennae are usual in the former groups whereas bipectinate species are associated with Myrtaceae feeders. Typical oenochromines are robust moths with hairy undersides and capable of rapid flight. Rosy tones are prevalent in the vestiture and the underside of the forewing may have a large purplish blotch. Mimicry of dead and decaying leaves is common in this group as exemplified by *Parepisparis* (Scoble and Edwards, 1990) and various other genera. Pupae in this group are usually glossy.

Oenochroma (21 described spp.) and *Arhodia* (*ca* 7 likely spp.) span the continent, although the former genus may be polyphyletic (Murillo-Ramos *et al.*, 2019). *O. quardrigramma* (Fig. 18.1.10) is from wet forest on the east coast whereas *O. pallida* and *O. cycnoptera* tolerate the arid zone. The unusual male of *O. alpina* is diurnal in the Australian Alps where its larvae graze shrubs of *Grevillea australis*. The common *O. vinaria* regularly exploits ornamental *Grevillea* and *Hakea* spp. in urban gardens and *O. turneri* has been reared from *Macadamia* trees in plantations.

The handsome *Gastrophora henricaria* is common in the eastern sclerophyll forests but rare at its western disjunct limit in the Mt Lofty Ranges; it is absent from Tasmania. *Parepisparis* (Fig. 18.2.1) is also eastern, extending from Victoria to Papua New Guinea, and sexual dimorphism can be considerable in this genus. *Cernia amyclaria* occurs in an arc from the Pilbara, across monsoonal northern Australia to extreme eastern Victoria. Also on *Eucalyptus*, a small *Arhodia* species in the southern semi-arid zone features extensive population outbreaks in mallee woodlands. This zone is also rich in species of *Dinophalus* associated with Proteaceae, which along with *Lissomma* also contributes to the fauna of the seasonally dry savannahs in the north. The slow-growing larvae of *Phallaria ophiusaria* are unusual in being polyphagous (Anderson, 1902; McFarland, 1988).

The genus *Sarcinodes* extends from north-east Queensland to north-east India and is diverse in Sundaland where it is hosted by tropical Proteaceae such as *Helicia*.

18.8.5.2 Oenochrominae (s.l.)

This group includes the large genus *Taxeotis* (*ca* 50 spp.), mainly found in the temperate to semi-arid zone. The cryptic adults rest on the ground amongst leaf litter during the day. Some lay unattached eggs but life histories are otherwise unrecorded.

18.8.6 Subfamily Larentiinae

The Larentiinae subsumes more than 6000 species globally and the monophyly of these, the carpet moths and their relatives seems secure (Õunap *et al.*, 2016; Murillo-Ramos *et al.*, 2019). The hindwing venation is distinctive for anastomosis between Sc + R_1 and Rs. Male secondary characters can be elaborate, including the presence of hair tufts or pencils on the body, legs and wings and one or more pairs of coremata in many genera.

Viidalepp (2011) summarized the morphological definitions of the larentiine tribes from a global perspective but noted the poor state of knowledge of the southern hemisphere taxa.

Fig. 18.2. Geometridae: larvae. 1. Oenochrominae – *Parepisparis* sp. Photo: Emily Noble. 2. Geometrinae: Hemitheini – *Chlorocoma dichloraria*. Photo: David Akers. 3. Geometrinae: Nemoriini – *Chlorodes boisduvalaria*. Photo: Simon Grove. 4. Geometrinae: Dysphaniini – *Dysphania numana*. Photo: Sarah Renneberg. 5. Larentiinae: Eupitheciini – *Microdes squamulata*. Photo: David Akers. 6. Ennominae: Boarmiini – *Pholodes sinistraria*. Photo: Nadine Gaffney. 7. Ennominae: Boarmiini – *Bracca matutinata*. Photo: Samuel Frankel. 8. Ennominae: Diptychini – *Mochlotona phasmatias*. Photo: Michael Charles. 9. Ennominae: Diptychini – *Mnesampela privata*. Photo: Paul George. 10. Ennominae: Diptychini – *Plesanemma fucata.* Photo: Franegan. 11. Ennominae: Diptychini – *Fisera eribola.* Photo: Craig Loechel. 12. Ennominae: Diptychini – *Paralaea* sp. Photo: Ian Moodie. 13. Ennominae: Diptychini – *Thalaina clara.* Photo: David Akers. 14. Ennominae: Diptychini – *Chlenias zonaea.* Photo: Kristi Ellingsen. 15. Ennominae: unplaced – *Xenomusa monoda.* Photo: Keith Martin-Smith.

Some groups in the Australian fauna can be assigned to existing tribes but others, such as the fern-feeding *Chaetolopha*, remain unallocated.

18.8.6.1 Trichopterygini

These are slender moths of distinctive appearance with most having rather small hindwings. Male sternite A2 is modified. *Episteira*, *Tympanota* and *Sauris* are Australian representatives of this worldwide group (Dugdale, 1980) which is absent from Tasmania. Many are associated with Sapindales as larvae (Schmidt, 2016), although *Sauris malaca* takes new growth of *Argyrodendron actinophyllum* (Malvaceae) (P. Hendry, pers. commun.) and an outlier population of *Tympanota perophora* is on a subalpine conifer, *Podocarpus lawrencei*.

18.8.6.2 Eupitheciini

This tribe includes the notoriously 'difficult' pug moths of the northern hemisphere. It includes numerous small geometrids whose larvae often feed on floral parts including immature fruits and seeds. Although diverse in the Holarctic, the Australian fauna is modest in size. The extensive Old World genus *Gymnoscelis* (19 spp. locally) is mainly tropical in Australia. *Pasiphilodes testulata* and *Phrissogonus laticostata* are very common moths in temperate areas and can be minor pests in pome orchards. *Sigilliclystis insigillata* spans the continent whereas *Chloroclystis filata* is Bassian; the yellow larva of the latter grazes Fabaceae flowers, including *Acacia*, on which it is cryptic. The temperate zone genus *Microdes* (11 spp.; Fig. 18.2.5), included in this tribe by Viidalepp (2011), reaches New Zealand. *Pasiphila* appears absent from Australia although it is diverse in New Zealand.

18.8.6.3 Melanthiini

This is a tribe close to the eupitheciines. They are mainly Holarctic but have Afrotropical and Indo-Australian representatives. *Collix* occurs in coastal Queensland and *Eccymatoge* in more temperate areas.

18.8.6.4 Asthenini

Xue and Scoble (2002) reviewed membership of this tribe. The female signum is distinctive for its arrangement of radiating thin spines. *Asthena* and *Poecilasthena* have larvae on Ericaceae and Proteaceae. *Eois* and *Minoa* have been recently readmitted to this group.

18.8.6.5 Xanthorhoini

The Xanthorhoini are a largely amphipolar tribe and diverse in mountains. Most males feature eversible coremata on A8 and this sternite is narrowed. The male genitalia usually possess a setose calcar and the costa of the valva projects distally (Schmidt, 2013). *Chrysolarentia* is the largest Australian genus and includes colourful diurnal species such as *C. perornata* (Fig. 18.1.11) in the Australian Alps and Tasmania. '*Xanthorhoe*' (*s.l.*) and *Epyaxa* are shared with New Zealand. Some of the larger species feature woody plants in their diet with *Acodia* and *Austrocidaria* on *Coprosma* (Rubiaceae). Xanthorhoine larvae often play dead if disturbed. *Hydrocotyle* (Araliaceae) and *Plantago* (Plantaginaceae) are common food plants in Australia. *Visiana* has been reared from *Urtica* (Urticaceae) (Schmidt, 2016).

18.8.6.6 'Hydriomenini'

The scope of this tribe in Australia remains uncertain. *Aponotoreas* has both diurnal and nocturnal species and is shared with New Zealand. *Melitulias* includes shrub-feeding species with androconial patches. Close to this genus is a complex of species lacking androconia and allocated by early authors to Holarctic genera such as *Hydriomena*.

Chaetolopha is a member of a distinctive austral clade associated with ferns in damp forests (Schmidt, 2002), including bracken (Schmidt, 2016). Its tribal affinities remain unknown.

18.8.7 Subfamily Ennominae

This, the largest and most diverse geometrid subfamily, is distributed globally and involves about 25 tribes accounting for more than half of all species richness in the Geometridae. Historically, ennomines were broadly recognized by the absence or reduction of vein M_2 in the hindwing. The male sternite A3 commonly bears a transverse comb of setae. The female ductus bursae is often sclerotized and fluted; a stellate signum is common (Holloway, 1993). Genetic evidence offers good support for the monophyly of Ennominae (Murillo-Ramos *et al.*, 2019).

Confidence in the monophyly of the tribes varies: many lack a single convincing apomorphy but are defined on clusters of broadly shared features. Half the ennomine tribes present in Australia are represented by a few outliers or 'edge of range' Asian genera in the Wet Tropics. Modern treatments of large elements of the austral fauna include Young (2006a, b, 2008a, b) for 'Nacophorini' (now Diptychini).

The first four tribes listed below are members of the informal 'boarmiine' lineage based, in part, on the presence of a fovea in the male forewing and a bifid pupal cremaster.

18.8.7.1 Abraxini

This is a near cosmopolitan group with minor representation in tropical Queensland: *Abraxas* (3 spp.), *Cassephyra* (1 sp.) and *Heterostegane* (1 sp.). A black and white wing pattern is common in the group, often contrasting with a yellow abdomen. *Abraxas* is a model for mimetic moths in Asia, such as the epicopeiid *Parabraxas*. This tribe is sister to the Macariini which together are related to the Boarmiini (Zheng *et al.*, 2022).

18.8.7.2 Eutoeini

The Eutoeini is a palaeotropical clade, possibly close to the Abraxini, with the male valvae strongly divided in most species. *Luxiaria*, *Nadagarodes*, *Probithia* and *Zeheba* contribute one or two species each to the fauna of north-east Queensland.

18.8.7.3 Boarmiini

This is an extensive cosmopolitan group with numerous embedded clades and more to be resolved. At least 37 genera are known to inhabit Australia. Evidence of rapid radiation within Boarmiini comes from polytomies in molecular phylogenies (Wahlberg *et al.*, 2010). Although monophyletic on genetic evidence, a morphological synapomorphy is lacking (Jiang *et al.*, 2017). A fovea is often present in the male forewing. Boarmiine eggs are typically elongated, with polygonal cells arranged in longitudinal rows and a soft chorion. Most larvae are mimics of live or dead twigs or petioles. Diagnostic characters include four lateral setae on the A6 proleg, an absence of extra prolegs, a uniordinal interrupted mesoseries crochet arrangement in first instar larvae and, in the mature larva, a biordinal uninterrupted mesoseries. A fleshy protuberance just dorsad to the spiracle on abdominal segment A2 is commonly present. Noteworthy Australian groups include *Scioglyptis* on *Acacia* and *Exocarpos* (Santalaceae) and various genera on Myrtaceae including *Gastrinodes* on *Eucalyptus*, *Euphronarcha* on *Leptospermum* and *Syneora* mainly on *Melaleuca*. *Lipogya* is common in the semi-arid zone. Polyphagous taxa include some *Cleora*, *Ectropis excursaria*, *Hyposidra talaca*, *Pholodes sinistraria* (Fig. 18.2.6) and *Zermizinga sinuata*, the latter with brachypterous females. Food plant specialists include *Phelotis* on *Exocarpos* and a small tropical Indo-Australian group which feed on basal plant groups, including *Milionia* (2 spp. locally) on conifers and *Craspedosis leucosticta* probably on *Piper* (Piperaceae). The handsome caterpillars of *Bracca* (Fig. 18.2.7) feed on primitive rainforest angiosperms including *Austrobaileyana* (*B. rotundata*; De Baar, 2008) (Austrobaileyaceae) and *Wilkiea huegeliana* (*B. matutinata*) (Monimiaceae). Other sympatric genera with a single representative in north Queensland include *Praesos* and *Kranandra*.

Widespread in the Old World, *Cleora* in Australia involves about 20 species of strong-flying moths and includes some with highly variable intraspecific phenotypes. A small ally of *Cleora displicata* extends inland to the semi-arid woodlands on *Acacia* but may be part of an unresolved larger complex including *C. injectaria*. The genus reaches islands in the south west Pacific such as Norfolk Island and Fiji; *C. illustraria* may have recently crossed Bass Strait to colonize northern Tasmania where it feeds on blackwood trees, *Acacia melanoxylon*.

Amblychia (2 spp.) and *Chorodna strixaria*, both outliers of genera richest in the Oriental tropics, are unusually large boarmiines. *Scioglyptis* (6+ spp.) includes the largest examples in the temperate zone, whereas *Phelotis cognata* and the mainly semi-arid *Lipogya* (4 spp.) are among the smallest.

18.8.7.4 Macariini

A widespread tribe, but probably of African origin, the macariines are medium-sized moths in which the male valva is deeply divided, a pair of horn-like setae often adorn the uncus (absent in *Isturgia* and *Paramelora*) and the chaetosemata are elongated in most members (Scoble and Krüger, 2002). Macariine eggs are distinctive. Their shape ranges from narrow and elongated to broader and subelliptical. The chorion is usually densely pitted or coralline and mesh-like, overlain by hexagonal, concave cells with moderately broad, elevated and double-ridged walls. The aeropyles are large, elevated domes with small openings and are absent from the middle of the top side. Although rather species poor here, the tribe includes some of the most common and widespread Australian geometrids. These are often associated with pioneer plants enriched in nitrogen such as *Acacia* and *Dodonaea* (Sapindaceae). *Parosteodes* is widespread on *Dodonaea* hosts whereas *Isturgia penthearia* is on *Acacia*. *Psilosticha mactaria*, which feeds on *Pultenaea* (Fabaceae), may belong here (Young, 2006a). *Paramelora zophodesma* is common inland in shrublands featuring *Acacia* and *Senna* (Fabaceae) but the broadly sympatric *Paramelora lynchota* is misplaced in this genus. *Chiasmia* is an extensive Old World genus on Fabaceae including *Acacia*; one species, *C. gratularia*, extends well inland.

18.8.7.5 Caberini

Casbia is the largest and most widespread caberine genus in Australia and is locally common in temperate woodlands and forests. *Petelia* occupies the more tropical latitudes. Many caberines feed on fast-growing Rhamnaceae such as *Pomaderris* and *Alphitonia*. The larvae usually rest at full stretch beneath the leaves.

18.8.7.6 Gonodontini

Gonodontis (5 spp. in Australia) is a sexually dimorphic, mesic group which extends to south-east Asia. The larva of *G. luteola*, featuring a prognathous head and a dorsal pair of diverging fleshy tentacles on segment A2, feeds on *Flindersia australis* (Hendry and Moss, 2011). Unusually, *G. stramenticea* occupies the eastern arid zone. The rare *Proboloptera* (1 sp.) is endemic to south-east Australia. Various unrelated woody plant genera are reported as hosts across the tribe.

18.8.7.7 Hypochrosini

The Hypochrosini are an Old World group, typically with the underside brightly coloured in red or yellow. Five genera reach north Queensland where *Corymica* is the most commonly seen. *Ctimene* (1 or 2 spp.), also in the Top End, is diurnal. Foodplants in South-east Asia include Fabaceae, Rubiaceae and Sterculiaceae.

18.8.7.8 Diptychini

This large tribe, as recently redefined (Sihvonen *et al.*, 2015; Murillo-Ramos *et al.*, 2019), accounts for about 70 genera in the Australian fauna and is especially rich in the south. Sclerophyllous Myrtaceae (*Eucalyptus*, *Corymbia*) and Fabaceae (*Acacia* and various mirbelioids), but generally not Proteaceae, are important food plants for this group.

Murillo-Ramos *et al.* (2019) folded the Lithinini into the Diptychini on genetic evidence. A near cosmopolitan group, including New Zealand (Weintraub and Scoble, 2004), 'lithinines' are often associated with ferns as food plants, including bracken (Holloway, 1987). *Idiodes* is the most common genus and dead-leaf mimicry is typical of the adults. Their larvae are cryptic on fern fronds but upon disturbance drop from their host plant and agitate or thrash violently if further provoked. The transfer to this tribe of *Acalyphes* and *Dirce* (Fig. 18.1.12), colourful diurnal geometrids superficially resembling Holarctic Archiearinae, was first proposed by Young (2006b, 2008b).

Morphological and biological revisions are available for *Thalaina* (Figs 18.1.13, 18.2.13), *Plesanemma* (Fig. 18.2.10), *Mnesampela* (Fig. 18.2.9) and *Paralaea* (Fig. 18.2.12) (McQuillan, 1981, 1984, 1985; McQuillan *et al.*, 2001), *Palleopa* (Young and McQuillan, 2001) and *Archephanes* (Young and McQuillan, 2003). Eggs and larvae of multiple genera, including *Fisera* (Fig. 18.2.11), were illustrated in monochrome by McFarland (1988) and in more detail by Young (2006b, 2008b). *Chlenias* (Fig. 18.2.14), *Paralaea* and *Mnesampela* include occasional outbreak species (Elliott *et al.*, 1980).

18.8.7.9 Plutodini

Of uncertain affinity, these are small but distinctive yellow moths with bold reddish-brown blotches which occupy the Palaeotropics. *Plutodes* (1 sp.) is sparingly represented in Australia.

18.8.7.10 Scardamiini

Two colourful genera of this small African and Indo-Australian tribe inhabit wet forests in north-east NSW and Queensland. *Aplochlora pisichroa* is green and *Scardamia* spp. (3 spp.) are shades of orange. Larvae are expected to be on Salicaceae such as *Flacourtia*.

18.8.7.11 Unassigned to tribe

Various ennomine genera are too poorly studied to allocate them to existing tribes with confidence. Some examples include *Coelocrossa*, *Neritodes*, *Picrophylla* and *Peridelias*. *Xenomusa* (Fig. 18.1.15) has two species in eastern wet forests with the colourful larva of *X. monoda* (Fig. 18.2.15) on *Hedycarya angustifolia* (Monimiaceae).

18.9 Frontiers

Lepidoptera have historically played an important role in the framing and testing of ideas in co-evolution, ecology and biogeography. As a biodiverse clade, Geometridae also recommend themselves as models for probing ecological questions at multiple scales, as they may be second only to butterflies in terms of ease of identification, sampling tractability and life history knowledge. Recent examples include an exploration of the role of elevation and food plants via the turnover in macromoths at intervals in the Eungella Range, central Queensland (Odell *et al.*, 2016) and global meta-studies which include Australian datasets (Beck *et al.*, 2017). Rainforest geometrids in Papua New Guinea were used as a model group to show that elevational gradients are important for maintaining high regional biodiversity and to allow for movement of species in response to climate change (Toko *et al.*, 2023). Geometrids are likely to be influential in food webs and new technologies are making it possible to document their importance as both herbivores and prey (e.g. Wenda *et al.*, 2023).

Australian geometrids also present a special case study in the assembly of a remote continental fauna, having evolved largely in isolation until the Sahul–Sunda collision created the opportunity for a biotic exchange with Asia about 20 mya. Floristically, there was strong asymmetry in the exchange in favour of eastwards migrations of Asian lineages into Australia (Crayn *et al.*, 2015) and the consequences of this are also visible in the Lepidoptera. However, with the exceptions of Holloway (1976, 1977, 1979), few studies linking vegetation communities to their local Geometridae have been undertaken to date. But it is an apparent paradox that the relatively modest diversity of looper moths in south-western Australia does not reflect the global hotspot status of the sympatric flora. As the need for habitat restoration rises as an environmental priority, such information will be necessary to guide good decision making.

The Australian fauna also offers an independent test-bed for evaluating new ideas such as the influence of geometrid phylogeny on trait characters, previously based on northern hemisphere evidence alone (e.g. Higginson *et al.*, 2012).

18.10 Conservation Issues

Geometrids are exposed to the same range of environmental threats to their survival as most other phytophagous insects. Their sensitivity to these challenges, however, varies among species and groups (New, 2023). Clarity in the alpha taxonomy in some groups is lacking but is necessary to propose good conservation management practices, as modelled by the study of Patrick *et al.* (2010) for a xanthorhoine species complex in New Zealand.

Rare and declining Lepidoptera are often associated with particular threatened habitats such as saltmarshes, mangroves and alpine meadows. Threatening processes can also act on species indirectly through the decline of their food plants. Major fire events, increasingly common, threaten native podocarps (Khan *et al.*, 2023) and other palaeo-austral plants which host rare and unusual geometrids. Habitat loss and fragmentation have degraded the prospects of species recolonizing habitat from nearby source populations. Land use changes are implicated in some cases (McQuillan *et al.*, 2024). Tasmania has several Geometridae formally listed as threatened but species in other states need to be formally assessed.

Like much of the Australian biota, moths are highly exposed to the impacts of megafires (Godfree *et al.*, 2021) and the vagaries of rampant 'hazard reduction burning', now the dominant paradigm in fire risk management, especially in the extensive periurban zone where fire-related anxiety is high. Further away, the immolation of formerly fire-resistant or sensitive vegetation types by increasingly frequent dry lightning (Styger *et al.*, 2018) is eroding insect-rich habitats and replacing them with simpler habitats more tolerant of frequent fire. Most Australian habitats are progressively drying out (Jones and Ricketts, 2024) as evaporation rates increase. Unfortunately, these trends and outcomes are running far ahead of our ability to understand the consequences for the Lepidoptera fauna.

Saltmarsh and mangrove geometrids are unable to move inland as sea level rises if their specialized food plants are bounded by infrastructure, coastal hardening or agricultural development. Extensive development on former coastal wetlands has resulted in local authorities applying broad-spectrum insecticides to vegetation for mosquito control.

The clearing of native vegetation proceeds apace in Australia at the present time and, beyond habitat loss and fragmentation, represents an irreversible loss of new and essential knowledge about the native fauna. Protected areas in Australia are biased away from fertile landscapes suitable for intensive agriculture and are, arguably, overrepresented on unproductive land (Engert *et al.* 2023).

Acknowledgements

We are grateful for improvements suggested by Dr Robert Hoare, Landcare Research, New Zealand. We thank the photographers who suppled us with images for the figures – individuals are named in the figure legends. Sarah Fayed assisted with the figures.

Note

The chapter author Catherine J. Byrne also publishes under the name C.J. Young.

References

Adeleye, M.A., Hopf, F., Haberle, S.G., Stannard, G.L., McWethy, D.B. *et al.* (2024). Landscape burning facilitated Aboriginal migration into Lutruwita/Tasmania 41,600 years ago. *Science Advances* 10, eadp6579.

Amoo, M., Salzmann, U., Pound, M.J., Thompson, N. and Bijl, P.K. (2022) Eocene to Oligocene vegetation and climate in the Tasmanian Gateway region were controlled by changes in ocean currents and p CO2. *Climate of the Past* 18, 525–546.

Anderson, E. (1902) Notes on the moths *Gastrophora henricaria* and *Phallaria ophiusaria*, Gn. *Victorian Naturalist* 19, 59–62.

Angas, G.F. (1847) *South Australia Illustrated*. Thomas M'Lean, London.

Ban, X, Jiang, N., Cheng, R., Xue, D. and Han, H. (2018) Tribal classification and phylogeny of Geometrinae (Lepidoptera: Geometridae) inferred from seven gene regions. *Zoological Journal of the Linnean Society* 184, 653–672.

Bazinet, A.L., Cummings, M.P., Mitter, K.T. and Mitter, C.W. (2013) Can RNA-Seq resolve the rapid radiation of advanced moths and butterflies (Hexapoda: Lepidoptera: Apoditrysia)? An exploratory study. *PLOS One* 8, e82615.

Beck, J., McCain, C.M., Axmacher, J.C., Ashton, L.A., Bärtschi, F. *et al.* (2017) Elevational species richness gradients in a hyperdiverse insect taxon: a global meta-study on geometrid moths. *Global Ecology and Biogeography* 26, 412–424.

Bendib, A. (2001) Dorsal sex pheromone glands in female Geometridae (Lepidoptera: Geometroidea): A new apomorphy of the family. *European Journal of Entomology* 98, 195–199.

Blackall, A.G., Mackay, D.A. and Whalen, M.A. (2023) Impact of reserve area on reproduction of a moth-pollinated *Stackhousia* Sm. (Celastraceae) species in a fragmented landscape. *Austral Ecology* 48, 2053–2075.

Bryceson, S.R., Hemming, K.T.M., Duncan, R.P. and Morgan, J. (2023) The contemporary distribution of grasses in Australia: A process of immigration, dispersal and shifting dominance. *Journal of Biogeography* 50, 1639–1652.

Byrne, M. (2008) Evidence for multiple refugia at different time scales during Pleistocene climatic oscillations in southern Australia inferred from phylogeography. *Quaternary Science Reviews* 27, 2576–2585.

Byrne, M. and Murphy, D.J. (2020) The origins and evolutionary history of xerophytic vegetation in Australia. *Australian Journal of Botany* 68, 195–207.

Byrne, M., Steane, D.A., Joseph, L., Yeates, D.K., Jordan, G.J. *et al.* (2011) Decline of a biome: evolution, contraction, fragmentation, extinction and invasion of the Australian mesic zone biota. *Journal of Biogeography* 38, 1635–1656.

Cardillo, M. and Pratt, R. (2013) Evolution of a hotspot genus: Geographic variation in speciation and extinction rates in *Banksia* (Proteaceae). *BMC Evolutionary Biology* 13, 155.

Carpenter, R.J., Jordan, G.J., Mildenhall, D.C. and Lee, D.E. (2011) Leaf fossils of the ancient Tasmanian relict *Microcachrys* (Podocarpaceae) from New Zealand. *American Journal of Botany* 98, 1164–1172.

Carpenter R.J., Jordan, G.J. and Hill, R.S. (2016) Fossil leaves of *Banksia*, Banksieae and pretenders: Resolving the fossil genus *Banksieaephyllum*. *Australian Systematic Botany* 29, 126–141.

Common, I.F.B. (1990) *Moths of Australia*. Melbourne University Press, Carlton, Victoria.

Cook, M.A. and Scoble, M.J. (1992) Tympanal organs of geometrid moths: A review of their morphology, function, and systematic importance. *Systematic Entomology* 17, 219–232.

Cook, M.A., Harwood, L.M., Scoble, M.J. and McGavin, G.C. (1994) The chemistry and systematic importance of the green wing pigment in emerald moths (Lepidoptera: Geometridae, Geometrinae). *Biochemical Systematics and Ecology* 22, 43–51.

Crayn, D.M., Costion, C. and Harrington, M.G. (2015) The Sahul–Sunda floristic exchange: dated molecular phylogenies document Cenozoic intercontinental dispersal dynamics. *Journal of Biogeography* 42, 11–24.

Das, S., Baumgartner, J.B., Esperon-Rodriguez, M., Wilson, P.D., Yap, J-Y.S. *et al.* (2019) Identifying climate refugia for 30 Australian rainforest plant species, from the last glacial maximum to 2070. *Landscape Ecology* 34, 2883–2896.

De Baar, M. (2008) Life history notes on *Bracca rotundata* and *Oxycophina theorina*. *Australian Entomologist* 35, 141–143.

Ding, W., Xu, H., Wu, Z., Hu, L., Huang, L. *et al.* (2023) The mitochondrial genomes of the Geometroidea (Lepidoptera) and their phylogenetic implications. *Ecology and Evolution* 13, e9813.

Dröllner, M., Barham, M., Kirkland, C.L., Danišík, M., Bourdet, J. *et al.* (2023) Directly dating Plio-Pleistocene climate change in the terrestrial record. *Geophysical Research Letters* 50, e2023GL102928.

Dugdale, J.S. (1980) Australian Trichopterygini (Lepidoptera: Geometridae) with descriptions of eight new taxa. *Australian Journal of Zoology* 28, 301–40.

Dugdale, J.S., Emmerson, A.W. and Hoare, R.J.B. (2023) *Declana* and *Ipana* (Insecta: Lepidoptera: Geometridae: Ennominae). *Fauna of New Zealand* 82, 1–120.

Elliott, H.J. and Bashford, R. (1978) The life history of *Mnesampela privata* (Guen.) (Lepidoptera: Geometridae), a defoliator of young eucalypts. *Journal of the Australian Entomological Society* 17, 201–204.

Elliott, H.J., Bashford, R. and Palzer, C. (1980) Biology of *Stathmorrhopa aphotista* Turner (Lepidoptera: Geometridae), a defoliator of *Eucalyptus* spp. in southern Tasmania. *Australian Forestry* 43, 81–86.

Engel, J., Woodhead, J., Hellstrom, J., White, S., White, N. and Green, H. (2020) Using speleothems to constrain late Cenozoic uplift rates in karst terranes. *Geology* 48, 755–760.

Engert, J.E., Pressey, R.L. and Adams, V.M. (2023) Threatened fauna protections compromised by agricultural interests in Australia. *Conservation Letters* 16, e12975.

Felder, R. and Rogenhofer, A. F. (1875) *Reise der österreichischen Fregatte Novara um die Erde* (Zool.). 2(Abt. 2) (5): pls 121–140, Wien.

Ghanavi, H.R., Chazot, N., Sanmartín, I., Murillo-Ramos, L., Duchêne, S. *et al.* (2025) Region-specific diversification dynamics and biogeographic history of one of the most diverse families of insects. *Systematic Entomology* 50, 206–220.

Godfree, R.C., Knerr, N., Encinas-Viso, F., Albrecht, D., Bush, D. *et al.* (2021) Implications of the 2019–2020 megafires for the biogeography and conservation of Australian vegetation. *Nature Communications* 12, 1023.

Goldfinch, G.M. (1929) Revision of Australian Geometridae (Lepidoptera). *Proceedings of the Linnean Society of New South Wales* 54, 378–407.

Goldfinch, G.M. (1944) Notes on Australian Boarmiidae and Oenochromidae (Lepidoptera) with descriptions of new species. *Proceedings of the Linnean Society of New South Wales* 69, 189-197.

Guenée, A. (1857) [1858] Uranides et Phalénites, Tome I. et II. In: Boisduval, J.-B.A.D. and Guenée, A. (eds) *Histoire Naturelle des Insectes. Species Général des Lépidoptères, 9: [I]-LVI + 1–514; 10: 1–584.* Roret, librairie-édit., Paris.

Guest, E. (1887) A classified list of Geometrina found around Balhannah, with notes on species. *Transactions and Proceedings of the Royal Society of South Australia* 9, 126–141.

Hendry, P. and Moss, J. (2011) Life history notes on the moth *Gonodontis luteola* (Turner, 1904) (Lepidoptera: Geometridae) and *Speiredonia spectans* (Guenée, 1852) (Lepidoptera: Noctuidae). *Metamorphosis Australia* 61, 10–13.

Hermsen, E., Gandolfo, M. and Zamaloa, M.C. (2012) The fossil record of *Eucalyptus* in Patagonia. *American Journal of Botany* 99, 1356–1374.

Hewish, M., Marriott, P., Edwards, T., Kallies, A. and Williams, S. (2014) *Moths of Victoria: Part 5, Satin Moths and Allies – Geometroidea (A).* Entomological Society of Victoria, Melbourne, Victoria.

Hewish, M., Marriott, P., Edwards, T., Kallies, A., Williams, S. and Byrne, C. (2016) *Moths of Victoria: Part 7, Bark Moths and Allies – Geometroidea (D).* Entomological Society of Victoria, Melbourne, Victoria.

Higginson, A.D., De Wert, L., Rowland, H.M., Speed, M.P. and Ruxton, G.D. (2012) Masquerade is associated with polyphagy and larval overwintering in Lepidoptera. *Biological Journal of the Linnean Society* 106, 90–103.

Holloway, J.D. (1976) *Moths of Borneo with Special Reference to Mt Kinabalu.* Malaysian Nature Society, Kuala Lumpur.

Holloway, J.D. (1977) *The Lepidoptera of Norfolk Island, their Biogeography and Ecology.* Series Entomologica, 13. W. Junk, The Hague, The Netherlands.

Holloway, J.D. (1979) *A Survey of the Lepidoptera, Biogeography and Ecology of New Caledonia.* Series Entomologica, 15. W. Junk, The Hague, The Netherlands.

Holloway, J.D. (1987) Bracken-feeding Geometridae in the genus *Idiodes* Guenée, 1857, and allied taxa in the tribe Lithinini (Lepidoptera). *Tinea* 12 (supplement), 242–248.

Holloway, J.D. (1994) The Moths of Borneo. Part 11. Family Geometridae. Subfamily Ennominae. *Malayan Nature Journal* 47, 1–309.

Holloway, J.D. (1996) The Moths of Borneo Part 9: Family Geometridae, subfamilies Oenochrominae, Desmobathrinae and Geometrinae. *Malayan Nature Journal* 49, 147–326.

Holloway, J.D. (1997) The Moths of Borneo Part 10: Geometridae: Sterrhinae, Larentiinae, Addenda to other subfamilies. *Malayan Nature Journal* 51, 1–242.

Jensen, L.M., Wallis, I.R. and Foley, W.J. (2015) The relative concentrations of nutrients and toxins dictate feeding by a vertebrate browser, the greater glider *Petauroides volans*. *PLOS One* 10, e0121584.

Jiang, N., Li, X., Hausmann, A., Cheng, R., Xue, D. and Han, H. (2017) A molecular phylogeny of the Palaearctic and Oriental members of the tribe Boarmiini (Lepidoptera: Geometridae: Ennominae). *Invertebrate Systematics* 31, 427–441.

Jones, R.N. and Ricketts, J.H. (2024) Comparing observed and projected changes in Australian fire climates. *Fire* 7, 113.

Jordan, G.J., Harrison, P.A., Worth, J.R.P., Williamson, G.J. and Kirkpatrick, J.B. (2015) Palaeoendemic plants provide evidence for persistence of open, well-watered vegetation since the Cretaceous. *Global Ecology and Biogeography* 25, 127–140.

Kang, C.K., Stevens, M., Moon, J.Y., Lee, S.I. and Jablonski, P.G. (2015) Camouflage through behavior in moths: the role of background matching and disruptive coloration. *Behavioural Ecology* 26, 45–54.

Khan, R., Hill, R.S., Liu, J. and Biffin, E. (2023) Diversity, distribution, systematics and conservation status of Podocarpaceae. *Plants* 12(5), 1171.

Knight, C.L. and Wilf, P. (2013). Rare leaf fossils of Monimiaceae and Atherospermataceae (Laurales) from Eocene Patagonian rainforests and their biogeographic significance. *Palaeontologia Electronica* 16(3), 26A.

Lamont, B.B., He, T., Milne, L.A. and Cowling, R.M. (2024) Out of Africa: linked continents, overland migration and differential survival explain abundance of Proteaceae in Australia. *Perspectives in Plant Ecology, Evolution and Systematics* 62, 125778.

Lees, D.C. and Smith, N.G. (1991) Food plant associations of the Uraniinae (Uraniidae) and their systematic, evolutionary, and ecological significance. *Journal of the Lepidopterists' Society* 45, 296–347.

Linnaeus, C. (1758) *Systema Naturae*, 10th edn. Laurentius Salvius, Stockholm, Sweden.

Lower, O.B. (1905) Descriptions of new Australian Lepidoptera, with synonymic notes. No XXIII. *Transactions of the Royal Society of South Australia* 29, 173–180.

Lower, O.B. (1918) The Lepidoptera of Broken Hill, New South Wales. IV. *Transactions of the Royal Society of South Australia* 42, 226–239.

Lucas, T.P. (1888) Contribution to a knowledge of the genus *Iodis*. *Proceedings of the Linnean Society of New South Wales* (Series 2) 3, 1263–1269.

Macphail, M.K. and Tornhill, A.H. (2016) How old are the eucalypts? A review of the microfossil and phylogenetic evidence. *Australian Journal of Botany* 64, 579–599.

Madden J.L. and Bashford R. (1977a) The life history and behaviour of *Chlenias* sp., a geometrid defoliator of Radiata Pine in Tasmania. *Journal of the Australian Entomological Society* 16, 371–378.

Madden J.L. and Bashford R. (1977b) Population biology of *Chlenias* sp., a geometrid defoliator of *Pinus radiata* in Tasmania. *Journal of the Australian Entomological Society* 16, 379–388.

Marriott, P. (2011) *Moths of Victoria: Part 3, Waves and Carpets – Geometroidea (A)*. Entomological Society of Victoria, Melbourne, Victoria.

Marriott, P. (2012) *Moths of Victoria: Part 4, Emeralds and Allies – Geometridae (B)*. Entomological Society of Victoria, Melbourne, Victoria.

McFarland, N. (1973) Some observations on the eggs of moths and certain aspects of first instar larval behaviour. *Journal of Research on the Lepidoptera* 12, 199–208.

McFarland, N. (1988) *Portraits of South Australian Geometrid Moths*. Allen Press, Lawrence, Kansas.

McQuillan, P.B. (1981) A review of the Australian moth genus *Thalaina* (Lepidoptera: Geometridae: Ennominae). *Transactions of the Royal Society of South Australia* 105, 1–23.

McQuillan, P.B. (1984) A new genus for *Chlenias fucata* Felder and Rogenhofer (Lepidoptera: Geometridae: Ennominae) and a new Tasmanian species. *Journal of the Australian Entomological Society* 23, 297–306.

McQuillan, P.B. (1985) A taxonomic revision of the Australian autumn gum moth genus *Mnesampela* Guest (Lepidoptera: Geometridae, Ennominae). *Entomologica Scandinavica* 16, 175–202.

McQuillan, P.B. (1986) Trans-Tasman relationships in the highland moth (Lepidoptera) fauna. In: Barlow, B.A. (ed.) *Flora and Fauna of alpine Australasia ages and origins*. CSIRO/Leiden: E.J. Brill, Australia, pp. 265–276.

McQuillan, P.B. (1996) The Tasmanian geometrid moths associated with the genus *Amelora* auctorum (Lepidoptera: Geometridae: Ennominae). *Invertebrate Taxonomy* 10, 433–506.

McQuillan, P.B. (2004) An overview of the Tasmanian geometrid moth fauna and its conservation status. *Journal of Insect Conservation* 8, 209–220.

McQuillan, P.B. and Edwards, E.D. (1996) Geometridae. In: Nielsen, E.S., Edwards, E.D. and Rangsi, V. (eds) *Checklist of the Lepidoptera of Australia*. Monographs on Australian Lepidoptera 4, CSIRO, Collingwood, Victoria, pp. 220–228, 355–357.

McQuillan, P.B., Taylor, R.J., Bereton, R.N. and Cale, P.G. (1998) Seasonal patterns of activity in geometrid moths (Lepidoptera: Geometridae) from a lowland and highland eucalypt forest in Tasmania. *Australian Journal of Entomology* 37, 228–237.

McQuillan, P.B., Young, C.J. and Richardson, A.M.M. (2001) A revision of the Australian moth genus *Paralaea* Guest (Lepidoptera: Geometridae: Ennominae). *Invertebrate Taxonomy* 15, 277–317.

McQuillan, P.B., Edwards, E.D. and Camilleri, J. (2023) Oswald Bertram Lower (1864–1925): A South Australian pioneer in the discovery of Australia's biodiversity. *Historical Records of Australian Science* 35, 1–15.

McQuillan, P.B, Schmidt, O., Byrne, C.J. and Warren, M. (2024) Rediscovery and systematics of the rare brown-caped carpet moth *Chrysolarentia excentrata* (Guenée, 1857 [1858]) comb. nov. (Lepidoptera: Geometridae: Larentiinae) in Tasmania. *Papers and Proceedings of the Royal Society of Tasmania* 158, 79–86.

Meyrick, E. (1888) Revision of Australian Lepidoptera. II. *Proceedings of the Linnean Society of New South Wales* (Series 2) 2, 835–928.

Meyrick, E. (1890) Revision of Australian Lepidoptera. III. *Proceedings of the Linnean Society of New South Wales* (Series 2) 4, 1117–1216.

Meyrick, E. (1891) Revision of Australian Lepidoptera IV. *Proceedings of the Linnean Society of New South Wales* (Series 2) 5, 791–879.

Meyrick, E. (1892) Revision of Australian Lepidoptera V. *Proceedings of the Linnean Society of New South Wales* (Series 2) 6, 581–678.

Miller, J.T., Murphy, D.J., Ho, S.Y.W., Cantrill, D.J. and Seigler, D. (2013) Comparative dating of *Acacia*: combining fossils and multiple phylogenies to infer ages of clades with poor fossil records. *Australian Journal of Botany* 61, 436–445.

Murillo-Ramos, L., Brehm, G., Sihvonen, P., Hausmann, A, Holm, S. *et al.* (2019) A comprehensive molecular phylogeny of Geometridae (Lepidoptera) with a focus on enigmatic small subfamilies. *PeerJ* 7, e7386.

Murillo-Ramos, L., Chazot, N., Sihvonen, P., Õunap, E., Jiang, N. *et al.* (2021a) Molecular phylogeny, classification, biogeography and diversification patterns of a diverse group of moths (Geometridae: Boarmiini). *Molecular Phylogenetics and Evolution* 162, 107–198.

Murillo-Ramos, L., Friedrich, E., Williams, S., Wahlberg, N., Brehm, G. and Sihvonen, P. (2021b) A morphological appraisal of the new subfamily Epidesmiinae (Lepidoptera: Geometridae) with an overview of all geometrid subfamilies. *Zoological Journal of the Linnean Society* 193, 1205–1233.

Murillo-Ramos, L., Twort, V., Wahlberg, N. and Sihvonen, P. (2023) A phylogenomic perspective on the relationships of subfamilies in the family Geometridae (Lepidoptera). *Systematic Entomology* 48, 618–632.

Narango, D.L., Tallamy, D.W. and Shropshire, K.J. (2020) Few keystone plant genera support the majority of Lepidoptera species. *Nature Communications* 11(1), 5751. Erratum: *Nature Communications* 2021 12(1), 1062.

New, T.R. (2023) *The Other Lepidoptera: Moth Conservation in Australia*. Springer, Cham, Switzerland.

Nokelainen, O., Silvasti, S.A., Strauss, S.Y., Wahlberg, N. and Mappes, J. (2024) Predator selection on phenotypic variability of cryptic and aposematic moths. *Nature Communications* 15, 1678.

Odell, E.H., Ashton, L.A. and Kitching, R.L. (2016) Elevation and moths in a central eastern Queensland rainforest. *Austral Ecology* 41, 133–144.

Õunap, E., Viidalepp, J. and Truuverk, A. (2016) Phylogeny of the subfamily Larentiinae (Lepidoptera: Geometridae): Integrating molecular data and traditional classifications. *Systematic Entomology* 41, 824–843.

Õunap E., Nedumpally V., Yapar E., Lemmon A.R. and Tammaru T. (2024) Molecular phylogeny of north European Geometridae (Lepidoptera: Geometroidea). *Systematic Entomology* 50, 32–67.

Parker, S.A. (1966) Albert S. Meek's collecting-locality on the Cape York Peninsula, 1898. *Emu* 66, 121–122.

Patrick, B., Hoare, R. and Rhode, B. (2010) Taxonomy and conservation of allopatric moth populations: A revisionary study of the *Notoreas perornata* Walker complex (Lepidoptera: Geometridae: Larentiinae), with special reference to southern New Zealand. *New Zealand Journal of Zoology* 37, 257–283.

Pierce, F.N. (1914) *The Genitalia of the Group Geometridae of the Lepidoptera of the British Islands: An Account of the Morphology of the Male Clasping Organs and the Corresponding Organs of the Female*. F.N. Pierce, Liverpool.

Pitkin, L.M., Han, H. and James, S. (2007) Moths of the tribe Pseudoterpnini (Geometridae: Geometrinae): a review of the genera. *Zoological Journal of the Linnean Society* 150, 334–412.

Pole, M.S. (2007) Early Eocene dispersed cuticles and mangrove to rainforest vegetation at Strahan-Regatta Point, Tasmania. *Palaeontologia Electronica* 10, 1–66.

Prout, L.B. (1910) Lepidoptera Heterocera, Fam. Geometridae, Subfam. Oenochrominae. *Genera Insectorum* 104. 120 pp.

Prout, L.B. (1933) The Indoaustralian Geometridae. In: Seitz, A. (ed.) (1920–1941) *The Macrolepidoptera of the World 12*. Verlag A. Kernen, Stuttgart, Germany, pp. 1–356.

Rajaei, H., Greve, C., Letsch, H., Stüning, D., Wahlberg, N. *et al.* (2015) Advances in Geometroidea phylogeny, with characterization of a new family based on *Pseudobiston pinratanai* (Lepidoptera, Glossata). *Zoologica Scripta* 44(4), 418–436.

Rajaei, H., Hausmann, A., Scoble, M., Wanke, D., Plotkin, D. *et al.* (2022) An online taxonomic facility of Geometridae (Lepidoptera), with an overview of global species richness and systematics. *Integrative Systematics* 5, 145–192.

Regier, J.C., Mitter, C., Zwick, A., Bazinet, A.L., Cummings, M.P. *et al.* (2013) A large-scale, higher-level, molecular phylogenetic study of the insect order Lepidoptera (moths and butterflies). *PLOS One* 8, e58568.

Rix, M.G., Edwards, D.L., Byrne, M., Harvey, M.S., Joseph, L. and Roberts, J.D. (2015) Biogeography and speciation of terrestrial fauna in the south-western Australian biodiversity hotspot. *Biological Reviews* 90, 762–793.

Sandiford, M., Wallace, M. and Coblentz, D. (2004) Origin of the *in situ* stress field in south-eastern Australia. *Basin Research* 16, 325–338.

Sarto i Monteys, V., Hausmann, A., Solórzano Kraemer, M.M., Hammel, J., Baixeras, J. *et al.* (2022) A new fossil inchworm moth discovered in Miocene Dominican amber (Lepidoptera: Geometridae). *Journal of South American Earth Sciences* 120, 104055.

Schmidt, O. (2002) A revision of the genus *Chaetolopha* Warren (Lepidoptera: Geometridae: Larentiinae) with a description of *Parachaetolopha*, gen. nov. *Invertebrate Systematics* 16, 703–733.

Schmidt, O. (2013) The structure and musculature of male terminalia in the tribe Xanthorhoini Pierce and related tribes (Lepidoptera: Geometridae: Larentiinae), with particular reference to the Palaearctic and Australian regions. *Zootaxa* 3721, 552–572.

Schmidt, O. (2016) Larval food plants of Australian Larentiinae (Lepidoptera: Geometridae) – a review of available data. *Biodiversity Data Journal* 4, e7938.

Scoble M.J. (ed.) (1999) *Geometrid Moths of the World – a Catalogue (Lepidoptera, Geometridae).* CSIRO Publishing, Collingwood, Victoria.
Scoble M.J. and Edwards E.D. (1990) *Parepisparis* Bethune Baker and the composition of the Oenochrominae (Lepidoptera: Geometridae). *Entomologica Scandinavica* 20, 371–399.
Scoble, M.J. and Krüger, M. (2002) A review of the genera of Macariini with a revised classification of the tribe (Geometridae: Ennominae). *Zoological Journal of the Linnean Society* 134, 257–315.
Sihvonen, P. and Kaila, L. (2004) Phylogeny and tribal classification of Sterrhinae with emphasis on delimiting Scopulini. *Systematic Entomology* 29, 324–358.
Sihvonen, P., Mutanen, M., Kaila, L., Brehm, G., Hausmann, A. and Staude, S.H. (2011) Comprehensive molecular sampling yields a robust phylogeny for geometrid moths (Lepidoptera: Geometridae). *PLOS One* 6(6), e20356.
Sihvonen, P., Staude, H.S. and Mutanen, M. (2015) Systematic position of the enigmatic African cycad moths: An integrative approach to a nearly century old problem (Lepidoptera: Geometridae, Diptychini). *Systematic Entomology* 40, 606–627.
Sihvonen, P., Murillo-Ramos, L., Brehm, G., Staude, H. and Wahlberg, N. (2020) Molecular phylogeny of Sterrhinae moths (Lepidoptera: Geometridae): Towards a global classification. *Systematic Entomology* 45, 606–634.
Slee, A. and Shulmeister, J. (2015) The distribution and climatic implications of periglacial landforms in eastern Australia. *Journal of Quaternary Science* 30, 848–858.
Smithers, C.N. (1998) A species list and bibliography of the insects recorded from Norfolk Island. *Technical Reports of the Australian Museum* 13, 1–55.
Steinbauer, M.J. and Matsuki, M. (2004) Suitability of *Eucalyptus* and *Corymbia* for *Mnesampela privata* (Guenée) (Lepidoptera: Geometridae) larvae. *Agricultural and Forest Entomology* 6, 323–332.
Steinbauer, M.J., McQuillan, P.B. and Young, C.J. (2001) Life history and behavioural traits of *Mnesampela privata* that exacerbate population responses to eucalypt plantations: Comparisons with Australian and outbreak species of forest geometrid from the Northern Hemisphere. *Austral Ecology* 26, 525–534.
Steinbauer, M.J., Östrand, F., Bellas, T.E., Nilsson, A., Andersson, F. *et al.* (2004) Identification, synthesis and activity of sex pheromone gland components of the autumn gum moth (Lepidoptera: Geometridae), a defoliator of *Eucalyptus*. *Chemoecology* 14, 217–223.
Styger, J., Marsden-Smedley, J. and Kirkpatrick, J. (2018) Changes in lightning fire incidence in the Tasmanian Wilderness World Heritage Area, 1980–2016. *Fire*, 1, 38.
Tautel, C. and Stüning, D. (2023) *Haemochrosis* gen. n., a new genus of Hypochrosini from Wallacea, with description of a new species from Sulawesi and the first observation of 'reflex bleeding' in the family Geometridae (Lepidoptera Geometridae Ennominae). *Antenor* 9, 111–129.
Thornhill, A.H., Crisp, M.D., Külheim, C., Lam, K.E., Nelson, L.A. *et al.* (2019) A dated molecular perspective of eucalypt taxonomy, evolution and diversification. *Australian Systematic Botany* 32, 29–48.
Tindale, N.B. (1961) A new species of *Chlenias* (Lepidoptera, Boarmiidae) on *Acacia aneura*, with some central Australian native beliefs about it. *Records of the South Australian Museum* 14, 131–140.
Toko, P.S., Koane, B., Molem, K., Miller, S.E. and Novotny, V. (2023) Ecological trends in moth communities (Geometridae, Lepidoptera) along a complete rainforest elevation gradient in Papua New Guinea. *Insect Conservation and Diversity* 16, 649–657.
Turner, A.J. (1904) Revision of Australian Lepidoptera. Family Geometridae. *Proceedings of the Royal Society of Victoria* 16, 218–284.
Turner, A.J. (1907) Revision of Australian Lepidoptera. III. *Proceedings of the Linnean Society of New South Wales* 31, 678–710.
Turner, A.J. (1910) Revision of Australian Lepidoptera. V. *Proceedings of the Linnean Society of New South Wales* 35, 555–653.
Turner, A.J. (1922) Australian Lepidoptera of the group Geometrites. *Transactions of the Royal Society of South Australia* 46, 225–294.
Turner A.J. (1926a) New and little known Tasmanian Lepidoptera. *Papers and Proceedings of the Royal Society of Tasmania* 1925, 81–117.
Turner A.J. (1926b) A revision of the Lepidoptera of Tasmania. *Papers and Proceedings of the Royal Society of Tasmania* 1925, 118–151.
Turner A.J. (1928) A revision of the Lepidoptera of Tasmania. Part II. *Papers and Proceedings of the Royal Society of Tasmania* 1927, 29–65.
Turner, A.J. (1929) Revision of Australian Oenochromidae (Lepidoptera). I. *Proceedings of the Linnean Society of New South Wales* 54, 463–504.

Turner, A.J. (1930) Revision of Australian Oenochromidae (Lepidoptera). II. *Proceedings of the Linnean Society of New South Wales* 55, 1–40.

Turner A.J. (1939) A second revision of the Lepidoptera of Tasmania. *Papers and Proceedings of the Royal Society of Tasmania* 1938, 57–115.

Viidalepp, J. (2011) A morphological review of tribes in Larentiinae (Lepidoptera: Geometridae). *Zootaxa* 3136, 1–44.

Wahlberg, N., Snäll, N., Viidalepp, J., Ruohomäki, K. and Tammaru, T. (2010) The evolution of female flightlessness among Ennominae of the Holarctic forest zone (Lepidoptera, Geometridae). *Molecular Phylogenetics and Evolution* 55, 929–938.

Walker, F. (1856) Noctuidae. *List of the specimens of lepidopterous insects in the collection of the British Museum*, 9, 1–252. British Museum (Natural History), London.

Wallis, I.R., Nicolle, D. and Foley, W.J. (2010) Available and not total nitrogen in leaves explains key chemical differences between the eucalypt subgenera. *Forest Ecology and Management* 260, 814–821.

Wang, H., Holloway, J.D., Wahlberg, N., Wang, M. and Nylin, S. (2019a) Molecular phylogenetic and morphological studies on the systematic position of *Heracula discivitta* reveal a new subfamily of Pseudobistonidae (Lepidoptera: Geometroidea). *Systematic Entomology* 44, 211–225.

Wang, H., Holloway, J.D., Wahlberg, N., Wang, M. and Nylin, S. (2019b) Molecular phylogenetic and morphological studies on the systematic position of *Heracula discivitta* reveal a new subfamily of Pseudobistonidae (Lepidoptera: Geometroidea). *Systematic Entomology* 44, 211–225.

Weintraub, J.D. and Scoble, M.J. (2004) Lithinini (Insecta: Lepidoptera: Geometridae: Ennominae). *Fauna of New Zealand* 49, 48.

Wenda, C., Nakamura, A. and Ashton, L.A. (2023) Season and herbivore defence trait mediate tri-trophic interactions in tropical rainforest. *Journal of Animal Ecology* 92, 466-476.

White, T.C.R. (2015) Senescence-feeders: A new trophic sub-guild of insect herbivores. *Journal of Applied Entomology* 139, 11–22.

Xue, D. and Scoble, M.J. (2002) A review of the genera associated with the tribe Asthenini (Lepidoptera: Geometridae: Larentiinae). *Bulletin of the Natural History Museum. Entomology Series* 71, 77–133.

Yamamoto, S. and Sota, T. (2007) Phylogeny of the Geometridae and the evolution of winter moths inferred from a simultaneous analysis of mitochondrial and nuclear genes. *Molecular Phylogenetics and Evolution* 44(2), 711–723.

Yan, Q., Korty, R., Zhang, Z. and Wang H. (2019) Evolution of tropical cyclone genesis regions during the Cenozoic era. *Nature Communications* 10, 3076.

Young, C.J. (2006a) Molecular relationships of the Australian Ennominae (Lepidoptera: Geometridae) and implications for the phylogeny of the Geometridae from molecular and morphological data. *Zootaxa* 1264, 1–147.

Young, C.J. (2006b) Descriptions of the eggs of some southern Australian Geometridae (Lepidoptera). *Zootaxa* 1287, 3–294.

Young, C.J. (2008a) Characterisation of the Australian Nacophorini using adult morphology, and phylogeny of the Geometridae based on morphological characters. *Zootaxa* 1736, 1–141.

Young, C.J. (2008b) Characterisation of the larvae of Australian Nacophorini. *Zootaxa* 1862, 1–74.

Young, C.J. and McQuillan, P.B. (2001) Redescription of the little-known Australian geometrid moth *Palleopa innotata* Walker, 1866 (Lepidoptera: Geometridae: Ennominae). *Insect Systematics and Evolution* 32, 263–278.

Young, C.J. and McQuillan, P.B. (2003) Redescription and life-cycle of *Archephanes zalosema* Turner, 1926 (Lepidoptera: Geometridae: Ennominae), a specialist on Winteraceae. *Insect Systematics and Evolution* 34, 81–94.

Zamaloa, M.C., Gandolfo, M.A. and Nixon K.C. (2020) 52 million years old *Eucalyptus* flower sheds more than pollen grains. *American Journal of Botany* 107, 1763–1771.

Zhang, Y., Huang, S.Y., Liang, D., Wang, H.S. and Zhang, P. (2020) A multilocus analysis of Epicopeiidae (Lepidoptera, Geometroidea) provides new insights into their relationships and the evolutionary history of mimicry. *Molecular Phylogenetics and Evolution* 149, 106847.

Zheng, X., Zhang, R., Yue, B., Wu, Y., Yang, N., and Zhou, C. (2022) Enhanced resolution of evolution and phylogeny of the moths inferred from nineteen mitochondrial genomes. *Genes* 13, 1634.

19 The Sphingidae: The Hawkmoths

Peter Hendry
Sheldon, Redland City, Queensland, Australia

Abstract

This chapter describes current knowledge of the hawkmoths, Sphingidae, with special focus on the Australian fauna. The chapter covers the number of species worldwide, early records and well-publicized species. The morphology is dealt with along with pests and conservation. All of the known Australian species, with notes on their distribution and host plants, are summarized in an Appendix.

19.1 Introduction

The moth family Sphingidae is one of ten families in the superfamily Bombycoidea (Zwick *et al.*, 2011). Worldwide, it contains 2053 taxa, comprising 1676 species and 377 subspecies in 204 genera in four subfamilies (extracted from Kitching, 2023, who recognizes the monotypic subfamily Langiinae found in parts of Asia). Only 4.3% of the world's taxa are found in Australia, 87 species (two of which are represented by two subspecies) in 31 genera. Twenty-four species are designated to subspecies level (Moulds *et al.*, 2020), in three subfamilies. The family is well represented in the scientific literature with at least 4160 titles published (Avon, 2016), more than a quarter of which relate directly or indirectly to the Australian fauna (Moulds *et al.*, 2020). In spite of this, there is still much to learn.

In 2020 the Australian Sphingidae were monographed by Max Moulds, James Tuttle and David Lane. The family thereby joined the very small number of Australian moth families for which modern monographic treatments exist. The work is also unique inasmuch as the Sphingidae are a relatively large and well-known family. Other monographs, inevitably, have focused on smaller families or taxa within families. The Moulds, Tuttle and Lane volume (2020) has been widely praised and has been described as 'one of the finest books on hawkmoths (perhaps indeed on any moths) ever produced' (Kitching and Edwards, 2020). The account presented here draws heavily on that work.

19.2 A Little History

Possibly the earliest record of a hawkmoth is in the Cave of the Trois-Frères in southwestern France, dated to 11–14,000 BCE. Howse (2010), after inverting a photograph from the cave, speculates in Lerio-Gourhan's *The Art of Prehis-*

Corresponding author: bevyjoy93@gmail.com

DOI: 10.1079/9781800627680.0019

toric Man in Western Europe that 'the drawings become unmistakable insects – hawk moths, perhaps ... there are also indications of hindwings, one of which appears to have eye spots on it'. The Stone Age artist may have drawn an eyed hawkmoth, *Smerinthus ocellatus*.

Of the more publicized species, *Xanthopan praedicta* was speculated about before it was discovered. When in 1862 Charles Darwin received a specimen of the orchid *Angraecum sesquipedale* from Madagascar, with a very long, slender, hollow floral spur, he predicted there would be, in Madagascar, a gigantic long-tongued moth capable of pollinating it (Minet *et al.*, 2021). Alfred Russel Wallace (1867) further speculated that, based on the fact that large sphinx moths of the tropics have proboscides nearly as long as the spur of *A. sesquipedale*, referencing that of *Macrosila* (now *Neococytius*) *cluentius* from South America at nine and one quarter inches (23.5 cm) and *Macrosila* (now *Xanthopan*) *morganii*, from tropical Africa at seven and a half inches (19 cm), these in principle could reach the nectar in the largest flowers of *A. sesquipedale*. Subsequently, Rothschild and Jordan (1903) described *Xanthopan morganii praedicta* from Madagascar, noting, '*X. morgani* [sic] *praedicta* can do for *Angraecum* what is necessary; we do not believe that there exists in Madagascar a moth with a longer tongue than is found in this sphingid'.

A recent chapter in this story was written when Minet *et al.* (2021) raised Rothschild and Jordans' *X. morganii praedicta* to full species status, *Xanthopan praedicta*.

The most well-known and extensively documented sphingid would have to be the death's-head hawkmoth which actually comprises three members of the genus *Acherontia*. With a skull-like pattern on its thorax, it has long been associated with death. Latreille (1805) wrote:

> The stain on its thorax, which roughly represents a skull, is seen by different people as the precursor of death. It appeared one year in Brittany, in quite large numbers, and as at that time there was an epidemic disease which caused many sick people to die, their mortality was attributed to it.

Moses Harris (1840) noted Latreille's comments and further quotes the author of *A Journal of a Naturalist*, who in turn references an unnamed correspondent in German Poland, who wrote that the death's-head hawkmoth:

> is a common creature, in the potato-fields of his village; it is regarded not as the creation of a benevolent being, but the device of evil spirits—spirits enemies to man—conceived and fabricated in the dark, and the very shining of its eyes is thought to represent the fiery element whence it is supposed to have proceeded. Flying into their apartments in the evening it at times extinguishes the light; foretelling war, pestilence, hunger, death to man and beast.

Also adding to its mystique is the fact it is capable of making a noise, first described by de Réaumur (1734) as 'a kind of cry, which could well have been taken for a lamentable funereal cry'. Harris (1840), summed up the thinking of the day, as follows: '[t]his noise has been the subject of much investigation'.

19.3 Morphology

The morphology of the Sphingidae is comprehensively covered and supported by well-illustrated diagrams in Moulds *et al.* (2020). Lachlan (2020) says of this work:

> Apart from the extremely detailed text it contains 43 beautifully executed figures of the adult body, wings, male and female genitalia, larva and pupa. The extraordinary number of named structural parts make this the most detailed overview of hawkmoth morphology since Rothschild and Jordan (1903). To my knowledge the figure of the complete female reproductive system is the first ever published for a hawkmoth.

19.3.1 Eggs

The eggs of hawkmoths (Fig. 19.1) are, generally, nearly spherical and slightly flattened, and in the case of the Australian endemic genera *Hopliocnema*, *Chelacnema* and *Synoecha*, are slightly elongated (Moulds *et al.*, 2020). The chorion is nearly smooth and the micropyle inconspicuous. They are usually pale green or yellow in colour and are nearly always laid singly on the underside of the leaves of the larval food plant. Exceptions are the *Hopliocnema* and *Chelacnema* that have grey eggs and *Coenotes* that lays eggs in tight clusters (Moulds *et al.*, 2020).

19.3.2 Larvae

Hawkmoth larvae (Figs 19.2–19.5), typical of lepidopteran larvae in their general form, are for the most part devoid of structural ornamentation in later instars, apart from a usually prominent caudal horn mid-dorsally on the eighth abdominal segment. Exceptions to this are found in the Australian endemic genus *Coequosa*, in which *C. triangularis* has only a very short caudal horn and then only in the first and second instars, and in *Hopliocnema brachycera* the first instar of which is generally lacking a caudal horn, although it may sometimes be present; in later instars, it is effectively lacking or reduced to a pointed nub (Moulds *et al.*, 2020). Abdominal spiracles are present on segments 1–8 and ventral prolegs on segments 3–6 with crochets arranged in a biordinal meso-series. The anal prolegs (or claspers) are on segment 10, along with the anal plate which always bears raised tubercles. The first instar larvae are slender, sometimes with inconspicuous long setae, and with (when present) a very long, slender caudal horn (Fig. 19.4). Mature larvae are stout, cylindrical and sometimes taper gradually towards the relatively small head (Fig. 19.5). Sometimes oblique lateral stripes and/or conspicuous eye-spots are present on the thorax or abdominal segments.

Fig. 19.1. *Theretra latreillii* egg. Photo: the author.

Fig. 19.2. *Psilogramma casuarinae* 5th larval instar on *Chionanthus ramiflorus*. Photo: the author.

Fig. 19.3. *Hippotion rosetta* 4th instar larva. Photo: the author.

Fig. 19.4. *Theretra latreillii* 1st instar larva. Photo: the author.

Fig. 19.5. *Eupanacra splendens* head of larva. Photo: the author.

Moulds *et al.* (2020) provide a key to the last instar larvae of Australian sphingids and Moulds (2024) to the first instar larvae.

19.3.3 Pupae

The distinctive pupae of hawkmoths are fairly stout, elongated and tapering at both ends, with the head being more rounded than the pointed rear and often only slightly narrower than the thorax. The head is dominated by the proboscis which is usually long, but can be very short in some genera. Although the appendages, including the proboscis, antennae and legs, are generally firmly fused (Fig. 19.6) in most Australian genera, the proboscis can be partly free within a separate sheath in a few instances (e.g. *Psilogramma*; Fig. 19.7) within the Sphinginae tribe Sphingini, except *Acherontia*, while the pupa of *Megacorma obliqua obliqua* is unknown. A free proboscis also applies to the pupa of the monotypic genus *Pergesa* (Macroglossinae) from parts of Asia, including India, China and the Philippines. As with the larvae, spiracles occur on the prothorax (either partly or entirely obscure) and on abdominal segments 1–8. Those on segment 1 are nearly always hidden by the edge of the forewing. The male pupae have a genital 'scar' on abdominal segment 9 while in the female it is on both segments 8 and 9. A clasper scar exists on segment 10 above the anal cavity and a spined cremaster is formed at the apex of segment 10. Moulds *et al.* (2020) provide a key to the pupae of Australian sphingids.

Fig. 19.6. *Hippotion rosetta* pupa. Photo: the author.

Fig. 19.7. *Psilogramma casuarinae* pupa showing 'free' proboscis. Photo: the author.

19.3.4 Adults

Adult hawkmoths (Figs. 19.9–19.15)) are medium-sized to large, stout-bodied moths with a wingspan 36–190 mm (Common, 1990). The rounded head is densely covered with short scales, but is sometimes crested. The large compound eyes contain as many as 30,000 ommatidia (Moulds *et al.*, 2020); chaetosemata and ocelli are absent. The antennae are short, becoming gradually thicker, but tapered near the tip which is usually hooked, mostly filiform but in some extralimital genera bipectinate. The unscaled proboscis is sometimes extremely long; in the case of the Neotropical *Amphimoea walkeri*, the world record holder, it measures up to 280 mm (Amsel, 1938, in Bauder *et al.*, 2014). However, in about one-fifth of hawkmoths worldwide it is very short (Miller, 1977, in Moulds *et al.*, 2020).

The maxillary palps are reduced to a single barely discernible segment. The three-segmented labial palps are usually closely appressed to the frons, with large basal and second segments but an almost inconspicuous third segment. At the base of the proboscis is a pair of very small lobe-like sensory structures known as pilifers; in some species these, along with a modified section 2 of the labial palps, are used for detecting ultrasonic sound as, unlike most large moths, hawkmoths have no hearing organ on the abdomen or thorax. Common (1990) describes the thorax as stout with large tegulae, and it may be smooth scaled or crested. The legs consist of five segments, the coxa, trochanter, femur, tibia and tarsus, to which are attached the pretarsal claws. The tibia has an epiphysis and in some species an apical projection, referred to by Rothschild and Jordan (1903) as a 'thorn' (Fig. 19.8). The mid- and hind-tibia carry a pair of apical spurs while the hind-tibia also carries a pair of mid-spurs, missing in some species. The 10 segmented abdomen in the male has segments 9 and 10 modified for sexual purposes while in the female segments 8–10 are modified for such purposes.

Most male hawkmoths have a pair of androconial tufts (hair pencils) on the anterolateral corner of sternite III, although these are missing in many Smerinthinae (Moulds *et al.*, 2020). *Gnathothlibus erotoides*, and possibly all *Gnathothlibus* species, also have androconial tufts on the foreleg coxa. The males of *Coequosa* have hidden anal tufts which they can display on cue (Fig. 19.9), possibly to disseminate pheromones to attract females although this, along with the use of the other forms of androconial tufts, has not been formally confirmed. The more prominent anal tufts in genera including *Macroglossum* and *Cephonodes* and to a lesser extent *Ambulyx*, found in both male and female, are associated with flight.

Fig. 19.8. *Cephonodes cunninghami* 'thorn' on foreleg. Photo: Dick Whitford.

The forewings are long and relatively narrow, and the hindwing is always much shorter. The frenulum in males is a single strong bristle which hooks behind a membranous retinaculum on Sc beneath the forewing; it is rarely rudimentary but even then, the frenulum base is still well developed. In females the frenulum is a group of several finer bristles, and the retinaculum is a tuft of projecting scales near Cu beneath the forewing (Common, 1990; Moulds *et al.*, 2020).

The venation as described by Moulds *et al.* (2020) is as follows: in the forewing the radius R and the radial sector veins Rs_1, Rs_2 and Rs_3 run closely parallel to each other, and the base of the median veins M_2 is always closer to the base of M_3 than to that of M_1. Anal veins 1A and 2A fuse to a single vein beyond their bases in the forewing and the posterior cubital vein CuP is absent. The forewing discal cell does not reach mid-length of the wing, and its distal margin (formed by the discocellular cross-veins) is straight or gently curved. In the hindwing, R crosses to the subcostal vein Sc near mid-length of the cell, and M_1 to CuA_2, each arising separately from the cell. At the apex of the entirely fused 1A + 2A in the hindwing, the margin protrudes a little, forming the tornus or anal angle, and as in the forewing, CuP is missing. The hindwing distal cell is always short.

The external male genitalia comprise mostly heavily sclerotized structures which provide important diagnostic characters for distinguishing species and genera and sometimes tribes (Moulds *et al.*, 2020). Common (1990) described the male and female genitalia as follows. In the male, the uncus may be single or bifid; the gnathos is sometimes present in the form of a single or pair of lobes and the valvae are usually

Fig. 19.9. (A,B) *Coequosa australasiae* males showing hair pencil display. Photos: the author.

Fig. 19.10. *Acosmeryx miskini*. Photo: the author.

broad, with a well-defined sacculus. A saccus is sometimes developed. The juxta is sclerotized, sometimes produced laterally into a pair of long processes. The aedeagus is cylindrical, with a coecum penis, and sometimes with sclerotized apical teeth or projections; cornuti are sometimes present. In the female genitalia, the ostium bursae may be centrally or asymmetrically placed, and flanked by a sclerotized sterigma. The posterior section of the ductus bursae may be strongly sclerotized, and the corpus bursae is usually membranous. One or, occasionally, two signa are nearly always present.

The internal reproductive system of the male, figured and described by Moulds *et al.* (2020) as 'a tangled mass of ducts of remarkable length occupying much of abdominal segments 4-8', comprises a fused pair of testes with the vasa deferentia each joining a long accessary gland before merging to continue as a long singular duct to the penis. The complete female reproductive system, with its complex associated ducts rarely illustrated, is detailed and figured by Moulds *et al.* (2020) who, among other things, explain how the ductus bursae is connected to the common oviduct by a singular tube, the ductus seminalis, that, intriguingly in many species, instead of going direct to the common oviduct, loops the ductus bursae to varying degrees in different genera, a phenomenon first noticed by Kitching (2002).

19.4 Natural Enemies of Hawkmoths

Under the heading *Natural Enemies*, Moulds *et al.* (2020) give an in-depth view of all kinds of hawkmoth enemies, covering all life stages, from eggs to adults. Pathogens, including viruses, bacteria and pathogenic fungi; predators, including birds, bats and invertebrates; endoparasites, including nematodes; ectoparasites including mites and midges; along with parasitoids. In this last category, Hymenoptera and Diptera are extensively covered, with parasitoids being recorded for 48 of the Australian taxa.

Mites in the family Otopheidomenidae are commonly associated with adult Sphingidae in the Neotropics and the genus *Prasadiseius* exclusively so (Prasad, 2011, in Moulds *et al.*, 2020). The only known Otopheidomenidae from Australia is *Nabiseius melinae* found in association with Hemiptera. Mite–sphingid adult associations in Australia are yet to be confirmed.

While there is no evidence of any parasitoid having a specific sphingid host, an analysis of the Sphingidae–parasitoid associations in Moulds *et al.*'s (2020) Appendix 1 shows there may be some relationships specific to life stages. Seventy-six percent of 98 species of parasitoids listed indicated an association with a particular life stage of the host. Using records where three or more concern the same family or genus of parasitoid shows that eggs are parasitized by Hymenoptera from four families: Encyrtidae (genus *Ooencyrtus*, 6 records); Eupelmidae (genus *Anastatus*, 5 records); Scelionidae (genus *Telenomus*, 9 records) and Trichogrammatidae (genus *Trichogramma*, 10 records). Hymenoptera were also recorded parasitizing the larval stage, but from families different from those attacking eggs. The families Braconidae (*Microplitis*), Chalcididae (*Brachymeria*), Eulophidae (*Euplectrus*) and Ichneumonidae (*Agrypon*) all attacked third, fourth or fifth instar sphingid larvae and Eulophidae (*Euplectromorpha*) was also recorded from a first instar larva.

Apart from a non-parasitoidal sarcophagid, the order Diptera is represented only by the Tachinidae. Sphingid larvae are affected by several tachinid genera. Those with three or more records are as follows: *Ceromya* (4), *Palexorista* (5) and *Winthemia* (5). Pupae are also affected by several tachinids, with the genus *Blepharipa* having 17 records. None of the genera associated with the pupae are listed for any other life stage, although this generalization is again based on very few numbers. Further research, with larger amounts of data, may prove this concept of host life stage specificity incorrect.

19.5 Hawkmoths as Pests

In many parts of the world, some hawkmoth larvae are regarded as significant pests of crops, usually referred to as 'hornworms' for obvious morphological reasons. In North America the tobacco hornworm (*Manduca sexta*) and the tomato hornworm (*M. quinqumaculata*) are widespread pests of Solanaceae including tomatoes, potatoes, aubergines and, not least, tobacco. Other species are occasional nuisance pests in horticulture (Tuttle, 2007).

No Australian species are regarded as significant pests although some are regarded as occasional nuisance species. The larvae of 21 species have been reported as feeding on cultivated plants (Moulds, 1981, 1984), while 11 species feed on some 70 species of garden ornamentals (Moulds, 1984). Their role as pests in Australia is probably outweighed by their usefulness as pollinators.

These observations notwithstanding, the 2013 declaration of the Western Australian Biosecurity and Agriculture Management Act 2007 placed the death's-head hawkmoths, *Acherontia atropos* and *A. styx*, on the prohibited organisms list even though they are not recorded from Australia. These, along with *A. lachesis*, are the only hawkmoth species considered pests (of beehives) as adults. In Australia *A. lachesis* is known only from Dauan Island in the Torres Strait and at this stage poses little threat.

19.6 Conservation

Although some regional studies have been conducted, for example on the Paraguayan hawkmoths (Smith, 2022), and individual studies, notably on the Hawaiian *Tinostoma smaragditis* (Heddle *et al.*, 2000), there is still much work to be done to evaluate the conservation status of most hawkmoths globally.

With no changes to the Sphingidae section since 2004, *The IUCN Red List of Threatened Species* lists four species. Both *Hyles hippophaes*, found in parts of Europe and Asia, and *Proserpinus proserpina*, found in parts of Europe and Asia, including the Middle East and northern Africa, have been downgraded from vulnerable to data deficient (World Conservation Monitoring Centre, 1996a, b). *Tinostoma smaragditis*, only known from Hawaii, was initially listed as extinct and was reassessed after its rediscovery by Adam Asquith in 1997. In 2002, *T. smaragditis* was noted as data deficient, being upgraded to endangered in 2000 (Heddle, 2004). The only other species on the Red List is *Euproserpinus wiesti*, from western parts of the United States which is listed as critically endangered (World Conservation Monitoring Centre, 1996c). Though not in the Red List, another Hawaiian species, *Manduca blackburni*, was listed as endangered in February 2000 by the US Fish and Wildlife Service.

The Australian fauna remain unassessed. *Amplypterus panopus panopus*, *Cypa decolor euroa*, *Daphnis hypothous crameri*, *Eurypteryx molucca*, *Macroglossum ungues ungues*, *Marumba timora* and *Megacorma obliqua obliqua* are known in Australia from single specimens. These may prove to be just vagrants. Other endemic species like *Hopliocnema lacunosa* are poorly represented in collections and could only be designated as data deficient.

19.7 The Australian Sphingidae

The list of Australian species that is the Appendix to this chapter largely follows the higher classification and nomenclature set out in Moulds *et al.* (2020), with the exception of the recent revision of *Gnathothlibus* by Eitschberger *et al.* (2024), in which *Gnathothlibus erotoides* Wallengren (1858a,b) has been reinstated, replacing the name *G. eras* (de Boisduval, 1832) in Australia. Eitschberger *et al.* (2024) state 'we are convinced that *G. eras* ... is an endemic species of Tahiti and has nothing to do with the representatives of the *G. erotus-eras* species group of Australia'. This has been reflected in the Appendix.

The Eitschberger *et al.* (2024) paper also described 11 new species of *Gnathothlibus* from Australia as follows. From Mt Glorious near Brisbane, Queensland, *G. gloriosa*, *G. lanei* and *G. pseudolanei* with a further unnamed species noted as *Gnathothlibus* spec. 6. From Buderim, Queensland, *G. buderimensis*. From Dauan Island, Torres Strait, *G. dauanensis* with a further three unnamed species noted as *Gnathothlibus* spec. 2, spec. 3 and spec. 4. From Hammond Island, Queensland, *G. hammondensis*. From elsewhere in Australia: *G. lachlani* from Kuranda, north Queensland; *G. maxmouldsi* from Cape Tribulation, north Queensland; *G. riottei* from Christmas Island; and, *G. wallengreni* and *G. queenslandi* from Christmas Creek, Queensland. The status of these species is questionable. Their distributions are very restricted and mostly noted as 'only known from type location' or, in the case of those from various islands, 'endemic to'. The possibility of having up to four distinct species of *Gnathothlibus* from an area the size of Mt Glorious does not seem plausible. Because of the high likelihood of these species being junior synonyms, they should be excluded from any checklist pending further investigation of their status. They are not included in the Appendix.

The following changes, however, are proposed by Ian Kitching (2023) based on a largely unpublished phylogeny.

- The genera *Imber* and *Coequosa* are in the subfamily Smerinthinae as *incertae sedis*.
- *Cypa* is in the informal 'Cypa genus group': the monophyly of the group remains to be demonstrated.
- The tribe Sphingulini is referred to as 'Australian Sphingulini' used as an informal taxon to contain the Australasian genera and to remove paraphyly of the tribe Sphingulini. Monophyly of the group remains to be demonstrated.
- *Cerberonoton severina* is in the informal grouping 'Xanthopan genus group'. This taxon is used to contain those genera previously referred to as tribe Cocytiina. Cocytiina cannot be used for this group as it is a junior homonym of *Cocytides* (de Boisduval, 1875), and a formal name has yet to be erected.
- *Leucomonia bethia* and the genus *Psilogramma* are in the grouping 'Psilogramma genus group'. This informal taxon is used to contain those genera that comprise the sister clade to Acherontiina (*Psilogramma*, *Macropoliana* and *Notonagemia*) and their relatives. It is used to preserve the current concept of Acherontiina and to prevent paraphyly of the subtribe Sphingina. It is provisionally diagnosed by a divided uncus; a new subtribe should be erected for the group in due course.
- The genera *Eupanacra* and *Gnathothlibus* are in the subtribe Macroglossini *incertae sedis*.

Information regarding distribution and host plants in the Appendix largely follows Moulds *et al.* (2020), though in summary form. That work should also be referred to for more detailed taxonomic information (on genera as well as species) as well as type localities. Many Australian species differ from each other in subtle ways. Again, Moulds *et al.* (2020) should be consulted for identification purposes, keys to species, notes on similar species and, most importantly, splendid illustrations of all stages of most species.

Additional information on distributions recorded since Moulds *et al.* (2020) is as follows.

- *Ambulyx wildei* (Fig. 19.11). Eastern parts of Indonesia (Kitching, 2023).
- *Eupanacra splendens splendens*. Moulds *et al.* (2020) give the distribution as Cape York peninsula to Brisbane, noting 'The only Brisbane record is from larvae found at Nathan (T. Brewer)'. However, recent social media exchanges (2020–2024), backed up with images, show it is well established at the Sunshine Coast through Brisbane to at least as far south as the Gold Coast. The author has often found eggs and larvae on *Epipremnum pinnatum* growing on his property at Redland City since 2022.
- *Eurypteryx molucca*. Moulds *et al.* (2020) note a single specimen from Boigu Island. A single final instar larva has since been reared through to adult, from Dauan Island, Torres Strait (Lane *et al.* 2021).
- *Theretra indistincta indistincta*. A range extension, from the far north corner of New South Wales to northern Sydney, was published by Moore (2021).
- *Macroglossum nubilum*. A recent find in the collection of John Moss was a specimen of *M. nubilum* from Lenthall's scrub via Torbanlea, Hervey Bay, Queensland, dated 2 January 1990. This extends the distribution given in Moulds *et al.* (2020), as north-east Queensland Iron Range to Townsville, by some 1100 km. Moulds *et al.* (2020) also noted three specimens in the South Australian museum labelled Brisbane which were considered erroneous. This specimen may add weight to the possible correctness of the Brisbane labels.

Fig. 19.11. *Ambulyx wildei*. Photo: the author.

Fig. 19.12. *Hyles livornicoides*. Photo: the author.

Additional information on host plants recorded since Moulds *et al.* (2020) is as follows. Asterisked species are exotics.

- *Acosmeryx anceus anceus*. *Saurauia andreana* (Actinidiaceae) (Moulds, 2024).
- *Agrius convolvuli*. **Ocimum basilicum* (Lamiaceae) (Moulds, 2024).
- *Cerberonoton severina*. *Polyalthia submontana* subsp. *sessiliflora* (Annonaceae) (Moulds, 2024).
- *Chelacnema ochra*. *Eremophila platythamnos* subsp. *exotrachys* (Moulds *et al.*, 2020 as *E. rubra* var. *exotrachys*).
- *Eurypteryx molucca*. *Alyxia spicata* (Apocynaceae) (Lane *et al.*, 2021).
- *Hippotion rosetta*. *Spermacoce stenophylla* (Rubiaceae) (Moulds, 2024).
- *Hyles livornicoides* (Fig. 19.12). *Commicarpus chinensis* (Nyctaginaceae) (Moulds *et al.*, 2020 as the synonym *Boerhavia chinensis*).
- *Psilogramma casuarinae*. *Chonanthus ramiflorus* (Lamiaceae) (new record, P. Hendry, unpublished).
- *Theretra latreillii*. **Spathiphyllum wallisii* (Araceae) (Moulds, 2024).
- *Theretra oldenlandiae*. *Spermacoce stenophylla* (Rubiaceae) (Moulds, 2024).

Note that within the checklist in Moulds *et al.* (2020, pp. 36–37), the names of two authors were applied incorrectly: Thon (1828) for *Theretra oldenlandiae* and Clark (1922) for *Daphnis dohertyi dohertyi*. However, within the body of the work the correct authors are shown: *Theretra oldenlandiae* (Fabricius, 1775) and *Daphnis dohertyi dohertyi* (Rothschild, 1897).

Fig. 19.13. *Psilogramma menephron nebulosa*. Photo: the author.

The Appendix to this chapter lists 87 species organized into 31 genera. Two of these species are further divided into two subspecies, both of which occur within Australian political boundaries. Of the 87 species, two are probably vagrants having been recorded only once, one from the Kimberley coast and one from the Darwin area. Four species (plus two subspecies) only occur on Christmas Island (or, in one case, the Cocos Keeling Islands) outside the biogeographic boundaries of Australia. A further seven species are New Guinea species only recorded from the far northern Torres Strait Islands of Boigu, Darnley and/or Dauan. This leaves 74 species as the 'core' fauna of the Australian continent (including Tasmania). Of these 74 species, 29 (39%) are endemic with a further 42 (57%) shared with the Moluccas, Sulawesi, New Guinea, the Solomons and associated islands – in other words, tropical areas of biogeographical 'Australasia' to our north.

We have information on at least some of the local larval food plants of all but four of these 74 'continental' species, making this by far the best known family of any Australian family of moths with a moderate number of members. This no doubt reflects the striking appearance and size of these moths, features which have led to a group of enthusiastic specialists working over many years.

Eight species of the continental hawkmoth fauna are truly polyphagous with larvae which exploit between five and 13 plant families. The remainder exploit at most four families of plants; most restrict themselves to a single plant family. For this set of more specialized species, feeding records exist across 28 plant families. For most plant families, only one or two feeding records are known. The exceptions are the Rubiaceae, Vitaceae, Apocynaceae, Araceae and Scrophulariaceae. Of these five families, the Rubiaceae dominate with a total of 23 species recorded feeding on members of the family (19 exclusively so). Eight species are recorded from various Vitaceae, and four each on the Apocynaceae, Araceae and Scrophulariaceae. The Rubiaceae specialists are all macroglossines.

The biogeographic aspects of these distribution and food plant records are discussed in more detail in Chapter 4.

Fig. 19.14. *Tetrachroa edwardsi*. Photo: the author.

Fig. 19.15. *Theretra nessus*. Photo: the author.

19.8 Host Plants: A Word of Warning

Extensive studies of the Australian hawkmoth food plants have been published (Moulds, 1981, 1984, 1998; Moulds *et al.*, 2020). Within all of these publications, there have been corrections to previous records. Some are misidentifications of the plant or moth species involved, while others stem from the fact that finding a larva on a plant does not constitute a host plant record. Under *Coenotes eremophilae*, Moulds (1984) warns:

> ... larvae of this species are frequently numerous and often defoliate their initial food source. During their search for additional food, larvae sometimes investigate plants they normally would not eat. Care should be taken by observers not to document such plants as larval food plants.

This is borne out in Moulds *et al.* (2020) where the record for *Agrius convolvuli* on *Abutilon oxycarpum* from Moulds (1981) is 'considered erroneous and was probably based on a wandering larva'.

A recent find of up to 40 eggs and first instar larvae on ten different plants of *Hymenosporum flavum* (Pittosporaceae) proved to be *Theretra latreillii*. Apart from a little epidermal chewing, *H. flavum* was not used as a food plant and a transfer to *Cayratia clematidea* proved to be a saving grace, confirming that finding eggs or larvae on a plant does not constitute a host plant record, even if there is evidence of plant material being consumed. They need to be bred through to an adult.

Acknowledgements

I thank Roger Kitching for his support, Max Moulds for his support, checking of the manuscript and suggested amendments, Daphne Bowden for proof reading the original manuscript and to my wife Beverly for her support and help with typing parts of the manuscript. A special thanks to Dick Whitford for allowing me to use and modify his photo of *Cephonodes cunninghami* (Fig. 19.8). All other photos are by Peter Hendry.

References

Avon, C. (2016) Sphingidae of the world Bibliography. Available at: www.academia.edu/27296167/Sphingidae_of_the_world_Bibliography_by_Christophe_Avon_2016 (accessed 5 October 2025).

Bauder, J.A.S., Warren, A.D. and Krenn H.W. (2014) Evolution of extreme proboscis lengths in Neotropical Hesperiidae (Lepidoptera). *Journal of Research on the Lepidoptera* 47, 65–71.

Common, I.F.B. (1990) *Moths of Australia*. Melbourne University Press, Melbourne, Victoria.

Curtis, J. (1837) *British entomology; being illustrations and descriptions of the genera of insects found in Great Britain and Ireland: containing coloured figures from nature of the most rare and beautiful species, and in many instances of the plants upon which they are found*. London.

De Boisduval, J.B.A.D. (1832) *Voyage de Découvertes de l'Astrolabe exécuté par Ordre du Roi, Pendant les Années 1826–1827–1828–1829, sous le Commandement de M.J. Dumont D'Urville. Faune entomologique de l'Océan Pacifique, avec l'illustration des insectes nouveaux recueillis pendant le Voyage. Part 1. Lépidoptères*. J. Tasta, Paris, pp. i–iv, 5–267.

Eitschberger, U., Malm, T. and Verdecia, V. (2024) Dritte arbeit zur revision des *Gnathothlibus erotus-eras*-artenkomplexes: die arten Australiens, einschließlich der Weihnachtsinseln, und von Tahiti (Lepidoptera, Sphingidae). *Neue Entomologische Nachrichten* 83, 1–197.

Harris, A.C. (1990) A specimen of *Cizara andeniae* (Lepidoptera: Sphingidae) captured in Naseby, central Otago. *New Zealand Entomologist* 13, 39–39.

Harris, M. (1840) *The Aurelian a Natural History of English Moths and Butterflies, together with the figures of their transformations and of the plants on which they feed, drawn, engraved, and coloured from the natural objects*. Henry G. Bohn, London.

Heddle, M.L. (2004) *Tinostoma smaragditis. The IUCN Red List of Threatened Species 2004*. e.T21913A9339981.

Heddle, M.L., Wood, K.R., Asquith, A., Gillespie, R.G. (2000) Conservation status and research on the fabulous Green Sphinx of Kaua'i, *Tinostoma smaragditis* (Lepidoptera: Sphingidae), including checklists of the vascular plants of the diverse mesic forests of Kaua'i, Hawai'i. *Pacific Science* 54, 1–9.

Howse, P. (2010) *Butterflies: Decoding Their Signs and Symbols*. Firefly Books, Richmond Hill Canada.

Kirby, W.F. (1883) *The Zoological Record for 1882*. Zoological Society, London.

Kirby, W.F. (1892) *A synonymic catalogue of Lepidoptera Heterocera (moths). Volume 1 Sphinges and Bombyces*. Gurney and Jackson, London.

Kitching, I.J. (2002) The phylogenetic relationships of Morgan's Sphinx, *Xanthopan morganii* (Walker), the tribe Acherontiini, and allied long-tongued hawkmoths (Lepidoptera: Sphingidae, Sphinginae). *Zoological Journal of the Linnean Society* 135, 471–527.

Kitching, I.J. (2023) *Sphingidae Taxonomic Inventory*. Available at: http://sphingidae.myspecies.info/ (accessed 25 February 2024).

Kitching, R.L. and Edwards, E.D. (2020) Books about Australian moths 1805–2020: An annotated bibliogrphy. *Australian Entomologist* 47, 119–154.

Lachlan, R.B. (2004) Description of a second species of *Gnathothlibus* Wallengren (Lepidoptera: Sphingidae) from Australia. *Australian Entomologist* 31, 111–118.

Lachlan, R.B. (2020) Book Review. *Hawkmoths of Australia. Identification, biology and distribution. Australian Entomologist* 47, 80–82.

Lane, D.A., Moulds, M.S. and Hopkinson, M. (2021) First account of the life history of the hawkmoth *Eurypteryx molucca* C. and R. Felder, 1874 (Lepidoptera: Sphingidae: Macroglossini). *Australian Entomologist* 48, 195–201.

Latreille, P.A. (1805). Famille Quatre-vingtieme. Sphingides; *sphingides; Histoire naturelle, générale et particulière, des Crustacés et des insects*, 14. F. Dufart, Paris.

Linnaeus, C. (1771) *Mantissa plantarum altera generum editionis VI. and specierum editionis II pars altera*. Laurentius Salvius, Stockholm, Sweden.

Minet, J., Basquin, P., Haxaire, J., Lees, D.C. and Rougerie, R. (2021) A new taxonomic status for Darwin's "predicted" pollinator: *Xanthopan praedicta* stat. nov. (Lepidoptera Sphingidae Sphinginae). *ANTEN-*

OR, Études de Lépidoptérologie Tropicale/Tropical Lepidopterology Studies, vol. 8, Fasc. 1 69–95. Alexanor, Paris.

Miskin, W.H. (1891) A revision of Australian Sphingidae. *Proceedings of the Royal Society of Queensland* 8, 1–28.

Moore, A.S. (2021) A review of southern range extensions for *Theretra indistincta indistincta* (Butler, 1877) (Lepidoptera: Sphingidae) in New South Wales. *Australian Entomologist* 48, 255–257.

Moore, F. (1882) *The Lepidoptera of Celyon*, vol. II. L. Reeves & Co., London.

Moritz, R.F.A., Kirchner, W.H. and Crewe, R.M. (1991) Chemical camouflage of the death's hawkmoth (*Acherontia atropos* L.) in honeybee colonies. *Naturwissenschsften* 78, 179–182.

Moulds, M.S. (1981) Larval foodplants of hawk moths (Lepidoptera: Sphingidae) affecting commercial crops in Australia. *General and Applied Entomology* 13, 68–80.

Moulds, M.S. (1984) Larval foodplants of hawk moths (Lepidoptera: Sphingidae) affecting garden ornamentals in Australia. *General and Applied Entomology* 16, 57–64.

Moulds, M.S. (1998) New larval food plants for Australian hawk moths (Lepidoptera: Sphingidae). *Australian Entomologist* 25, 13–22.

Moulds, M.S. (2024) Key to first and second instar larvae together with new foodplant records for Australian hawkmoths (Lepidoptera: Sphingidae). *Atalanta* 55, 296–304.

Moulds, M.S., Tuttle, J.P. and Lane, D.A. (2020) *Hawkmoths of Australia: Identification, Biology and Distribution*. Monographs on Australia Lepidoptera 13. CSIRO Publishing, Melbourne, Victoria.

Rothschild, L.W. and Jordan, K. (1903) A revision of the lepidopterous family Sphingidae. *Novitates Zoologicae* 9 (suppl.), 1–972.

Smith, P. (2022) Diversity, distribution patterns and preliminary conservation assessment of the Paraguayan hawkmoths (Lepidoptera: Sphingidae) *Journal of Insect Conservation* 26, 327–335.

Tuttle, J. P. (2007). *The Hawk Moths of North America: A Natural History of the Sphingidae of the United States and Canada*. Wedge Entomological Research Foundation, Washington, DC.

Wallace, A.R. (1867) Creation by law. *Quarterly Journal of Science* 4, 471–488.

Wallengren, H.D.J. (1858a) Nya fjäril-slägten. *Öfversigt af Kongelige Vetenskaps-Akademiens Förhandlingar* 15, 75–84, 135–143, 209–215.

Wallengren, H.D.J. (1858b) Lepidoptera, species novas descripsit. In: *Kongliga Svenska Fregatten Eugenies resa omkring jorden: Under befäl af C.A. Virgin, åren 1851–1853*. P.A. Norstedt, Stockholm, Sweden, pp. 351–390.

World Conservation Monitoring Centre (1996a) *Hyles hippophaes. The IUCN Red List of Threatened Species 1996*. e.T10542A3198676.

World Conservation Monitoring Centre (1996b) *Proserpinus proserpina. The IUCN Red List of Threatened Species 1996*. e.T18366A8153516.

World Conservation Monitoring Centre (1996c) *Euproserpinus wiesti. The IUCN Red List of Threatened Species 1996*. e.T8373A12908327.

Zolotuhin, V.V. and Ryabov, S.A. (2012).*The Hawkmoths of Vietnam*. Korporatsiya Tekhnologiy Prodvizheniya, Ulyanovsk, Russia.

Zwick, A., Regier, J.C., Mitter, C. and Cummings, M.P. (2011) Increased gene sampling yields robust support for higher level clades within Bombycoidea (Lepidoptera). *Systematic Entomology* 36, 31–43.

APPENDIX The Australian sphingid fauna. Note that all references to naming authorities may be found in Moulds *et al.* (2020)

TAXON	DISTRIBUTION		LARVAL FOOD PLANTS
	AUSTRALIA	EXTRA-LIMITAL	
SMERINTHINAE: Ambulycini			
***Ambulyx* (Westwood, 1847)**			
dohertyi queenslandi (Clark, 1928)	Cape York to Townsville, Qld	–	Unknown
wildei (Miskin, 1891) (Fig. 19.11)	Cooktown to Tully, Qld	Moluccas, New Guinea	Unknown
***Amplypterus* (Hübner, 1819)**			
panopus panopus (Cramer, 1779)	Darwin (one known), NT	India, China, SE Asia	Unknown
SMERINTHINAE: Smerinthini			
***Imber* (Moulds *et al.*, 2010)**			
tropicus (Moulds, 1983)	Broome, WA to NE Qld	–	*Brachychiton spp. (Sterculiaceae)*
***Coequosa* (Walker, 1856)**			
australasiae (Donovan, 1805) (Fig. 19.9a,b)	E Qld to NE Vic.	–	
triangularis (Donovan, 1805)	SE Qld to NE Vic.	–	Banksia, Grevillea, Hakea etc. (Proteaceae)
***Cypa* (Walker, 1865)**			
*decolor euroa (*Rothschild & Jordan,1903)	Dauan I. (one known), Qld	New Guinea	Unknown
SMERINTHINAE: Sichiini			
***Marumba* (Moore, 1882)**			
timora (Rothschild & Jordan, 1903)	Kimberleys (one known), WA	Lesser Sundas, indonesia	Unknown
SPHINGINAE: Sphingulini			
***Chelacnema* (Moulds *et al.*, 2020)**			
ochra (Tuttle *et al.*, 2012)	S WA, NT, SA	–	*Eremophila* spp. (Scrophulariaceae)
***Coenotes* (Rothschild & Jordan, 1903)**			
arida (Moulds & Melichar, 2014)	Central WA,, NT	–	*Ptilotus obovatus* (Amaranthaceae)
eremophilae (Lucas, 1891)	Inland Australia, widespread	–	Polyphagous (<13 herbaceous families)
***Hopliocnema* (Rothschild & Jordan, 1903)**			
brachycera (Lower, 1897)	Mainland, widespread	–	*Eremophila longifolia* (Scrophulariaceae)
lacunosa (Tuttle *et al.*, 2012)	Inland S WA	–	*Eremophila longifolia* (Scrophulariaceae)(in captivity)
***Synoecha* (Rothschild & Jordan, 1903)**			
marmorata (Lucas, 1891)	Arid Qld & NSW	–	*Eremophila mitchellii* (Scrophulariaceae)

Continued

TAXON	DISTRIBUTION AUSTRALIA	EXTRA-LIMITAL	LARVAL FOOD PLANTS
***Tetrachroa* (Rothschild & Jordan, 1903)**			
edwardsi (Olliff, 1890) (Fig. 19.14)	Coastal & inland Qld & NSW	–	*Parsonsia* spp. (Apocynaceae)
SPHINGINAE: Sphingini: Sphingina			
***Cerberonoton* (Zolotuhin & Ryabov, 2012)**			
severina (Miskin, 1891)	Torres Strait to SE Qld	Tanimbar, NG; Solomons	Various Annonaceae & Bignoniaceae
***Leucomonia* (Rothschild & Jordan, 1903)**			
bethia (Kirby, 1877)	Northern half of Australia	–	*Clerodendron floribundum* (Lamiaceae)
***Psilogramma* (Rothschild & Jordan, 1903)**			
argos (Moulds and Lane, 1999)	Broome WA; Darwin, NT; Coen to Toomba N Qld		
casuarinae (Walker, 1856)	Cape York, Qld to central Vic.	New Guinea	Polyphagous (<6 families0
discistriga discistriga (Walker, 1856)	Christmas Island	S & SE Asia to Philippines	Unknown
exigua (Brechlin *et al.*, 2010)	Broome WA to trpical N Qld	–	*Clerodendron floribundum* (Lamiaceae), Vitex spp., Jasminum sp. (Oleaceae)
maxmouldsi (Eitschberger, 2001)	SE Qld, NE NSW	–	*Clerodendron floribundum* (Lamiaceae)
menephron nebulosa (Butler, 1876) (Fig. 19.13)	Torres Strait to NE NSW	New Guinea	*Spathodea campanulata* (Bignoniaceae), *Chionanthus ramiflorus* (Oleaceae)
papuensis (Brechlin, 2001)	Torres Strait to Eungella, Qld	Aru Is, New Guinea	Various Bignoniaceae, Gentianaceae, Lamiaceae
penumbra (Lane *et al.*, 2011)	Kimberleys, Top End of NT	–	Unknown
SPHINGINAE: Sphingini: Acherontiina			
***Acherontia* (Laspeyres, 1809)**			
lachesis (Fabricius, 1798)	Dauan Is only, Qld	S, SE and W Asia	Solanaceae (extralimitally 29 families)
***Agrius* (Hübner, 1819)**			
convolvuli (Linnaeus, 1758)	Mainland Australia & Tas.	Widespread in the Old World	Various Convolvulaceae, *Ocimum basilicum* (Lamiaceae)
godarti (Macleay, 1826)	Mainland Aust. exc. SW WA	–	Various Convolvulaceae, *Salsola* sp.(Amaranthaceae)

Continued

TAXON	DISTRIBUTION		LARVAL FOOD PLANTS
	AUSTRALIA	EXTRA-LIMITAL	
***Magacorma* (Rothschild & Jordan, 1903)**			
obliqua obliqua (Walker, 1856)	Dauan I, Qld (one only)	S & SE Asia	Spathodea campanulata, Oroxylum indicum (Bignoniaceae) (probable)
MACROGLOSSINAE: Hemarini			
***Cephonodes* (Hübner, 1819)**			
australis (Kitching & Cadiou, 2000)	Coastal, Kimberleys to NSW	–	*Gardenia* sp., *Larsenaikia* sp, *Pavetta* spp., *Psychotris, Psydrax* (Rubiaceae)
cunninghami (Cramer, 1777) (Fig. 19.8)	Christmas I to Brisbane	Timor, New Guinea; Philippines	Various Rubiaceae
janus (Miskin, 1891)	NT to SE Qld	–	*Aidia racemosa, Atractocarpus i., Psydrax odoprata* (Rubiaceae)
kingii (Macleay, 1826)	Widespead, non-arid mainland	–	Various Rubiaceae
MACROGLOSSINAE: Macroglossini, Macroglossina			
***Angonyx* (de Boisduval, 1875)**			
papuana papuana (Rothschild & Jordan, 1903)	Torres Strait to Mission B., Qld	New Guinea	*Strychnos minor* (Loganiaceae)
***Cizara* (Walker, 1856)**			
ardeniae (Lewin, 1805)	E Qld to NE Vic.	Vagrant to New Zealand	*Coelospermum* sp., *Coprosma* spp., *Gynochthodes* sp. (Rubiaceae)
***Daphnis* (Hübner, 1819)**			
dohertyi dohertyi (Rothschild, 1897)			
hypothous crameri (Eitschberger & Melichar, 2010)	Christmas Island (one only)	S & SE Asia (Oriental Region)	Unknown (extralimitally Apocynaceae, Rubiaceae)
moorei (Macleay, 1866)	Broome, WA to Brisbane, Qld	New Guinea; Solomons	*Breonia chinensis, Guettarda speciosa, Nauclea orientalis* (Rubiaceae)
placida placida (Walker, 1856)	N WA to S NSW, coastal	S & SE Asia; Melanesia	Various Apocynaceae, *Alangium* (Cornaceae)
protrudens protrudens (Felder & Felder, 1874)	Torres Strait to tropical Qld	Moluccas; NG; Solomons	*Timonius timon* (Rubiaceae)
***Eupanacra* (Cadiou & Holloway, 1989)**			
splendens splendens (Rothschild, 1894)	Cape York to SE Qld	New Guinea	Various exotic Araceae, *Raphidophora australasica* (Araceae)
***Eurypteryx* (Felder & Felder, 1874)**			
molucca (Felder & Felder, 1874)	Boigu & Dauan Is, Torres Strait	Moluccas; NG; Philippines	*Alyxia spicata* (Apocynaceae)

Continued

TAXON	DISTRIBUTION AUSTRALIA	DISTRIBUTION EXTRA-LIMITAL	LARVAL FOOD PLANTS
***Gnathothlibus* (Wallengren, 1858a)**			
australiensis (Lachlan, 2004)	Kimberleys, WA, to NE NSW	–	*Melastoma* spp., (Melastomataceae)
erotoides (Wallengren 1858b)	Kimberleys, WA to Victoria	SE Asia; Pacific Is	Polyphagous (<5 families favouring Vitaceae, Rubiaceae)
***Macroglossum* (Scopoli, 1777)**			
alcedo (de Boisduval, 1832)	Torres Strait to Innisfail, Qld	New Guinea; Aru; Tanimbar	*Hodgkinsonia frutescens, Psychotria* sp. (Rubiaceae
corythus corythus (Walker, 1856)	Cocos Keeling Islands	S & SE Asia	*Guettarda speciosa, Morinda citrifolia* (Rubiaceae)
corythus approximans (Lucas, 1891)	Darwin, NT; Torres Strait to Mackay, Qld	New Guinea	*Coelospermum reticulatum, Guettarda speciosa, Morinda citrifolia* (Rubiaceae)
dohertyi doddi (Clark, 1922)	Torres Strait to Cardwell, Qld	Moluccas; New Guinea	*Myrmecodium beccarii, M. platytyrea, M. tuberosa* (Rubiaceae)
errans (Walker, 1856)	NT to Wollongong, NSW	Tanimbar, NG; Solomons	Various Rubiaceae (<7 genera)
joannisi (Rothschild & Jordan, 1903)	NT; Torres Strait Coen to Kuranda, Qld	–	*Memecylon* spp., (Melastomataceae)
melas melas (Rothschild & Jordan, 1903)	Dauan Island, Torres Strait, Qld	Seram, Aru, N G, Louisiade Archipelago.	Unknown
micacea micacea (Walker, 1856)	Torres Strait , E Qld to Armidale, NSW	New Guinea.	*Strychnos psilosperma* (Loganiaceae)
nubilum (Rothschild & Jordan, 1903)	Iron Range to Townsville, N Qld	N G, Aru & Woodlark Is	*Coelospermum paniculatum*; *Morinda citrifolia* (Rubiaceae)
papuanum (Rothschild & Jordan, 1903)	Torres Strait to Mackay E Qld	New Guinea.	**Spermacoce latifolia*, *Pentas lanceolata* (Rubiaceae)
prometheus lineata (Lucas, 1891)	Koolan Is, WA, N NT, Torres Str to Burleigh, E Qld	New Guinea & adjacent islands.	*Gynochthodes jasminoides*; *Morinda citrifolia* (Rubiaceae)
prometheus prometheus (de Boisduval, 1875)	Christmas Island	SE Asia & south-western India	*Gynochthodes macrophylla*; *Morinda citrifolia* (Rubiaceae)
queenslandi (Clark, 1927)	Dauan Island to Kuranda, Qld	–	*Psychotria spp* (Rubiaceae)
rectans (Rothschild & Jordan, 1903)	Top End NT & Torres Strait to Ayr Qld	Sulawesi, NG , Kai & Tanimbar Islands	*Psychotria poliostemma* (Rubiaceae)
tenebrosa (Lucas, 1891)	Torres Strait to Yeppoon Qld	Indonesia, New Guinea, Solomon Is	*Coelospermum paniculatum* var. *syncarpum* (Rubiaceae)

Continued

TAXON	DISTRIBUTION AUSTRALIA	EXTRA-LIMITAL	LARVAL FOOD PLANTS
ungues ungues (Rothschild & Jordan, 1903)	Christmas Island	Java east to Tanimbar Island	*Paederia* (Rubiaceae)
vacillans (Walker, 1865)	Far north WA, NT, Torres Strait to Brisbane, Qld	NG, east Indonesia & Timor	*Strychnos lucida* (Loganiaceae)
***Nephele* (Hübner, [819)**			
hespera (Fabricius, 1775)	Torres Strait to Laur, N Qld	Iran, Afghanistan, China, Thailand, Indonesia & NG	*Carissa* spp. (Apocynaceae).
subvaria (Walker, 1856)	Northern WA, NT, most of Qld, to Sydney, NSW	New Caledonia	*Carissa* spp. (Apocynaceae).
***Pseudangonyx* (Eitschberger, 2010)**			
excellens (Rothschild, 1911)	Torres Strait to Kuranda, Qld	Aru & New Guinea	Unknown
***Zacria* (Haxaire & Melichar, 2003)**			
vojtechi (Haxaire & Melichar, 2003)	Southern half WA, southern parts of NT	–	*Psydrax attenulata*, *P. rigidula* (Rubiaceae)
MACROGLOSSINAE: Macroglossini, Acosmerygina			
***Acosmeryx* (de Boisduval, 1875)**			
anceus anceus (Stoll, 1781)	Cape York to north of Townsville, Qld	NG, Indonesia, Sulawesi	*Cayratia clematidea* (Vitaceae) *Saurauia andreana* (Actinidiaceae)
cinnamomea (Herrich-Schäffer, 1869)	North WA, Top End NT, eastern Qld, NSW to Vic	Papua, NG	*Cayratia* spp, *Cissus* spp. & **Vitis* spp. (Vitaceae)
miskini (Murray, 1873) (Fig. 19.10)	Windsor Tableland, Qld to Wollongong, NSW	–	*Cayratia clematidea*, *Cissus spp & *Vitis* spp. *(Vitaceae)*
MACROGLOSSINAE: Macroglossini, Choerocampina			
***Hippotion* (Hübner, 1819)**			
boerhaviae (Fabricius, 1775)	Christmas Island, eastern Qld to the NSW	Asia, Philippines, Indonesia, NG, Solomons, Vanuatu, New Caledonia.	*Curculigo ensifolia* (Hypoxidaceae) **Pentas lanceolata* (Rubiaceae)
brennus (Stoll, 1782)	Torres Strait Islands to Brisbane	Moluccas & Seram, Indonesia, NG, Solomon Is	*Hibbertia scandens* (Dilleniaceae), (4 genera of Rubiaceae)

Continued

TAXON	DISTRIBUTION AUSTRALIA	DISTRIBUTION EXTRA-LIMITAL	LARVAL FOOD PLANTS
celerio (Linnaeus, 1758)	Australia including Tas. & offshore islands	Europe, Africa, Middle East, Asia, China, Philippines, Indonesia, NG, New Caledonia, Fiji & Samoa	Various plants in the families (Apocynaceae) (Araceae) (Balsaminaceae) (Begoniaceae) (Bignoniaceae) (Convolvulaceae) (Dilleniaceae) (Nyctaginaceae) (Polygonaceae) (Rubiaceae) (Vitaceae)
johanna (Kirby, 1877)	Iron Range, Qld to Brunswick Heads, NSW	New Guinea	*Persicaria decipiens* (Polygonaceae), **Pentas lanceolata* (Rubiaceae)
rosetta (Swinhoe, 1892)	Cocos-Keeling & Christmas Is, north WA, Top End NT, Torres Strait, Qld to NSW	Asia, Taiwan, Philippines, Indonesia, NG, Solomons & New Caledonia	*Nepenthes spp* (Nepenthaceae) (6 geners of Rubiaceae).
scrofa (de Boisduval, 1832)	Throughout Australia including Tasmania	New Caledonia & Vanuatu	Families (Balsaminaceae) (Onagraceae) (Polygonaceae) (Rubiaceae) (Vitaceae)
velox (Fabricius, 1793)	Cocos-Keeling & Christmas Is, north WA, north NT, Torres Strait, Qld, to Sydney, Nsw, ACT, Norfolk Is	Asia, Philippines, Indonesia, NG, several Pacific Islands east to Samoa.	**Bougainvillea spectabilis*; *Pisonia aculeata*, *P. grandis*, *P. umbellifera* (Nyctaginaceae)
***Hyles* (Hübner, 1819)**			
livornicoides (Lucas, 1892) (Fig. 19.12)	Throughout mainland Australia	–	(Chenopodiaceae) (Nyctaginaceae) (Portulacaceae) (Vitaceae) (Zygophyllaceae)
***Theretra* (Hübner, 1819)**			
celata celata (Butler, 1877)	North WA, north NT, Torres Str. to Canberra, ACT	New Guinea to the Solomon Islands	*Dendrocnide* spp.; *Pipturus* (Urticaceae); *Cayratia* spp.; *Cissus spp* (Vitaceae)
indistincta indistincta (Butler, 1877)	North WA, north NT, Torres Strait to Sydney, NSW	New Guinea, the Moluccas and Sulawesi	*Cayratia* spp; *Cissus* spp.; *Leea indica*; **Vitis vinifera* (Vitaceae);
inornata (Walker, 1865)	North WA, north NT, Torres Strait to Gold Coast, Qld	–	*Ampelocissus frutescens*; *Cissus* spp.; *Clematicissus opaca* (Vitaceae).
insularis insularis (Swinhoe, 1892)	Dauan and Darnley Islands in Torres Strait	North and South Moluccas, New Guinea.	Unknown

Continued

TAXON	DISTRIBUTION		LARVAL FOOD PLANTS
	AUSTRALIA	EXTRA-LIMITAL	
latreilli (Macleay, 1826)	North WA, north NT, Torres Str. to Wollongong, NSW	Papua New Guinea and the Solomon Islands.	(Araceae) (Balsaminaceae) (Lythraceae) (Onagraceae) (Polygonaceae) (Rubiaceae)(Vitaceae)
lucasii (Walker, 1856)	Christmas Island	Much of Asia	(Actinidiaceae) (Araceae) (Balsaminaceae) (Bignoniaceae) (Lythraceae) (Meliaceae) (Onagraceae) (Rutaceae) (Sapindaceae) (Vitaceae)
margarita (Kirby, 1877)	North half of Australia, eastern south to Vic	–	*Boerhavia dominii*, *B. pubescens* (Nyctaginaceae)
nessus nessus (Drury, 1773) (Fig. 19.15)	Christmas Island, Torres Strait Islands to Sydney	Asia, NG, Solomons, introduced to Hawaii	*Alocasia brisbanensis* (Araceae); *Dioscorea* spp. (Dioscoreaceae).
oldenlandiae (Fabricius, 1775)	Christmas Island, north Aus., eastern south to Tas	Afghanistan, Asia, New Guinea	(Araceae) (Balsaminaceae) (Bignoniaceae) (Dilleniaceae) (Lecythidaceae) (Onagraceae) (Rubiaceae) (Vitaceae)
queenslandi (Lucas, 1891)	Jardine River, Qld to Wallaga Lake, NSW	Papua New Guinea	*Dendrocnide* spp.; *Pipturus argenteus* (Urticaceae)
silhetensis intersecta (Butler, 1876)	North WA, north NT, Torres Strait to NE NSW	Sulawesi, Moluccas, NG, Philippines, Guam, Solomons, Vanuatu, New Caledonia & Fiji.	*Typhonium flagelliforme* (Araceae); *Ludwigia* spp (Onagraceae); *Hedyotis* sp., **Pentas lanceolata* (Rubiaceae); *Cyratia acris*, *C. clematidea* (Vitaceae).
tryoni (Miskin, 1891)	Cooktown, Qld to Sydney, NSW	Philippines, Sulawesi, Moluccas, NG, Solomons	*Alocasia* spp.; **Colocasia*; **Cyrtosperma*; **Remusatia;* **Zantedeschia* (Araceae)
turneri (Lucas, 1891)	North WA, north NT, Torres Strait to Byfield, Qld	–	*Dioscorea bulbifera*, *D. transversa* (Dioscoreaceae) *Cayratia trifolia* (Vitaceae)

* indicates non-native species

20 The Noctuoidea: An Overview

Roger L. Kitching
School of Environment and Science, Griffith University, Nathan, Queensland, Australia

Abstract

To date, the largest of all lepidopteran superfamilies, the Noctuoidea, contains at least 60,000 species globally. Following recent phyogenetic analyses, seven families are recognized, all of which occur in Australia. Of the three basal 'trifid' families, the newly recognized Scranciidae have but a single known Australian representative. The Notodontidae are represented by 80 species and the Oenosandridae by eight. The Oenosandridae are one of the few families wholly endemic to Australia. The remaining four 'quadrifid' families, the Euteliidae, Noctuidae, Erebidae and Nolidae, make up the bulk of a very rich fauna. The Erebidae and Noctuidae in particular each contain many hundreds of known species and an equally large or larger numnber awaiting discovery and/or description. Many species, especially of the Noctuidae, have emerged as crop and pasture pests.

20.1 Introduction

The Noctuoidea is the largest of all lepidopteran superfamilies with as many as 70,000 described species globally, of which about 1500 species are of economic importance (Zhang, 1994; [1]Kitching and Rawlins, 1999; Zahiri *et al.*, 2011). The Noctuoidea form a well-established monophyletic grouping within the 'Macrolepidoptera' in turn within the massive clade Obtectomera (see Chapter 3). According to Kawahara *et al.* (2019), they diverged from other clades of macromoths in the late Cretaceous (*ca.* 90 mya) and diversified widely in the Palaeogene, possibly driven by bat predation (see below). The few known noctuoid fossils, often only questionably placed within lower taxa, are mostly from Eocene, Miocene and Pleistocene strata (Sohn *et al.*, 2012). The identity of the Cretaceous fossil egg from North America described by Lawrence Gall and Bruce Tiffney (1983) has been challenged (Kristensen and Skalski, 1999) but if this identity is accepted, it supports the late Cretaceous origin posited by Kawahara *et al.*'s (2019) dated phylogenomic tree.

As might be expected of such a massive taxon as the Noctuoidea, the seven families and about 70 recognized subfamilies within the superfamily are biologically and ecologically diverse. General accounts of the biology of constituent families may be found in Scoble (1992), Common (1990) and Kitching and Rawlins (1999) (among others). Many species have free-living larvae feeding, variously, on a vast range of hosts from lichens and algae to forest trees. Others may form webs and live colonially, or construct leaf shelters. A few are detritivores, even predators. Adults are overwhelm-

Corresponding author: r.kitching@griffith.edu.au
[1]see endnote

DOI: 10.1079/9781800627680.0020

ingly nectar feeders, although some species in the Calpinae can feed by piercing fruit, a behaviour which appears to have led to some species adopting blood- and tear-sucking habits (Bänziger, 2021; Zilli *et al.*, 2024).

The major families now included in the Noctuoidea (Notodontidae, Erebidae, Nolidae, Euteliidae and Noctuidae) have worldwide distributions. The Oenosandridae are restricted to Australia and the Scranciidae to the Old World.

20.2 Taxonomic History of the Superfamily

The Noctuoidea have had a long and tortuous taxonomic history. Holloway (2011), among others, presents the taxonomic history of the group in some detail and clarity. The recent availability of a growing series of papers based on molecular phylogenomics is gradually bringing insight and agreement to the field although key questions remain (see, e.g., Holloway, 2011; Zahiri *et al.*, 2011, 2012). The root name of the noctuoid clade was first coined by Pierre Latreille in 1809 for his concept of the nominate family Noctuidae. Other key family-level designations followed in the 19th century (Erebidae Leach, 1815; Notodontidae Stephens, 1829; Nolidae Bruand, 1847; Euteliidae Grote, 1882). In 1991 American entomologist James Miller added the Oenosandridae and Scranciidae (initially as the notodontine tribe Scranciini, redefined and raised to family level by St Laurent *et al.*, 2023a).

The fundamental division of the noctuoid families into two groups based on forewing venation was initially devised by Hampson (1898–1913). Those families with three branches to the cubitus vein of the forewing are referred to as 'trifid'. This groups the Notodontidae, Oenosandridae and the newly elevated Scranciidae although these three families do not form a monophyletic group (see below). The remaining families appear with four branches to the cubitus vein in the forewing and are referred to, accordingly, as the 'quadrifids'. These are the Erebidae, Nolidae, Euteliidae and Noctuidae. These four families do appear in modern phylogenies as a monophyletic clade.

The seminal paper by Zahiri *et al.* (2011) marked a new era in classification of the superfamily based on phylogenomic analyses examining multiple ribosomal and nuclear genes. Their 2011 paper identified six clades with well-established monophylies. Perhaps the most dramatic outcome (from a lepidopterist's viewpoint) was the resurrection of the family Erebidae (Leach, 1815) to include a large set of subfamilies, including the very familiar Arctiinae and Lymantriinae, both of which had previously been considered families. These six clades have subsequently gained acceptance as defining the noctuoid families. As noted above, a seventh family, the Scranciidae, was separated from among the Notodontidae by St Laurent *et al.* (2023a).

The Noctuoidea are clearly monophyletic under both morphological and molecular analyses. The defining morphological features are tympanal organs on the metathorax with associated structures on the adjacent abdominal segments. Kitching and Rawlins (1999) discuss this structure at length.

As a hearing organ, the tympanum perceives ultrasound and has been credited as a co-evolved adaptation to prey echolocation by insectivorous bats (Fenton and Fullard, 1981; Spangler, 1988; Corcoran and Conner, 2012; Ter Hofstede and Ratcliffe, 2016). As a general correlate of the highly successful radiation of both the Noctuoidea and the Microchiroptera, this idea still has credibility but as dated phylogenies push back the origin of the Noctuoidea into the Cretaceous (Kawahara *et al.*, 2019) – that is, prior to the origin of the micro-bats – alternative explanations for the original purpose of the tympanum in the Noctuoidea are needed.

The existence of small insectivorous pterosaurs, such as the Anurognathidae, throughout the Mesozoic are likely to have driven the evolution of avoidance mechanisms in their prey. The hearing ability of the earliest noctuoids could have been part of this avoidance behaviour although there is no evidence of echolocation in pterosaurs (Bestwick *et al.*, 2018; David Britton, pers. commun.).

Suggestions that the tympanum may have originally existed to receive mating signals then come into play (Conner, 1987; Skals *et al.*, 2003). Most authors have suggested that this form of within-species communication evolved as a secondary use of the tympanal system (Zha *et al.*, 2009) although this seems predicated on an *a priori* assumption of the bat defence hypothesis about tympanal origins. The role of auditory communication in mating and lekking behaviour in the noctuid subfamily Agaristinae is discussed at length in Chapter 22. The absence of any Cretaceous fossils of noctuoids, apart from Gall and Tiffney's (1983) debatable egg, makes it currently impossible to know the tympanal state (or purpose) in the earliest noctuoids.

20.3 Phylogeny

The six lineages identified in their 2011 paper by Reza Zahiri and colleagues have been confirmed and elaborated on subsequently (Zahiri *et al.*, 2012; Regier *et al.*, 2016; St Laurent *et al.*, 2023a), gradually adding more gene regions and species to the ongoing analyses.

The fundamental division between the two (now three) trifid families and the four quadrifid clades is well established for the superfamily. The exact relationship among the basic trifid lineages (Scranciidae, Oenosandridae, Notodontidae) and the quadrifids remains a work in progress. The most recent discussion on the topic is that of St Laurent *et al.* (2023a) as part of their elevation of the scranciid clade to full family status. In attempting to resolve the position of the Scranciidae, they posit seven alternative phylogenetic hypotheses among the four clades (accepting the quadrifids as a well-established mega-clade), each of which arises from one or more of their analyses based on nucleotide or amino acid residues and on a range of inference frameworks plus their underlying assumptions concerning data completeness and informativeness. In an exhaustive and complex statistical analysis, they showed that only two of these seven topologies represented significantly unbiased outcomes. The first of these showed the Scranciidae as sister to all other Noctuoidea, with the Oenosandridae as sister to the notodontids and quadrifids. The second 'credible' topology again places the scranciids as basal to all other noctuoids. Oenosandrids emerge as sister to the notodontids and the [Oenosandridae + Notodontidae] clade is sister to the quadrifines. Of these two configurations, they show the first (Fig. 20.1) to be much more likely than the second. The work continues.

Relationships among the four quadrifid families remain challenging. In their initial molecular analysis, Zahiri *et al.* (2011) hypothesized

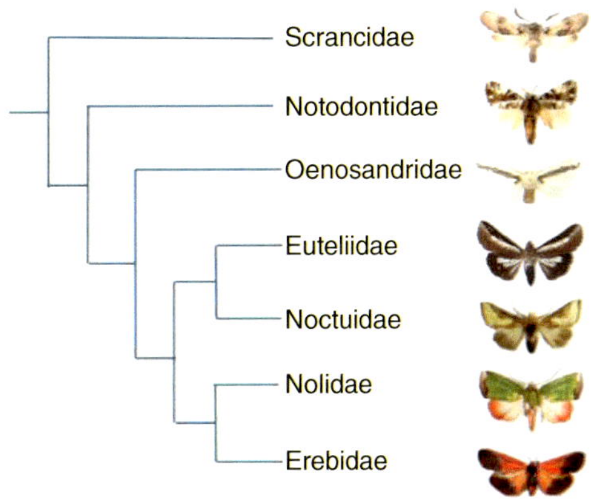

Fig. 20.1. One of several hypotheses of noctuoid phylogeny. Redrawn and modified from St Laurent *et al.* (2023a).

two paired, monophyletic, groups: Euteliidae with Noctuidae and Nolidae with Erebidae. In later work (Zahiri *et al.*, 2012), this hypothesis was modified to show the Nolidae as sister to the other three families. The [Euteliidae + Noctuidae] pairing was retained, sister to the Erebidae. In contrast, Regier *et al.*'s (2016) analysis hypothesized Erebidae as sister to the other three families and the Noctuidae as sister to a Euteliidae/Nolidae pairing. Finally, to date, Zahiri *et al.* (2023) in their discussion of the Euteliidae returned to their original hypothesis with two clades each containing two sister families. They concluded:

> Relationships among the four families within the quadrifid Noctuoidea ... are not clear, although they form a strongly supported ... monophyletic group ... with Euteliidae sister to Noctuidae and Nolidae sister to Erebidae in both ML [Maximum Likelihood] and BI [Bayesian Inference] analyses. (Zahiri et al., 2023)

This latest phylogenetic hypothesis will be tested further as gene sequences and the number of species sampled continue to increase.

20.4 The Australian Fauna and its Global Context

20.4.1 The Scranciidae

As discussed above, the basal noctuoid Scranciidae were elevated to family status from among the Notodontidae by St Laurent *et al.* (2023a). Those authors provide a full diagnosis for the new family. For adults and in summary, they state: '... the combination of the elongate broad, accessory cell, absence of cteniophores and frequent absence of socii is diagnostic of Scranciidae'. The 'accessory cell' is a feature of the forewing venation. 'Cteniophores' are membranous outpushings of the pleura on each side of the fourth abdominal sternite and 'socii' refers to paired lobes on the uncus of the male genitalia (definitions from Scoble, 1992; see also definitions and illustrations in Holloway, 1983).

For larvae they write: 'Scranciid caterpillars are unnmistakeable having both reduced prolegs on abdominal segments three and four and stemapodiform anal prolegs with extendable filaments on the tenth abdominal segment'. Here 'stemapodiform' refers to the modification of the anal prolegs to produce long extendable filaments.

The only known Australian scranciid is *Lasioceros aroa* (Fig. 20.2). This species is included in the Australian checklist (Edwards, 1996b) as a notodontine notodontid but *Lasioceros* is one of the 22 current genera listed by St Laurent *et al.* (2023a) making up the family globally. Australian distribution records (Atlas of Living Australia, www.ala.org.au/) suggest it is a coastal tropical species and ranges from the extreme north-west of Western Australia, through the coastal Northern Territory to Cape York and southwards along the Queensland Pacific Coast to about Mackay. Other records exist from Papua New Guinea. Other members of the genus occur in South-east Asia as far as Sulawesi.

I can find no host plant records for *Lasioceros* spp. Other genera of Scranciidae apparently favour Euphorbiaceae but a range of other plant families have been recorded as larval hosts (St Laurent *et al.*, 2023a).

Fig. 20.2. *Lasioceras aroa*, the only Australian scrancid. Photo: Dominic Funnell.

20.4.2 Oenosandridae

The endemic Australian family Oenosandridae – referred to as 'snub moths' – was separated from the Notodontidae: Thaumatopoeinae by Miller (1991) based on his examination of *Oenosandra boisduvalii* (Fig. 20.3). He wrote: '[t]he species exhibits a seemingly contradictory set of characters'. These contradictory and defining characters include trifid venation in the forewing, socii in the male genitalia and a typanum configuration resembling that of the quadrifids rather than the Notodontidae.

Fig. 20.3. *Oenosandra boisduvalii,* an example of the Australian endemic family Oenosandridae. Photo: Kristi Ellingsen.

As discussed earlier, molecular analyses have confirmed the Oenosandridae as a separate clade (or perhaps clades) within the Noctuoidea. Zahiri *et al.* (2011) show the Oenosandridae as sister to the four quadrifid clades with the Notodontidae as sister to the [Oenosandridae + quadrifid families]. The more recent work of St Laurent *et al.* (2023a) suggests the Oenosandridae are sister to all Noctuoidea (including the Notodontidae) except the Scranciidae.

Subsequent to Miller's recognition of the family, other genera have been added to the taxon, resulting in eight described species now being listed as oenosandrids in the Australian checklist: *Dicerateucha* (1 species), *Discophlebia* (5 species), *Nycteropa* (1 species) and *Oenosandra* (1 species). No molecular analyses to date have included more than *Oenosandra* and *Discophlebia*. The monophyly of the family remains to be demonstrated unequivocally (Zahiri *et al.*, 2012; St Laurent *et al.*, 2023a). All four genera are restricted to Australia and most distribution records are from the southern half of the continent.

There are host plant records for *O. boisduvalii* and *D. catocalina* (https://lepidoptera.butterflyhouse.com.au/oeno/oenosandridae.html). In both cases the larvae feed on *Eucalyptus* leaves, suggesting that the family has diverged and radiated (modestly) with this dominant element of the Australian flora.

20.4.3 The Notodontidae

The Notodontidae are a moderately sized family of larger noctuoids generally referred to as prominents and include species of economic importance. They are distributed globally and include more than 4700 described species. The infra-family classification has recently been resolved in a thorough work by St Laurent *et al.* (2025) based on extensive phylogenomics as well as morphological characteristics, particularly of larvae. These authors identify 21 subfamilies globally, of which 11 are present in Australia. They examined 150 species in their analyses but still suggest the Australian and New Guinea faunas were underrepresented. Nevertheless, with the aid of parallel, unpublished local analyses they were able to assign almost all Australian genera to well-supported subfamilies. They suggest that the Australian genera *Gallaba* (Lusurinae) and *Archigargetta* (unassigned) are 'enigmatic' and worthy of further study.

Notodontids have very distinctive larvae with a variety of ornamentation and 'highly irregular' structures. Larval feeding preferences include groups particularly associated with the Salicaceae, Passifloraceae, Juglandaceae, Fabaceae and Poaceae (St Laurent *et al.*, 2023a, 2023b).

The Australian checklist (Edwards, 1996a) names 80 species of Australian notodontids. The online list created by Don Herbison-Evans and Stella Crossley (https://lepidoptera.butterflyhouse.com.au/noto/notodontidae.html) adds an additional species of *Trichiocercus* to this number. The single species of Australian scranciid should be removed from these lists. Key works of particular relevance to the Australian fauna, in addition to those of St Laurent and colleagues, are Turner (1922), Kiriakoff (1968), Holloway (1983) and Schintlmeister (2013).

Host plant records exist for about 40% of the named Australian species (https://lepidoptera.butterflyhouse.com.au/noto/notodontidae.html). Feeding records extend across 17 plant families but by far the most prominent among these are the Myrtaceae (16 moth species) followed by the Mimosaceae (6 species) and Proteaceae (4 species). The close association of the speciose notodontid genus *Epicoma* with genera of Myrtaceae implies a strong Australian co-radiation.

20.4.4 The Euteliidae

The Euteliidae along with the Stictopteridae were recognized as sister families separate from the Noctuidae (*s.l.*) by Mitchell *et al.* (2006). The two groupings had been recognized as sister taxa by earlier workers (Holloway, 1985; Kitching, 1987) who identified a number of morphological synapomorphies (see below). Subsequent molecular analyses have recognized the two taxa as making up a well-supported monophyletic clade within the quadrifid Noctuoidea, leading them to be recognized currently as two subfamilies of the family Euteliidae (so called because of naming priority over Stictopteridae) (Zahiri *et al.*, 2011, 2012, 2023). Globally the family comprises 520 species organized into 34 genera (van Nieukerken *et al.*, 2011), making it the smallest of the four quadrifid noctuoid families. The Euteliidae are the sister family to the Noctuidae (*s.s.*) as now recognized, with an estimated divergence time of 53 mya in the early Eocene (Zahiri *et al.*, 2023).

A large suite of synapomorphies defines the family. These include:

- a reduced female frenulum
- a clypeofrons with scaling only at the margins
- hindlegs shorter than midlegs
- a double structure to the counter-tympanal hood
- and several genitalic characters in both males and females (Zahiri *et al.*, 2023).

These authors also define the two subfamilies morphologically.

The molecular phylogeny presented by Zahiri *et al.* (2023) defines the Stictopterinae as basal to the Euteliinae. Two tribes within the Stictopterinae are recognized.

The Australian euteliid fauna comprises about 35 named species in 13 genera spread across the two subfamilies (https://bie.ala.org.au/). The Herbison-Evans and Crossley website (https://lepidoptera.butterflyhouse.com.au/euteliidae/euteliidae.html) lists nine local food plant records. Zahiri *et al.* (2023) suggest that plants belonging to the Malpighiales were probably the ancestral food plants of the family and, indeed, the two Australian host plant records from the Stictopterinae are species of Clusiaceae. They note a more recent global bias among euteliid food plants towards the Anacardiaceae and four of the eight Australian records are from that plant family. At least two species of Australian euteliines have been recorded feeding on Myrtaceae (*Pataeta carbo* on *Eucalyptus* and *Callistemon*, *Targalla delatrix* on *Eugenia* spp.). Several species are minor pests of tropical fruit.

20.4.5 The Noctuidae

As discussed earlier, the Noctuidae – the 'owlet moths' – emerge as a strongly supported monophyletic clade within the quadrifid noctuoids in all recent molecular analyses (see in particular Zahiri *et al.*, 2013, Keegan *et al.*, 2021). Even after the many species previously included in the catch-all 'Noctuidae' (*s.l.*) are variously reassigned to the other three quadrifid groups, the Noctuidae (*s.s.*) remain one of the largest lepidopteran families with more than 12,000 described species globally and, according to estimates, thousands awaiting description (Kitching and Rawlins, 1999; van Nieukerken *et al.*, 2011; Keegan *et al.*, 2021). The Noctuidae are indicated as the sister group to the Euteliidae from which they diverged in the Eocene, diversifying dramatically in the Miocene. Current phylogenies identify 26 subfamilies although some may not be monophyletic – especially the nominate Noctuinae (Keegan *et al.*, 2021). I return to this point below.

Despite the well-established family monophyly, the Noctuidae show few if any unequivocal synapomorphies. The hindwing venation has received considerable attention (discussed at length by Zahiri *et al.*, 2013). Many recognized subfamilies show clear 'trifine' hindwing venation where the M2 vein is usually reduced or vestigial whereas the cubital vein appears to be three-branched (hence 'trifine' – not to be confused with 'trifid', a condition of the forewing). Other subfamilies, however, do not show the M2 reduction and have been called 'pseudoquadrifine'.

The most recent published phylogeny of the family is that of Keegan *et al.* (2021). They sampled 341 genera across the 20 (of 21) subfamilies recognized at that time. Their selection of target species greatly increased the geographic scope of sampling compared with earlier studies although the African continent remained substantially undersampled. They examined eight protein-coding genes in this study and clarified the status of many subfamilies and described two as new. The most derived subfamily, however, the Noctuinae, and some currently recognized 'adjacent' clades (Eriopinae, Bryophilinae, Phosphilini), remain systematically unresolved. The authors say: 'Consistent with the findings of Mitchell *et al.* (2006) and Regier *et al.* (2016), we found the Noctuinae to be an unnatural assemblage'.

The Noctuinae, of course, remain the largest of currently recognized subfamilies and are still undersampled in molecular analyses. There is ongoing research seeking phylogenetic hypotheses among the species usually regarded as noctuines. Sisson *et al.* (2023), for example, have examined the putative tribe Noctuini (which includes such key pest genera as *Spodoptera*, *Agrotis*, *Euxoa* and *Ochropleura*), finding, in their words, only 'equivocal support' for the monophyly of the tribe as a whole and for the two subtribes within it.

In such a vast assemblage of species as the Noctuidae, seeking general patterns of host plant use is a challenging task. This is discussed at length by Zahiri *et al.* (2013). They draw some conclusions based on available data suggesting woody plants represent the basal larval association within the family (indeed, within the entire superfamily) and that several lineages that are basal within their phylogeny retain this woody plant association. These include the Dilobinae, Raphiinae, Pantheinae, Acronictinae and several tribes within the Noctuinae. Other more derived subfamilies have retained specialized feeding preferences: Agaristinae on the Vitaceae, Dyopsinae on the Urticaceae, Bagisarinae on the Malvaceae, and so on. These associations represent broad preferences rather than unvarying patterns. Other more derived subfamilies have switched to herbs and grasses. These include the Heliothinae, many Noctuinae, Amphipyrinae and Metoponinae. Many of these have become highly polyphagous with food plants coming from many plant families. Here too, the tendency to become agricultural and pasture pests has arisen many times (see below).

The Australian Noctuidae comprise about 400–500 named species (Edwards, 1996b) and the 'usual' rump of many species awaiting attention. Various attempts have been made to match the known Australian genera against currently recognized subfamilies and attempts can be found on both the Atlas of Living Australia website (https://bie.ala.org.au/species/https://biodiversity.org.au/afd/taxa) and the Herbison-Evans/Crossley lists (https://lepidoptera.butterflyhouse.com.au/noctuidae/noctuidae.html). The moving target nature of global subfamily designations makes these assignments challenging. Keegan *et al.* (2021), for instance, made no attempt to separate the majority of the Australian Acronictinae/Amphipyrinae complex which they did not place within their phylogeny, leaving them unresolved as in the Edwards (1996b) checklist. The Australian fauna probably contains representatives of 11 or 12 noctuid subfamilies although this includes at least one, the Cucullinae, where the only Australian representative, *Neogalea sunia*, was introduced as a biocontrol agent for the shrubby weed *Lantana*.

The exception to this dearth of information is the subfamily Heliothinae which has been comprehensively monographed by Marcus Matthews (1999). Matthews recognized 38 Australian heliothines in five genera. Eight of these species he described as new.The focus on the Heliothinae, of course, reflected the interest in the group for economic reasons, with two species, *Helicoverpa armigera* and *H. punctigera*, being among Australia's most damaging pests of agriculture (see Chapter 32). It is no surprise, therefore, that Matthews' authoritative study was funded largely by various industry funding bodies. Biogeographic aspects of this study are discussed in Chapter 4.

Known larval food plants of the Australian Noctuidae are listed in the Herbison-Evans/Crossley database. A substantial majority of species have no recorded food plant so generalizations must be qualified. Some smaller subfamilies appear to follow the global patterns. Our only dyopsine is recorded feeding on Urticaceae, aedines on Convolvulaceae. Other lineages (or putative lineages) are much more generalist in their eating habits. Species listed as acronictines (probably also including amphipyrines) have been recorded from over 23 families of angiosperms. Eustrotiines, heliothines, plusiines and noctuines are similarly catholic in their preferences. Occasional 'exceptional' feeding habits have been noted. A few acronictines feed on ferns, eustrotiines on lichens and carnivory (principally on scale insects) appears very occasionally in both the Acontiinae and Noctuinae. The highly polyphagous nature of many species underpins their roles as major and minor pests of forestry, agricultures, pastoralism and horticulture (see below).

The chapter by Michael Braby in this volume describes the current state of knowledge of the Australian species of a the noctuid subfamily Agaristinae (Chapter 22). Other chapters present accounts of *Helicoverpa* species (Heliothinae) (Chapter 32) and *Agrotis infusa* (Noctuinae) (Chapter 37).

20.4.6 The Nolidae

Until relatively recently, the currently recognized noctuoid family Nolidae existed as a scattering of subfamilies within the Noctuidae (and the now subsumed Arctiidae). After widespread discussions in the community of noctuidologists, Jeremy Holloway (1998) reviewed the morphology of four of these groups, concluding they were best regarded as a single monophyletic family, the Nolidae. This hypothesis was confirmed and expanded by Kitching and Rawlins (1999) who included nine subfamilies within the Nolidae. This expanded concept of the Nolidae was then used in Holloway's (2003) treatment of the large Bornean fauna.

The Nolidae, in this expanded sense, emerged as one of the four monophyletic quadrifid lineages within the molecular analyses of Zahiri *et al.* (2011). The 11 putative nolids included in their analyses sorted into six subfamilies. Their analyses did not include any representatives of the Afridinae, Collomeninae or Risobinae. A more comprehensive analysis followed. Zahiri *et al.* (2013) sampled 82 nolids, examining eight separate gene regions. They identified eight 'strongly supported' monophyletic subfamilies, one of which – the Diphtherinae – they described as new, although again two putative subfamilies were not included in their samples (the Central and North American Afridinae and the Asian Camptolominae).

The Nolidae, sometimes referred to as 'tuft moths', as defined above, are estimated to contain 1700 species globally (van Nieukerken *et al.*, 2011). The principal defining morphological synapomorphy is the boat-shaped chrysalis with a vertical, anterior exit slit (Kitching and Rawlins, 1999).

The Australian nolid fauna comprises about 180 named species (https://bie.ala.org.au/species/) organized into over 80 genera. Some genera remain to be assigned to subfamilies. The database of Herbison-Evans and Crossley records food plants for representatives of 25 Australian nolid genera. Overall, 20 different plant families are recorded ranging from canopy trees to grasses and sedges. Several genera of noline and one chloephorine have been recorded feeding on Myrtaceae. One species of *Nola* was reared on lichen. Chapter 31 presents an account of the biology of the *Uraba lugens* (Nolinae), the gum leaf skeletonizer.

20.4.7 The Erebidae

This huge family was resurrected by Zahiri *et al.* (2011) to describe the largest monophyletic clade within their molecular phylogeny of the Noctuoidea. The authors chose Leach's (1815) name rather than the equally venerable Arctiidae or Herminiidae to avoid confusion with more recent usages of these latter two names (Holloway, 2011; Zahiri *et al.*, 2012). Globally the Erebidae number about 25,000 named species and an incalculable number awaiting discovery and/or description (Zahiri *et al.*, 2012). Sisson *et al.* (2025) present an account of the history and current state of the systematics of the family.

The monophyly of the Erebidae now seems beyond doubt and the taxa which belong within it are now well established. Both of these conclusions are based on more and more thorough molecular analyses. Nevertheless, as Zahiri *et al.* (2012) point out, 'the family Erebidae is difficult to define morphologically in a manner that can be applied in a dichotomous key'. A variety of morphological characters unite subsets of the now emerging subfamilies but there are notable exceptions even within these groupings. Zahiri *et al.* (2012) discuss these similarities and differences at length.

The most recent molecular analyses define clades which equate to 18 subfamilies within the Erebidae which are 'moderately to well supported' (Zahiri *et al.*, 2012). Many of these are familiar as groupings previously recognized with either family or subfamily status with morphological definitions. These include well-known groupings such as the Arctiinae, Lymantriinae, Herminiinae, Hypeninae and Aganainae. The Scoliopteryginae emerge as the basal group although the 'tightness' of the phylogeny implies a rapid diversification close to the base of the subfamily.

The Australian fauna now placed within the Erebidae includes more than 1300 named species with many more recognized but awaiting descriptions. No doubt many more await discovery. Representatives of 16 of the 18 subfamilies defined by Zahiri *et al.* (2012) make up this fauna. The dominant subfamilies are the nominate Erebinae (about 230 named species), the Arctiinae (*ca.* 250 species) and the Hypeninae (over 100 species).

Connecting subfamilies with preferred larval host plants is always challenging, especially for larger subfamilies. This is complicated because we have reliable host plant records for only a small fraction of the Australian erebids. Where there are multiple records for particular subfamilies, some patterns do appear. The basal Scoliopterygidae seem to have radiated with Malvaceae, the Aganainae with Moraceae, and the Calpinae with the Menospermataceae. In contrast, records for the Hypeninae, Lymantriinae and Erebinae span a huge range of plant families from canopy trees to weedy herbs. Unusual larval feeding records include conifers (Hypeninae, Lymantriinae), fungi (Hypeninae), leaf litter (Herminiinae) and lichens (lithosiine arctiines).

Two further chapters in this work examine erebids in more detail: Chapter 21 discusses the Arctiinae and Chapter 30 the fruit-piercing moths (Calpinae).

20.5 Economic Importance

The Noctuoidea contain a range of species which have become major pests of agriculture (see Chapter 23). Some of these have very widespread distributions beyond Australia and some, certainly, have originated overseas, no doubt imported with foodstuffs. A perhaps surprising number, however, are endemic to Australia and have switched to readily available crop plants. The polyphagous nature of many noctuids has no doubt facilitated this process.

The large majority of noctuoid species recorded as major pests are from the family Noctuidae but there are a few exceptions. The Thaumatopoeinae (Notodontidae) provide us with *Ochrogaster* species, a serious nuisance pest of humans and livestock (see Chapter 29) A few species of nolids have become pests of economic importance in Australia. The gum leaf skeletonizer (*Uraba lugens*) is a major pest of plantation eucalypts (see Chapter 31). Some Malvaceae-feeding species of *Earias* are pests of cotton. Within the Erebidae, the fruit-piercing moths (Calpinae, *Eudocima* spp.) stand out as pests of tropical fruit where it is the adults' feeding activities that cause loss rather than the larvae (see Chapter 30). Within the Noctuidae, however, well over 30 species cause major economic damage in Australia. These range from specific problem species such as the grapevine moth *Phalaenoides glycinae* (Agaristinae) and the chevron cutworm *Diarsia intermixta* (Noctuinae), a pest of turnips in Tasmania, through to the real villains attacking a wide range of crops across huge areas of the continent. Species of *Agrotis*, *Helicoverpa*, *Heliothis*, *Leucania*, *Proteuxia* and *Spodoptera* (among others) cost the agricultural, pastoral and horticultural industries untold millions each growing season. The often draconian chemical and other management interventions add to this cost by often unquantified environmental impacts. Detailed accounts of *Helicoverpa* species (Chapter 32) and *Agrotis infusa* (Chapter 37) appear elsewhere in this volume.

Acknowledgements

This chapter has benefitted greatly from critical and constructive comments by David Britton, the late Jeremy Holloway, Scott Miller and Ryan St Laurent. I thank them sincerely for this input.

Note

[1]The author wishes to make clear that the many references to 'Kitching' in the literature of the Noctuoidea and other taxa recognize the important and ongoing contributions of Dr Ian J. Kitching of the Natural History Museum, London. Dr I.J. Kitching is an esteemed colleague but no relative of the chapter author.

References

Bänziger, H. (2021) *Vampire Moths: Behaviour, Ecology and Taxonomy of Blood-Sucking* Calyptra. Natural History Publications, Kota Kinabalu, Malaysia.

Bestwick, J., Unwin, D.M., Butler, R.J., Henderson, D.M. and Purnell, M.A. (2018) Pterosaur dietary hypotheses: A review of ideas and approaches. *Biological Reviews* 93, 2021–2048.
Bruand, T. (1847) Catalogue systématique et synonymique des Lépidoptères du département du Doubs. *Mémoires de la Société d'Émulation du Doubs* 2(Ser.1), 97–124.
Common, I.F.B. (1990) *Moths of Australia*. Melbourne University Press, Carlton, Victoria.
Conner, E.E. (1987) Ultrasound: Its role in the courtship of the arctiid moth, *Cycnia tenera*. *Experientia* 43, 1029–1031.
Corcoran, A.J. and Conner, W.E. (2012) Sound strategies: The 65-million-year-old battle between bats and insects. *Annual Review of Entomology* 57, 21–39.
Edwards, E.F. (1996a) Notodontidae. In: Nielsen, E.S., Edwards, E.D. and Rangi, T.V. (eds) *Checklist of the Lepidoptera of Australia*. Monographs on Australian Lepidoptera, 4. CSIRO Publishing, Collingwood, Victoria, pp. 272–274.
Edwards, E.F. (1996b) Noctuidae. In: Nielsen, E.S., Edwards, E.D. and Rangi, T.V. (eds) *Checklist of the Lepidoptera of Australia*. Monographs on Australian Lepidoptera, 4. CSIRO Publishing, Collingwood, Victoria, pp. 291–333.
Fenton, M.B. and Fullard, J.H. (1981) Moth hearing and the feeding strategies of bats. *American Scientist* 69, 266–275.
Gall, L.F. and Tiffney, B.H. (1983) A fossil noctuid egg from the Late Cretaceous of eastern North America. *Science* 219, 507–509.
Grote, A.R. (1882) *New Check List of North American Moths*. Privately printed.
Hampson, G.F. (1898–1913) *Catalogue of the Lepidoptera Phalaenae in the British Museum*. British Museum (Natural History), London, pp. 1–13.
Holloway, J.D. (1983) The Moths of Borneo: Family Notodontidae. *Malayan Nature Journal* 37, 1–107.
Holloway, J.D. (1985) The Moths of Borneo: Family Noctuidae: Subfamilies Euteliinae, Stictopterinae, Plusiinae, Pantheinae. *Malayan Nature Journal* 38, 157–317.
Holloway, J.D. (1998) The classification of the Sarrothripinae, Chloephorinae, Camptolominae and Nolinae as the Nolidae (Lepidoptera: Noctuoidea). *Quadrifina* 1, 247–276.
Holloway, J.D. (2003) *The Moths of Borneo, Part 18, Nolidae*. Southdene Sdn. Bhd, Kuala Lumpur, Malaysia.
Holloway, J.D. (2011) The Moths of Borneo: Families Phaudidae, Himantopteridae and Zygaenidae: Revised and annotated checklist. *Malayan Nature Journal* 63, 1–548.
Kawahara, A.Y., Plotkin, D., Espeland, M., Meusemann, K., Toussaint, E.F.A. *et al.* (2019) Phylogenomics reveals the evolutionary timing and pattern of butterflies and moths. *Proceedings of the National Academy of Sciences* 116, 22657–22663.
Keegan, K.L., Rota, J., Zahiri, R., Zilli, A., Wahlberg, N. *et al.* (2021) Towards a stable global Noctuidae (Lepidoptera) taxonomy. *Insect Systematics and Diversity* 5, 1–24.
Kitching, I.J. (1987) Spectacles and silver Y's: A synthesis of the systematics, cladistics and biology of the Plusiinae (Lepidoptera: Noctuidae. *Bulletin of the British Museum (Natural History) (Entomology)* 54, 153–234.
Kitching, I.J. and Rawlins, J.E. (1999) The Noctuoidea. In: Kristensen, N.P. (ed.) *Lepidoptera, Butterflies and Moths* 1, *Handbook of Zoology* 4(1). De Gruyter, Berlin & New York, pp. 355–401.
Kiriakoff, S.G. (1968) Lepidoptera Familia Notodontidae. In: Wytsman, P. (ed.) *Genera Insectorum*, 217C. Edition Mercurites, Brussels, Belgium, pp. 1–269.
Kristensen, N.P. and Skalski, A.W. (1999) Phylogeny and palaeontology. In: Kristensen, N.P. (ed.) *Lepidoptera, Butterflies and Moths* 1, *Handbook of Zoology* 4(1). De Gruyter, Berlin & New York, pp. 7–25.
Latreille, P.A. (1806–1809) *Histoire Naturelle Générale et particulière des Crustacés et des Insectes*. F. Dufart, Paris.
Leach, W.E. (1815) Entomology. In: *Edinburgh Encyclopaedia*. Brewster, Edinburgh, Scotland, p. 134.
Matthews, M. (1999) *Heliothine Moths of Australia: A Guide to Pest Bollworms and Related Noctuid Groups*, Monographs on Australian Lepidoptera, 7, CSIRO Publishing, Collingwood, Victoria.
Miller, J.S. (1991) Cladistics and classification of the Notodontidae (Lepidoptera: Noctuoidea) based on larval and adult morphology. *Bulletin of the American Museum of Natural History* 204, 1–230.
Mitchell, A., Mitter, C. and Regier, J.C. (2006) Systematics and evolution of the cutworm moths (Lepidoptera: Noctuidae): Evidence from two protein-coding genes. *Systematic Entomology* 31, 21–466.
Regier, J.C., Mitter, C., Mitter, K., Cummings, M.P., Bazinet, A.L. *et al.* (2016) Further progress on the phylogeny of Noctuoidea (Insecta: Lepidoptera) using an expanded gene sample. *Systematic Entomology* 42, 82–93.
Schintlmeister, A. (2013) *World Catalogue of Insects. Volume 2. Notodontidae and Oenosandridae (Lepidoptera)*. Brill, Leiden, The Netherlands.

Scoble, M.J. (1992) *The Lepidoptera: Form, Function and Diversity*. British Museum (Natural History), London.
Sisson, M.S., Flom, M.J., Crossley, J.L. and Simmons, R.B. (2023) A molecular phylogeny of Noctuini (Lepidoptera: Noctuidae: Noctuinae). *Insect Systematics and Diversity* 7, 1–10.
Sisson, M.S., Dowdy, N.J., Fisher, M.L., Gall, L.F., Goldstein, P.Z. *et al.* (2025) Erebidae systematics: Past, present, and future – Progress in understanding a diverse lepidopteran lineage. *Insect Systematics and Diversity* 9.3: 5.
Skals, N., Plepys, D. and Löfstedt, C. (2003) Foraging and mate-finding in the silver Y moth, *Autographa gamma*, (Lepidoptera: Noctuidae) under the risk of predation. *Oikos* 102, 351–357.
Sohn, J.-C., Labandeira, C., Davis, D. and Mitter, C. (2012) An annotated catalogue of fossil and subfossil Lepidoptera (Insecta: Holometabola) of the world. *Zootaxa* 3286, 1–132.
Spangler, H.G. (1988) Moth hearing, defense and communication. *Annual Review of Entomology* 33, 59–81.
Stephens, J.F. (1829) *The Nomenclature of British Insects*. Baldwin & Cradoc, London.
St Laurent, R.A., Goldstein, P.Z., Miller, S.E. and Robbins, R.K. (2023a) Hidden in plain sight: Phylogenomics reveals a new branch on the Noctuoidea tree of life. *Systematic Entomology* 49, 1–21.
St Laurent, R.A., Goldstein, P.Z., Miller, J.S., Markee, A., Staude, H.S. *et al.* (2023b) Phylogenetic systematics, diversification and biogeography of Cerurinae (Lepidoptera: Notodontidae) and a description of a new genus. *Insect Systematics and Diversity* 7, 1–25.
St Laurent, R.A., Goldstein, P.Z., Prada-Lara, L., Schintlmeister, A., Miller, S.E. *et al.* (2025) Phylogenomics of prominent moths (Lepidoptera: Notodontidae): A subfamily-level reclassification. *Smithsonian Contributions to Zoology* 657, 1–133.
Ter Hofstede, H.M. and Ratcliffe, J.M. (2016) Evolutionary escalation: The bat-moth arms race. *Journal of Experimental Biology* 219, 1589–1602.
Turner, A.J. (1922) Revision of Australian Lepidoptera: Saturniidae, Bombycidae, Eupterotidae, Notodontidae. *Proceedings of the Linnean Society of New South Wales* 47, 348–390.
van Nieukerken, E.J., Kaila, L., Kitching, I., Kristensen, N.P., Lees, D.C. *et al.* (2011) Order Lepidoptera Linnaeus, 1758. In: Zhang, Z.Q. (ed.) Animal biodiversity: An outline of higher-level classification and survey of taxonomic richness. *Zootaxa*, pp. 212–221.
Zahiri, R., Kitching, I.J., Lafontaine, J.D., Mutanen, M., Kaila, L. *et al.* (2011) A new molecular phylogeny offers hope for a stable family level classification of the Noctuoidea (Lepidoptera). *Zoologica Scripta* 40, 158–173.
Zahiri, R., Holloway, J.D., Kitching, I.J., Lafontaine, J.D., Mutanen, M. and Wahlberg, N. (2012) Molecular phylogenetics of Erebidae (Lepidoptera: Noctuoidea). *Systematic Entomology* 37, 102–124.
Zahiri, R., Lafontaine, J.D., Holloway, J.D., Kitching, I.J., Schmidt, B.C. *et al.* (2013) Major lineages of Nolidae (Lepidoptera: Noctuoidea) elucidated by molecular phylogenetics. *Cladistics* 29, 337–359.
Zahiri, R., Holloway, J.D., Rota, J., Schmidt, B.C., Pellinen, M.J. *et al.* (2023) Evolutionary history of Euteliidae (Lepidoptera: Noctuoidea). *Systematic Entomology* 48, 445–462.
Zha, Y., Chen, Q. and Lei, C. (2009) Ultrasonic hearing in moths. *Annales de la Société Entomologique de la France* 45, 145–156.
Zhang, B.C. (1994) *Index of Economically Important Lepidoptera*. CABI International, Wallingford, UK.
Zilli, A., Barbut, J., Dorward, L.J. and Lees, D.C. (2024) The bird tear-drinking moths of the genus *Hemiceratoides* (Lepidoptera: Erebidae). *Zoological Journal of the Linnean Society*, 202, 1–22.

21 The Arctiinae: The Tiger, Lichen and Wasp Moths

David R. Britton[1]* and Georgina E. Binns[2]
[1]*Elizabeth Macarthur Agricultural Institute, NSW Department of Primary Industries and Regional Development, Menangle, NSW, Australia;* [2]*School of Natural Sciences, Macquarie University, Macquarie Park, NSW, Australia*

Abstract
The Arctiinae (tiger, lichen and wasp moths) are a conspicuous and fascinating component of the Australian moth fauna. This subfamily is diverse, often aposematic and chemically defended, and frequently polyphagous. Despite the larger Arctiini being relatively well known, there is almost nothing known of life histories of the tribes Lithosiini and Syntomini, even though the adults of these moths are ubiquitous in most populated areas of Australia. This chapter highlights features and knowledge gaps for the Australian Arctiinae fauna, including the current state of knowledge of taxonomy, diversity, life histories, distribution and ecology, especially in relation to chemical and behavioural ecology. It provides direction for future researchers interested in these moths.

21.1 Introduction

The Arctiinae are a diverse but relatively discrete and recognizable subfamily within the Erebidae (see Chapter 20). The subfamily name (Arctos = bear or bear-like) refers to the conspicuous secondary setae which are found in many arctiine larvae but are especially well developed in the 'woolly bear' larvae of moths in the tribe Arctiini. In many species these secondary setae are plumose and, while not unique to the subfamily, this provides a useful character for larval identification. The adults, and in some cases the larvae, are often aposematically coloured, hence the common name 'tiger moths' applied to larger species in the Arctiini. Adults have metepisternal tymbal organs for production of ultrasonic sound, along with thoracic tymbals for hearing; analogous organs occur in other groups of Lepidoptera, but the structure and placement of the arctiine examples are apomorphic for the group (Scoble, 1992; Kitching and Rawlins, 1998). Another unique feature of the subfamily is the laterally paired invaginated dorsal pheromone glands in adult females (Bendib and Minet, 1998).

Species in the subfamily that feed on vascular plants are often polyphagous, and few species appear to have specialized food plant requirements, although life histories are lacking for most Australian species, especially those in the tribes Lithosiini and Syntomini (Robinson *et al.*, 2023).

Arctiini have prothoracic glands which can produce distasteful and odorous secretions, presumably to deter predation, with most Lithosiini and Syntomini appearing to have similar glands (Fig. 21.1A) (Kitching and Rawlins, 1998). Pyrazines, produced from prothoracic glands

*Corresponding author: david.britton@dpi.nsw.gov.au

DOI: 10.1079/9781800627680.0021

Fig. 21.1. (A) Reflex 'bleeding' of defence secretions in *Amata nigriceps* (Syntomini), Yarramundi, NSW. Photo: G.E. Binns. (B) Males of *Euchromia creusa* (Arctiini) with coremata expanded while imbibing pyrrolizidine alkaloids from senescent rattlepod plants (*Crotalaria* sp., Fabaceae). Photo: D. Britton, Kangu Beach, Bougainville. (C) A drawer of *Ardices curvata* (Arctiini) specimens demonstrating variability in wing patterns. Photo: Heather Poole, © 2025 Australian Museum.

and used as an odour defence, have also been found in Australian species of *Amata* in the tribe Syntomini (Rothschild *et al.*, 1984). Aposematic colouration is common in the subfamily, presumably associated with chemical defence. Mimicry, both Batesian and Müllerian, also occurs (Chiocchio *et al.*, 2021). The visual appearance of these moths is often augmented by auditory advertising and other behaviours which may deter potential predators (Conner *et al.*, 2008). The acquisition and storage of chemical defences may be a basal characteristic of the Arctiinae (Weller *et al.*, 1999). The behaviour and chemical ecology of the ornate arctiid *Utetheisa ornatrix* have been studied in detail in North America (Conner, 2009) and have shown how these chemical defences are inextricably linked to reproductive behaviour in this moth.

21.2 Taxonomic History in Australia

Australian publications on the Arctiinae before Zahiri *et al.* (2011, 2012) give the subfamily full family status. Turner (1940) provided the first review of the Australian arctiine fauna, since this work there have been relatively few new Australian species described for the subfamily and no new genera. Common (1990) provided a summary of the Australian arctiine fauna in *Moths of Australia*, some of which is updated here, and Edwards (1996) provided a checklist of valid species. The most significant recent taxonomic changes affecting the Arctiinae have been from extralimital lepidopterists working to the north of Australia, most notably J.D. Holloway's monumental series *The Moths of Borneo*. The generic concepts used in Turner's revision relied heavily on wing venation, among other external characters (Turner, 1940). Holloway (1988, 2001, 2011), in revising the group for the fauna of Borneo, made several decisions that include genera with Australian representatives. His revisions included examination of genitalia but he did not have the time, or possibly the access, to examine all the affected Australian species to confirm their status. Britton (2010) updated the Edwards checklist to include Holloway's changes where it was unambiguous.

Several Australian Arctiini were included in the paraphyletic genus *Spilosoma* (Edwards, 1996). Dubatolov (2005) reviewed some of the Australian Arctiini, resurrecting *Ardices* for *A. curvata*, *A. glatignyi* and *A. canescens*. Two other species are placed in the genus *Spilarctia*: *S. erythrastis* and *S. nobilis* (ABRS, 2025). Dubatolov (2004) placed Australian species which had been in the genera *Aloa* and *Amsacta* into the genus *Paramsacta*, including *P. marginata* *P. costalis* and *P. gangara*.

Tribal classification in the Arctiinae is stable, with three tribes recognized, the Arctiini, Lithosiini and Syntomini, with all tribes represented in the Australian fauna (Zahiri *et al.*, 2012). Subtribal classifications are not as well defined and, for the bulk of Australian Lithosiini and elsewhere, there have been relatively few valid structures established below tribe and above genus, none of which have been tested for the Australian fauna (Palting and Moore, 2022).

21.3 Tribe Arctiini

Australia is disproportionally depauperate in Arctiini, with only 34 described species (ABRS, 2025). The Australian species have strong affinities with faunas to the north of Australia, with many represented by genera or species that also occur in Papua New Guinea, Indonesia and South-east Asia (Holloway, 1988, 2011; de Vos, 2024). Adults are medium to large moths, ranging in size from around 13 mm wingspan for species of *Phaos* to 34 mm for *Amerila crokeri*. Endemic genera include *Ardices* (three species) *Cheliosea* (one species) and the montane genus *Phaos* (two species). This last species at least appears to be allied to the New Zealand tiger moths *Metacrias* (three species) (Patrick *et al.*, 2003; Dubatolov, 2005).

Species of *Ardices* are common moths in south-east and south-west Australia. *Ardices curvata* is found between south-east Queensland and Victoria with a few records from South Australia and Tasmania. It is a strikingly patterned orange and black species; the extent and placement of the black markings are highly variable, even within a single brood (Fig. 21.1C). This extreme variation led Butler to describe no less than five distinct synonyms based on a series of *A. curvata* specimens available in the British Museum (Natural History Museum) (Butler, 1878). The other two species of *Ardices* have white wings with black markings. *Ardices glatignyi* is the only species in the genus found in Western Australia and is the only *Ardices* regularly found in Tasmania and South Australia. *A. canescens* has a similar distribution to *A. curvata*, again with a few records from South Australia and Tasmania. All three species have typical 'woolly bear' larvae which are found feeding on a broad range of introduced and native herbs and woody plants.

Both described species of *Phaos* are found in montane subalpine and alpine regions in south-eastern Australia where the males are diurnally active. As with *Metacrias* in New Zealand, the females are micropterous and flightless. *Phaos aglaophara* is found in disjunct alpine regions along the Dividing Range from southern New South Wales through to Victoria where the larvae feed on daisies (Asteraceae) (Marriott, 2009). *P. interfixa* is a variable species restricted to Tasmania where it appears to form isolated altitudinally clinal populations between coastal and subalpine/

alpine habitats. Higher altitude populations previously were assigned to the synonym *P. acmena*. Larvae have been found feeding on prostrate herbs, especially Asteraceae, including introduced and native *Taraxacum* sp. (Common, 1990; P. McQuillan, pers. commun.). Larvae of both species appear to remain close to their birthplace; this behaviour, their preference for higher altitudes on the mainland, and having flightless females restrict the ability of populations to disperse to more favourable habitats, making *Phaos* vulnerable to changing climate and land usage.

Cheliosea cosmeta is a drab species that occurs in the drier inland regions of southern mainland Australia. The life history does not appear to have been recorded.

The large cosmopolitan genus *Utetheisa* is represented in Australia by five species. The subgenus *Utetheisa* is represented by three very similar and somewhat variable species of 'salt and pepper' moths: *U. lotrix*, *U. pulchelloides* and *U. pectinata*. Males of *U. pulchelloides* can be separated from *U. lotrix* by their dentate antennae and the presence of a fold in the hindwing containing hair-like scent scales (Common, 1990; Marriott, 2009; de Vos, 2024). Females of both species have filiform antennae, but those of *U. pulchelloides* are somewhat thicker. Males of *U. lotrix* have moniliform antennae and lack the fold in the hindwing. Although *U. lotrix* and *U. pulchelloides* are distributed through most of the Old World tropics, *Utetheisa pectinata* is a regional endemic that is found from islands in the Arafura Sea and the northern coastline of the Northern Territory.

Adults of *U. lotrix* and *U. pulchelloides* are day-flying and can sometimes undergo massed seasonal movements (D. Britton, pers. obs.). Larval host plants include many species which are considered toxic to other herbivores. Larvae of *U. lotrix* feed on rattlepods *Crotalaria* spp. (Fabaceae) and those of *U. pulchelloides* on introduced pasture weeds such as *Heliotropium* spp. (Heliotropiaceae) and *Echium plantagineum* (Boraginaceae) (Common, 1990).

The other two Australian species in *Utetheisa* are very different in appearance and their black and white wing patterns could be confused with tiger moths in the genus *Nyctemera*. *Utetheisa aegrotum* is found in north-east Queensland, and *U. pellax* occurs in northern Australia, Papua New Guinea, Vanuatu and Indonesia (West Papua).

The large genus, *Nyctemera*, is represented by three species, with only *N. amicus* endemic to Australia. This species is found over most of temperate southern Australia through to the subtropics; the blue, black and orange larvae feed on Asteraceae, notably introduced weedy *Senecio*, along with native species in the same genus (Common, 1990). The similar *N. balulus* has a broad distribution from India across South-east Asia to Samoa in the Pacific, preferring more northerly latitudes to *N. amicus*. Larval host plants again include herbaceous Asteraceae. *Nyctemera secundiana* is a junior synonym of *N. balulus* (de Vos, 2024); Australian specimens designated as '*N. secundiana*' can usually be assigned to *N. amicus*. *Nyctemera luctosa* is a paler species found throughout Papua New Guinea, Indonesia (West Papua and Ambon) and the Philippines through to Vanuatu. In Australia it is restricted to north-east Queensland.

Spilarctia is another large Old World genus that is represented by two species found in north-east Queensland: *S. erythrastis* and *S. nobilis*. Nothing is published on their life history.

Eight species belonging in *Amerila* are found in Australia. The adult moths are the largest Australian Arctiinae and have earned the common name of 'frothers' based on the conspicuous froth produced from their prothoracic glands when the moths are threatened. At least five of the Australian species also occur in Papua New Guinea and beyond, and most have a northerly distribution within Australia. The most widespread species in Australia is *A. crokeri* with records as far south as Nelson Bay, NSW. The endemic species *A. rubripes* has a similar distribution within Australia. Its larvae are notable for feeding on the introduced rubber vine *Cryptostegia grandiflora* (Apocynaceae) (Zborowski and Edwards, 2007) although it is doubtful if they exert any control of this serious weed.

Argina astrea is a common species in northern Australia. The orange and black adults are often spotted fluttering in vegetation during the day but are also active at night. This widely distributed species is found from east Africa throughout South-east Asia and on many Pacific islands; the conspicuous larvae are most often found feeding on rattlepods *Crotalaria* spp. (Fabaceae).

The strikingly coloured day-flying species in the genus *Euchromia* previously have been grouped with the Syntomini, but recent phylogenies have found strong evidence that the genus belongs in the Arctiinae (Przybyłowicz *et al.*, 2019). The genus has an Old World distribution from Africa through tropical Asia, Melanesia and as far east as Fiji; the three Australian species are found only in the tropical northern regions. Adults of *Euchromia creusa* have been observed imbibing pyrrolizidine alkaloids from senescing or withered plant material in Bougainville, Papua New Guinea (D. Britton, pers. obs.). Both the moth and the plants also occur in northern Queensland (Fig. 21.1B).

21.4 Tribe Lithosiini

The Lithosiini contain the bulk of diversity and endemicity of Australian Arctiinae, with 220 described species (ABRS, 2025). Members of the tribe commonly are referred to as lichen moths, and the few known life histories indicate that larvae predominantly feed on encrusting lichen and algae. An alternative common name is 'footman', in reference to those species that hold their wings wrapped around the body lengthwise, resembling the liveried household servants of the same name, as seen in the alternating footman *Tigrioides alterna* (Fig. 21.2D).

Adults are diverse in appearance, ranging from small (*ca* 5 mm wingspan) to medium-sized moths (*ca* 25 mm wingspan). Many of the smaller species superficially resemble species of microlepidoptera from other families. Most species are active at night, although some aposematic species of *Asura* mimic (or are mimicked by) moths in the genus *Amata* and are active during the day when both are flying. Adults are often sexually dimorphic, male adults usually have larger pectinations on their antennae and, in some species, well-developed scale patches and other glandular structures on the wings and abdomen, possibly for dissemination of sex pheromones. The male of *Calamidia hirta* has greatly elongated labial palpi compared with the female. Wing shape and maculation can differ dramatically between sexes in some genera, such as *Calamidia* and *Macaduma*. Females frequently have a large tuft of scales on the tip of the abdomen. Male moths often bear extensive patches of modified scales on their wings, as in species of *Teulisna* and many other genera. The males of *Chamaita barnardi* bear bizarre, clubbed structures on the hind legs.

The exact function of many of these sexually dimorphic characters is not known, but may relate to broadcast of glandular secretions, or in some cases the scales may even be deciduous and shed as part of reproductive behaviour, as seen in the American scarlet wasp moth *Cosmosoma myrodora* (Arctiinae: Arctiini) which, during courtship, releases 'flocculent' scales containing pyrrolizidine alkaloids (Conner *et al.*, 2000).

Larval mandibles are modified to match their encrusting (lichenous) diet with a mola, a uniquely flattened and heavily sclerotized area on the inner margin of the mandibles which they use to grind through tough lichen thalli (Jacobson and Weller, 2002; Palting and Moore, 2022). Larval apomorphies include the mandibular mola and the arrangement of labral setae, both of which support the monophyly of the tribe (Bendib and Minet, 1999). As with the Arctiini, there appears to be dietary acquisition and sequestration of defensive compounds (Hesbacher *et al.*, 1995), although this remains to be demonstrated for Australian taxa.

Diversity of the tribe is greatest in the northern wet tropics in Australia, with many genera shared with extralimital regions to the north, but the tribe is also notable for having a significant southerly endemic radiation with several speciose genera, most notably *Thallarcha* and *Termessa* (Fig. 21.2A–C). The Australian fauna has many undescribed species and potentially new genera awaiting description throughout the continent, and some described species in the 'brown' Australian genera such as *Thermeola*, *Lambula* and *Scoliacma* can be difficult to identify to species on habitus alone.

Although Holloway (1988, 2001, 2011) helped refine some of the broader generic concepts which include Australian species, there are still issues with defining genera, some of which will require additional integrative studies with fauna to the north of Australia. One example is the genus *Asura* which, in earlier publications, included a very large group of species distributed from India to South-east Asia and through to Australia. The type species is the Australian *A. cervicalis*. Edwards (1996) treated *Asura* in a much more restrictive sense for Australian species, and Holloway (2001, 2011) further defined the genus using genitalic characters to exclude almost all the extralimital species, whereas de Vos (2024) still includes Papuan species in the genus, differing from Holloway's narrower concept. The genus as it currently stands in Australia is almost certainly polyphyletic based on the few known larvae. The *Asura* '*cervicalis*' group have sparsely hirsute larvae and the *Asura* '*lydia*' group have heavily hirsute larvae (D. Britton, pers. obs.). *Asura polyspila* appears to be closely related to *A. lydia*, with extraordinarily hairy larvae which may even be the hairiest Lepidoptera of all (Fig. 21.2E).

Australian species currently included in the genus *Eilema* are almost certainly misplaced

Fig. 21.2. Lithosiini. (A) The favoured footman *Termessa gratiosa*, a common lichen moth in the Sydney basin. (B) Speckled footman *Termessa catocalina*. (C) Snowy footman *Termessa nivosa*. (D) Alternating footman *Tigrioides alterna*. Images: D. Britton, Wedderburn, NSW. (E) The densely hirstute larva of *Asura polyspila* consuming an earthstar *Geastrum triplex* (Geastraceae). Photo: Steve Pearson, Airlie Beach, Queensland. (F) Male and female of *Anestia ombrophanes* mating on the outside of the female cocoon. The shed skin from the final instar and empty pupal exuvia remain inside. The female laid infertile eggs before mating. As with many Lithosiini, the cocoon is formed by interweaving the secondary larval setae from the final instar to make a protective shelter. Photo: D. Britton.

under the current concept for this genus (Dubatolov and Zolotuhin, 2011).

Many species of Lithosiini are rare in collections, and have extremely localized distributions, so additional targeted collecting will be required to support a major revision.

The biggest knowledge gap for Lithosiini in general relates to life histories. The larvae of only a few species are known in Australia, and hosts and life histories are fragmentary at best for most species, whether Australian or otherwise. The reasons for this are in part related to their diet of lichens, algae and fungi, but also because most larvae shelter during the day and only feed at night. Vascular plants are often readily identifiable in the field by lepidopterists or can be collected and pressed for later identification, and there are regional field guides to enable identification. Lichens, algae and related organisms, on the other hand, are a mystery to lepidopterists compared with vascular plants; field guides don't exist for most of Australia and expertise is hard to find. Apart from photography, how do you collect and voucher a lichen? How do you take material back for rearing larvae? Given the worldwide diversity of Lithosiini, there are relatively few published examples where larvae have been unambiguously associated with named hosts; most host records are unnamed lichens, mosses and algae. For Australia, Ferguson (2018) described and reared late instar larvae of *Termessa shepherdi* on lichen sourced from the field locality of the larvae but did not identify which lichens were utilized beyond describing them as foliose.

Lichens are symbionts consisting of a fungus, often an ascomycete, in a relationship with a photobiont, usually algal but sometimes a cyanobacterium. The hyphae of the fungal component provide protective layers, food storage and attachment to the substrate, while the algal/cyanobacterial component is capable of photosynthesis, providing the lichen with carbohydrates. As well as physical protection, the fungal component can synthesize chemicals which are often unique to lichens and often species specific. These compounds are significant in relation to lichenivory.

There is considerable diversity in lichen forms. Common lichen forms are crustose (flat two-dimensional against the substrate), foliose (sheet-like, but lobed and rising slightly above the substrate) and fruticose (branching away from the substrate). The outer part of lichens exposed to the environment is referred to as the cortex, formed from a thin layer of dense fungal hyphae. Under the cortex is the photobiont layer and under the photobiont is a thicker layer of loosely woven hyphae referred to as the medulla. The base of the lichen is often fastened to the substrate by strands of hyphae referred to as rhizines. The Australian National Botanic Gardens has produced an excellent online resource for Australian lichens (ANBG, 2024), and the Australian Biological Resources Study has checklists and identification resources available online (ABRS, 2024).

Observations of extralimital lichen moth species indicate that larvae often feed on the cortex to expose the photobiont layer which is preferentially consumed. They will usually avoid the medulla which may contain toxic compounds (Hesbacher *et al.*, 1995). The larvae of a single species may vary their feeding strategy depending on availability of hosts. Moskowitz and Westphal (2002) observed larvae of the North American *Hypoprepia fucosa* feeding on algae growing on concrete but were also able to observe the same species in captivity feeding on an epiphytic species of lichen, where the larvae removed the fungal component of the lichen to feed on the algal cortex underneath while avoiding consumption of the medulla.

The few species which have been reared in laboratory experiments show that host choice is related to the presence of secondary metabolites, but the relationship is not a simple one. In host choice experiments, species of *Katha* (as '*Eilema*' species) preferred *Melanelia exasperata*, a lichen species that does not contain polyphenolic secondary metabolites, and larvae developed faster and grew bigger on this host (Pöykkö and Hyvärinen, 2003). In free-choice experiments, however, they also foraged on other phenolic-containing lichen species if they were present, even though they were not as nutritionally suitable (Pöykkö and Hyvärinen, 2003), indicating that perhaps these phenolics may have a role outside of nutrition for larvae. Pöykkö *et al.* (2005) manipulated five species of lichens experimentally by removing secondary metabolites and presenting them to larvae of *Katha deplana* (as '*Eilema depressum*'), demonstrating that two species of these lichens are toxic to larvae in their natural state but were acceptable hosts once treated, whereas another lichen was tolerated even with secondary metabolites present. It is likely that these secondary metabolites are effective at preventing larvae from using some lichen species as hosts but may contribute to unpalatability of the moths.

Although there remains a focus on lichens as a dietary source, Wagner (2009) notes that many species of Lithosiini could be regarded as generalist algae feeders, just that they must consume the fungal component of the lichens to reach that food source; the lichens determine host specificity rather than the algal component.

It is likely that some Australian species in the tribe may not feed on lichen at all. Palting and Moore (2022) in their molecular subtribal phylogeny of the western hemisphere Lithosiini noted that species in the exotic subtribe Clemensiina may be specialist algivores. The extremely hairy larvae of *Asura polyspila* (Fig. 21.2E) have been observed feeding on the collared earthstar *Geastrum triplex* (Geastraceae), a cosmopolitan ground-dwelling saprophytic fungus (S. Pearson, pers. commun.), so fungi may also be dietary items for some species.

Published host records for Australian species sometimes include vascular plant species. These should be treated with caution as they may be based on collection of late instar larvae from plants where they may have been grazing on epiphytic algae and lichen, rather than feeding on the putative host. One example of this is *Palaeosia bicosta*, a species that is very common as larvae and adults in introduced *Pinus radiata* plantations (Moore, 1962, 1963, 1972; Britton, 1994). Moore (1963) notes that large numbers of *Asura cervicalis* larvae have on occasion defoliated *Ficus rubiginosa*, and that this species also feeds on *Pinus radiata*.

Cocoons of Lithosiini are often formed from the secondary setae of the larvae woven together with silk. Species in the genus *Cyana* demonstrate the most elaborate version of this behaviour, including the endemic species *Cyana meyricki* (Monteith, 2008; Leong, 2010). *Cyana* larvae weave an elegant and sparse mesh-like cocoon, suspending themselves on a fine silk harness in the middle of this structure for pupation. These mesh cocoons may provide protection from pupal parasitoids and smaller predators simply by making it difficult for probing ovipositors and mouth parts to reach the pupa, but they may also mitigate the extremes of hot and wet tropical climates (Hawes, 2016a). The empty cocoons of some Thai species of *Cyana* are utilized by spiders (Hawes, 2016b).

Micropterous (small or reduced wings) females are known for a few species, most notably in the genus *Anestia* (Fig. 21.2F). Females of these are often encountered on the walls of buildings or on fence lines where they are found clinging to their cocoon. The genus currently contains two species separated on male wing patterns but most collections contain intermediate specimens, suggesting that there is a single variable species, *A. ombrophanes*. The evolutionary drivers for development of micropterous females may relate to resource allocation to increase egg production. The resources that otherwise would have been used for wing development and associated musculature can be allocated towards egg production. Wingless females can have heavier abdomens as they no longer need to become airborne (Heppner, 1991).

Females of *Castulo plagiata* have never been collected and they are suspected to be flightless or may not be attracted to light; males of the species occur at higher altitudes in NSW, the ACT and Victoria, and closer to sea level in Tasmania. *Castulo doubledayi* has females that are capable of flight, and which are often collected with males at light. *Phenacomorpha bisecta* is a widespread but rarely collected species that also has flightless females.

21.5 Tribe Syntomini

The tribe Syntomini was erected by Herrich-Schäffer in 1846 (Herrich-Schäffer, 1843–1856), but the syntomines became a tribe of the Arctiinae and were separated from Ctenuchinae ('handmaidens' and allies) and Euchromiini when the subfamily was reclassified, firstly using morphological characteristics (Kitching and Rawlins, 1998) and then with molecular methods (Zahiri *et al.*, 2012). Only three genera in this tribe are found in Australia, the speciose *Amata* (39 described species), *Ceryx* (2 species) and *Eressa* (7 species). These Old World genera are also found throughout Madagascar, Africa and Asia (Przybyłowicz *et al.*, 2019).

The adult moths can be identified by their long and narrow forewings (*ca* 20–30 mm wingspans), much shorter hindwings, and long, narrow and striped abdomens. Most of the tribe have thin, pectinate antennae, with exceptions found in *Eressa* males where the majority have lightly plumose antennae. The Australian Syntomini possess a functional proboscis and will actively feed at flowers as adults; the larvae are predominantly polyphagous but have been noted to feed on different plant families, such as Chenopodiaceae and dead eucalypt leaves (Common, 1990). *Amata* larvae have been successfully raised on dandelion (Asteraceae) leaves (Binns *et al.*, 2022) and decaying garden rose petals (G.E. Binns, pers. obs.). The Syntomini are widely distributed in Australia and are found in all states and territories except Tasmania.

Amata species (previously in the Ctenuchinae) are notoriously difficult to identify due to their cryptic morphology among species and their vague original descriptions. Turner (1905) attempted to revise the Australian Syntomini and noted around 25 synonyms in what he referred to as *Syntomis* (=*Amata*), some of which are frustratingly still present in online resources and published checklists, adding to the confusion. Attempts to tease apart *Amata* species in NSW, using molecular data and genital morphology, found that *A. nigriceps* is likely to be a complex of three to four subspecies, most of which are living sympatrically along the east coast (G.E. Binns, unpublished data). This genus desperately needs a thorough revision and accompanying identification key, and this is also true for non-Australian *Amata* (Przybyłowicz *et al.*, 2019). *Eressa* also deserves some taxonomic attention as at least one of the species, *Eressa angustipenna*, has several features that deviate from its congeners, including thin, serrate antennae and much reduced hindwings (less than a quarter of the size of the forewing; Fig. 21.3D), which likely place it either in *Amata* or within its own genus.

The individuals in this tribe are recognizably hymenopteran mimics and are commonly known as wasp moths. The moths are either black or dark brown, with yellow, orange or reddish abdominal stripes and wing spots, or transparent wing spot windows in *Ceryx* species and some *Amata*. These bright colours contrast with the dark brown or black ground colours, and are almost certainly aposematic in function. Some male *Amata*, however, possess iridescent blue on their abdominal segments (e.g. *A. antitheta*, Fig. 21.3A), or amongst the black scales of their wings (e.g. *A. nigriceps*, Fig. 21.3B) which may facilitate mate recognition.

Although pyrazines have been detected in some Australian *Amata* (Rothschild *et al.*, 1984), little is currently known about the chemical ecology of Australian syntomines, and whether they are sequestering secondary plant defence compounds or synthesizing their own, or both. Recent metabolomic work on sex differences in chemical defences in *A. nigriceps* in New South Wales found several plant defence compounds, including glycine linoleamide, a toxin employed by plants as a defence against herbivory (Weiss *et al.*, 2018), and plant alkaloids such as imperialine, from the genus *Fritillaria* (Liliaceae). The latter can cause spasms and cardiac arrest in humans (Al-Snafi, 2019) and was found exclusively in female *A. nigriceps* moths (Binns *et al.*, 2025).

Wasp moths are predominantly diurnal, although they will often come to light traps at night. These moths are often described as being 'not strong fliers' but recent research has dubbed this behaviour as 'nonchalant flight', advertising the unpalatability of these moths to potential predators (Dowdy and Conner, 2019). Males will display lekking behaviour in large numbers during mating seasons, although no evidence of male territoriality has been observed. Female moths lay eggs covered with densely packed scales shed from the terminal abdominal segment, creating a spongy nest which may protect the eggs from predation and parasitoids.

Larvae are generally moderately hairy and brown; the larval stage lasts about 40 days in summer, and *Amata* have been recorded to overwinter as larvae. *A. aperta* larvae remained in larval form for over 280 days in a pilot rearing experiment under glasshouse conditions (Fig. 21.3C; Binns *et al.*, 2022). As with many lithosiines, syntomines also use larval setae to create a cocoon around their pupal case, most likely as protection from parasitoids.

Species of *Amata* appear to be models for other Arctiinae, both diurnal and nocturnally active species, with many species of *Asura* (Lithosiini) having similar orange and black colours. A striking example of this mimicry complex is the rare lichen moth species *Pseudophanes melanoptera* (Lithosiini) (Fig. 21.3E) which closely resembles *Amata bicolor* (Fig. 21.3F). The few known specimens of *P. melanoptera* have been collected

Fig. 21.3. Syntomini. (A) *Amata antitheta*. Male was hand raised on rose petals; female was caught in Cairns, QLD, in 2019. (B) *Amata nigriceps*. Macro-image of this male specimen shows the iridescent blue that can be seen at certain angles, due to the shape and structure of the scales. (C) Final instar *Amata aperta*. Larvae were raised from eggs hatched in April 2020. This glasshouse pilot experiment was to determine what food source could be used to raise the moths, either a 'natural' diet of rose petals or artificial diet of 'moth porridge'. It was unknown at the time that *Amata* overwinter as larvae and this individual was 264 days old when its portrait was taken. (D) '*Eressa*' *angustipenna*, Wyoming, NSW. Images: G.E. Binns. (E) Male lectotype of *Pseudophanes melanoptera*, Kuranda, QLD (Lithosiini, ANIC). Image: D. Britton. (F) *Amata bicolor* (Syntomini), Jungle Creek Kuranda, QLD. Image: Cynthia Chan, © 2025 Australian Museum.

in the vicinity of Kuranda, Queensland, a region which is also a stronghold of *A. bicolor*. This mimic is very rare in collections compared with the model, and until relatively recently was only known from the type series collected by F.P. Dodd in 1906. David Rentz has collected specimens in Kuranda, and there is a recent observation from nearby Julatten. Perhaps the excellent mimicry has fooled collectors as well?

21.6 Future Directions

The Australian Arctiinae offer some inspiring options for researchers and natural historians. They can provide accessible models for studying the interactions between larval and adult diet, defence mechanisms and sexual selection. Preliminary evidence suggests that at least some Australian species utilize different families of defensive chemicals compared with their overseas relatives. The current lack of knowledge of host plants and adult feeding behaviours hinders research into the source of these chemicals, so documenting the basic natural history of these moths is another important step towards understanding their chemical ecology. The interactions between arctiines and their potential endemic predators, in particular Australian insectivorous bats, would be another fascinating area to research. For example, many of the 'brown' Australian Lithosiini which do not appear visually aposematic may well be advertising their distasteful nature acoustically when active at night, but this has yet to be observed and documented.

The subfamily and component tribes of the Arctiinae are readily recognizable, but a significant amount of work is required to better outline the beta and alpha taxonomy for Australian representatives. Comparing results with the current work on the moth fauna of West Papua (Irian Jaya) by Rob de Vos and others would allow for a better understanding of how Australian Arctiinae have evolved and radiated in relation to the complex fauna of South-east Asia. Unlike some of the megadiverse noctuoid lineages, the Australian arctiine fauna represents a manageable group for this type of systematic research.

Acknowledgements

We thank Matt Shaw and Russell Cox for providing access to the Australian Museum entomology collection, and Cynthia Chan and Heather Poole for imaging specimens from this collection. Steven and Allison Pearson allowed us to use their images of *Asura polyspila* from Airlie Beach. Peter McQuillan provided insight into the diversity and life history of Tasmanian *Phaos*. Peter Mackay reviewed a draft of this chapter.

References

ABRS (2024) Lichen Databases and Online Resources. Available at: www.dcceew.gov.au/science-research/abrs/online-resources/lichen (accessed 1 May 2024).

ABRS (2025) Australian Faunal Directory. Australian Biological Resources Study, Canberra. Available at: https://biodiversity.org.au/afd/home (accessed March 2025).

Al-Snafi, A.E. (2019) *Fritillaria imperialis* – A review. *IOSR Journal of Pharmacy* 9, 47–51.

ANBG (2024) Australian National Botanic Gardens. Australian Lichens. Available at: www.anbg.gov.au/lichen/index.html (accessed May 2024).

Bendib, A. and Minet, J. (1998) Female pheromone glands in Arctiidae (Lepidoptera): evolution and phylogenetic significance. *Comptes Rendus de l'Academie des Sciences Paris* 321, 1007–1014.

Bendib, A. and Minet, J. (1999) Lithosiinae main lineages and their possible interrelationships. 1. Definition of new or resurrected tribes (Lepidoptera: Arctiidae). *Annales de la Société Entomologique de France* 35, 241–263.

Binns, G.E., Hämäläinen, L., Kemp, D.J., Rowland, H.M., Umbers, K.D. and Herberstein, M.E. (2022) Additive genetic variation, but not temperature, influences warning signal expression in *Amata nigriceps* moths (Lepidoptera: Arctiinae). *Ecology and Evolution* 12, e9111.

Binns, G.E., Hämäläinen, L., Rowland, H.M., Caputi, L., Kunert, M. *et al.* (2025) Sex differences in warning signals and composition of chemical defences in *Amata nigriceps* moths (Lepidoptera: Arctiinae). *Royal Society Open Science* 12, 242186.

Britton, D.R. (1994) Nutritional Ecology of *Chlenias* spp. (Geometridae) on an Introduced conifer *Pinus radiata* D. Don: The Effects of Choosing a Non-Native Host Plant in Lepidoptera. MSc thesis, La Trobe University, Bundoora, Victoria.

Britton, D.R. (2010) Subfamily Arctiinae. Australian Faunal Directory. Australian Biological Resources Study, Canberra. Available at: https://biodiversity.org.au/afd/taxa/Arctiinae (accessed March 2025).

Butler, A.G. (1878) Description of new Lepidoptera of the group Bombycites, in the collection of the British Museum. *Proceedings of the Zoological Society of London* 1878, 381–388.

Chiocchio, A., Arduino, P., Cianchi, R., Canestrelli, D. and Zilli, A. (2021) Unveiling cryptic diversity among Müllerian co-mimics: Insights from the Western Palaearctic *Syntomis* moths (Lepidoptera: Erebidae: Arctiinae). *Organisms Diversity & Evolution* 21, 119–133.

Common, I.F.B. (1990) *Moths of Australia*. Melbourne University Press, Carlton, Victoria.

Conner, W.E. (2009) *Utetheisa ornatrix*, the Ornate Arctiid. In: Conner, W.E. (ed.) *Tiger Moths and Woolly Bears. Behavior, Ecology, and Evolution of the Arctiidae*. Oxford University Press, New York, pp. 1–10.

Conner, W.E., Boada, R., Schroeder, F. and Eisner, T. (2000) Chemical defense: Bestowal of a nuptial alkaloidal garment by a male moth on its mate. *Proceedings of the National Academy of Sciences USA* 97, 14406–14411.

Conner, W.E., Hristov, N.I. and Barber, J.R. (2008) Sound strategies: Acoustic aposematism, startle, and sonar jamming. In: Conner, W.E. (ed.) *Tiger Moths and Woolly Bears. Behavior, Ecology, and Evolution of the Arctiidae*. Oxford University Press, New York, pp. 177–191.

De Vos, R. (2024) The Tigermoths (Lepidoptera: Erebidae, Arctiinae) of Papua Indonesia. The Papua Insects Foundation. Available at: www.papua-insects.nl/insect%20orders/Lepidoptera/Erebidae/Arctiinae/Arctiinae%20list.htm (accessed March 2025).

Dowdy, N.J. and Conner, W.E. (2019) Nonchalant flight in tiger moths (Erebidae: Arctiinae) is correlated with unpalatability. *Frontiers in Ecology and Evolution* 7, 480.

Dubatolov, V.V. (2004) A new genus is established for *Bombyx lineola* Fabricius, 1793, with systematic notes on the genus *Aloa* Walker, 1855 (Lepidoptera, Arctiidae). *Atalanta* 35, 403–413.

Dubatolov, V.V. (2005) On the status of the Australian genus *Ardices* Walker, 1855 with the description of a new subgenus for *A. curvata* Donovan, 1805 (Lepidoptera, Arctiidae). *Atalanta* 36, 173–179.

Dubatolov, V.V. and Zolotuhin, V.V. (2011) Does *Eilema* Hübner, [1819] (Lepidoptera, Arctiidae, Lithosiinae) present one or several genera? *Euroasian Entomological Journal* 10, 367–380.

Edwards, E.D. (1996) Arctiidae. In: Nielsen, E.S., Edwards, E.D. and Rangsi, T.V. (eds) *Checklist of the Lepidoptera of Australia*. Monographs on Australian Lepidoptera, 4. CSIRO Publishing, Canberra, ACT, pp. 278–286.

Ferguson, D.J. (2018) Description of the larva of *Termessa shepherdi* Newman (Lepidoptera: Erebidae: Arctiinae: Lithosiini) from southeastern New South Wales. *Australian Entomologist* 45, 245–249.

Hawes, T.C. (2016a) Cocoon orientation by *Cyana* moths on lichen-host trees. *Oriental Insects* 50, 111–118.

Hawes, T.C. (2016b) Spiders appropriate empty *Cyana* moth cocoons for nest armour. *Natural History Bulletin of the Siam Society* 61, 127–130.

Heppner, J.B. (1991) Brachyptery and aptery in Lepidoptera. *Tropical Lepidoptera* 2, 11–40.

Herrich-Schäffer, G.A.W. (1843–1856) *Systematische Bearbeitung der Schmetterlinge von Europe, zugleich als Text. Revision und Supplement zu Jacob Hübner's Sammlung europäische Schmetterlinge.* G.J. Manz, Regensberg, Germany.

Hesbacher, S., Giez, I., Embacher, G., Fiedler, K., Max, W. *et al.* (1995) Sequestration of lichen compounds by lichen-feeding members of the Arctiidae (Lepidoptera). *Journal of Chemical Ecology* 21, 2079–2089.

Holloway, J.D. (1988) *The Moths of Borneo (Part 6): Family Arctiidae, Subfamilies Syntominae, Euchromiinae Arctiinae; Noctuidae misplaced in Arctiidae (Camptoloma, Aganainae)*. Southdene Sdn. Bhd., Kuala Lumpur, Malaysia.

Holloway, J.D. (2001) The Moths of Borneo: Family Arctiidae, subfamily Lithosiinae. *Malayan Nature Journal* 55, 279–486.

Holloway, J.D. (2011) The Moths of Borneo: Families Phaudidae, Himantopteridae and Zygaenidae; revised and annotated checklist. *Malayan Nature Journal* 63, 1–548.

Jacobson, N.L. and Weller, S.J. (2002) *A Cladistic Study of the Arctiidae (Lepidoptera) by using Characters of Immatures and Adults*. Entomological Society of America, Lanham, MD.

Kitching, I.J. and Rawlins, J.E. (1998) The Noctuoidea. In: Kristensen, N.P. (ed.). *Lepidoptera, Moths and Butterflies. Volume 1: Evolution, Systematics, and Biogeography*. Handbuch der Zoologie/Handbook of Zoology. Volume IV Arthropoda: Insecta Part 35. i–x. Walter de Gruyter, Berlin, Germany, pp. 355–401.

Leong, T.M. (2010) Cocoon and pupa of the moth *Cyana perornata* (Walker, 1854) from Singapore, with a deduction of its larval identity (Lepidoptera: Arctiidae: Lithosiinae). *Nature in Singapore* 3, 175–181.

Marriott, P. (2009) *Moths of Victoria. Part 2 Tiger Moths and Allies. Noctuoidea* (A). Entomological Society of Victoria, Melbourne, Victoria.

Monteith, G. (2008) The mystery of the arctiid moth, *Cyana meyricki* Rothschild – An Insect Houdini. *Entomological Society of Queensland News Bulletin* 36, 237–243.

Moore, K.M. (1962) Insect attack on *Pinus* spp. Forestry Commission of N.S.W., Division of Forest Management, Research Note 12, 1–15.

Moore, K.M. (1963) Observations on some Australian forest insects. 16. Lepidoptera attacking *Pinus* spp. in New South Wales. *Proceedings of the Linnean Society of New South Wales* 88, 340–360.

Moore, K.M. (1972) Observations on some Australian forest insects. 30. A supplementary list of insects attacking *Pinus* spp. in New South Wales. *Australian Zoologist* 17, 59–64.

Moskowitz, D.P. and Westphal, C. (2002) Notes on the larval diet of the Painted Lichen Moth *Hypoprepia fucosa* Hübner (Arctiidae: Lithosiinae). *Journal of the Lepidopterists Society* 56, 289–290.

Palting, J.D. and Moore, W. (2022) Molecular phylogeny of lichen tiger moths (Lepidoptera, Erebidae, Arctiinae, Lithosiini): A contribution toward classifying Western Hemisphere genera. *ZooKeys* 1108, 119–139.

Patrick, B., Patrick, H. and Edwards, E. (2003) Tiger moths and pheromones. *The Weta* 25, 14–17.

Pöykkö, H. and Hyvärinen, M. (2003) Host preference and performance of lichenivorous *Eilema* spp. larvae in relation to lichen secondary metabolites. *Journal of Animal Ecology* 72, 383–390.

Pöykkö, H., Hyvärinen, M. and Bačkor, M. (2005) Removal of lichen secondary metabolites affects food choice and survival of lichenivorous moth larvae. *Ecology* 86, 2623–2632.

Przybyłowicz, Ł., Lees, D.C., Zenker, M.M. and Wahlberg, N. (2019) Molecular systematics of the arctiine tribe Syntomini (Lepidoptera, Erebidae). *Systematic Entomology* 44, 624–637.

Robinson, G.S., Ackery, P.R., Kitching, I., Beccaloni, G.W. and Hernández, L.M. (2023) HOSTS – a Database of the World's Lepidopteran Hostplants. Available at: https://data.nhm.ac.uk/dataset/hosts (accessed 6 October 2025).

Rothschild, M., Moore, B.P. and Brown, W.V. (1984) Pyrazines as warning odour components in the monarch butterfly, *Danaus plexippus*, and in the moths of the genera *Zygaena* and *Amata* (Lepidoptera). *Biological Journal of the Linnean Society* 23, 375–380.

Scoble, M.J. (1992) *The Lepidoptera. Form, Function and Diversity*. Natural History Museum Publications and Oxford University Press, Oxford.

Turner, A.J. (1905) Revision of Australian Lepidoptera. II. *Proceedings of the Linnean Society of New South Wales* 29, 832–859.

Turner, A.J. (1940) A revision of the Australian Arctiidae (Lepidoptera). *Proceedings of the Royal Society of Queensland* 51, 51–131.

Wagner, D.L. (2009) The immature stages: Structure, function, behavior, and ecology. In: Connor, W.E. (ed.) *Tiger Moths and Woolly Bears. Behavior, Ecology, and Evolution of the Arctiidae*. Oxford University Press, New York, pp. 31–53.

Weiss, L.C., Albada, B., Becker, S.M., Meckelmann, S.W., Klein, J. *et al.* (2018) Identification of *Chaoborus* kairomone chemicals that induce defences in *Daphnia*. *Nature Chemical Biology* 14, 1133–1139.

Weller, S.J., Jacobson, N.L. and Conner, W.E. (1999) The evolution of chemical defences and mating systems in tiger moths (Lepidoptera: Arctiidae). *Biological Journal of the Linnean Society* 68, 557–578.

Zahiri, R., Kitching, I.J., Lafontaine, J.D., Mutanen, M., Kaila, L. *et al.* (2011) A new molecular phylogeny offers hope for a stable family level classification of the Noctuoidea (Lepidoptera). *Zoologica Scripta* 40, 158–173.

Zahiri, R., Holloway, J.D., Kitching, I.J., Lafontaine, J.D., Mutanen, M. and Wahlberg, N. (2012) Molecular phylogenetics of Erebidae (Lepidoptera, Noctuoidea). *Systematic Entomology* 37, 102–124.

Zborowski, P. and Edwards, E.D. (2007) *A Guide to Australian Moths*. CSIRO Publishing, Collingwood, Victoria.

22 The Agaristinae: The Day Moths

Michael F. Braby
Australian National Insect Collection, CSIRO, Canberra, ACT, Australia, and Division of Ecology and Evolution, Research School of Biology, Australian National University, Acton, ACT, Australia

Abstract
The Agaristinae, a subfamily of the Noctuidae, exhibit an array of life history tactics. Adults may be diurnal, crepuscular or nocturnal. They may exhibit audible sound production; cryptic, mimetic and warningly coloured wing patterns; aposematic larvae; and univoltine and bivoltine generations with protracted pupal dormancy. Many agaristines are day flying, hence the common name 'day moth'. The diurnal species frequently feed at flowers, use visual cues for mating, often have swollen, club-like antennae and have bright, aposematic wing colour patterns. This chapter presents an up-to-date synthesis of the taxonomy, classification, diversity, distribution, biology, ecology and economic importance of the Australian agaristine fauna, together with an overview of their life cycle and seasonality, adult activity, mating behaviour and acoustic communication, larval food plants, aposematism and chemical defence.

22.1 Introduction

The Agaristinae are noctuid moths exhibiting a fascinating array of life history traits, including adults with diurnal, crepuscular and nocturnal activity; audible sound production; cryptic, mimetic and warningly coloured wing patterns; aposematic larvae; and univoltine and bivoltine generations with protracted pupal dormancy. Many agaristines are day flying, hence the common name 'day moth'. The diurnal species frequently feed at flowers, use visual cues for mating, often have swollen, club-like antennae, and have bright, aposematic wing colour patterns – classic warning colours consisting of contrasting patches, spots or bands of white, yellow, orange or occasionally red against a black or dark brown ground colour (Kitching, 1984; Common, 1990; Nielsen and Common, 1991). In many ways their behaviour and appearance are similar to butterflies. Indeed, the spectacular painted day moth or Joseph's coat moth, *Agarista agricola*, was originally described by Donovan (1805) as a butterfly in the genus *Papilio* (Papilionoidea)! The life histories and larval food plants of *Agarista agricola*, *Eutrichopidia latina*, *Phalaenoides glycinae*, *P. tristifica*, *P. behri* and *Cruria donowani*, all conspicuous species around Sydney or Melbourne, have been well known for over 100 years, described and often lavishly illustrated with colour plates (e.g. Fig. 22.1) (Lewin, 1805; Scott, 1864, 1891; McCoy, 1878; French, 2013). French's manuscript, incidentally, remained unpublished for over 100 years.

Since those pioneering studies, a great deal of knowledge has accumulated on the taxonomy,

Corresponding author: michael.braby@anu.edu.au

DOI: 10.1079/9781800627680.0022

Fig. 22.1. Historical colour plate depicting the adults, immature stages (larva, pupa, cocoon) and larval food plants of three common species of *Phalaenoides* around Melbourne (after McCoy, 1878). Shown are *P. tristifica* on *Epilobium tetragonum* (figs 1–4); *P. behri* on *Amyema* sp. (figs 5–8); and *P. glycinae* on *Vitis vinifera* (figs 9–13).

classification, diversity, distribution, biology, ecology and economic importance of the Australian agaristine fauna. This chapter provides an up-to-date synthesis of these topics, as well as an overview of their life cycle and seasonality, adult activity, mating behaviour and acoustic communication, larval food plants, aposematism and chemical defence.

22.2 Taxonomy and Classification

The Agaristinae were traditionally considered a separate family closely related to Noctuidae (e.g. Hampson, 1901, 1910; Jordan, 1912; Kiriakoff, 1977). In contrast, Strand (1912) and Turner (1920) treated Agaristinae as a subfamily of Noctuidae, but universal adoption of this taxonomic rank did not occur until the 1980s following publication of a major checklist and revision of the North American Noctuidae (Franclemont and Todd, 1983) and Ian Kitching's (1984) global review of the higher classification of Noctuidae.

The Agaristinae are widely regarded as monophyletic (Kiriakoff, 1977; Kitching, 1984; Holloway, 1989) but the relative importance of characters considered to delineate the group has varied among workers. Hampson (1901, 1910), for example, considered the filiform structure of the antennae to be the primary character for distinguishing agaristines from other noctuids, whereas Jordan (1912) and Turner (1920) found little structural evidence to diagnose agaristines, although both acknowledged that they comprised a distinct natural group. Jordan (1912) included genera with filiform, ciliate and bipectinate antennae in his revision and classification of the Indo-Australian fauna. Kiriakoff (1963, 1977) noted that agaristines are a very homogeneous group morphologically; he observed that they possess enlarged tympanal organs similar to some other noctuids, but are distinguished by the presence of a paired vesicle ('bulla abdominalis') at the base of the abdomen, a synapomorphy he considered to be a reliable character for differentiating the lineage.

Agaristines are informally classified as 'trifine' noctuids in which vein M_2 of the hindwing is absent (Kitching, 1984), the base of the male abdomen has a pair of complex hair pencils or brush organs (scent organs which disseminate pheromones; Edgar *et al.*, 1978), and the male genitalia have a costal process associated with the harpe (Holloway, 1989). Phylogenies based on molecular data support the 'trifine' classification (Mitchell *et al.*, 1997, 2000, 2006). Recent, large-scale phylogenetic analysis of Noctuidae and Noctuoidea suggest Agaristinae are most closely related to either Oncocnemidinae (Mitchell *et al.*, 2006; Keegan *et al.*, 2021) or Acronictinae and Amphipyrinae (Regier *et al.*, 2016; Li *et al.*, 2024).

Strand (1912) recognized just over 500 valid species of Agaristinae referred to 137 genera in his global checklist of the fauna, but a more recent estimate of Scoble (1992) was 300 species in the subfamily. In Australia, Common (1990) recognized 40 species referred to 21 genera. More recent publications indicate there are now 45 described Australian species in 18 genera (Edwards, 1996; Braby *et al.*, 2018; Braby and Heath, 2024; Wragg and Braby, 2025) (Table 22.1). The genus *Cremnophora*, containing the species *C. angasii*, was added to the Australian agaristine fauna by Nye (1975), which was followed by Common (1990) and Edwards (1996), but this proposal was rejected by Wragg and Braby (2025) on phylogenetic grounds. The morphology of the male genitalia, frons, and wing pattern elements suggests *Cremnophora* probably belongs in the Amphipyrinae. Edwards (1996) revised the status of the following four taxa to full species: *Cruria latifascia*, *Cruria tropica*, *Idalima butleri* and *Idalima tasso*. Edwards (1996) also synonymized the genus *Prostheta* under *Hecatesia* and treated its type species, *Prostheta acrypta*, as a junior synonym of *Hecatesia exultans*. Braby *et al.* (2018) recognized *Cruria darwiniensis* from the Top End of the Northern Territory as a species distinct from *Cruria tropica* from northern Queensland. Braby and Heath (2024) added the genus *Sarbanissa* and species *S. diana* from Christmas Island to the Australian fauna. More recently, Wragg and Braby (2025) synonymized *Agaristodes* with *Apina*, *Comocrus* with *Phalaenoides* and *Periscepta* with *Idalima* based on cladistic analyses of morphological characters (Fig. 22.2).

There are at least a further eight undescribed species of agaristine (M.F. Braby, unpublished data), which suggests the Australian fauna is likely to contain at least 53 species once the taxonomic inventory is complete (Table 22.1).

Kiriakoff (1977) provisionally classified the Agaristinae from the Palaearctic, Oriental and Australian regions into two groups and 13 informal subgroups based on features of the male

Table 22.1. Summary of the geographical distribution, adult activity and larval food plants of the Australian Agaristinae. Classification follows Wragg and Braby (2025). Distribution data are primarily from Hampson (1901, 1910), Jordan (1912), Kiriakoff (1977), Marriot *et al.* (2017), Braby *et al.* (2018), McQuillan *et al.* (2019) and specimens in the ANIC and the author's private reference collection. Adult activity data are based largely on unpublished observations of the author, supplemented with records in the literature (see References). Larval food plant data are derived from Braby (2025).

Species	Distribution	Adult activity	Larval food plants
Leucogonia ekeikei (Bethune-Baker, 1906)	Australia (QLD), New Guinea	nocturnal	
Leucogonia cosmopis (Lower, 1897)	Australia (WA, NT, QLD)	nocturnal	
Zalissa stichograpta Turner, 1943	Australia (QLD)	nocturnal	
Zalissa catocalina Walker, 1865	Australia (QLD), New Guinea	nocturnal	Vitaceae
Zalissa pratti (Bethune-Baker, 1906)	Australia (QLD), New Guinea	nocturnal	
Sarbanissa diana Sugi, 1996	Australia (Christmas Island), Java, Bali	nocturnal	
Hecatesia fenestrata Boisduval 1828	Australia (QLD, NSW, ACT, VIC, TAS, SA)	diurnal, crepuscular	Lauraceae
Hecatesia sp. 'Arnhem Land'	Australia (NT)	diurnal	Lauraceae
Hecatesia thyridion Feisthamel, 1839	Australia (NSW, VIC, SA, WA)	diurnal, crepuscular	Lauraceae
Hecatesia exultans Walker, [1865]	Australia (VIC, SA, WA)	diurnal	Lauraceae
Hecatesia sp. 'Amata'	Australia (WA, NT, QLD)	diurnal	Lauraceae
Argyrolepidia aequalis (Walker, 1864)	Australia (QLD), New Guinea, Solomon Islands	nocturnal, possibly crepuscular	Vitaceae
Argyrolepidia thoracophora (Turner, 1920)	Australia (QLD)	nocturnal	
Argyrolepidia fracta (Rothschild, 1899)	Australia (QLD), New Guinea	nocturnal, possibly crepuscular	
Coenotoca subaspersa (Walker, [1865])	Australia (QLD, NSW, VIC)	diurnal, nocturnal	Vitaceae
Coenotoca unimacula (Lower, 1903)	Australia (QLD)	nocturnal	Vitaceae
Mimeusemia centralis (Rothschild, 1896)	Australia (NT, QLD)	nocturnal	Vitaceae, Menispermaceae
Mimeusemia econia Hampson, 1900	Australia (WA, QLD), Sulawesi, Bali	nocturnal	
Mimeusemia simplex (T.P. Lucas, 1891)	Australia (QLD), New Guinea, Maluku		
Mimeusemia sp. 'Wet Tropics'	Australia (QLD)	nocturnal	
Periopta ardescens (Butler, 1884)	Australia (WA, NT, QLD)	crepuscular, nocturnal	Vitaceae
Periopta diversa (Walker, [1865])	Australia (WA, NT, QLD)	diurnal	Rubiaceae
Ipanica cornigera (Butler, 1886)	Australia (WA, NT, QLD, NSW, VIC, SA), New Guinea, Tanimbar and adjacent islands	nocturnal	Nyctaginaceae
Apina callisto (Angas, 1847)	Australia (NSW, ACT, VIC, TAS, SA, WA)	Diurnal	Asteraceae, Crassulaceae, Fabaceae, Geraniaceae, Lamiaceae, Malvaceae, Plantaginaceae, Polygonaceae
Apina feisthamelii (Herrich-Schäffer, [1853])	Australia (NSW, ACT, VIC, TAS)	Diurnal	Onagraceae, Rubiaceae

Phalaenoides tristifica (Hübner, 1818)	Australia (NSW, ACT, VIC, TAS, SA)	Diurnal	Onagraceae
Phalaenoides glycinae Lewin, 1805	Australia (QLD, NSW, ACT, VIC, TAS, SA, WA)	Diurnal	Vitaceae, Dilleniaceae, Onagraceae, Balsaminaceae, Asteraceae
Phalaenoides sp. 'Wet Tropics'	Australia (QLD)	diurnal, possibly crepuscular	
Phalaenoides behri (Angas, 1847)	Australia (WA, NT, QLD, NSW, ACT, VIC, SA)	diurnal	Loranthaceae
Radinocera vagata (Walker, 1865)	Australia (WA, NT, QLD), New Guinea, Tanimbar and adjacent Islands	nocturnal	Vitaceae
Radinocera sp. 'N Qld'	Australia (QLD)	nocturnal	
Radinocera sp. 'sandstone'	Australia (WA, NT)	crepuscular, nocturnal	Vitaceae
Radinocera maculosa (Rothschild, 1896)	Australia (QLD, NSW)	nocturnal	
Agarista agricola (Donovan, 1805)	Australia (NT, QLD, NSW), New Guinea, West Timor	diurnal	Vitaceae
Platagarista macleayi (W.J. Macleay, 1864)	Australia (QLD, NSW, VIC)	diurnal	Dilleniaceae
Eutrichopidia latina (Donovan, 1805)	Australia (QLD, NSW, ACT, VIC, TAS, SA)	diurnal	Dilleniaceae, Haloragaceae
Eutrichopidia sp. 'Wet Tropics'	Australia (QLD)	diurnal	Dilleniaceae
Burgena varia (Walker, 1854)	Australia (QLD, NSW), New Guinea, Maluku	diurnal	Vitaceae
Cruria synopla Turner, 1903	Australia (QLD, NSW, VIC)	diurnal	Araceae
Cruria latifascia Jordan, 1912	Australia (QLD)	diurnal	
Cruria tropica (T.P. Lucas, 1891)	Australia (QLD)	diurnal	
Cruria darwiniensis (Butler, 1884)	Australia (NT)	diurnal	Araceae
Cruria donowani (Boisduval, 1832)	Australia (WA, NT, QLD, NSW, ACT, VIC, SA), East Timor	diurnal	Vitaceae, Dilleniaceae, Nyctaginaceae
Cruria kochii (W.J. Macleay, 1866)	Australia (QLD), New Guinea	diurnal	
Cruria epicharita Turner, 1911	Australia (QLD)	diurnal	
Idalima affinis (Boisduval, 1832)	Australia (QLD, NSW, ACT, VIC)	diurnal	Dilleniaceae
Idalima metasticta Hampson, 1910	Australia (WA, NT)	diurnal	Dilleniaceae
Idalima polysticta (Butler, 1875)	Australia (NSW, ACT, VIC, SA)	diurnal	Dilleniaceae
Idalima butleri (Swinhoe, 1892)	Australia (WA)	diurnal	
Idalima sp. 'Arnhem Land'	Australia (NT)	diurnal	Dilleniaceae
Idalima aethrias (Turner, 1908)	Australia (WA, NT, QLD)	diurnal	Dilleniaceae
Idalima leonora (Doubleday, 1846)	Australia (WA, NT)	diurnal	Dilleniaceae
Idalima tasso (Jordan, 1912)	Australia (QLD)	diurnal	

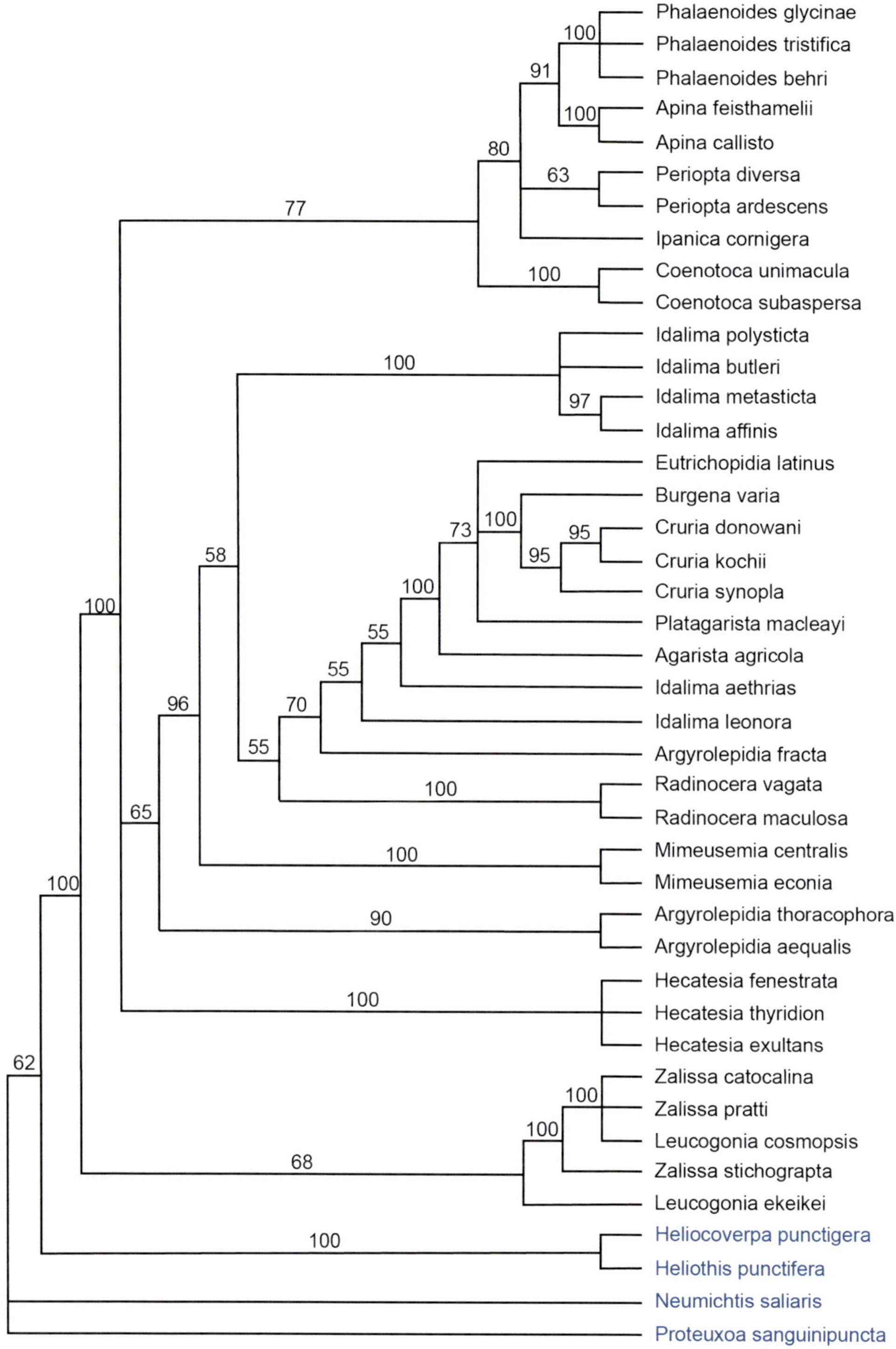

Fig. 22.2. Cladistic analysis based on 50% majority rule tree using morphological characters (n = 39 characters; 13,232 trees; tree length= 257; over 90 islands of length 257; CI = 0.409; RI = 0.687) (after Wragg and Braby, 2025). Numbers above the nodes are the percentage of trees in which that node occurs. Outgroup taxa are highlighted in blue.

genitalia. His first group (Group A) contained the Asian genus *Episteme* (in Subgroup I), whereas his second group (Group B) contained all the other genera (in Subgroups II–XIII). The Australian genera were classified into eight of these subgroups. There has been no explicit phylogenetic analysis to test the validity of Kiriakoff's hypothesis. Wragg and Braby (2025) found that although there was evidence of a close relationship between *Burgena* and *Cruria* (supporting Subgroup III), other subgroups were found to be incongruent, such as the delineation and classification of species assigned to *Phalaenoides* (in Subgroups IV and VI).

Although Wragg and Braby's (2025) phylogeny (Fig. 22.2) was only partially resolved, several patterns were evident which differ from systematic classifications proposed by Kiriakoff (1977) and Edwards (1996). First, the genera *Leucogonia* and *Zalissa* were closely related and these genera, together with *Sarbanissa*, probably comprise the sister group to the rest of the Australian agaristines. All members of this group have a simple, tubular vesica, are cryptically patterned and nocturnal compared with the brightly coloured and diurnal species, which have more elaborate vesica shapes, usually with curved diverticula. Second, several genera (*Hecatesia, Argyrolepidia, Coenotoca, Mimeusemia, Periopta, Ipanica, Apina*) comprised a 'three-barred' group in which most members have three bars (antemedial, medial, postmedial) on the forewing. The constituent species have an expanded or usually simple, tubular vesica, sometimes with a small, straight or spiralled diverticulum, and longitudinal thoracic markings. A large proportion of these species have the forewing drab or with a reddish-brown ground colour, some with iridescent markings, and are often nocturnal or crepuscular. The remaining genera (*Radinocera, Agarista, Platagarista, Eutrichopidia, Burgena, Cruria, Idalima* and *Phalaenoides*) have the diverticula curved and sometimes the vesica is also curved, or permanently everted and sclerotized (as in *Burgena* and *Cruria*). Nearly all of these species are brightly coloured and diurnal.

22.3 Diversity and Distribution

Agaristines have a pantropical distribution (Hampson, 1901, 1910; Jordan, 1912; Strand, 1912; Kiriakoff, 1977; Kitching, 1984). They are thought to have originated in the eastern Palaearctic during the Coenozoic, with a palaeotropical expansion and differentiation in the New World giving rise to a Nearctic and Neotropical element (tribe Alypiini), followed by a westward expansion and radiation in Africa and Madagascar giving rise to an Afrotropical element (tribe Heracliini) (Kiriakoff, 1977). The Agaristinae of the Oriental region are considered to be a more recent extension of the Palaearctic faunal element (tribe Agaristini) (Kiriakoff, 1977). Of 18 genera recognized in Australia (Table 22.1), a large proportion (seven genera, i.e. 39%) is endemic to the continent and its surrounding islands (i.e. *Hecatesia, Coenotoca, Periopta, Apina, Phalaenoides, Platagarista,* and *Idalima*). Most other Australian genera extend to New Caledonia (*Eutrichopidia*), mainland New Guinea and its adjcaent islands (*Leucogonia, Zalissa, Argyrolepidia, Ipanica, Radinocera*) or mainland New Guinea and East Timor and/or Indonesia (*Agarista, Burgena, Cruria*) (Hampson, 1901, 1910; Jordan, 1912; Kiriakoff, 1977; Common, 1990). Only *Sarbanissa* and *Mimeusemia* occur more widely in South-east Asia, extending to India, Indochina, China, Taiwan and Japan (Kiriakoff, 1977; Wei *et al.*, 2019). At the species level, a very high proportion (39 species out of 53, i.e. 74%) is endemic to Australia (Table 22.1).

Despite high levels of endemism, Kiriakoff (1977) noted the lack of a distinct Australian faunal element, stating the 'Australian genera hardly differ from those of the continental and insular East Indies'. Accordingly, the Australian fauna is considered to be a subset of the Oriental fauna, which in turn is a southern extension of the Palaearctic fauna (Kiriakoff, 1977). If this hypothesis is correct, then the Australian Agaristinae are relatively recent, arriving from the Oriental region perhaps during the Miocene (or possibly earlier) when opportunities for dispersal (by island hopping) became more favourable following collision of the Australian continental plate with the South-east Asian plate.

Although agaristines occur in all biomes and major vegetation communities across the continent, most species occur in the tropical regions of northern Australia and along the eastern coast where they occur predominantly in rainforest, monsoon vine thicket, savannah woodland and eucalypt heathy woodland. Only a few species are restricted to the temperate latitudes. *Idalima butleri* is endemic to heathy woodlands of the south-western corner of Western Australia, whereas its sister species *I. polysticta* is restricted to a limited area in south-eastern Australia where it occurs in eucalypt woodland with an understorey of heath. *Hecatesia thyridion* and *H. exultans* both have disjunct distributions with populations in south-western Western Australia and in south-eastern Australia (Victoria, South Australia) where they typically occur in drier habitats, including mallee woodland, mallee heath, heathland or *Banksia* open woodland. *Apina feisthamelii* is restricted to subalpine grasslands in Tasmania and mainland south-eastern Australia (ACT, NSW, Victoria). *Apina callisto* and *Phalaenoides tristifica* are also restricted to temperate habitats, with the former species occurring in grasslands and pastures across southern Australia, and the latter species widespread in south-eastern Australia, favouring swamplands up to 1900 m where it may co-occur with *A. feisthamelii*. *Ipanica cornigera* and *Phalaenoides behri* are the only species to extend widely into the arid zone of central Australia. On a state-by-state basis, Queensland has the highest species richness with 39 species, followed by New South Wales and Western Australia (18 species each). In contrast, relatively few species occur in South Australia (11 species), the Australian Capital Territory (10) and Tasmania (6) (Table 22.1).

22.4 Biology

22.4.1 Life cycles and seasonality

The life histories and biology of the Australian Agaristinae have been well documented for *Phalaenoides glycinae* (McCoy, 1878; Scott, 1891; Cordingley, 1980a,b, 1981b); *P. tristifica* (McCoy, 1878; Scott, 1891; Wragg, 1999); *Apina callisto* (Larcombe, 1914; Humphries, 1950; Lewis, 1985, 1989); and, *A. feisthamelii* (Wragg, 1999). There is fragmentary information for several other species (Scott, 1864, 1891; McCoy, 1878; Moss, 2010; Braby, 2011, 2015; French, 2013).

In general, eggs are disc shaped or semi-spherical with faint radial ribs. They are initially green but within a day change to yellow-brown with small reddish or orange-brown markings and thus are well camouflaged. Eggs are laid singly. A few species lay their eggs directly on the larval food plant, but many oviposit on objects adjacent to the plant, such as faces of rocks, soil, twigs, branches or leaves of other plants. In *A. callisto*, females lay their eggs in patches of bare earth, often in highly disturbed areas. The eggs of all species develop directly and hatch soon afterwards. The larvae of agaristines are without secondary setae, aposematically coloured, and with the posterior abdominal segments enlarged or truncated to form a hump (Common, 1990; Nielsen and Common, 1991). Larvae of the four well-studied species noted above all have six larval instars. Larvae generally feed on the plant on which their eggs were laid, but in *A. callisto* larvae shift their feeding preferences as they develop, with early instars (instars II, III) preferring the introduced annual, *Arctotheca calendula* (Asteraceae), and later instars (instars IV–VI) favouring the native perennial, *Plantago antarctica* (Plantaginaceae) (Lewis, 1989).

Larval development is rapid, typically being completed within a few weeks or months. In *P. glycinae*, mean larval development varies with temperature, from 18 days at 27°C to 61 days at 15°C (Cordingley, 1980a), whereas larvae of *A. callisto*, which feed during winter, take up to 3 months to complete development (Lewis, 1985, 1989). When ready to pupate, larvae may wander significant distances away from the food plant to find suitable pupation sites. Cordingley (1980a) estimated that *P. glycinae* larvae may wander for 2 days at an average velocity of 0.64 m/min, indicating that they are capable of travelling up to a maximum distance of a kilometre. Pupation occurs in soil, soft wood of rotting logs, wood fragments in the soil, bark at the base of trees or even inside fence posts. The cocoon consists of a chamber or cell composed of particles of soil or wood in which the larva pupates. In *A. callisto*, the larvae seek a bare patch of dry earth to excavate an elaborate hole in the ground, which takes 8–10 h or even longer to construct (Larcombe, 1914; Humphries, 1950). The entrance of the hole is then sealed before the larva constructs the cocoon, which takes several weeks to make.

The pupae of agaristines are various shades of brown, sclerotized and often rugose (Common, 1990; Nielsen and Common, 1991). Duration of the pupal stage of directly developing generations is usually around 2–3 weeks, but all species that have been reared undergo pupal diapause during the temperate winter or tropical dry season, with the dormancy period of some species lasting for 1 or even 2 years (Braby, 2011; Braby *et al.*, 2018). The exception to this pattern is *A. callisto*, which feeds as a larva during winter and pupates in August–September, spending up to 8 months over the hottest time of year in the pupal stage (Fig. 22.3A) (Larcombe, 1914; Humphries, 1950).

Apina callisto and *A. feisthamelii* are both univoltine, with a short flight season lasting approximately 1–2 months (Figs 22.3, 22.4) (Lewis, 1985, 1989; Wragg, 1999). Adults of *A. callisto* fly only in autumn (early April to early May) with a peak in abundance that lasts for about a week in early to mid-April in the ACT (Fig. 22.3B), whereas *A. feisthamelii* flies only in spring (mid-October to mid-December) with a peak in abundance in early November (Fig. 22.4). Both species occur in open grasslands: *A. callisto* occurs widely in lowland temperate grasslands and pastures, whereas *A. feisthamelii* is restricted to subalpine frost hollows at high altitudes (>1100 m in the ACT). In contrast, both *P. tristifica* (Fig. 22.4) and *P. glycinae* (Figs 22.5, 22.6) have a longer flight season, with two discrete generations annually (bivoltine) but the second (summer) generation is far more abundant than the first (spring) generation (Cordingley, 1980b; Wragg, 1999). *Phalaenoides glycinae* may have a partial third generation at some locations, but the offspring rarely survive due to food shortfall (Cordingley, 1980b).

The phenology of the other agaristine species has been less well studied but a univoltine or bivoltine life cycle seems to be characteristic of the Australian fauna. In the tropics of northern Australia, the adults of most species are highly seasonal and appear to be univoltine, partially bivoltine or bivoltine – often they peak in abundance during the build-up to the early wet season following the pre-monsoon storms, which may be followed by sporadic occurrences later in the wet season and/or early dry season (Braby *et al.*, 2018). The larvae of many of these tropical species (e.g. *Hecatesia* spp., *Periopta* spp., *Radinocera* spp., *Agarista agricola*, *Cruria* spp.) are highly seasonal and occur only during the wet season. The larvae of these species specialize on the new leaf growth of the food plants, which are often short-lived seasonal perennial vines or shrubs, seasonally deciduous shrubs or short-lived annuals or seasonal herbs (Braby *et al.*, 2018). In contrast, four tropical species of *Idalima* in the Top End have a longer flight season and although they peak in abundance during the early wet season, they probably have two or more generations annually, with the larvae feeding on perennial shrubs (Braby *et al.*, 2018). The mistletoe-feeding *P. behri* has a long flight period in both tropical and temperate zones, and is probably multivoltine, with three annual generations in northern Australia and at least two generations annually in southern Australia.

22.4.2 Adult activity

Agaristines are mostly either nocturnal or diurnal but some are also crepuscular, flying just before or after sunset (Table 22.1). Eight Australian genera (*Leucogonia*, *Zalissa*, *Sarbanissa*, *Argyrolepidia*, *Coenotoca*, *Mimeusemia*, *Ipanica*, *Radinocera*) are nocturnal, and these taxa are regularly observed or captured at night by light traps. As noted earlier, all of these taxa are cryptically patterned or have the forewing with a drab reddish-brown ground colour, although a few have iridescent blue markings, such as *Argyrolepidia aequalis* and *A. fracta*. The remaining ten genera (*Hecatesia*, *Periopta*, *Apina*, *Phalaenoides*, *Agarista*, *Platagarista*, *Eutrichopidia*, *Burgena*, *Cruria*, *Idalima*) are all diurnal; they are brightly coloured and often have contrasting colours of white, yellow, orange or red against a black or dark brown ground colour. I have observed that *Coenotoca subaspersa* is an exception to this rule, being both diurnal (flying from late morning to mid-afternoon) and nocturnal (regularly attracted to light traps). Females of two species of *Hecatesia* (*H. fenestrata*, *H. thyridion*) are occasionally collected at light traps at night, but it is not clear if this genus is also nocturnal. The diurnal species generally fly only during sunshine, although some tropical species will fly when overcast provided conditions are hot. Many of the diurnal species fly in the early to mid-morning and/or mid- to late afternoon and are generally not active at midday, especially when conditions are hot. During cooler weather, however, male activity in some species may continue until

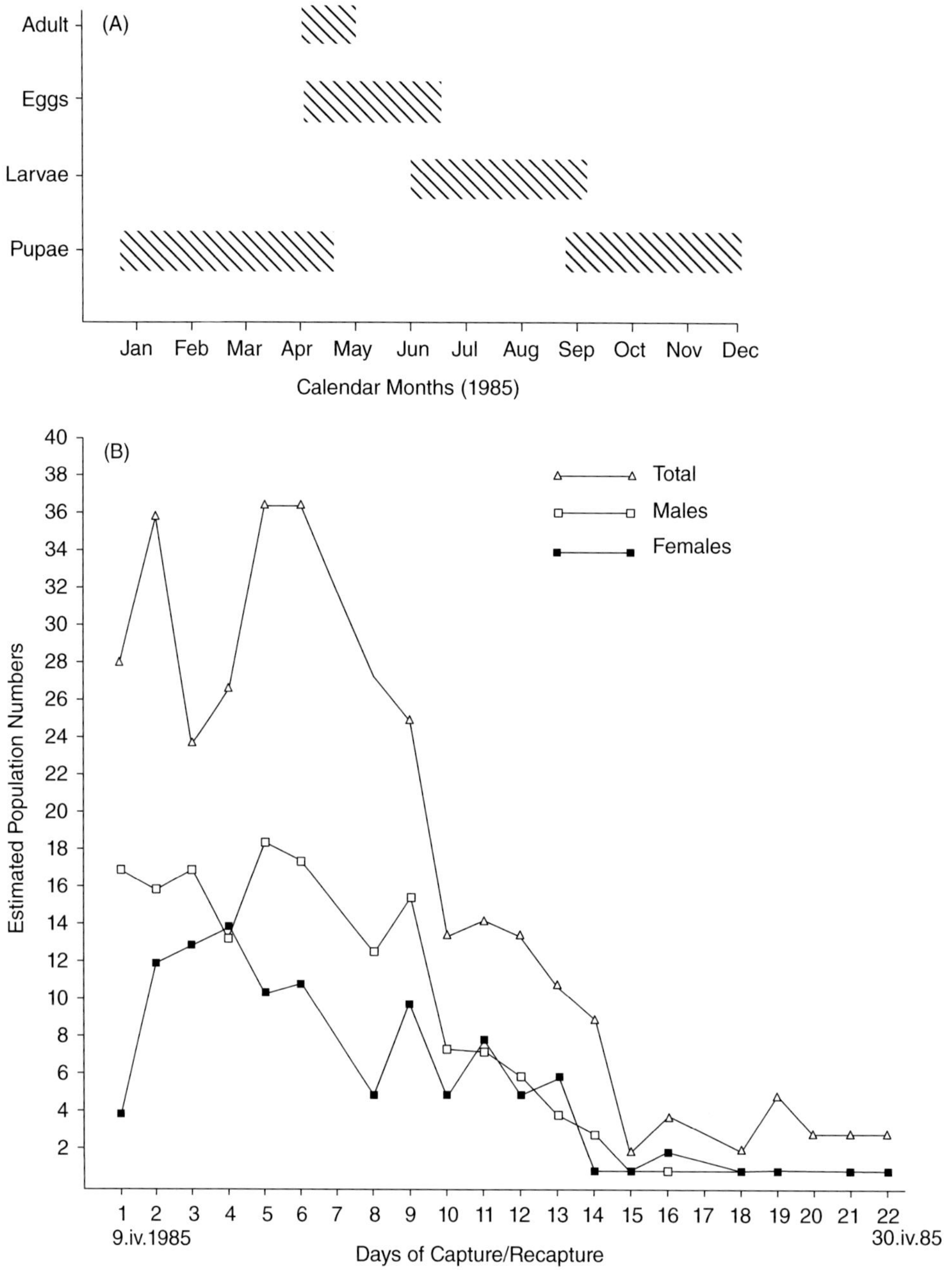

Fig. 22.3. Biology of *Apina callisto*. (A) Life cycle showing phenology of the adult and immature stages. (B) Seasonal changes in population size of adults during the 1985 flight season (after Lewis, 1985). Moths were monitored at a site in Canberra, ACT (CSIRO Division of Entomology, Black Mountain Laboratories), with daily estimates of population size calculated using the Manly–Parr method based on a capture–mark–recapture study from 9–30 April. The flight season commenced on 4 April and ceased on 7 May 1985.

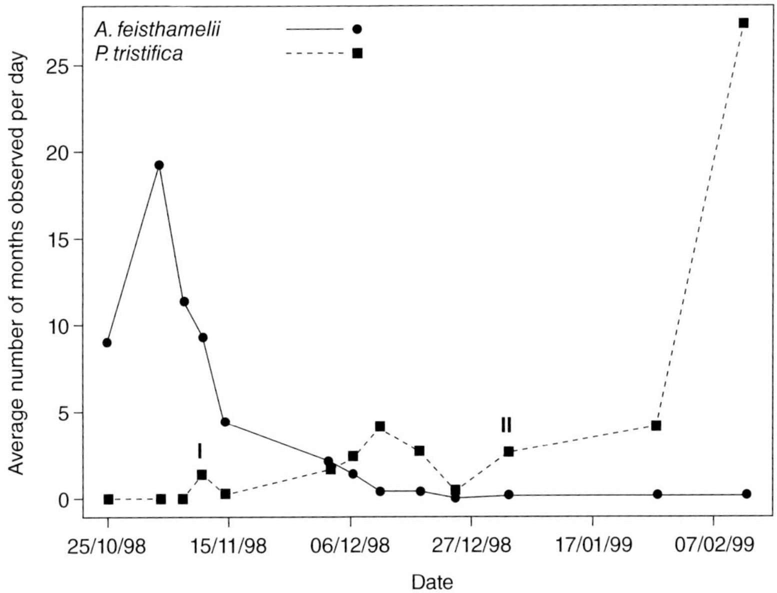

Fig. 22.4. Seasonal changes in relative abundance of adult *Apina feisthamelli* and *Phalaenoides tristifica* during the 1998–1999 flight season at Mt Clear, Namadgi National Park, ACT (after Wragg, 1999). Numbers I and II indicate the start of the first (spring generation) and second (summer generation) flight periods of *P. tristifica*, respectively. Adult moths were sampled approximately twice a week (25 Oct 1998–2 Jan 1999) and then fortnightly (10 Jan–13 Feb 1999) using a transect count method, which involved walking a transect (1800 m long × 10 m wide × 5 m in front of the observer) over a period of 2 h. Counts were conducted during sunny or partially cloudy conditions and repeated two or three times each day (0830–1030, 1130–1330, 1530–1730 h AEDT) and then averaged for each daily sample.

around midday (McFarland, 1976). The two species of *Apina* are an exception and fly throughout much of the day, typically after 1000 h, but their flight periods are restricted to the cooler months of the year (spring or autumn) (Lewis, 1985; Wragg, 1999). All species of *Hecatesia* fly during the afternoon, and in *H. thyridion* (Alcock *et al.*, 1989) and *H. fenestrata* (Common, 1990) activity may continue until dusk. *Hecatesia exultans* is unusual in that adults are active from mid-morning to mid-afternoon (i.e. 1000–1500 h AWST), mainly during winter (June–August) in northern parts of the species' range in western Western Australia (Alcock and Bailey, 1995) or late winter–early spring (August–October) in western Victoria (F. Douglas, pers. commun.).

The diurnal agaristines typically fly with a fast and erratic flight; many species fly in the understorey or lower canopy but some such as *Hecatesia exultans*, *Hecatesia* sp. 'Amata', *Apina* spp., *Periopta diversa*, *Phalaenoides tristifica* and *Idalima metastica* fly very close to ground level. Agaristines invariably settle head downwards, often low down on a tree trunk or leaf of a shrub, in a tectiform position – a roof-like resting position in which the forewings completely cover the hindwings. Because agaristines can hear sound, they are very difficult to approach when settled and easily take flight, often from even the slightest disturbance.

24.4.3 Mating behaviour and acoustic communication

Mate location and territorial behaviour have been reported for only a few of the diurnal-flying species. McFarland (1976) first described hilltopping behaviour in *P. behri* and noted that such behaviour was rare in diurnal moths, but otherwise similar to that reported for many butterflies and other insects (Shields, 1967; Scott, 1968; Alcock, 1987; Braby, 2000). In this species,

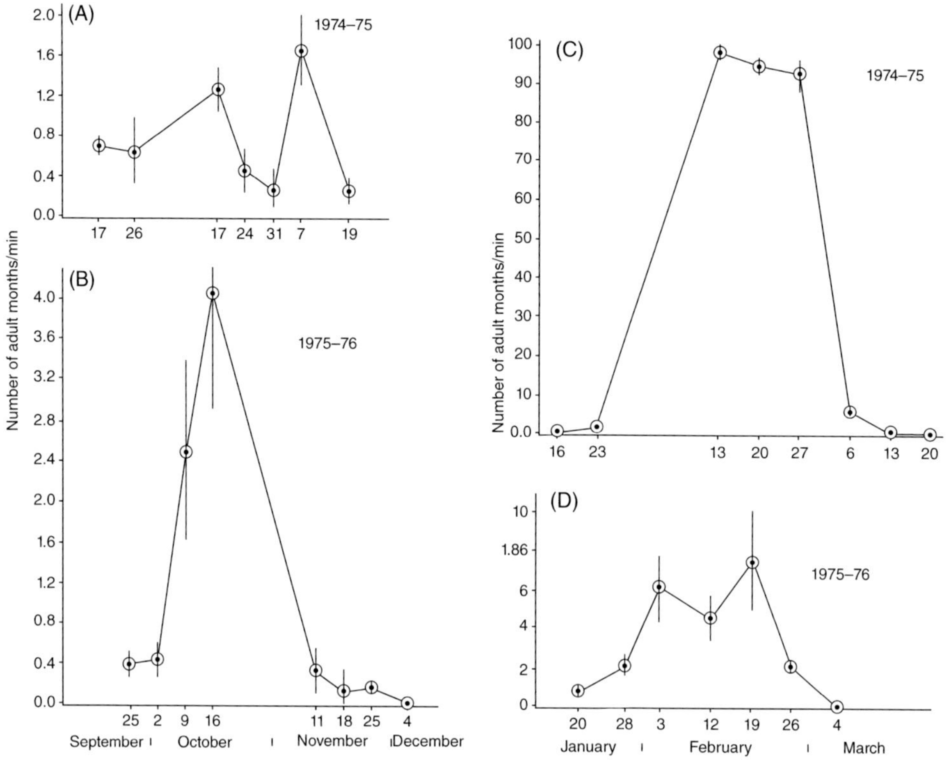

Fig. 22.5. Seasonal variation in the number of adults of *Phalaenoides glycinae* in a commercial vineyard 70 km SE of Adelaide, SA. (A, B) First (spring generation) flight period. (C, D) Second (summer generation) flight period (after Cordingley, 1980b). Adult day moths were sampled during two seasons: (A, C) 1974–1975 and (B, D) 1975–1976. Shown are the number of adults (± SE) per sampling unit. Adults were counted approximately each week using an area count method of 30 m^2 (3 m × 10 m) for 15 min/1 h. For each sample, three separate subsamples were made, including moths that were flying (5 mins), those resting on vines (5 mins), and those feeding on flowering weeds growing along the edge of the vineyard (5 mins). Used with permission from John Wiley & Sons.

males visit landmarks (summits of hills) where they show strong site fidelity, spending long periods patrolling, and even returning to, the same hilltop, occasionally alighting to rest on a particular shrub; they are territorial, engaging in site defence by chasing off intruders, such as rival males. These hilltops are visited by females for mating. Males of *P. behri* typically hilltop during the early to mid-morning depending on temperature (McFarland, 1976), but they may also hilltop during the late afternoon. Although hilltopping is less well documented in other diurnal agaristines (Jelinek *et al.*, 1994; Moss, 1999a; Wainer and Yen, 2000), I have observed that similar mate location behaviour occurs regularly in seven species: *Hecatesia fenestrata* (dusk), *H. thyridion* (dusk), *Phalaenoides glycinae* (mid-morning to late afternoon), *Agarista agricola* (early to mid-morning), *Eutrichopidia latina* (early to mid-morning but sometimes continuing to late morning or early afternoon depending on temperature), *Eutrichopidia* sp. 'Wet Tropics' (early to mid-morning), and *Cruria donowani* (late morning to early afternoon for subsp. *C. d. donowani* in eastern Australia). When hilltopping, males of *A. agricola* and *E. latina* frequently perch on foliage with wings outstretched (i.e. 180° or slightly tectiform), occasionally patrolling light gaps (openings) in the canopy, but if two or more males are present, they fly rapidly and engage in aggressive aerial combat behaviour. In contrast, *P. glycinae* males rarely settle when hilltopping and patrol rapidly around the canopy of trees for short periods before departing.

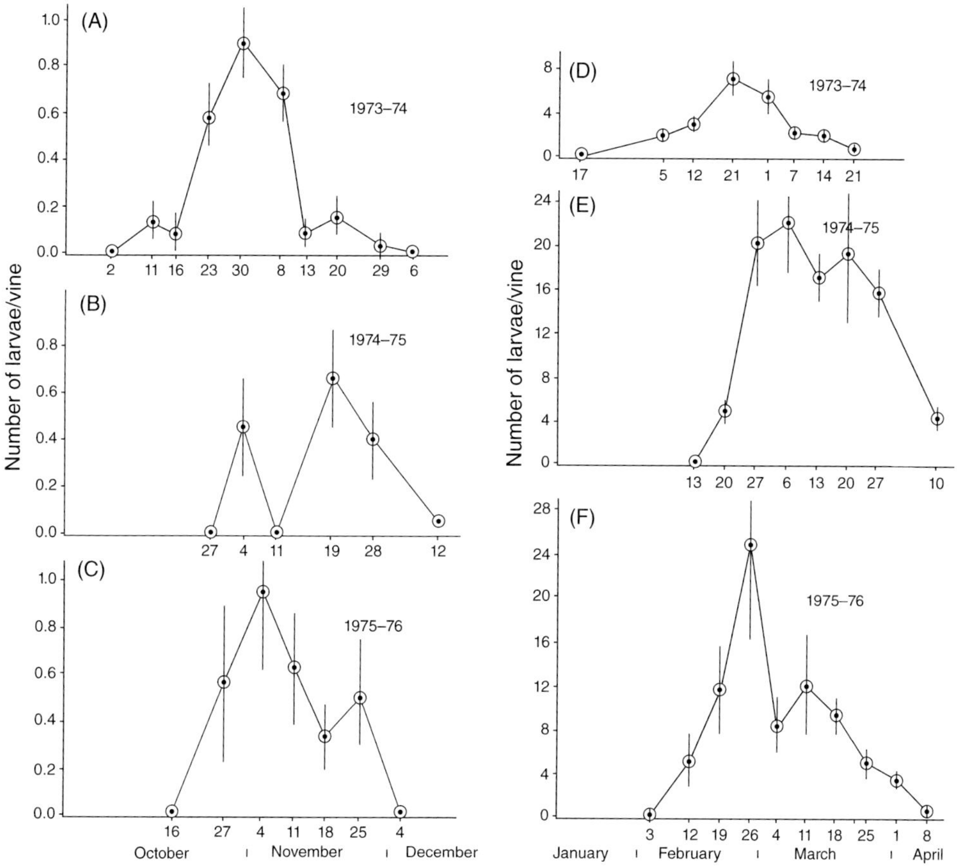

Fig. 22.6. Seasonal variation in the number of larvae of *Phalaenoides glycinae* in a commercial vineyard 70 km SE of Adelaide, SA. (A–C) First (spring) generation. (D–F) Second (summer) generation (after Cordingley, 1980b). Larvae were sampled during three seasons: (A, D) 1973–1974; (B, E) 1974–1975; and (C, F) 1975–1976. Shown are the number of larvae (± SE) per sampling unit. Larvae were counted approximately each week using a stratified method involving 40 vines (arranged in five rows, each with eight vines); on each sampling occasion five vines (one from each row) were sampled randomly. Used with permission from John Wiley & Sons.

In addition to hilltopping, other mating tactics are used. For example, males of *P. glycinae* also patrol around the canopy of prominent trees near the breeding areas to establish mating territories, especially during the mid- to late afternoon (M.F. Braby, unpublished data). Males of most species of *Cruria*, including *C. synopla*, *C. latifascia*, *C. darwiniensis*, *C. donowani* and *C. kochii*, establish territories that appear to be used for mate location. Typically, several males will congregate in a small open area within the forest or woodland; within these arenas they perch (head downwards) within 0.5–2.0 m of the ground on the trunks of particular trees or foliage (or sometimes people if they are standing still!) while waiting for receptive females (Grund, 2010; Grimshaw, 2020). Occasionally, they make short patrolling flights before returning to the same site to perch, which are defended aggressively against conspecific males. The same males use these arenas on consecutive days, at least for 12 days in *C. donowani* (Grimshaw, 2020), suggesting *Cruria* employs a lek mating system.

In contrast, males of *Apina* spp. patrol the breeding areas and rarely settle; they fly rapidly close to ground level (within *ca.* 0.2–1.0 m) searching for unmated females, which remain settled on the ground (Lewis, 1985; Wragg,

1999). The males of these species are the only Australian agaristines to possess bipectinate antennae, so presumably the females release a pheromone to attract males whilst settled. Males of the tropical *Periopta diversa* employ similar patrolling tactics to *Apina*, flying within 0.5 m above the ground in open disturbed grassy areas to locate unmated females, but they are active for only 1–2 h in the early morning just after sunrise (M. F. Braby, unpublished data).

Another component of mate location and courtship behaviour is the ability of males of several diurnal species to produce audible sound. The hearing organs are located on each side of the metathorax (Scoble, 1992). Acoustic communication is rare in the Lepidoptera (Scoble, 1992; Alcock and Bailey, 1995), but at least nine Australian agaristines communicate acoustically, including *Hecatesia* (5 species), *Platagarista macleayi*, *Eutrichopidia* (2 species) and *Agarista agricola*. The mechanism of sound production differs in all four genera and it is most elaborate in *Hecatesia*, which are commonly known as 'whistling moths' (Bailey, 1978; Alcock *et al.*, 1989; Common, 1990; Nielsen and Common, 1991; Alcock, 1995). In this genus, the males possess a pair of conspicuous knob-like structures ('castanets') near the costa of the forewing, below which is a hyaline corrugated membranous area that acts as a resonant chamber (Bailey, 1978). A percussive sound is produced when the castanets strike each other as the wings are beaten together in an upright position above the thorax in rapid succession. The song comprises a series of clicks or chirps, with each chirp composed of a group of rapid sound pulses produced by several wingbeats (Alcock *et al.*, 1989; Alcock and Bailey, 1995), and it is emitted either during a special display flight in a relatively small open area (*H. fenestrata*, *H. thyridion*, *Hecatesia* sp. 'Arnhem Land') or when perched on low vegetation (*H. exultans*, *Hecatesia* sp. 'Amata'). In *H. fenestrata*, *H. thyridion* and *Hecatesia* sp. 'Arnhem Land' calling males typically fly rapidly close to the ground in a figure of eight, with the height varying among species, from below 0.5 m (*H. thyridion*) to 1–2 m (*Hecatesia* sp. 'Arnhem Land') or 2–3 m (*H. fenestrata*). The call is audible in most species at a distance of up to 10 m or more (Common, 1990), except in *H. exultans* which is ultrasonic (30 kHz) and for most humans can only be detected by an ultrasound detector (Alcock and Bailey, 1995). The structure of the resonant chamber of *Hecatesia* is similar to that of *Pemphigostola synemonistis* from Madagascar, except in this species the sound is produced by rubbing the midleg (which bears a series of ribs or 'file' on the tibia and tarsus) along a groove on the ventral surface of the forewing; the groove has a series of ribs along the base of the cubitus (Scoble, 1992).

In male *Platagarista*, the basal segment of the tarsus of the hindleg has a series of transverse ridges and grooves which form a 'file', and the wing membranes between the basal veins of both wings are semi-hyaline and enlarged (curved), which possibly function as resonators. A loud stridulating sound is produced in flight when the file of the hind tarsus is rubbed against the underside of the veins (Common, 1990). In this species, the males fly rapidly in a series of tight horizontal spirals or small circles at a height of 2–3m above the ground in more open areas, although the height at which they fly varies and sometimes they may fly in the canopy of treetops. Typically, males are most active during early to mid-morning (D.R. Britton and E.P. Beaver, pers. commun.), and sometimes large congregations may be encountered as they patrol along tracks, crests and cliff tops of sandstone escarpments (Moss, 1999b).

The method of sound production in *Eutrichopidia* appears to be similar to *Platagarista*, except the hind tarsi lack the file and only the hindwing has the intervening flexed membrane between the basal veins; the sound comprises a single stridulation or series of rapid stridulations in flight and is audible to the human ear at close range, but it is produced only when males are engaged in hilltopping behaviour, typically in the early or mid-morning (M.F. Braby and F. Douglas unpublished data). The mechanism of sound production in *Agarista agricola* has not been determined; the sound comprises a pulse of several rapid clicks in flight and only occurs when males are hilltopping in the early morning (M.F. Braby, unpublished data). It is possible that sound is produced by rubbing the hindlegs against the undersurface of the hindwing.

Acoustic signalling in *Hecatesia* is emitted only during courtship when males display (call) within small territories comprising open patches of habitat for mating over long periods (Alcock *et al.*, 1989; Alcock and Bailey, 1995). The males are highly aggressive and defend their territories against intruders. In *H. exultans*, territories are defended through a contest behaviour known as 'buzz bumping' in which they amplify the frequency, duration and intensity of the call in an attempt to ram the opponent male (Alcock and Bailey, 1995). The territories are often close together and thus comprise a lek mating system. Sexually receptive females enter the lek specifically to mate. Copulation is brief: in *H. thyridion* it lasts only 50–135 sec (Alcock *et al.*, 1989), whereas in *H. exultans* it varies from 5 to 11 mins (Alcock and Bailey, 1995). In *H. exultans*, male acoustic signalling occurs from mid-morning to mid-afternoon (Alcock and Bailey, 1995), whereas in *H. thyridion* it typically occurs during late afternoon during sunny weather (Alcock *et al.*, 1989) but it also occurs when males are hilltopping at dusk (M.F. Braby and F. Douglas, unpublished data). Presumably, sound production in *Platagarista*, *Eutrichopidia* and *Agarista* has a similar function to *Hecatesia* in courtship and reproduction, especially since it occurs only in males. In *Eutrichopidia* and *Agarista*, sound production seems to occur only when males are hilltopping.

Within the mating territory, *H. thyridion* males have also been observed to exhibit 'vegetation rubbing' in which they appear to mark scents (volatile chemicals) on certain plants (Alcock *et al.*, 1989). The significance of this behaviour is unclear, but the abdominal brush organs of *Phalaenoides glycinae* have been found to release two pheromones (2-phenylethanol, 2-phenylethanolacetate) (Edgar *et al.*, 1978) that are widely believed to be used in courtship to attract females. Edgar *et al.*'s (1978) survey of the presence of abdominal brush organs suggests most, if not all, male Australian agaristines possess these structures, including the three named species of *Hecatesia*.

22.5 Ecology

22.5.1 Larval food plants

A critical review of the larval food plants of the Australian Agaristinae was recently compiled by Braby (2025). Associations between food plants and moths have been documented for two-thirds of the species (34 out of 53 or 64%) and almost all of the genera (16 out of 18 or 89%) (Table 22.1).

There are several interesting patterns with respect to diet breadth from the perspective of the moths. First, the most overwhelming pattern is the high level of specialization at the generic level. At least ten genera are monophagous (i.e. feeding on one plant family), with six of these (*Zalissa*, *Argyrolepidia*, *Coenotoca*, *Radinocera*, *Agarista*, *Burgena*) specializing on Vitaceae. Two genera (*Platagarista*, *Idalima*) appear to be specialists of Dilleniaceae, whereas the two other genera feed on other plants, with *Hecatesia* on Lauraceae and *Ipanica* on Nyctaginaceae. Three genera are oligophagous (i.e. feeding on two or three plant families) and these involve different combinations of plant families: *Mimeusemia* on Vitaceae and Menispermaceae; *Periopta* on Vitaceae and Rubiaceae; and *Eutrichopidia* on Haloragaceae and Dilleniaceae.

The second striking pattern is that three genera (*Apina*, *Phalaenoides*, *Cruria*) are highly polyphagous (i.e. feeding on three or more plant families). The genus *Apina* is recorded to feed on no less than ten families (Asteraceae, Crassulaceae, Fabaceae, Geraniaceae, Lamiaceae, Malvaceae, Onagraceae, Plantaginaceae, Polygonaceae, Rubiaceae) and many of these are not used by other agaristines. *Phalaenoides* is recorded from six families (Asteraceae, Balsaminaceae, Dilleniaceae, Loranthaceae, Onagraceae, Vitaceae), whereas *Cruria* is recorded from four families (Araceae, Dilleniaceae, Nyctaginaceae, Vitaceae), although food plant data are known for only three of the seven species in that genus (Table 22.1).

From the perspective of the Angiospermae, the patterns of food plant usage indicate that, despite high levels of specialization at species and generic levels, Australian Agaristinae as a whole feed on a broad set of 19 plant families in 16 orders and eight higher informal groups (clades) (Table 22.1). Most of these plant taxa are not closely related (APG IV, 2016). Even when records for the highly polyphagous *Apina callisto* are excluded, the diet breadth of the remaining 15 genera of agaristines is still relatively broad, with 12 plant families in 12 orders in eight informal groups (Table 22.1). A robust phylogeny of the angiosperms indicates that most of these 12 plant families are not closely related and, moreover, each of these families is classified in a different order (APG IV, 2016).

The lack of a well-resolved global phylogeny of Agaristinae precludes analyses of deep evolutionary patterns of host usage and determination of the ancestral host, but Vitaceae (Vitales) are the most widely exploited family (used by 12 agaristine species in ten genera), followed by Dilleniaceae (Dilleniales) (used by 11 agaristine species in five genera), a family which is not closely related to Vitaceae. In contrast, four families are each used by only two agaristine genera: Nyctaginaceae (by *Cruria* and *Ipanica*), Rubiaceae (by *Apina* and *Periopta*), Onagraceae (by *Apina* and *Phalaenoides*), and Asteraceae (by *Apina* and *Phalaenoides*). The remaining six families (Lauraceae, Araceae, Menispermaceae, Haloragaceae, Loranthaceae, Balsaminaceae) are each exploited by a single genus of Agaristinae (Table 22.1). These patterns are similar to the Oriental fauna in which the families Vitaceae and Dilleniaceae appear to be the most frequently used plants by Asian agaristines, with a further ten families used infrequently (Kitching, 1984; Sugi, 1987; Holloway, 1989; Robinson *et al.*, 2002). In South Africa, the larvae of 11 species feed predominantly on Vitaceae and rarely on other plants (Staude *et al.*, 2023).

The use of Lauraceae (Laurales) by *Hecatesia* is interesting because they belong in the magnoliid clade, which represents an early lineage of 'primitive' plants (APG IV, 2016). No other agaristine is known to feed on magnoliids in Australia. The exploitation of *Cassytha* (7 species) by *Hecatesia* appears to be the only example of a lineage in the Australian agaristine fauna in which there has been radiation on a particular plant genus/family, with up to five species of *Hecatesia* endemic to Australia specializing on *Cassytha* (Braby, 2025).

Even without a well-supported phylogeny of the Agaristinae, the pattern of host usage is clear. Despite high levels of monophagy, the Australian agaristines as a whole are exploiting a large set of distantly related plants. Such a pattern could only have been generated by frequent host shifts and repeated colonization rather than through a process of co-speciation or co-evolution. In other words, there is no evidence of phylogenetic conservatism in the Australian Agaristinae, and related species of moths do not feed on closely related species of plants at the higher taxonomic levels.

22.5.2 Aposematism and chemical defence

The larvae of agaristines are invariably warningly coloured ('aposematic') and the adults of many species are diurnal and also aposematic. Moreover, the adults of some species, such as *Platagarista*, release a strong odour when captured. These traits suggest the moths are distasteful or unpalatable to most predators and chemically defended (poisonous) through the sequestration of secondary plant compounds from their larval food plants or from ingested components that are used as precursors for chemical defensive compounds. Indeed, when larvae are molested they regurgitate a green fluid from their mouth as an additional defence (Humphries, 1950; Cordingley, 1980a). Late instar larvae of *P. glycinae* and *P. behri* not only regurgitate but adopt an 'aggressive' pose when provoked or threatened by rearing their head backwards to display the bright orange-coloured thoracic legs; this behaviour is most pronounced in *P. behri*. In *P. glycinae*, Cordingley (1980a) observed that the larvae also evert a prosternal gland – a ventral eversible thoracic organ, which is a defensive structure that releases volatile chemicals when everted (James *et al.*, 2012). Scott (1891) noted that the larvae of *P. glycinae* are not attacked by birds, even when in high abundance, apparently due to their defensive fluid regurgitations or by some other protective means, although Cordingley (1981a) observed larvae to be eaten by cuckoos (Horsfield's bronze cuckoo, pallid cuckoo) and the Australian magpie. French (2013) also commented that all birds except cuckoos avoid the larvae of *P. glycinae*. Similarly, larvae of *A. callisto*, which remain fully exposed on the ground, do not seem to be eaten by bird predators (Humphries, 1950), although F. Douglas (pers. commun.) once observed larvae being eaten by a fan-tailed cuckoo.

Very little is known of the toxins or secondary plant compounds of the two major plant families (Vitaceae, Dilleniaceae) exploited by Agaristinae. Raffauf's (1996) comprehensive survey of alkaloids, one of the largest groups of secondary plant compounds, found little evidence for the presence of alkaloids in Vitaceae and Dilleniaceae, and none in the relevant genera for Nyctaginaceae, Onagraceae and Loranthaceae, although positive detections were found in some genera of Rubiaceae (*Asperula*, *Oldenlandia*), Araceae (*Alocasia*, *Typhonium*), Menispermaceae and Lauraceae (*Cassytha*). Recent phytochemical studies of *Leea indica* (Vitaceae), a larval food plant of *Agarista agricola*, showed the presence of various secondary plant metabolites, including alkaloids, flavonoids, glycosides, phenolics, saponins, steroids, tannins and terpenoids as major bioactive components (Khuniad *et al.*, 2022). Similarly, a review of the phytochemistry of Dilleniaceae identified the presence of flavonoids, terpenoids, lignoids, phenolic derivatives and other secondary compounds, although the distribution of these components in *Hibbertia*, a major genus exploited by Australian Agaristinae with at least 14 species used (Braby, 2025), remains unknown (Lima *et al.*, 2014).

Interestingly, a study of *Hecatesia exultans* and its larval food plant, *Cassytha* spp. (Lauraceae) which contain aporphine alkaloids, found that the larvae (and adults) of this moth did not sequester these toxic compounds and thus were palatable to potential predators (Talianchich *et al.*, 2003). Predator avoidance in this species apparently relies on escape tactics involving rapid and erratic flight. It is unclear if this evasive defence mechanism in *H. exultans* extends more generally to the genus because adults of the four other species of *Hecatesia* are highly conspicuous, having bright orange patches on the hindwings and a brilliant orange-red abdomen, which is obvious in flight and 'flame'-like in colour. It is also possible that mimicry (e.g. Batesian) may also be involved, especially in *Hecatesia* sp. 'Amata' which resemble distasteful species of *Amata* (Erebidae: Arctiidae) moths.

22.6 Economic Importance

The only Australian agaristine of economic importance is the native vine day moth or vine moth, *Phalaenoides glycinae*, which has become a pest of commercial grapevines, *Vitis vinifera* (Vitaceae), in the wine-producing areas of south-eastern Australia (McCoy, 1878; Froggatt, 1907; French, 1916; Fenner, 1961; Jones, 1967). *Apina callisto* naturally occurs in grasslands but, following European settlement, it has adapted to pastures where its larvae feed on a large suite of introduced broad-leaved annual and perennial weeds (Lewis, 1989). However, it does not appear to be of any economic significance, although in some years the larvae may be extremely abundant.

The biology and ecology of *P. glycinae* in vineyards have been studied extensively by Cordingley (1980a, b, 1981a, b). Interestingly, before the introduction of cultivated grapevines and establishment of commercial vineyards, McCoy (1878) noted that in Victoria the common herb *Pseudognaphalium luteoalbum* (Asteraceae) was the preferred food plant of *P. glycinae*. However, following establishment of grapevines, the moth switched hosts to *V. vinifera*, increased in abundance and completely abandoned its former reliance on Asteraceae. Most damage inflicted on vines occurs during the second (summer) generation (Cordingley, 1980b), and high numbers of larvae may defoliate vines, causing the plants to produce new leaves rather than grapes (Cordingley, 1980a). For small-scale vineyards and residential and periurban gardens, physical hand removal of larvae is an effective method to prevent leaf damage and loss. This method is not practical for large-scale commercial vineyards, and four main strategies of control have been proposed to manage caterpillars: (i) chemical, (ii) biological, (iii) pheromonal and (iv) integrative.

Chemical control involves application of an insecticide to the foliage. Historically, azinphos, carbaryl and lead arsenate have been used (Fenner, 1961; Jones, 1967), but current registered poisons in Australia include carbamates, synthetic pyrethroids, Indoxacarb, and various products containing strains of *Bacillus thuringiensis* (e.g. Dipel®) (APVMA, 2025). Biological control involves the use of natural enemies to regulate numbers of caterpillars; Cordingley (1981a) identified several important candidate predators and parasitoids, including the predatory shield bug *Oechalia schellenbergii* (Hemiptera: Pentatomidae) and the parasitoid flies *Winthemia* sp. and *Exorista* sp. (Diptera: Tachnidae). Other natural enemies (e.g. parasitoid Hymenoptera) and disease (e.g. granulosis virus) were considered less effective. The Indian or common myna (*Acridotheres tristis*) was specifically introduced into Australia by the Acclimatisation Society to control larvae of *P. glycinae* (as well as locusts), but this attempt failed because the birds do not attack and eat the larvae (McCoy, 1878). As noted above, bird species in general rarely eat the larvae of *P. glycinae* (and those of other agaristines) so they are not an effective biological control agent of this moth. The application of pheromone control to disrupt mating behaviour, based on scent chemicals released by the abdominal brush organs (Edgar *et al.*, 1978), does not appear to have been investigated further.

An integrative pest management approach, which may offer great potential, involves a combination of chemical and biological control methods. The idea is based around the principles of regenerative agriculture in which patches of native bushland are retained or restored among commercial vineyards. These biodiverse, non-grape-producing areas perform a range of vital ecosystem services of benefit to the vineyards, including predation by natural enemies, pollination and nitrogen fixation, thereby reducing reliance on pesticides and herbicides.

22.7 Future Research

A robust phylogenetic hypothesis of Agaristinae is still lacking at both national and global levels. A well-resolved phylogeny would provide a systematic framework for a more informed classification based on evolutionary relationships and help resolve the status of several genera. In particular, the monophyly of *Leucogonia*, *Zalissa*, *Argyrolepidia* and *Idalima* is doubtful (Fig. 22.2), and these genera require taxonomic revision. Both *Leucogonia* and *Argyrolepidia* extend into New Guinea and/or South-east Asia (Indonesia) so inclusion of the Papuan and Oriental species is a requisite for determining the monophyly of these genera and may provide information on the affiliations of these groups (Wragg and Braby, 2025).

In addition to the eight undescribed species (Table 22.1), the taxonomic status and distribution of *Mimeusemia simplex* are not clear and require investigation. This species is known only from the holotype male from Queensland, but the exact location from which the specimen was collected remains a mystery. Jordan (1912) considered *M. simplex* to be polytypic, containing five other subspecies from Maluku and mainland New Guinea and adjacent islands and, although most of these clearly belong to *Mimeusemia*, examination of Lucas' type of *Agarista simplex* in the South Australian Museum suggests that *M. simplex* may well be specifically distinct from these Indonesian/New Guinea taxa. A phylogenetic hypothesis would also help determine the ancestral host plants and trace the evolution of host plant usage.

It remains to be determined if Agaristinae are indeed unpalatable, as their warning colour patterns suggest. It is likely that the larvae of most species are toxic to predators, but the mechanism of unpalatability is not fully understood. The role of secondary plant compounds (e.g. sequestration of alkaloids and other metabolites) in chemical defence of Agaristinae requires further study, especially for the key plant families Vitaceae and Dilleniaceae. The significance of the abdominal brush organs requires further research to determine if the pheromones produced are used in courtship (Edgar *et al.*, 1978; Zha *et al.*, 2009). A phylogenetic hypothesis of Agaristinae would also determine the evolutionary pattern of adult activity. It is assumed that nocturnal behaviour is the ancestral (plesiomorphic) state and that diurnal behaviour is the derived (apomorphic) state; however, it is not clear if diurnality evolved once or more than once and if there has been reversal to nocturnality in some taxa.

Finally, noctuids are characterized by the presence of tympanal organs on the thoracic segments that allow them to hear ultrasounds, most likely a defence against the echolocation of predatory insectivorous bats (Rydell *et al.*, 1995). The diurnal agaristines appear to have taken this key innovation further by using the typanum for intraspecific communication, specifically for females to receive mating signals by courting males (Scoble, 1992; Zha *et al.*, 2009). However, the evolutionary patterns of sound production are not clear: has acoustic communication evolved once or multiple times across the subfamily? There are at least four different mechanisms of sound production among the Australian genera of *Hecatesia*, *Platagarista*, *Eutrichopidia*, and *Agarista*, suggesting acoustic communication has evolved independently, but the relationships of these four genera are not well understood. Moreover, the mechanism of sound production in *Hecatesia* is elaborate and complex, yet the Neotropical agaristine *Darceta* appears to have evolved the same mechanism. Again, it is not clear if these two genera are closely related, and if acoustic signalling evolved once from a common ancestor, or if the same mechanism has evolved independently in each of these biogeographic regions.

Acknowledgements

I thank Fabian Douglas, Sharyn Wragg, Ethan Beaver and David Britton for biological information. David Britton and Fabian Douglas kindly reviewed a draft of this chapter.

References

Alcock, J. (1987) Leks and hilltopping in insects. *Journal of Natural History* 21, 319–328.

Alcock, J. (1995) Blowing the whistle on mating moths. *Australian Geographic* 38, 22–23.

Alcock, J. and Bailey, W.J. (1995) Acoustical communication and the mating system of the Australian whistling moth *Hecatesia exultans* (Noctuidae: Agaristinae). *Journal of Zoology* 237, 337–352.

Alcock, J., Gwynne, D.T. and Dadour, L.R. (1989) Acoustic signaling, territoriality, and mating in whistling moths, *Hecatesia thyridion* (Agaristidae). *Journal of Insect Behavior* 2, 27–37.

APG IV (2016) An update of the Angiosperm Phylogeny Group classification for the orders and families of flowering plants: APG IV. *Botanical Journal of the Linnean Society* 181, 1–20.

APVMA (2025) Australian Pesticides and Veterinary Medicines Authority. Public Chemicals Registration Information System. Available at: https://portal.apvma.gov.au/pubcris?p_auth=q029h4ZJ&p_p_id=pubcrisportlet_WAR_pubcrisportlet&p_p_lifecycle=1&p_p_state=normal&p_p_mode=view&p_p_col_id=column-1&p_p_col_pos=2&p_p_col_count=4&_pubcrisportlet_WAR_pubcrisportlet_javax.portlet.action=search (accessed 24 March 2025).

Bailey, W.J. (1978) Resonant wing systems in the Australian whistling moth *Hecatesia* (Agarasidae [sic], Lepidoptera). *Nature* 272, 444–446.

Braby, M.F. (2000) *Butterflies of Australia. Their Identification, Biology and Distribution*. CSIRO Publishing, Collingwood, Melbourne.

Braby, M.F. (2011) New larval food plant associations for some butterflies and diurnal moths (Lepidoptera) from the Northern Territory and eastern Kimberley, Australia. *The Beagle, Records of the Museums and Art Galleries of the Northern Territory* 27, 85–105.

Braby, M.F. (2015) New larval food plant associations for some butterflies and diurnal moths (Lepidoptera) from the Northern Territory and Kimberley, Australia. Part II. *Records of the Western Australian Museum* 30, 73–97.

Braby, M.F. (2025) Review of larval food plant associations of the Agaristinae (Lepidoptera: Noctuidae) in Australia. *Austral Entomology* 64(2), e70010, 1–17.

Braby, M.F. and Heath, P.M. (2024) *Sarbanissa diana* Sugi, 1996 (Lepidoptera: Noctuidae) newly recorded from Australia. *Australian Entomologist* 51, 179–182.

Braby, M.F., Franklin, D.C., Bisa, D.E., Williams, M.R., Williams, A.A.E. *et al.* (2018) *Atlas of Butterflies and Diurnal Moths in the Monsoon Tropics of Northern Australia*. ANU Press, Canberra, ACT.

Common, I.F.B. (1990) *Moths of Australia*. Melbourne University Press, Collingwood, Victoria.

Cordingley, C.L. (1980a) Biology of the grape vine moth, *Phalaenoides glycinae* Lewin (Lepidoptera: Agaristidae). *Journal of the Australian Entomological Society* 19, 181–187.

Cordingley, C.L. (1980b) Seasonal abundance of the grape vine moth, *Phalaenoides glycinae* Lewin (Lepidoptera: Agaristidae) and distribution of eggs and larvae in a vineyard. *Journal of the Australian Entomological Society* 19, 189–200.

Cordingley, C.L. (1981a) Factors causing mortality of the grape vine moth *Phalaenoides glycinae* Lewis (Lepidoptera: Agaristidae). *Journal of the Australian Entomological Society* 20, 33–38.

Cordingley, C.L. (1981b) Reproductive biology of the grape vine moth, *Phalaenoides glycinae* Lewin (Lepidoptera: Agaristidae). *Journal of the Australian Entomological Society* 20, 25–31.

Donovan, E. (1805) *An Epitome of the Natural History of the Insects of New Holland, New Zealand, New Guinea, Otaheite, and other Islands in the Indian, Southern, and Pacific Oceans: including the figures and descriptions of one hundred and fifty-three species of the more splendid, beautiful, and interesting insects, hitherto discovered in those countries, and which for the most part have not appeared in the works of any preceding author*. F.C. & J. Rivington, London.

Edgar, J.A., Cockrum, P.A. and Carrodus, B.B. (1978) Male scent-organ chemicals of the vine moth, *Phalaenoides glycinae* Lew. (Agaristidae). *Specialia Experientia* 37, 861–862.

Edwards, E.D. (1996) Noctuidae. In: Nielsen, E.S., Edwards, E.D. and Rangsi, T.V. (eds) *Checklist of the Lepidoptera of Australia*. Monographs on Australian Lepidoptera, 4. CSIRO Publishing, Collingwood, Victoria, pp. 291–336, 370–380.

Fenner, T.L. (1961) Vine moth. *Journal of Agriculture, South Australia* 65, 73.

Franclemont, J.G. and Todd, E.L. (1983) Noctuidae. In: Hodges, R.W. (ed.) *Checklist of the Lepidoptera of America North of Mexico*. Classey, London, pp. 120–159.

French, C. (1916) Insect pests of the fruit, flower and vegetable garden. *Journal of the Department of Agriculture, Victoria* 14, 433–438.

French, C. (2013) *A Handbook of the Destructive Insects of Victoria, with Notes on the Methods of Prevention and Extirpation. Part IV*. Department of Primary Industries, Melbourne, Victoria.

Froggatt, W.W. (1907) *Australian Insects*. William Brooks & Company, Sydney, NSW.

Grimshaw, P. (2020) Observations and notes on Donovan's day moth (*Cruria donowani*) from Mt Crosby, Brisbane, 19th March to 2nd May 2020. *Metamorphosis Australia. Magazine of the Butterfly and Other Invertebrates Club* 98, 20–25.

Grund, R. (2010) New range extensions and other data for selected butterflies and day flying moths from coastal areas of the far westcoast, northern Eyre Peninsula and mid north regions of South Australia. *Victorian Entomologist* 40, 36–38.

Hampson, G.F. (1901) *Catalogue of the Lepidoptera Phalaenae in the British Museum. Volume III. Catalogue of the Arctiadae (Arctianae) and Agaristidae in the Collection of the British Museum*. British Museum (Natural History), London.

Hampson, G.F. (1910) *Catalogue of the Lepidoptera Phalaenae in the British Museum. Volume IX. Catalogue of the Noctuidae in the Collection of the British Museum*. British Museum (Natural History), London.

Holloway, J.D. (1989) *The Moths of Borneo. Part 12. Family Noctuidae, Trifine Subfamilies: Noctuidae, Heliothinae, Hadeninae, Acronictinae, Ampipyrinae, Agaristidae*. *Malayan Nature Journal* 42, 57–226.

Humphries, L. (1950) Tommy the tunneller. *Wild Life: Australian Nature Magazine* 12, 104–107.

James, D.G., Hebert, V. and LePage, J. (2012) The prosternal gland in Pacific Northwest butterfly larvae with preliminary chemical analyses of emissions. *Journal of the Lepidopterists' Society* 66, 137–142.

Jelinek, A., Britton, D.R. and New, T.R. (1994) Conservation of a 'threatened butterfly community' at Mount Piper, Victoria. *Memoirs of the Queensland Museum* 33, 115–120.

Jones, E.L. (1967) Pests of the grape vine. *Agricultural Gazette of New South Wales* 78, 708–714.

Jordan, H.E.K. (1912) Family Agaristidae. In: Seitz, A. (ed.) The Macrolepidoptera of the World: A Systematic Account of all the Known Macrolepidoptera. Division II: The Macrolepidoptera of the Indo-Australian fauna. Volume 11: Noctuiform Phalaense. Alfred Kernen Verlag, Stuttgart, Germany.

Keegan, K.L., Rota, J., Zahiri, R., Zilli, A., Wahlberg, N. *et al.* (2021) Toward a stable global Noctuidae (Lepidoptera) taxonomy. *Insect Systematics and Diversity* 5, 1–24.

Khuniad, C., Nahar, L., Ritchie, K.J. and Sarker, S.D. (2022) Therapeutic potential of *Leea indica* (Vitaceae). *Journal of Natural Products Discovery* 1, 1–19.

Kiriakoff, S.G. (1963) The tympanic structures of the Lepidoptera and the taxonomy of the order. *Journal of the Lepidopterists' Society* 17, 1–6.

Kiriakoff, S.G. (1977) Lepidoptera Noctuiformes. Agaristidae I (Palaearctic and Oriental genera). *Das Tierreich* 97, 1–180.

Kitching, I.J. (1984) An historical review of the higher classification of the Noctuidae (Lepidoptera). *Bulletin of the British Museum of Natural History (Entomology)* 49, 153–234.

Larcombe, E.E. (1914) Notes on *Alpina callisto*, the crow's-foot caterpillar. *Australian Naturalist* 3, 203–205.

Lewin, J.W. (1805) *Prodromus Entomology. Natural History of Lepidopterous Insects of New South Wales*. T. Bensley, London.

Lewis, M.J. (1985) Systematics and Biology of *Apina callisto*. Unpublished Honours thesis, Division of Botany and Zoology, Australian National University, Canberra, ACT.

Lewis, M.J. (1989) Shift in food plant preference with developmental stage by the larvae of *Apina callisto* Angas (Lepidoptera: Noctudiae). *General and Applied Entomology* 21, 5–10.

Li, X., Breinholt, J.W., Martinez, J.I., Keegan, K., Ellis, E.A. *et al.* (2024) Large-scale genomic data reveal the phylogeny and evolution of owlet moths (Noctuoidea). *Cladistics* 40, 21–33.

Lima, C.C., Lemos, R.P.L. and Conserva, L.M. (2014) Dilleniaceae family: An overview of its ethnomedicinal uses, biological and phytochemical profile. *Journal of Pharmacognosy and Phytochemistry* 3, 181–204.

Marriot, P., Edwards, E.D., Kallies, A., Hewish, M. and Williams, S. (2017) *Moths of Victoria. Part 8: Night Moths and Allies. Noctuoidea (B)*. Entomological Society of Victoria, Melbourne, Victoria.

McCoy, F. (1878) *Natural History of Victoria. Prodromus of the Zoology of Victoria; or, Figures and Descriptions of the Living Species of all Classes of the Victorian Indigenous Animals*. Volume 1, Decade I. George Robertson, Melbourne, Victoria.

McFarland, N. (1976) Hilltopping and defence behaviour in a diurnal agaristid moth. *Australian Entomological Magazine* 3, 25–29.

McQuillan, P.B., Forrest, J.A., Keane, D. and Grund, R. (2019) *Caterpillars, Moths and their Plants of Southern Australia*. Butterfly Conservation South Australia Inc., Adelaide, South Australia.

Mitchell, A., Cho, S., Regier, J.C., Mitter, C., Poole, R.W. and Matthews, M. (1997) Phylogenetic utility of *Elongation Factor-1a* in Noctuoidea (Insecta: Lepidoptera): The limits of synonymous substitution. *Molecular Biology and Evolution* 14, 381–390.

Mitchell, A., Mitter, C. and Regier, J.C. (2000) More taxa or more characters revisited: Combining data from nuclear protein-encoding genes for phylogenetic analysis of Noctuoidea (Insecta: Lepidoptera). *Systematic Biology* 49, 202–224.

Mitchell, A., Mitter, C. and Regier, J.C. (2006) Systematics and evolution of the cutworm moths (Lepidoptera: Noctuidae): Evidence from two protein-coding nuclear genes. *Systematic Entomology* 31, 21–46.

Moss, J.T. (1999a) Insect observations in the Granite Belt – Part 1. *Butterfly and Other Invertebrates Club Newsletter* 13, 14–16.

Moss, J.T. (1999b) Insect observations in the Granite Belt – Part 2. *Butterfly and Other Invertebrates Club Newsletter* 15, 10–12.

Moss, J.T. (2010) Life history notes on the day-flying moth *Cruria synopla* Turner, 1903 and its distinction from *C. donowani* (Boisduval, 1832) (Lepidoptera: Noctuidae). *Metamorphosis Australia. Magazine of the Butterfly and Other Invertebrates Club* 59, 4–8.

Nielsen, E.S. and Common, I.F.B. (1991) Lepidoptera. In: Naumann, I.D. (ed.) *The Insects of Australia*. Melbourne University Press, Carlton, Victoria, pp. 817–915.

Nye, I.W.B. (1975) Noctuidea (part): Noctuidae, Agaristidae, and Nolidae. In: Nye, I.W.B. (ed.) *The Generic Names of Moths of the World*. British Museum (Natural History), London, pp. 1–571.

Raffauf, R.F. (1996) *Plant Alkaloids: A Guide to their Discovery and Distribution*. Food Products Press, New York.

Regier, J.C., Mitter, C., Mitter, K., Cummings, M.P., Bazinet, A.L. *et al.* (2016) Further progress on the phylogeny of Noctuoidea (Insecta: Lepidoptera) using an expanded gene sample. *Systematic Entomology* 42, 82–93.

Robinson, G.S., Ackery, P.R., Kitching, I.J., Beccaloni, G.W. and Hernández, L.M. (2002) *Hostplants of the Moth and Butterfly Caterpillars of America North of Mexico*. Memoirs of the American Entomological Institute vol. 69. American Entomological Institute, Gainesville, FL.

Rydell, J., Jones, G. and Waters, D. (1995) Bats and hearing moths: who are the winners? *Oikos* 73, 419–424.

Scoble, M.J. (1992) *The Lepidoptera: Form, Function, and Diversity*. Oxford University Press, Oxford.

Scott, A.W. (1864) *Australian Lepidoptera and Their Transformations, Drawn from the Life, by Harriet and Helena Scott; with Descriptions, General and Systematic by A.W. Scott., M.A., Ash Island, Hunter River, New South Wales. Vol.* 1. John van Voorst, London.

Scott, A.W. (1891) *Australian Lepidoptera and their transformations, by the late Alexander Walker Scott, with illustrations drawn from the life by his daughters, Harriet Morgan and Helena Forde. Vol.* 2. *Edited and revised by Arthur Sidney Olliff and Helena Forde*. Australian Museum, Sydney, NSW.

Scott, J.A. (1968) Hilltopping as a mating mechanism to aid the survival of low density species. *Journal of Research on the Lepidoptera* 7, 191–204.

Shields, O. (1967) Hilltopping. *Journal of Research on the Lepidoptera* 6, 69–178.

Staude, H., Picker, M. and Griffiths, C. (2023) *Southern African Moths & their Caterpillars*. Pelagic Publishing, London.

Strand, E. (1912) Noctuidae: Agaristinae. *Lepidopterorum Catalogus* Part 5, 1–82.

Sugi, S. (1987) *Larvae of Larger Moths in Japan*. Kodansha, Tokyo, Japan.

Talianchich, A., Bailey, W.J. and Ghisalberti, E.L. (2003) Palatability and defense in the aposematic diurnal whistling moth, *Hecatesia exultans* Walker (Lepidoptera: Noctuidae: Agaristinae). *Australian Journal of Entomology* 42, 276–280.

Turner, A.J. (1920) Revision of the Australian Noctuidae. *Transactions of the Royal Society of South Australia* 44, 120–189.

Wainer, J.W. and Yen, A.L. (2000) A survey of the butterfly fauna at the Paps Scenic Reserve, Mansfield, Victoria. *Victorian Naturalist* 117, 131–140.

Wei, F-H., Kishida, Y. and Wang, M. (2019) A review of the genus *Sarbanissa* Walker, 1865 in China, with description of a new species (Lepidoptera: Noctuidae: Agaristinae). *Zootaxa* 4648, 354–370.
Wragg, S.A. (1999) *Biology of Agaristodes feisthamelii, Phalaenoides tristifica and their close relatives, with a phylogenetic investigation of the Australian Agaristinae (Noctuidae: Lepidoptera).* Unpublished Honours thesis, Department of Botany and Zoology, Australian National University, Canberra, ACT.
Wragg, S.A. and Braby, M.F. (2025) Systematics and phylogenetic relationships of the Australian Agaristinae (Lepidoptera: Noctuidae) based on morphology. *Records of the Australian Museum* 77, 137–173.
Zha, Y., Chen, Q. and Lei, C. (2009) Ultrasonic hearing in moths. *Annales de la Société Entomologique de la France* 45, 145–156.

23 Moths as Pests in Australia

Myron P. Zalucki[1]*, Helen F. Nahrung[2] and Mark Schutze[3]

[1]*School of the Environment, The University of Queensland, Brisbane, Queensland, Australia;* [2]*Forest Research Institute, University of the Sunshine Coast, Sippy Downs, Queensland, Australia;* [3]*Biosecurity Queensland, Queensland Department of Primary Industries, Brisbane, Queensland, Australia*

Abstract

This short chapter introduces the following nine chapters, each of which deals, at the species level, with moths that are pests in Australia. Here we place these in the context of the entire fauna. Surprisingly, there is currently no comprehensive list of pest species of moths and whether they are native or introduced. Our scouring of the literature, databases and websites suggests *ca* 290, of which about 80 are invasive from elsewhere. Some 67% of the accidentally introduced non-native species are pests, compared with 1.7% of the native fauna, which probably reflects the availability of introduced, widely cultivated crop hosts. The moths identified as pests from our searches represent 33 families, but most come from just ten families. What is apparent for most of these families – Erebidae, Noctuidae, Crambidae, Pyralidae, Tineidae, Hepialidae and Limacodidae – is that pest species are proportionately higher than might be expected from the families' diversity in the moth fauna as a whole. We suspect this reflects resource use in these species and what we humans consider to be of value.

23.1 Introduction

When first contemplating this introduction to pest moths in Australia, we asked some deceptively simple questions. How many species of moth in Australia are considered pests? Why are they considered pests? How many of these 'pest' species are native and how many are invasive?

23.2 Why is Something Considered a Pest?

This is the slightly easier question of the set posed. The answer is relatively simple – it is an anthropogenic value judgement. Usually, a species reaches this status because it impacts negatively upon something we consider valuable. Our health and the wellbeing of domesticated mammals tend to be high on the value chain. There are a surprising number of moths that in the caterpillar and/or adult stages can cause medical and veterinary problems (Battisti *et al.*, 2024). Yet, unless the symptoms are dramatic, they tend to be ignored. *Ochrogaster lunifer*, the deceptively innocuous-sounding bunny-tail moth, is an exception as it causes horses to abort (see Chapter 29).

Few people appreciate finding holes in their favourite garments caused by the larvae of clothes moths. In Australia these are likely to be the invasive species *Tinea pellionella* or *Tineola bisselliella*. Similarly in the kitchen pantry, various species, again introduced, may spoil stored food such as

*Corresponding author: m.zalucki@uq.edu.au

DOI: 10.1079/9781800627680.0023

rice, flour, nuts and pasta. The culprits are probably the Indian meal moth (*Plodia interpunctella*), the Mediterranean flour moth (*Ephestia kuehniella*) and/or the tropical warehouse moth (*Ephestia cautella*). With all these household pests, by the time you notice them they are well established and difficult to eliminate.

The insects that damage stored food products do not only do so in the home. David Rees (Chapter 28, this volume) provides an account of the moth species that damage food in storage and processing facilities – key parts of the supply chain. These pests require constant vigilance and add to production and consumer costs.

Before food and other commodities make it into storage, moths directly reduce growth and yield of staple food (corn, maize, potatoes, rice, wheat), fibre (cotton, flax, hemp), fruit (apples, oranges, peaches, pears, plums, etc.), nuts (almond, macadamia, pecan), vegetables (Brassicae, capsicum, tomatoes), forage (lucerne, forage brassica) and forestry (hardwoods and softwoods) – and the all-important grapes. How much damage moths cause to these commodities depends on their population ecology and in part on their management.

Note that virtually all the commodities we grow for food and fibre in Australia are 'invasive' species in the sense that they have been deliberately introduced and, perhaps not surprisingly, some of their major pests too are invasives from elsewhere. Nevertheless, there are many native/endemic species that are significant 'pests'.

This connects with our other simple questions: how many species of moth in Australia are considered pests? How many are 'native'? And, how many are invasive? It turns out these simple questions are not easy to answer! There is no single comprehensive list of moth species that are considered pests, or for any other group of insects for that matter, perhaps with the exception of the well-studied beetles, the Scarabaeidae that afflict sugar cane (see Allsopp, 2010).

23.3 The Australian Pest Fauna

The Australian Plant Pest Database (APPD), at least in name, appears promising as a resource. It is essentially a (restricted-access) catalogue of the label data of museum specimens of 'plant-feeding' insects held in major state and federal collections in Australia. After pulling out all the Lepidoptera records from the APPD and eliminating the five butterfly families that occur in Australia, there were 58,436 accessions recorded and another 4837 from the Queensland database (QDPC). Removing overseas records, quarantine interceptions, specimens with a '?' beside them, those identified to family or genus only, and duplicates yielded 2690 species. The names of these species and families were checked against the taxonomy used in the GBIF database – some were obvious spelling mistakes and others simply could not be located against the Atlas of Living Australia (ALA) and Google, resulting in 2436 allegedly pest species of moth in Australia. At almost ten times the number of pest moths estimated by Common (1990), it became clear that 'plant pest' in the APPD was a misnomer and most records simply represent plant associations, rather than pests. Indeed, for nearly all the species there was no information on what plant the species damaged to justify the pest label.

To narrow our list down to species considered 'pests', we next reviewed compilations by Robinson and Nielsen (1987), Bailey (2007), Gerozisis *et al.* (2008), Malipatil *et al.* (2019), Battisti *et al.* (2024), Brockerhoff *et al.* (2023) and the websites that list 'pests' in Australia. Websites were those maintained by the Australian Government (2024), the Coffs Harbour Butterfly House (2025), CSIRO (2025), Business QLD (2025) and PADIL (2025). This yielded a total of 287 species of moths that are referred to as 'pests' by someone.

We then attempted to allocate the native/invasive status of these species using similar literature and distributional searches. Native species are easy to determine if they are only found in Australia. Of the 287 pests, however, approximately 80 are introduced, presumably by human endeavour (Table 23.1). In similar manner, at least 12 native Australian species have been donated to the rest of the world. Of course, many more of these pest insects may be invasive in the biogeographic sense, having entered when the Australian continental shelf collided with South-east Asia some 25 mya and during the periods when changing sea levels formed land bridges on and off the Australian continent. Others may have arrived post-1788, by hitchhiking on ships or potentially as natural migrants whose establishment was facilitated by the introduction of host plants.

Table 23.1. Breakdown by family of moth species identified as pests, classified as native and non-native and number introduced as weed biological control agents. Species considered major pests by someone (see text for details) and covered in the book are named. * includes one species that is described as cosmopolitan.

Family	Non-native	Native	Biological control	Major pest species (in bold – species included in this book)
Anthelidae		6		
Bedelliidae	1			
Bucculatricidae		1	2	
Coleophoridae	1			*Coleophora alcyonipennella* (Kollar, 1832)
Cosmopterigidae	1			*Ithome lassula* Hodges, 1962
Cossidae		3		***Endoxyla cinereus* (Tepper, 1890)**
Crambidae	12*	22	5	*Deanolis sublimbalis* Snellen, 1899; *Maruca vitrata* (Fabricius, 1787); *Nacoleia octasema* Meyrick, 1886; *Nymphula depunctalis* (Guenée); *Sceliodes cordalis* (Doubleday, 1843); *Scirpophaga innotata* (Walker, 1863); *Spoladea recurvalis* (Fabricius, 1775)
Depressariidae		1	1	
Erebidae	17	43	3	***Eudocima aurantia* (Moore, 1887); *E. jordani* (Holland, 1900); *E. phalonia* (Clerck, 1764); *E. materna* (Linnaeus, 1767); *E. salaminia* (Cramer, 1777)**; *Mocis alterna* Walker, 1858
Eupterotidae		1		
Euteliidae	1	1		
Gelechiidae	5	3	2	***Sitotroga cerealella* (Olivier, 1789)**; *Phthorimaea operculella* (Zeller, 1873)
Geometridae		14	6	
Gracillariidae	1	1	2	
Hepialidae		11		*Oncopera intricata* Walker, 1856; *O. rufobrunnea* Tindale, 1933; *Oxycanus antipoda* (Herrich-Schäffer)
Incurvariidae		1		*Perthida glyphopa* Common, 1969
Lasiocampidae		1		
Limacodidae		10		
Lyonetiidae			1	
Noctuidae	11	28	2	***Agrotis infusa* Boisduval, 1835**; *A. munda* Walker; *Conogethes punctiferalis* Guenée, 1854; ***Helicoverpa armigera* (Hübner, 1808); *H. punctigera* Wallengren, 1860**; *Persectania ewingii* Westwood, 1839: *Spodoptera exempta* (Walker, 1856); *S. frugiperda* J.E. Smith, 1797; *Tiracola plagiata* (Walker, 1857)
Nolidae		6		***Uraba lugens* Walker, 1863**
Notodontidae		6		***Ochrogaster lunifer* Herrich-Schaffer, 1853**
Oecophoridae	2	2		
Plutellidae	1			***Plutella xylostella* (Linnaeus, 1767)**
Praydidae	1			

Continued

Table 23.1. Continued.

Family	Non-native	Native	Biological control	Major pest species (in bold – species included in this book)
Psychidae		4		
Pterophoridae	1*	1	4	*Sphenarches anisodactylus* (Walker, 1864)
Pyralidae	14	13	5	*Apomyelois ceratoniae* (Zeller, 1839); *Corcyra cephalonica* (Stainton, 1866); ***Ephestia elutella* (Hübner, 1796); *E. kuehniella* Zeller, 1879**; *Etiella behrii* (Zeller, 1848); *Galleria mellonella* (Linnaeus, 1758); ***Plodia interpunctella* (Hübner, 1813)**
Saturniidae		2		
Sesiidae	1	1	4	
Sphingidae		1		
Tineidae	10	1		
Tortricidae	3	17	5	***Cydia pomonella* (Linnaeus, 1758); *Epiphyas postvittana* (Walker, 1863); *Grapholita molesta* (Busck, 1916)**; *Merophyas divulsana* (Walker, 1863)
Xyloryctidae		3		
TOTALS	83	204	42	

Attaching any one of these categories unequivocally to the 287 target species is often challenging. Some resolution will be reached when species level, comprehensive and dated phylogenies of the species in Australia can be compared with their overseas relatives. This has been done for the Australian dacine fruit flies (Starkie *et al.*, 2022). At this stage many of the moth species that are considered native are also widespread outside Australia.

In addition, there are the species of moths deliberately added to the Australian fauna as biological control agents of weeds (Table 23.2). Weeds are plants we do not like and once again are usually invasive. People demand the right to grow whatever they like in their gardens and some 27,500 of these exotic plants have been brought into Australia (Groves *et al.*, 2005). Of these, 10% will escape and become naturalized and 10% of these will go on to be considered weeds needing control (Williamson and Fitter, 1996). Many of these are candidates for classic biological control using herbivorous insects. In a sense, these are insects we would like to behave as pests of introduced plants! The classic textbook success story is *Cactoblastis cactorum* introduced to control prickly pear (see Chapter 17) but closer scrutiny suggests accidentally introduced disease may also have contributed (Zalucki, 2015). The list of moth species introduced for weed biological control is well documented and accessible (ibiological catalog, 2025; Winston *et al.*, 2025). A similar resource for pests would be invaluable.

Of the 12,074 named moths in Australia, some 160 are introduced – 1.3% of the fauna (updated from Mally *et al.*, 2022). Some 67% of the accidentally introduced non-native species are pests, compared with 1.7% of native species, and probably reflects the availability of introduced, widely cultivated crop host plants. The 287 moths identified as pests from our searches represent 33 families, but 83% of the pests come from just ten families. In descending order, these families are: Erebidae (21% of pest moths: 8% of all moths), Noctuidae (14% of pest moths: 5% of all moths), Crambidae (12% of pest moths: 9% of all moths), Pyralidae (9% of pest moths: 4% of all moths), Tortricidae (7% of pest moths: 7% of all moths), Geometridae (5% of pest moths: 14% of all moths), Hepialidae (4% of pest moths: 1% of all moths), Tineidae (4% of pest moths: 2% of all moths), Limacodidae (4% of pest moths: 1% of all moths) and Gelechiidae (3% of pest moths: 4% of all moths).

What is apparent for most of these families– Erebidae, Noctuidae, Crambidae, Pyralidae, Tineidae, Hepialidae and Limacodidae – is that pests are proportionately higher than might be expected from the families' diversity in the moth fauna as a whole, in some cases by a factor of four. We suspect this reflects resource use in these species. The proportion of invasive pests is high in the Erebidae (28%), Noctuidae (28%), Crambidae (35%), Pyralidae (52%) and Tineidae (91%). In families with pests that are all native, their pest status again reflects host use. The Hepialidae, for example, feed on Poaceae and are pests of pastures. Given the importance of the sheep and cattle industries, they achieve pest status, albeit often described as localized and irregular. The Limacodidae are of medical importance and the proportion of 'pests' reflects the high value we place on ourselves. On the other hand, Australia's two most speciose moth families – Geometridae (14% of the fauna) and Oecophoridae (19%) – account for only 5% and 1% of pests, respectively, with only two invasive pest species (both Oecophoridae) between them. Geometridae are often found on Myrtaceae (see Chapter 18) and only a few make it as pests. The Oecophoridae are largely detritus feeders, again on myrtaceaous leaf litter (see Chapter 12).

Very few species make it to 'major' pest status (Table 23.1). A number of these are covered in this book. One of the major introduced pests is the diamondback moth, *Plutella xylostella* (see Chapter 25). The species has been spread globally and is a major pest of brassicas, although a close look at this species' host plants suggests it is polyphagous, having been recorded on 11 plant families (Zalucki *et al.*, 2021). Nevertheless, it is viewed as a brassica specialist and in 2012 cost the world economy $5–7 billion in this host group alone (Zalucki *et al.*, 2012). With inflation, the figure is likely larger today. The species' claim to fame is that it has developed resistance to every pesticide that has been used against it. As Furlong points out in Chapter 25, the species can be well managed with introduced parasitoids and very basic 'soft' integrated pest management (IPM) practices. The challenge is always getting growers to buy in to the approach. Interestingly, in Australia there is an endemic, *Plutella australiana*,

Table 23.2. Species of moth by family deliberately introduced and established to control various species of plant considered to be weeds.

Family	Species
Bucculatricidae	*Bucculatrix ivella* Busck
Bucculatricidae	*Bucculatrix parthenica* Bradley
Crambidae	*Euclasta gigantalis* Viette
Crambidae	*Niphograpta albiguttalis* (Warren)
Crambidae	*Orphanostigma haemorrhoidalis* (Guenée)
Crambidae	*Samea multiplicalis* (Guenée)
Crambidae	*Xubida infusellus* (Walker)
Depressariidae	*Agonopterix umbellana* (Fabricius)
Erebidae	*Eublemma amoena* (Hübner)
Erebidae	*Hypena laceratalis* Walker
Erebidae	*Tyria jacobaeae* (Linnaeus)
Gelechiidae	*Aristotelia ivae* Busck
Gelechiidae	*Evippe* sp. #1
Geometridae	*Chiasmia assimilis* (Warren)
Geometridae	*Comostolopsis germana* Prout
Geometridae	*Digrammia pallidata* (Packard)
Geometridae	*Eueupithecia cisplatensis* Prout
Geometridae	*Eueupithecia vollonoides* Hausmann
Geometridae	*Leuciris fimbriaria* (Stoll)
Gracillariidae	*Dialectica scalariella* (Zeller)
Gracillariidae	*Neurostrota gunniella* (Busck)
Lyonetiidae	*Leucoptera spartifoliella* (Hübner)
Noctuidae	*Neogalea sunia* (Guenée)
Noctuidae	*Notioplusia illustrata* (Guenée)
Pterophoridae	*Hellinsia balanotes* (Meyrick)
Pterophoridae	*Lantanophaga pusillidactylus* (Walker)
Pterophoridae	*Platyptilia isodactyla* (Zeller)
Pterophoridae	*Wheeleria spilodactylus* (Curtis)
Pyralidae	*Cactoblastis cactorum* (Berg)
Pyralidae	*Hypocosmia pyrochroma* Jones
Pyralidae	*Macrorrhinia endonephele* (Hampson)
Pyralidae	*Melitara junctolineella* Hulst
Pyralidae	*Tucumania tapiacola* Dyar
Sesiidae	*Carmenta mimosa* Eichlin & Passoa
Sesiidae	*Carmenta* sp. (Beutenmüller) nr *ithacae*
Sesiidae	*Chamaesphecia mysiniformis* (Boisduval)
Sesiidae	*Pyropteron doryliformis* (Ochsenheimer)
Tortricidae	*Cochylis atricapitana* (Stephens)
Tortricidae	*Crocidosema lantana* Busck
Tortricidae	*Epiblema strenuana* (Walker)
Tortricidae	*Platphalonidia mystica* (Razowski & Becker)
Tortricidae	*Tortrix* sp.

which is likely not well recognized and may be an occasional pest.

The earworms (*Helicoverpa* spp.) (see Chapter 32) are perhaps the standout pests of field crops in Australia. There are two main offending species, both polyphagous – you name it, they probably feed on it (Cunningham and Zalucki, 2014). One, *H. armigera*, is now almost global (Kriticos *et al.*, 2015) and the other, *H. punctigera*, is endemic to Australia and a few nearby islands, but could spread overseas (Oliveira *et al.*, 2021). Both are key pests of cotton. *H. armigera* develops resistance to insecticides rapidly, whereas *H. punctigera* has not, to date, been recorded with field resistance, and both are the main reason the cotton industry embraced genetically modified cotton varieties that express Bt toxins. In other crops, and there are many, growers must be content with IPM in its various guises (Zalucki *et al.*, 2009) although area-wide management approaches are developing, necessitated by the need to manage resistance (Downes *et al.*, 2017; Ives *et al.*, 2017). There is a third species (*H. assulta*) that was once a major pest of the now defunct tobacco industry in Australia but can damage other crops in the Solanaceae.

The major fruit-feeding pests in the Tortricidae (Chapter 27) and Erebidae (Chapter 30) are an interesting contrast. The tortricid pests *Cydia pomonella*, *Grapholita molesta* and *Epiphyas postvittana* cause damage, often cosmetic, directly to fruiting structures in their caterpillar stages. The former two are invasive and the light brown apple moth (*Epiphyas postvittana*) is an endemic we have donated to the world. These are pests of fruits typically grown in temperate climates (apples, pears, plums, etc.) and tend to be restricted to such climates. A group of erebid species in the genus *Eudocima* (*E. aurantia*, *E. jordani*, *E. phalonia*, *E. maternal*, *E. salaminia*) (Chapter 30) afflict fruit production in the tropics by feeding directly on fruit structures of a range of plants (carambola, banana, citrus, fig, guava, kiwifruit, longan lychee, mango, stonefruit, persimmon, papaya) in the adult stage. The proboscis is hardened, pointed and barbed. The species can outbreak and are migratory.

Forest pests covered include the gumleaf skeletonizer, *Uraba lugens* (Chapter 31) and the jarrah leaf miner, *Perthida glyphopa* (Chapter 24). As the common names imply, they have very different feeding habits but as is common in forest systems, both show outbreak dynamics and can significantly impact their host trees. The gumleaf skeletonizer has the added bonus of being urticating.

The processionary caterpillars, *Ochrogaster lunifer* (Chapter 29), present the difficult research and management problems faced when dealing with a cryptic species complex. At least three species are confirmed in the taxon (Mather *et al.*, 2019) and many more remain to be diagnosed. The taxon is a minor pest of trees and shrubs but poses its greatest problem to the thoroughbred horse-stud industry – ingestion of setae causes mares to abort their foals – and in humans causes various medical reactions, probably greatly underestimated (Battisti *et al.*, 2024). If many of the pests are indeed species complexes, then how many species of moth in Australia are indeed pests?

The giant woodborer, *Endoxyla cinereus* (Chapter 26), is again a minor pest of its eucalypt host, in no small part the result of cockatoo damage as the birds rip apart the trees seeking the moth's caterpillars. As with all species, particularly the endemic ones, once you look closely it is a key part of that intricate web we call life on Earth. How many threads can we really afford to lose before the fabric is ruined?

Acknowledgements

Lin Weifan provided invaluable help to MPZ with initial data searches and checking of species names. Thanks to Mike Day and Rachel Winston for assisting with the weed biological control database

References

Allsopp, P.G. (2010) Integrated management of sugarcane whitegrubs in Australia: An evolving success. *Annual Review of Entomology* 55, 329–49.

Australian Government (2024) Pests, Diseases and Contaminants of Horticultural Commodities. Available at: www.agriculture.gov.au/sites/default/files/documents/pests-diseases-contaminants-horticultural-commodities.pdf (accesssed 7 October 2025).

Bailey, P.T. (ed.) (2007) *Pests of Field Crops and Pastures. Identification and Control.* CSIRO Publishing, Collingwood, Victoria.

Battisti, A., Walker, A.A., Uemura, M., Zalucki, M.P., Brinquin, A.-S. *et al.* (2024) Look but do not touch: The occurrence of venomous species across Lepidoptera. *Entomologia Generalis* 44, 29–39.

Brockerhoff, E.G., Gresham, B., Meurisse, N., Nahrung, H.F., Perret-Gentil, A. *et al.* (2023) Pining away and at home: Global utilisation of *Pinus radiata* by native and non-native insects. *NeoBiota* 84, 137–167.

Business Qld (2025) Available at: www.business.qld.gov.au/industries/farms-fishing-forestry/agriculture/biosecurity/plants/insects/horticultural (accessed 7 October 2025).

Butterfly House (2025) Available at: https://lepidoptera.butterflyhouse.com.au/moths.html (accessed 7 October 2025).
Common, I.E.B. (1990) *Moths of Australia*, Melbourne University Press, Carlton, Victoria.
CSIRO (2025) Pests. Available at: www.csiro.au/en/research/animals/pests (accessed 7 October 2025).
Cunningham, J.P. and Zalucki, M.P. (2014) Understanding heliothine (Lepidoptera: Noctuidae) pests: what is a host plant? *Journal of Economic Entomology* 107, 881–896.
Downes, S., Kriticos, D., Parry, H., Paull, C., Schellhorn, N. and Zalucki, M.P. (2017) Perspective on management of *Helicoverpa armigera*: transgenic Bt cotton, IPM, and landscapes. *Pest Management Science* 73, 485–492.
Gerozisis, J., Hadlington, P. and Staunton, I. (2008) *Urban Pest Management in Australia*. UNSW Press, Sydney, NSW.
Groves, R.H., Boden, R. and Lonsdale, W.M. (2005) *Jumping the Garden Fence: Invasive Garden Plants in Australia and their Environmental and Agricultural Impacts*. CSIRO report prepared for WWF Australia. WWF Australia, Sydney, NSW.
Ibiolcontrol Catalog (2025) Available at: www.ibiocontrol.org/catalog/ (accessed 7 October 2025).
Ives, A., Paull, C., Hulthen, A., Downes, S., Andow, D. *et al.* (2017) Spatio-temporal variation in landscape composition may speed resistance evolution of pests to Bt crops. *PLOS One* 12(1), e0169167.
Kriticos, D.J., Ota, N., Hutchison, W.D., Beddow, J., Walsh, T. *et al.* (2015) The potential distribution of invading *Helicoverpa armigera* in North America: Is it just a matter of time? *PLOS One* 10(7), e0119618.
Malipatil, M.B., Williams, D.G. and Semeraro, L. (2019) *Pests of Pome and Stone Fruit and their Predators and Parasitoids. A Pocket Guide*. Agriculture Victoria, Melbourne, Victoria.
Mally, F.R., Turner, R.M., Blake, R.E., Fenn-Moltu, G., Bertelsmeier, C. *et al.* (2022) Moths and butterflies on alien shores: global biogeography of non-native Lepidoptera. *Journal of Biogeography* 49, 1455–1468.
Mather, A., Zalucki, M.P., Farrell, J., Perkins, L.E. and Cook, L.G. (2019) The Australian processionary caterpillar, *Ochrogaster lunifer* Herrich-Schäffer (Lepidoptera: Notodontidae), comprises cryptic species. *Austral Entomology* 58, 816–825.
Oliveira, DeM.R.C., Zalucki, M.P., Pastori, P.L. and Kriticos, D.J. (2021) Current and future potential distributions of *Helicoverpa punctigera* (Lepidoptera: Noctuidae): Is this the next FAW? *Bulletin of Entomological Research* 112, 119–130.
PADIL (2025) Available at: www.padil.gov.au/ (accessed 7 October 2025).
Robinson, G.S. and Nielsen, E.S. (1987) The pest species of 'Tinea' (clothes moths) occurring in Australia (Lepidoptera: Tineidae). *General and Applied Entomology* 19, 45–48.
Starkie, M.L., Cameron, S.L., Krosch, M.N., Phillips, M.J., Royer, J.E. *et al.* (2022) A comprehensive phylogeny helps clarify the evolutionary history of host breadth and lure response in the Australian Dacini fruit flies (Diptera: Tephritidae). *Molecular Phylogenetics and Evolution* 172, 107481.
Williamson, M. and Fitter, A. (1996) The varying success of invaders. *Ecology* 77, 1661–1666.
Winston, R.L., Schwarzlander, M., Hinz, H.L., Day, M.D., Cock, M.J.W. and Julien, M.H. (eds) (2025) *Biological Control of Weeds: A World Catalogue of Agents and Their Target Weeds*. USDA Forest Service, Forest Health Technology Enterprise Team. Available at: www.ibiocontrol.org/catalog/ (accessed 26 March 2025).
Zalucki, M.P. (2015) From natural history to continental scale perspectives: An overview of contributions by Australian entomologists to applied ecology – a play in three acts. *Austral Entomology* 54, 231–245.
Zalucki, M.P., Adamson, D. and Furlong, M.J. (2009) The future of IPM: whither or wither? *Australian Journal of Entomology* 48, 85–96.
Zalucki, M.P., Shabbir, A., Silva, R., Adamson, D., Liu, S.-S. and Furlong, M.J. (2012) Estimating the economic cost of one of the world's major insect pests, *Plutella xylostella*: Just how long is a piece of string? *Journal of Economic Entomology* 105, 1115–1129.
Zalucki, J.M., Heckel, D.G., Wang, P., Kuwar, S., Vassão, D.G. *et al.* (2021) A generalist feeding on Brassicaceae: It does not get any better with selection. *Plants* 10, 954.

24 The Jarrah Leaf Miner: *Perthida glyphopa*

Allan J. Wills
Biodiversity and Conservation Science, Department of Biodiversity, Conservation and Attractions, Manjimup, Western Australia, Australia

Abstract
The jarrah leaf miner *Perthida glyphopa* Common (Lepidoptera: Incurvariidae) is a small, 'primitive' moth endemic to south-western Australia. The leaf-mining larvae of the species can cause extensive defoliation of the forest-dominant jarrah tree *Eucalyptus marginata*. Outbreaks and spread of the leaf miner may well have been associated with postcolonial human environmental disturbance and disruption of earlier Aboriginal fire regimes. Damaging populations of jarrah leaf miner have abated in the 21st century from their extent in the 1980s, a period coincident with increasing temperatures, diminishing rainfall and increasing forest leaf area index in south-western Australia.

24.1 Introduction

The leaf-mining larvae of the small moth, the jarrah leaf miner, *Perthida glyphopa* (Lepidoptera: Incurvariidae) can be serious defoliators of jarrah *Eucalyptus marginata* in Western Australia, affecting growth rate (Mazanec, 1974). Damage to the leaves of jarrah was first attributed to the leaf miner by Newman and Clark (1926) who referred to it as *Tinea* sp. in the family Tineidae. Common (1969) recognized the misplacement within this taxon and described the jarrah leaf miner, assigning it as the type species of a new genus *Perthida* in the Incurvariinae (Incurvariidae). Molecular phylogeny of non-ditrysian lepidopteran lineages by Regier *et al.* (2015) supports the assignment to Incurvariidae and its relationships with other genera of leaf miners within this family, such as *Vespina* from east Asia (Hirowatari *et al.*, 2021) and *Basiluera* and *Simacauda* from Chile (Nielsen and Davis, 1981; Heckford *et al.*, 2023).

24.2 The Overall Status of the Species

Early 'outbreaks' of damage by the jarrah leaf miner (Fig. 24.1) were associated with the vicinity of ports in south-western Australia (Newman and Clark, 1926; Abbott *et al.*, 1999a, b), leading to speculation that the species was an introduced exotic. Exotic incursions are not unknown for Incurvariidae. *Simacauda dicommatias* from Chile, for example, is established in Cornwall, England (Heckford *et al.*, 2023). Port areas were early centres of ecological disturbance by colonial settlers in Western Australia, including

Corresponding author: allan.wills@dbca.wa.gov.au

DOI: 10.1079/9781800627680.0024

Fig. 24.1. Jarrah leaf miner damage showing blotch mines and holes left after larval excision. Photo: Allan Wills.

felling and coppicing of mature trees and disruption of Noongar (Aboriginal) burning regimes (Abbott, 2003) – habitat modifications which could account for initiation of outbreaks by promoting young foliage which is favoured for oviposition sites by the moth. Newman and Clark (1926) noted similar damage on *Eucalyptus gomphocephala*, *E. rudis*, *E. salubris* and *E. transcontinentalis*. The latter two species of trees were inland and distant from the coastal occurrences of jarrah leaf miner. Investigation of early presence of the characteristic damage on herbarium specimens led to the species being considered endemic but restricted in distribution at the time of colonization of Western Australia. The species was formally described by Ian Common in 1969.

By the mid-20th century, the species was widespread and abundant in south-western Australia (Abbott *et al.*, 1999a, b). Other Western Australian eucalypts on which similar leaf miners occur have been listed by Wallace (1970): *E. camaldulensis*, *E. todtiana*, *E. gomphocephala*, *E. diversicolor*, *E. loxophleba*, *E. redunca* var. *elata*, *E. patens*, *E. jacksoni*, *E. occidentalis* and *E. erythrocorys*. The moth is also sometimes present on the exotic plantation species *E. globulus*. Given the widely dispersed and localized ranges of some of these eucalypts, often overlapping the range of *E. marginata*, it is not clear whether occurrences of damage on other species represent a wide host range or the presence of several species of leaf miner. The leaf miner causing damage to *E. rudis* is known to be genetically distinct and isolated from *P. glyphopa* (Mahon *et al.*, 1982).

24.3 The Occurrence and Distribution of the Species

Common (1969) and Wallace (1970) described the distribution of jarrah leaf miner as south-western Australia, south from the Murchison River. Evidence of damage characteristic of *Perthida* spp. was noted on herbarium specimens of *E. rudis* collected in 1901 between the Moore and Murchison Rivers. Abbott *et al.* (1999b) examined 38 herbarium sheets of *E. marginata* from north of the Moore River and none carried evidence of *P. glyphopa*, whereas two of seven sheets carried evidence of damage on *E. rudis* pre-1920 from the same area, perhaps indicating only that the *Perthida* sp. associated with *E. rudis* is present north of the Moore River. Likewise, *P. glyphopa* was absent from 16 herbarium specimens of *E. marginata* collected from east of the main jarrah forest and north of Cranbrook. A 1983 map of 'infestation' of jarrah forest by jarrah leaf miner (Fig. 24.2) shows that leaf damage of some extent was detected in extensive areas of jarrah forest west of the Albany Highway and on the coastal plain to the north of Bunbury. Evidence of successful excision of larvae from jarrah leaves is known from the coastal plain as much as 40 km north of Wanneroo, north of Perth, and to within 10 km of Northam, east of Perth, and Cranbrook and Tambellup, north of Mt Barker (Abbott, 1992; Abbott *et al.*, 1999a, b).

24.4 Basic Life History and Natural History of the Species

The life history of the jarrah leaf miner was described and illustrated in detail by Mazanec (1983). The following account draws heavily on Mazanec's description of the life history.

24.4.1 Eggs

Eggs are deposited under the lower epidermis of host leaves. The embryo becomes visible within the egg 2–3 weeks after oviposition, developing into a larva with a light brown sclerotized head moving freely within the egg. Eggs are laid into leaves in April to May.

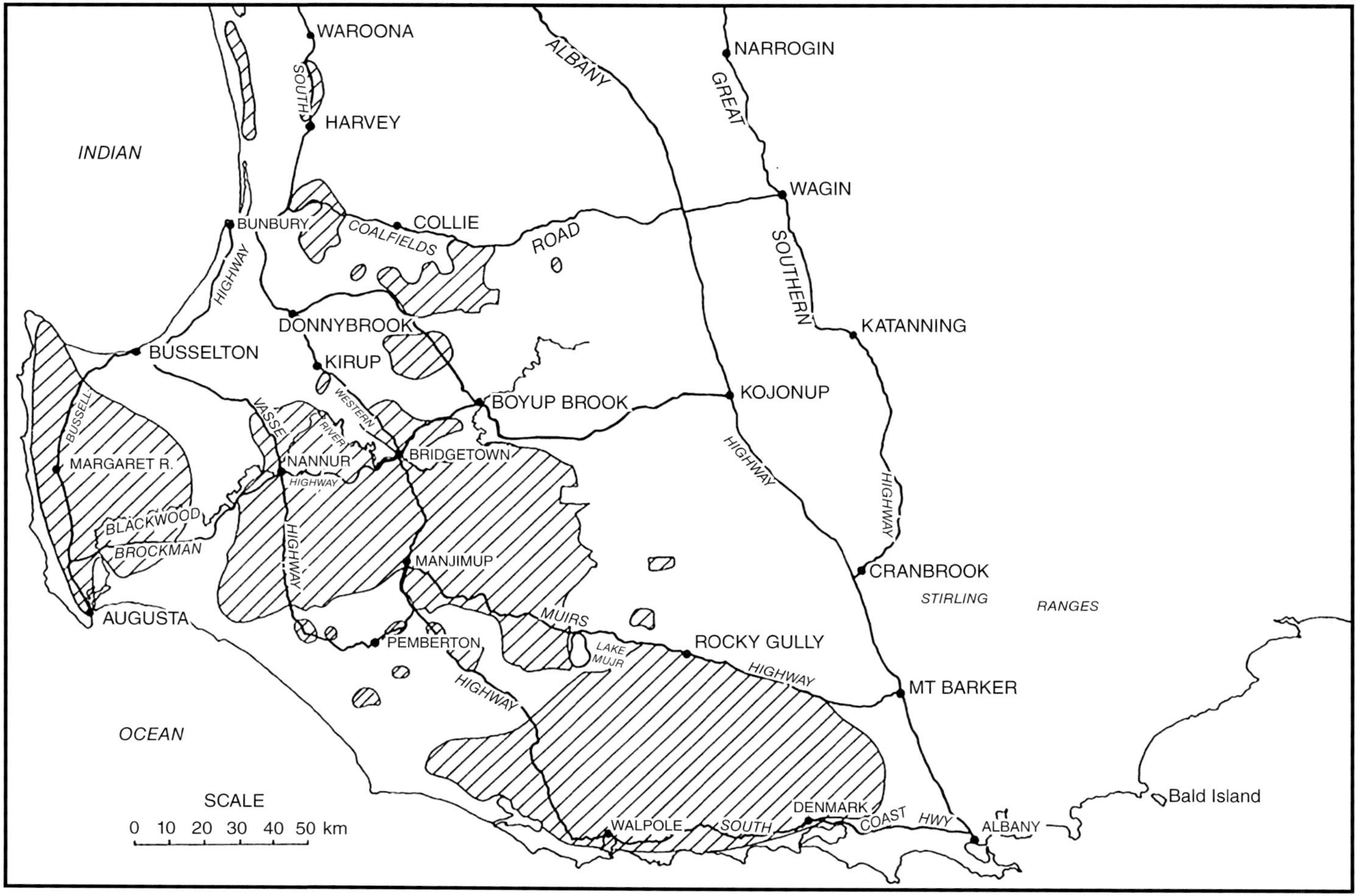

Fig. 24.2. Extent of damaging 'infestation' of Western Australian jarrah forest by jarrah leaf miner in 1983. The ranges of jarrah and jarrah leaf miner extend beyond these hatched areas. Reproduced with permission from Abbott (1992).

24.4.2 Larvae

Larvae hatch in late May. First instar larvae hatch and consume the chorion of the egg and then begin to feed within the leaf. Larvae avoid veins and oil glands, excavating an irregular mine by eating out all the tissue including the contents of epidermal cells on both sides of the leaf, leaving some of the side walls and outer walls covered by the cuticle. The second and subsequent instars eat the thin veins and chew through the oil glands without ingesting the oil. The larvae transfer the oil to their faecal pellets at the opposite end of the mine, keeping the feeding area of the mine free of oil and waste.

Larvae feed until August–September in warmer areas and September–October in cooler areas. On completion of the feeding phase, the mature larva constructs the cell in which it will spend the summer buried in the soil. The cell is constructed by an intricate cutting and fastening of oval-shaped discs from the leaf cuticle. The final excision of the cell is completed at night. The released cell containing the larva leaves a hole in the mine (Fig. 24.1). Reaching the ground, the larva buries itself up to 2.2 cm deep in the soil before daylight. More deeply buried cells are more likely to survive. Having buried itself in the soil, the larva withdraws into the cell and seals it. It remains in aestivating diapause within the cell for the next 6–7 months.

24.4.3 Pupae

During the prepupal stage of 1–2 days, there is longitudinal expansion of segments and lateral narrowing of the first abdominal segment accompanied by elongation of the larva. The pupal stage itself lasts about 37 days in February–March. Ecdysis and emergence occur at the soil surface.

24.4.4 Adults

In the cooler forest areas moths emerge in April, whereas in the warmer coastal areas, emergence is in April–May. Males begin emerging before females and are more numerous (Mazanec and Justin, 1994). Female moths have two ovaries consisting of four polytrophic ovarioles. At emergence, each ovariole contains approximately equal numbers of mature and immature eggs, allowing oviposition immediately after mating. Mazanec and Justin (1994) found an average of 142 eggs per newly emerged female.

24.5 The Phenology of the Species

Protandrous emergence of jarrah leaf miner moths, a higher ratio of males (Mazanec, 1984) and the tendency of males to occupy the axils of leaves and forks of twigs awaiting passing females ensure that mating occurs soon after emergence of females in April and May (Mazanec and Justin, 1986). Some jarrah trees appear to be resistant to either oviposition or development of mines (Mazanec, 1985). Ovipositing jarrah leaf miners do not discriminate between resistant leaves unfavourable to development of the mines and susceptible leaves where mines are more frequently successful (Abbott *et al.*, 1994).

Chronic defoliation arising from jarrah leaf miner infestation of susceptible trees leads to loss of older leaves and a predominance of younger leaves in affected canopies. These young leaves (*ca* 6 months old) are the preferred oviposition sites compared with leaves 1–2 years old (Wallace, 1970; Mazanec and Justin, 1986). Accordingly, jarrah leaf miner might be considered a flush feeder (*cf.* Gely *et al.*, 2020) whereby outbreaks leading to defoliation promote preferred oviposition sites. The majority of new leaf material is produced by *E. marginata* in November to January (Davison and Tay, 1989), at the time when larvae are aestivating in soil and litter (Table 24.1). This window of opportunity for the host to produce new leaves without defoliation means that the host is rarely killed by defoliation, even though defoliation to some extent may be repeated seasonally for decades (Wills *et al.*, 2004) leading to chronically affected trees expressing recession of crowns (Abbott *et al.*, 1995).

Mazanec described the immature stages and life history of a series of parasitoids of jarrah leaf miner (Mazanec, 1988, 1990a–c). The parasitoids attack in sequence: *Chrysonotomyia* sp. C (Eulophidae), *Enytus* sp. (Ichneumonidae), then *Chrysonotomyia* sp. A (Eulophidae) and

Table 24.1. Phenology of plant host *Eucalyptus marginata*, leaf miner *Perthida glyphopa* and two parasitoids of *P. glyphopa*, *Enytus* sp. and *Chrysonotomyia* sp. C. *Euclayptus marginata* phenology based on Mazanec (1989) and Davison and Tay (1989). Leaf miner and parasitoid phenologies based on Mazanec (1988, 1990a,b).

Month	Host *E. marginata*	Leaf miner *P. glyphopa*	Parasitoid *Chrysonotomyia* sp. C (Hymenoptera: Eulophidae)	Parasitoid *Enytus* sp. (Hymenoptera: Ichneumonidae)
April		Emergence of moths, mating and oviposition	Pupae, adults	Pupae, feeding larvae, adults
May		Emergence of moths, mating and oviposition	Pupae, adults	Pupae, feeding larvae, adults
June		Larval feeding commences	Adults, one or two eggs deposited within host	Pupae, feeding larvae, adults
July		Larval feeding	Slow maturation of larvae feeding within living host	Adults, eggs
August		Larval feeding, excision of larval cells in warmer areas, burial in soil and aestivation	Slow maturation of larvae within living host.	Adults, eggs, feeding larvae within host
September	Shedding of severely damaged leaves	Larval feeding, excision of larval cells, burial in soil and aestivation	Slow maturation of larvae within living host	Feeding larvae within host
October	Shedding of severely damaged leaves	Excision of larval cells, burial in soil and aestivation	Slow maturation of larvae within living host. Aestivating larvae in host cell	Feeding larvae within host
November	New leaves produced	Aestivation	Aestivating larvae in host cell	Feeding larvae within host
December	New leaves produced	Aestivation	Aestivating larvae in host cell	Feeding larvae within host
January	New leaves produced	Aestivation	Aestivating larvae in host cell	Feeding larvae within host
February		Pupation	Aestivating larvae in host cell	Feeding larvae within host
March		Pupation	Pupae	Feeding larvae within host

Diaulomorpha sp. (Eulophidae). The phenologies of *Chrysonotomyia* sp. C and *Enytus* sp. are detailed in Table 24.1 based on Mazanec (1988, 1990a).

24.6 Management of the Species

From colonial times to the present, jarrah has been an important timber of commerce in Western Australia. The chronic nature of population outbreaks and their threat to long-term timber production through loss of diameter increment (Mazanec, 1974) motivated experiments with insecticides to control jarrah leaf miner populations. These were initially small scale, using implantation of dimethoate into tree trunks (Wallace, 1966) and spraying of individual trees, then were extended to an aerial spraying experiment using dimethoate covering about 40 hectares (Van Didden, 1967). Fortunately, these experiments were not extended to a wider area of forest. Research towards an ecological understanding of the jarrah leafminer was initiated by CSIRO with the work of Murray Wallace (Wallace, 1970) and then Zdenek (Zan) Mazanec (Mazanec, 1981, 1983).

A second method of control was the use of prescribed fire. Fire creates a flush of new leaves and leads to an increase in the number of eggs deposited in leaves over that expected from the number of females surviving the fire (Mazanec, 1981), indicating immigration from source populations outside the burnt area. The scale of the fire in relation to source populations in unburnt area is important, with moths capable of flying more than a kilometre (Mazanec and Justin, 1986). Migration can obscure localized effects of fire and timber harvesting on numbers of mines per leaf and mine success rates (Abbott *et al.*, 1993). Severity of the fire is also critical, with aestivating larvae and eggs and larvae in leaves being able to survive low-severity fires. Even for severe fires at scales of 5000–7000 ha, timing of fire is critical, with an autumn fire leading to a decrease in the percentage of larvae excising from leaves the following spring and an increase in the number of mines per leaf the following year (Abbott *et al.*, 1993, 1999a, b). A spring fire led to an increase in the percentage of mines excising in the following spring (Abbott *et al.*, 1993, 1999a, b). Thus, large-scale outbreaks render autumn fire a tool of only temporary benefit for control of leaf miner populations.

Damaging populations of jarrah leaf miner have abated in the 21st century from their extent in the 1980s (Fig. 24.2), a period coincident with increasing temperatures, diminishing rainfall and increasing leaf area index in forests of south-western Australia (Liu *et al.*, 2019; Philip and Yu, 2020). Mazanec (1981) described relationships between antecedent rainfall and leaf index, and among mean daily temperature in April, leaf index, and eggs deposited per female. Leaf area index cannot be considered independent of leaf miner populations because of the feedback between damage and leaf area index. Mazanec, however, published the bulk of his population monitoring data in a key factor analysis (Mazanec, 1996) and this mortality data can be compared with regional rainfall anomalies over the monitoring period (Fig. 24.3). Clearly, mortality increased under drier conditions. This supports a hypothesis that jarrah leaf miner populations have declined over extensive areas of their range due to adverse rainfall conditions which affect host foliage phenology at similarly broad scales, and possibly in combination with a suite of parasitoids and predators (Mazanec, 1987, 1988, 1990a–c, 1996).

24.7 Future Research

The absence of a spread from the coastal plain outbreak into the northern jarrah forest adjacent to Perth seems perplexing, given the opportunity for this to happen early over a short distance from the coastal plain around Perth, and given the apparent spread of outbreaks into the Manjimup and Collie areas from the south-eastern range of jarrah (Abbott *et al.*, 1999a, b). However, leaf miners are sensitive to regional differences in biotic and abiotic conditions (Maunsell *et al.*, 2015, 2016). Genomic studies might reveal cryptic diversity in this apparently widespread leaf miner species, perhaps responding to subtle regional variation in its host (Wheeler and Byrne, 2006).

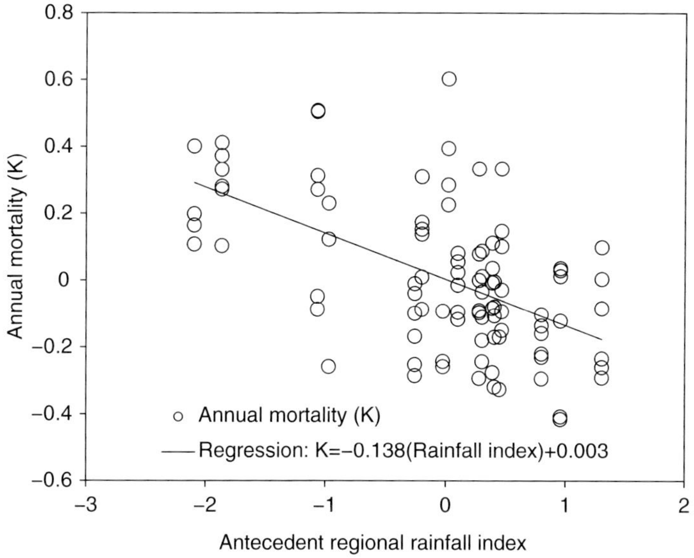

Fig. 24.3. Mortality data from Mazanec (1996) compared with the antecedent regional rainfall anomaly. Mortality (K) = log (eggs laid year 1) – log (eggs laid year 2). Rainfall index = average of normalized annual rainfall for Rocky Gully, Manjimup and Bridgetown for antecedent year. R^2 = 0.30.

Precolonial Noongar fire regimes are imprecisely known. The apparent rarity of severe fires in the precolonial dendrochronological record (Burrows *et al.*, 1995), the tendency of high fire frequency to create discontinuity in fuels capable of supporting fire (Burrows *et al.*, 2021), and pre- and early colonial records of Noongar fires (Ian Abbott, 2003, pers. commun.) lead to speculation that Noongar fire regimes resulted in relatively fine-scale mosaics of different fuel ages. These mosaics likely allowed for low-severity surface fires which did not affect leaf age demography of tree crowns in forest landscapes dominated by mature trees (Burrows *et al.*, 1995; Wills, 2024). Burning to implement fine-grain fuel age mosaics by fire management authorities has been largely experimental in south-western Australia to date (Burrows *et al.*, 2021). Investigation of effects of fine-grain mosaics could be informative in explaining responses of jarrah leaf miner populations (and arboreal fauna generally) to fire in low-severity conditions.

The effects of climate change on leaf miners in south-western Australia are largely unknown. The contraction of outbreaks due to an ongoing drying trend affecting the abundance of preferred food is suggested as an effect of climate change. Is a broader suite of locally endemic leaf miners negatively affected by drying trends imposed by climate change in south-western Australia? It should be noted that jarrah leaf miner must have been very uncommon before colonization by Europeans in Western Australia as there is no evidence for the presence of jarrah leaf miner on herbarium sheets of the host before 1878 (Abbott *et al.*, 1999).

Accurate methods of censusing life stages of populations are essential for documenting the effects of biotic and abiotic factors on population change and how cryptic species interact. Mazanec trapped excising larvae and emergent adults in funnel traps of standard size and counted oviposition per unit leaf area to census populations of jarrah leaf miner at relatively high population densities (Mazanec, 1978). While Mazanec used the pheromone-producing glands of female jarrah leaf miner moths to attract male moths in small-scale experiments, identification of the active substances in the pheromone has not been carried out to date. Sticky traps using sex pheromones as an attractant and light trapping using UV light are useful for detecting presence of leaf miners at low population densities (e.g. Mangrio *et al.*, 2023).

References

Abbott, I. (1992) Records of outbreaks of defoliating insects in jarrah forests, south-west Western Australia, from 1960 to 1990. *Technical Report (Western Australia. Department of Conservation and Land Management)* 28, 1–29.

Abbott, I. (2003) Aboriginal fire regimes in south-west Western Australia: Evidence from historical documents. In: Abbott, I. and Burrows, N. (eds) *Fire in Ecosystems of South-west Western Australia: Impacts and Management*. Backhuys Publishers, Leiden, The Netherlands, pp. 119–146.

Abbott, I., Van Heurck, P. and Burbidge, T. (1993) Ecology of the pest insect jarrah leaf miner (Lepidoptera) in relation to fire and timber harvesting in Jarrah Forest in Western Australia. *Australian Forestry* 56, 264–275.

Abbott, I., Van Heurck, P., Burbidge, T. and Wills, A. (1994) Factors influencing the performance of jarrah leafminer (Lepidoptera) within stands of jarrah forest of Western Australia *Australian Forestry* 57, 165–170.

Abbott, I., van Heurck, P., Burbidge, T. and Wills, A. (1995) Cutting out the leafminer. *Landscope* 11, 43–47.

Abbott, I., Wills, A. and Burbidge, T. (1999a) Reinfestation of *Eucalyptus marginata* ground coppice by jarrah leafminer after scorch by autumn or spring fires. *Australian Forestry* 62, 160–165.

Abbott, I.A., Wills, A. and Burbidge, T. (1999b) Historical incidence of *Perthida* leafminer species (Lepidoptera) in southwest Western Australia based on herbarium specimens. *Australian Journal of Ecology* 24, 144–150.

Burrows, N.D., Ward, B. and Robinson, A.D. (1995) Jarrah forest fire history from stem analysis and anthropological evidence. *Australian Forestry* 58, 7–16.

Burrows, N., Stephens, C., Wills, A. and Densmore, V. (2021) Fire mosaics in south-west Australian forest landscapes. *International Journal of Wildland Fire* 30, 933–945.

Common, I.F.B. (1969) A new genus *Perthida* for the Western Australian jarrah leaf miner *P. glyphopa* sp. n. and *Tinea phoenicopa* Meyrick (Lepidoptera: Incurvariidae). *Australian Journal of Entomology* 8, 126–130.

Davison, C.M. and Tay, F.C.S. (1989) Phenology of *Eucalyptus marginata* on sites infested with *Phytophthora cinnamomi*. *Australian Journal of Botany* 37, 193–206.

Gely, C., Laurance, S.G.W. and Stork, N.E. (2020) How do herbivorous insects respond to drought stress in trees? *Biological Reviews* 95, 434–448.

Heckford, R.J., Beavan, S.D. and Lees, D.C. (2023) *Simacauda dicommatias* (Meyrick, 1931) (Lepidoptera: Incurvariidae), a South American species established in Cornwall, new to the British Isles and unknown in mainland Europe, with first details of its biology. *Entomologist's Gazette* 73, 193–218.

Hirowatari, T., Yagi, S., Ohshima, I., Huang, G.H. and Wang, M. (2021) Review of the genus *Vespina* (Lepidoptera, Incurvariidae) with two new species from China and Japan. *Zootaxa* 4927, 209–233.

Liu, N., Harper, R.J., Smettem, K.R.J., Dell, B. and Liu, S. (2019) Responses of streamflow to vegetation and climate change in southwestern Australia. *Journal of Hydrology* 572, 761–770.

Mahon, R.J., Miethke, P.M. and Mahon, J.A. (1982) The evolutionary relationships of three forms of the jarrah leaf miner, *Perthida glyphopa* (Common) (Lepidoptera: Incurvariidae). *Australian Journal of Zoology* 30, 243–250.

Mangrio, G.Q., Gilal, A.A., Rajput, L.B. and Gabol, A.H. (2023) Performance of pheromone and light traps in monitoring and management of tomato leafminer, *Tuta absoluta* (Lepidoptera: Gelechiidae). *Journal of the Saudi Society of Agricultural Sciences* 22, 288–297.

Maunsell, S.C., Kitching, R.L., Burwell, C.J. and Morris, R.J. (2015) Changes in host–parasitoid food web structure with elevation. *Journal of Animal Ecology* 84, 353–363.

Maunsell, S.C., Burwell, C.J., Morris, R.J., McDonald, W.J., Edwards, E.D. *et al.* (2016) Elevational turnover in the composition of leaf miners and their interactions with host plants in Australian subtropical rainforest. *Austral Ecology* 41, 238–247.

Mazanec, Z. (1974) Influence of jarrah leaf miner on the growth of jarrah. *Australian Forestry* 37, 32–42.

Mazanec, Z. (1978) A sampling scheme for estimating population density of the jarrah leaf miner, *Perthida glyphopa* (Lepidoptera: Incurvariidae). *Australian Journal of Entomology* 17, 275–285.

Mazanec, Z. (1981) Environmental conditions influencing the population density of the jarrah leafminer, *Perthida glyphopa* (Lepidoptera: Incurvariidae). In: *Eucalypt Dieback in Forests and Woodlands*. Proceedings of a conference, CSIRO Division of Forest Research, Canberra, ACT, August 1980, pp. 140–146.

Mazanec, Z. (1983) The immature stages and life history of the jarrah leafminer, *Perthida glyphopa* Common (Lepidoptera: Incurvariidae). *Australian Journal of Entomology* 22, 101–108.

Mazanec, Z. (1984) Studies on the sex ratio in the jarrah leafminer *Perthida glyphopa* (Lepidoptera: Incurvariidae). *Australian Journal of Entomology* 23, 69–74.

Mazanec, Z. (1985) Resistance of *Eucalyptus marginata* to *Perthida glyphopa* (Lepidoptera: Incurvariidae). *Australian Journal of Entomology* 24, 209–221.

Mazanec, Z. (1987) Natural enemies of *Perthida glyphopa* Common (Lepidoptera: Incurvariidae). *Australian Journal of Entomology* 26, 303–308.

Mazanec, Z. (1988) Immature stages and life history of *Chrysonotomyia* sp. C (Hymenoptera: Eulophidae), a parasitoid of the jarrah leafminer. *Australian Journal of Entomology* 27, 279–292.

Mazanec, Z. (1990a) Immature stages and life history of *Enytus* sp. (Hymenoptera: Ichneumonidae), a parasitoid of *Perthida glyphopa* Common (Lepidoptera: Incurvariidae). *Australian Journal of Entomology* 29, 57–66.

Mazanec, Z. (1990b) The immature stages and biology of *Chrysonotomyia* sp. A (Hymenoptera: Eulophidae), a parasitoid of *Perthida glyphopa* Common (Lepidoptera: Incurvariidae). *Australian Journal of Entomology* 29, 139–146.

Mazanec, Z. (1990c) The immature stages and life history of *Diaulomorpha* sp.(Hymenoptera: Eulophidae), a parasitoid of *Perthida glyphopa* Common (Lepidoptera: Incurvariidae). *Australian Journal of Entomology* 29, 147–159.

Mazanec, Z. (1996) *Key Factor Analysis of Population Data for* Perthida glyphopa *Common (Lepidoptera: Incurvariidae).* Division of Entomology Technical Report No, 67. CSIRO, Canberra, ACT.

Mazanec, Z. and Justin, M.J. (1986) Oviposition behaviour and dispersal of *Perthida glyphopa* Common (Lepidoptera: Incurvariidae). *Australian Journal of Entomology* 25, 149–160.

Mazanec, Z. and Justin, M.J. (1994) Fecundity and oviposition by *Perthida glyphopa* Common (Lepidoptera: Incurvariidae). *Australian Journal of Entomology* 33, 223–234.

Newman, L.J. and Clark, J. (1926) The jarrah leaf miner, *Tinea* sp. *Australian Forestry Journal* 9, 95–99.

Nielsen, E.S. and Davis, D.S. (1981) A revision of the Neotropical Incurvariidae *s.str.,* with the description of two new genera and two new species (Lepidoptera: Incurvarioidea). *Steenstrupia* 7, 25–57.

Philip, P. and Yu, B. (2020) Interannual variations in rainfall of different intensities in South West of Western Australia. *International Journal of Climatology* 40, 3052–3071.

Regier, J.C., Mitter, C., Kristensen, N.P., Davis, D., van Nieukerken, E.J. *et al.* (2015) A molecular phylogeny for the oldest (nonditrysian) lineages of extant Lepidoptera, with implications for classification, comparative morphology and life-history evolution. *Systematic Entomology* 40, 671–704.

Van Didden G.W. (1967) Aerial Spraying of Jarrah Leafminer in W.A. Available at: https://library.dbca.wa.gov.au/Journals/080044/080044-05.015.pdf (accessed 7 October 2025).

Wallace, M.M.H. (1966) Trunk implantation of dimethoate to control jarrah leaf miner and psyllids in eucalypts. *Australian Forestry* 30, 131–138.

Wallace, M.M.H. (1970) The biology of the jarrah leaf miner, *Perthida glyphopa* Common (Lepidoptera: Incurvariidae). *Australian Journal of Zoology* 18, 91–104.

Wheeler, M.A. and Byrne, M. (2006) Congruence between phylogeographic patterns in cpDNA variation in *Eucalyptus marginata* (Myrtaceae) and geomorphology of the Darling Plateau, south-west of Western Australia. *Australian Journal of Botany* 54, 17–26.

Wills, A.J. (2024) Listening to the seasons. *Landscope* 39, 36–41.

Wills, A.J., Burbidge, T.E. and Abbott, I. (2004) Impact of repeated defoliation on jarrah (*Eucalyptus marginata*) saplings. *Australian Forestry* 67, 194–198.

25 The Diamondback Moths: *Plutella xylostella* and *Plutella australiana*

Michael J. Furlong
School of the Environment, University of Queensland, Brisbane, Queensland, Australia

Abstract

Plutella xylostella is the most significant pest of brassicaceous crops worldwide and is estimated to cost up to US$6.9 billion a year in yield losses and management expenses. It is an important pest in Australia where it was recently shown to exist in sympatry with the newly discovered cryptic species, *Plutella australiana*. *P. xylostella* is widely distributed throughout the country but the full extent of the distribution of *P. australiana* is not as well understood, although it occurs throughout eastern Australia, from far north Queensland, through coastal New South Wales, Victoria, southern South Australia, southern West Australia and Tasmania. Both species share many weedy hosts throughout their distributions, and both occur frequently in canola crops but *P. australiana* appears to be absent from Brassica vegetable crops, perhaps reflecting its susceptibility to insecticides. Both species persist on weedy species in the Brassicaceae when crop plants are absent. Although *P. xylostella* and *P. australiana* can hybridize and produce viable offspring in the laboratory, it is not known if this occurs in nature but genetic studies to date have not found evidence of widespread hybridization. As elsewhere in the world, *P. xylostella* is attacked by a wide range of predatory natural enemies in Australia, including many different species of Araneae, Neuroptera, Hemiptera, Coleoptera and Formicidae, and local complexes have the capacity to suppress populations in vegetable crops. The parasitoid fauna attacking *P. xylostella* in Australia is dominated by species introduced as classical biological control agents including the larval-pupal parasitoid *Diadegma semiclausum* (Hymenoptera: Ichneumonidae). The native larval parasitoids *Apanteles ippeus* (Braconidae) and *Diadegma rapi* (Ichneumonidae) are less effective against *P. xylostella* than *D. semiclausum*.

25.1 Introduction

In Australia 'diamondback moths' refer to the cryptic species *Plutella xylostella* and *Plutella australiana* (Plutellidae), which are the only confirmed members of the genus on the continent. The common name of these small moths (adults are approximately 10 mm long) derives from the beige-coloured diamond pattern that forms along the back of adult insects by markings on the hind margins of the folded forewings (Fig. 25.1). The pattern is more prominent in males, and it can be indistinct or even absent in females. *Plutella xylostella* is a significant global pest of Brassica crops (Furlong *et al.*, 2013; Li *et al.*, 2016) with one of the most extensive distributions of all Lepidoptera (Talekar and Shelton, 1993; Furlong *et al.*, 2013). While *P. xylostella* was first formally recorded in Australia more than 140 years ago (Tryon, 1889), *P. australiana*

Corresponding author: m.furlong@uq.edu.au

DOI: 10.1079/9781800627680.0025

Fig. 25.1. Adult male *P. xylostella* (approximately 10 mm in length) showing the beige-coloured diamond pattern formed by markings on the hind margins of the folded forewings. Author's own figure.

was first described a little over a decade ago (Landry and Hebert, 2013); it is probably native and its pest status remains to be determined (Perry, 2019).

25.2 Distinguishing Between *P. xylostella* and *P. australiana*

The existence of a second *Plutella* species in Australia was discovered serendipitously in an ambitious project to barcode the CO1 gene of the Lepidoptera in the Australian National Insect Collection (Hebert *et al.*, 2013). A subsequent study compared CO1 sequences between putative *P. xylostella* specimens (n = 397) from Asia, Australia, Europe and North America and showed that the samples separated into two clusters with 8.6% sequence divergence (Landry and Hebert, 2013). One cluster consisted of individuals from all four continents and represented true *P. xylostella* (average intraspecific divergence = 0.7%) while the other was composed of insects collected only in Australia (average intraspecific divergence = 0.1%) and was subsequently described as *P. australiana* by Landry and Hebert (2013).

Although external characters do not allow separation of these cryptic species, Landry and Hebert (2013) describe morphological differences in genitalia between the taxa. Previously Justus and Mitchell (1999) provided a comprehensive description of the reproductive systems of male and female *P. xylostella* from different regions of the world. They described highly specific configurations of the genitalia of both sexes in different populations and suggested that simple and exact fits between the aedeagus and female bursal duct helped maintain reproductive isolation within the species. Interestingly, they noted modifications to the aedeagal and bursal structures in the Australian population – before *P. australiana* was recognized. According to Landry and Hebert (2013), in male *P. australiana*, the genitalia are narrower than in *P. xylostella* and the most visible difference between the species is the shape of the vinculum-saccus. In *P. australiana* this is long and slender with a slight medial constriction and a protruded and inflated anterior apex whereas in *P. xylostella* it is shorter, broader and deeply concave. Despite these differences, following examination of additional specimens, Kariyawasam (2018) concluded that these characteristics of males are not distinct enough to separate the species reliably. In female *P. australiana*, the abdominal sternum 7 has a flat heart-shaped melanized area surrounding the antrum which has a short, curved tubular projection. In *P. xylostella*, the area of sternum 7 surrounding the antrum is bordered by a markedly raised pair of folds which form two conical projections bracing the longer tubular projection of the antrum. These structures have been confirmed as reliable diagnostic features by Kariyawasam (2018) and are consistent with historical descriptions of three female *P. xylostella* specimens from New Zealand (Dugdale, 1973; Kariyawasam, 2018), suggesting that further exploration for *P. australiana* beyond Australia is warranted.

25.3 Distributions and Host Plant Relationships

Historically, *Plutella xylostella* was widely considered to have originated in the Mediterranean region of Europe (Talekar and Shelton, 1993). More recent studies have proposed that the pest originated in southern Africa (Kfir, 1998) or Asia (Juric *et al*., 2017) but the most comprehensive and detailed genetic study to date posits that its origin is South American (You *et al*., 2020). Interestingly, You *et al*. (2020) found that their putative basal South American lineage was most closely associated with *P. australiana*, suggesting that the Australian endemism of *P. australiana* may need re-evaluation if additional cryptic species of *P. xylostella* are found elsewhere in the world. This non-diapausing species is highly migratory (Chapman *et al*., 2002). It has the capacity to reach extreme latitudes and, provided that suitable host plants are available, it can become seasonally abundant in regions of the world where environmental conditions preclude its year-round persistence (Furlong *et al*., 2013). In Australia, *P. xylostella* is most abundant in coastal areas but it can persist seasonally in all but the hottest and most arid areas in the central and central-west parts of the country (Furlong and Zalucki, 2017).

The economic importance of *P. xylostella* is defined by its pest status on vegetable and oilseed crops in the Brassicaceae, especially those in the genus *Brassica* (Ahuja *et al*. 2010). The pest also utilizes weed species in this family as important refugia when crop plants are absent (Sarfraz *et al*., 2006; Kahuthia-Gathu *et al*., 2009). In Australia these hosts include wild turnip (*Rapistrum rugosum*), wild radish (*Raphanus raphanistrum*), wild mustard (*Rhamphospermum arvense*), sand rocket or Lincoln weed (*Diplotaxis tenuifolia*), sea rocket (*Cakile maritima*) and Ward's weed (*Carrichtera annua*) (Perry *et al*., 2018). Some of these species can also serve as important 'green bridges' that allow *P. xylostella* populations to persist, or even build up, before cultivation of crop hosts (Perry, 2019) and they can also facilitate the local persistence of primary parasitoids (Furlong and Zalucki, 2007).

Outside the Brassicaceae, *P. xylostella* has been reported to utilize *Capparis sandwichiana* (Capparaceae) in Hawaii (Robinson and Sattler, 2001) and host associations with Australian members of this genus, which occur in drier inland areas of eastern Australia, should be investigated. Similarly, *P. xylostella* has been recorded feeding on members of the Cleomaceae (Landry and Hebert, 2013) and the Tropaeolaceae (Sarfraz *et al*., 2006) and both families, along with approximately 200 additional species of Brassicaceae, many of which are native, occur in Australia (Australian Virtual Herbarium, 2025). Although there are several unreliable reports of *P. xylostella* host associations in older literature (see Kariyawasam, 2018), there are verified recurring reports of its close association with peas (*Pisum sativum*, Fabaceae) (Thorsteinson, 1960; Löhr and Rossbach, 2004). The alleles allowing survival on *P. sativum* are not fixed (Henniges-Janssen *et al*., 2011), indicating the pest's significant genetic plasticity. Such short-term host shifts can allow *P. xylostella* to exploit enemy free space (Rossbach *et al*., 2008) and the host plant relationships of the pest need to be investigated more thoroughly.

Following its discovery, surveys based on adult moth specimens showed that *P. australiana* was distributed throughout eastern Australia, from far north Queensland, through coastal New South Wales and into eastern South Australia (Landry and Hebert, 2013). Subsequent light trap surveys confirmed the presence of adult moths in Tasmania (Kariyawasam, 2018). When larvae (Fig. 25.2) were collected from *Brassica* vegetable crops (broccoli, cabbage, cauliflower, kale and swede) from sites south-west of Sydney in New South Wales, south-east Queensland, Hobart in Tasmania and Nelson in New Zealand, *P. australiana* was only recovered from two proximate sites in NSW and it always occurred in sympatry with *P. xylostella* (Kariyawasam, 2018). At one of these sites, *P. australiana* accounted for 79% of larvae collected from a cabbage crop and, at the other, it accounted for a similar proportion of larvae collected from weedy field mustard (*Brassica rapa*) growing in a kale crop, but all larvae collected from kale plants were *P. xylostella*.

In an extensive survey of *Plutella* spp., Perry *et al*. (2018) collected larvae from commercial *Brassica* vegetable crops, canola and brassicaceous weeds at 75 sites across Australia. *P. xylostella* was found at all sites and although *P. australiana* was recovered from all states except Tasmania, it represented only *ca*. 10% of individuals collected. *P. australiana* was never recovered from *Brassica* vegetable crops but it was found in sympatry with *P. xylostella* in a single kale crop and in approximately one-third of all canola crops sampled (Perry *et al*., 2018). Similarly, the two species occurred in sympatry in stands of wild turnip, wild radish and sand rocket but *P. australiana* was never found on sea rocket or Ward's weed. It did occur on African mustard (*Brassica tournefortii*) in the absence of *P. xylostella* at one site in southern

Fig. 25.2. Fourth instar *P. xylostella* larva (approximately 10 mm long) feeding on a cabbage (*Brassica oleracea*) leaf. Author's own figure.

Queensland (Perry *et al.*, 2018). In complementary laboratory studies, *P. australiana* could complete development on sand rocket (Perry *et al.*, 2018) but the host status of the remaining non-crop plants from which larvae were recovered requires confirmation and the search for further host plants both within and beyond the Brassicaceae should continue.

25.4 Reproductive Isolation of Species and Genetic Structure of Populations

In the laboratory, *P. australiana* and *P. xylostella* can hybridize but interspecific mating success is more likely between *Wolbachia*-infected female *P. australiana* and male *P. xylostella* than in reciprocal crosses (Perry *et al.*, 2018). Progeny of all crosses were highly male biased but back-crosses of these hybrid males with females of either species resulted in viable offspring, indicating that interspecies gene flow is possible (Perry *et al.*, 2018). Whether hybridization occurs in nature is not known, but whole-genome analysis of multiple *P. xylostella* and *P. australiana* individuals found no evidence of widespread introgression between the species (Ward and Baxter, 2018).

Australian populations of *P. xylostella* have low levels of genetic diversity compared with populations from elsewhere (Endersby *et al.*, 2006), possibly indicating a small initial founding population. A nuclear genetic study confirmed the low genetic diversity in Australian *P. xylostella* populations while simultaneously showing significant genetic diversity in *P. australiana* populations (Perry *et al.*, 2018). This later work may explain previous studies (Pichon *et al.*, 2006; Roux *et al.*, 2007) that reported greater genetic diversity in Australian *P. xylostella* than Endersby *et al.* (2006) as the existence of the more genetically diverse cryptic species was not known at the time. Despite the lack of genetic diversity and structure in *P. xylostella* and the likely capacity of the species to move vast distances in Australia (Furlong *et al.*, 2008), it is unlikely that a completely randomly interbreeding panmictic population exists (Saw *et al.*, 2006; Perry *et al.*, 2020). For example, when host plants are abundant *P. xylostella* shows little proclivity to migrate and movement is typically restricted to short-range, trivial flight (Mo *et al.*, 2003) which can lead to isolation and genetic changes because of local selection for insecticide resistance (Endersby *et al.*, 2008).

Genetic evidence suggests that *P. xylostella* in Australia originates from either Europe (Juric *et al.*, 2017) or Asia (Saw *et al.*, 2006) and that it and the native *P. australiana* experienced allopatric divergence. The mechanisms that maintain reproductive isolation between the species now that they are in sympatry are not understood but they might involve cytoplasmic incompatibility mediated by extremely high rates of *Wolbachia* infection in *P. australiana* (Perry *et al.*, 2018) or compromised mating because of morphological incongruences (Justus and Mitchell, 1999).

25.5 Natural Enemies

Plutella xylostella is attacked by a wide range of natural enemies and due to its pest status many of these have been the subject of significant research to identify effective pathogens, predators and parasitoids for control programmes (Gurr *et al.*, 2018). In Australia, studies have documented fungal pathogens (Vickers *et al.*, 2004), predators (e.g. Furlong *et al.*, 2004a) and parasitoids (Goodwin, 1979; Furlong and Zalucki, 2007) attacking *P. xylostella* in horticultural (Furlong *et al.*, 2004b) and canola crops (see Furlong *et al.*, 2008) but there are no reports of granulosis, nuclear polyhedrosis or cypro viruses, all of which can be important natural enemies of the pest elsewhere in the world (Grzywacz *et al.*, 2010). Significantly, much of this work was conducted before the existence of *P. australiana* was recognized and the relationships between natural enemies attacking it and those attacking *P. xylostella* are unknown.

Various molecular methods have been employed to demonstrate that *P. xylostella* is attacked by a wide range of predatory arthropods in horticultural crops. In South Australia species of Nabidae (Heteroptera) and Lycosidae (Araneida) have been recorded as important predators of *P. xylostella* in *Brassica* crops (Ma *et al.*, 2005; Hosseini *et al.*, 2008). In Queensland more extensive molecular studies show that many different spiders (members of the Araneidae, Clubionidae, Lycosidae, Linyphiidae, Oxypodidae, Salticidae, Tetragnathidae and Thomisidae), Neuroptera (*Micromus* sp.), Heteroptera (Lygaeidae, Miridae (*Deraeocoris* sp.) and Nabidae), Coleoptera (Coccinellidae (*Hippodamia variegata*), Staphylinidae), Dermaptera (Labiduridae (*Labidura truncata*, *Nala lividipes*)) and Hymenoptera (Formicidae) feed on *P. xylostella* in *Brassica* crops (Murtiningsih, 2015) and that local predator complexes can significantly reduce *P. xylostella* populations (Furlong *et al.*, 2014). Much less is known about the complexes of predatory arthropods in canola-growing regions of Australia, but spiders are abundant in Australian grain crops (Whitehouse and Lawrence, 2001). They are likely important predators of both *P. xylostella* and *P. australiana* in these agro-ecosystems but this requires investigation.

The hymenopteran parasitoid fauna attacking *P. xylostella* in Australia is dominated by species introduced as classical biological control agents and their distributions, dynamics and impacts on host populations in horticultural crops are well understood (Table 25.1 and see, for example, Goodwin, 1979; Furlong and Zalucki, 2007). Foremost amongst these is the larval-pupal parasitoid *Diadegma semiclausum* (Hymenoptera: Ichneumonidae) which was imported from New Zealand alongside the pupal parasitoid *Diadromus collaris* (Hymenoptera: Ichneumonidae) in the 1940s, where they had been introduced from the United Kingdom (Furlong *et al.*, 2013). Both species readily established in Australia (Wilson, 1960; Yarrow, 1970) and each is now widely distributed throughout *Brassica* vegetable and canola-growing regions (Table 25.1). Of more limited importance is the less well distributed larval-pupal parasitoid *Oomyzus sokolowskii* (Hymenoptera: Eulophidae) and the pupal parasitoid *Brachymeria phya* (Hymenoptera: Chalcididae) (Table 25.1; Furlong and Zalucki, 2007). The origin of these species is not known but *O. sokolowskii* may have been one of the five unidentified species of *P. xylostella* parasitoids from Spain, Sri Lanka, India and China liberated in Western Australia between 1903 and 1909 (Wilson, 1960).

Similarly, the larval parasitoids *Apanteles ippeus* (Braconidae) and *Diadegma rapi* (Ichneumonidae) are also reported to have more limited impacts on *P. xylostella* populations than the introduced *D. semiclausum* (Yarrow, 1970; Goodwin, 1979; Hamilton, 1979; Furlong and Zalucki, 2007). Both *A. ippeus* (Yarrow, 1970) and *D. rapi* (Goodwin, 1979; Azidah *et al.*, 2000) are considered indigenous to Australia but their native hosts have not been determined. The parasitoid fauna associated with *P. australiana* has not been investigated and it is possible that this species is the native host for either or both of these parasitoids. Venkatraman (1964) reported contemporaneous parasitism of *P. xylostella* by *D. semiclausum* and *D. rapi* over an extended period in NSW and showed that parasitism by *D. semiclausum* was consistently significantly higher than that by *D. rapi*. *P. australiana* is more abundant in NSW than elsewhere in Australia (Perry *et al.*, 2018) and while it is not known if any of the hosts recorded as *P. xylostella* by Venkatraman (1964) were *P. australiana*, this is a possibility.

The relationships among the parasitoid species introduced for the biological control of *P. xylostella* with *P. australiana* and those

Table 25.1. Distribution and provenance of major hymenopteran primary parasitoids reported attacking *P. xylostella* in Australia. Adapted from Furlong *et al.* (2008).

Host stage	Parasitoid	Provenance	Known distribution
Egg	*Trichogrammatoidea bactrae* Nagaraja (Trichogrammatidae)	Uncertain	Queensland
	Trichogramma pretiosum Riley (Trichogrammatidae)	United States, introduced for *Helicoverpa* sp.	Queensland
Larva	*Apanteles ippeus* Nixon (Braconidae)	Uncertain, likely indigenous	New South Wales, Queensland, South Australia, Tasmania, Victoria, Western Australia
	Diadegma rapi (Cameron) (Ichneumonidae)	Uncertain, likely indigenous	New South Wales, Victoria, South Australia, Tasmania, Western Australia
	Diadegma semiclausum (Hellén) (Ichneumonidae)	United Kingdom via New Zealand	New South Wales, Queensland, South Australia, Tasmania, Victoria, Western Australia
	Oomyzus sokolowskii (Kurdjumov) (Eulophidae)	Uncertain	Queensland, Tasmania, Western Australia
Pupa	*Brachymeria phya* (Walker) (Chalcididae)	Uncertain	New South Wales, Queensland, South Australia, Victoria
	Diadromus collaris (Gravenhorst) (Ichneumonidae)	United Kingdom via New Zealand	New South Wales, Queensland, South Australia, Tasmania, Victoria, Western Australia

between *P. xylostella* and the putatively native parasitoids require clarification. Such work would likely explain much of the confusion that has surrounded perceived *P. xylostella*/parasitoid relationships in Australia since classical biological control programmes began in the 1940s (Table 25.1).

25.6 Pest Status and Management

Plutella xylostella is the most serious pest of brassicaceous crops worldwide (Furlong *et al.*, 2013; Li *et al.*, 2016). Zalucki *et al.* (2012) estimated that the combined yield loss and management expenses for *P. xylostella* cost the world economy up to US$5 billion annually (US$6.9 billion in 2024). In Australia these combined costs across the vegetable and canola industries were estimated at US$0.20 billion (US$0.27 billion in 2024). Unlike *P. xylostella*, *P. australiana* is not a pest of vegetable crops and while it does pose a threat to canola production, its significance as a pest remains to be determined (Perry *et al.*, 2018; Perry, 2019).

Despite its significant pest status, *P. xylostella* is usually considered to be an induced pest that can be suppressed by its natural enemies, provided it has been reunited with key species in its invasive range (Furlong *et al.*, 2013). Typically, growers immediately resort to synthetic chemical insecticides to control *P. xylostella* (Furlong *et al.*, 2013) and this, coupled with the pest's unrivalled capacity to evolve resistance to these products, has had catastrophic effects. There are 1099 reported cases of *P. xylostella* resistance to a total of 104 different insecticides worldwide; 28 of these cases are in Australia, where resistance to six different synthetic insecticides has been recorded (Mota-Sanchez and Wise, 2025).

Conversely, *P. australiana* is more susceptible to insecticides (Perry *et al.*, 2018) and this may contribute to its apparent exclusion from *Brassica* vegetable crops. Although the evolution of insecticide resistance poses a serious threat to the sustainable management of *P. xylostella*, of equal or possibily greater concern is the phenomenon of pest resurgence that synthetic insecticides induce by the destruction of natural enemies (Furlong *et al.*, 2013). *P. xylostella* can be managed economically and sustainably in Australian vegetable crops by the integration of selective organic insecticides and natural enemies (Furlong *et al.*, 2004a; Thurman and Furlong, 2025), but the sustained adoption of integrated pest management approaches in Australia remains as elusive as elsewhere in the world (Furlong *et al.*, 2013).

References

Ahuja, I., Rohloff, J. and Bones, A.M. (2010) Defence mechanisms of Brassicaceae: implications for plant-insect interactions and potential for integrated pest management. In: Lichtfouse, E., Hamelin, M., Navarrete, M. and P. Debaeke (eds) *Sustainable Agriculture*, 2. Springer, New York, pp. 623–670.

Australasian Virtual Herbarium (2025) Available at: https://avh.chah.org.au/ (accessed 26 April 2025).

Azidah, A.A., Fitton, M.G. and Quicke, D.L.J. (2000) Identification of the *Diadegma* species (Hymenoptera: Ichneumonidae, Campopleginae) attacking the diamondback moth, *Plutella xylostella* (Lepidoptera: Plutellidae). *Bulletin of Entomological Research* 90, 375–389.

Chapman, J.W., Reynolds, D.R., Smith, A.D., Riley, J.R., Pedgley, D.E. and Woiwod, I.P. (2002) High-altitude migration of the diamondback moth *Plutella xylostella* to the UK: A study using radar, aerial netting, and ground trapping. *Ecological Entomology* 27, 641–650.

Dugdale, J.S. (1973) The genus *Plutella* (Hyponomeutidae) in New Zealand and the family position of *Circoxena* (Lepidoptera). *New Zealand Journal of Science* 16, 1009–1023.

Endersby, N.M., McKechnie, S., Ridland, P. and Weeks, A. (2006) Microsatellites reveal a lack of structure in Australian populations of the diamondback moth, *Plutella xylostella* (L.). *Molecular Ecology* 15, 107–118.

Endersby, N.M., Ridland, P.M. and Hoffmann, A.A. (2008) The effects of local selection versus dispersal on insecticide resistance patterns: longitudinal evidence from diamondback moth (*Plutella xylostella* (Lepidoptera: Plutellidae)) in Australia evolving resistance to pyrethroids. *Bulletin of Entomological Research* 98, 145–157.

Furlong, M.J. and Zalucki, M.P. (2007) Parasitoid complex of diamondback moth in south-east Queensland: first records of *Oomyzus sokolowskii* (Hymenoptera: Eulophidae) in Australia. *Australian Journal of Entomology* 46, 167–175.

Furlong, M.J. and Zalucki, M.P. (2017) Climate change and biological control: the consequences of increasing temperatures on host–parasitoid interactions. *Current Opinion in Insect Science* 20, 39–44.

Furlong, M.J., Shi, Z.H., Liu, Y.Q., Guo, S.J., Lu, Y.B. *et al.* (2004a) Experimental analysis of the influence of pest management practice on the efficacy of an endemic arthropod natural enemy complex of the diamondback moth. *Journal of Economic Entomology* 97, 1814–1827.

Furlong, M.J., Shi, Z.H., Liu, S.S. and Zalucki, M.P. (2004b) Evaluation of the impact of natural enemies on *Plutella xylostella* L. (Lepidoptera: Yponomeutidae) populations on commercial *Brassica* farms. *Agricultural and Forest Entomology* 6, 311–322.

Furlong, M.J., Ju, K.H., Su, P.W., Chol, J.K., Il, R.C. and Zalucki, M.P. (2008) Integration of endemic natural enemies and *Bacillus thuringiensis* to manage insect pests of *Brassica* crops in North Korea. *Agriculture, Ecosystems and Environment* 125, 223–238.

Furlong, M.J., Wright, D.J. and Dosdall, L.M. (2013) Diamondback moth ecology and management: problems, progress, and prospects. *Annual Review of Entomology* 58, 517–541.

Furlong, M.J., Rowley, D.L., Murtiningsih, R. and Greenstone, M.H. (2014) Combining ecological methods and molecular gut-content analysis to investigate predation of a lepidopteran pest complex of *Brassica* crops. *Entomologia Experimentalis et Applicata* 153, 128–141.

Goodwin, S. (1979) Changes in numbers in the parasitoid complex associated with the diamond-back moth, *Plutella xylostella* (L.) (Lepidoptera), in Victoria. *Australian Journal of Entomology* 27, 981–989.

Grzywacz, D., Rossbach, A., Rauf, A., Russell, D.A., Srinivasan, R. and Shelton, A.M. (2010) Current control methods for diamondback moth and other *Brassica* insect pests and the prospects for improved management with lepidopteran-resistant Bt vegetable brassicas in Asia and Africa. *Crop Protection* 29, 68–79.

Gurr, G.M., Reynolds, O.L., Johnson, A.C., Desneux, N., Zalucki, M.P. *et al.* (2018) Landscape ecology and expanding range of biocontrol agent taxa enhance prospects for diamondback moth management. A review. *Agronomy for Sustainable Development* 38, 23.

Hamilton, J.T. (1979) Seasonal abundance of *Pieris rapae* (L.), *Plutella xylostella* (L.) and their diseases and parasites. *General and Applied Entomology* 11, 59–66.

Hebert, P.D.N., de Waard, J.R., Zakarov, E.V., Prosser, S.W.J., Sones, J.E. *et al.* (2013) A DNA 'Barcode Blitz': Rapid digitization and sequencing of a natural history collection. *PLOS One* 8, e68535.

Henniges-Janssen, K., Reineke, A., Heckel, D.G. and Groot, A.T. (2011) Complex inheritance of larval adaptation in *Plutella xylostella* to a novel host plant. *Heredity* 107, 421–423.

Hosseini, R., Keller, M.A., Schmidt, O. and Baker, G. (2008) Predators associated with crop pests of *Brassica* spp. in South Australia. In: Shelton, A.M., Collins, H.L., Zhang, Y.J. and Wu, Q.J. (eds) *Proceedings of the Fifth International Workshop on Management of Diamondback Moth and Other Crucifer Pests*. China Agricultural Science and Technology Press, Beijing, China, pp. 338–344.

Juric, I., Salzburger, W. and Balmer, O. (2017) Spread and global population structure of the diamondback moth *Plutella xylostella* (Lepidoptera: Plutellidae) and its larval parasitoids *Diadegma semiclausum* and *Diadegma fenestrale* (Hymenoptera: Ichneumonidae) based on mtDNA. *Bulletin of Entomological Research* 107, 155–164.

Justus, K.A. and Mitchell, B.K. (1999) Reproductive morphology, copulation, and inter-populational variation in the diamondback moth, *Plutella xylostella* (L.) (Lepidoptera: Plutellidae). *International Journal of Insect Morphology and Embryology* 28, 233–246.

Kahuthia-Gathu, R., Löhr, B., Poehling, H.M. and Mbugua, P.K. (2009) Diversity, distribution and role of wild crucifers in major cabbage and kale growing areas of Kenya. *Bulletin of Entomological Research* 99, 287–297.

Kariyawasam, H.K.T.N. (2018) Taxonomy, Distribution and Pest Status of *Plutella* Species (Lepidoptera: Plutellidae) in Australia and New Zealand. MAppSc thesis, Queensland University of Technology.

Kfir, R. (1998) Origin of the diamondback moth (Lepidoptera: Plutellidae). *Annals of the Entomological Society of America* 91, 164–167.

Landry, J.F. and Hebert, P.D.N. (2013) *Plutella australiana* (Lepidoptera, Plutellidae), an overlooked diamondback moth revealed by DNA barcodes. *ZooKeys* 327, 43–63.

Li, Z., Feng, X., Liu, S., You, M. and Furlong, M.J. (2016) Biology, ecology, and management of the diamondback moth in China. *Annual Review of Entomology* 61, 277–296.

Löhr, B. and Rossbach, A. (2004) Diamondback moth, *Plutella xylostella* (L.), on peas in Kenya: impact of the host shift on the pest and its parasitoid. In: Endersby, N.M. and Ridland, P.M. (eds) *Proceedings of the Fourth International Workshop on the Management of Diamondback Moth and Other Crucifer Pests*. The Regional Institute, Melbourne, Victoria, pp. 133–139.

Ma, J., Li, D., Keller, M., Schmidt, O. and Feng, X. (2005) A DNA marker to identify predation of *Plutella xylostella* (Lep., Plutellidae) by *Nabis kinbergii* (Hem., Nabidae) and *Lycosa* sp. (Aranaea, Lycosidae). *Journal of Applied Entomology* 129, 330–335.

Mo, J., Baker, G., Keller, M. and Roush, R. (2003) Local dispersal of the diamondback moth [*Plutella xylostella* (L.)] (Lepidoptera: Plutellidae). *Environmental Entomology* 32, 71–79.

Mota-Sanchez, D. and Wise, J.C. (2025) The Arthropod Pesticide Resistance Database. Michigan State University. Available at: www.pesticideresistance.org (accessed 25 April 2025).

Murtiningsih, R. (2015) Predation of Cabbage Pests: A Combined Ecological and Molecular Approach. PhD thesis, School of Biological Sciences, University of Queensland.

Perry, K.D. (2019) The Colonisation of Canola Crops by the Diamondback Moth, *Plutella xylostella* L., in Southern Australia. PhD thesis, University of Adelaide.

Perry, K.D., Baker, G.J., Powis, K.J., Kent, J.K. Ward, C.M. and Baxter, S.W. (2018) Cryptic *Plutella* species show deep divergence despite the capacity to hybridize. *BMC Evolutionary Biology*. 18, 77.

Perry, K.D., Keller, M.A. and Baxter, S.W. (2020) Genome-wide analysis of diamondback moth, *Plutella xylostella* L., from *Brassica* crops and wild host plants reveals no genetic structure in Australia. *Scientific Reports* 10, 12047.

Pichon, A., Arvanitakis, L., Roux, O., Kirk, A.A., Alauzet, C. *et al.* (2006) Genetic differentiation among various populations of the diamondback moth, *Plutella xylostella* (Lepidoptera: Yponomeutidae). *Bulletin of Entomological Research* 96, 137–144.

Robinson, G.S. and Sattler, K. (2001) *Plutella* in the Hawaiian Islands: Relatives and host-races of the diamondback moth (Lepidoptera: Plutellidae). *Bishop Museum Occasional Papers* 67.

Rossbach, A., Löhr, B. and Vidal, S. (2008) Interspecific competition between *Diadegma semiclausum* Hellen and *Diadegma mollipla* (Holmgren), parasitoids of the diamondback moth, *Plutellla xylostella* (L.), feeding on a new host plant. *Bulletin of Entomological Research* 98, 135–143.

Roux, O., Gevrey, M., Arvanitakis, L., Gers, C., Bordat, D. and Legal, L. (2007) ISSR–CR: Tool for discrimination and genetic structure analysis of *Plutella xylostella* populations native to different geographical areas. *Molecular Phylogenetics and Evolution* 43, 240–250.
Sarfraz, M., Dosdall, L. M. and Keddie, B. A. (2006) Diamondback moth–host plant interactions: implications for pest management. *Crop Protection* 25, 625–639.
Saw, J., Endersby, N.M. and McKechnie, S.W. (2006) Low mtDNA diversity among widespread Australian diamondback moth *Plutella xylostella* (L.) suggests isolation and a founder effect. *Insect Science* 13, 365–373.
Talekar, N.S. and Shelton, A.M. (1993) Biology, ecology, and management of the diamondback moth. *Annual Review of Entomology* 38, 275–301.
Thorsteinson, A.J. (1960) Host selection in phytophagous insects. *Annual Review of Entomology* 5, 193–218.
Thurman, J.H. and Furlong, M.J. (2025) Biocontrol of diamondback moth (*Plutella xylostella*) in organic crops: Spatial and seasonal dynamics. *Agriculture, Ecosystems and Environment* 385, 109567.
Tryon, H. (1889) *Report on Insect and Fungus Pests*, vol. 1. Queensland Department of Agriculture, Brisbane, Queensland.
Venkatraman, T.V. (1964) Experimental studies in superparasitism and multiparasitism in *Horogenes cerophaga* (Grav.) and *Hymenobosmina rapi* (Cam.), the larval parasites of *Plutella maculipennis*. *Indian Journal of Entomology* 27, 1–32.
Vickers, R.A., Furlong, M.J., White, A. and Pell, J.K. (2004) Initiation of fungal epizootics in diamondback moth populations within a large field cage: Proof of concept for auto-dissemination. *Entomologia Experimentalis et Applicata* 111, 7–17.
Ward, C.M. and Baxter, S.W. (2018) Assessing genomic admixture between cryptic *Plutella* moth species following secondary contact. *Genome Biology and Evolution* 10, 2973–2985.
Whitehouse, M.E. and Lawrence, L. (2001) Are spiders the perfect predator? *Australian Cotton Grower* 22, 30–33.
Wilson, F. (1960) A *Review of the Biological Control of Insects and Weeds in Australia and Australian New Guinea*. Technical Communication No. 1. Commonwealth Agricultural Bureaux. Commonwealth Institute of Biological Control, Ottawa, Canada.
Yarrow, W.H.T. (1970) Parasites of *Plutella xylostella* (L.) in south-eastern Queensland. *Queensland Journal of Agricultural and Animal Sciences* 27, 321–324.
You, M., Ke, F., You, S., Wu, Z., Liu, Q. *et al.* (2020) Variation among 532 genomes unveils the origin and evolutionary history of a global insect herbivore. *Nature Communications* 11, 2321.
Zalucki, M.P., Shabbir, A., Silva, R., Adamson, D., Liu, S.S. and Furlong M.J. (2012) Estimating the economic cost of one of the world's major insect pests, *Plutella xylostella*: Just how long is a piece of string? *Journal of Economic Entomology* 105, 1115–1129.

26 The Giant Wood Moth: *Endoxyla cinereus*

Jessa H. Thurman
CSIRO, Agriculture & Food, Dutton Park, Queensland, Australia

Abstract

The giant wood moth (*Endoxyla cinereus*) is the heaviest recorded moth in the world and plays complex roles as an ecosystem engineer, cultural resource and forestry pest. This chapter explores this moth's unique life history, which includes a multi-year developmental period completed inside the trunks of eucalypts. The ecological significance of this species is also considered based on its boring habits, abundance and large geographic range. Its cultural significance to Indigenous Australians is briefly highlighted with reference to other large species of *Endoxyla*. Finally, future directions for research projects relevant to the giant wood moth are highlighted.

26.1 Introduction

The giant wood moth, *Endoxyla cinereus* (Tepper 1890), is one of the largest and most abundant species of wood moths (Cossidae) in Australia (see Chapter 15). Found across coastal regions of northern Queensland to South Australia, the giant wood moth is a species of significant ecological, cultural and economic importance. It holds the title for the heaviest moth in the world, with a recorded female weighing 31.2 grams (Beccaloni, 2010), but it is not the largest species in its genus by wingspan, a distinction held by *E. macleayi* which may be even heavier than *E. cinereus* (Edwards, 1996).

The large size of the giant wood moth is made possible by its long larval development, the majority of which is spent inside the trunks or branches of *Eucalyptus* trees. Larvae go through two main development phases which can cumulatively last for 2 years with some suspected to last up to 3 years (Dodd, 1916). Similar to ghost moths (Hepialidae) like New Zealand's puriri moth (*Aenetus virescens*) (Grehan, 1983), the larvae of the giant wood moth feed on the callus tissue formed from undifferentiated, rapidly dividing cells that grow from the cambium layer under the bark of trees. This nutritious food resource paired with their protective tunnel allows the larvae to survive for long periods and grow to large sizes (15 cm long and 3 cm wide).

The tunnels formed by larvae also create habitat in the form of tree hollows for other animals, and the larvae serve as a food resource for other animals, including yellow-tailed black cockatoos (*Zanda funerea*). It is culturally important,

Corresponding author: jessa.thurman@csiro.au

DOI: 10.1079/9781800627680.0026

like many other species of *Endoxyla*, to Indigenous Australians who harvested the larvae, commonly referred to as witchetty grubs, from their host trees (Tepper, 1890; Dodd, 1916; Thurman, 2022). At the same time, the moth's wood-boring larval habits have made it a significant pest in *Eucalyptus* plantations, where it can significantly reduce yields.

The giant wood moth can be distinguished from other *Endoxyla* species based on its large size, with females reaching wingspans of 25cm (Edwards, 1996). There are four other species of large *Endoxyla* which also feed in eucalypt branches on the east coast of Australia which can be confused with the more common *E. cinereus*. The giant wood moth can be distinguished from these other large *Endoxyla* species by the dark-coloured inverted heart-shaped thoracic blotch and overall grey coloration with lightly marked wings (Fig. 26.1). Unlike *E. affinis* and *E. magnifica*, which have plain thoraces, or *E. macleayi*, which has well-marked wings and broader hindwings, *E. cinereus* has subtler wing markings and lacks the distinctive antennae colours of the other species.

The giant wood moth was originally described as *Cossus cinereus* by John Gottlieb Otto Tepper in 1890 from a specimen collected in Adelaide, South Australia, and a few years later as *Xyleutes boisduvali* by Rothschild (1896), based on material collected from the Burdekin River region in Queensland. Currently, it is recognized as *Endoxyla cinereus* and has a broad distribution from tropical north Queensland, along the east coast of Australia and down to South Australia. This overlaps with the range of its key host trees, including the flooded gum (*Eucalyptus grandis*), forest red gum (*E. tereticornis*), Sydney blue gum (*E. saligna*) and river red gum (*E. camaldulensis*). These host trees are commonly found in riparian, sclerophyll and wet sclerophyll forests; however, they are also seen within parklands and urban environments where the moth has also been documented. Given this extensive distribution, future taxonomic work should assess whether populations of the giant wood moth across this range represent a single species or multiple 'cryptic' forms. Such efforts may lead to the identification of additional synonyms or the reinstatement of previously synonymized names, though this will require careful evaluation of historical records, as early taxonomic work on *Endoxyla* is notably inconsistent.

Fig. 26.1. A recently emerged giant wood moth (*Endoxyla cinereus*) dries its wings on the trunk of a flooded gum (*Eucalyptus grandis*) in St Lucia, Queensland. Its pupal case (not pictured) is lodged in the emergence hole which exposes the chamber where the larva fed for nearly 2 years. Queensland. Photos by J.H. Thurman.

26.2 Life History

26.2.1 Eggs and larval dispersal

Oviposition by the giant wood moth has not been observed in the wild, but in controlled conditions, females extend their ovipositor up to 5 cm and lay their eggs in large to very large multilayered clusters (1000–10,000 eggs). One female (Fig. 26.2A) was observed by the author to lay over 100,000 eggs in clusters (Fig. 26.2B) over the course of 3 days. Each egg is *ca* 1 mm long and opaque yellow, with a thin membrane suggesting that they are susceptible to desiccation. It is most likely that the female uses her long ovipositor to lay the clusters of eggs under bark in the wild. Previous studies have also suggested that the eggs are covered in a glutinous

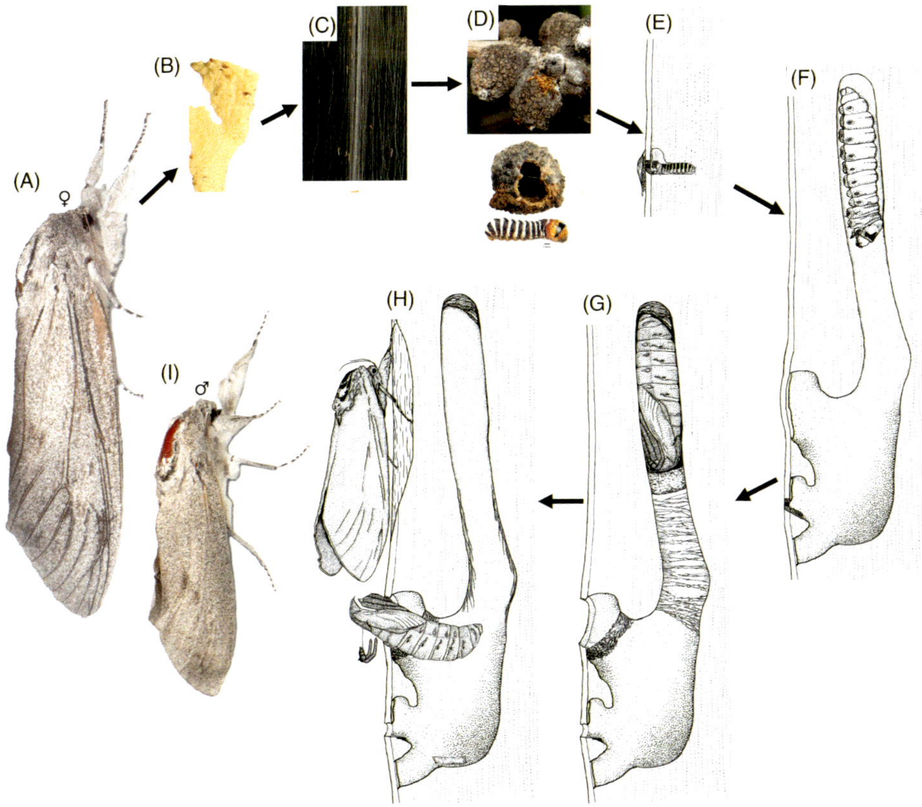

Fig. 26.2. The full lifecycle of the giant wood moth (*Endoxyla cinereus*). (A) Adult females lay (B) over 100,000 eggs. After hatching and sheltering, (C) larvae silk disperse. (D) Larvae enter the opening of scale insect galls (*Apiomorpha* spp.) and feed on the gall tissue and scale insect. (E) A larva moves to the trunk or branch of a *Eucalyptus* tree to excavate a tunnel. (F) The larva feeds on the callus tissue under the bark and further excavates its tunnel. (G) The larva prepares to pupate by opening its tunnel with an exit hole, clearing the lower initial entrance hole, lining its tunnel with mucus-covered silk, and then plugging itself in the top of the tunnel to pupate. (H) The pupa pushes through these barriers and emerges from the pupal case through the exit hole. (I) An adult male was reared from egg to adulthood over a period of roughly 2 years from the female pictured in (A) by J.H. Thurman. Photos by J.H. Thurman, illustrations by both J.H. Thurman and Geoff Thompson.

secretion (Fearn and Leong, 2022), but this has not been observed by the author with captive moths.

The dark purple and white stripes of first instar larvae become visible within the egg 10 days after being laid, with hatching occurring 12 days after laying. Upon hatching, larvae consume their egg casing and then shelter underneath their egg cluster, layering tunnels of silk over the egg mass. This sheltering and clustering of hatched larvae can continue for 5 days until a mass dispersal event occurs (Fig. 26.2C). This mass dispersal has been observed numerous times (Dodd, 1916; Monteith, 1991) and is likely to be triggered by environmental factors, although which such factors are not clear. During dispersal, larvae continue to lay silk and as thousands of larvae disperse, sheets of silk are formed over surfaces. Larvae will climb adjacent vegetation before silk dropping to disperse by the wind. It is thought that the majority of larvae die during this period.

26.2.2 Larval feeding in galls – larval phase 1

For over a century, the later larval stages of the giant wood moth have been studied in trees, but it has also been known that there is a missing larval phase between their dispersal and establishment in the tree trunk or branch (Illidge and Quail, 1903; Dodd, 1916). Recently, the author reared larvae through this 'mystery phase', showing that they spend their time feeding within the galls of scale insects, *Apiomorpha* spp. (Hemiptera: Eriococcidae) (Fig. 26.2D). These galls can be formed on the same host trees (e.g. *Eucalyptus tereticornis*) within which moth larvae later establish in the trunks or branches. Gall feeding is a transitionary phase for the larvae, similar to that seen in ghost moths like *Aenetus* spp. which feed in leaf litter before establishing in trees (Grehan, 1983).

After wind dispersing, early instar larvae must find an active gall of a female scale insect (*Apiomorpha* spp.). The early instar larvae (Fig. 26.2C) are small enough to enter the apical orifice – the natural opening of the gall. Larvae have been observed feeding on gall tissue and the scale insect itself. Each larva may feed in multiple scale insect galls before being large enough to establish in the tree trunk or branch. This feeding period is expected to vary from just over 2 months to 8 months. The 2-month feeding period is based on observations by the author of captive reared larvae, whereas the second, longer estimate is based on observations of when adults are present for egg laying, and when larvae are observed establishing in host trees.

During this gall-dependent phase, larvae grow from roughly 1.5 mm long to 25 mm long, maintaining their purple and white stripes with a sclerotized head capsule and prothoracic segment that is orange and black (Fig. 26.2D). Together, these colours may exist as temporary pseudo-aposematic coloration, although the larvae are not toxic and lack defences. This signal may aid larvae exposed when they move from the gall into the trunk or branch of their host tree for the next phase of their larval development.

26.2.3 Larval establishment in the host tree – larval phase 2

This larval phase is better understood than what I have called phase 1, and is the longest period of development for the moth. When about 25 mm long, individual larvae will migrate to the trunk or branch of a eucalypt, likely the same species which was host to the galls that the earlier stages utilized. This migration occurs during the day and establishment has been observed to occur about mid-day. Larvae have been observed moving their head back and forth while climbing up tree trunks, likely sensing which establishment location may be most suitable. The time of day may be key for the questing larvae to locate areas of the trunk or branch most exposed to sunlight for temperature regulation or its influence on the regrowth of callus tissue upon which the larva relies.

Once a suitable location is found to establish itself, the larva begins to spin a silken hammock over its body, attaching it to the bark. This hammock is laid in a circular pattern with criss-crossing strands over the back of the larvae. Once the hammock is formed, the larva is in a C-shape, and removes strips of bark which are incorporated into the silk. It rotates, removing strips in a circular pattern until the larva itself is no longer visible. It continues this excavation until a groove with two narrow sides under the bark is formed. Overnight this tunnel is completed and sealed off with the entry blocked by compacted silk. Within 2 days, a hole is cut from the centre of this compacted silk through which strips of bark and wood, alongside pellets of frass, are discarded. The silken hammock resembles more of a pad once all the strips of bark are incorporated, and it eventually falls away from the tree (Fig. 26.2E).

Younger phase 2 larval stages can be difficult to detect once the larval pad has fallen away, but older stages (Fig. 26.2F) are made more obvious by the discarded frass and strips of bark which accumulate near the base of the tree, as well as by the swollen appearance around the tunnel's entrance. In this tunnel, the larva is consuming the callus tissue which regenerates under the bark. This tunnel is slowly expanded with a larger feeding area under the bark and a tunnel that penetrates back into the heart wood, forming a 'J' shape (Fig. 26.2F). After a year, this tunnel can be up to 8 cm into the tree and 15 cm long. This penetrating tunnel into the heartwood of the tree is later used as a pupation chamber. By this stage, larvae have also lost their purple stripes and have a cream-coloured body. Their sclerotized first thoracic segment maintains the same pattern and is orange-yellow with black markings.

26.2.4 Pupation and adult emergence

As larvae approach pupation, they modify their tunnel structure by first clearing their lower entrance hole (originally sealed with compacted silk) and creating a larger exit hole for adult emergence (Fig. 26.2G). This pupation chamber is then blocked by 5 cm of mucus-covered silk and a plug composed of silk and strips of bark. Once this future pupation chamber is formed, the prepupal larva is suspected to take a few days before pupating. This suspicion is based on observations of its parasitoid, *Virgulibracon endoxylaphagus*, which attacks prepupal larvae for several days after this pupation chamber has been formed (Fig. 26.3B). In due course pupation occurs inside this chamber and lasts approximately 56 days (ranging from 31 to 83 days).

Adults emerge primarily from late spring to early summer (November to January), and emergence days are likely influenced by environmental triggers such as rainfall. Adults also emerge during the day, usually around 3.30pm, inflating their wings and drying them near the exit tunnel. Adult emergence is most obviously detected by the presence of the pupal case which remains lodged in the exit tunnel (Fig. 26.2H). Their grey coloration allows the moths to blend in with the grey tree bark. Overall development time does not appear to vary between the sexes, even though females (Fig. 26.2A) are much larger than males (Fig. 26.2I). Adults only live for a few days, with females surviving for 5 days and males over a week. Neither sex has mouthparts in their adult form, relying on stored fats accumulated during their extended larval period to find a mate and reproduce.

26.3 Ecology

26.3.1 Host trees and forestry impacts

The giant wood moth exhibits a close ecological relationship with its host trees (primarily *Eucalyptus* spp.). It feeds on and completes most of its larval and its pupal stages inside the tree, and the tree it selects likely influences the moth's overall survival. Smaller trees, for instance, prevent the larva from having deep tunnels, making it more susceptible to parasitism and predation. The larva also needs the cambium to regenerate and provide it with nutrition as it slowly expands its feeding chamber under the bark. Studies from forestry plantations indicate that the giant wood moth prefers smooth-barked eucalypts, such as *Eucalyptus grandis*, *E. saligna* and *E. tereticornis* (Lawson, 2003; Listyanto *et al.*, 2010). Some forestry studies also report a higher incidence of *E. cinereus* in fertilized and irrigated trees (Wylie and Peters, 1993; Lawson *et al.*, 2002) whereas others indicate a preference for stressed trees (Carne *et al.*, 1974; Listyanto *et al.*, 2010), highlighting potential geographic and environmental influences on host tree interactions.

The larval stage of the giant wood moth can have a significant impact on tree health, primarily through the excavation of large tunnels within the heartwood over multiple years (McInnes and Carne, 1978; Griffiths *et al.*, 2004; King and Lawson, 2005). These tunnels weaken the structural integrity of trees, increasing the likelihood of wind damage, breakage and mortality. In addition, secondary damage is inflicted by yellow-tailed black cockatoos (*Zanda funerea*) (Fig. 26.3A), which tear away bark and wood to extract larvae, further exposing the tree's core and exacerbating decay (Carne *et al.*, 1974; Dickinson *et al.*, 2003). In forestry plantations, particularly those of *Eucalyptus grandis*, this combination of larval and cockatoo foraging damage has resulted in significant economic losses (McInnes and Carne, 1978), and has led to alternative species of *Eucalyptus* being grown to avoid moth damage.

In some cases, tree mortality induced by larvae may help maintain open-canopy conditions favourable for the regeneration of host tree species, while also creating habitat features such as tree hollows (see below).

26.3.2 Predation and parasitism

The late-stage larvae of the giant wood moth serve as a critical food resource for various

Fig. 26.3. (A) Yellow-tailed black cockatoo (*Zanda funerea*) excavating the larval tunnel of the giant wood moth (*Endoxyla cinereus*) in Mt Archer, Queensland. Photo by Peter Odekerken. (B) The parasitoid wasp (*Virgulibracon endoxylaphagus*) parasitizes the prepupa of the giant wood moth (*Endoxyla cinereus*), using its long ovipositor to access the pupation chamber. Photo by J.H. Thurman.

predators and hosts for parasitoids. To minimize exposure to these natural enemies, the larvae construct retreating tunnels within the tree, providing them with relative protection (Fig. 26.2F,G). Despite this defence, certain predators and parasitoids are still capable of accessing the larvae. Yellow-tailed black cockatoos (*Zanda funerea*), for example, use their strong beaks to excavate deep tunnels and retrieve the larvae (Fig. 26.3A). Larvae of the giant wood moth are generally at their largest from August to September, a season which overlaps with the yellow-tailed black cockatoo's nesting season (Dickinson *et al.* 2003). Excavations by the cockatoos are frequently observed during this time, with groups excavating from trees within the same general area. This overlap in phenology suggests that these large larvae are likely an important food source supporting the bird's reproduction.

The giant wood moth is also targeted by a specialized parasitoid wasp, *Virgulibracon endoxylaphagus* (Hymenoptera: Braconidae) described by Quicke and Ingram in 1993. This wasp has one of the longest recorded ovipositors (5 cm) relative to its body size (Fig. 26.3B). This parasitoid uses its long ovipositor to 'drill' through the wood and parasitize the prepupal stage of the giant wood moth within its pupation chamber (Monteith, 1991; Quicke and Ingram, 1993). Although much of the life history of this parasitoid remains poorly understood, it is believed to be a specialist, parasitizing only the giant wood moth. Observations by the author have suggested that *V. endoxylaphagus* has a single generation each year, overwintering as larvae and then emerging as adult wasps from August to November. Each parasitized larva of the giant wood moth can support over 100 wasp larvae, which are likely from multiple female wasps. Gatherings of female wasps can be observed outside one tunnel of a giant wood moth, with each wasp taking its turn to use the same drill site and insert its ovipositor. The depth of the pupation chamber has likely been the largest selection factor acting on *V. endoxylaphagus*, explaining why this species has such a long ovipositor.

Other specialized predators and parasitoids of the giant wood moth may be further identified through continued ecological research, especially since a significant proportion of the moth's offspring do not survive to adulthood (Dodd, 1916).

26.3.3 Tree hollow formation

The giant wood moth functions as an ecosystem engineer by creating tree hollows and potentially influencing forest succession dynamics and the maintenance of biodiversity. Tree hollows are a key habitat feature used by many Australian animals including birds, bats, possums and frogs (Goldingay, 2009; Newton *et al.*, 2022) as well as a variety of invertebrates. As larvae, the giant wood moth bore into the heartwood of eucalypts, with a larger feeding chamber located under the bark. From larval pupation, these chambers are exposed with the formation of their exit holes, and frogs, including the bleating tree frog (*Litoria dentata*), have been documented using this space (Mo, 2015). Other invertebrates also use these hollows including the red triangle slug (*Triboniophorus graeffei*), and several species of termites, ants and cockroaches.

Perhaps of more importance, the tunnels formed by the giant wood moth can accelerate the formation of larger tree hollows used by birds and other vertebrates, as they create a large initial chamber which is further expanded by termites, fungal decay or other natural processes. Records of this transition are rare as they require decades of observation, but rainbow lorikeets (*Trichoglossus moluccanus*) and scaly-breasted lorikeets (*T. chlorolepidotus*) have been observed to use these wood moth tunnels which were further modified by termites and fungal decay (G.B. Monteith, pers. commun., 2020). Previous studies on the formation of tree hollows have focused on the role of termites and fire (Woolley *et al.*, 2018; Rochelmeyer *et al.*, 2025), but future research should also consider the role of the giant wood moth and other insect borers, such as ghost moths (Hepialidae) and longhorn beetles (Cerambycidae). Just as woodpeckers initiate tree hollows in the Americas (Cockle *et al.*, 2012), these insects may play a comparable role in Australian forests. Given its abundance and the size of its tunnels, the giant wood moth in particular likely modifies habitat structure through tree hollow formation in ways that influence broader forest dynamics.

26.4 Cultural Significance

The giant wood moth is culturally important, like many other species of *Endoxyla*, to Indigenous Australians throughout its distribution (Tepper, 1890; Dodd, 1916; Thurman, 2022). The larvae of large species of *Endoxyla*, including *E. cinereus*, are nutritionally rich (Miller *et al.*, 1993) and an integral part of traditional life, especially in arid habitats (Tindale, 1953; Turpin and Si, 2017).

Wood moth larvae are commonly called ‘witchetty grubs’ with various spellings, but this term has been contemporarily applied to any insect larvae used by Indigenous Australians. The popular common name originates from the Arabana people in South Australia, who harvest ‘mako witjuti’ or the larva of a wood moth (*Endoxyla* sp.) from ‘witjuti’ or the wattle shrub (*Acacia kempeana*) (Tindale, 1953). This naming convention reflects the traditional knowledge of Indigenous Australians and their ecological understanding of this species. Future collaboration could highlight the ecologies of other species of large wood moths which are currently poorly documented.

26.5 Future Research

Several potentially productive areas for future research have been flagged in this chapter and others emerge on reflection.

As indicated, it is possible that *Endoxyla cinereus* is a complex of recently evolved ‘cryptic’ species. Current methods of molecular analysis of museum specimens could resolve this hypothesis, and several specimens have been sequenced for CO1 already. Whether such entities are structured geographically or by the species of their host trees (or both) poses further consequential questions.

The roles of *E. cinereus* as a pest of forestry plantations and as an ecosystem engineer creating tree hollows and, thereby, opportunities for other elements of the forest ecosystem are, from a human standpoint, contradictory. Critical comparisons using standard survey methods in ecosystems that have different management histories might quantify the roles of the giant wood moth and its ‘engineering’ activities. Fundamentally, do the conservation values of the moth outweigh the costs of its pestiferous behaviour and does this balance change across the forest landscape?

The prominence of witchetty grubs in Indigenous Australian culture and popular versions of that culture warrant a multidisciplinary study to sort out exactly which moth species are involved and where. This may be aided by genetic work as moths of cultural relevance are usually harvested as larvae and difficult to identify to species. The extent and interpretation of Indigenous natural history knowledge of the species would also be both interesting and revealing.

Lastly, the behaviour and sociobiology of the wood moths' parasitoid are intriguing. That several females assembled to parasitize the same moth larvae and used the same oviposition borehole raises questions of their relatedness, social hierarchy, the extent of intraspecific competition among parasitoid larvae, and many other questions that careful study and molecular analysis might elucidate.

References

Beccaloni, G. (2010) Giant wood moth. In: *Big Bugs: Life-Size*. Natural History Museum, London.

Carne, P.B., Greaves, R.T.G. and McInnes, R.S. (1974) Insect damage to plantation-grown eucalypts in north coastal New South Wales, with particular reference to Christmas beetles (Coleoptera: Scarabaeidae). *Australian Journal of Entomology* 13, 189–206.

Cockle, K.L., Martin, K. and Robledo, G. (2012) Linking fungi, trees, and hole-using birds in a Neotropical tree-cavity network: Pathways of cavity production and implications for conservation. *Forest Ecology and Management* 264, 210–219.

Dickinson, G.R., Huth, J.R. and Lawson, S.A. (2003) Foraging behaviour of the yellow-tailed black cockatoo for larvae of the giant wood moth in hardwood plantations in south-east Queensland. *Sunbird* 33, 99–107.

Dodd, F.P. (1916) IV. Notes on the great Australian Cossidae. In: Oberthür, C., Houlbert, C. and Dodd, F.P. (eds) *Contributions à l'Etude des Grands Lépidoptères d'Australie*. Imprimerie Oberthür, Rennes, France, pp. 33–37.

Edwards, E.D. (1996) Cossidae. In: Nielsen, E.S., Edwards, E.D. and Rangsi, T.V. (eds) *Checklist of the Lepidoptera of Australia, Monographs on Australian Lepidoptera*, 4. CSIRO Publishing, Collingwood, Victoria, pp. 119–121.

Fearn, S. and Leong, D. (2022) Ecological observations on the giant wood moth *Endoxyla cinereus* (Tepper, 1890) (Lepidoptera: Cossidae: Zeuzerinae) in southeast Queensland. *Records of the Queen Victoria Museum and Art Gallery* 121, 1–20.

Goldingay, R.L. (2009) Characteristics of tree hollows used by Australian birds and bats. *Wildlife Research* 36, 394–409.

Grehan, J.R. (1983) Larval establishment behaviour of the borer *Aenetus virescens* (Lepidoptera: Hepialidae) in live trees. *New Zealand Entomologist* 7, 413–417.

Griffiths, M., Wylie, R., Lawson, S., Pegg, G. and McDonald, J. (2004) Known or potential threats from pests and diseases to prospective tree species for high value timber plantings in northern Australia. Prospects for High-value Hardwood Timber Plantations in the 'Dry' Tropics of Northern Australia. Proceedings of a Workshop, Mareeba, October, pp. 1–15.

Illidge, R. and Quail, A. (1903) Australian woodboring Cossidae–*Endoxyla macleayi* Scott; *E. boisduvalii* Roths; *Culama expressa* Lucas; with incidental reference to other species. *Proceedings of the Royal Society of Queensland* 17, 161–174.

King, J. and Lawson, S. (2005) Insect pests and diseases of rainforest timber species grown in plantations. In: Erskine, P.D., Lamb, D. and Bristow, M. (eds) *Reforestation in the Tropics and Subtropics of Australia using Rainforest Tree Species*. Rural Industries Research and Development Corporation, Canberra, ACT, pp. 114–128.

Lawson, S.A. (2003) *Susceptibility of Eucalypt Species to Attack by the Giant Wood Moth* (Endoxyla cinerea) *in Queensland*. Hardwoods Queensland Report, 7. Department of Primary Industries, Brisbane, Queensland.

Lawson, S.A., Wylie, F.R., Wylie, R.L. and Ryan, P. (2002) *Longicorn beetles (Phoracantha spp.) and giant wood moths* (Endoxyla *spp.*)*: Emerging threats in tropical and subtropical eucalypt plantations of Queensland, Australia*. Forestry Research Support Programme for Asia and the Pacific, Bangkok, Thailand.

Listyanto, T., Glencross, K., Nicholls, J. D., Schoer, L. and Harwood, C. (2010) Performance of eight eucalypt species and interspecific hybrid combinations at three sites in northern New South Wales, Australia. *Australian Forestry* 73, 47–52.

McInnes, R.S. and Carne, P.B. (1978) Predation of cossid moth larvae by yellow-tailed black cockatoos causing losses in plantations of *Eucalyptus grandis* in north coastal New South Wales. *Australian Wildlife Research* 5, 101–121.

Miller, J.B., James, K.W. and Maggiore, P.M.A. (1993) *Tables of Composition of Australian Aboriginal Foods*. Aboriginal Studies Press, Canberra, ACT.

Mo, M. (2015) Frogs using cossid moth galleries: An opportunity in flooded gum plantations in the New South Wales North Coast bioregion. *Australian Zoologist* 37, 501–507.

Monteith, G.B. (1991) The life and times of the giant wood moth. *Wildlife Australia* 28, 8–10.

Newton, J.P., Bateman, P.W., Heydenrych, M.J., Mousavi-Derazmahallah, M. and Neville, P. (2022) Home is where the hollow is: Revealing vertebrate tree hollow user biodiversity with eDNA metabarcoding. *Environmental DNA* 4, 1078–1091.

Quicke, D.L.J. and Ingram, S.N. (1993) Braconine wasps of Australia. *Memoirs of the Queensland Museum* 33, 299–336.

Rochelmeyer, E., Morphy, B.P., Levick, S.R. Stobo-Wilson, A. and. Richards, A.E. (2025) The influence of fire and termites on tree hollow development in an Australian tropical savanna. *Forest Ecology and Management* 577, 122403.

Rothschild, W. (1896) Descriptions of some new species of Lepidoptera, with remarks on some previously described forms. *Novitates Zoologicae* 3, 600–603.

Tepper, J.G.O. (1890) *Common Native Insects of South Australia: A Popular Guide to South Australian Entomology. Part II: Lepidoptera, or Butterflies and Moths*. E.S. Wigg and Sons, London.

Thurman, J.H. (2022) Beyond the pest: life history, ecology and ethnoentomology of the giant wood moth (*Endoxyla cinereus*). *Austral Ecology* 47, 733–747.

Tindale, N.B. (1953) On some Australian Cossidae including the moth of the witjuti (witchety) grub. *Transactions of the Royal Society of South Australia* 76, 56–65.

Turpin, M. and Si, A. (2017) Edible insect larvae in Kaytetye: Their nomenclature and significance. *Journal of Ethnobiology* 37, 120–140.

Woolley, L.-A., Murphy, B.P., Radford, I.J., Westaway, J. and Woinarski, J.C.Z. (2018) Cyclones, fire and termites: The drivers of tree hollow abundance in northern Australia's mesic tropical savanna. *Forest Ecology and Management*, 419–420, 146–159.

Wylie, F.R. and Peters, B.C. (1993) Insect pest problems of eucalypt plantations in Australia. *Australian Forestry* 56, 358–362.

27 Tortricid Pests: Codling Moth, Light-brown Apple Moth and Their Relatives

Mike Elias
Australian National Insect Collection, CSIRO, Black Mountain, ACT, Australia

Abstract

Tortricid moths are some of the most economically significant horticultural pests. *Cydia pomonella* (the codling moth) and *Epiphyas postvittana* (the light-brown apple moth) are two of the most damaging of the tortricid species present in Australia. Both primarily affect apples but also attack other fruits such as pears, walnuts and citrus. Despite targeting similar crops, their origins and behaviours differ. Codling moth, native to Europe, has spread almost globally, establishing itself in most major apple-growing regions. Western Australia is one apple-growing region in which codling moth is notably absent. In contrast, light-brown apple moth is native to Australia and has become invasive in several other areas, most notably New Zealand and California.

Codling moth larvae are endoparasites of plants and burrow into fruit, causing significant damage and making them unmarketable. The species has multiple generations per year in warm climates, contributing to persistent infestations. It is arguably the most destructive pest on apples in the world. Although it does opportunistically feed on other economically significant fruits and nuts, it has a strong preference for apples. In contrast, light-brown apple moth is highly polyphagous, feeding on over 500 plant species. Its larvae cause damage by eating leaves and buds of their host, reducing yield and occasionally attacking the surface of the fruit, making it unmarketable. It will do significant damage in apple orchards if present in large numbers and while it is not nearly as significant a pest on apples as codling moth, it is a significant pest on a much broader range of crops. The spread of these pests has led to strict quarantine measures to prevent further infestations in unaffected regions.

Environmental factors play a crucial role in the development and activity of both species. Warmer temperatures accelerate their life cycles, allowing them to produce multiple generations per season. Codling moth development is closely linked to temperature thresholds, while light-brown apple moth is influenced by seasonal availability of host plants. Their adaptability to different climates makes them difficult to control, requiring region-specific management strategies.

27.1 Introduction

Tortricidae are a very large family of moths containing over 9000 described species (Brown, 2005). They are mostly found in temperate regions or in the alpine regions of the tropics (Meijerman and Ulenberg, 2000). Many species of tortricid moths have affinities with apple trees as larval hosts (Chapman, 1973). Indeed, the family includes more apple pests than any other family (Chapman, 1973), some of which are among the most economically damaging of all apple pests (Meijerman and Ulenberg, 2000).

Corresponding author: michael.elias@csiro.au

DOI: 10.1079/9781800627680.0027

Overall, the Tortricidae ('leafroller moths') are one of the most economically important families of the Lepidoptera. There are several tortricid pests in Australia which cause serious damage to the Australian economy. The lucerne leafroller, *Merophyas divulsana*, is a serious pest on lucerne, lettuce and various herbs (Patrick and Dugdale, 1994). Other species are pests of cotton (*Crocidosema plebejana*) and macadamia (*Cryprophlebia ombrodelta*) (Common, 1990). Elsewhere, the omnivorous leafroller, *Platynota stultana*, is an exotic pest known to cause economic damage on kiwifruit, capsicum, citrus, cotton, apple, plum, peach, pears, blackberry, raspberry and grapes. This chapter, however, will focus on the two major apple pests *Cydia pomonella* (the codling moth) and *Epiphyas postvittana* (the light-brown apple moth – 'LBAM'). Both are major pests on apples but are also able to do significant damage to a range of other fruits and nuts such as other pome fruit, some stone or citrus fruits, walnuts and avocadoes. Although both pests are known to damage similar crops, they have very different origins and ecologies.

Cydia pomonella (commonly known as codling moth) was first described by Linnaeus in the 10th edition of *Systema Naturae* (1758–59) as *Phalaena pomonella*. It is one of hundreds of species in the genus *Cydia*, which contains several significant pest species but none quite as economically damaging as *Cydia pomonella*. Despite the large number of species in the genus, codling moth is rarely confused with other species, perhaps because it has a fairly distinctive ecological impact in apple orchards. The common name codling moth is probably derived from the English apple variety, codling apples, which used to be a common variety of cooking apple. Due to its economic impact, codling moth is one of the most studied species in the field of agricultural entomology.

Epiphyas postvittana is a species which has been variously placed in several different genera. It was first described as *Teras postvittana* by Walker in 1863 who created a second species, *Teras basialbana*, which turned out to be a synonym of *T. postvittana*. Since then, *E. postvittana* and its various junior synonyms have been placed in six different genera before being moved to the genus *Epiphyas* by Common (1961). The genus *Epiphyas* currently houses a few dozen species, all from Australia. Light-brown apple moth is a highly invasive economic pest.

These two tortricid species are of significant economic importance in Australia and globally. They increase the cost of fruit production (Suckling *et al.*, 2014; Kadoić Balaško *et al.*, 2020) and have significant cost to trade due to biosecurity precautions taken to prevent the spread of these pests (Wearing *et al.*, 2001; Aquino *et al.*, 2008).

27.2 Occurrence And Distribution

27.2.1 Codling moth

This species is one of the most significant pests of apple orchards worldwide, causing severe economic losses due to direct fruit damage and the cost of control measures (Fig. 27.1). Native to Europe, the moth has now spread to apple-growing regions across North and South America, Australia, New Zealand and parts of Asia. According to Froggatt (1901), it was first recorded in Australia – in Tasmania – in the 1850s, later spreading to New Zealand (in the 1860s) and the Australian mainland (in 1885 in Victoria, 1887 in New South Wales and 1889 in Queensland). One can suppose this mirrored the developing apple industry. Elsewhere, it is a pest of walnuts (Caprile and Vossen, 2011).

27.2.2 Light-brown apple moth

Unlike codling moth, the light-brown apple moth is an Australian native originally restricted to the cooler regions of south-eastern Australia, including Tasmania (Geier and Briese, 1988) (Fig. 27.2). Supremely polyphagous across a wide range of native and exotic plant families (Geier and Briese, 1988, record 122 genera of larval food plants across 56 plant families), it rapidly became established as a pest of great economic importance. In Australia, the LBAM is a major pest affecting fruit and vegetable crops, causing economic losses due to direct damage and the cost of management. LBAM larvae feed on leaves, buds and fruit surfaces, leading to reduced yield, downgraded produce quality and increased susceptibility to secondary infections, such as fungal pathogens. Damage levels vary depending on environmental conditions but in the absence of effective pest control measures, fruit damage has been recorded

Fig. 27.1. The codling moth, *Cydia pomonella*. Photo: Mike Elias.

Fig. 27.2. The light brown apple moth, *Epiphyas postvittana*. Photo: Mike Elias.

at levels between 5% and 20% in some Australian orchards (Manktelow *et al.*, 2005). The moth's presence in exported produce also raises phytosanitary concerns, leading to additional costs in order to meet quarantine standards.

27.3 Life Histories

27.3.1 Codling moth

Codling moths lay between 30 and 70 eggs per female (Caprile and Vossen, 2011). These eggs are deposited on either the surface of the fruit or on the leaves of host plants such as apples and pears. Codling moth eggs are small, measuring approximately 1 mm in length, disc-shaped and opaque white when first laid (Williams, 2000; Caprile and Vossen, 2011). As they approach hatching, the eggs darken and become translucent, revealing the developing larva inside. The incubation period for codling moth eggs varies depending on environmental conditions, particularly temperature. In optimal conditions, eggs hatch within 6–14 days. Warmer temperatures accelerate development, while cooler conditions can extend the incubation period beyond 2 weeks (Neilsen and Dunley, 2002).

Upon hatching, codling moth larvae are less than 3 mm in length and initially cream-coloured with a dark brown head (Wiman and Stoven, 2022). Newly emerged larvae begin searching for an entry point into the host fruit. They bore into the maturing fruit, often entering near the calyx or stem. As they feed, they create tunnels through the fruit's flesh and into the seed cavity, leaving behind a distinctive trail of frass near the oviposition site. The larval stage consists of five instars, all of which are dedicated fruit borers. These larvae primarily consume the seeds and inner flesh of the fruit, preferring to remain deep within their feeding site for protection. As they progress through instars, larvae grow to approximately 12–20 mm in length before exiting the fruit to pupate (Caprile and Vossen, 2011). The duration of the larval stage is typically 3–4 weeks but can vary depending on temperature and host availability (Witzgall *et al.*, 2008).

Once mature, codling moth larvae exit the fruit and seek out a protected location to pupate. This may include bark crevices, leaf litter or soil near the base of the host tree (Caprile and Vossen, 2011; Wiman and Stoven, 2022). The pupation process takes place within a silken cocoon, where the larva undergoes metamorphosis into an adult moth. The pupal stage typically lasts between 10 days to 3 weeks, but environmental conditions play a significant role in its duration. Under cool conditions, pupation may be extended, while warm weather accelerates development (Williams, 2000). Some codling moths overwinter in the pupal stage, remaining dormant in their cocoons until favourable temperatures return in spring (Neilsen and Dunley, 2002; Wiman and Stoven, 2022).

Adult codling moths are small, measuring 12–15 mm in length, with a wingspan of approximately 20 mm. They exhibit a greyish overall coloration, composed of fine bands of various shades of grey and brown scales. The wing tips may contain patches of differently coloured scales, and studies have shown that the pigmentation of these scales can be influenced by the larval diet (Tadic, 1957). Adult moths emerge from their pupal cases and become active primarily during dusk and nighttime hours. Mating occurs shortly after emergence, with females releasing pheromones to attract males. Egg laying typically begins within a few days of mating, with females depositing eggs individually on suitable host surfaces (Caprile and Vossen, 2011).

Codling moths can complete multiple generations per year, with 2–3 generations being typical in temperate regions (Witzgall *et al.*, 2008; Brunner, 2018a, b). In warmer climates, as many as four generations may be observed. The life cycle duration is highly temperature dependent, with the entire process taking approximately 6–8 weeks in optimal conditions.

27.3.2 Light brown apple moth

Female LBAMs lay egg masses on the upper surfaces of smooth-leaved host plants. Females begin laying eggs approximately 6–10 days after emerging from the pupal stage. Each female can lay between 300 and 1500 eggs throughout her lifetime, depending on environmental conditions (Lewis and Hodges, 2010). These eggs are pale blue-green and scale-like, with a slightly pebbled surface. Each egg mass contains between 10 and 170 eggs, although typical clusters contain 10–60 eggs. The eggs measure approximately 0.8–0.95 mm in length (Lewis and Hodges, 2010). The incubation period is highly dependent on temperature and can range from 5 to 30 days. Warmer conditions accelerate development, while cooler temperatures prolong the hatching process. Eggs laid in the summer may hatch in as little as 5 days, whereas those laid in autumn may take close to a month to develop.

Upon hatching, larvae emerge as small, pale yellow-green caterpillars with light brown heads. Early-instar larvae typically feed on the underside of leaves and create a fine webbing for protection. Some larvae migrate across host plants and may suspend themselves from silk threads to reach new feeding sites (Lewis and Hodges, 2010). LBAM larvae progress through six developmental instars, during which they grow from an initial length of approximately 1 mm to a mature size of 10–18 mm. As they develop, their coloration deepens to a medium green with a darker green central stripe and lighter side stripes (Danthanarayana, 1975). The larval stage usually lasts 6–8 weeks but under ideal conditions, it can be as short as 3 weeks. Larvae are highly responsive to disturbances, wriggling vigorously and retreating into shelters, or dropping to the ground to avoid predation (Espinosa and Hodges, 2009). Older larvae construct protected nests by rolling leaves or spinning silk between adjacent leaves, flowers, fruits or berries. These silk shelters provide protection while they feed (Espinosa and Hodges, 2009). In some cases, larvae may enter fruit bunches, causing significant agricultural damage. Overwintering typically occurs at the second to fourth larval instar, with the larvae sheltering in plant debris or within host plants. Once warmer weather resumes, they resume feeding and development (Lewis and Hodges, 2010).

When larvae reach maturity, they enter the pupal stage, which takes place inside their feeding shelters. Initially, pupae are green but they quickly harden into brown chrysalises. Pupae are typically 10–15 mm long, with female pupae being slightly larger than males. This stage lasts between 1 and 4 weeks, depending on temperature. Higher temperatures shorten the pupal stage, while cooler conditions prolong it (Danthanarayana, 1975). Pupation usually occurs within rolled leaves or in silken cocoons spun between two leaves. In some cases, pupation may take place within fruit clusters or beneath plant debris. Once pupation is complete, the adult moth emerges (Lewis and Hodges, 2010).

Adult LBAMs are small moths, approximately 10 mm in length when at rest with folded wings. They are sexually dimorphic, with males having distinct brown markings on their forewings while females appear more uniformly light brown. Females can be identified by a characteristic dark mark on the hind margin of each forewing (Danthanarayana, 1975). LBAM adults are primarily nocturnal, with peak flight activity occurring for a few hours after sunset and before sunrise. Males are attracted to females by sex pheromones released shortly after emergence. While most adults do not travel far, some individuals have been observed flying up to 600 metres in search of mates (Espinosa and Hodges, 2009). The adult moths have relatively short lifespans, with most surviving only 2–3 weeks.

The duration of the LBAM life cycle is significantly influenced by temperature and seasonality. In warmer regions, the complete cycle from egg to adult may take only 6 weeks, allowing for multiple generations per year. In cooler climates, development slows considerably, resulting in only two generations annually (Espinosa and Hodges, 2009).

27.4 Phenology

27.4.1 Codling moth

Temperature is the primary factor governing codling moth development. The rate of development at each life stage is dictated by accumulated heat units, measured in degree-days (DD), with a lower developmental threshold of approximately 10°C (Barnes, 1991). Codling moths develop faster under higher temperatures, with increased voltinism in warmer regions (Wearing *et al.*, 2001). Degree-day models are widely used to predict emergence and optimize control measures, such as pesticide application and mating disruption (Riedl *et al.*, 1976).

Microclimatic conditions within orchards also play a role in modulating the phenology of codling moths. Orchard structure, such as tree canopy density, influences temperature regulation and consequently development rates (Kührt *et al.*, 2006). In shaded areas, codling moth development can be delayed, affecting the synchrony of adult emergence with host fruit availability (Knight, 2014).

Photoperiod is another key environmental cue that influences diapause induction. As daylight hours shorten in late summer, codling moths enter diapause to overwinter as larvae within protective cocoons (Riedl and Croft, 1978). The interaction between temperature and photoperiod determines the proportion of individuals entering diapause versus those continuing development into additional generations. Warmer autumns, associated with climate change, may disrupt this balance, leading to prolonged activity and increased overwintering survival (Witzgall *et al.*, 2008).

Codling moth populations exhibit significant regional variation in phenology due to differences in climate, altitude and orchard management practices. In cooler temperate regions such as northern Europe and parts of North America, codling moths typically complete one or two generations per year (univoltine or bivoltine life cycle) (Barnes, 1991). In contrast, in warmer regions such as southern Europe,

California and parts of Australia, codling moths can be multivoltine, producing 3–5 generations annually (Wearing *et al.*, 2001). Altitude plays a crucial role in determining local population dynamics. Higher-elevation orchards experience lower average temperatures, which slow development rates and reduce the number of generations per season (Kührt *et al.*, 2006). Similarly, orchards located near water bodies may experience milder microclimates, influencing emergence timing and population build-up (Foster *et al.*, 1991).

Regional phenological patterns necessitate localized pest management strategies. In areas with a single annual generation, control measures may focus on a single well-timed intervention whereas in warmer climates, repeated applications of insecticides or pheromone-based mating disruption may be required throughout the season (Knight, 2014). Degree-day models must be calibrated to specific regions to provide accurate predictions of emergence and oviposition periods (Riedl *et al.*, 1976).

The impact of climate change on codling moth phenology is an area of active research. Rising temperatures are expected to increase the number of generations per year in many regions, extending the period of fruit vulnerability to infestation (Witzgall, 2008). Shifts in seasonal temperature patterns could also alter diapause induction, leading to changes in overwintering survival rates and earlier spring emergence (Riedl and Croft, 1978).

27.4.2 Light-brown apple moth

The phenology of the LBAM is shaped by a combination of environmental factors, including temperature, photoperiod, humidity and host plant availability. These factors influence developmental rates, reproductive cycles and seasonal population dynamics, leading to regional variability in the number of generations per year and the overall population structure.

Temperature is the primary driver of LBAM development, affecting the duration of each life stage and the number of generations per year. Research has established a lower development threshold of approximately 7°C, with optimal development occurring between 20°C and 25°C (Danthanarayana, 1975). Degree-day models are widely used to predict key phenological events, such as egg hatch and adult emergence, enabling more precise pest management timing.

Temperature fluctuations can impact the survival and fecundity of *E. postvittana*. At lower temperatures, developmental rates slow significantly, leading to extended generation times and prolonged larval or pupal stages (Wearing *et al.*, 1991). In contrast, excessively high temperatures can reduce egg viability and larval survival, thereby limiting population growth in warmer regions (Danthanarayana, 1983).

Day length plays a crucial role in the reproductive behaviour and diapause induction of *E. postvittana*. In cooler regions, shorter day lengths may trigger diapause in late-season generations, allowing the species to survive through unfavourable winter conditions (Wearing *et al.*, 1991). In areas with milder winters, however, diapause induction is less common, leading to more continuous population cycles.

Photoperiod can also influence mating behaviour and fecundity. Studies suggest that longer daylight hours correlate with increased reproductive activity and oviposition rates, whereas shorter photoperiods can delay mating and reduce overall egg production (Foster and Howard, 1998).

Host plant phenology significantly impacts the population dynamics of *E. postvittana*, as larvae require young, tender plant tissues for optimal development. Seasonal availability of suitable host plants determines larval survival rates and reproductive success (Brockerhoff *et al.*, 2010). In orchards and cultivated environments, populations tend to peak when fresh vegetative growth is available, aligning with the phenology of key host plants. The presence of alternative host plants in surrounding landscapes can support populations even when primary crop hosts are not available, potentially leading to sustained pest pressure throughout the year (Lo *et al.*, 2000; Suckling and Brockerhoff, 2010).

Humidity levels play an important role in egg viability and larval survival. High humidity conditions generally favour population growth by preventing egg desiccation and improving larval development rates. Conversely, arid conditions can suppress population expansion by increasing mortality rates in early life stages (Danthanarayana, 1975). Moisture availability can also influence pupation success, particularly in drier climates where pupae are more susceptible to desiccation. Microclimatic conditions within orchards, such as irrigation or dense foliage cover, can create localized environments that enhance survival rates, allowing populations to persist even in otherwise unfavourable conditions (Foster *et al.*, 1991).

In temperate regions, *E. postvittana* typically completes 2–3 generations per year, with overwintering occurring primarily in the larval or pupal stages. Adult emergence begins in early spring, coinciding with the resumption of host plant growth (Wearing *et al.*, 1991). Seasonal temperature fluctuations in these regions strongly influence developmental rates and generation timing, often resulting in peaks of adult activity during late spring and summer. Population dynamics in temperate areas are further modulated by natural enemies and environmental conditions that affect overwintering success (Brockerhoff *et al.*, 2010; Feng *et al.*, 2015).

In warmer climates, where temperatures remain favourable year round, *E. postvittana* exhibits continuous breeding with overlapping generations. The absence of a pronounced winter period prevents the need for diapause, allowing for sustained population growth (Brockerhoff *et al.*, 2010). In these regions, population regulation is primarily driven by host plant availability and environmental stressors such as excessive heat or drought. The complex population structure in subtropical and tropical areas complicates pest management, as moth activity is less predictable than in temperate environments (Markwick *et al.*, 2013).

Extreme temperatures and low humidity levels constrain *E. postvittana* populations in arid and semi-arid regions. Although natural populations may struggle under such conditions, irrigation and the presence of cultivated crops can create microclimates that facilitate moth survival and reproduction (Danthanarayana, 1975). In these environments, moth populations are typically lower but can exhibit periodic outbreaks following favourable climatic events, such as increased rainfall or extended warm periods. Adaptive behavioural responses, such as seeking shaded microhabitats or modifying reproductive timing, can help populations persist despite challenging conditions (Foster and Howard, 1998).

Degree-day accumulation models provide a valuable tool for predicting the phenology of *E. postvittana*. These models estimate developmental milestones based on accumulated thermal units, allowing for more precise pest forecasting and management (Riedl *et al.*, 1976; Wearing *et al.*, 1991). Region-specific calibrations of degree-day models improve their predictive accuracy, taking into account variations in local climate conditions and population dynamics. The integration of these models into pest management programmes enables growers to optimize control measures, reducing reliance on broad-spectrum insecticides while enhancing the effectiveness of targeted interventions.

27.5 Overall Status and Economic Impacts

27.5.1 Codling moth

The codling moth (*Cydia pomonella*) is one of the most significant pests of apple orchards worldwide, causing severe economic losses due to direct fruit damage and the cost of control measures. Native to Europe, the moth has now spread to apple-growing regions across North and South America, Australia, New Zealand and parts of Asia. It primarily infests apples but also attacks pears, quinces and walnuts. Due to its high reproductive capacity and ability to develop resistance to insecticides, the codling moth remains a persistent threat to apple production.

The codling moth is responsible for substantial economic losses in apple orchards. The larvae burrow into fruit, making them unmarketable for fresh consumption and reducing their value for processing. Damage rates vary by region and management practices but in untreated orchards, infestation levels can reach 80% or higher, leading to devastating losses (Witzgall *et al.*, 2008). Even in well-managed orchards, minor infestations can still result in significant economic burdens due to the high cost of preventive measures.

The global apple industry is valued at approximately US$90 billion. In the United States alone, apple growers spend over $100 million each year on pest management programmes targeting the codling moth (Jones *et al.*, 2013). These costs include insecticide applications, mating disruption techniques, monitoring systems and postharvest quarantine treatments to meet export requirements.

In Australia, where apples contribute significantly to the horticultural economy, the economic impact of the codling moth is equally significant. The Australian apple industry is valued at approximately AU$600 million per year, and losses due to codling moth infestation can account for up to 5% of total production (Suckling *et al.*, 2014). In response, growers invest heavily in integrated pest management (IPM) strategies to mitigate the pest's impact, with control costs reaching up to AU$250 per hectare per season (Knight, 2007).

The control of codling moth involves a combination of chemical, biological and cultural strategies, all of which contribute to high production costs. Conventional control relies on multiple insecticide applications per season, particularly organophosphates, carbamates and newer selective insecticides such as insect growth regulators and neonicotinoids. Resistance development, however, has forced growers to adopt alternative control methods, including mating disruption, sterile insect techniques and biological control agents.

Mating disruption, which uses pheromone dispensers to interfere with moth reproduction, has gained popularity as an environmentally friendly alternative. Although effective, the initial investment is high, costing between $150 and $250 per hectare (Witzgall *et al.*, 2008). Additionally, the effectiveness of this method depends on the size of the treated area and the pest population levels, making it less suitable for smaller orchards or regions with high infestation pressure.

Biological control measures, such as the introduction of parasitoid wasps like *Mastrus ridens*, have shown promise in reducing codling moth populations (Suckling *et al.*, 2014). However, these methods require extensive research and implementation costs, and their success varies depending on environmental conditions and orchard management practices.

Codling moth infestations pose significant challenges for apple exporters, as many countries impose strict quarantine restrictions on imported fruit. Apples infested with codling moth larvae are often rejected at borders, leading to trade disruptions and financial losses. For instance, the presence of codling moth in Australian apple exports has resulted in increased scrutiny from trading partners such as China and Japan, requiring growers to implement costly pre-export treatments and certification processes (Jones *et al.*, 2013).

In New Zealand, strict phytosanitary regulations have led to increased adoption of non-chemical treatments, such as controlled atmosphere storage and irradiation, to ensure compliance with export standards. While these measures help maintain market access, they add to the overall cost of production and reduce profit margins for growers.

One of the greatest challenges in codling moth management is the development of resistance to commonly used insecticides. Resistance to organophosphates was first documented in the 1980s and since then, resistance to pyrethroids, neonicotinoids and insect growth regulators has emerged in various apple-growing regions (Knight, 2007). This has necessitated the development of more sustainable control strategies, including the use of microbial insecticides like *Cydia pomonella granulovirus* (CpGV) and genetically modified apple varieties with increased pest resistance.

Despite ongoing research, the codling moth remains a persistent problem for apple growers. The high cost of pest management, combined with the risk of resistance and export restrictions, underscores the need for continued investment in innovative control methods. Sustainable approaches, such as improved orchard sanitation, enhanced biological control and advanced monitoring technologies, will be crucial in minimizing economic losses while ensuring long-term viability for the apple industry.

27.5.2 Light-brown apple moth

The light brown apple moth is a polyphagous pest native to Australia, known to feed on over 500 plant species, including commercially significant crops such as apples, citrus, grapes, pears and stone fruits (Varela *et al.*, 2010). Beyond Australia, the LBAM has established itself in regions like New Zealand, Hawaii and parts of North America, notably California. The pest's ability to infest a wide variety of plants makes it a significant concern for agricultural industries worldwide.

Efforts to manage the LBAM involve substantial investments in pest control measures, including monitoring, chemical treatments, and biological control methods. In Australia, growers implement IPM strategies that combine natural predators, pheromone traps, and selective insecticides to control LBAM populations. Maintaining effective control, however, remains challenging due to the moth's high reproductive potential and ability to develop resistance to insecticides.

New Zealand has experienced significant issues with the LBAM, particularly during the 1960s to 1980s when it caused extensive damage to fruit crops. In apple orchards, LBAM infestations resulted in crop losses ranging from 8% to 26%, with peak damage reaching 48% in some instances (Collyer and van Geldermalsen, 1975). The high damage levels led to increased pesticide applications, which in turn contributed to the development of resistance in LBAM populations. Resistance to organochlorine insecticides was first reported in New Zealand in the early 1960s, highlighting the need for alternative pest control strategies (Collyer and van Geldermalsen, 1975).

The economic burden of the LBAM in Australia is significant. Studies estimate that the pest causes AU$21.1 million in annual losses due to reduced production and pest management expenses. This equates to approximately 1.3% of the gross fruit value of major crops such as apples, pears, oranges and grapes (Varela *et al.*, 2010). These figures highlight the critical need for effective control strategies to minimize economic damage.

The economic impact of the LBAM extends beyond Australia, with the pest imposing significant costs on agricultural industries in regions where it has become established. In California, the estimated impact on fruit production costs has been projected to exceed US$100 million annually. Applying the Australian damage estimate of 1.3% to California's 2005 gross value of US$5.4 billion for apples, pears, oranges and grapes suggests a potential annual production loss of US$70.2 million (Varela *et al.*, 2010). These figures underscore the financial burden that LBAM presents to growers and the broader economy.

Since the 1980s, New Zealand growers have successfully managed the LBAM through a combination of natural control methods and selective insecticides. The widespread use of organophosphate insecticides, which were previously applied 6–9 times per season on pome fruit crops, has significantly declined. Over the last decade, the use of organophosphates has dropped by 97% and the overall frequency of insecticide applications has decreased by approximately 50% (Manktelow *et al.*, 2005). These reductions reflect the success of integrated pest management strategies and the increased reliance on biological control agents.

The overuse of broad-spectrum insecticides has historically contributed to resistance issues in LBAM populations, particularly in regions with intensive agricultural activity. In both Australia and New Zealand, the reliance on chemical controls has necessitated the adoption of more sustainable pest management approaches. IPM strategies now emphasize the use of biological control agents, such as parasitoid wasps and tachinid flies, in combination with selective insecticides and cultural practices. These approaches have proven effective in maintaining LBAM populations at manageable levels while reducing the environmental and economic costs associated with pesticide use.

27.6 Management

Codling moth (*Cydia pomonella*) and light brown apple moth (*Epiphyas postvittana*) are two significant pests affecting apple orchards and other horticultural crops worldwide. Effective management strategies are crucial to reducing economic losses and maintaining sustainable agricultural practices. An IPM approach, which incorporates multiple control techniques, is recommended for the long-term suppression of these pests (Balasko *et al.*, 2020). Codling moth is primarily an apple pest which is sometimes found on other fruit, while light brown apple moth is more of a generalist.

Effective pest management begins with comprehensive surveillance to detect and quantify infestations. Monitoring tools are essential for establishing action thresholds and determining the necessity of intervention. Pheromone traps, which attract male moths using synthetic sex pheromones, provide critical data on population densities and flight activity patterns (Gut and Brunner, 1998). This information helps growers time their management actions to maximize effectiveness and minimize unnecessary pesticide applications. Additionally, degree-day models can help predict pest emergence based on accumulated heat units, allowing for more precise treatment timing.

Cultural practices are fundamental to disrupting the life cycles of both codling moth and LBAM. Regular removal of infested fruit from trees and orchard floors eliminates breeding sites and reduces pest populations. Pruning and thinning branches improves air circulation and sunlight penetration, creating unfavourable conditions for egg laying and larval development. Additionally, bagging fruit offers a physical barrier that prevents female moths from laying eggs on the surface of developing apples, thereby reducing infestations. Cover cropping and interplanting with repellent plant species can also contribute to pest suppression by discouraging oviposition.

Biological control plays a crucial role in the management of these pests. *Trichogramma* wasps, parasitoids that attack codling moth and LBAM eggs, have been shown to significantly reduce population densities (Mills *et al.*, 2000). Other natural enemies, such as the parasitoid *Dolichogenidea tasmanica*, target larvae of the LBAM (Lo and Blank, 1992). Entomopathogenic viruses, such as the codling moth granulovirus (CpGV) and the *Epiphyas postvittana* granulovirus (EpGV), provide effective biopesticide options that are highly specific to their target pests and environmentally sustainable (Lacey *et al.*, 2008; Suckling *et al.*, 2014). Additionally, fungal pathogens like *Beauveria bassiana* have shown potential in controlling both species under controlled conditions.

Despite the increasing emphasis on sustainable management strategies, chemical pesticides remain a primary tool for controlling codling moth and light brown apple moth. Insecticides must be applied at the correct developmental stage to be effective. For codling moth, treatments are most effective when applied after egg laying but before larvae burrow into fruit, as chemical control is largely ineffective once larvae are protected inside developing fruit (Knight, 2007) This is not the case for light brown apple moth. Unlike codling moth, LBAM are primarily leaf feeders and do not seek out fruit but attack them incidentally (Geier and Briese, 1980). The incidence of LBAM attacking fruit in an orchard is directly proportional to their population density (Geier and Briese, 1980).

Resistance to commonly used insecticides has been reported in codling moth populations worldwide (Reyes *et al.*, 2009; Balasko *et al.*, 2020). To mitigate resistance, growers must rotate chemical classes and integrate non-chemical control methods into their pest management programs (IRAC, 2021). Insect growth regulators (IGRs), which interfere with moulting and development, provide an additional tool for managing populations while reducing harm to beneficial organisms.

Mating disruption is a highly targeted approach that reduces successful reproduction in pest populations. This method involves the deployment of synthetic pheromone dispensers throughout orchards, which interfere with male moths' ability to locate females. Studies have demonstrated that mating disruption can significantly reduce codling moth damage when implemented correctly (Witzgall *et al.*, 2010). Similarly, the use of pheromone-based disruption has proven effective against light brown apple moth. Combining this method with other control strategies can enhance overall efficacy and reduce reliance on chemical insecticides.

The sterile insect technique (SIT) involves the release of sterilized male moths, which compete with wild males for mates. This results in unsuccessful reproduction and a gradual decline in the pest population. The SIT has been successfully employed in various regions as part of area-wide pest suppression programmes (Bloem *et al.*, 2007). Early trials have demonstrated the effectiveness of the SIT in controlling codling moth populations, with promising results for its potential application against light brown apple moth as well (Suckling *et al.*, 2014; APAL, 2023). However, implementation of the SIT requires extensive co-ordination and logistical support to maintain effective population suppression over large areas.

Preventing the spread of these pests to new regions is a key management goal. Although codling moth is well established in many apple-growing regions, certain areas, such as

Western Australia, remain free from infestation. Maintaining area freedom requires strict quarantine measures, including inspection of imported fruit and plant materials, and the implementation of regulatory controls to prevent accidental introductions. Similarly, the light brown apple moth has been subject to rigorous containment efforts in regions where it poses a threat to agriculture and native ecosystems. Public awareness and compliance with movement restrictions play a crucial role in preventing further spread.

References

Apple and Pear Australia Limited (APAL) (2023) Sterile insect technology for control of codling moth. Available at: https://apal.org.au/sterile-insect-technology-for-control-of-codling-moth/ (accessed November 2024).

Aquino, T., Bamboo, B., Kimes, K., Nurseries, G., Aptos, C.A. *et al.* (2008) *Light Brown Apple Moth (LBAM). Implications for California Agriculture*. Phytosanitary Advisory no 12-2007. California Department of Food and Agriculture, Sacramento, CA.

Balasko, J., Smith, R. and Taylor, P. (2020) Strategies for sustainable codling moth control. *Journal of Pest Management* 35, 67–82.

Barnes, M.M. (1991) Codling moth occurrence, host race formation, and damage. In: van der Geest L.P.S. and Evenhuis H.H. (eds) *Tortricid Pests: Their Biology, Natural Enemies and Control*. Elsevier Science, Amsterdam, The Netherlands, pp. 313–327.

Bloem, S., Carpenter, J.E. and Dorn, S. (2007) Area-wide control of codling moth using sterile insect technique. *Entomologia Experimentalis et Applicata* 122, 239–246.

Brockerhoff, E.G., Barratt, B.I.P., Beggs, J.R., Fagan, L.L., Kay, N. *et al.* (2010) Impacts of exotic invertebrates on New Zealand's indigenous species and ecosystems. *New Zealand Journal of Ecology* 34, 158–174.

Brown, J.W. (2005) Tortricidae (Lepidoptera). In: *World Catalogue of Insects 5*. Apollo Books, Stenstrup, Denmark.

Brunner, J. (2018a) *Codling Moth: Biology and Management*. Washington State University, Seattle, WA.

Brunner, J.F. (2018b) Management of codling moth: Strategies for control in apple orchards. *Journal of Fruit Pest Management* 12, 87–102.

Caprile, J. and Vossen, P. (2011) *Codling Moth, Pest Notes 7412, 1–6*. University of California Extension Service, Davis, CA.

Chapman, P.J. (1973). Bionomics of the apple-feeding *Tortricidae*. *Annual Review of Entomology*, 18, 73–96.

Collyer, E. and van Geldermalsen, M. (1975) Light brown apple moth in New Zealand. *New Zealand Journal of Zoology* 2, 1–15.

Common, I.F.B. (1961) The generic position of the Australian light-brown apple moth (Lepidoptera: Tortricidae). *Proceedings of the Linnean Society of New South Wales* 86, 177–182.

Common, I.F.B. (1990). *Moths of Australia*. Melbourne University Press, Carlton, Victoria.

Danthanarayana, W. (1975) The bionomics, distribution, and host range of the light brown apple moth, *Epiphyas postvittana* (Walker) (Tortricidae). *Australian Journal of Zoology* 23, 419–437.

Danthanarayana, W. (1983) Population ecology of the light brown apple moth, *Epiphyas postvittana* (Lepidoptera: Tortricidae). *Journal of Animal Ecology* 52, 1–33.

Espinosa, A. and Hodges, R. (2009) *Epiphyas postvittana*. *Bugwood Wiki*. Available at: https://wiki.bugwood.org/Epiphyas_postvittana (accessed 8 October 2025).

Feng, Y., Wratten, S., Sandhu, H. and Keller, M. (2015) Host plants affect the foraging success of two parasitoids that attack light brown apple moth *Epiphyas postvittana* (Walker) (Lepidoptera: Tortricidae). *PLOS One* 10, e0124773.

Foster, S.P. and Howard, A.J. (1998) Influence of stimuli from *Camellia japonica* on ovipositional behaviour of generalist herbivore *Epiphyas postvittana*. *Journal of Chemical Ecology* 24, 1251–1275.

Foster, S.P., Howard, A.J. and Harris, M.O. (1991) Behavioral manipulation methods for insect pest management. *Annual Review of Entomology* 36, 375–400.

Froggatt, W.W. (1901) *The Codling Moth (Carpocapsa pomonella, Linn.)*. *New South Wales Agricultural Gazette*, Miscellaneous Publication 521, 1–9.

Geier, P.W. and Briese, D.T. (1980) The population dynamics of *Epiphyas postvittana* in Australian orchards. *Australian Journal of Ecology* 5, 155–167.

Geier, P.W. and Briese, D.T. (1988) The light-brown apple moth, *Epiphyas psotvittana* (Walker): a native leafroller fostered by European settlement. In: Kitching, R.L. and Jones, R.E. (eds) *The Ecology of Pests: Some Australian Case Histories*. CSIRO, Melbourne, Victoria, pp. 130–155.

Gut, L.J. and Brunner, J.F. (1998). Pheromone-based management of codling moth (Lepidoptera: Tortricidae) in Washington apple orchards. *Journal of Agricultural Entomology* 15, 387–406.

Insecticide Resistance Action Committee (IRAC) (2021) Maintaining Effective Control Strategies for Codling Moth, *Cydia pomonella*. Available at: https://irac-online.org/documents/cydia-pomonella-irm-poster/ (accessed 4 March 2025)

Jones, V.P., Hilton, R., Brunner, J.F., Bentley, W.J., Alston, D.G. *et al.* (2013) Predicting the emergence of the codling moth, *Cydia pomonella* (Lepidoptera: Tortricidae), on a degree-day scale in North America. *Pest Management Science* 69, 1393–1398.

Kadoić Balaško, M., Bažok, R., Mikac, K.M., Lemic, D. and Pajač Živković, I. (2020). Pest management challenges and control practices in codling moth: a review. *Insects* 11, 38.

Knight, A.L. (2007) Insecticide resistance in codling moth: challenges and solutions. *Pest Management Science* 63, 800–810.

Knight, A.L. (2014) Adjusting the phenology model of codling moth (Lepidoptera: Tortricidae) in Washington state apple orchards. *Environmental Entomology* 36, 1485–1493.

Kührt, U., Samietz, J., Höhn, H. and Dorn, S. (2006) Modelling the phenology of codling moth: Influence of habitat and thermoregulation. *Agriculture, Ecosystems and Environment* 117, 29–38.

Lacey, L.A., Thomson, D., Vincent, C. and Arthurs, S.P. (2008) Codling moth granulovirus: A biopesticide success story. *IOBC Bulletin* 31, 93–108.

Lewis, C.A. and Hodges, A.C. (2010) *Light Brown Apple Moth,* Epiphyas postvittana *(Walker) (Insecta: Lepidoptera: Tortricidae)*. Department of Entomology and Nematology, UF/IFAS Extension. Available at: https://edis.ifas.ufl.edu/publication/IN841 (accessed 8 October 2025).

Linnaeus, C. (1758–59) *Systema Naturae per Regna tria Naturae*, 10th edn. Laurentius Salvius, Stockholm, Sweden.

Lo, P.L. and Blank, R.H. (1992) The role of parasitoids in controlling light brown apple moth populations. *New Zealand Journal of Ecology* 16, 123–131.

Lo, P.L., Suckling, D.M., Bradley, S.J., Walker, J.T.S., Shaw, P.W. and Burnip, G.M. (2000) Factors affecting feeding site preferences of lightbrown apple moth, *Epiphyas postvittana* (*Lepidoptera: Tortricidae*), on apple trees in New Zealand. *New Zealand Journal of Crop and Horticulatural Science* 28, 235–243.

Manktelow, D., Stevens, P.S., Walker, J.T.S., Gurnsey, S., Park, N.M. *et al.* (2005) *Trends in Pesticide Use in New Zealand, 2004*. Technical Report, Ministry of Environment, Wellington, New Zealand.

Markwick, N.P., Poulton, J., Espley, R.V., Rowan, D.D., McGhie, T.K. *et al.* (2013) Red-foliaged apples affect the establishment, growth, and development of the light brown apple moth, *Epiphyas postvittana*. *Entomologia Experimentalis et Applicata* 146, 261–275.

Meijerman, L. and Ulenberg, S.A. (2000) *Arthropods of Economic Importance: Eurasian Tortricidae*. ETI, Amsterdam, The Netherlands

Mills, N., Pickel, C., Mansfield, S., McDougall, S., Buchner, R. *et al.* (2000) Mass releases of wasps can reduce damage from codling moth. *California Agriculture*, 54, 22–25.

Neilsen, D.G. and Dunley, J.E. (2002) Effect of temperature on codling moth development. *Journal of Economic Entomology* 95, 270–275.

Patrick, B. and Dugdale, J. (1994) Australian lucerne leafroller, *Merophyas divulsana* (Lepidoptera: Tortricidae), in New Zealand. *New Zealand Journal of Zoology* 21, 305–308.

Reyes, M., Franck, P., Olivares, J., Margaritopoulos, J., Knight, A. and Sauphanor, B. (2009) Worldwide variability of insecticide resistance mechanisms in the codling moth, *Cydia pomonella* L.(Lepidoptera: Tortricidae). *Bulletin of Entomological Research*, 99, 359–369.

Riedl, H. and Croft, B.A. (1978) Effects of photoperiod and effective temperature on the seasonal phenology of the codling moth (Lepidoptera: Tortricidae). *Canadian Entomologist* 110, 455–470.

Riedl, H., Croft, B.A. and Howitt, A.J. (1976) Forecasting codling moth phenology based on pheromone trap catches and physiological-time models. *Canadian Entomologist* 108, 449–460.

Suckling, D.M. and Brockerhoff, E.G. (2010) Invasion biology, ecology, and management of the light brown apple moth. *Annual Review of Entomology* 55, 285–306.

Suckling, D.M., Stringer, L.D., Baird, D.B., Butler, R.C., Sullivan, T.E.S. *et al.* (2014) Light brown apple moth (*Epiphyas postvittana*) (Lepidoptera: Tortricidae) colonization of California. *Biological Invasions* 16, 1851–1864.

Tadic, M. D. (1957) *The Biology of the Codling Moth (Carpocapsa pomonella* L.*) as a Basis for its Control*. University of Belgrade, Belgrade, Serbia.

Varela, L.G., Walker, J.T.S., Lo, P.L. and Rogers, D.J. (2010) New Zealand lessons may aid efforts to control light brown apple moth in California. *California Agriculture* 64, 6–12.

Walker, G. (1863) *List of Specimens of Lepidopterous Insects in the Collection of the British Museum. Crambites and Tortricites*. British Museum (Natural History), London.

Wearing, C.H., Thomas, W.P., Dugdale, J.S. and Danthanarayana, W. (1991) Tortricid pests of pome and stone fruits, Australian and New Zealand species. In: van der Geest L. and Evenhuis, H.H. (eds), *Tortricid Pests: Their Biology, Natural Enemies and Control*. Elsevier, Amsterdam, The Netherlands, pp. 453–472.

Wearing, C.H., Hansen, J.D., Whyte, C., Miller, C.E. and Brown, J. (2001) The potential for spread of codling moth (Lepidoptera: Tortricidae) via commercial sweet cherry fruit: A critical review and risk assessment. *Crop Protection* 20, 465–488.

Williams, D. (2000) *Codling Moth, Priority Pest Insects and Mites*. Agriculture Victoria, Melbourne, Victoria.

Wiman, N. and Stoven, H. (2022) Apple pests. In: Pscheidt, J.W. and Ocamb, C.M. (eds) *Tree Fruit Crops: Pacific Northwest Pest Management Handbooks*. Oregon State University, Corvallis, OR.

Witzgall, P., Stelinski, L., Gut, L. and Thomson, D. (2008) Codling moth management and chemical ecology. *Annual Review of Entomology* 53, 503–522.

Witzgall, P., Kirsch, P. and Cork, A. (2010) Sex pheromones and their impact on pest management. *Journal of Chemical Ecology* 36, 80–100.

28 Moths of Stored Products

David Rees
Gungahlin, ACT, Australia

Abstract

Today, important pests of stored products, including moths present in Australia, are a subset of the species now found worldwide, the origins of which are mostly obscure. The most important pests of stored product are members of the Pyralidae and Gelechiidae. They can be significant pests of cereal grains but most are better known as pests of processed foodstuffs. In addition, members of the Oecophoridae and Tineidae can be found in storages, mainly as scavengers on crop residues. The most important moth species, *Plodia*, *Cadra* and *Ephestia*, commonly infest finished processed products. Any presence of these creatures on a finished product is effectively a total loss in terms of being able to sell the product.

28.1 Introduction

About 10,000 years ago human populations started to transition from hunter-gatherer societies to ones relying on settled agriculture, based on domesticated animals and production of a variety of crops. From this grew towns and cities and the need to store large quantities of cereals, pulses and other edible dried material from one season to the next. Such accumulations would have been at risk of attack by a range of pests and over time, a number of species, mostly beetles but also moths, started infesting these stored commodities. Some thrived and became significant pests and were spread widely by trade. Archaeology provides evidence of this happening in historic times, especially for beetles as the bodies survive well when buried. Beetle pests of stored grain such as *Oryzaephilus*, *Tribolium* and *Sitophilus*, for example, appear to have arrived in Britain in Roman times (AD 49–410) (Smith *et al.*, 2020).

In Australia, prior to European settlement Indigenous people hunted and gathered food, but some also grew and stored tubers and grass seeds (Gerritsen, 2010). It is unclear if such materials were attacked by storage pests. By the time of European discovery and settlement of Australia, however, stores of ships arriving here were plagued with storage pests. Joseph Banks in his journey on the *Endeavour* in 1769 wrote (Banks, 1771):

> Our bread indeed is but indifferent, occasiond by the quantity of Vermin that are in it. I have often seen hundreds nay thousands shaken out of a single bisket. We in the Cabbin have however an

 Australian Moths: A Natural History (ed. R.L. Kitching)
DOI: 10.1079/9781800627680.0028

> easy remedy for this by baking it in an oven, not too hot, which makes them all walk off, ... They are of 5 kinds, ...

Today, important pests of stored products, including moths present in Australia, are a subset of the species now found worldwide, the origins of which are mostly obscure. Many are likely to have arrived with the First Fleet or soon after, given the conditions of those days as described by Joseph Banks.

28.2 Moths of Stored Products

The most important families as pests of stored product are members of the Pyralidae and Gelechiidae. They can be significant pests of cereal grains, but most are better known as pests of processed foodstuffs. In addition, members of the Oecophoridae and Tineidae can be found in storages, mainly as scavengers on crop residues (Rees, 2004).

28.2.1 Pyralidae

In Australia, members of the genera *Cadra*, *Corcyra*, *Ephestia*, *Plodia* and *Pyralis* are present as pests of stored products. These have been spread worldwide as a result of trade.

28.2.1.1 Cadra *spp.*, Ephestia *spp.*, Plodia interpunctella

Members of these genera are important pests of a wide range of stored products of plant origin, including milled, processed and manufactured foodstuffs. *Ephestia* species include the moths known as cacao, tobacco or warehouse moths. *Cadra* spp. include the almond moths, fig moths and others. *Plodia interpunctella* is the Indian meal moth. These common names can be misleading and particular species may infest a range of substances beyond their eponyms. These moths are frequently encountered in mills, food and animal feed processing facilities, grocery, pet food and feed stores and households.

Adults of *Cadra* and *Ephestia* species are similar in appearance. When fresh, forewings (7–14 mm long) are grey, marked with vague darker wavy lines running across the wing. The head has labial palps curving upwards. Identification to species is difficult by external examination of specimens and is best done by dissection and examination of the male genitalia (Ferguson, 1987; Mound, 1989). The forewings of *Plodia interpunctella* (7–9 mm long) are distinctively bicoloured, cream with the apical half reddish-brown, with the labial palps pointing forwards. Larvae of *Cadra* and *Ephestia* species are similar in appearance; full grown, they are about 15–20 mm long, white or pink. The bases of hairs on the body are marked with a black spot. Larvae of *Plodia* are of similar size, creamy white without pigmented spots at the bases of hairs (Rees, 2004)

In warm temperate to tropical environments *Cadra cautella* (Fig. 28.1) is a major pest of mills and related facilities. In temperate regions, it tends to be replaced by *Ephestia kuehniella*. In southern Australia, *C. cautella* is typically found in indoor locations, often inside food-processing plants, whereas *E. kuehniella* is more likely to be found in cooler, unheated locations (Rees and Wright, 1998). Both species can be pests of bulk and bagged stored grain and oilseeds and in bulk commodities they attack the surface layers. *Ephestia elutella* (Fig. 28.2) can be found in mills, and in the past was associated with tobacco processing (Champ, 1966). *C. figulilella* (Fig. 28.3) and *C. callidella* are minor pests, typically of dried fruits, infestations of which may begin in the field prior to harvest. They can be found in Australia in vineyards containing drying crop residues (Rees and Wright, 1988).

In Australia, *Plodia interpunctella* is the most commonly encountered moth pest of milled, manufactured and packaged dried foodstuffs throughout the manufacturing, distribution and retail chain. It is one of the most important pests of stored products in terms of its economic impact and is the most frequently encountered pest of such products in Australian homes (Rees, 2004).

Adult moths of these genera are short-lived (days to a couple of weeks), depending on temperature, but can lay up to about 200 eggs. Eggs of *Plodia* are sticky and adhere to food or surfaces such as packaging, machinery or the fabric of the storage structure when laid, while those of *Ephestia* and *Cadra* are laid loose among food, packaging (including plastics) and so forth. Adult moths are typically inactive during the day and can be found resting on vertical surfaces,

Fig. 28.1. Pyralidae: *Cadra cautella*. Photo: CSIRO.

Fig. 28.2. Pyralidae: *Ephestia elutella*. Photo: CSIRO.

head upwards. They become active at dusk and dawn, especially in still, warm conditions. Adult moths are able to follow odour trails from attractive food sources from a distance, which will vary depending on ambient temperature (in warm mean temperatures >25°C flight is more easily sustained) and the strength of the odour trail. An exact figure of the distance these insects can fly is hard to calculate as there are many variables. Marked individual *Plodia* adults, for example, have been recovered in flight traps over distances of up to 276 m from a release point (Campbell and Mullen, 2004). Tests in a flight mill show that adult *E. kuehniella* can fly a mean distance of 24–34 km in a 24 h period. Moths are capable of flying 207 m in a single bout when pheromones are present; this drops to 41 m when food is present nearby (Abshire, 2024). Female moths emit pheromones to attract males. These pheromones have been synthesized

Fig. 28.3. Pyralidae: *Cadra figulilella*. Photo: CSIRO.

and are available for trapping and monitoring populations of these moths (Rees and Wright, 1988; Arbogast *et al.*, 2000).

Newly hatched larvae forage to find food and will initially enter the product via cracks and crevices within the commodity or processed food. When fully grown, larvae will often wander and become obvious and may exit foodstuff in order to pupate. Reproduction is continuous under ideal conditions, with the life cycle for *C. cautella* and *P. interpunctella* taking as little as a month under warm humid conditions (30°C, 75% r.h. and breeding can occur between about 15–35°C). In temperate regions, reproduction may slow or stop during the coolest part of the year, when larvae may enter diapause.

Infestations of *Plodia*, *Cadra* and *Ephestia* species are visually characterized by copious amounts of webbing produced by larvae. This binds the product together, preventing product flow. In addition, there is contamination of the commodity with faecal pellets, larval and pupal skins, and living and dead individuals. Loss due to contamination and binding of the product greatly outweighs the loss caused by direct consumption. When feeding on whole grain, larvae preferentially consume the germ and bran.

Mills and food-processing facilities contain many sites that are at risk of infestation by these moths. These include the storage of ingredients, the processing and production line itself, and the general fabric of the facility, in which food residues may accumulate, often in conditions conducive to the rapid development of pest populations. Although it is important to ensure that ingredients are not infested before processing, such earlier infestations may not be the main cause of moth populations found in finished milled and processed food products. Most inputs are subjected to processes that, by their nature, are highly 'insecticidal' such as milling, steaming, rolling, extruding and baking. Finished products are, however, highly susceptible to infestation once processing is complete. Moth larvae infesting the fabric of the facility may contaminate the product once past the last processing step and are simply packed into the product.

Weighing and packing is often the last process a product undergoes and is an especially vulnerable point for infestation. Very careful attention needs to be paid to this equipment to ensure it is kept clean and infestation free. The act of processing grains and similar products always produces a quantity of dust and fragments, which can blow about and collect on machinery surfaces often in inaccessible places as well as residues accumulating in voids and dead spots within machines. If conditions are suitable, such accumulations can sustain populations of moths and other pests unless they are removed frequently.

Packaged goods spend months in distribution and retail systems before final consumption.

During this time, they may be stored in conditions suitable to the development of larvae already present. They are also at risk of infestation from populations of moths residing nearby. Package seals can be defective at time of manufacture and damage to packages can accumulate over time as a result of handling and transport. Tiny (<4 mm long) newly hatched moth larvae are able to enter though very small defects in a package. Such damage may not be obvious to casual observation but commonly can be found as damage or imperfections in the package seal. They can enter via holes of just 0.45 mm in diameter (Mohandass *et al.*, 2007).

Adult moths, especially *Plodia*, may fly in from nearby general urban or suburban environments. Warehouse and retail facilities are busy places with goods and people coming and going and may be open and lit to the outside at times when moths can be most active, especially on calm warm evenings when temperatures are high enough to allow moths to fly readily such as in summer in southern Australia and all year round in subtropical and tropical regions.

Goods on pallets are often shrink wrapped, which may offer good protection if tightly wound and assuming it remains undamaged, and a suitable liner is placed on the pallet to restrict access from below. In contrast, loose wrapping over goods on a pallet as a weather/dust protector may make the situation worse, as it may act effectively as a trap. Moths will typically land on a place of interest and will usually walk upwards, between the plastic and the cartons, where they may become trapped and over time many may accumulate, mate and lay eggs.

Infested packages are often identified by frass and other debris leaking out from obvious holes made in them, typically 2–3 mm in diameter. Such obvious damage is typically exit holes made by mature larvae looking for a place to pupate. If an infested package is stored compressed directly next to others, the larva may be able to bite its way from one package to the next. Pupation typically occurs in a crack or crevice nearby. New packaging, such as cardboard and multi-layer paper sacks stored close to production lines, can become infested by wandering larvae in search of places to pupate.

In the home, partly used and poorly resealed packages of infestable goods can be found in many a pantry. These are at high risk of being infested by adult moths moving about on warm evenings as previously discussed, as well as by moths already present in the building. When an infested package is found, it is usually the manufacturer that is blamed, without taking account of the extent of infestation risk within the home. Populations of these moths in the home can usually be managed by a combination of stock control, long-term storage of infestable materials in insect-proof containers, such as screw-topped or clip-sealed jars and containers, as well as regular cleaning of the storage space to avoid the build-up of spillages in which infestations often start. Freezing of sealed packages of newly bought materials followed by the above methods may additionally manage infestation risk.

28.2.1.2 Pyralis farinalis

Adult *Pyralis farinalis* (Fig. 28.4), meal moths, are distinctive relative to other 'storage' moths. The fore- and hindwings are marked with two wavy white lines going across them. On the forewing they delineate basal and apical purplish brown coloration with a central band of pale brown. The forewing is broad and about 10 mm in length. Larvae are creamy white in colour and grow to about 25 mm long (Rees, 2004).

The species is found worldwide but mostly in temperate regions. In Australia, it is found widely in the more humid settled areas of the southern half of Australia, from south-east Queensland to south-west Western Australia. It is a minor pest compared with *C. cautella* and *P. interpunctella*. It is most likely to be encountered in aged crop residues and crop waste, especially if these are damp. In domestic situations, it can often be encountered in compost bins and in sheds and barns where animal feed (hay, straw and the like) is kept. It is unlikely to be encountered on the processing line in well-run mills, processing facilities and in retail premises with a rapid turnover of stock.

28.2.1.3 Aphomia (Corcyra) cephalonica

Adult *Aphomia cephalonica*, rice moths, are similar in general appearance to *Cadra* and *Ephestia* species, being grey-brown. However, the forewing of *Aphomia* lacks the vague darker markings seen in the other two genera and is more tapered towards the tip. Forewings measure 8–15 mm, and females are larger than males. Labial palps in the female are long and point forward, in the male they are short and hidden by scales. Larvae are white in colour and grow to 15–20 mm long. Abdominal spiracles are obviously thickened on the rear side (Mound, 1989).

The species is known as a pest of mills and food-processing facilities, much like *C. cautella*, especially in tropical climates where it can be a major pest of the rice industry. Historically, it known as a pest of bagged grain in tropical Queensland but replaced in importance by *C. cautella* in southern Queensland (Champ, 1966). There appears to be little modern data as to the pest status of this insect in Australia.

28.2.2 Gelechiidae

Sitotroga cerealella (Fig. 28.5), the so-called rice grain moth, is unlikely to be confused with any other moth commonly associated with stored grain on account of its size, general appearance when fresh and lifestyle. It is the only member of its family that is an important pest of stored grain. Adult moths are smaller than other pest species, with forewings about 5 mm long, light

Fig. 28.4. Pyralidae: *Pyralis farinalis*. Photo: CSIRO.

Fig. 28.5. Gelechiidae: *Sitotroga cereallela*. Photo: CSIRO.

greyish brown in colour and heavily fringed with hairs. Labial palps of the adult are long and strongly curved upwards, as is typical for this family. Unlike other moth pests of stored products, larvae of this species are internal feeders on whole cereal grain. Up to 150 eggs are laid by a female on seed heads or in crevices between grains. On hatching, larvae bore into a grain, excavating a cavity as they feed, and remain concealed within the grain. True legs of these larvae are reduced in size and prolegs are vestigial. Before pupating, the larva constructs a silk cover over an exit hole made in the grain. Upon hatching, the adult moth pushes its way through this to exit the grain. Unlike other major moth pests, this species does not attack milled or processed grain products.

Sitotroga cerealella is a pest of grains including wheat, barley, rice, maize, millet and sorghum. It is found worldwide in the tropics, subtropics and warmer parts of temperate regions. The species may overwinter in temperate regions as larvae inside stored grains as well as in crop residues left in the paddock. Infestations can begin in the field as crops mature before harvest. Crops with open heads, such as sorghum, and crops, such as maize, which can be left in the field to dry prior to harvest are at particular risk. Grain stored in bags is more at risk of damage than grain stored in bulk where only the surface layers are usually open to attack, due to the inability of adult moths to penetrate into the grain bulk (Rees, 2004). Damage caused to intact grains can permit other pests, including other moths, which cannot feed easily on whole grains, to attack the produce. While present in Australia, modern harvesting, bulk handling and bulk storage methods have greatly reduced its impact as a pest. There appears to be little data as to the current status of this pest in Australia.

28.2.3 Oecophoridae

Two species of oecophorid moths are known to attack stored products in Australia: *Endrosis sarcitrella* (Fig. 28.6), the white-shouldered house moth, and *Homannophila pseudospretella*, the brown house moth. Both are introduced and are found worldwide, mostly in temperate regions. *E. sarcitrella* is distinctive in appearance, having its head, thorax and base of forewings covered with bright white scales. The forewings, 6–10 mm long, have an obvious pattern of light and darker patches. *H. pseudospretella* is a drab brown moth with forewings that are 6–12 mm in length, brown with several dark spots in the centre (Rees, 2004). Both species have long upward curving palps, a feature which will easily separate them from the pyralids with which they can be found.

Both species are minor pests of stored products, being typically found in aged residues and

Fig. 28.6. Oecophoridae: *Endrosis sarcitrella*. Photo: CSIRO.

in situations of poor hygiene, compost bins and the like. Both are present in urban and suburban environments in southern Australia. Larvae of *H. pseudospretella* can feed on wool and as such can potentially attack carpets.

28.2.4 Tineidae

Larvae of members of the Tineidae feed wholly or mostly on dried material of animal origin. Members of the genera *Tinea* and *Tineola* are pests of products containing natural fibres such as wool. *Nemapogon granella* can be found infesting dried material of plant origin.

Unlike other stored product moths, the head of an adult tineid moth is covered in rough scales that give it a hairy appearance, palps are short, do not point upwards and may not be easily visible. Moths of *Tinea* and *Teneola* are small, dull, light to dark brown moths, while forewings of *N. granella* are obviously mottled grey/white and about 8 mm long (Rees, 2004).

Nemapogon granella, the European grain worm (or moth), is a minor pest of damp and aged residues of stored grains and their products. It has been known to attack aged mouldy corks of wine bottles held in storage – collectors beware! It is present in south-eastern Australia, probably as a result of introduction from Europe, and is a minor pest. In contrast, *Tineola* and *Tinea* spp. can be significant pests of woollen garments and carpets in domestic premises in Australia. Larvae of *Tinea* spp. spend their lives encased in a silken case, which they carry about and do not leave until they emerge as adults. Larvae of *Nemapogon* and *Tineola* leave lots of webbing and silken tunnels as they feed.

28.3 Conclusion

Relative to other major pests of stored products, especially beetles, the amount of material moths destroy directly is relatively small. Pest control and safe storage as applied for beetles in modern storage will manage moths in or directly on the raw material, and this comes at a cost. Australian environmental conditions of relatively high temperatures and mild to warm winters mean that pest pressure, without effective control, remains high all year round. Moth infestations produce a lot of webbing, frass and other detritus and this can not only contaminate products but also block machinery, leading to costly downtime, loss of production, damage and repairs. Downtime is a major cost, as many production facilities are run at or near capacity.

The most important moth species, *Plodia*, *Cadra* and *Ephestia*, commonly infest finished processed products. Any presence of these creatures on a finished product is effectively a total loss in terms of being able to sell the product. It is actually worse than that, given reputational and other flow-on issues. These can be critical, especially for small businesses establishing a product in the market. For retailers in Australia, especially those in the more or less subtropical and tropical zones anywhere north of a line from Perth to Sydney, the ambient pest pressure from the general environment is likely to be significant all year round.

Processed supplies to commercial customers, such as flour to bakeries, are typically supplied on an as-needed basis with very short 'use by' time frames, as time in storage increases risk of postproduction infestations. Major retailers like to turn stock over quickly and this is one of the reasons why. What is hard to manage is what happens in households, where infestable goods are kept, often in part-used opened packets, for indeterminate lengths of time, often in conditions highly suitable for pest population growth. Most people are repelled by caterpillars in their muesli or other foodstuffs and will discard them on sight. For some, the presence of insects and/or fragments of insects can be allergenic and can raise a range of other issues, including mental health.

References

Abshire, J., Harman, R., Bruce, A., Gillatte, S., Maille, J.M. *et al.* (2024) Flight capacity and behavior of *Ephestia kuehniella* (Lepidoptera: Pyralidae) in response to kairomonal and pheromonal stimuli, *Environmental Entomology* 53, 567–576.

Arbogast, R.T., Kendra, P.E., Mankin, R.W. and McGovern, J.E. (2000) Monitoring insect pests in retail stores by trapping and spatial analysis. *Journal of Economic Entomology* 93, 1531–154.

Banks, J. (1771) *Series 03. 02: Volume 2: Joseph Banks – Endeavour Journal, 15 August 1769 – 12 July 1771*. State Library of New South Wales. Available at: https://collection.sl.nsw.gov.au/intro-modal (accessed 8 October 2025).

Campbell, J. and Mullen, M. (2004) Distribution and dispersal behavior of *Trogoderma variabile* and *Plodia interpunctella* outside a food processing plant. *Journal of Economic Entomology* 9, 1455–1464.

Champ, B. (1966) Insect and mites associated with stored products in Queensland, *Queensland Journal of Agricultural and Animal Sciences* 23, 155–175.

Ferguson, D.C. (1987) Adult moths (Lepidoptera). In: Gorham, J.R. (ed.) *Insects and Mite Pests in Food: An Illustrated Key*. USDA Agricultural Handbook No. 655. United States Department of Agriculture, Washington DC, pp. 231–244.

Gerritsen, R. (2010) Evidence for indigenous Australian agriculture. *Australian Science* July/August, 35–37.

Mohandass, S., Arthur, F.H., Zhu, K.Y. and Throne, J.E. (2007) Biology and management of *Plodia interpunctella* (Lepidoptera: Pyralidae) in stored products. *Journal of Stored Products Research* 43, 302–311.

Mound, L. (1989) *Common Insect Pests of Stored Food Products, A Guide to their Identification*. British Museum (Natural History), London.

Rees, D. (2004) *Insects of Stored Products*. CSIRO Publishing, Collingwood, Victoria.

Rees, D. and Wright, E.J. (1988) Using pheromone traps to monitor pest populations. In: *Stored Grain in Australia: Proceedings of the Postharvest technical Conference, Canberra, 26–29 May*. CSIRO Entomology, Canberra, ACT, pp. 181–187.

Smith, D., Hill, G., Kenward, H. and Allison, E. (2020) Development of synanthropic beetle faunas over the last 9000 years in the British Isles. *Journal of Archaeological Science* 115, 105075.

29 Processionary Caterpillars: *Ochrogaster lunifer*, A Cryptic Species Complex

Juliane Farrell[1], Kai Hart[1], Nicole McMullen[1], Mizuki Uemura[1], Lynda Perkins[1], Andrew Walker[2], Lyn Cook[1], Andrea Battisti[3] and Myron P. Zalucki[1]*

[1]School of the Environment, University of Queensland, Brisbane, Queensland, Australia; [2]Institute of Molecular Biology, University of Queensland, Brisbane, Queensland, Australia; [3]Department of Agronomy, Food, Natural Resources, Animals and Environment, University of Padova, Padova, Italy

Abstract

As is the case with most widespread moth species, research on *Ochrogaster lunifer* has been localized, sporadic and piecemeal. The weight of evidence summarized here points to *Ochrogaster lunifer* being a species complex. Since the species is of medical and veterinary importance, it is likely that systematic research efforts will continue and the species' limits will be resolved as this will help sort out the ecology and inform management. We can only speculate how many cryptic species complexes await discovery in Australia's rich moth fauna.

29.1 Introduction

Every Australian autumn, long chains of hairy caterpillars in head-to-tail processions can be seen filing across open ground, paths and roads where frequently they are squashed by foot traffic, bikes and cars. These processions attract the attention of the public and media when they occur in populated areas. Invariably questions are asked if such processions are unusual, associated with climate change or other confounding factors, and are the caterpillars dangerous? The processions happen every year but are just more likely to be noticed during caterpillar outbreaks. Regarding danger, yes: it is best to look but not touch.

The creator of these processions is currently known as *Ochrogaster lunifer*, also referred to as the Australian processionary caterpillar, bag shelter moth and bunny-tail moth, and is a lepidopteran insect of medical and veterinary significance, a group whose species number is likely to be greatly underestimated (Battisti *et al.*, 2024). Contact with the minute urticating true setae can cause pruritic dermatitis, localized pain and erythema, ophthalmia nodosa, paediatric osteomyelitis and, in extreme cases, anaphylaxis (Southcott, 1987; Bockxmeer and Green, 2013; Randell *et al.*, 2017). Arguably the most important impact from these processionary caterpillars, however, is equine amnionitis and foetal loss in horses (EAFL) (Carrick *et al.*,

*Corresponding author: m.zalucki@uq.edu.au

DOI: 10.1079/9781800627680.0029

2014). Ingestion of caterpillar exuviae and the attached urticating setae causes a range of outcomes from focal mucoid placentitis to abortion or perinatal death (Cawdell-Smith *et al.*, 2012, 2013; Todhunter *et al.*, 2014a, b), although the precise aetiology is not understood.

Long before cities and horse studs developed in Australia, to the First Nations people of the southern coast of Queensland the processions of larvae seen in autumn signalled the coming of the mullet season (Bartholomew, 2022). The lines of caterpillars indicated that mullet fish were leaving streams to the south and making their way to Queensland to spawn. Alas, too little is documented about Indigenous people's associations with and uses of *O. lunifer*. In Central Australia, Arrernte and other desert people used the outer silk layers of the bag-shelters as a dressing with a poultice of emu bush (*Eremophila longifolia*) for burns and deep wounds (Walsh *et al.*, 2014). The Arrernte believe that the more larvae in a procession during March–April, the colder will be the days during the oncoming winter. A suggested explanation for this association could be that more rain in the hot wet season results in larger numbers and/or size of larvae, and greater soil moisture retention along with clear skies in the cooler dry season results in more frequent frosty days (Alice Springs Field Naturalists Club, 2023). Greater transfer of Indigenous knowledge to non Indigenous peoples would fertilize cross-cultural understanding and better inform scientific research going forward.

29.2 Overall Status

If you were to open a drawer labelled '*Ochrogaster lunifer* Herrich-Schäffer 1855 (Notodontidae: Thaumetopoeinae)' in the Lepidoptera section of the Australian National Insect Collection (ANIC), you would be struck by the diversity of the adult moths. Although they look similar, there are nuanced differences in their morphology that correlate with collection site, host and nest type. Currently, the subfamily Thaumetopoeinae comprises 106 species in 20 genera (Schintlmeister, 2013; Basso *et al.*, 2017). Taxonomic work on the subfamily indicates three major clades (Basso *et al.*, 2017; Battisti *et al.*, 2017); the Australian genera – *Ochrogaster* with a single recognized species and *Epicoma* with 20 species (Global Biodiversity Information Facility, www.gbif.org/) – form the first clade. The second is made up of African genera and the third clade is a conglomeration of African, Asian and European genera (Battisti *et al.*, 2017).

The type specimen for *O. lunifer* is unknown and there are numerous synonyms of the currently defined species, most of which are in the collection of the British Museum of Natural History (BMNH): *Teara contraria* (BMNH), *Teara interrupta* (BMNH), *Poecilocampa leucopyga* (BMNH), *Darala cinctifera* (BMNH), *Ochrogaster ruptimacula* (BMNH), *Teara erebodes* (South Australian Museum, Adelaide), *Marane rubricorpus* (BMNH), *Teara discistriga* (BMNH) and *Adelaidina dispar* (South Australian Museum, Adelaide).

Ochrogaster lunifer has long been suggested to be a cryptic species complex based on the variety of nesting habits, host use and distribution patterns (Common, 1990; Floater, 1996a; Floater and Zalucki, 1999; and below). Genetic analyses suggest there are at least two reproductively isolated species in this complex, and probably many more (Mather *et al.*, 2019), complicating any account of the biology and ecology of the 'species'. Here, we proceed by summarizing what is known about *O. lunifer* in terms of the associations between host plant, geographic location and nesting form.

At least five nesting forms have been recognized (Perkins *et al.*, 2016) and there are probably more. Various workers have studied the taxon in their local area. Quantitative data can be found in van Schagen *et al.* (1992a, b) for canopy nesters in south-west Western Australia (see also Mills, 1950, 1951a, b), while Floater (1996a–c, 1997a, b, 1998), and Floater and Zalucki (1999, 2000) have provided an extensive account of ground-nesting populations in coastal south-eastern Queensland. Perkins *et al.* (2023, 2024) studied a trunk-nesting form in south-east Queensland and Farrell *et al.* (2023) record canopy-nesting forms in eucalypts and acacias in south-east Queensland and northern New South Wales. Steinbauer (2019) provided information from southern Australia. Some aspects of *O. lunifer* biology are shared among different populations, but there is little data currently available comparing their biology and ecology.

29.3 Occurrence and Distribution

Based predominantly on light trap sampling of adults, *O. lunifer* has been collected Australia-wide (Fig. 29.1A). At face value this suggests the species has a wide climatic niche, but also assumes *O. lunifer* is a single species. Where collection records include details of host plant and nest form, a more complex picture emerges (Fig. 29.1B).

With few exceptions, the canopy-nesting form in eastern Australia (New South Wales, Queensland, Victoria) appears confined to inland areas, although in the Hunter Valley all three nesting forms occur in close proximity, sometimes on the same host (see Table 29.1), with the trunk- and ground-nesting forms being more common in the coastal areas (Fig. 29.1B). The Great Dividing Range which separates the eastern Australian coastal areas from inland areas appears to be the boundary between distinct *O. lunifer* populations. In South Australia, *O. lunifer* is less common, possibly because the dominant eucalypt is the mallee, a low-growing multi-stemmed plant that appears not to be a preferred host. In Victoria, acacias, mallee and other eucalypts with loose bark that often hangs in strips from the trunk dominate the environment and *O. lunifer* appears not to use these species as hosts, possibly due to difficulties navigating the loose bark. In Western Australia, *O. lunifer* have mainly been recorded on acacias but there are few records from there or the Northern Territory.

29.4 Basic Life History and Natural History

Adults have non-functional mouth parts and are not strong flyers. After mating, females deposit eggs in layered masses, cover them in the scales of the anal tuft and die shortly afterwards. Ground nesters oviposit at the base of their host and the covering scales are white (Fig. 29.2A, B). Canopy and trunk nesters lay on small branches in the canopy and the egg mass appears golden (Fig. 29.2C). This egg mass colour difference is due to the piliform scales being golden coloured in canopy nesters, whereas they are white in ground-nester females. The lamellar scales are white in all nesting forms (Perkins *et al.*, 2023).

First instars, at least in *Acacia* ground nesters (Floater, 1997a), and most likely all forms, are non-feeding and construct the early nest structure by fluffing the scale mass before moulting. Second instars emerge from the scale mass at the base of the host and then travel in a single

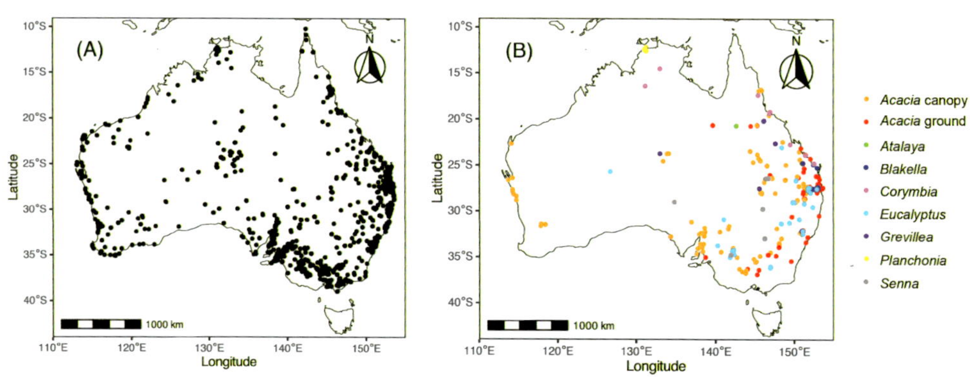

Fig. 29.1. Occurrence data for *Ochrogaster* from (A) all records available in the Atlas of Living Australia (www.ala.org.au,) and (B) records of nests where the nest type and host are identifiable, combining our own data with observations in iNaturalist Australia (https://inaturalist.ala.org.au/home). Nests are coloured by host plant genus as these may represent distinct species of *Ochrogaster*, and those on *Acacia* are further divided by nest type (canopy or ground) because these do represent distinct species. There is variation in nest types on some other hosts, such as *Corymbia*, but there are too few data currently available to confidently assign host–nest type combinations.

Table 29.1. Some life history attributes of *Ochrogaster lunifer* recorded from different nesting forms and hosts.

Host	Nest form; egg mass colour	Number of eggs Range, Median, Mean± SD, (sample size)	Mean Egg size (mm) ± SD, (sample size)	Pupal size (mm) mean, Range, mean± SD (sample size)	Time in pupal stage (d) ±SD,	Number of instars
Eucalyptus spp.	canopy; gold	50-383, 117 (241)	1.5±0.08, (105)	21.9, 19.6-24.1, (514)	46.95 ± 5.01	7
Acacia spp.	canopy; gold	24-329				6*
Blakella tessellaris	trunk; gold	154-343, 180 (128) **294 ± 28 (4)**	1.6 ± 0.06 (64) **1.5 ± 0.14 (209)**	24.3, 21.8-26.8, (128) **♀:27±1.5 (56)** **♂:22±1.3 (41)**	49.5 ± 5.88	**8**
Acacia spp.	ground; white	38-536 137 (85) *268±95 (24)*	1.4±0.05, 70	22.3, 19.9-24.7, (417) *♀:24.4±1.33 (25)* *♂:21.1±1.20 (29)*	45.4 ± 3.67	*8*
Allocasuarina littoralis	ground; white	69-461 (6)				

Data from Farrell (2023), (plain text), Floater (1996) (italics), and Perkins *et al.* (2023) (bold), Van Schagen *et al.* (1992ab) (*). Values in brackets are sample sizes when known

file, leaving a silk trail, to the youngest leaves in the outer canopy where they feed gregariously (Fig. 29.2D, E) and then return to the nest – given their size, a truly epic journey. It is at this stage that most nests fail as larvae never make it back (Floater and Zalucki, 1999). If they survive this first journey, they will make this ascent daily to feed and return to the nest, except when moulting. In the early instars (II to III–IV) feeding is during the day, switching to night feeding in the later instars (IV–VIII). Moulting takes place in the expanding nest. If multiple egg masses are laid on the same plant, which is common (Floater and Zalucki, 2000), the nests typically coalesce into a larger structure (Fig. 29.2G). The silk trail, which is assiduously followed, suggesting a trail pheromone, becomes obvious on the trunk (Fig. 29.2H) although thigmotaxis also plays a major role in maintaining processions (Steinbauer, 2009). The distribution, structure and abundance of the offensive setae are detailed in Perkins *et al.* (2016, 2019).

Larvae remain gregarious throughout their development. If the host is defoliated (Fig. 29.2F), they will process in search of another to complete development and may feed on other food plants (Farrell *et al.*, 2023). In early autumn when larvae become mature, the colony will process at sunrise in search of a pupation site, and this is known as a prepupation procession (Uemura *et al.*, 2020). Before processing, the colony forms a cluster and larvae move amongst each other in a ball (M. Uemura, personal observation). Eventually, a leader of the procession emerges from the cluster and remaining larvae follow. The leader is typically a larger female and the procession travels on average 17.4 m h^{-1} orienting either north or south (Uemura *et al.*, 2020). The larva uses one of the stemmata that is optimized for sky light polarization vision to orientate in the environment (Uemura *et al.*, 2021). The procession travels on average 40 m to a potential pupation site which is usually in a shaded area of the environment (Uemura *et al.*, 2020). If the site reached at the end of the procession is unsuitable for pupation, the colony frequently restarts the prepupation procession the following sunrise, and distances of a few hundred metres for the total procession have been recorded.

29.4.1 Eggs

Egg mass size can be highly variable (Table 29.1), reflecting location, adult size and larval diet, and averages 200–300 eggs. In general, egg size is more similar in canopy- and trunk-nesting forms, with ground nesters having smaller eggs (Table 29.1), although there are differences

Fig. 29.2. *Ochrogaster lunifer*. (A) Egg mass of ground-nesting form on an *Acacia*. (B) Early-stage ground nest with third instar larvae emerging to ascend and feed. (C) Egg mass of a canopy nester. (D) Late instar larvae of canopy-nesting forms (dark brown) with ground nesters (grey). (E) Ground-nesting larvae processing to find pupation sites. (F) Ground nest at the base of an acacia. (G) Canopy nest in an acacia. (H) 'Tree-hugging' nest form on a *Blakella*. Authors' own photographs.

correlated with host species (Farrell, 2023). In ground nesters, egg masses are often clumped on the same tree (Floater and Zalucki, 2000), with 22 egg masses counted on single *Acacia* spp. (M. Zalucki, personal observation).

29.4.2 Larvae

Instar number may differ between nesting forms and host plants (Table 29.1), but data is lacking in this area. To date, studies have utilized *Acacia*- and eucalypt-dwelling caterpillars, but van Schagen *et al.* (1992) studied the canopy-nesting form in Western Australia whereas Floater (1996) studied the *Acacia* ground-nesting form in Queensland, and Perkins *et al.* (2023) a trunk nester on *Blakella* (formerly *Corymbia* (partim)).

One of the more striking morphological differences between canopy- and ground-nesting forms can be seen in colour differences (Fig. 29.1D). Late instar canopy-nesting caterpillars are brown in appearance whereas ground nesters are grey. Larvae are social throughout their development, co-operate in nest building and may have rudimentary division of labour, as has been suggested in pine processionary moths, *Thaumetopoea pityocampa* (Uemura *et al.*, 2023), but this has not been studied in Australian species.

29.4.3 Pupae

Pupal size differs between nesting form populations. Trunk nesters ($\bar{x}$ = 24.3 mm, 21.8–26.8, n = 128) were the largest on average followed by ground nesters ($\bar{x}$ = 22.3 mm, 19.9–24.7, n = 417) and canopy nesters ($\bar{x}$ = 21.9 mm, 19.6–24.1, n = 514). The latter two did not differ because variability was high for all three nesting forms. Females are much larger than males (Table 29.1).

29.4.4 Adults

Adult size (reflecting pupal size; Table 29.1) and wing polymorphism are highly variable. The colour of above-ground nesters is lighter than ground nesters in both sexes.

29.5 Host Relationships

Females of ground nesters on acacias have been observed to alight frequently on foliage in the canopy, and then walk backwards down the trunk to the base where they deposit their egg mass in layers, finally covering them in scales from their anal tuft. Host records are often based on presence of nests but these may be the result of either oviposition and larval survival *in situ*, or larvae moving among plants. To be a true 'host', females need to lay eggs, immatures need to complete development and produce viable adults that replace the previous generation all on the same species of plant. These requirements are rarely fully tested (but see Perkins *et al.*, 2024) and the actual data available are presence records that at best approximate food plants as opposed to true hosts (Farrell *et al.*, 2023) (Table 29.2).

Hosts have been recorded in eight plant families, with most (55%) in the Fabaceae, mainly *Acacia* spp. (37 of 39 records) and 22% in the Myrtaceae (16 *Eucalyptus* spp. and three *Blakella* and two *Corymbia* spp.). The remainder generally only have a few records per family: Casuarinaceae (3), Lecythidaceae (1), Pinaceae (1), Pittosporaceae (1), Proteaceae (2), Sapindaceae (3). Female egg laying and subsequent early instar larvae have a relatively narrow host range (Perkins *et al.*, 2024), but after defoliation larger larvae are more polyphagous and will feed on other hosts. The host status of Casuarinaceae needs to be confirmed as large larvae will certainly feed on plants, and eggs have been observed, but not early instars. Transplants are more likely to be successful with large larvae (Farrell *et al.*, 2023). In terms of broad nesting forms, most hosts have only one nesting type: ground (44%), canopy (28%) and trunk (6%), but some hosts have various combinations and in three cases all three nesting forms have been recorded on the one host species.

29.6 Species Phenology and Dynamics

As far as is known, the species is univoltine (Fig. 29.3) throughout its range.

Adults emerge in spring, often over a protracted period perhaps influenced by earlier rainfall events, sometimes sporadically. They are

Table 29.2. Nesting forms (canopy, trunk, ground) recorded for host food plants of *Ochrogaster lunifer*, common names, distribution and attribution included.

Genus species	Common name	Distribution	Nest form
Family: Fabaceae			
Acacia acuminata[2]	Jam wattle	WA	Canopy, ground
Acacia anceps [3]	Port Lincoln wattle	WA, SA, Qld	Ground
Acacia aneura[15]	Mulga	NSW, SA	Canopy, ground
Acacia aulacocarpa[3]	Hickory wattle	Qld, NSW	Ground
Acacia bivenosa[1]	Dune wattle	Qld, NT, WA	Ground
Acacia cardiophylla[8]	Wyalong wattle	NSW	Ground
Acacia concurrens [3]	Black wattle	Qld	Canopy, ground
Acacia crassicarpa[9]	Northern wattle	Qld, PNG	Unreliable data
Acacia chrystotricha[14]	Bellinger River wattle	NSW	Ground
Acacia decora[1]	Western silver wattle	Qld, NSW, Vic	Ground
Acacia disparrima[1]	Salwood, hickory	Qld, NSW	Ground
Acacia fimbriata[3]	Fringed wattle	Qld, NSW	Ground
Acacia georginae[10]	Georgina gidgee	Qld, NT, SA	Ground
Acacia glaucocarpa[7]	Hickory wattle	Qld, NSW	Ground
Acacia hakeoides[1]	Western black wattle	NSW, Vic, SA, WA	Canopy
Acacia hamiltoniana[1]	Hamilton's wattle	NSW	Ground
Acacia harpophylla[1]	Brigalow	Qld, NSW	Ground
Acacia holosericea[7]	Strap wattle	Qld, NT, WA	Ground
Acacia homalophylla[11]	Yarran	NSW	Canopy
Acacia implexa[3]	Lightwood wattle	Qld, NSW, Vic	Canopy, ground
Acacia iteaphylla[1]	Willow-leaf wattle	SA, NSW, WA	Ground
Acacia leiocalyx [3]	Black wattle	Qld, NSW	Ground
Acacia longifolia[12]	Sydney golden wattle	Qld, NSW, Vic, SA	Not recorded
Acacia maidenii [3]	Maiden's wattle	Qld, NSW, Vic	Ground
Acacia mangium[9]	Brown salwood	Qld, PNG	Not recorded
Acacia melanoxylon[1]	Australian blackwood	Qld, NSW, Vic, Tas, SA	Ground
Acacia murrayana[1]	Murray's wattle	Qld, NSW, Vic, SA, NT, WA	Canopy
Acacia neriifolia[1]	Pechy, silver wattle	Qld, NSW	Ground
Acacia oswaldi[6]	Boree, umbrella wattle	Qld, NSW, SA, WA, NT	Canopy
Acacia pendula[16]	Weeping Myall	Qld, NSW, Vic, SA, WA	Canopy
Acacia podalyriifolia[7]	Mount Morgan wattle	Qld, NSW, Vic, SA	Ground
Acacia prominens[17]	Gosford wattle	NSW, Vic	Ground
Acacia pycnantha[15]	Golden wattle	Vic, SA, Qld, NSW	Ground
Acacia rostellifera[2]	Summer scented wattle	WA	Canopy
Acacia salicina[1]	Cooba, sally wattle	Qld, NSW, Vic, SA, NT	Canopy
Acacia shirleyi[1]	Lancewood	Qld, NT	Canopy
Acacia stenophylla[1]	River Cooba, River Myall	Qld, NSW, Vic, SA, NT	Canopy, trunk, ground
Senna artemisioides[15]	Rosewood/bullock-bush, *Cassia*	Qld, NSW, Vic, SA, WA, NT	Canopy, trunk, ground
Leucaena leucocephala[1]	Tree lucerne, river tamarind	Qld; introduced	Ground
Family: Myrtaceae			
Blakella citriodora[1]	Lemon-scented gum	Qld, NSW	Trunk
Blakella tessellaris[7]	Moreton Bay ash	Qld	Canopy, trunk, ground
Blakella torelliana	Cadaghi	Qld, NSW	Canopy
Corymbia intermedia[3]	Pink bloodwood	Qld, NSW	Ground
Corymbia terminalis[13]	Bloodwood	NT	Canopy
Eucalyptus albens[14]	White box	Qld, NSW, Vic, SA	Not recorded
Eucalyptus blakelyi[1]	Blakely's red gum	Qld, NSW, Vic	Trunk, ground
Eucalyptus calycogona[1]	Gooseberry mallee	Vic, SA, WA	Canopy

Continued

Table 29.2. Continued.

Genus species	Common name	Distribution	Nest form
Eucalyptus conica[1]	Fuzzy box	Qld, NSW	Canopy
Eucalyptus coolabah[1]	Coolibah	Qld, NSW, SA, NT, WA	Canopy
Eucalyptus crebra[1*]	Narrow-leaved ironbark	Qld, NSW, Vic	Canopy
Eucalyptus dawsonii[1]	Slatey gum	NSW	Trunk, ground
Eucalyptus melanophloia[1]	Silver-leaved ironbark	Qld, NSW	Canopy, trunk
Eucalyptus microcarpa[12]	Grey box	Qld, NSW, Vic, SA	Trunk
Eucalyptus moluccana[1]	Grey, gum-topped box	Qld, NSW	Trunk
Eucalyptus oleosa[1]	Red, acorn mallee	Qld, NSW, Vic, SA, WA	Canopy
Eucalyptus orgadophila[1]	Mountain coolibah	Qld	Canopy
Eucalyptus populnea[1]	Brimble box	Qld, NSW	Canopy
Eucalyptus scoparia[12]	Wallangara white gum	Qld, NSW	Trunk
Eucalyptus sideroxylon[1]	Red ironbark	Qld, NSW, Vic	Ground
Eucalyptus tereticornis[12]	Forest red gum	Qld, NSW, Vic, SA, WA	Canopy
Family: Casuarinaceae			
Allocasuarina littoralis[1]	She-oak	Qld, NSW	Ground
Casuarina cunninghamiana[1]	River she-oak	NSW, Vic, NT	Ground
Casuarina sp.[1]	She-oak	Qld, NSW	Trunk, ground
Family: Proteaceae			
Grevillea striata[3]	Beefwood	WA, NT, Qld, NSW, SA	Ground
Persoonia cornifolia[3]	Broad-leaved geebung	Qld, NSW	Ground
Family: Pinaceae			
Pinus elliotti[3]	Slash pine	Qld, NSW	Ground
Family: Lecythidaceae			
Planchonia careya[13]	Cocky apple	Qld, NT, WA	Canopy
Family: Sapindaceae			
Acer rubrum[14]	Red maple	Introduced	Not recorded
Alectryon oleifolius canescens[5]	Boonaree, inland rosewood	Qld, NSW, Vic, SA, WA	Ground
Alectryon oleifolius[5]	Rosewood	Qld, NSW, Vic, SA, NT, WA	Canopy
Family: Pittosporaceae			
Pittosporum angustifolium[1]	Gumbi gumbi	Qld, NSW, Vic, SA, NT, WA	Ground

Sources: [1]Farrell (2023); [1*]M.P. Zalucki (pers. obs.); [2]Jenkins (1962); [3]Floater (1996); [4]Steinbauer and Mitchell (2016); [5]Steinbauer (2009); [6]Steinbauer (2019); [7]Perkins *et al.* (2023); [8]Chisholm (1938); [9]Wylie *et al.* (1998); [10]L. Cook (pers. obs.); [11]Chisholm (1929); [12]Major (2011); [13]Whelan (2010); [14]Herbison-Evans and Crossley (2022); [15]McFarland (1979); [16]Froggatt (1923); [17]Froggatt (1896).

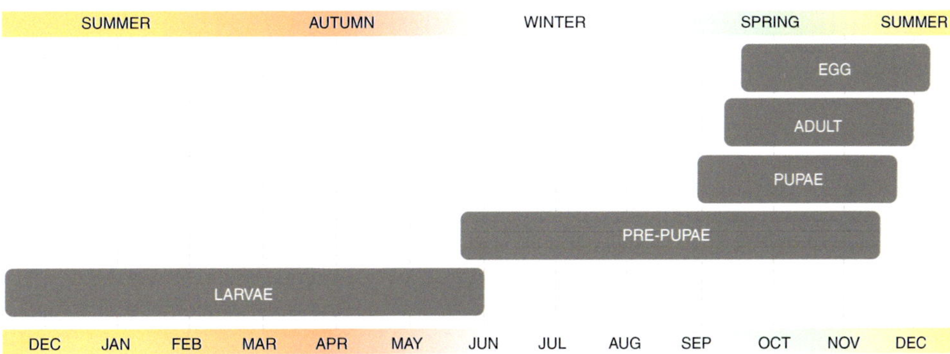

Fig. 29.3. Occurrence of *Ochrogaster* developmental stages over the year.

short-lived, so males and females must emerge synchronously to produce fertilized eggs. Larvae feed from late spring or early summer until late autumn. In general, ground nesters process to prepupation sites earlier, and canopy and trunk nesters later. Indeed, trunk nesters may remain in nests for protracted periods after apparently completing development (Perkins *et al.*, 2024). How synchronization is achieved is not clear, or how it might vary with putative species and geographic locations.

The abundance of the taxon varies greatly over time and space and likely depends on the distribution of hosts and egg-laying behaviour. There can be localized outbreaks leading to defoliation of host plants (Floater and Zalucki, 2000). In open, homogeneous habitats dominated by acacias, more egg batches were laid on high-quality larger trees, which is likely to increase population growth. In habitats containing more diverse species of trees, more egg batches were laid on low-quality but highly apparent trees, which may represent an adaptation to reduce population growth and stabilize population dynamics. The aggregation of batches on small apparent trees in diverse habitats led to 'outbreaks' on these trees year after year, even when population levels were low, while site-wide outbreaks were rare (Floater and Zlucki, 2000). Battisti *et al.* (2023) described a similar mechanism in pine processionary moths.

In thoroughbred horse-breeding areas where ornamental plantings of hosts are common, such as the Hunter Valley region (NSW), farm outbreaks vary from year to year and are associated with increased cases of EAFL (Carrick *et al.*, 2014; J. Farrell and M. Zalucki, personal observation).

29.7 Natural Enemies and the Defensive Function of Setae

Antipredator defence mechanisms can be chemical, morphological or behavioural in nature (see Chapter 6) and are typically categorized by whether they function before (primary defence) or after (secondary defence) being discovered by a predator. Primary defence mechanisms, such as crypsis, aposematism and mimicry, reduce or avoid detection, encounter and approach by predators (Edmunds, 1974; Ruxton *et al.*, 2004). In contrast, secondary defence mechanisms activated after discovery deter the enemy from pursuing an attack (Edmunds, 1974; Ruxton *et al.*, 2004). Uemura *et al.* (2017) examined the defences of *O. lunifer* caterpillars towards ants and found evidence for a chemical repellent associated with setae. For the eggs, their arrangement in masses offered some protection to inner eggs (Uemura *et al.*, 2019), as did their covering of anal true setae of the adult females.

The long setae which give larvae their characteristic hairy appearance can act as a physical barrier and visual deterrent. The smaller true setae are thought to act as a secondary defence against homeothermic animals (Battisti *et al.*, 2011) and are the causative agents of urticaria and life-threatening immune reactions in other processionary moths (Moneo *et al.*, 2015), and EAFL (Cawdell-Smith *et al.*, 2012). In *O. lunifer* true setae are hollow, containing a venom consisting of numerous putative toxins from the secapin, odorant-binding protein, and arylphorin protein families with the insect immune modulator, Diedel (Walker *et al.*, 2023). The molecular modes of action of these molecules are unknown, but a highly similar secapin peptide from the venom of the related oak processionary caterpillar, *Thaumetopoea processionea*, was found to be an agonist of the pain- and heat-associated neuronal receptor TRPV1 (Seldeslachts *et al.*, 2024).

It is likely that the venom of processionary caterpillars has defensive capabilities on both naïve and previously exposed predators. Venom of *Thaumetopoea pityocampa* causes degranulation of mast cells through mechanisms independent of the allergen-mediating immunoglobulin E (IgE) (Lamy *et al.*, 1983) and hence would be expected to cause irritation in naïve predators. Individuals previously sensitized to *Thaumetopoea* spp. are likely to have additional inflammation mediated by IgE (Lamy *et al.*, 1983). However, the corresponding data for *O. lunifer* is currently lacking.

Regardless of the mechanism, the inflammatory symptoms caused by true setae typically do not occur immediately, and therefore their usefulness during a confrontation is questionable as predators are unlikely to be repelled before the caterpillars are killed or eaten. Moreover, it is unlikely for the predator to be able to connect any of the symptoms that may eventuate back to the original attack given the delayed time frame of its effect. Possibly, these deviations from the typical defensive adaptations of arthropods are related to the social nature of *O. lunifer* and have evolved due to selection pressures similar to inclusive fitness, kin selection or colony selection (Moneo *et al.*, 2015). Similar apparent contradictions in venom biology, including lethal sting autotomy in bees and wasps (Hermann, 1971) and induction of long-lasting allodynia (lowered pain thresholds) by ant venom peptides (Eagles *et al.*, 2022), occur amongst social Hymenoptera.

Given their extensive array of defensive features, *O. lunifer* have been assumed to be subject to relatively little predation. To date, mainly insect natural enemies have been documented feeding on the various life stages (Table 29.3). Of these recorded enemies, most are parasitoids: at least nine species of Tachinidae across all nesting forms and seven species of Hymenoptera across five families (Table 29.3). Invertebrate egg mass predators such as dermestid beetles have been reported to be damaging to their population sizes (Floater and Zalucki, 1999). Overall, the egg and neonate stages appear to be the most vulnerable to invertebrate predators and host plant effects, whereas later instar larvae, pupae and adults are more prone to attacks by parasitoids (Floater, 1996b; Floater and Zalucki, 1999) and birds. *Cacomantis pyrrhophanus*, *Cuculus pallidus*, *Chysococcyx basalis*, *Ch. plagosus*, *Petroica multicolor*, *Colluricincla harmonica*, *Oriolus sagittatus* and *Cracticus nigrogulari* (Young, 1973; Mackey, 1985) have been reported taking large larvae from eucalypt canopies.

There is a suggestion that different nesting forms on different hosts may have different suites of natural enemies (Table 29.3). This is quite apparent for the egg parasitoid *Anastatus fuiligispina*. Levels of egg parasitism of ground nesters on acacias were very high (95%) (Uemura *et al.*, 2019) but none were found in 24 egg masses on trunk nesters on *Blakella tessellaris* in the same habitat (Perkins *et al.*, 2023).

29.8 Management of *Ochrogaster lunifer*

Defoliation and damage of acacias and ornamental eucalypts rarely warrant intervention other than for aesthetic reasons. Of more concern are the health risks to humans caused by setae of larvae during outbreaks. Similarly, high-value thoroughbred horse studs can suffer increased rates of EAFL during outbreak seasons. In these cases, management is necessary to prevent exposure to setae. It is not known what a sufficient dose is to cause adverse outcomes for pregnant mares (see Perkins *et al.*, 2019), so a zero-tolerance approach is taken.

Management in these situations is not complicated once you know the biology and ecology of the species involved. Removal has been the most practical and effective approach to date. The initial stage is searching the subject property for the presence of *O. lunifer* and trees and shrubs that are potential host plants. This needs to be undertaken by a suitably trained and experienced person. Ground nesters lay white egg masses that look like cotton wool balls at the base of a host tree trunk. The white egg masses are easily detected during the egg-laying season in October–December. Often two or three searches up to Christmas/New Year will identify all the potential nests of ground nesters. The egg masses or early-stage nests are more easily removed than a much larger later-stage nest of several hundred larvae.

Egg masses, larvae and nest material including shed exoskeletons need to be removed, bagged and destroyed when detected to minimize the risk of mares, and in some circumstances humans, inadvertently contacting this material. Such material is very light and can easily be dispersed in the wind so collection and transport need to be avoided on windy days and should be done with care and wearing appropriate personal protection. Destruction may be by deep burial or by burning. Eggs of canopy and trunk nesters are more difficult to locate, and a cherry picker may be necessary. Canopy and trunk nests become more obvious once they are larger, and particular care is needed not to rupture nests during removal.

Generally, a few years and rounds of inspections are needed to reduce populations greatly but the approach works (Carrick *et al.*, 2014). Where there is the potential for *O. lunifer* to reinfest a property from neighbouring areas, monitoring is an annual requirement. The education and training of on-site personnel greatly assist the long-term success of management programmes.

Table 29.3. Natural enemies (parasitoids and predators) recorded attacking *Ochrogaster lunifer* life stages (eggs, eggs/first instar larvae (L1), larvae) arranged by nesting form (canopy, trunk, ground) and host (*Eucalyptus*, *Blakella*, *Acacia*). Within a nesting form column, hosts are distinguished by shades of grey. No colour indicates both host genera.

	Nest and host type			Natural enemies		
	Canopy nests	Trunk nests	Ground nests			
	Eucalyptus, Acacia	*Eucalyptus, Blakella*	*Eucalyptus, Acacia*	Order	Family	Functional group
Eggs	*Anastatus fuligispina*		*Anastatus fuligispina*	Hymenoptera	Eupelmidae	Parasitoid
			Elachertus sp.	Hymenoptera	Eulophidae	
Eggs/L1	*Dermestes ater*	*Dermestes ater*	*Dermestes ater*	Coleoptera	Dermestidae	Predator
	Dermestes frischii		*Dermestes frischii*	Coleoptera	Dermestidae	
			Trogoderma apicipenne	Coleoptera	Dermestidae	
	Titanoceros thermoptera			Lepidoptera	Pyralidae	
Larvae	*Anamastax* sp.	*Anamastax* sp.		Diptera	Tachinidae	Parasitoid
	Carcelimyia sp.	*Carcelimyia dispar*	*Carcelimyia dispar*	Diptera	Tachinidae	
		Blepharipa sp.		Diptera	Tachinidae	
	Exorista sp. 1	*Exorista* sp.		Diptera	Tachinidae	
	Exorista sp. 2			Diptera	Tachinidae	
	Microtropesa violacescens	*Microtropesa violacescens*		Diptera	Tachinidae	
	Palexorista sp.		*Palexorista* sp.	Diptera	Tachinidae	
	Winthemia sp.	*Winthemia* sp.		Diptera	Tachinidae	
	Unidentified spp.			Diptera	Tachinidae	
	Meteorus luteus			Hymenoptera	Braconidae	
			Dirhinus sp.	Hymenoptera	Chalcididae	
			Paramesius sp.	Hymenoptera	Diapriidae	
		Unidentified sp. 1		Hymenoptera	Diapriidae	
	Eurytoma sp. 1			Hymenoptera	Eurytomidae	
	Eurytoma sp. 2			Hymenoptera	Eurytomidae	
	Eriborus sp.			Hymenoptera	Ichneumonidae	
	Lycosa sp.			Araneae	Lycosidae	Predator
			Dindymus circumcinctus	Hemiptera	Pyrrhocoridae	
			Pristhesancus plagipennis	Hemiptera	Reduviidae	
			Unidentified sp. (ants)	Hymenoptera	Formicidae	
Legend:	*Acacia*	*Eucalyptus*	*Blakella*		Both host species	

Sources: Farrell (2023), Floater (1996a), Floater and Zalucki (1999), Perkins *et al.* (2023, 2024), Uemura *et al.* (2019), van Schagen *et al.* (1992b).

Another management tool is the removal of host trees, as was done for the monospecific eastern tent caterpillar in the USA (Fitzgerald, 1995). Removal of all hosts on a property is probably not possible given the extensive host list (Table 29.2) but may be an option for high-risk areas. Isolating pregnant mares from paddocks where *O. lunifer* may be present is also an option.

29.9 Key Outstanding Questions

Species like *Ochrogaster lunifer* challenge our interpretation of ecology and biology. It is either polyphagous, polymorphic and plastic for a wide range of characters and has a wide climate niche or, more likely, is a complex of host-specialist species with unique but related attributes. In either case, many characters are likely to vary due to localized adaptation, and these need to be recorded and understood if we are to make predictions of population changes and manage health and veterinary risks. The likely numerous species in the complex may have unique venoms and toxins associated with their defensive setae, perhaps adapted to deal with different suites of natural enemies that occur in different habitats. The general conclusion is 'one size does not fit all'. We suspect many biological and ecological studies of moths suffer from poor definition and identification of species limits. Just how diverse is Australia's moth fauna?

Acknowledgements

Much of the work in south-east Queensland and northern NSW has been supported by the thoroughbred racing industry, ARC-Linkage and ARC discovery grants, as well as funds from the European Union's Horizon 2020 Program for Research & Innovation under grant agreement no.771271 'HOMED' and UQ.

References

Alice Springs Field Naturalists Club (2023) Newsletter, May. Available at: https://alicefieldnaturalists.org.au/23_05.pdf (accessed 8 October 2025).

Bartholomew, K. (2022) Processionary Caterpillar a Welcome Sign Ahead of Annual Mullet Season. ABC News, Sunshine Coast. Available at: www.abc.net.au/news/2022-04-30/processionary-caterpillars-march-through-south-east-queensland/101021888 (accessed 8 October 2025).

Basso, A., Negrisolo, E., Zilli, A, Battisti, A. and Cerretti, P. (2017) A total evidence phylogeny for the processionary moths of the genus *Thaumetopoea* (Lepidoptera: Notodontidae: Thaumetopoeinae). *Cladistics* 33, 557–573.

Battisti, A., Holm, G., Fagrell, B., and Larsson, S. (2011) Urticating hairs in arthropods – their nature and medical significance. *Annual Review of Entomology*, 56, 203–220.

Battisti, A., Larsson, S. and Roques, A. (2017) Processionary moths and associated urtication risk: global change-driven effects. *Annual Review of Entomology* 62, 323–342.

Battisti, A., Hódar, J.A., Hernańdez, R. and Larsson, S. (2023) Aggregative oviposition varies with density in processionary moths – Implications for insect outbreak propensity. *Ecological Entomology* 48, 102–111.

Battisti, A., Walker, A.A., Uemura, M., Zalucki, M.P., Brinquin, A.-S. *et al.* (2024) Look but do not touch: The occurrence of venomous species across Lepidoptera. *Entomologia Generalis* 44, 29–39.

Bockxmeer, van J.J. and Green, J. (2013) Paediatric osteomyelitis after exposure to toxic *Ochrogaster lunifer* moth. *Medical Journal of Australia* 199, 331–332.

Carrick, J.B., Perkins, N.R. and Zalucki, M.P. (2014) Causes of abortion in Australia (2005–2012) – proportion of cases due to equine amnionitis and foetal loss (EAFL). *Journal of Equine Veterinary Science* 34, 212–214.

Cawdell-Smith, A.J., Todhunter, K.H., Anderson, S.T., Perkins, N. R. and Bryden, W.L. (2012) Equine amnionitis and fetal loss: mare abortion following experimental exposure to processionary caterpillars (*Ochrogaster lunifer*). *Equine Veterinary Journal* 44, 282–288.

Cawdell-Smith, A.J., Todhunter, K.H., Perkins, N.R. and Bryden, W.L. (2013) Exposure of mares to processionary caterpillars (*Ochrogaster lunifer*) in early pregnancy: An additional dimension to equine amnionitis and fetal loss. *Equine Veterinary Journal* 45, 755–760.

Chisholm, E.C. (1929) Birds of the East Bogan District, county of Flinders, New South Wales. *Emu – Austral Ornithology* 29, 143–151.

Chisholm, E.C. (1938) The birds of Barellan, New South Wales with botanical and other notes. *Emu – Austral Ornithology* 37, 301–313.

Common, I.F.B. (1990). *Moths of Australia*. Melbourne University Press, Collingwood, Victoria.

Eagles, D.A., Saez Natalie, J., Krishnarjuna, B., Bradford J.J., Chin, Y.K.-Y. *et al.* (2022). A peptide toxin in ant venom mimics vertebrate EGF-like hormones to cause long-lasting hypersensitivity in mammals. *Proceedings of the National Academy of Sciences* 119, e2112630119.

Edmunds, M. (1974) *Defence in Animals: a Survey of Anti-predator Defences*. Longman, Harlow, UK.

Farrell, J.F. (2023) The Biology, Ecology, and Management of Processionary Caterpillars Ochrogaster lunifer, Primary Agents of Equine Amnionitis and Foetal Loss (EAFL). PhD thesis, University of Queensland, Brisbane, Australia.

Farrell, J.F., Zalucki, M.P. and Battisti, A. (2023) Host specificity in canopy nesting forms of *Ochrogaster lunifer*: The older children do not care? *Insects* 14, 420.

Fitzgerald, T.D. (1995) *The Tent Caterpillars*. Cornell University Press, Ithaca, New York.

Floater, G.J. (1996a) Life history comparisons of ground- and canopy-nesting populations of *Ochrogaster lunifer* Herrich-Schäffer (Lepidoptera: Thaumetopoeidae): Evidence for two species? *Australian Journal of Entomology* 35, 223–230.

Floater, G.J. (1996b) The Brooks Dyar rule and morphometrics of the processionary caterpillar, *Ochrogaster lunifer* Herrich-Schäffer (Lepidoptera: Thaumetopoeidae). *Australian Journal of Entomology* 35, 271–278.

Floater, G.J. (1996c) Estimating movement of the processionary caterpillar, *Ochrogaster lunifer* Herrich-Schäffer (Lepidoptera: Thaumetopoeidae) between discrete resource patches. *Australian Journal of Entomology* 35, 279–283.

Floater, G.J. (1997a) Spatial Dynamics of a Gregarious Herbivore. PhD thesis, University of Queensland, Brisbane, Australia.

Floater, G.J. (1997b) Rainfall, nitrogen and host plant condition: consequences for the processionary caterpillar, *Ochrogaster lunifer* (Thaumetopoeidae). *Ecological Entomology* 22, 247–255.

Floater, G.J. (1998) Tuft scales and egg protection in *Ochrogaster lunifer* Herrich-Schäffer (Lepidoptera: Thaumetopoeidae). *Australian Journal of Entomology* 37, 34–39.

Floater, G.J. and Zalucki, M.P. (1999) Life tables of the processionary caterpillar *Ochrogaster lunifer* Herrich-Schäffer at local and regional scales. *Australian Journal of Entomology* 38, 330–339.

Floater, G.J. and Zalucki, M.P. (2000) Habitat structure and egg distributions in the processionary caterpillar *Ochrogaster lunifer*: lessons for conservation and pest management. *Journal of Applied Ecology* 37, 87–99.

Froggatt, W.W. (1896) On the bag shelters of lepidopterous larvae of the genus *Teara*. *Proceedings of the Linnean Society of New South Wales* 21, 258–260.

Froggatt, W.W. (1923) *Forest Insects of Australia*. Forestry Commissioners of New South Wales, Sydney, NSW.

Hermann, H.R. (1971) Sting autotomy, a defensive mechanism in certain social Hymenoptera. *Insectes Sociaux* 18, 111–120.

Jenkins, C.F.H. (1962) Insect pests of forests, 2: Tussock moths and bag-shelter moths. *Journal of Agriculture, Western Australia* 3, 2–4.

Lamy, M., Vincendeau, P., Ducombs, G. and Pastureaud, M. H. (1983) Irritating substance extracted from the *Thaumetopoea pityocampa* caterpillar mechanism of action. *Experientia* 39, 299–299.

McFarland, N. (1979) Annotated list of larval food plant records for 280 species of Australian moths. *Journal of the Lepidopterist's Society (Supplement)* 33, 1–72.

Mackey, A. P. (1985) Moth species eaten by pied butcher birds. *Australian Journal of Entomology* 24, 93–94.

Major, R. (2011) Notice of final determination. *NSW Government Gazette* 4833.

Mather, A., Zalucki, M.P., Farrell, J., Perkins, L.E. and Cook, L.G. (2019) The Australian processionary caterpillar, *Ochrogaster lunifer* Herrich-Schäffer (Lepidoptera: Notodontidae), comprises cryptic species. *Austral Entomology* 58, 816–825.

Mills, M.B. (1950) Observations on processionary caterpillars. *Western Australian Naturalist* 2, 84–87.

Mills, M.B. (1951a) Bag shelter caterpillars and their habits. *Western Australian Naturalist* 3, 61–67.

Mills, M.B. (1951b) Bag shelter caterpillars and their habits. *Western Australian Naturalist* 3, 84–92.

Moneo, I., Battisti, A., Dufour, B., García-Ortiz, J.C., González-Muñoz, M. *et al.* (2015) Medical and veterinary impact of the urticating processionary larvae. In: Roques, A. (ed.) *Processionary Moths and Climate Change: An Update.* Springer, Dordrecht, The Netherlands, pp. 359–410.

Perkins, L.E., Zalucki, M.P., Perkins, N.R., Cawdell-Smith, A.J., Todhunter, K.H. *et al.* (2016) The urticating setae of *Ochrogaster lunifer*, an Australian processionary caterpillar of veterinary importance. *Medical and Veterinary Entomology* 30, 241–245.

Perkins, L.E., Cribb, B.W., Pagendam, D. and Zalucki, M.P. (2019) Variation in morphology and air-borne dispersal of the urticating apparatus of *Ochrogaster lunifer*, an Australian processionary caterpillar, and implications for livestock and humans. *Journal of Insect Science* 19, 1–8.

Perkins, L.E., Uemura, M. and Zalucki, M.P. (2023) A trunk-nesting form of the processionary caterpillar *Ochrogaster lunifer* (Lepidoptera: Notodontidae) restricted to a single host species *Corymbia tessellaris* (Myrtaceae), with some comparisons to the ground-nesting form. *Austral Entomology* 62, 246–256.

Perkins, L E., Uemura, M., Zalucki, M.P. and Cook, L.G. (2024). Host specificity of two co-occurring nesting-forms of the bag-shelter moth *Ochrogaster lunifer* (Lepidoptera: Notodontidae) is congruent with there being cryptic species. *Austral Entomology* 63, 406–412.

Randell, D., Fett, M., Jenner, H., Frances, S. and Shanks, D. (2017) Caterpillar dermatitis in military cadets. *Journal of Military and Veterans' Health* 25, 7–10.

Ruxton, G.D., Sherratt, T.N. and Speed, M.P. (2004) *Avoiding Attack: The Evolutionary Ecology of Crypsis, Warning Signals and Mimicry.* Oxford University Press, Oxford.

Seldeslachts, A., Undheim, E.A.B., Vriens, J., Tytgat, J. and Peigneur, S. (2024) Exploring oak processionary caterpillar induced lepidopterism (part 2): Ex vivo bio-assays unmask the role of TRPV1. *Cellular and Molecular Life Sciences* 81, 281.

Schagen, J.J. van, Hobbs, R.J. and Majer, J.D. (1992a) Defoliation of trees in roadside corridors and remnant vegetation in the Western Australian wheatbelt. *Journal of the Royal Society of Western Australia* 75, 75–81.

Schagen, J. van, Majer, J.D. and Hobbs, R.J. (1992b) Biology of *Ochrogaster lunifer* Herrich-Schaeffer (Lepidoptera: Thaumetopoeidae), a defoliator of *Acacia acuminata* Bentham, in the Western Australian wheatbelt. *Australian Entomological Magazine* 19, 19–24.

Schintlmeister, A. (2013). *Notodontidae and Oenosandridae (Lepidoptera). World Catalogue of Insects.* Brill, Leiden, The Netherlands.

Southcott, R.V. (1987) Moths and butterflies. In: Covacevich, J., Davie, P. and Pearn, J. (eds) *Toxic Plants and Animals: A Guide for Australia*. Queensland Museum, Brisbane, Queensland, pp. 243–257.

Steinbauer, M.J. (2009) Thigmotaxis maintains processions of late-instar caterpillars of *Ochrogaster lunifer*. *Physiological Entomology* 34, 345–349.

Steinbauer, M.J. (2019) Wing pattern polyphenism in two behavioural forms of *Ochrogaster lunifer* (Lepidoptera: Notodontidae). *Austral Entomology* 58, 432–442.

Steinbauer, M.J. and Mitchell, M. (2016) Explaining phenotypic and behavioural variability in *Ochrogaster lunifer* (Lepidoptera: Notodontidae) from an evolutionary ecology perspective. *F1000Research* 5, 351 (poster).

Todhunter, K.H., Cawdell-Smith, A.J., Bryden, W. L., Perkins, N.R., and Begg, A.P. (2014a) Processionary caterpillar setae and equine fetal loss: 1. Histopathology of experimentally exposed pregnant mares. *Veterinary Pathology* 51, 1117–1130.

Todhunter, K.H., Cawdell-Smith, A.J., Bryden, W.L., Perkins, N.R., and Begg, A.P. (2014b) Processionary caterpillar setae and equine fetal loss: 2. Histopathology of the fetal-placental unit from experimentally exposed mares. *Veterinary Pathology* 51, 1131–1142.

Uemura, M., Perkins, L.E., Zalucki, M.P. and Cribb, B.W. (2017) Predator–prey interaction between greenhead ants and processionary caterpillars is mediated by chemical communication. *Animal Behaviour* 129, 213–222.

Uemura, M., Perkins, L., Battisti, A. and Zalucki, M.P. (2019) Egg mass structure of the processionary caterpillar *Ochrogaster lunifer* (Lepidoptera: Notodontidae): is the outer egg layer sacrificed for attack by the egg parasitoid *Anastatus fuligispina* (Hymenoptera: Chalcidoidea: Eupelmidae)? *Austral Entomology* 58, 810–815.

Uemura, M., Perkins, L., Zalucki, M.P. and Battisti, A. (2020) Movement behaviour of two social urticating caterpillars in opposite hemispheres. *Movement Ecology* 8, 3807.

Uemura, M., Zalucki, M.P. and Battisti, A. (2021) Behavioural plasticity and tree architecture shapes tent and foraging locations of pine processionary larval colonies Behaviour of pine processionary colonies. *Entomologia Generalis* 41, 121–136.

Uemura, M., Zalucki, M.P. and Battisti, A. (2023) Large male caterpillars are the primary builders: exploring tent construction and foraging behaviour in gregarious pine processionary caterpillar. *Insects* 14, 829.

Walker, A.A., Perkins, L.E., Battisti, A., Zalucki, M.P. and King, G.F. (2023) Proteome of urticating setae of *Ochrogaster lunifer*, a processionary caterpillar of medical and veterinary importance, including primary structures of putative toxins. *Proteomics* 23, e2300204.
Walsh, F., Christophersen, P. and McGregor, S. (2014) Indigenous perspectives on biodiversity. In: Morton, S., Sheppard, A. and Lonsdale, M. (eds) *Biodiversity: Science and Solutions for Australia*. CSIRO Publishing, Melbourne, Victoria, pp. 81–199.
Whelan, P. (2010) *Stinging and Itchy Caterpillars in the Top End of the Northern Territory*. Department of Health & Families, NT Government Report, Darwin, NT.
Wylie, R., Floyd, R., Elliot, H., Khen, C.V., Intachat, J. *et al.* (1998) Insect pests of tropical acacias: a new project in southeast Asia and northern Australia. In: Turnbull, J.W., Crompton, H. R. and Pinyopusarerk, K. (eds) *Recent Developments in Acacia Planting.* Proceedings of an International Workshop held in Hanoi, Vietnam, 27–30 October 1997. ACIAR Proceedings No. 82, pp. 234–239.
Young, H. (1973) Birds noted eating processionary caterpillars. *Australian Bird Watcher* 5, 79.

30 The Fruit-piercing Moths: *Eudocima* Species – Identities, Distribution, Ecology and Pest Status

Donald P.A. Sands

CSIRO Health and Biosecurity, Brisbane, Queensland, Australia, and School of the Environment, University of Queensland, St Lucia, Queensland, Australia

Abstract

Seven species of fruit-piercing moths (*Eudocima* spp., Erebidae, Calpinae) occur in Australia: *Eudocima phalonia*, *E. salaminia*, *E. materna*, *E. iridescens*, *E. jordani*, *E. cocalus* and *E. aurantia*. Fruit-piercing moths are distinguished from other Erebidae by the presence of modified and broadened subapical spines of the proboscis, enabling both sexes to penetrate the firm skin of ripening tropical and subtropical fruit. *Serrodes campana* is also known to pierce fruit. *E. phalonia*, *E. salaminia* and *E. materna* are the most important pests in the Northern Territory, Queensland and north-eastern New South Wales. Larvae of Australian *Eudocima* spp. feed on rainforest vines (Menispermaceae), whereas in Papua New Guinea (PNG) and Pacific islands, larvae of *E. fullonia* also feed on *Erythrina* spp. (Fabaceae). Investigations on the natural enemies of *E. fullonia* in the Pacific Region, identified potential biological control agents from PNG, including two egg parasitoids: *Telenomus lucullus* (Scelionidae) and an undescribed *Ooencyrtus* sp. (*papilionis* Ashmead, species group, Encyrtidae). Both parasitoids were released on Pacific islands, achieving moderate levels of control, but were insufficiently host specific for introduction into Australia.

30.1 Introduction

More than 50 species of fruit-piercing moths (*Eudocima* Billberg 1820, Erebidae, Calpinae) occur worldwide. They are pests of horticulture in South-east Asia (Banziger, 1982; Zaspel and Branham, 2008), India (Baptist, 1945; Singh *et al.*, 2019), the Philippines (Zilli and Hogenes, 2002; Leroy *et al.*, 2021), the South Pacific islands (Cochereau, 1977; Sands and Liebregts, 2006), South America (Vargas-Fonseca *et al.*, 2020) and Australia (Tryon, 1924: Sands and Schotz, 1989; Fay and Halfpapp, 2006). Mostly referred to *Othreis* Hübner, 1823, in the Australian literature (e.g. Sands, 1987a; Sands and Schotz, 1991, 1992), Edwards (1996) synonymized *Othreis* and 18 other genera of fruit-piercing moths with *Eudocima*. The identities of fruit-piercing moths and four newly described *Eudocima* spp. were subsequently reviewed by Zilli and Hogenes (2002).

In Australia, both sexes of fruit-piercing moths (*Eudocima* spp.) cause damage by piercing fruit. Destructive piercing species are *E. phalonia*, *E. salaminia*, *E. materna*, *E. jordani*, *E. cocalus* and *E. aurantia*. The uncommon *E. iridescens* is not considered a pest species. An additional moth, *Serrodes campana*, has been reported piercing fruit in northern Queensland (Fay and Halfpapp, 2006). The most widely distributed and

Corresponding author: donsands2@gmail.com

DOI: 10.1079/9781800627680.0030

important pest species, *E. phalonia*, occurs in Africa, India, southern China, several Indo-Pacific countries, Australia, the western islands of the South Pacific (Fig. 30.1) and as far east as Hawaii (Baptist, 1945; Waterhouse and Norris, 1987; Leroy *et al.*, 2021). The Australian species of fruit-piercing moths also occur in Papua New Guinea (PNG), where they are not so abundant and not considered pests (Sands and Liebregts, 2006). In New Caledonia, Cochereau (1972a, b, 1974a, b, 1977) conducted extensive studies on *E. phalonia*, its parasitoids and predators, and provided the basis for the comprehensive review by Waterhouse and Norris (1987) of the ecology, distribution, food plants, pest status and natural enemies of this moth in the Pacific Region. Subsequently, Sands (1987a, 1996), Sands and Schotz (1989), Fay and Halfpapp (1991a, b) and Fay (1994) undertook studies on fruit-piercing moths in Australia, aiming towards a better understanding of the identities, ecology, pest status and control measures for the moths.

Most recently, Leroy *et al.* (2021) reviewed the distribution, ecology, pest status and management of *E. phalonia* in countries to the north of Australia and provided a synthesis of data and publications on this moth with the focus on their pest status. Sands (1987a) and Sands and Schotz (1989, 1992) focused on identifying the pest species of fruit-piercing moths in Australia and in the Pacific Region, distinguishing piercing species from fruit-sucking moths, and investigating natural enemies and their potentials as biological control agents.

Natural enemies of *Eudocima* spp. are few in species and uncommon in Australia, leading to an abundance of moths and consequential damage to crops. Moreover, aposematically coloured mature larvae of the moths deter vertebrate predation. They are likely to be toxic from feeding as larvae on Menispermaceae, vines known to be rich in alkaloids (Blanchfield *et al.*, 1993; Hungerford *et al.*, 1998). Chemical control of fruit-piercing moths in orchards is difficult to

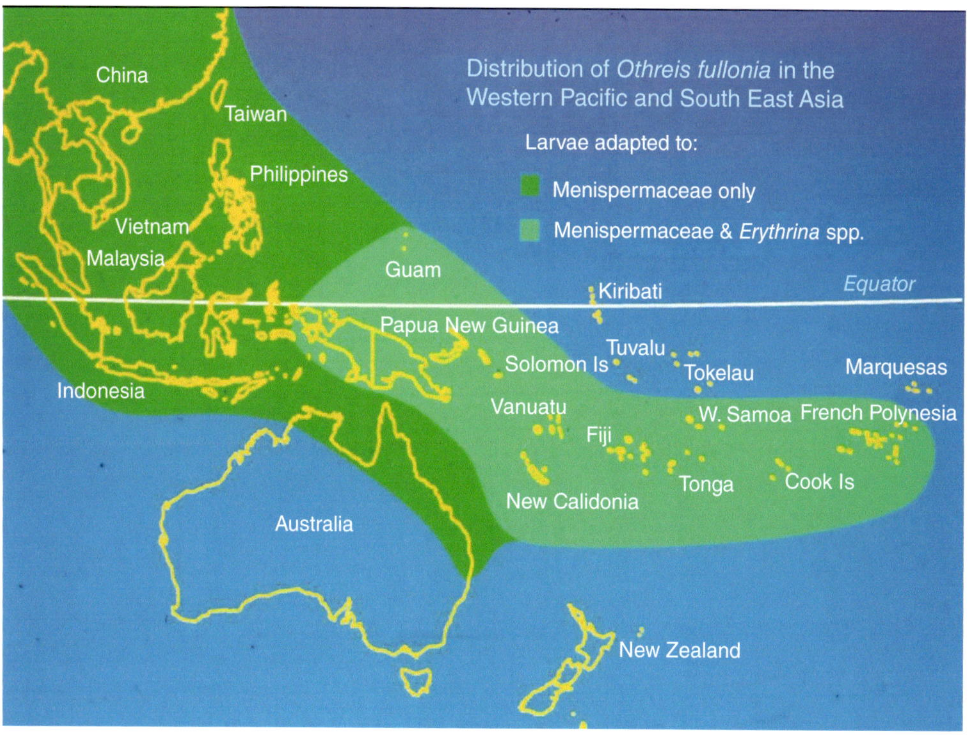

Fig. 30.1. Distribution of *Eudocima phalonia* (= *O. fullonia*) and food plants in the western Pacific Region, now including Hawaii.

achieve, as pesticides cannot be applied to fruit close to harvest due to human health hazards associated with residues (Waterhouse and Norris, 1987).

30.2 The Australian Fruit-piercing Moths

Seven species of *Eudocima* are dealt with in this account. Details of taxonomy, synonymy, similar species, global distributions and pest status are presented for each species in the Appendix to this chapter. Details of known larval food plants are in Table 30.1 and of known natural enemies in Table 30.2.

Fruit-piercing moths can damage pre-ripened, ripening or ripe fruit by piercing the skin with the tip of their sclerotized proboscides and then penetrating tissues to abrade pulp with the barbed apex, to extract the juice. The most widely distributed species in northern and eastern Australia are *E. phalonia* (Fig. 30.3A–C), *E. salaminia* (Fig. 30.2, Fig. 30.3D–F) and *E. materna* (Fig. 30.3G–I). Four species, *E. jordani* (Fig. 30.4A–C), *E. cocalus* (Fig. 30.4D–F), *E. aurantia* (Fig. 30.3J–L) and *E. iridescens* (Fig. 30.4G–I), are restricted to tropical rainforest areas of northern Queensland (Fay and Halfpapp, 2006; Zborowski and Edwards, 2007). With the exception of the uncommon species *E. iridescens*, both sexes of other *Eudocima* spp. cause substantial damage to fruit.

Table 30.1. Food plants of *Eudocima* spp. (Menispermacee) and seasonal areas of moth occurrence.

		Areas of occurrence	
Larval food plants*	*Eudocima* spp.	Perennial	Seasonal (October–May)
Carronia multisepalea	*Eudocima phalonia*	N/A	Qld: Gympie – Tweed River; NSW: Qld border – Dorrigo
Echinostephania aculeata	*Eudocima phalonia*	N/A	Qld: Gympie – Clarence River
Hypserpa decumbens	*Eudocima phalonia*	N.T.; Qld: Bamaga –Mackay	Qld: Rockhampton – Tweed River; NSW: Tweed River – Pt Macquarie
Legnephora moorei	*Eudocima phalonia*	N.T.; Qld: Bamaga – Mackay	Qld: Rockhampton – Tweed River
Legnephora moorei	*Eudocima materna*	Qld: Bamaga – Mackay	NSW: Tweed River – Richmond River
Pachygone ovata	*Eudocima phalonia*	Qld: Cape York – Sarina	Qld: Rockhampton – Bundaberg
Parapachygone longifolia	*Eudocima iridescens*	N.E. Qld: Bamaga – Paluma	N/A
Pleogyne australis	*Eudocima phalonia*	N.T; Qld: Cape York – Mackay	Qld: Rockhampton –Tweed River; NSW: Qld border – Richmond River
Pycnarrhena novoguineensis	*Eudocima iridescens*	N.E. Qld: Bamaga –Paluma	N/A
Sarcopetalum harveyanum	*Eudocima phalonia*	Qld: Cape York – Mackay	Qld: Rockhampton – Tweed River; NSW: Qld border – Coffs Harbour
Stephania bancroftii	*Eudocima phalonia*	Qld: Forty Mile Scrub, Little Mulgrave	N/A
Stephania japonica japonica	*Eudocima salaminia*	N.T; Qld: Cape York – NSW Border	Qld: south to Tweed River

Continued

Table 30.1. Continued.

Larval food plants*	*Eudocima* spp.	Areas of occurrence	
		Perennial	Seasonal (October–May)
Stephania japonica all varieties	*Eudocima phalonia*	N.T; Qld: Cape York – Sarina	Qld: Rockhampton – Tweed River; NSW: Tweed River – Coffs Harbour
S. japonica var. *discolor*	*Eudocima phalonia*	N.T; Qld: Cape York – Sarina	NSW: Tweed River – Pt Macquarie
S. japonica (var. *japonica*)	*Eudocima phalonia*	N.T; Qld: Cape York – Sarina	Qld: Rockhampton – Tweed River; NSW: Tweed River – Coffs Harbour

*Other Menispermaceae listed in Bostock and Holland (2010) are occasional food plants in northern Queensland.

Table 30.2. Arthropod natural enemies of fruit-piercing moths in Australia.

Natural enemy*	*Eudocima* host(s)	Stage attacked
Parasitoids		
Echthromorpha agrestoria (Ichneumonidae)	*E. salaminia*	Larva/pupa
Euplectrus melanocephalus (Eulophidae)	*E. salaminia, E. materna, E. phalonia*	Larva
Ooencyrtus sp. (nr. *cochereau*) (Encyrtidae)	*E. salaminia*	Egg
Ooencyrtus spp. (*papilionis* group), 2 spp. (Encyrtidae)	*E. salaminia, E. materna, E. phalonia*	Egg
Teleonomus sp. (Scelionidae)	*E. salaminia, E. materna, E. phalonia*	Egg
Anastatus sp. (Eupelmidae) (?hyperparasitoid)	*E. salaminia*	Egg
Trichogramma chilonis (Trichogrammatidae)	*E. salaminia, E. materna*	Egg
Trichogramma sp. (Trichogrammatidae)	*E. salaminia*	Egg
Exorista sorbillans (Tachinidae)	*E. salaminia*	Larva/pupa
Predators		
Pheidole megacephala (Formicidae)	*E. salaminia, E. phalonia*	Egg
Myrmecia gulosa (Formicidae)	*E. salaminia*	Larva
Oecophylla smaragdina (Formicidae)	*E. phalonia*	Larva
Iridomyrmex glaber (Formicidae)	*E. salaminia*	Egg
Forcipomyia sp. (Ceratopogonidae)	*E. salaminia*	Larva
Gminatus wallengreni (Reduviidae)	*E. salaminia*	Larva
Pristhesancus plagipennis (Reduviidae)	*E. salaminia*	Larva
Oechalia schellenbergii (Pentatomidae)	*E. salaminia*	Larva

*Sands and Broe (1991); Sands (1996); D. Sands and W. Liebregts, unpublished; H. Fay, pers. commun.)

Fig. 30.2. Three fruit-piercing moths, *E. salaminia*, piercing tomatoes.

The classification of fruit-piercing moths in the family Erebidae by Zahiri *et al.* (2012) is followed here, and the recognized species of *Eudocima* have been listed by Zilli and Hogenes (2002) and Zaspel and Branham (2008). Except for the species' name for *Eudocima phalonia* (= *Eudocima fullonia* in Edwards, 1996), the genera, species, synonymies and combinations used here follow the taxonomic arrangements of Edwards (1996), Zilli and Hogenes (2002) and Zaspel and Branham (2008). In studies preceding Edwards (1996), fruit-piercing moths were mostly included in the genera *Othreis* Hübner, *Eudocima* Billberg, *Khadira* Moore and *Rhytia* Hübner (Kitching, 1984). All were subsequently combined as *Eudocima* spp. by Edwards (1996).

These genera historically were variously included in subfamilies of the Noctuidae – Ophiderinae and Othreinae. Edwards (1996) placed *Eudocima* in the subfamily Catocalinae, regarding

Fig. 30.3. *Eudocima* spp., cosmopolitan pest species. (A–C) *E. phalonia*. A. Male upperside, B. female upperside, C. male underside. (D–F) *E. salaminia*. D. Male upperside, E. female upperside, F. male underside. (G–I) *E. materna*. G. Male upperside, H. female upperside, I. male underside. (J–L) *E. aurantia*. J. Male upperside, K. female upperside, L. male underside.

Fig. 30.4. Rainforest adapted *Eudocima* spp. (A–C) *E. jordani*. A. Male upperside, B. female upperside, C. male underside. (D–F) *E. cocalus*. D. Male upperside, E. female upperside, F. male underside. (G–I) *E. iridescens*. G. Male upperside, H. female upperside, I. male underside.

Ophiderinae and Othreinae as junior synonyms of that name. Zahiri *et al.* (2012) applied rigorous molecular phylogenetic analysis to the entire noctuoid assemblage, introducing a much more restricted hypothesis for the Noctuidae and elevating the Erebinae (previously a further junior synonym of the Catocalinae according to Edwards, 1996) to full family status within which the subfamily Calpinae was recognized. The fruit-piercing moths now reside within this erebine subfamily (see Chapter 20 for an extended discussion of these changes).

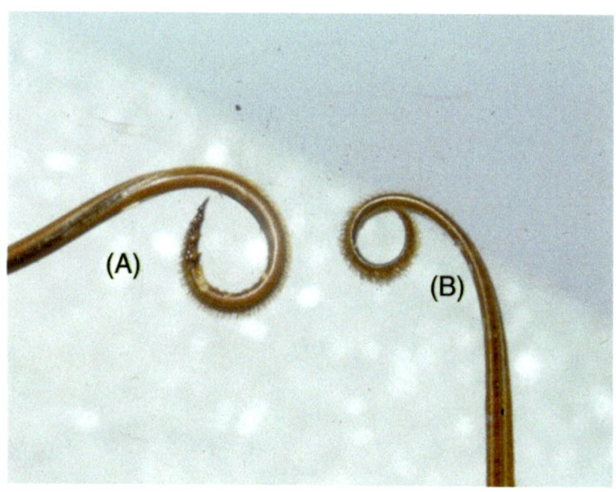

Fig. 30.5. Moth proboscides. (A) Proboscis of *E. salaminia*. (B) Proboscis of fruit-sucking moth, *Calyptra* sp.

In addition to adult morphology, Australian fruit-piercing moths (*Eudocima* spp.) are distinguishable from other Erebidae by the modified, heavily sclerotized tip and broadened spines or 'tearing hooks', near the apex of the proboscis of both sexes (Figs 30.5A, 30.6). Several other genera of large erebid moths (e.g. *Serrodes* Guenée, *Ophiusa* Ochsenheimer, *Phyllodes* Boisduval, *Thyas* Hubner) are referred to as 'fruit-sucking moths'. *Ophiusa parcemaculata*, for example, does not pierce intact ripening fruit but feeds on wounds in fruit, overripe or firm fruit damaged by other organisms (Cocking *et al.*, 2022). *Phyllodes imperialis* (and subspecies) is also considered to be a fruit-sucking moth, and has not been observed piercing firm fruit (Sands, 2012a, b). The apical tip of the proboscis of the southern subspecies *P. imperialis smithersi* is not heavily sclerotized as in the piercing *Eudocima* spp., despite *P. i. smithersi* having a row of broadly based *ventral* spines on the proboscis somewhat similar to the *dorsal* 'tearing hooks' of fruit-piercing moths. Also, unlike *Eudocima* spp., the proboscis of *P. i. smithersi* has fine setae on the dorsal surface near the apex. Confirming the identities of fruit-sucking moths and distinguishing the species from fruit-piercing moths (*Eudocima* spp.) involved experiments exposing each moth species to intact fruit at various stages of ripeness, and fruit scarred or damaged by other organisms, and examining for evidence of piercing (D.P.A. Sands unpublished).

Species of *Calyptra* (Erebidae: Calpinae), taxonomically related to *Eudocima* (Zaspel and Branham, 2008), have abundant, slender pointed spines on the proboscis (Fig. 30.5B) which is not prominently broadened at the base as in *Eudocima* spp. (Figs 30.5A, 30.6). *Calyptra minuticornis* is said to pierce fruit in South-east Asia (Banziger, 1968) but although known from Australia (Edwards, 1996; Hughes and Moss, 2012), this species has not been recorded piercing fruit in areas where it occurs in Queensland and northern New South Wales. In South-east Asia several *Calyptra* spp., including *C. minuticornis*, are also hematophagous, referred to as 'vampire moths' (Zaspel *et al.*, 2012). These moths are known to pierce the skin of large vertebrates to feed on blood (Banziger, 1979). Other *Calyptra* spp. feed on secretions of the eyes of large animals (Banziger, 1990). In Papua New Guinea, *C. minuticornis* was frequently observed feeding at night on secretions from the eyes and wounds of cattle (D.P.A. Sands and P. Bakker, unpublished). Such feeding on vertebrate blood or secretions of eyes of large animals by *C. minuticornis* has not been reported in Australia.

Fig. 30.6. Proboscis of *E. salaminia* (scanning electron microscope; CSIRO Biennial Report 14, 1985–87).

30.3 Damage To Fruit

In Australia, the wide range of tropical, subtropical and temperate fruit damaged by fruit-piercing moths includes oranges (Figs 30.7, 30.8), mandarins, grapefruit, carambolas, kiwifruit, bananas, apples, stone fruit, mangoes, persimmons, to-

Fig. 30.7. *Eudocima phalonia*. Female moth piercing a pre-ripe orange.

matoes, pawpaw, custard apples, litchis, longans and figs (Waterhouse and Norris, 1987). Both sexes of adult fruit-piercing moths damage pre-ripened or ripening fruit by piercing the skin with the tip of their sclerotized proboscides, then penetrating the tissues to macerate the pulp to extract juice. Fermenting wounds following damage attract flies, birds, etc., and rotting follows when fungi and micro-organisms invade damaged tissues, causing fruit fall (Waterhouse and Norris, 1987; Fay, 1991; Fig. 30.8). Piercing by the moths often provides the volatiles and preconditions attractive to fruit flies to oviposit, and infected fruit is easily overlooked during packing. It can destroy intact fruit in contact in containers when ready for marketing (Sands and Schotz, 1989).

Both sexes of fruit-piercing moths seek fruit soon after they have emerged from pupae and pierce fruit to obtain sugars. The sex ratio of moths captured in orchards varies according to fruit variety, ripeness and isolation of the orchard. The feeding behaviour is distinctive: after penetration by the proboscis, the moth moves its head up and down to varying depths, so that the tip is spiralled within the fruit and, with the tearing hooks surrounding the apex of the proboscis, macerates the pulp to produce a cavity 5 mm or more within the tissues (Fig. 30.6). Sands and Schotz (1987), working with *E. phalonia* and *E. salaminia*, found no difference in the longevity of moths fed 10% sucrose solutions and moths fed ripe banana. Feeding frequency averaged 1–3 piercings each night on one or more fruit. Young moths (<8 days post-eclosion) pierced fruit more frequently than older moths (10–30 days), and female moths required more food (to obtain sugars) from fruit than males, in order to mature oocytes. A well-fed female *E. phalonia* could potentially (based on mature oocytes) deposit up to 600 eggs in a lifetime of *ca.* 30 days; 530 eggs were deposited by a single female *E. phalonia* in one night (Sands and Schotz, 1987). One (laboratory-reared) moth deposited 1200 eggs when fed supplementary sugars.

Fig. 30.8. *Eudocima phalonia*. Cavity in orange abraded by moth proboscis.

30.4 Life Histories

Early life history studies on *E. phalonia* were made by Cochereau (1977) in New Caledonia. Banziger (1982) studied this and several other species in Thailand. The life histories of important Australian species, *E. phalonia*, *E. salaminia* and *E. materna*, were discussed by Tryon (1924).

30.4.1 Food plants

Tryon (1924) noted that all *Eudocima* spp. feed on rainforest vines (Menispermaceae). Studies on *E. phalonia* by Waterhouse and Norris (1987) confirmed this preference. Sands and Schotz (1987, 1991), Fay and Halfpapp (1993), Fay (1996) and Sands and Chan (1996) subsequently noted that most species of *Eudocima* had only one or two species of larval food plants (Table 30.1). In Papua New Guinea, the Solomon Islands, Micronesia and most South Pacific islands, the larvae of 'biotypes' of *E. phalonia* also feed on *Erythrina* spp. (Fabaceae) (Cochereau, 1974a; Waterhouse and Norris, 1987; Sands and Liebregts, 2006). Several species of the plant are grown for domestic and decorative use, or as fence posts (e.g. *Erythrina variegata*).

In Australia, larvae of *E. phalonia* have only been observed feeding on Menispermaceae (Table 30.1) and are not known to feed naturally on *Erythrina* spp. (Sands and Chan, 1996). Fruit-piercing moths do not overwinter south of about Rockhampton, Queensland, and studies on the effects of temperatures on the immature stages by Sands *et al.* (1991) indicated that *Eudocima* spp. were basically tropically adapted species, breeding throughout the year in northern Queensland with some species migrating south in spring, to breed only in the warmer months in south-eastern Queensland and northern New South Wales.

The pest species of fruit-piercing moths that are widely distributed throughout the Indo-Pacific Region may have originated in countries to the north-west of Australia and gained entry in historic times to Australia via mainland New Guinea. The extensive distribution of *E. phalonia* may in part be attributed to the migratory capabilities of adults, its polyphagous larvae feeding on various species of menisperm vines, and geographical biotypes that also feed on *Erythrina* spp. (Fabaceae) in several Pacific countries and mainland New Guinea.

In Australia, the widely distributed vine *Tinospora smilacina*, the endemic food plant of several *Eudocima* spp. in Australia (Table 30.1), undoubtedly influences the abundance of two pest species, *E. materna* and *E. phalonia*. This deciduous vine is abundant in dry and wet rainforests, and moist eucalypt forests in Queensland and Northern Territory. In dry seasons, when leafless and not supporting moth larvae, the vine, unlike other Menispermaceae, can survive basal severance or cool-burning around the vine near ground level. It can produce long trailing roots from living stems to aid regrowth. Another serious pest species, the monophagous *E. salaminia*, is dependent on the abundant vine *Stephania japonica*. The abundance of these two vines, *T. smilacina* and *S. japonica*, is considered to have considerable influence on the abundance and distribution of the three pest species, *E. materna*, *E. phalonia* and *E. salaminia*.

30.4.2 Oviposition

Females of *Eudocima* spp. mostly oviposit beneath the subterminal leaves or on young petioles of their food plants. Eggs are deposited singly or in small groups, rarely in large batches, depending on species. The large eggs (<1.1 mm) of *E. salaminia* are deposited beneath the leaves of the food plant *S. japonica*, or sometimes on the upper surface, on the bark of trees and twigs of other plants growing nearby. In the Pacific islands, *E. phalonia* may lay large batches of eggs, especially when deposited on *Erythrina* spp. Eggs of both species hatch in 3–4 days in summer or 20 days or longer during the cooler months. The young green larvae of *E. salaminia* consume the eggshell in warm weather but leave much of it intact when it is cool, before searching for young, terminal growth of the food plant. First instar larvae seek soft-textured leaves to begin feeding but not on the terminal shoot. First instar larvae are very prone to starvation if leaves are too firm for initial incisions by their mandibles. First instar larvae of both species are great wanderers and can search for food of suitable quality for up to 3 days before dying if food is not located.

30.4.3 Postovum development

The duration of immature development of *E. salaminia* and *E. phalonia* depends on temperatures, with development of egg to adult taking about 35 days in summer and 80 days in late autumn. There are five or occasionally six larval instars. First instar larvae are green (Fig. 30.9A), instars 2–3 are black with white and orange markings, and instars 4–6 are grey, black or brown (Fig. 30.9B, D). Larvae of *E. phalonia* are sometimes green with extensive brightly coloured marks of white, cream, orange or blue (Fig. 30.9 C). Mature larvae leave the food plant to pupate and spin loosely woven cocoons among the leaves of neighbouring shrubs 2 m or more from the ground, where they pupate. Unlike some other Erebidae, there is no evidence for pupal diapause in *Eudocima* spp., and low winter temperatures in subtropical areas resulted in mortality of pupae.

Adults emerge from cocoons soon after dusk and remain quiescent for about 2 h. Moths mate within 24–72 h of emergence and then seek fruit to pierce for the first meal. In experiments with *Eudocima* spp. in the field and in cages (using methods of Kumar and Lal, 1983), mating of *E. salaminia* was only observed when moths were held in shaded areas, but shade was not a requirement for mating for *E. phalonia* or *E. materna*. The pre-oviposition period for *E. salaminia* was 3 days in summer but up to 5 days in *E. phalonia* (Sands and Schotz, 1987).

Fig. 30.9. Larvae of *E. phalonia*. (A) First instar. (B) Fourth instar. (C, D) Green and dark forms, aposematic defensive pose.

30.5 Seasonal Abundance and Food Plant Quality

Of the most important species in northern and eastern Australia, *E. phalonia*, *E. materna* and *E. salaminia*, *E. phalonia* is the most abundant in tropical Queensland, *E. materna* in the Northern Territory and *E. salaminia* in the more temperate areas of southern Queensland and northern New South Wales. *E. phalonia* and *E. salaminia* occasionally occur as far south as Wollongong, NSW, but have not been sufficiently abundant there to cause significant damage to fruit. In northern Queensland *E. jordani* and *E. aurantia* also damage fruit (Fay, 2002). *E. jordani* is mostly a pest in lowland, high-rainfall areas north from Townsville, whereas *E. phalonia* was the main destructive species on a farm surrounded by rainforest at submontane Kuranda in 1986, where about 30 tropical and subtropical fruits were grown (Sands and Schotz, 1987). *E. materna* and *E. aurantia* are sometimes abundant and destructive in late summer and autumn as far south as Rockhampton on the central Queensland coast. *E. materna* is not usually abundant in subtropical areas south of Rockhampton despite its ability to migrate into Victoria and the Australian Capital Territory in some seasons. In the Northern Territory and northern Western Australia, *E. materna* is the predominant pest species. Levels of 'importance' were based on: (i) the range of fruit attacked, (ii) the abundance of the moths, and (iii) the distribution of the breeding habitats. In south-eastern Queensland attack by fruit-piercing moths varies from season to season (Huber *et al.*, 1998) but, on lychees and carambolas for example, 40% or more of the fruit may be unmarketable from moth damage and total losses sometimes occur if nets are not used to protect the trees.

In tropical regions of Queensland, north from about Paluma (19°S), the five *Eudocima* spp. reproduce persistently when climatic conditions are favourable, when the food plant understories are undisturbed and not replaced by weeds, and if food plants are producing soft terminal growth suitable for feeding by moths' early instars. The older, firmer leaves of food plants were found to be unpalatable except to advanced instars. Growing tips were avoided by early instars in favour of expanded soft leaves at nodes 2–5 from the growing apices. Studies with *E. phalonia* and *E. salaminia* (D.P.A. Sands, unpublished) indicated the nutrient-rich, soft apical leaves of *Stephania japonica* were higher (by dry weight) in nitrogen concentrations than the lower, older and firmer leaves. Survival of first to third instar larvae increased when fed softer leaves than when fed firmer, older leaves, and resulted in more rapid larval development times to pupation.

The leaf toughness of food plants is an important major factor affecting larval survival of fruit-piercing moth species (Sands and Brancatini, 1991), although toughness varied with host plant, temperatures and seasonal growth. The common food plant *Tinospora smilacina*, for example, an important food plant for several moth species (Table 30.1), is deciduous and loses its leaves during dry periods and, mainly in winter and spring, and after rains the vines can produce massive growth of new, soft leaves, promoting larval survival. Other food plant species may become dormant in winter, only producing leaves that are too tough for first instar larvae.

In northern Queensland, Fay and Halfpapp (1993) noted that tropical rainforest, coastal and dry tropical rainforests containing three forms of the vine *T. smilacina* supporting *E. phalonia* and *E. materna*, mostly between November and March for the former species and January to June for the latter. Fay and Halfpapp (1993) noted that *E. cocalus* utilized two forms of *T. smilacina* and preferred rainforest, habitats to which *E. jordani* and *E. aurantia* were restricted for breeding. In the drier inland habitats in northern Queensland, only *E. materna* and *E. phalonia* occurred and only during a limited season. There was temporal separation for the two species when utilizing *T. smilacina* in these habitats. Fay and Halfpapp (1993) noted that the larvae of *E. phalonia*, *E. jordani* and *E. aurantia* fed on more than one species of menisperm vine, whereas the host plants of *E. materna* and *E. cocalus* were always vines of the genus *Tinospora*. The abundance of moths correlated temporally with the presence of larvae on the main host *T. smilacina*, and the differential habitats and host plants utilized by the moths were associated with variations in their seasonal activity, importance and local pest status.

30.6 Migrations of Fruit-piercing Moths

In subtropical regions of south-eastern Queensland and northern NSW, plant communities supporting breeding of *E. phalonia*, *E. materna* and *E. salaminia* are mostly lowland and submontane subtropical rainforest, dry rainforest and wet sclerophyll forest. For the monophagous *E. salaminia*, wet sclerophyll forest may also support the food plant *S. japonica* in areas from about Port Macquarie, NSW, north to south-eastern Queensland.

In most years (to 2005) both sexes of adult *E. phalonia* and *E. salamina* were observed migrating in late spring or early summer, from perennial breeding sites in northern Queensland to areas of south-eastern Queensland and far northern New South Wales. The abundance of individuals migrating varied from year to year and according to seasonal rainfall. The densities of south-eastern arrivals appeared to relate to favourable weather conditions in the northern breeding sites (Table 30.1) supporting abundant growth of food plants, and high moth densities developing in the preceding months.

Hundreds of moths have occasionally been observed migrating and, on one occasion, huge numbers flying from north to south were observed at the top of a building in Brisbane City (the late M. DeBaar, pers. commun.). On arrival in the south-eastern regions, the condition of migrating *E. phalonia* and *E. salaminia* was remarkably 'fresh' considering the long journeys, with little wing damage or fading on specimens examined. Migrations of other *Eudocima* spp. have not been reported but occasional appearances of *E. materna* in southern states (southern NSW, South Australia and the Australian Capital Territory) occur in areas where no Menispermaceae food plants are present to support breeding, indicating that this species migrates well beyond the range of its breeding sites. In early studies by CSIRO using radar of insects crossing Torres Strait, fruit-piercing moths were noted among the many other Lepidoptera observed.

The occurrence of *E. phalonia* south from Mackay, Queensland, to northern NSW is seasonal in most years, with adults first appearing in late spring months following their annual migrations from further north. Increases in abundance follow from breeding locally through the warmer months from December until about May each year, with build-up of moth populations from breeding on several local menisperm vines, mostly *Legnephora moorei*, *Tinospora smilacina*, *Sarcopetalum harveyanum* and *Stephania japonica* subsp. *timoriensis*. The absence of adults and immature stages of *E. phalonia* in winter in southern regions of eastern Australia has been attributed to the failure of larvae to survive the cool and dry months of subtropical regions. Adults and immature stages reappear after the migrating adults arrive and commence breeding on local food plants from late October or November, through until about May (Huber *et al.*, 1998).

30.7 Natural Mortality

In south-eastern Queensland and northern New South Wales, cold snaps in autumn months, with temperatures dropping below 5°C at night for more than 10 days, can kill the immature stages of *E. phalonia* and *E. salaminia*. Scarcity of soft leaves at the time of larval eclosion from eggs is also a common cause of mortality.

Natural enemies of fruit-piercing moth *E. phalonia* in Australia and the Pacific region were referred to by Waterhouse and Norris (1987) and Sands (1996), following the extensive studies in New Caledonia by Cochereau (1972a, b, 1973, 1974a, b, 1977). In Australia, the natural enemies of *E. phalonia*, *E. salaminia* and *E. materna*, the three most important pest species in Queensland and the Northern Territory, were investigated by Sands (1996). These studies indicated that parasitoids played little part in affecting the abundance of the moths or their status as pests in Australia. In Queensland, at least one *Trichogramma* spp. (Trichogramatidae) was recovered from eggs of *Eudocima* spp. The egg parasitoids appeared to be very seasonal and present only when egg densities were relatively abundant. Another abundant egg parasitoid, *Ooencyrtus* sp. (*papilionis* sp. group) (Encyrtidae), occasionally appeared from parasitized eggs of *E. phalonia* in northern Queensland. This species was closely related (*viz. papilionis* species group) to several other *Ooencyrtus* spp. occurring on Pacific Islands (Cochereau, 1972b, 1974a, b, 1977; Waterhouse and Norris, 1987). In Papua New Guinea, two egg parasitoids (*Ooencyrtus* sp. and *Telenomus lucullus*) were abundant and could potentially influence the abundance of *E. phalonia* in PNG (Sands and Papacek, 1993) (Table 30.2).

Larval parasitoids of *Eudocima* spp. were generally uncommon and only one species, *Euplectrus melanocephalus* (Hymenoptera, Eulophidae), was found attacking larvae of *Eudocima* spp. (Jones and Sands, 1999) (Fig. 30.10). *E. melanocephalus* is related to its congener, *platyhypenae*, a parasitoid of *E. phalonia* from New Caledonia and Fiji (Maddison, 1982), and *Euplectrus maternus*, a parasitoid of *Eudocima* spp., from India (Muniapan *et al.*, 2004; Magar *et al.*, 2017). In northern Queensland *E. melanocephalus* was not sufficiently abundant to reduce the build-up in fruit-piercing moth densities, where it parasitized few moth larvae in rainforest areas. *Echthromorpha*

Fig. 30.10. *Euplectrus melanocephalus*. (A) Adult wasp. (B) Gregarious parasitoids on larvae of *Eudocima iridescens*. (C) Close-up of parasitoid larva.

agrestoria (Ichneumonidae), a wasp thought likely to be specific to *Eudocima* spp., was only rarely encountered attacking larvae of *E. salamina* (Table 30.2).

Predators of eggs and young larvae of *Eudocima* spp. included ants, *Pheidole megacephala* and *Tetramorium* sp., and a haemolymph-sucking midge, *Forcipomyia* sp. (Diptera: Ceratopogonidae) commonly observed on mature larvae of *E. salaminia*. No pupal parasitoids emerged from pupae but diseased, dead and decomposing pupae sometimes produced larvae, pupae and adults of unidentified flies (mostly Sarcophagidae).

30.8 Seasonal Occurrence and Abundance of Moths

In northern Queensland, the increasing abundance of moths in the warmer months of spring follows flushing of subterminal leaves of the larval food plants (Table 30.1), providing soft leaves of the food plants that support first instar larvae, and with few predators (mainly ants) to attack the eggs and young larvae. Following increases in abundance of two species, *E. phalonia* and *E. salaminia*, in spring in northern Queensland, both species migrate to southern Queensland in most years. Before migrating, the build-up in abundance of moths in tropical regions occurs north of Townsville, where their numbers are enhanced by moist warm weather, the nutritious, soft leaves of their food plants, particularly *T. smilacina*, and scarcity of natural enemies attacking the immature stages. High moth densities in northern Queensland appear to be the trigger for the migrations of moths that head south in late spring and early summer (Huber *et al.*, 1998). After migrating, and following their arrival in southern Queensland and northern NSW, the moths seek fruit crops on which to feed, and female moths search for food plants on which to deposit eggs, starting the first summer generations in the southern regions.

In south-east Queensland in moist summer and autumn months, large numbers of moths breed on local food plants (Table 30.1), where up to four southern generations continue to damage fruit wherever it is available. During dry summers and autumns, moth numbers are fewer and not so destructive, due to the high mortality of their first instars that fail to survive on food plants with tough leaves and with lower levels of nutrients (mainly nitrogen). As cooler weather develops in autumn in southern Queensland and northern NSW, larval feeding and immature development are slowed and by about June each year, in the absence of diapausing stages, mortality results and no stages overwinter or survive. The annual occurrence of moths in southern regions of Queensland and northern New South Wales is dependent on arrivals of adults in spring and subsequent local generations.

Species of fruit-piercing moths are pests of a wide range of commercial fruits, and can cause crop losses of more than 50% (Fay and Halfpap, 1993). In Australia, the food plants of fruit-piercing moths, vines in the family Menispermaceae, are represented by 13 genera and 24 species (Bostock and Holland, 2010) which occur through coastal and subcoastal regions of north and eastern Australia and on several Torres Strait Islands. The density and diversity of these vines are greatest in the tropics of north-east Queensland. Of these menisperm vines, about 60% support fruit-piercing moth larvae, ranging from host-specific species of moths to polyphagous species. *Tinospora smilacina* and *Stephania japonica* (and its subspecies) are the most widely distributed vines with forms or varieties that vary according to the plant communities they occupy, from tropical rainforests to semi-deserts. The abundance and distribution of *T. smilacina* (Huber *et al.*, 1998) and *S. japonica* are major factors contributing to the abundance and distribution of the fruit-piercing moths, and their migrations and the seasonal damage they cause to fruit.

With predicted increases in global temperatures, the distribution of fruit-piercing moths is likely to extend south, beyond areas currently climatically suitable for survival and development of immature stages, for example *E. phalonia* and *E. salaminia* (Sands *et al.*, 1991), and where their food plants (for example, *S. japonica*) already occur.

30.9 Biological Control Programmes

Biological control agents for *E. phalonia* were investigated by Cochereau (1974b) in New

Caledonia, where several egg parasitoids (*Ooencyrtus* spp.) and a tachinid fly, *Winthemia caledoniae*, were identified as natural enemies of *E. phalonia*. In Fiji, Kumar and Lal (1983) conducted studies on the host parasitoids and seasonal abundance of *E. phalonia*. Biological control of *E. phalonia* was proposed by Waterhouse and Norris (1987) and subsequently several natural enemies from the Pacific region were evaluated as potential agents (Sands and Broe, 1991; Sands *et al.*, 1993). In Papua New Guinea, Sands (1996) identified two potential biological control agents of *E. phalonia* – the egg parasitoids *Telenomus lucullus* (Scelionidae) and an *Ooencyrtus* sp. (*papilionis* species group) (Encyrtidae). The host specificity of *T. lucullus* and *Ooencyrtus* sp. was studied in the CSIRO quarantine facility by exposing the parasitoids to eggs of a range of taxonomically related erebid moths, then checking for evidence of development by immature stages of the parasitoids in the moth eggs. Those studies found that *T. lucullus* was a specific parasitoid of eggs of *Eudocima* spp. but the *Ooencyrtus* sp. developed on several non-target species when their eggs were in the presence of the food plants of *E. fullonia*.

A series of host range tests exposing the parasitoids to related erebid moths showed that *T. lucullus* and *Ooencyrtus* sp. (*papilionis* species group) were adequately host-specific for release on western Pacific islands and subsequently, both parasitoids were released in Samoa, Tonga, Fiji and the Cook Islands (Sands, 1992). *T. lucullus* and *Ooencyrtus* sp. became established in Tonga and Fiji, and *T. lucullus* became established in Samoa and the Cook Islands (Sands and Broe, 1991; Sands, 1996; Sands and Liebregts, 2006). After establishment of both parasitoids in Samoa and Tonga, levels of egg parasitism increased dramatically, declines in the abundance of eggs and adult moths followed, and reductions in damage to fruit have been observed in Fiji and Cook Islands (Sands and Liebregts, 1992).

In Australia, host specificity studies with non-target moths indicated that, although the two parasitoids were host-specific for *Eudocima* spp., one of these *Eudocima* species, *iridescens*, was not considered a pest species and accordingly should not be targeted in control programmes. The decision not to release those parasitoids was made to avoid affecting indigenous non-target hosts taxonomically related to the pest species (Sands and Liebregts, 2006). This study was one of the first examples in Australia where exotic agents were tested for possible impacts on non-target indigenous insects. The larval parasitoid *Euplectrus maternus*, of Indian origin, was released on Guam in 1998–1999 in an attempt to control *E. phalonia* but by 2004 the parasitoids had failed to become established (Muniappan *et al.*, 2004).

Knowledge of the likely origin of *E. phalonia* and of most other *Eudocima* spp. (Leroy *et al.* (2021) has alleviated concerns sometimes expressed when evaluating the host range of a biological control agent to target a native pest species (Sands, 1987b; Sands and Van Driesche, 2000; Van Lenteren *et al.*, 2006). Such concerns are no longer relevant to biological control of the important pest *E. phalonia*, a species considered to have originated in Africa (Leroy *et al.*, 2021).

30.10 Pest Management and Control

Early methods recommended by Tryon (1924) for controlling fruit-piercing moths ranged from 'destroying the native food plants' (e.g. in Tryon, 1924) to '... destroy the entire woody vegetation where such food plants may be expected to exist', or exposing poison baits such as bananas impregnated with arsenic. These methods would have had serious impacts on non-target fauna and pose other unacceptable non-target risks.

Fruit-piercing moths are difficult to control due to their attack on fruit that is close to harvest, when the use of insecticides is considered ineffective or unsafe (Waterhouse and Norris, 1987). Baptist (1944) suggested many methods to control the moths, including (i) attracting moths into traps with poisoned baits, (ii) smoking orchards to disguise the attractive volatiles from fruit, (iii) deterrent sprays containing repellent volatiles, (iv) early harvest of fruit, (v) hand-capture using nets, (vi) bagging and screening fruit, (vii) repelling moths from orchards using lights, and (viii) destruction of larval food plants. In Thailand eradication of food plants was suggested by Banziger (1982) and this method for controlling *E. fullonia* was successfully implemented in a small valley in New Caledonia by Cochereau (1972b, 1977).

Illumination barriers near fruit crops evaluated by Fay and Halfpap (1991b, 1995) were said to help reduce damage by the moths. Physical control by netting fruit trees has been the most successful method to control the moths, by preventing access by moths as well as bats and birds. This method was shown to be the most effective way of reducing damage by fruit-piercing moths (Fay, 1988, 2002; Fay and Halfpapp, 1990, 1991a,b; Sands, 1996).

Acknowledgements

The studies on fruit-piercing moths referred to here formed part of CSIRO's Division of Entomology programme, *Pests of Horticulture*. Special thanks go to Wilco Liebregts for his participation in experimental studies on insects on Pacific Islands, to Dr Harry Fay (Qld Department of Primary Industries) for collaboration with research, and Veronica Brancatini, Rose Broe, Richard Chan, Martina Schotz and Peter Van der Graaf (CSIRO) for their participation in experimental studies on the moth. I am most grateful to the late Doug Waterhouse, Ted Edwards and Ian Common, and to Brian Fletcher, Tim New and Marianne Horak for many helpful discussions and advice given to me throughout this project. You Ning Su (ANIC, CSIRO) kindly prepared the images for the figures of mounted *Eudocima* spp. Much of the work was supported by the Horticultural Research and Development Corporation, the Australian Centre for International Agricultural Research (ACIAR) and the CRC for Tropical Pest Management.

References

Banziger, H. (1968) Preliminary observations on a skin-piercing blood sucking moth (*Calyptera eustrigata* (Hems.)) (Lep. Noctuidae) in Malaya. *Bulletin of Entomological Research* 58, 159–163.

Banziger, H. (1979) Skin-piercing blood-sucking moths II: Studies on a further three adult *Calyptra*/*Calpe*/ sp. (Lepid., Noctuidae). *Acta Tropica* 36, 23–37.

Banziger, H. (1982) Fruit-piercing moths. (Lep., Noctuidae) in Thailand: A general survey and some new perspectives. *Mitteilungen der Schweizerischen Entomologischen Gesellschaft* 55, 213–240.

Banzier, H. (1990) Moths with a taste for tears. *New Scientist* 1744, 48–51.

Baptist, B.A. (1944) The fruit piercing moth (*Othreis fullonia* L.) with special reference to its economic importance. *Indian Journal of Entomology* 6, 1–13.

Blanchfield, J.T., Kitching W., Sands, D.P.A., Thong Y.H., Kennard C.H.L, *et al.* (1993) Alkaloids from Australian *Stephania* (Menispermaceae) species. *Natural Products Letters* 3, 305–312.

Bostock, P. and Holland, A.E. (eds) (2010) *Census of the Queensland Flora 2010*. Queensland Herbarium, Department of Environment and Resource Management, Brisbane, Queensland.

Brou, V.A. (1994) New records of two fruit-piercing moths encountered in Louisiana and other southern states. *Southern Lepidoptera News* 16, 39–40.

Brou, V.A. Jr. and Zilli, A. (2016) An overlooked sibling of the fruit-piercing moth *Eudocima phalonia* (Linnaeus, 1763) from Africa (Lepidoptera, Erebidae, Calpinae). *Zootaxa* 4109, 391–399.

Cochereau, P. (1972a) Population dynamics of a fruit sucking moth, *Othreis fullonia* Clerck (Lepidoptera, Noctuidae) in New Caledonia. In: *Abstracts: Proceedings of the 14th International Congress of Entomology*. 22–30 August. Australian National University, Canberra, ACT, pp. 210–211.

Cochereau P. (1972b) La lutte biologique dans le Pacifique. Works and documents of the ORSTOM, Paris, France. *Série Biologie* 16, 89–104.

Cochereau P. (1973) Contrôle biologique naturel des papillons piqueurs de fruits. *Fruits* 28, 367–375.

Cochereau P. (1974a) Les papillons piqueurs de fruits et leurs commensaux. Works and documents of the ORSTOM, Paris, France. *Série Biologie* 22, 3–42.

Cochereau P. (1974b) Complexe parasitaire en Nouvelle-Calédonie d'un papillon piqueur de fruits: *Othreis fullonia* Clerck (Lepidoptera, Noctuidae, Catocalinae). Works and documents of the ORSTOM, Paris, France. *Série Biologie* 25, 35–65.

Cochereau, P. (1977) Biologie et ecologie des populations en Nouvelle-Caledonie d'un papillon piquer de fruiits: *Othreis fullonia* (Clerck). *Cahiers ORSTOM Serie Biologie* 71, 322.

Cocking, G., Bond, S. and Edwards, T. (2022) *Moths in the A.C.T.* Glen Cocking, Canberra, ACT.

Common, I.F.B. (1990) *Moths of Australia*. Melbourne University Press, Carlton, Victoria.

CSIRO (1987) *Division of Entomology, Biennial Report 1985–87*, CSIRO Canberra, ACT.

Duncan, S. (2011) Connecting the bush in the sunshine coast Hinterland. *Bush Hands*, Sunshine Coast Regional Council, 5.

Edwards, E.D. (1996) Noctuidae. In: Nielsen, E.S., Edwards, E.D. and Rangsi, T.V. (eds) *Checklist of the Lepidoptera of Australia*. Monographs on Australian Lepidoptera 6. CSIRO Publishing, Collingwood, Victoria, pp. 291–333.

Fay, H.A.C. (1988) Fruit-piercing moths (F. Noctuidae) in tropical Australia: Aspects of their ecology and potential for their control. In: *Proceedings of the XVIII International Congress of Entomology*, Vancouver, B.C., Canada, 3–9 July 1988.

Fay, H.A.C. (1991) *Ecology and Control of Fruit-piercing Moths and Leaf-eating Beetles in Developing Strategies for their Management in Tropical Tree Crops*. Rural Industries Research and Development Report, Project DAQ-47A. Queensland Department of Primary Industries, Mareeba, Queensland.

Fay, H.A.C. (1994) The relative acceptabilities of three Australian Menispermaceae as food plants for larvae of the fruitpiercing moth, *Othreis fullonia*. *Entomologia Experimentalis et Applicata* 72, 67–75.

Fay, H.A.C. (1996) Evolutionary and taxonomic relationships between fruit-piercing moths and Menispermaceae. *Australian Systematic Botany* 9, 227–233.

Fay, H.A.C. (2002) Fruitpiercing moths and fruitspotting bugs: Intractable pests of tree fruits in a reduced-insecticide environment. In: *Proceedings of the International Symposium on Tropical and Subtropical Fruits*; Cairns, Australia. 26 November–1 December 2000, pp. 485–493.

Fay, H.A.C. and Halfpapp, K.H. (1987) Activity of adult fruit-piercing moths and the host-plant associations of their larvae. In: *4th Workshop of Tropical Agricultural Entomology*, Darwin, 10–15 May, 1987.

Fay, H.A.C. and Halfpapp, K.H. (1990) Potential methods for control of fruit-piercing moths (Lepidoptera: Noctuidae) in tropical Australia. In: *Proceedings of the First Asia-Pacific Conference of Entomology*. Chiang Mai, Thailand, 8–13 November 1989. Entomology and Zoology Association of Thailand, Bangkok, pp. 204–208.

Fay, H.A.C. and Halfpapp, K.H. (1991a) Feeding attractants and light barriers as control measures for fruit-piercing moths – progress in their evaluation. In: *Proceedings of the 5th Workshop of Tropical Entomology*, Townsville, Australia 1–5 July.

Fay, H.A.C. and Halfpapp, K.H. (1991b) Utilization of A-fenispermaceae by larvae of five species of fruit-piercing moth in North Queensland. In: *Proceedings of the 5th Workshop of Tropical Entomology*, Townsville, Australia 1–5 July.

Fay, H.A.C. and Halfpapp, K.H. (1993) Differential habitat affinities of five species of fruitpiercing moths (Lepidoptera: Noctuidae) in their utilization of *Tinospora smilacina* Benth as a larval host plant in north Queensland. *Australian Journal of Ecology* 18, 451–462.

Fay, H.A.C. and Halfpapp, K.H. (1995) Reducing fruit-piercing moth damage in orchards with illumination barriers. In: *Proceedings of the Workshop on Fruit-piercing Moths*, Brisbane, Australia. 16–21 January.

Fay, H.A.C. and Halfpapp, K.H. (1999) Activity of fruit-piercing moths, *Eudocima* spp. (Lepidoptera: Noctuidae), in north Queensland crops: Some effects of fruit type, locality and season. *Australian Journal of Entomology* 38, 16–22.

Fay, H.A.C. and Halfpapp, K.H. (2006) Fruit maturity and soundness relevant to feeding choice by fruit-piercing moths (Lepidoptera: Noctuidae) in citrus crops in northeast Australia. *International Journal of Pest Management* 52, 317–324.

Herbison-Evans, D. and Crossley, S. *Eudocima fullonia* (Clerck [1874] Fruit-Piercing Moth. Available at: http://lepidoptera.butterflyhouse.com.au/calp/fullonia.html (accessed January 2025).

Huber, P., Fay, H.A., Sands, D.P.A. and Seymour, J. (1998) Seasonality of fruit-piercing moths in northern Queensland with special reference to the principal pest species, *Othreis fullonia* (Clerck) (Lepidoptera: Noctuidae). In: Zalucki, M.P., Drew, R.A.I., and White, G.G. (eds) *Proceedings of 6th Australian Applied Entomology Research Conference*, 29 September to 2 October, University of Queensland, Brisbane, pp. 87–93.

Hughes, L. and Moss, J. (2012) Fruit-piercing moths – night raiders! *Metamorphosis Australia* 67, 4–9.

Hungerford, N.L., Sands, D.P.A. and Kitching, W. (1998) Isolation and structure of some constituents of the Australian medicinal plant, *Tinospora smilacina* ('snakevine'). *Australian Journal of Chemistry* 51, 1103–1111.

Jones, P. and Sands, D.P.A. (1999) *Euplectrus melanaocephalus* Girault (Hymenoptera: Eulophidae), an ectoparasitoid of larvae of fruit-piercing moths (Noctuidae: Catocalinae) from northern Queensland. *Australian Journal of Entomology* 37, 377–381.

Kitching, I.J. (1984) An historical review of the higher classification of the Noctuidae: Lepidoptera). *Bulletin of the British Museum (Entomology)* 54, 143–234.

Kumar, K. and Lal, S.N. (1983) Studies on the biology, seasonal abundance and host-parasite relationship of fruit sucking moth *Othreis fullonia* (Clerck) in Fiji. *Fiji Agricultural Journal* 45, 71–77.

Leroy, L., Mille, C. and Fogliani, B. (2021) The common fruit-piercing moth in the Pacific Region: a survey of current status of a significant worldwide economic pest, *Eudocima phalonia* (Lepidoptera: Erebidae), with a focus on New Caledonia. *Insects* 12, 117.

Maddison, P.A. (1982) *Fruit-piercing Moths*. South Pacific Commission Advisory Leaflet 14. South Pacific Commission, Noumea, New Caledonia.

Magar, P.N., Kulkarni, S.R. and Chandele, A.G. (2017) Natural parasitoids of fruit-piercing moth, *Eudocima* spp. *Journal of Biological Control* 31, 10–12.

Miller, D. and Brou, V.A. (2013). A new U.S. state record for a tropical fruit-piercing moth in the family Erebidae Leach. *Southern Lepididopterist's News* 35, 27–28.

Muniappan, R., Bamba, J.P., Cruz, J. and Reddy, G.V.P. (2004) Biology, rearing and field release on Guam of *Euplectrus maternus*, a parasitoid of the fruit-piercing moth, *Eudocima fullonia*. *BioControl* 49, 537–551.

Sands, D.P.A. (1987a) The natural history of fruit-piercing moths in southeastern Queensland. *Bulletin of the Entomological Society of Queensland* 15, 19–25.

Sands, D.P.A. (1987b) (i) Procedures for collection and identification of arthropod and weed pests and their likely natural enemies, (ii) Practical problems in host specificity testing; (iii) Packaging and shipping live and dead insects. In: Ferrar P., and Stechmann, D.H. (eds) *Proceedings of Workshop on Biological Control in the South Pacific*, Tonga, 17–25 October 1985, pp. 221–223, 246–247.

Sands, D.P.A. (1992) Australian sponsored biological control projects in the South Pacific. In: *Biological Control towards a Safer Environment. Proceedings of the International Conerence on Biological Control in Tropical Agriculture*, Genting Highlands, Malaysia, 1991.

Sands, D.P.A. (1996) Natural enemies and prospects for biological control of fruit-piercing moths. In: Welsh, A. and Ferguson, J. (eds) *Proceedings of the 4th National Lychee Seminar including longans*, Australian Lychee Grower's Association, Yeppoon, Queensland, 26–28 September, pp. 110–117.

Sands, D.P.A. (1997) The 'safety' of biological control agents: assessing their impact on beneficial and other non-target hosts. *Memoirs of the Museum of Victoria* 56, 611–616.

Sands, D.P.A. (1998) Guidelines for testing host specificity of agents for biological control of arthropod pests. In: Zalucki, M.P., Drew, R.A.I., and White, G.G. (eds) *Pest Management-Future Challenges, Proceedings of 6th Australian Applied Entomology Research Conference*, University of Queensland, Brisbane, 29 September to 2 October, pp. 556–560.

Sands, D.P.A. (2012a) Review of Australian *Phyllodes imperialis* Druce (Lepidoptera: Erebidae) with description of a new subspecies from subtropical Australia. *Australian Entomologist* 39, 281–292.

Sands, D.P.A. (2012b) The southern pink underwing moth. In: Curtis, L.K., Dennis, A.J., McDonald, K.R.M., Kyne, P.M. and Debus, S.J.S. (eds) *Queensland's Threatened Animals*. CSIRO Publishing, Collingwood, Victoria, pp. 38–39.

Sands, D.P.A. and Brancatini, V. (1991) A portable penetrometer for measuring leaf toughness in insect herbivory studies. *Proceedings of the Entomological Society of Washington* 93, 786–788.

Sands, D.P.A. and Broe, R. (1991) Prospects for biological control of the fruit-piercing moth, *Othreis fullonia* (Clerck) (Lepidoptera: Noctuidae). In: Johnson, M.W., Ulman, D.E. and Vargo, A. (eds) *Proceedings 1st Agricultural Development in the American Pacific ('ADAP') Crop Protection Conference*, University of Hawaii, Honolulu, May 18–19 1989, pp. 155–158.

Sands, D.P.A. and Chan, R.C. (1996) Survivorship of Australian *Othreis fullonia* on *Erythrina variegata*: hypotheses for development of host-plant biotypes in the Pacific. *Entomologia Experimentalis et Applicata* 80, 145–148.

Sands, D.P.A. and Liebregts, W.J.M.M. (1992) Establishment of *Telenomus* sp. (Hymenoptera: Scelionidae) in Western Samoa, for biological control of *Othreis fullonia* (Clerck) (Lepidoptera: Noctuidae).

In: Schreiner, I. and Nafus, D. (eds) *Proceedings of the Second ADAP Crop Protection Conference*, University of Guam, Mangilao, Guam, May 29–30 1990, pp. 16–21.

Sands, D.P.A. and Liebregts, W.J.M.M. (2006) Biological control of fruit-piercing moth (*Eudocima fullonia* [Clerck]) (Lepidoptera: Noctuidae) in the Pacific: Exploration, specificity, and evaluation of parasitoids. In: Hoddle, M.S. (ed.) *Proceedings of the Second International Symposium on Biological Control of Arthropods*, Davos, Switzerland, 12–16 September 2005. United States Department of Agriculture, Forest Service, Morgantown, WV, pp. 267–276.

Sands, D.P.A. and Papacek, D. (1993) Specificity requirements of exotic agents for biological control of arthropod pests. In: Corey, A.A., Dall, D.J. and Milne, W.M. (eds) *Pest Control and Sustainable Agriculture*. CSIRO, Melbourne, Victoria, p. 495.

Sands, D.P.A. and Schotz, M. (1987) Ecology of fruit-piercing moths in subtropical Australia. In: Strictland, G.R. (ed.) *Tropical Agricultural Entomology, Proceedings of a Workshop*, Darwin, Northern Territory, 10–15 May, pp. 47–51.

Sands, D.P.A. and Schotz, M. (1989) Advances in research on fruit-piercing moths of subtropical Australia. In: Batten, D. (ed.) *Proceeding 4th Australasian Conference Tree and Nut Crops, 14–20 August 1988*. Exotic Fruit Growers Association, Lismore, NSW, pp. 378–383.

Sands, D.P.A. and Schotz, M. (1991) Ecology of fruit-piercing moths in subtropical Australia. In: Magallona, E.D. (ed.) *Proceedings of the 11th International Congress of Plant Protection*. Manila, Philippines, 5–9 October 1987, pp. 229–232.

Sands, D.P.A. and Schotz, M. (1992) Ecology of fruit-piercing moths in subtropical Australia. In: Strictland, G.R. (ed.) *Tropical Agricultural Entomology, Proceedings of the 4th Workshop*, Darwin, Northern Territory, 10–15 May 1987, pp. 47–51.

Sands, D.P.A. and Van Driesche, R.G. (2000) Evaluating the host range of agents for biological control of arthropods: rationale, methodology and interpretation. In: Van Driesche, R.G., Heard, T.A., McClay, A.S. and Reardon, R. (eds) *Proceedings: Host Specificity Testing of Exotic Arthropod Biological Control Agents: the Biological Basis for Improvement in Safety*. Forest Health Technology Enterprise Team 1999–01, USDA Forest Service, Morgantown, WV, pp. 69–83.

Sands, D.P.A. Schotz, M. and Bourne, A.S. (1991) Effects of temperature on development and seasonality of *Eudocima salaminia* (Cramer) (Lepidoptera: Noctuidae) in eastern Australia. *Bulletin of Entomological Research* 81, 291–296.

Sands, D.P.A., Liebregts, W.J.M.M. and Broe, R. (1993) Biological control of the fruit-piercing moth, *Othreis fullonia* (Clerck) (Lepidoptera: Noctuidae) in the Pacific. *Micronesica* Suppl. 4: 25–31.

Singh, N., Ahmad, J., Kaustubh, K. and Joshi, R. (2019). New distributional records of *Eudocima* Billberg, 1820 (Lepidoptera: Erebidae: Calpinae) from Gangetic Plains (India). *Records of the Zoological Survey of India* 119, 238–245.

Tryon, H. (1924) Orange-piercing moths (Fam. *Ophiderinae*). *Queensland Agricultural Journal* 21, 385–396.

Van Lenteren, J.C., Cock, M.J.W., Hoffmaster, T.S. and Sands, D.P.A. (2006) Host specificity in arthropod biological control: Methods for testing and interpretation of the data. In: Bigler, F., Babendreier, B. and Kuhlmann, U. (eds) *Environmental Impact of Invertebrates for Biological Control of Arthropods, Methods and Risk Assessment*. CAB International, Wallingford, UK, pp. 38–63.

Vargas-Fonseca, S., Correa-Carmona, Y., Montes-Rodríguez, J.M., Calero-Mejía, H. and Zilli, A. (2020) Synopsis of fruit-piercing moths of the genus *Eudocima* (Lepidoptera, Erebidae) from Colombia. *ZooKeys* 953, 85–104.

Waterhouse, D.F. and Norris K.R. (1987). *Biological Control Pacific Prospects*. Australian Centre for International Agricultural Research (ACIAR), Melbourne, Victoria.

Zahiri, R., Holloway, J.D., Kitching, I.J., Lafontaine, J.D., Mutanen, M. and Wahlberg, N. (2012) Molecular phylogenetics of Erebidae (Lepidoptera, Noctuoidea). *Systematic Entomology* 37, 102–124.

Zaspel, J.M. and Branham, M.A. (2008) World checklist of tribe Calpini Lepidoptera: Noctuidae: Calpinae. *Insecta Mundi* 47, 1–15.

Zaspel, J.M., Zahiri, R., Hoy, M.A., Janzen, D., Weller, S.J. and Wahlberg, N. (2012) A molecular phylogenetic analysis of the vampire moths and their fruit-piercing relatives (Lepidoptera: Erebidae: Calpinae). *Molecular Phylogenetics and Evolution* 65, 786–791.

Zborowski, P. and Edwards, E.D. (2007) *A Guide to the Australian Moths*. CSIRO, Collingwood, Victoria.

Zilli, A. and Hogenes, W. (2002) An annotated list of the fruit-piercing moth genus *Eudocima* Billberg, 1820 (*sensu* Poole) with descriptions of four new species (Lepidoptera: Noctuidae, Catocalinae). *Quadrifina* 5, 153–207.

Appendix – The Australian species of fruit-piercing moths: taxonomy, synonymy, similar species, distribution and pest status

Details of recorded larval food plants for each species are given in Table 30.1. Known natural enemies for each species are listed in Table 30.2. The contents of these tables are discussed in the main text.

Genus ***Eudocima* Billberg, 1820.** Type species: *Phalaena salaminia* Cramer, 1777 (Zaspel and Branham 2008).

***Eudocima salaminia* (Cramer, 1777), 'owlet moth' (Figs 30.2, 30.3D–F)**
Synonymy: Edwards (1996)

Eudocima salaminea Hübner, 1823 (misspelt)
Ophideres atkinsoni (Scott, 1890)

Similar species: *Eudocima nigricilia* (Prout, 1924) from mainland New Guinea. The uninterrupted terminal band of the hindwing of *E. nigricilia* distinguishes this species from *E. salaminia.*
Distribution. India, China, SE Asia, South Pacific islands (east to Samoa), Papua New Guinea; Australia: NT, Qld; seasonally south-eastern Queensland and northern NSW and occasional migrants reaching areas south to Sydney.
Pest status. E. salaminia is one of the most important pest species in eastern Australia. In south-eastern Queensland the fruit damaged includes carambolas, soursop, lychees, persimmons, oranges, mandarins, pawpaws, stone fruit, longans, mangoes, kiwifruit, tomatoes and aubergine. Indigenous species attacked by *E. salaminia* include palms, *Ficus* spp., lillypillies (*Syzigium* spp.), *Solanum* spp. and *Tabernaemontana pandacaqui* Lam. (Sands and Schotz, 1987). *E. salaminia* is said to be an uncommon species in India (Baptist, 1945).

***Eudocima phalonia* (Linnaeus, 1763), 'common fruit-piercing moth' (Figs 30.1, 30.3A–C, 30.9)**
Synonymy. Edwards (1996), Zill and Hogenes (2002), Zaspel and Braham (2008)

Phalaena phalonia (Linnaeus, 1763) (*Phalaena Noctua*)
Phalaena fullonia (Clerck, [1764])
Eudocima cocalus (Cramer, 1777)
Phalaena fullonica (Linnaeus, 1767) (*Phalaena Noctua*)
Noctua dioscoreae Fabricius, 1775
Ophideres dioscorae Moore, 1877
Noctua dioscoriae Hamson, 1894
Noctua dioscoriae Warren, 1913
Noctua dioscoriae Zhu and Chen, 1963
Phalaena pomona Cramer, 1776
Phalaena princeps Boisduval, 1832
Ophideres obliterans Walker, 1858
Ophideres fullionica Pinhey, 1975

Taxonomic identity. The generic and species names used for *Eudocima phalonia* (Linnaeus, 1763) have been subjected to uncertainties relating to the original descriptions and validity. The name *fullonia* was considered a synonym of *fulonica* Linn., by Zilli and Hogenes (2002) and Zaspel and Branham (2008). Zilli and Hogenes (2002) added *Eudocima phalonia* (Linnaeus, 1763) comb. n. '... must be used for the species known as *Eudocima fullonia* (Clerck, [1764]) and relegated *fullonia* to synonymy'. However, Edwards (1996), when he reviewed *Eudocima* spp. with synonyms and the origin of the name, *E. fullonia*, indicated the name *phalonia* had been applied as a genus for another species and was therefore unavailable for this taxon. Without attempting to resolve this discrepancy, the species name applied here for *E. phalonia* follows that of Zilli and Hogenes (2002). In Australia the most widespread and important pest species, *E. phalonia*, was considered by Waterhouse and Norris (1987) to have originated from the Indo-Malaysian region.
Similar species. Eudicima lequeuxi (Brou and Zilki) from Africa, and *E. euryzona* (Hampson) from Madagascar. This led Leroy *et al.* (2021) to suggest Africa was probably the origin of *E. phalonia*, based on morphology of these similar species occurring in Africa.
Distribution. Waterhouse and Norris (1987) noted this moth occurred in more than 67 countries across Asia, Africa and south-east China, Indonesia, Papua New Guinea and islands of the Pacific (Fig. 30.1). In Australia *E. phalonia* occurs in Northern Territory and Queensland, and seasonally in south-eastern

Queensland and New South Wales to the Hunter River and occasionally as far south as Wollongong. The occurrence of *E. phalonia* south from Mackay, Queensland, to northern NSW is seasonal in most years, with adults first appearing in late spring months following their annual migrations from further north. Increases in abundance follow from breeding locally through the warmer months from December until about May each year, with build-up of moth populations from breeding on several local menisperm vines, mostly *Legnephora moorei*, *Tinospora smilacina*, *Sarcopetalum harveyanum* and *Stephania japonica* subsp. *timoriensis*. The absence of adults and immature stages of *E. phalonia* in winter in southern regions of eastern Australia has been attributed to the failure of larvae to survive the cool and dry months of subtropical regions. Adults and immature stages reappear after the migrating adults arrive and commence breeding on local food plants from late October or November through until about May (Huber *et al.*, 1998).
Adult variation. In Australian specimens both sexes of *E. phalonia* vary in wingspan, colour and the extent of markings of the forewing. In host-range experiments (D.P.A. Sands, unpublished), variation in wingspan was attributed to species of food plant, the moisture content and concentration of nitrogen (% N dry weight) in leaves, leaf toughness and temperatures during larval development. Specimens of *E. phalonia* from Samoa differed somewhat in adult morphology when compared with specimens from Australia and Papua New Guinea, notable by their paler colour and different forewing markings (D.P.A. Sands, unpublished). The mature larvae varied in background colour from black and brown to green.
Pest status. Eudocima phalonia is the most serious pest species of fruit-piercing moth in Australia. In south-eastern Queensland the fruit most frequently damaged by the moths are carambolas, soursop, lychees, persimmons, citrus (mainly oranges and mandarins) and pawpaws (Sands and Schotz, 1987). Other fruit damaged include stone fruit, longans, mangoes, kiwifruit and guavas. One of the most seriously affected fruits is lychee, and *E. phalonia* can easily penetrate the hard shell of the fruit. The fruit of indigenous species attacked by the moths include palms, figs (*Ficus* spp.), lillypillies (*Syzigium* spp.), *Solanum* spp. and the shrub *Tabernaemontana pandacaqui* (Sands and Schotz, 1987).
Life history. In Australia larvae of *E. phalonia* are only known to feed on species of vines of the family Menispermaceae. In Papua New Guinea (PNG) and on Pacific islands to the east, larvae of *E. phalonia* feed on several *Erythrina* spp. (Fabaceae) as well as Menispermaceae (including *Stephania japonica* in Tonga) whenever these vines are present. In New Guinea and on Pacific islands, the interaction of *E. phalonia* with *Erythrina* spp. and the Menispermaceae as hosts is thought to be due to the similar alkaloids they possess (i.e. the tetracyclic *Erythrina* type), which are found in certain species of the plant family Menispermaceae (Fay and Halfpapp, 1987; Blanchfield *et al.*, 1993). Differences in alkaloids associated with certain menisperm genera may have contributed to the development of host plant specificity of some *Eudocima* spp.

***Eudocima materna* (Linneaus, 1767), 'dot-underwing piercing moth' (Fig. 30.3G–I)**
Eudocima materna (Linneaus, 1767) (*Phalaena Noctua*).
Synonymy: Edwards 1996, Zaspel and Braham (2008)
Noctua hybrida Fabricius, 1775
Ophideres chalcogramma Walker, 1865
Similar species. The morphologically similar *Eudocima apta* (Walker, 1858), from South America from Brazil, Argentina and Mexico (Zilli and Hogenes, 2002). *E. apta* occasionally migrates into the USA from South America (Brou, 1994; Miller and Brou, 2013).
Distribution. Africa, India (Sing *et al.*, 2019), Asia, Papua New Guinea, Australia. *E. materna* occurs in northern Western Australia south to Carnarvon (Common, 1990), the Northern Territory, eastern Queensland (common north from Gympie) and (uncommonly) north-eastern NSW. It occasionally migrates from its northern breeding areas to Victoria and the Australian Capital Territory (Cocking *et al.*, 2022).
Habitat. In Northern Territory, *E. materna* is an important pest species wherever its principal larval food plant, *Tinospora smilacina* and varieties, are abundant. *E. materna* breeds on this deciduous vine in dry vine thicket, beach forest, monsoon forest, gallery forest, lowland and upland rainforest and occasionally open woodland (Fay and Halfpapp, 1999). Larvae of *E. materna* occasionally feed on *Legnephora moorei* (F. Muell.) Miers (Common, 1990) in rainforest.
Pest status. Eudocima materna is a serious horticultural pest, particularly in fruit-growing areas of northern Australia and north-eastern Queensland as far south as Mackay. Its abundance in south-eastern Queensland varies with seasons and the phenology of production of leaves on its deciduous food plant, *T. smilacina*. During favourable climatic conditions, this moth has been known to migrate great distances from its breeding areas (Common, 1990).

***Eudocima aurantia* (Moore, 1877) (Fig. 30.3J–L)**
Ophideres aurantia Moore, 1777
Synonymy: Edwards (1996), Zaspel and Braham (2008)
Adris rutilis Moore, 1881
Morphologically similar species. None recognized.
Distribution. Sri Lanka, China, India, Andaman Islands, Sundaland, Sulawesi, Seram, Indonesia, Papua New Guinea. In Australia, *E. aurantia* occurs in eastern Queensland from Cape York Peninsula to Townsville (Common, 1990) and occasionally Mackay.
Habitat. *E. aurantia* reproduces in coastal and subcoastal woodland and rainforest communities of eastern Queensland, wherever the food plant *Tinospora smilacina* occurs. Moths oviposit anywhere on the vines, usually only when young leaves are present. The larvae leave the vine to pupate in loose cocoons they construct in leaves of nearby shrubs.
Pest status. A major cosmopolitan pest in Queensland, mainly north of Townsville but occasionally further south following favourable climatic conditions.

***Eudocima iridescens* (Lucas, 1894) (Fig. 30.4G–I)**
Ophideres iridescens Lucas, 1894.
Synonymy: Edwards (1996)
Ophideres pyrocrana Turner, 1908
Morphologically similar species: None recognized.
Distribution. Australia and PNG, with a record from New Zealand (noted on Wikipedia). Occurs in tropical rainforest areas of Queensland from Cape York Peninsula to Rollingstone and Paluma, with occasional records from migrations into south-eastern Queensland.
Habitat. *E. iridescens* is restricted to lowland tropical rainforest communities of northern Queensland. Moths oviposit low on the vines when young, soft leaves are present. Larvae leave the vine to pupate and construct cocoons in leaves of nearby rainforest plants.
Pest status. An uncommon species in northern Queensland, *E. iridescens* is considered an occasional pest where orchards have been established close to rainforest.

***Eudocima jordani* (Holland, 1900) (Fig. 30.4A–C)**
Ophideres jordani (Holland, 1900)
Synonymy: None recognized.
Morphologically similar species. None recognized.
Distribution. Papua New Guinea, northern Queensland.
Habitat. *E. jordani* reproduces in lowland tropical rainforest and wetland communities in northern Queensland. Moths oviposit low on vines when young, soft leaves are present. Larvae leave the vine to pupate and construct cocoons in leaves of nearby rainforest plants.
Pest status. A common species in northern Queensland, in and near tropical rainforests. *E. jordani* is at times a pest of fruit in orchards when crops are close to rainforest.

Eudocima cocalus (Cramer, 1777) (Fig. 30.4D–F)
Phalaena cocalus Cramer, 1777 (*Phalaena* (*Noctua*)
Synonymy: Edwards (1996), Zilli and Hogenes (2002)
Ophideres crepidolata Lucas, 1894;
Noctua maculata Weber, 1801
Ophideres plana Walker
Rhytia crepidolata Lucas, 1894
Ophideres crocale Pagenstacher, 1896
Morphologically similar species: *Eudocima tradawayi* Zilli and Hogenes, known from the Philippines.
Distribution. India, Indonesia, Malaysia, Philippines, China, Papua New Guinea, northern Queensland.
Habitat. *E. cocalus* reproduces in lowland tropical rainforest in northern Queensland. Moths oviposit on and near the vines when young, soft leaves are present. Larvae leave the vine to pupate and construct cocoons in leaves of nearby rainforest shrubs.
Pest status. An uncommon species in northern Queensland, in and near tropical rainforests. *E. cocalus* is a pest of fruit in orchards when crops are close to rainforest.

31 The Gumleaf Skeletonizer Moth: *Uraba lugens*

Mark A. Elgar and Kathryn B. McNamara*
School of BioSciences, University of Melbourne, Parkville, Victoria, Australia

Abstract

The gumleaf skeletonizer, *Uraba lugens*, is a nolid moth probably best known for its pest status in New Zealand and its potential to cause extensive damage to *Eucalyptus* plantations in Australia. As a pest, research on *U. lugens* has focused on how to predict and manage these destructive outbreaks. The species, however, has intriguing natural history traits that remain poorly understood and has proved to be a very useful 'model' species for investigating neglected ideas and contemporary concepts in sexual communication. Here we describe the natural history of this species, including recent research on aspects of its evolutionary ecology, chemical signalling and phenotypic plasticity.

31.1 Introduction

The gumleaf skeletonizer moth *Uraba lugens* is an Australian member of the Nolidae, that can cause extensive damage to *Eucalyptus* plantation crops in Australia and is an invasive pest in New Zealand. As a pest, research on *U. lugens* has focused on how to manage these destructive outbreaks. This chapter describes the natural history of this species, drawing on two excellent reviews (Campbell, 1962; Berndt and Allen, 2010) as well as descriptions of research on various aspects of evolutionary ecology, including chemical signalling and phenotypic plasticity. The species has intriguing natural history traits that remain poorly understood and has proved to be a very useful 'model' species for investigating neglected ideas and contemporary concepts in sexual communication.

Adult *U. lugens* do not have any remarkable physical features or colour patterns. Adults are small (body length about 10 mm) with a mostly grey colouration, and their wings (25–30 mm) are characterized by a few dark grey patterns. Adults are sexually dimorphic: females are typically larger and heavier (Fig. 31.1A), while males possess feathery, bipectinate antennae that are structurally more complex than the simple filiform antennae of females (Fig. 31.1B). The eggs of *U. lugens* are laid in characteristic lines, and the larvae are light green in colour and covered in dense, urticating hairs. The species is widely distributed in Australia and is designated a pest because of the large population outbreaks that can damage *Eucalyptus* host plants in both natural and managed forests. It was introduced into the North Island of Aotearoa New Zealand, where there is now an established population.

Unlike adults, the larvae have an unusual and remarkable visual attribute. The larvae of fourth or older instar *U. lugens* usually retain the head capsule of their exoskeleton when they moult (McFarland, 1978; Berndt and

*Corresponding author: mcnamara@unimelb.edu.au

DOI: 10.1079/9781800627680.0031

Fig. 31.1. Images of (A) adult female; (B) adult male; (C) larva with retained head capsules; (D) larvae on a skeletonized leaf. Sources: (A,B) Birgit E. Rhode, Landcare Research New Zealand Ltd (CC BY 4.0), (C,D) Donald Hobern (CC BY 2.0).

Allen, 2010), with the larvae accumulating a new headcase with each successive moult (Fig. 31.1C). This has led to the popular nickname 'the mad hatterpillar moth', although this may be a little unfair to the original, eccentric character. Accumulating head capsules with successive moults appears to be uncommon taxonomically, as it has been documented for only four other species of moths, including *Rhynchopala argentalis* in India, *Roeselia togatulalis* and *R. nitida* in Europe, and *Mimerastria mandshuriana* in Japan, all of which belong to the Nolidae (McFarland, 1978). The evolutionary significance of this remarkable larval characteristic is not clear.

31.2 Taxonomy

Originally described as *Uraba lugens* by Francis Walker in 1863 from a single specimen from the British Museum, the species' nomenclature has had some changes before settling on the original. Three years after his original description, Walker redescribed the species as *Coesa viduella*, and the species accumulated several other synonyms (Berndt and Allen, 2010). The moth is placed within the family Nolidae, now firmly established as a 'full' family within the Noctuoidea (see Chapter 20). Edwards (1996), in the *Australian Checklist*, notes: 'populations in different parts of Australia differ in characteristics of their biology. All populations are here included under the one specific name until a thorough revision is undertaken'. We await such a revision.

There is one other congeneric species in Australia, *Uraba deplanana*, also described by Walker from a single specimen collected from Tasmania and held by the British Museum. The holotype has been lost, however, and no further information about *U. deplanana* has been published.

31.3 Ecology

The gumleaf skeletonizer moth is endemic to Australia, with records throughout the continent except for the Northern Territory. According to the *Atlas of Living Australia* (www.ala.org.au), most records are found south of Maryborough (Queensland) on the eastern seaboard and Kalbarri (roughly 600 km north of Perth) on the western seaboard (Fig. 31.2A). Interestingly, there are records of *U. lugens* in north Queensland from Townsville north to Cooktown, but no records between Maryborough and Townsville. Thus, the species thrives in a range of climatic zones, from wet tropical regions in Queensland, through warm and dry regions of the continent to the cool, subalpine conditions of Tasmania (Berndt and Allen, 2010).

The larvae of *U. lugens* is polyphagous, with the larvae feeding on a remarkably broad range of host plants, including over 100 species of *Eucalyptus* and many other species within the Myrtaceae (Berndt and Allen, 2010). However, there is little evidence that the distribution of *U. lugens* is constrained by the distribution of these food plants (Kriticos *et al.*, 2007; Fig. 31.2B).

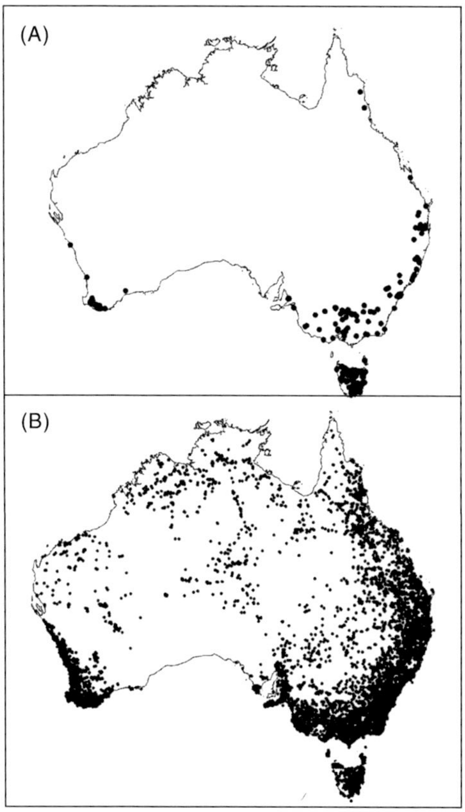

Fig. 31.2. Records of *U. lugens* (A) and distribution of the foodplants of *Uraba lugens* (B). Modified from Kriticos *et al.* (2007).

Populations of *U. lugens* may be univoltine or bivoltine and although this variation is thought to be determined by climatic conditions, other factors may also contribute (Berndt and Allen, 2010). In South Australia and Victoria, the bivoltine populations have a winter generation that starts around April–May and lasts roughly 8 months, whereas the summer generation starts in December and lasts approximately 4 months (Campbell, 1962; Allen, 1990). Campbell (1962) identified univoltine populations in the 'highland' regions of Victoria, while Farr (2002) reported both univoltine and bivoltine populations in the south of Western Australia.

Adults of *U. lugens* are thought to be poor fliers (Campbell, 1962; Morgan and Cobbinah, 1977), and so dispersal by adults is likely to be confined to short distances. There may be limited movement during the larval stages: late instar larvae may leave depleted food plants in search of other plants. The larvae usually leave the food plant to pupate on the ground among nearby vegetation. This can be facilitated by producing silk threads to lower themselves, a process that may allow them to be dispersed by the wind.

The larvae of *U. lugens* have several anti-predator traits (see below) but nevertheless have many natural enemies, including 22 species of primary parasitoids that attack the larvae and pupae, including tachinid flies, and braconid, chalcid, eulophid and ichneumonid wasps (Berndt and Allen, 2010). Berndt and Allen (2010) provide a comprehensive list of predators that include Erythraeidae mites that prey on eggs, and jumping spiders (Salticidae), pentatomid predatory bugs, nabid damsel bugs and lacewings (Chrysopidae) that prey on larvae. There are no accounts of vertebrate predators (Berndt and Allen, 2010).

31.4 Life History

31.4.1 Eggs and larvae

Eggs maintained at 30°C hatch in 1–4 days, while those at lower temperatures (10°C) hatch in 4–10 days (Allen and Keller, 1991). The ideal temperature for eggs is between 15–25°C, with a large proportion of eggs failing at lower and higher temperatures (Allen and Keller, 1991). The number of larval instars to pupation is influenced by temperature, with 10–14 instars at 15°C and 6–9 instars at 25°C (Allen and Keller, 1991). Field observations in Western Australia indicated that larval development varies across different host plant species, with larvae feeding on *Corymbia calophylla*, *C. ficifolia*, *C. citriodora* and *E. nicholii* developing comparatively rapidly and those feeding on *E. caesia* developing more slowly (Farr, 2002).

31.4.1.1 Larval foraging behaviour

The first instar larvae emerge synchronously from a cluster of eggs and forage gregariously close to the egg cases, which may provide some shelter (Campbell, 1962). The young larvae feed on the upper and lower epidermis, pallisade tissue and spongy mesophyll of the leaf, but tend to avoid the oil cells and veins. As a result, the leaf is 'skeletonized' (Fig. 31.1D) – and hence the source of the species' common name. This feeding strategy may allow the larva to avoid consuming eucalypt toxins. The larvae lose their gregarious foraging habit by the fifth or sixth instar, when they typically forage individually or in small groups (Campbell, 1962). These larger larvae will consume portions of the leaf blade, rather than just skeletonizing it (Campbell, 1962).

In a novel experiment, Leonard *et al.* (2016) provided evidence that larval foraging behaviour is affected by exposure to wind. Larvae of *U. lugens* forage openly on the leaf, where they may be exposed to desiccation (Hassell and Southwood, 1978) and being dislodged from the leaf (Yamazaki, 2011), with the risk of both being higher when there is a wind. By placing older, solitary larvae on small (60 cm) plants of *E. tereticornis* in the laboratory, Leonard *et al.* (2016) compared the movement of larvae over 2 h when they were or were not exposed to wind (~3 m/s), generated by pedestal fans. A smaller proportion of larvae remained on the leaves when exposed to wind, and these larvae were more likely to be found on the leeward than on the windward side of the plant. A larger proportion of larvae moved to the petiole or branch when exposed to wind, and roughly 10% of the larvae exposed to wind fell off the leaves within 2 h. Although the study revealed an intriguing effect of wind on the movement and location of larvae, the biological significance of the experiment remains a little unclear because the majority (roughly 70%) of the larvae remained on the leaves. Perhaps stronger winds are required to elicit more widespread responses.

Larvae of *U. lugens* are sensitive to the risk of predation. In laboratory experiments, Low *et al.* (2014) simulated predator attacks by gently pinching the body of the larvae with soft forceps, intended to simulate the bite of an arthropod or bird. The reaction of the larvae seemed to depend on the frequency of the attack, with relatively fewer larvae continuing to feed if they had been repeatedly attacked. Further, larvae that had been attacked were also less likely to resume foraging within 3 h, compared with larvae that had not been attacked.

Larvae have distinct morphological antipredator adaptations: they are pale green (matching their leaf background) and are entirely covered in venom-containing, urticating hairs. Numerous lepidopterans within the Zygaenoidea, Papilionoidea, Noctuoidea, Lasiocampoidea and Bombycoidea possess defensive setae, ranging from small, detachable hairs to much larger, hollow spines filled with venom (Battisti *et al.*, 2011). Human contact with the hairs of the larvae of *U. lugens* can cause a painful reaction, referred to as erucism (Derraik, 2006). Perhaps the apparent absence of vertebrate predators of *U. lugens* (Berndt and Allen, 2010) reflects the efficacy of urticating hairs as an antipredator defence.

31.4.1.2 Larval head-dress

Low *et al.* (2016) examined the significance of the larval head-dress in a field experiment, in which groups of larvae with different proportions of individuals with and without head-dresses were exposed to natural enemies. There was qualitatively greater survival of larval groups in which all individuals had head-dresses compared with those whose head-dresses had been experimentally removed, although this difference was not statistically significant. Larvae with head-dresses, however, had a significantly greater survival rate than those without head-dresses in the mixed groups comprising equal numbers of larvae with and without head-dresses (Low *et al.*, 2016). This latter result, suggesting the head-dresses may have some deterrence role, may be more informative because the treatment provided natural enemies with a choice of victim, which was not necessarily possible in the homogeneous groups. Although it is not clear how the head-dress provides protection to the larvae, this study provides a valuable and intriguing insight into this curious and uncommon trait.

31.4.2 Adult reproductive strategies: females

31.4.2.1 Sex pheromones

Adult female *U. lugens* attract potential male mates by releasing a sex pheromone, which males use to locate the females (Johnson *et al.*, 2017a). This signalling or 'calling' behaviour is unambiguous, and characterized by the female expanding her wings to reveal the tip of her abdomen and an everted gland containing the sex pheromone. Gibb *et al.* (2008) characterized the chemical composition of female sex pheromones by excising the pheromone glands from calling adult females and analysing the contents using gas chromatography–mass spectrometry. Three compounds were identified: (10E,12Z)-hexadecadien-1-yl acetate (E10,Z12–16:Ac), (10E,12Z)-hexadecadien-1-ol (E10,Z12–16:OH) and (Z)-11-hexadecen-1-yl acetate (Z11–16:Ac). Subsequent trapping trials in Queensland and Auckland, NZ, indicated that the two major components E10,Z12–16:Ac and E10,Z12–16:OH are sufficient, and necessary, for trapping males (Gibb *et al.* 2008).

31.4.2.2 Female calling behaviour

Several lines of evidence highlight that calling behaviour in *U. lugens* is plastic in its timing and intensity, responding adaptively to sociosexual cues in the environment. Under laboratory conditions, females begin calling shortly after the onset of scotophase, that is, the onset of dark, with the greatest proportion of females calling around 6 h into the scotophase, and a sharp decline in calling subsequently (Pham *et al.*, 2021).

Conventional wisdom and theoretical models predict that virgin females should increase their investment in calling behaviour as they age to maximize their likelihood of attracting a male and avoid dying unmated, referred to as 'mating failure' (see Umbers *et al.*, 2015). Critically, however, male *U. lugens* can mate once a day only (Pham *et al.*, 2020), and so there is a greater availability of receptive males, and thus likelihood of female mating, earlier in the scotophase. Accordingly, in contrast with theory and other moths, older virgin females were far less likely to call and called for shorter periods than younger virgin females. Instead, older females commenced calling earlier in the scotophase, perhaps to avoid competition with younger females whose sex pheromone is more attractive to males (Johnson *et al.*, 2017a; Pham *et al.*, 2021). Similarly, Pham *et al.* (2021) demonstrated that female *U. lugens* adjust their calling behaviour adaptively to reduce the risk of mating failure due to local competition with other females. Females reared as larvae at high density (providing a cue of future local mate competition) called earlier in the scotophase compared with females reared as larvae at low densities.

31.4.3 Adult reproductive strategies: males

31.4.3.1 Sexual selection and male reproductive traits

Like many moths, adults of *U. lugens* are sexually dimorphic in overall body size, with females typically larger and heavier than males. They are also dimorphic in their antennal structure, where males have feathery, bipectinate antennae and females have simple, filiform antennae. It is widely held that complex antennae in the Lepidoptera have evolved to allow males to detect minute amounts of sex pheromone that females release to reveal their location, but comparative studies provide little support for this view (Symonds *et al.*, 2012). Indeed, most moth species are sexually monomorphic in antennal structure. Nevertheless, recent research on *U. lugens* provides support for Charles Darwin's surprisingly neglected insight – that sexual selection will favour 'organs of sense' that improve the capacity of males to detect and encounter receptive females (Darwin, 1871; Elgar *et al.*, 2019). In a field experiment, either one or two females of *U. lugens* were placed in delta traps, which showed that males with longer antennae were better able to detect the solitary female (Johnson *et al.*, 2017a). Assuming the concentration of pheromone emerging from traps with a single female is less than from traps with two females, these data suggest that males with relatively larger antennae will detect females more rapidly, as Darwin predicted.

31.4.3.2 Cost of sexual signalling

Darwin's insight also highlights the potential costs of sexual signalling via pheromones. Research on chemical signalling, and particularly sex pheromones in insects, has tended to focus primarily on proximate questions – identifying the chemical composition and the underlying mechanisms in their production. The conventional wisdom is that the costs of pheromone production are trivial (Cardé and Baker, 1984; Fromhage *et al.*, 2016). Pham *et al.* (2022) ask whether female signal (pheromone) production and male signal reception (antennal microstructure) are costly in *U. lugens*, using an experimental immune challenge. The density of antennal sensilla of adult males that had been immune challenged as last instar larvae was lower than their unchallenged counterparts. Similarly, the pheromone of adult females that had been immune challenged as last instar larvae was less attractive to males than their unchallenged counterparts. This study also highlights the impact that parasite and pathogen attack can have on the ability of males and females of this species to detect and attract mates, and why variation in these signalling traits is maintained despite strong directional selection.

31.4.4 Copulation

The courtship and mating behaviour of *U. lugens* is not widely documented. Campbell (1962) notes that the male and female face opposite directions when *in copula*, with the wings of the male partially covering those of the female. Campbell (1962) suggests that copulation takes place shortly after the start of scotophase, but it is likely that copulation can begin later in the night since unmated females continue to call for six or more hours into the scotophase (Pham *et al.*, 2021). However, pairs are unlikely to begin copulation much later in the scotophase because the duration of copulation is lengthy, up to 4 h (Pham *et al.*, 2021).

One reason why so few studies document the courtship and copulation in *U. lugens* is because it is difficult to observe this behaviour in the laboratory (Pham *et al.*, 2021). We found a much higher proportion of studies of mating behaviour in moth species where males have simple antennae than in moth species where males have complex antennae (H.T. Pham *et al.*, unpublished). Our supporting experiment demonstrated that in *U. lugens*, the ability of males to locate a calling female was possible only under certain conditions, in which air flow was maximized, potentially allowing the establishment of a pheromone concentration gradient for males to follow.

31.4.5 Oviposition

Females of *U. lugens* are capital breeders (*sensu* Stephens *et al.*, 2009) and can produce large numbers of eggs during their lifetime. In a Western Australian population, dissected females that had been reared in the laboratory from late instar larvae had a mean of 404 ± 13 eggs, with a range of 14–677 eggs per female (n = 75) (Farr, 2002). Unsurprisingly, this measure of female fecundity strongly and positively correlated with pupal weight. Clutches of eggs are laid in lines, and the number of eggs within a clutch can range from 28 to 344, with a mean of 104 ± 5 (n = 103; Farr, 2002). Morgan and Cobbinah (1977) report that females prefer to lay eggs on maturing or mature leaves that are blue green to dark green in colour and largely ignore younger leaves.

Uraba lugens is polyphagous and yet female ovipositional choices are not necessarily in the best interests of the larvae. At the arboretum of the Waite Agricultural Research Institute of the University of Adelaide, South Australia, oviposition was recorded more than once on 174 species of *Eucalyptus* and two species of *Angophora* (Cobbinah, 1985). Interestingly, there is considerable variation in the survival of larvae across these host plant species. Other studies similarly report that females lay eggs on numerous plant species that are not suitable hosts for the larvae, which are unlikely to survive beyond the fourth instar (Morgan and Cobbinah, 1977; Cobbinah,

1985; Farr, 2002). Berndt and Allen (2010) found that of the 103 myrtaceous tree species that have been reported as host trees, less than 45 could be considered good hosts. It is not clear why females make these ovipositional 'mistakes', nor what impact they have on their reproductive success.

31.4.6 Larval environment and variation in adult traits

31.4.6.1 Host plants and nutrition

The nutritional quality of host plants can have a profound impact on the development of larvae and thus adult survival and reproductive output of phytophagous insects (Moreau *et al.*, 2017). The quality of the larval host plant can also affect life history traits and reproductive success of adult *U. lugens*. Pham *et al.* (2024) investigated the impact of host plant nutritional environment on larval performance and key adult reproductive traits. Specifically, they reared larvae on either *Eucalyptus camaldulensis* or *E. moorei*, which differed in their carbon and nitrogen leaf content. Adult males and females reared on *E. moorei* had longer wing lengths than those raised on *E. camaldulensis*. Larval diet also had an impact on adult male but not female longevity, with males raised on *E. camaldulensis* living longer than those raised on *E. moorei*. In a subsequent experiment, Pham *et al.* (2024) raised larvae on *E. moorei* host plants which were treated with commercial fertilizer or not, resulting in differences in nitrogen but not carbon leaf content. Adult females reared on fertilized plants were larger than those reared on non-fertilized plants, but there was no difference in male size. However, both males and females reared on fertilized plants lived longer than those on non-fertilized plants.

Pham *et al.* (2024) also documented the impact of larval diet on male and female reproductive traits. Diet did not have an impact on the attractiveness of female sex pheromones: Y-maze olfactometer assays indicated that males showed no preference for females reared on any of the different diets. Diet did affect male traits, however: adult males reared on *E. moorei* had longer antennae than those reared on *E. camaldulensis*, but the density of olfactory sensilla was greater for males reared on *E. moorei* than *E. camaldulensis*. Larger males had larger testes, but testes size was not significantly different across the two host plants. Interestingly, male antennal length, sensilla density and testes size were not significantly different across the fertilized and non-fertilized *E. moorei* host plants.

Combined, these data suggest that larval diet has different effects on male and female reproductive strategies. For males, host plant species can impact male investment in traits associated with mate search that have consequences for mating success (Johnson *et al.*, 2017a). In contrast, although host plant species did not affect the attractiveness of female sex pheromones, it did affect female size and longevity, which may have consequences for egg production and securing a mate. Clearly, *U. lugens* offers promising opportunities to investigate how male and female reproductive strategies interact with nutritional requirements (Pham *et al.*, 2024).

31.4.6.2 The social environment

Remarkably, the larval social environment similarly can influence adult traits, as reported for several moths, including the Indian meal moth, *Plodia interpunctella* (Gage, 1995), the armyworm, *Pseudaletia separata* (He and Miyata, 1997) and the almond moth, *Cadra cautella* (McNamara *et al.*, 2010). The larval social environment can predict key aspects of the adult environment that can in turn affect mating strategies. For example, larvae developing in relatively dense populations can anticipate that there will be a large adult population, and thus potentially strong competition for mating partners. Consistent with this prediction, adult male *U. lugens* raised in dense larval populations had larger testes and smaller antennae and wings than those raised in sparser larval populations (Johnson *et al.*, 2017b). A different pattern emerged for adult females raised in dense larval populations, which invested more heavily in signalling activity and whose pheromone was more attractive to males than females raised in sparse larval populations (Pham *et al.*, 2020).

31.5 Pest Status

The larvae of *U. lugens* are sometimes responsible for extensive defoliation of several species of *Eucalyptus* following population outbreaks. Most of the earlier research effort on *U. lugens* attempted to understand the factors leading to these outbreaks and devise ways of managing their potentially destructive consequences. Collett and Fagg (2010) document an extensive survey of mixed species eucalypt forests in East Gippsland, Victoria, from 2003 to 2006. During this time, there were two defoliation events, which Collett and Fagg (2010) argue were separate events caused by different species, *Uraba lugens* and cup moths *Doratifera* spp. The latter was responsible for much higher levels of defoliation (71%) than *U. lugens* (23%), although in both instances there was considerable variation across sites (Collett and Fagg, 2010). Collett and Fagg (2010) suggest that low rainfall may trigger these events: the drier conditions may reduce the risk of fungal infections, especially for the pupae, a view that is echoed by both Campbell (1962) and Elliott *et al.* (1998). Modelling by Wills and Farr (2017) similarly suggests that a reduction in groundwater and moisture conditions during the egg and early larval phases are important in determining an outbreak.

The impact of these outbreaks on the survival of affected trees is less clear. Pinkard *et al.* (2011) suggest that defoliating outbreaks are not necessarily lethal for trees generally but do contribute to physiological stress. Wills *et al.* (2004) found that in jarrah *E. marginata* forests, the outbreaks are transient and the foliage of defoliated trees is eventually replenished with little, if any, mortality. As Common (1990) notes, *U. lugens* is not among the most destructive of lepidopteran pests in Australia.

In contrast, *U. lugens* is considered a serious exotic pest of plantation and amenity *Eucalyptus* species in New Zealand. Extensive laboratory trials that investigated ovipositional preferences and larval survival on New Zealand plants suggest that there is little risk of *U. lugens* feeding on trees other than those within the Myrtaceae (Withers *et al.*, 2011). Control strategies in New Zealand have included the use of insecticides, pheromone trapping and introducing natural enemies. Mansfield *et al.* (2006) report that pyrethroids and spinosad (that is, a natural mixture of bacterially generated spinosyn A and spinosyn B) are the most effective insecticides. *Bacillus thuringiensis kurstaki* Berliner (*Btk*) was less effective against the younger skeletonizing larvae compared with old larvae that chewed the entire leaf. More broadly, the effectiveness of these insecticides also varied with the host plant. Of course, it is important to recognize that these methods will create a strong selection pressure favouring resistance traits in the local population (e.g. Barzman *et al.*, 2015) and so they are useful in the short term only. Although synthesized sex pheromones attract adult males (Gibb *et al.*, 2008) and may be useful in tracking population numbers (Suckling *et al.*, 2005), it is not yet established whether this could be an effective means of controlling *U. lugens* outbreaks.

There has been more extensive research on the potential impact of the braconid parasitoid *Cotesia urabae*. Females of this small (2.5–3.2 mm long) wasp oviposit into the larvae of *U. lugens*, and the developing wasp larvae take 2–3 weeks to emerge from the caterpillar and pupate alongside the now deceased host (Allen, 1990). The species was introduced into New Zealand in 2011 as a biological control agent protecting commercial eucalyptus plantations and ornamental trees (Avila *et al.*, 2013). Established populations have now been confirmed in several locations from Hawkes Bay in the south to Northland in the north (Avila *et al.*, 2023).

As with all species that acquire the status of 'pest', the focus of research tends to be on control. Like many forest insects, population outbreaks in *U. lugens* and our ability to predict them with any certainty are limited by the lack of consistent long-term data (>30 years) and, more fundamentally, our understanding of the biology of the species, including how foraging, courting, mating, ovipositing and dispersal behaviour contribute to fluctuations in population size. While these 'natural history' insights are frequently fascinating, if properly addressed they can often enhance our understanding and ability to better control the species' population sustainably. For example, recognizing the factors that influence female sex pheromone production and male perception may help inform mating disruption strategies with synthetic odours.

Acknowledgements

We thank the Australian Research Council for financial support of our research on *Uraba lugens*.

References

Allen, G.R. (1990) *Uraba lugens* Walker (Lepidoptera: Noctuidae): Larval survival and parasitoid biology in the field in South Australia. *Journal of the Australian Entomological Society* 29, 301–312.

Allen, G.R. and Keller, M.A. (1991) *Uraba lugens* (Lepidoptera: Noctuidae) and its parasitoids (Hymenoptera: Braconidae): Temperature, host size, and development. *Environmental Entomology* 20, 458–469.

Avila, G.A., Berndt, L.A. and Holwell, G.I. (2013) First releases and monitoring of the biological control agent *Cotesia urabae* Austin and Allen (Hymenoptera: Braconidae). *New Zealand Entomologist* 36, 65–72.

Avila, G.A., MacDonald, F.H., Hunt, S.K., Santos, K., Alavi, M. *et al.* (2023) Field assessment of parasitism on the New Zealand endemic *Nyctemera annulata* by *Cotesia urabae*: Can non-target impacts be ruled out? *Biological Control* 178, 105136.

Barzman, M., Bàrberi, P., Birch, A.N.E., Boonekamp, P., Dachbrodt-Saaydeh, S. *et al.* (2015) Eight principles of integrated pest management. *Agronomy for Sustainable Development* 35, 1199–1215.

Battisti, A., Holm, G., Fagrell, B. and Larsson, S. (2011) Urticating hairs in arthropods: Their nature and medical significance. *Annual Reviews in Entomology* 56, 203–220.

Berndt, L.A. and Allen, G.R. (2010) Biology and pest status of *Uraba lugens* Walker (Lepidoptera: Nolidae) in Australia and New Zealand. *Australian Journal of Entomology* 49, 268–277.

Campbell, K.G. (1962) The biology of *Roeselia lugens* (Walk.), the gumleaf skeletonizer moth, with particular reference to the *Eucalyptus camaldulensis* Dehn. (river red gum) forests of the Murray Valley Region. *Proceedings of the Linnean Society of New South Wales* 87, 316–338.

Cardé, R. and Baker, T. (1984) Sexual communication with pheromones. In: Bell, W. and Cardé, R. (eds) *Chemical Ecology of Insects*. Chapman and Hall, London, pp. 355–383.

Cobbinah, J.R. (1985) The gum leaf skeletonizer, *Uraba lugens*, and its hosts. Possible selection of strains of insects that are able to feed on resistant trees. *Insect Science and its Application* 6, 291–299.

Collett, N.G. and Fagg, P.C. (2010) Insect defoliation of mixed-species eucalypts in East Gippsland. *Australian Forestry* 73, 81–90.

Common, I.F.B. (1990) *Moths of Australia*. Melbourne University Press, Melbourne, Victoria.

Darwin, C. (1871) *The Descent of Man and Selection in Relation to Sex*. Murray, London.

Derraik, J. (2006) Erucism in New Zealand: Exposure to gum leaf skeletoniser (*Uraba lugens*) caterpillars in the differential diagnosis of contact dermatitis in the Auckland region. *New Zealand Medical Journal* 119, 1241.

Edwards, E.D. (1996) Noctuidae (Notes). In: Nielsen, E.S., Edwards, E.D. and Rangsi, T.V. (eds) *Checklist of the Lepidoptera of Australia*. Monographs in Australian Lepidoptera 4. CSIRO Publishing, Collingwood, Victoria, pp. 370–380.

Elgar, M.A., Johnson, T.L. and Symonds, M.R.E. (2019) Sexual selection and organs of sense: Darwin's neglected insight. *Animal Biology* 69, 63–82.

Elliott, H.J., Ohmart, C.P. and Wylie, F.R. (1998) *Insect Pests of Australian Forests: Ecology and Management*. Inkata Press, Melbourne, Victoria.

Farr, J.D. (2002) Biology of the gumleaf skeletoniser, *Uraba lugens* Walker (Lepidoptera: Noctuidae), in the southern jarrah forest of Western Australia. *Australian Journal of Entomology* 41, 60–69.

Fromhage, L., Jennions, M. and Kokko, H. (2016) The evolution of sex roles in mate searching. *Evolution* 70, 617–624.

Gage, M.J.G. (1995) Continuous variation in reproductive strategy as an adaptive response to population-density in the moth *Plodia interpunctella*. *Proceedings of the Royal Society* B, 261, 25–30.

Gibb, A.R., Suckling, D.M., Fielder, S., Bunn, B., Jamieson, L.E. *et al.* (2008) Major sex pheromone components of the Australian gum leaf skeletonizer *Uraba* lugens: (10E,12Z)-Hexadecadien-1-yl Acetate and (10E,12Z)-Hexadecadien-1-ol. *Journal of Chemical Ecology* 34, 1125–1133.

Hassell, M.P. and Southwood, T.R.E. (1978) Foraging strategies of insects. *Annual Review of Ecology, Evolution and Systematics* 9, 75–98.

He, Y. and Miyata, T. (1997) Variations in sperm number in relation to larval crowding and spermatophore size in the armyworm, *Pseudaletia separata*. *Ecological Entomology* 22, 41–46.

Johnson, T.L., Symonds, M.R.E. and Elgar, M.A. (2017a) Sexual selection on receptor organ traits: Younger females attract males with longer antennae. *Science of Nature* 104, 44.

Johnson, T.L., Symonds, M.R.E. and Elgar, M.A. (2017b) Anticipatory flexibility: Larval population density in moths determines male investment in antennae, wings and testes. *Proceedings of the Royal Society B*, 284, 20172087.

Kriticos, D.J., Potter, K.J.B., Alexander, N.S., Gibb, A.R. and Suckling, D.M. (2007) Using a pheromone lure survey to establish the native and potential distribution of an invasive Lepidopteran, *Uraba lugens*. *Journal of Applied Ecology* 44, 853–863.

Leonard, R.J., McArthur, M. and Hochuli, D.F. (2016) Exposure to wind alters insect herbivore behaviour in larvae of *Uraba lugens* (Lepidoptera: Nolidae). *Austral Entomology* 55, 242–246.

Low, P.A., McArthur, M. and Hochuli, D.F. (2014) Dealing with your past: Experience of failed predation suppresses caterpillar feeding behaviour. *Animal Behaviour* 90, 337–343.

Low, P.A., McArthur, M. and Hochuli, D.F. (2016) Head capsule stacking by caterpillars: Morphology complements behaviour to provide a novel defence. *PeerJ* 4, e1714.

Mansfield, S., Withers, T.M., Gous, S.F., Potter, K.J.B., Kriticos, D.J. *et al.* (2006) Potential of selective insecticides for managing *Uraba lugens* (Lepidoptera: Nolidae) on eucalypts. *Journal of Economic Entomology* 99, 780–789.

McFarland, N. (1978) Retention of cast head capsules by some nolid immatures in four Old World countries. *Journal of Research on the Lepidoptera* 17, 209–217.

McNamara, K.B., Elgar, M.A. and Jones, T.M. (2010) Adult responses to larval population size in the almond moth, *Cadra cautella*. *Ethology* 116, 39–46.

Moreau, J., Desouhant, E., Louapre, P., Goubault, M., Rajon, E. *et al.* (2017) How host plant and fluctuating environments affect insect reproductive strategies? In: Sauvion, N., Thiéry, D. and Calatayud, P.A. (eds) *Insect–Plant Interactions in a Crop Protection Perspective*. Elsevier, London, pp. 259–288.

Morgan, F.D. and Cobbinah, J.R. (1977) Oviposition and establishment of *Uraba lugens* (Walker), the gum leaf skeletoniser. *Australian Forestry* 40, 44–55.

Phạm, H.T., McNamara, K.B. and Elgar, M.A. (2020) Socially cued anticipatory adjustment of female signalling effort in a moth. *Biology Letters* 16, 20200614.

Phạm, H.T., McNamara, K.B. and Elgar, M.A. (2021) Age-dependent chemical signalling and its consequences for mate attraction in the gumleaf skeletonizer moth *Uraba lugens* Walker. *Animal Behaviour* 173, 207–213.

Phạm, H.T., Elgar, M.A., van Lieshout, E. and McNamara, K.B. (2022) Experimental immune challenges reduce the quality of male antennae and female pheromone output. *Scientific Reports* 12, 3578.

Phạm, H.T., McNamara, K.B. and Elgar, M.A. (2024) Diet-dependent reproductive investment in gumleaf skeletonizer moths, *Uraba lugens*. *Physiological Entomology* 49, 157–166.

Pinkard, E.A., Battaglia, M., Roxburgh, S. and O'Grady, A.P. (2011) Estimating forest net primary production under changing climate: Adding pests into the equation. *Tree Physiology* 31, 686–699.

Stephens, P.A., Boyd, I.L., McNamara, J.M. and Houston, A.I. (2009) Capital breeding and income breeding: Their meaning, measurement, and worth. *Ecology* 90, 2057–2067.

Suckling, D.M., Gibb, A.R., Dentener, P.R., Seldon, D.S., Clare, G.K. *et al.* (2005) *Uraba lugens* (Lepidoptera: Nolidae) in New Zealand: Pheromone trapping for delimitation and phenology. *Journal of Economic Entomology* 98, 1187–1192.

Symonds, M.R.E., Johnson, T.L. and Elgar, M.A. (2012) Pheromone titre and composition, population abundance, body size, and the evolution of elaborate antennae in moths. *Ecology and Evolution* 2, 227–246.

Umbers, K.D.L., Symonds, M.R.E. and Kokko, H. (2015) The mothematics of female pheromone signaling: Strategies for aging virgins. *American Naturalist* 185, 417–432.

Wills, A.J. and Farr, J.D. (2017) Gumleaf skeletoniser *Uraba lugens* (Lepidoptera: Nolidae) larval outbreaks occur in high rainfall Western Australian jarrah (*Eucalyptus marginata*) forest after drought. *Austral Entomology* 56, 424–432.

Wills, A.J., Burbidge, T.E. and Abbott, I. (2004) Impact of repeated defoliation on jarrah (*Eucalyptus marginata*) saplings. *Australian Forestry* 67, 194–198.

Withers, T.M., Potter, K.J.B., Berndt, L.A., Forgie, S.A., Paynter, Q.E. *et al.* (2011) Risk posed by the invasive defoliator *Uraba lugens* to New Zealand native flora. *Agricultural and Forest Entomology* 13, 99–110.

Yamazaki, K. (2011) Gone with the wind: Trembling leaves may deter herbivory. *Biological Journal of the Linnean Society* 104, 738–747.

32 The Earworms, Bollworms and Budworms: *Helicoverpa* spp.

Peter C. Gregg[1]* and Gary P. Fitt[2]

[1]*School of Environmental and Rural Sciences, University of New England, Armidale, NSW, Australia;* [2]*CSIRO Health and Biosecurity, Brisbane, Queensland, Australia*

Abstract

Helicoverpa armigera (Hübner) and *H. punctigera* (Wallengren) are the most prominent Australian species in the Heliothinae, a subfamily of the Noctuidae that includes many major pests worldwide. They are similar morphologically and have similar life histories, but there are important ecological differences including their host ranges, propensity for diapause and migration strategies. These differences have major implications for their management as agricultural pests. They are among the most intensively studied of all Australian insects and Australian research has contributed significantly to innovative strategies for these and similar pests including insecticide and transgenic resistance management and integrated pest management.

32.1 Introduction: The Genus *Helicoverpa* in Context

The subfamily Heliothinae (Noctuidae) is widely distributed in tropical, subtropical and temperate regions of the world (Mitter *et al.*, 1993). Most areas have one or more species that cause serious losses to crops. In Australia, species in the genus *Helicoverpa*, especially *H. armigera* and *H. punctigera*, are the major heliothine pests (Zalucki *et al.*, 1986; Matthews, 1999). In northern Australia *H. assulta* (Guenée) is an occasional pest of solanaceous crops (Matthews, 1999). While it does not appear to be as abundant in Australia as it is in east Asia, it may be more common than is often believed, as its adults and larvae are superficially similar to those of *H. armigera*.

Helicoverpa armigera and *H. punctigera* are known by a wide range of common names, depending on the crop they infest. Corn earworm and cotton bollworm for *H. armigera* and native budworm for *H. punctigera* are the most widely used. Collectively, the two species were estimated in 1997 to cause annual losses in Australia of between A$159 million and A$328 million (Adamson *et al.*, 1997). More recent information on aggregated losses is not available but Murray *et al.* (2013) estimate annual damage and control costs in grains alone totalling A$53.2 million.

As the common name suggests, *H. punctigera* is endemic to Australia. Its distribution is limited to the Australian mainland, adjacent islands (Holloway, 1977) and, as transient migrants, New Zealand (Fox, 1978). In contrast, *H. armigera* is widely distributed globally. It is likely that *H. armigera* was present in Australia before European settlement. Caterpillars fitting

*Corresponding author: pgregg49@bigpond.com

DOI: 10.1079/9781800627680.0032

its description were recorded in colonial times but the two species were not separated until the work of Common (1953), so it is not possible to differentiate these two species in early records.

The two species are morphologically similar as adults, with wingspans of 25–45 mm. Their colour varies from light fawn to dark grey-brown, with lighter colours reflecting higher rearing temperature (Hill, 1993). Hindwings are pale, with a broad dark outer margin. *H. armigera* can be distinguished from *H. punctigera* by the presence of a pale patch within this margin (Fig. 32.1) and by differences in the genitalia (Common, 1953; Kirkpatrick,1961). Pupae are morphologically similar but can be distinguished by the spacing of cremaster spines (Kirkpatrick, 1961). Larvae vary widely in colour from pale yellow-green through brown to almost black, depending in part on the host plant (Yamasaki *et al.*, 2009). They have several longitudinal lines and short setae borne on dark raised tubercles and can be distinguished from other noctuids by the sharply downward angling of the terminal abdominal segments. Separating the larvae of the two species is problematic, but larger larvae of *H. armigera* can often be identified by the presence of a dark saddle marking on the fourth segment and white hairs around the head. Eggs and small larvae cannot be reliably separated on morphological criteria but genetic differences are revealed by electrophoresis (Daly and Gregg, 1985), ELISA (Trowell *et al.*, 2000) or DNA sequencing (McKechnie *et al.*, 1993; Scott *et al.*, 2004; Walsh, 2016).

Helicoverpa armigera and *H. punctigera* have similar life histories in laboratory culture. They have generation times of about 5 weeks and fecundities of 1000–2000 eggs per female at optimum temperatures of 25–30°C (Zalucki *et al.*, 1986). Although they may not always be achieved under field conditions, these relatively *r*-selected characteristics give them potentially high intrinsic rates of increase compared with many other lepidopterans. Although the two species are similar in these respects, there are important differences in their ecological adaptations, especially regarding host plant range, diapause and migratory strategies. These differences have major implications for their pest

Fig. 32.1. (Top) *Helicoverpa armigera* adult and larva. (Bottom) *Helicoverpa punctigera* adult and larvae. Photos: P. Gregg.

status and the pest management tactics which can be deployed against them (Fitt, 1989; Fitt and Cotter, 2004).

32.2 *Helicoverpa armigera*

32.2.1 Taxonomy

Helicoverpa armigera was first described as *Noctua armigera* by Hübner (1805). Other names have included *Chloridea armigera*, *Heliothis armigera* and *Heliothis armiger*. Hardwick (1965) proposed the genus name *Helicoverpa* and separated *H. armigera* from the New World species *H. zea*, with which it can hybridize. *H. armigera* has about a dozen common names depending on the host plant it is found on (Zalucki *et al.*, 1986), the most common being cotton bollworm, corn earworm and tomato fruitworm.

32.2.2 Distribution

Helicoverpa armigera is a cosmopolitan species and the most globally widespread of all the *Helicoverpa* spp. It is a major agricultural pest throughout central and south-eastern Asia, the Middle East, Africa, southern Europe, Australia, New Zealand and the Pacific Islands (Kriticos *et al.*, 2015).

The native range is the Old World (Asia, Africa, Europe) and Australasia (Tay *et al.*, 2017), with Hardwick (1965) considering its indigenous region to be between 40°N and 40°S. Its range appears to have subsequently expanded throughout wider Asia, Europe, Africa and Oceania. The most noteworthy recent expansion of its distribution is throughout South America. After a first report in Brazil in 2013 (Tay *et al.*, 2013), *H. armigera* has rapidly spread throughout the country and to neighbouring countries such as Paraguay, Argentina, Bolivia and Uruguay (Arnemann *et al.*, 2019).

Helicoverpa armigera is currently present in at least 128 countries (Fig. 32.2). It is not yet established in northern America (Gonçalves *et al.*, 2019) but modelling suggests the potential for incursion into continental USA and further introgression with *H. zea* (Kriticos *et al.*, 2015).

Within Australia, *H. armigera* is widely distributed in eastern and northern Australia, but known from only a handful of specimens in the south-west (Matthews, 1999). Most records are coastal and from cropping regions of eastern and northern Australia, but extensive sampling in more arid inland Australia shows the species

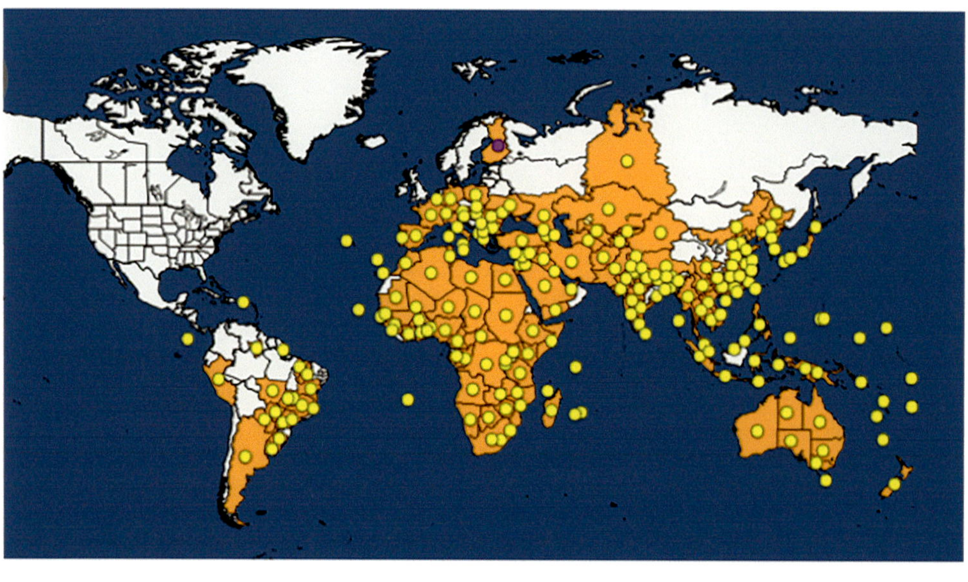

Fig. 32.2. Global distribution of *Helicoverpa armigera* (EPPO Global Database, 2020).

widely but irregularly distributed in low numbers (Gregg *et al.*, 2019).

32.2.3 Seasonal dynamics

There are considerable differences in the seasonal dynamics of *H. armigera* and *H. punctigera* in cropping areas of eastern Australia (Wardhaugh *et al.*, 1980; Wilson, 1983; Fitt, 1989). In southern Queensland and much of New South Wales, *H. armigera* has 4–5 generations per season and passes the winter in a pupal diapause (Fitt, 1989; Fitt and Daly, 1990). During the summer subpopulations develop through a first generation on spring weeds, winter legumes and early sown sorghum or sunflower. Low-density populations can also occur on extensive areas of wheat. In spring and early summer, populations are relatively synchronized but later in the summer, considerable overlap of generations occurs. The next 2–3 generations may occur on cotton, sorghum, sunflowers, maize and legumes. Hence, at any one time the *H. armigera* population occurs as a complex mosaic of subpopulations developing on many host plants, at slightly different rates, with differing levels of mortality and fitness effects on future fecundity and with considerable adult movement linking all subpopulations (Dillon *et al.*, 1996).

32.2.4 Host use and host selection

In Australia alone, *H. armigera* has been recorded on over 100 plant species from 36 families (Matthews, 1999). Hosts include nearly all the major field crops in Australia: cotton, sorghum, maize, sunflowers, chickpeas, lupins and most other legumes, lucerne, tobacco and wheat. Horticultural crops such as tomatoes, lettuce, capsicum and sweetcorn, and flowers such as chrysanthemums and roses suffer high rates of damage (Matthews, 1999). There are also numerous hosts among introduced weeds, pasture plants and native flora (Gregg *et al.*, 2019).

In contrast to *H. punctigera*, *H. armigera* feeds on both dicotyledonous and monocotyledonous plants and is more associated with cultivated crops than with native host plants. Its dynamics are largely determined by spatial and temporal patterns in agricultural landscapes. There is some evidence for geographic variation in host plant preference (Firempong and Zalucki, 1990a) but the underlying population genetics of host range are yet to be examined.

Host selection in the field is primarily determined by ovipositing females, though older larvae will move if the current host becomes unsuitable (Zalucki *et al.*, 1986). There is a hierarchy of oviposition preference displayed by females (Firempong and Zalucki, 1990a; Butter and Singh, 1996; Jallow and Zalucki, 1996; Cunningham and Zalucki, 2014). Moths were provided with nine different plant species from eight families (Firempong and Zalucki, 1990a). The top three and lowest three ranking plant species were the same but there was variation in the ranking of host plants between populations from different geographic locations. In some cases the most popular host was the most abundant crop in the area, suggesting the possibility of host races. In three of the five cases, however, the most popular plant species was also the plant from which the tested individuals had been collected, which could have influenced adult host preference.

Other factors can affect this hierarchy, such as plant abundance (Cunningham *et al.*, 2001), plant height (Sequeira and Moore, 2003) and phenology (Firempong and Zalucki, 1990b), previous experience (Cunningham *et al.*, 1998) and current egg load (Jallow and Zalucki, 1998).

Female moths are thought to use various physical and chemical cues during oviposition to determine the suitability of the host plant (Fitt, 1991; Jallow *et al.*, 1999). Hairy or rough surfaces are thought to be preferred for egg laying (Zalucki *et al.*, 1986). Plant phenology is also important. Peak egg laying is concurrent with peak flowering in many crops (Zalucki *et al.*, 1986) so it may be that flowering plants represent a higher quality food source for offspring than pre-flowering plants. This seems likely as larvae feed preferentially on fruiting bodies once they become available (Matthews, 1999).

There is evidence of learning in host selection and nectar foraging behaviour (Cunningham *et al.*, 1999, 2001). Experience with a host increased the likelihood of that host being selected for oviposition at a later date (Cunningham *et al.*, 1998). Similarly, experience with a flowering host species increases the probability of that species being selected for nectar foraging and the efficiency of foraging on that host (Cunningham *et al.*, 1998). It may be these factors in conjunction that cause high levels of oviposition on flowering plants in the field.

Females lay eggs singly or in groups of two or three, on flowers, fruiting bodies, growing tips and leaves. Although neonates will eat leaves, flowers and fruiting bodies are preferred feeding sites and larvae will move to these sites as they become available. In this way, high levels of economic damage to crops can be caused by relatively low numbers of larvae (Matthews, 1999).

First instar larvae are highly mobile within and between plants, with the capacity to drop off unsuitable hosts and to balloon on silk to other plants. This stage of the life cycle is the most vulnerable and subject to the highest mortality rates (Perovic *et al.*, 2008).

A number of studies have examined the behaviours and survivorship of early instar *Helicoverpa* larvae on various crops. This is the most vulnerable life stage and larvae show a range of complex behaviours and adaptations to optimize their feeding and movement within and between plants (Johnson and Zalucki, 2005; Perkins *et al.*, 2009). Luong *et al.* (2019, 2022) document the capacity of first instar *H. armigera* larvae to detect differences in suitability of a host plant and 'drop off' the plant, presumably to then wander in search of something more suitable. However, they suffer very high mortality when exposed on soil surfaces in most field cropping situations (Perovic *et al.*, 2008).

32.2.5 Long-distance movement

A major factor contributing to the pest status of *H. armigera* is its ability to spread via long-range migration (Jones *et al.*, 2019) and international trade (Tay *et al.*, 2017). Although long-distance movement certainly occurs in *H. armigera* in Australia (Daly and Gregg, 1985), it is probably a less frequent occurrence than in *H. punctigera* (Farrow and Daly, 1987; Walden, 1995) and populations of this species may be influenced more by local cropping and climatic conditions (Fitt, 1989; Maelzer and Zalucki, 2000). It is almost certainly a facultative migrant (Daly and Gregg, 1985; Farrow and Daly, 1987) and, being more closely associated with cropping areas than *H. punctigera*, *H. armigera* may rarely experience the widespread deterioration of breeding habitat needed to induce regular mass movements. Nevertheless, the absence of significant genetic differentiation between widely separated subpopulations of *H. armigera* (Daly and Gregg, 1985; Weeks *et al.*, 2010) suggests that gene flow occurs over much of eastern Australia at least. The general issue of migration in this species and others is further discussed in Chapter 7.

32.2.6 Pesticide resistance

Helicoverpa armigera has consistently developed resistance to pesticides (Forrester *et al.*, 1993; Downes *et al.*, 2017) with major episodes of resistance development to DDT, synthetic pyrethroids, carbamates and endosulfan during the 1960s to 1980s. For these reasons, *H. armigera* has been the focal species for management strategies deployed to manage resistance risk for new pesticides (Forrester *et al.*, 1993) and for transgenic crops (Downes *et al.*, 2017).

Selection for resistance to pesticides occurs whenever they are used, but the rate at which resistance genes increase in frequency is determined by the balance between selection and dilution through gene flow from other subpopulations (Daly, 1993). Cotton, by virtue of its long growing season, will support 2–3 generations of *H. armigera* and selection for resistance may be significant. Individual sowings of most other crops support only a single generation of *Helicoverpa* before senescing and so impose only one generation of selection per crop. However, the potential for multiple or staggered plantings of these crops over the season could result in selection of several successive generations. For example, the use of synthetic pyrethroids (SPs) on sorghum (often at low rates for sorghum midge control) is thought to have contributed greatly to the evolution of SP resistance in *H. armigera* (Daly and Murray, 1988). Inadvertent selection of a large part of the *H. armigera* population on extensive areas of sorghum, combined with intense selection on cotton where SPs were heavily used, led to the appearance of SP resistance within 5 years of their release in Australia (Forrester *et al.*, 1993).

In the case of transgenic Bt cotton varieties deployed in Australia, an ecologically based resistance management strategy was deployed from first commercial use in 1996 and continues today (Roush *et al.*, 1998). Compliance with the resistance management strategy for Bt cotton required a shift in focus from pest control at the level of individual fields or farms towards a co-ordinated area-wide landscape approach. Resistance genes are certainly present in the *H. armigera* populations and small changes in gene frequency have occurred through selection, but these have not resulted in crop damage due to Bt resistance.

32.3 *Helicoverpa punctigera*

32.3.1 Taxonomy

The species now known as *Helicoverpa punctigera* was described by Wallengren in 1860 as *Heliothis punctigera*. It has also been referred to as *Heliothis punctiger* (Matthews, 1999). Common (1953) separated *H. punctigera* from three other species in the genus *Heliothis*, including *H. armigera*. Before this date, records confused the species, with many early records of *H. armigera* likely to be *H. punctigera*. The mitogenome of *H. punctigera* has been characterized, and phylogenetic analysis indicates that the species is closer to the American *H. gelotopoeon* and *Heliothis subflexa* than it is to *H. armigera* (Walsh *et al.*, 2019). Whether judged by allozyme variation (Daly and Gregg, 1985) or mitochondrial DNA (McKechnie *et al.*, 1993), genetic distances among widely spaced populations are low and there is little evidence of population structure – the species is probably effectively panmictic.

32.3.2 Distribution

Helicoverpa punctigera is widespread through mainland Australia (Fig. 32.3A) and has been recorded on distant offshore islands (Holloway, 1977; Farrow, 1984) and transiently in New Zealand (Fox, 1978). Oliveira *et al.* (2022) modelled the potential worldwide distribution of *H. punctigera* based on climatic suitability and concluded that it has the potential to be a serious invasive pest in many tropical and subtropical regions.

Unlike *H. armigera*, *H. punctigera* is widespread and frequently very abundant in semi-arid inland Australia (Fig. 32.3B; Gregg *et al.*, 2019). Localities within these areas are not permanently inhabited (Fig. 32.3C) but can support large populations when rainfall, especially in autumn and winter, germinates annual host plants. Gregg *et al.* (2019) recognized six habitats within these areas, and quantified potential *H. punctigera* abundance as sandy deserts > floodplains > mulga, grasslands and saltbush > stony downs.

32.3.3 Seasonal dynamics

Helicoverpa punctigera eggs hatch in 2–8 days, and larvae develop through six instars over 13–45 days, with non-diapausing pupal development taking 10–38 days (Kirkpatrick, 1962). Development of all stages is strongly temperature dependent and models describing these relationships are provided by Allsopp *et al.* (1991). In the Namoi Valley of northern New South Wales, there are at least three and sometimes four generations per season, though later generations are often small. Optimum temperatures for development are around 35°C, but constant temperatures exceeding 38°C are lethal to eggs (Qayyum and Zalucki, 1987). Larvae can survive short exposures of up to 49°C, with the short-term thermal limit for pupae and adults being around 47°C and 46°C respectively (Bawa *et al.*, 2021a).

Helicoverpa punctigera has a facultative winter diapause in the pupal stage, induced by temperatures below 28°C and declining day length experienced in the larval stage (Cullen and Browning, 1978). Pupal development can be extended for several months, but winter diapause is much weaker in *H. punctigera* than it is in *H. armigera*, and this has been used to explain the scarcity of overwintering pupae of *H. punctigera* relative to *H. armigera* in northern NSW (Fitt and Daly, 1990). In contrast to *H. armigera*, it is usually early season populations (spring/early summer), originating from immigrants or early local breeding on non-crop vegetation, which cause

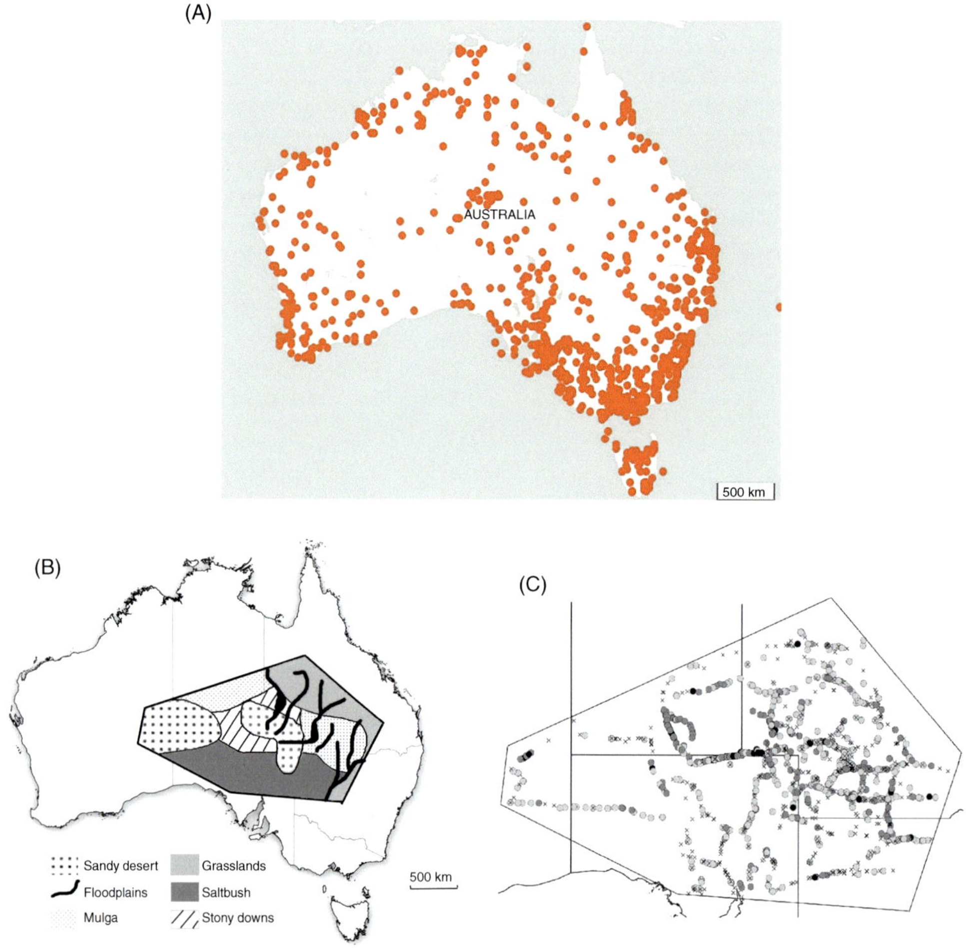

Fig. 32.3. *Helicoverpa punctigera*. (A) Distribution (records from Atlas of Living Australia (https://bie.ala.org.au/species/https://biodiversity.org.au/afd/taxa/3c441d79-1941-4e86-a351-4c08afe5e12b, accessed 30 August 2024). (B) Major habitat types in semi-arid inland Australia. (C) Inland abundance 1987–2017, from Gregg *et al*. (2019). Crosses indicate negative samples, circles positive samples, the darker shades indicate higher numbers.

the most damage to crops. Later season generations are often of less concern, possibly because of the impact of insecticides and natural enemies and of emigration from cropping areas.

32.3.4 Host use and host selection

Helicoverpa punctigera is highly polyphagous, with eggs and larvae being recorded from at least 233 plant species in 40 families including many crops and non-crop plants (Zalucki *et al.*, 1994; Gregg *et al.*, 2019). However, not all plants on which immature stages are found are good hosts and larval mobility may be important in relocation to better hosts (Cunningham *et al.*, 2011). The best hosts among non-crop plants are in the families Asteraceae and Fabaceae and to a lesser extent Malvaceae. Prominent species include *Rhodanthe floribunda*, *R. charsleyae*, *Senecio gregorii*, *Polycalymna stuartii* and *Ixiolaena brevicompta* (all Asteraceae) and *Cullen cinereum* (Fabaceae) (Walter and Benfield, 1994; Gregg *et al.*, 2019; Jones *et al.*, 2021).

Studies on ovipositional preferences and potential host plant races, comparable with those in *H. armigera*, are lacking for *H. punctigera*. Among native hosts, when legumes were found near daisies, they had more larvae if both were flowering, but this was reversed if only the daisy was flowering. Larval performance on the flowers and leaves of three inland hosts was strongly correlated with nitrogen content (Gregg *et al.*, 2016). Walter and Benfield (1994) and Jones *et al.* (2021) identified several species they termed 'primary hosts' for *H. punctigera*, which were preferred for both oviposition and larval selection, although Gregg *et al.* (2019) questioned the relevance of the primary host concept in areas where vegetation composition shows extreme fluctuations, such as inland non-cropping regions.

Among crops, *H. punctigera* was a major pest of cotton before the introduction of transgenic (*Bt*) cotton (Wilson and Greenup, 1977) but is now only a minor pest of that crop (Wilson *et al.*, 2013). Major crops now affected include grain legumes, field peas, canola and a range of horticultural crops (Ridland *et al.*, 1993; Clement *et al.*, 2000; Gu *et al.*, 2007). In contrast to *H. armigera*, *H. punctigera* rarely affects cereal and coarse grain crops. It was thought to be restricted to host plants with the C3 photosynthetic pathway, but more recently significant numbers of moths trapped in cropping areas have stable carbon isotope signatures which indicate development on C4 plants (Baker *et al.*, 2019). Laboratory studies suggest saltbushes and other chenopods as potential C4 hosts (Bawa *et al.*, 2021b,c), although few larvae have been found on them in the field.

32.3.5 Long-distance movement

The intermittent occurrence of *H. punctigera* on offshore islands and in New Zealand and its frequently sudden appearance in temperate parts of Australia, along with the molecular evidence of high gene flow, suggest that *H. punctigera* is highly migratory. It has been detected by radar flying in fast nocturnal winds at altitude in central western NSW (Drake and Farrow, 1985) and across Bass Strait to Tasmania (Drake *et al.*, 1981). Farrow and Daly (1987) suggested that *H. punctigera* was more migratory than *H. armigera*, and more than the American heliothines *Helicoverpa zea* and *Heliothis virescens*, and it has been characterized as an obligate migrant (Jones *et al.*, 2019) (see also Chapter 7).

One well-established migration pathway is from western Queensland and central Australia to the cropping areas of the eastern wheat belt in early spring (Gregg, 1995; Gregg *et al.*, 1995). This coincides with the desiccation of ephemeral host plants in the inland following the arrival of hot dry weather, and is facilitated by strong northerly to north-westerly winds ahead of cold fronts (Gregg *et al.*, 2001). In inland Western Australia a corresponding pathway brings *H. punctigera* to the cropping areas towards the coast, on north-easterly winds following the passage of high pressure (Walden, 1995). Other migration pathways, including return migration from cropping areas to the inland, remain to be determined.

32.3.6 Pesticide resistance

In contrast to *H. armigera*, *H. punctigera* has rarely developed resistance to any of the wide variety of insecticides which have been used against it (Gunning and Easton, 1994; Bird and Walker, 2019). Similarly, despite the existence of recessive alleles conferring resistance to the Cry-2Ab and Vip3a toxins in transgenic *Bt* cotton in previously unselected populations (Downes *et al.*, 2009, 2010), no control failures have been reported. The contrast with *H. armigera* has been attributed to dilution of resistance alleles by frequent immigration from the inland, along with limited overwintering in cropping areas (Fitt and Daly, 1990; Fitt, 1994), although the potential for development of resistance remains a consideration in the cotton industry's strategy for managing resistance to transgenic cotton (Wilson *et al.*, 2018).

32.4 Pest Management of *Helicoverpa* in Australia

Crops vary widely in their ability to tolerate *Helicoverpa* damage. Fitt (1997) provided an analysis of the costs of control across the major field crops where *Helicoverpa* is the major insect pest (cotton, sorghum, tomato, sweet corn, chickpeas, field peas, navy beans, mung beans). Dollar costs recorded then are not relevant today but the relativities between crops remain informative. Among those crops, cotton accounted for 65% of the *Helicoverpa* control costs. *Helicoverpa* represented the target for 80% of the pesticide sprays, although transgenic *Bt* cotton has dramatically changed that situation.

Cotton is particularly susceptible because of its long growing season during which 3–4 generations of *Helicoverpa* (*punctigera* and *armigera*) may attack the crop, and being a valuable crop, the value of damage is high. By contrast, sorghum is attacked only once during development, with only 50% of the total hectares requiring even one spray in most years. Three crops – cotton, tomato and sweet corn – have persistently high levels of damage from *Helicoverpa* and require considerable control. In sorghum and corn, *H. armigera* is the sole *Helicoverpa* species present whereas in all other crops, both *H. punctigera* and *H. armigera* may occur. Of the legumes, field peas and chickpeas are spring-flowering crops grown most extensively in southern, and more recently in western, Australia where they receive sporadic damage from populations of predominantly *H. punctigera*. On average, these legume crops may require only a single pesticide application.

32.4.1 Integrated pest management

Although *H. armigera* and *H. punctigera* are regarded as major pests in many crops, they are rarely the only pests present. Dealing with a suite of pests in a sustainable way requires an integrated pest management (IPM) approach.

Unfortunately, in many crops use of pesticides has been the main response to *Helicoverpa* management, particularly as economic damage can be sustained very quickly. Nonetheless, there have been real attempts to apply IPM in some horticultural and field cropping situations. Brier *et al.* (2008) outline the challenges and options for IPM in grain production systems of north-eastern Australia where *Helicoverpa* is an important pest. Holloway *et al.* (2008) provide similar perspectives for southern grain production systems. Cotton is a good example of a progressive development and application of IPM principles and highlights the challenges and complexities of utilizing IPM with pests like *Helicoverpa* (Fitt, 1989), but the comments below could be applied in many other crops.

Integrated pest management seeks to enhance sustainability and minimize pesticide use by including a broad range of tactics such as pest-resistant varieties, conservation and augmentation of beneficial insect populations, use of selective and short residual insecticides, recognition of the compensatory capacity of the plant and various cultural control practices. Downes *et al.* (2017) integrate many of the IPM tactics outlined below into the co-ordinated area-wide landscape management paradigm now applied in the era of transgenic crops. Below is a list of the key IPM components utilized in Australian cotton production. Wilson *et al.* (2004) provide more detail.

32.4.1.1 Pest-tolerant varieties: the foundation

The crop itself is the template on which a broad range of interactions between pests and their environment are formed and is thus the starting point for IPM. Cotton plants have naturally occurring biochemical and morphological defences against pests and a high capacity to compensate for pest damage. Several conventional host plant resistance traits have been incorporated into modern varieties. These include morphological and chemical traits which reduce the development of pest populations. (Sadras, 1995; Sadras and Fitt, 1997).

In 1996 a quantum step forward in pest resistance was achieved through genetically modified cottons expressing the delta-endotoxin genes from *Bacillus thuringiensis* subsp. *kurstaki* (*Bt*). These transgenic cottons are a good platform for IPM (Fitt and Wilson, 2000) as they reduce the need to control *H. punctigera*, the primary early-season cotton pest in Australia, thereby reducing disruption to beneficial insects caused by insecticide use, and conserving and maximizing beneficial insect activity.

32.4.1.2 Use of trap crops to concentrate pest populations

Spring chickpea crops can be used to capture eggs from *H. armigera* moths that emerge from overwintering diapause (Sequeira, 2001). These moths are potential carriers of genes for pesticide resistance from one season to the next (Daly and Fitt, 1990). Trap crops are a means to concentrate *H. armigera* populations into a limited area where they can be destroyed using biopesticides.

32.4.1.3 Conservation, enhancement and integration of natural enemies into pest management

Cotton fields typically harbour a rich diversity of arthropods. In Australia, up to 450 different species have been recorded in unsprayed fields (L. Wilson, unpublished) and a significant proportion of these are beneficials. It is striking that the key beneficial groups in cotton are similar in many parts of the world (Hearn and Fitt, 1992).

While predators and parasites are important components of IPM systems, there are often limitations in their capacity to control heliothines. These pests are highly mobile, highly fecund, well adapted to exploit diverse cropping systems (Fitt, 1989, 1994) and capable of explosive infestations of crops. Consequently, an important area of research has been to identify means to conserve, augment or manipulate beneficial populations. Conservation of natural enemies requires considerable ecological understanding of their seasonal phenology, habitat and prey requirements. The majority of predators are generalists, able to sustain populations on a diversity of prey types. Predator abundance can be readily monitored and estimates of abundance utilized in decision making through a predator/prey ratio which indicates when predators are sufficiently abundant to have impact (Mensah, 2002a, b). Mechanisms to encourage beneficial insects include the use of nursery crops such as perennial lucerne crops which are attractive to many insects and offer a permanent habitat on-farm, so providing some buffer against the unpredictability of natural populations. Brier *et al.* (2008) and Holloway *et al.* (2008) discuss a range of options to manage beneficial insects in grain cropping systems. Rizvi *et al.* (2022) outline the options for broader scale habitat manipulation to maintain and enhance beneficials in horticultural crops and situations.

32.4.1.4 Rigorous and regular crop sampling

Regular crop sampling for pests, plant damage and beneficial insects is critical for effective IPM (Deutscher and Wilson, 1999a). Regular sampling means that decisions to delay control can be monitored and action taken if the situation changes while the pest population can still be effectively controlled with selective insecticides.

32.4.1.5 Use of combined pest and damage thresholds

Thresholds are an essential tool to ensure that insecticides are only applied if economic loss is reasonably expected to occur (Deutscher and Wilson, 1999b). They have been established for several grain crops and cotton. By integrating compensatory responses into thresholds, it is possible to identify those situations where pests may have exceeded a threshold but the crop will recover without loss, thereby avoiding use of an insecticide to prevent non-economic damage (Wilson *et al.*, 2003).

32.4.1.6 Preferential use of selective insecticides

New-generation insecticides are much more selective than the older suite of organophosphate, carbamate and pyrethroid pesticides which characterized Australian crops many years ago. Many new compounds and chemical groups (e.g. spinosad, emamectin, indoxacarb, pymetrozine, diafethiuron and methoxyfenozide) together with biologicals (NPV virus and Bt sprays) provide powerful IPM tools as they are less disruptive to beneficial populations.

32.4.1.7 Semiochemicals

Pheromones have been identified for both *H. armigera* and *H. punctigera* (Rothschild, 1978; Rothschild *et al.*, 1982). They consist of straight chain isomeric aldehydes in the C_{14} to C_{16} range. Lures and traps have been developed for monitoring purposes (Gregg and Wilson, 1991). However, their usefulness is limited by a poor correlation between trap catches and the numbers and proportions of each species in the field (Daly and Fitt, 1993). Similarly, mating disruption has had little success, mainly due to immigration of females that have mated outside treated areas (Betts *et al.*, 1993).

Bisexual attract-and-kill, using plant volatile compounds that attract both sexes, offers more promise. Gregg *et al.* (2018) describe a range of opportunities for use of attract-and-kill approaches in pest management and Gregg *et al.* (2022) provide specific examples for use of an attractant derived from a specific blend of plant volatiles combined with a killing agent (commercialized as Magnet™) as part of resistance management strategies for transgenic cottons. They described a 'preventive approach', where attract-and-kill targets the cohort of moths which lays the last generation of eggs on cotton which would likely result in an overwintering population of diapausing pupae. This novel approach has the advantage of directly reducing the numbers of larvae which would otherwise be subjected to selection pressure and require cultivation ('pupae busting').

32.4.1.8 Genetically modified cottons in IPM

Genetically engineered cottons expressing the delta-endotoxin genes from *Bt* offer great potential to dramatically reduce pesticide dependence for control of the major lepidopteran pests and consequently offer real opportunities as a component of sustainable and environmentally acceptable IPM systems.

Bt cotton varieties expressing a single Cry 1Ac protein were first registered in Australia in 1996 (INGARD®). Fitt (2003) provides a comprehensive assessment of the impact of *Bt* cotton in Australia over the first 6 years of commercial use. Since then, the technology has advanced to Bollgard® 3 cotton varieties which expressed three independently acting *Bt* genes. The most consistent 'winner' from Bt cottons has been the environment, with 80–90% reductions in pesticide loads.

Bt cotton varieties are not perceived as 'magic bullets' for pest control in Australia. Instead, they are viewed broadly as an opportunity to address environmental concerns about cotton production and as a foundation on which to build IPM systems which incorporate a range of biological and cultural tactics (Fitt, 2000; Wilson *et al.*, 2004). Research has shown little effect of *Bt* cottons on non-target species, including non-lepidopterous pests, beneficial insects and other canopy-dwelling and soil-dwelling species (Fitt, 2000; Fitt and Wilson, 2000). Survival of beneficials is markedly higher than in conventional sprayed cotton, and they provide control for some secondary pests, particularly those that are induced pests in sprayed cotton (e.g. mites and aphids).

32.4.1.9 Effective resistance management strategies

Australia has a long history of problems with the evolution of pesticide resistance, but also a world-leading position in effective resistance management (Forrester *et al.*, 1993). In cotton, all new selective pesticides are incorporated in the Insecticide Resistance Management Strategy (IRMS) developed under the auspices of the industry-based Transgenic and Insecticide Management Strategies (TIMS) committee.

Destruction of diapausing pupae of *H. armigera* that are a potential reservoir of resistance genes (Fitt and Daly, 1990) is a core non-insecticidal component of both the IRM and IPM strategies. Growers are advised to sample cotton stubble for overwintering pupae to determine which fields require control and to prioritize those that do. Fields which have grown *Bt* cotton require mandatory cultivation to eliminate regrowth and destroy any potential resistant pest populations. In some situations, however, this becomes a real challenge, particularly where the winter is wet and cultivation cannot be undertaken. In that situation, the attract-and-kill strategy outlined above is an option.

The major challenge to sustainable use of transgenic *Bt* cottons is the risk that target pests, particularly *H. armigera*, may evolve resistance to the various *Bt* proteins (Cry 1Ac, Cry IIA, Vip3A). A pre-emptive resistance management strategy was implemented to accompany the release of the transgenic strategy, based on the use of structured refuges to maintain susceptible individuals in the population. This seeks to take advantage of the polyphagy and local mobility of *H. armigera* by utilizing gene flow to counter selection in transgenic crops. Growers can choose among a range of crops which can be grown as a refuge and several studies have quantified their relative value for producing *Bt*-susceptible *Helicoverpa* moths (Baker *et al.*, 2008, 2016; Baker and Tann, 2014).

In contrast, extensive natural refuges effectively nullify the resistance risk in *H. punctigera*. Indeed, *H. punctigera* provides an excellent natural example of the capacity of the refuge strategy to reduce resistance risk.

Key elements of the *Bt* cotton resistance management strategy are:

- effective refuges on each farm growing *Bt* cotton
- defined planting windows for *Bt* cotton to avoid late planted crops that may be exposed to abundant *H. armigera* late in the growing season
- mandatory cultivation of *Bt* crops and crop residues after harvest to destroy most overwintering pupae of *H. armigera*
- removal of volunteer Bt plants
- defined spray thresholds for *Helicoverpa* to ensure any survivors in the crops are controlled
- monitoring of *Bt* resistance levels in field populations.

Over the past 25 years the details of the strategy have evolved as progressively improved insect-tolerant *Bt* varieties have been deployed.

32.4.1.10 Extension and implementation of IPM

Achieving effective implementation requires a consistent and coherent communication and extension effort supported by decision support systems for use by consultants or growers. Since the inception of the SIRATAC system in the 1980s, computer-based decision support systems (DSS) have been a feature of the Australian cotton industry. The CottonLOGIC suite of DSS provides desktop support for the key elements of IPM and a benchmark against which decisions can be compared (Deutscher and Bange, 2003).

The final critical factor in gaining support for IPM systems came from favourable economic analyses of IPM. Hoque *et al.* (2000) analysed grower information and showed that the 'soft' IPM approach generally had equal or higher gross margins than a more traditional approach using 'hard', more disruptive chemicals. The difference was attributed to higher beneficial insect populations in the fields managed with more selective insecticides. This analysis was critical in providing economic credibility for the IPM approach.

32.5 Conclusion

Helicoverpa armigera and *H. punctigera* are undoubtedly the most intensively studied of all Australian insects. A search of the Web of Science Core Collection identified over 6000 papers published on them since 1987 alone. This reflects their status as among the most damaging pests in Australia and, in the case of *H. armigera*, in the world. It also reflects the fascination they have held for insect ecologists over many years as strongly r-selected, highly migratory and highly polyphagous insects. They have served as model organisms for the development of IPM strategies within Australia, and Australian research has arguably led the world in the development of such strategies for many crops, and for the management of resistance to both conventional insecticides and transgenic crops.

References

Adamson, D., Thomas, G. and Davis, E. (1997) *An Economic Estimate of Helicoverpa's effect on Australian Agricultural Production*. CRC for Tropical Pest Management, Brisbane, Queensland.

Allsopp, P.G., Daglish, G.J., Taylor, M.F.J. and Gregg, P.C. (1991) Measuring development of *Heliothis* species. In: Zalucki, M.P. (ed.) *Heliothis: Research Methods and Prospects*. Springer-Verlag. New York, pp. 90–101.

Arnemann, J.A., Roxburgh, S., Walsh, T., Guedes, J., Gordon, K. *et al.* (2019) Multiple incursion pathways for *Helicoverpa armigera* in Brazil show its genetic diversity spreading in a connected world. *Scientific Reports* 9(1), 19380.

Baker, G.H. and Tann, C.R. (2014) Refuge crop performance as part of the Bt resistance management strategy for *Helicoverpa* spp. (Lepidoptera: Noctuidae) in Australian cotton production systems. *Austral Entomology* 53, 240–247.

Baker, G.H., Tann, C.R. and Fitt, G.P. (2008) Production of *Helicoverpa* spp. (Lepidoptera, Noctuidae) from different refuge crops to accompany transgenic cotton plantings in eastern Australia. *Australian Journal of Agricultural Research* 59, 723–732.

Baker, G.H., Parry, H.R. and Tann, C.R. (2016) Pigeon pea refuge crops are likely to provide patchy delivery of *Helicoverpa* (Lepidoptera: Noctuidae) within Bt cotton production regions in eastern Australia. *Austral Entomology* 55,439–448.

Baker, G.H., Tann, C.R., Verwey, P. and Lisle, L. (2019) Do the plant host origins of *Helicoverpa* (Lepidoptera: Noctuidae) moth populations reflect the agricultural landscapes within which they are caught? *Bulletin of Entomological Research* 109, 1–14.

Bawa, S.A., Gregg, P.C., del Soccoro, A.P., Miller, C. and Andrew, N.R. (2021a) Estimating the differences in critical thermal maximum and metabolic rate of *Helicoverpa punctigera* (Wallengren) (Lepidoptera: Noctuidae) across life stages. *PeerJ* 9, e12479.

Bawa, S.A., Gregg, P.C., del Soccoro, A.P., Miller, C. and Andrew, N.R (2021b) Oviposition preferences of the native budworm, *Helicoverpa punctigera* (Wallengren), on C3 and C4 plants. *Crop and Pasture Science* 72, 939–946.

Bawa, S.A., Gregg, P.C., del Soccoro, A.P., Miller, C. and Andrew, N.R (2021c) Survival of immature *Helicoverpa punctigera* (Wallengren) (Lepidoptera: Noctuidae) on some C4 and C3 plants. *Austral Entomology* 60, 791–797.

Betts, M.D., Gregg, P.C., Fitt, G.P. and MacQuillan, M.J. (1993) A field trial of mating disruption for *Helicoverpa* spp. in cotton. In: Corey, S., Dall, D.F.J. and Milne, W.M. (eds) *Pest Control and Sustainable Agriculture*. CSIRO, Melbourne, Victoria, pp 298–300.

Bird, L.J. and Walker, P.W. (2019) Baseline susceptibility of *Helicoverpa punctigera* (Lepidoptera: Noctuidae) to Indoxacarb, Emamectin Benzoate, and Chlorantraniliprole. *Journal of Economic Entomology* 112, 818–826.

Brier, H.B., Murray, D.A.H., Wilson, L.J., Nicholas, A.H., Miles, M.M. *et al.* (2008) An overview of integrated pest management (IPM) in north-eastern Australian grain farming systems: Past, present and future prospects. *Australian Journal of Experimental Agriculture* 48, 1574–1593.

Butter, N.S. and Singh, S. (1996) Ovipositional response of *Helicoverpa armigera* to different cotton genotypes. *Phytoparasitica* 24, 97–102.

Clement, S.L., Wightman, J.A., Hardie, D.C., Bailey, P., Baker, G. and McDonald, G. (2000) Opportunities for integrated management of insect pests of grain legumes. In: Knight, R. (ed.) *Linking Research and Marketing Opportunities for Pulses in the 21st Century. Current Plant Science and Biotechnology in Agriculture*. Springer, New York, pp. 467–480.

Common, I.F.B. (1953) The Australian species of *Heliothis* (Lepidoptera:Noctuidae) and their pest status. *Australian Journal of Zoology* 1, 319–344.

Cullen, J.M. and Browning, T.O. (1978) The influence of photoperiod and temperature on the induction of diapause in pupae of *Heliothis punctigera*. *Journal of Insect Physiology* 24, 595–560.

Cunningham, J.P. and Zalucki, M.P. (2014) Understanding Heliothine (Lepidoptera: Heliothinae) pests: What is a host plant? *Journal of Economic Entomology* 107, 881–896.

Cunningham, J.P., Jallow, M.F.A., Wright, D.J. and Zalucki, M.P. (1998) Learning in host selection in *Helicoverpa armigera* (Hübner) (Lepidoptera: Noctuidae). *Animal Behaviour* 55, 227–234.

Cunningham, J.P., Zalucki, M.P. and West, S.A. (1999) Learning in *Helicoverpa armigera* (Lepidoptera: Noctuidae): New look at the behaviour and control of a polyphagous pest. *Bulletin of Entomological Research* 89, 201–207.

Cunningham, J.P., West, S.A. and Zalucki, M.P. (2001) Host selection in phytophagous insects: a new explanation for learning in adults. *Oikos* 95, 537–543.

Cunningham, J.P., Lange, C., Walter, G. and Zalucki, M.P. (2011) Host selection behavior in desert caterpillars. *Entomologia Experimentales et Applicata* 141, 1–7.

Daly, J.C. (1993) Ecology and genetics of insecticide resistance in *Helicoverpa armigera* – interactions between selection and gene flow. *Genetica* 90, 217–226.

Daly, J.C. and Fitt, G.P. (1990) Monitoring for pyrethroid resistance in relation to body weight in adult *Helicoverpa armigera* (Hübner) (Lepidoptera: Noctuidae). *Journal of Economic Entomology* 83, 705–709.

Daly, J.C. and Fitt, G.P. (1993) The need to validate monitoring techniques for insect populations. In: Corey, S., Dall, D.J. and Milne, W.M. (eds) *Pest Control and Sustainable Agriculture*. CSIRO, Canberra, ACT, pp. 414–416.

Daly, J.C. and Gregg, P.C. (1985) Genetic variation in *Heliothis* in Australia: Species identification and gene flow in the two pest species *H.armigera* (Hübner) and *H. punctigera* Wallengren (Lepidoptera: Noctuidae). *Bulletin of Entomological Research* 75, 169–184.

Daly, J.C. and Murray, D.A.H. (1988) Evolution of resistance to pyrethroids in *Heliothis armigera* (Hubner) (Lepidoptera, Noctuidae) in Australia. *Journal of Economic Entomology* 81, 984–988.

Deutscher, S. and Bange, M. (2003) Advancement in computerised decision support systems for Australian cotton systems. *Proceedings of the World Cotton Research Conference* 3, March, Cape Town, South Africa, pp. 865–872.

Deutscher, S. and Wilson, L. (1999a) *Insect and Damage Sampling for Cotton*. IPM Guidelines Supporting Document, 1. Australian Cotton Cooperative Research Centre, Narrabri, NSW, Australia.

Deutscher, S. and Wilson L.J. (1999b) *Insect Thresholds*. IPM Guidelines Supporting Document, vol. 8. Australian Cotton Cooperative Research Centre, Narrabri, NSW, Australia.

Dillon, M.L., Fitt, G.P., Hamilton, J.G. and Rochester, W.A. (1996) A simulation model of wind-driven dispersal of *Helicoverpa* moths. *Ecological Modelling* 86, 145–150.

Downes, S., Parker, T. and Mahon, R. (2009) Frequency of alleles conferring resistance to the *Bacillus thuringiensis* toxins Cry1Ac and Cry2Ab in Australian populations of *Helicoverpa punctigera* (Lepidoptera: Noctuidae) from 2002 to 2006. *Journal of Economic Entomology* 102, 733–742.

Downes, S., Parker, T. and Mahon, R. (2010) Incipient resistance of *Helicoverpa punctigera* to the Cry2Ab Bt toxin in Bollgard II cotton. *PLOS One* 5(9), e12567.

Downes, S., Kriticos, D., Parry, H., Paull, C., Schellhorn, N. and Zalucki, M.P. (2017) A perspective on management of *Helicoverpa armigera*: Transgenic Bt cotton, IPM, and landscapes. *Pest Management Science* 73, 485–492.

Drake, V.A. and Farrow, R.A. (1985) A radar and aerial trapping study of an early spring migration of moths (Lepidoptera) in inland New South Wales. *Australian Journal of Ecology* 10, 223–235.

Drake, V.A., Helm, K.F. and Readshaw, J.L. (1981) Insect migration across Bass Strait during spring: a radar study. *Bulletin of Entomological Research* 71, 449–466.

EPPO Global Database (2020) *Helicoverpa armigera* (HELIAR). Available at: https://gd.eppo.int/taxon/HELIAR/distribution (accessed 10 February 2025).

Farrow, R.A. (1984) Detection of transoceanic migration of insects to a remote island in the Coral Sea, Willis Island. *Australian Journal of Ecology* 9, 253–272.

Farrow, R.A. and Daly, J.C. (1987) Long-range movements as an adaptive strategy in the genus *Heliothis* (Lepidoptera: Noctuidae): A review of its occurrence and detection in four pest species. *Australian Journal of Zoology* 35, 1–24.

Firempong, S. and Zalucki, M.P. (1990a) Host plant preferences of populations of *Helicoverpa armigera* (Hübner) (Lepidoptera, Noctuidae) from different geographic locations. *Australian Journal of Zoology* 37, 665–673.

Firempong, S. and Zalucki, M.P. (1990b) Host plant selection by *Helicoverpa armigera* (Hübner) (Lepidoptera, Noctuidae) – role of certain plant attributes. *Australian Journal of Zoology* 37, 675–683.

Fitt, G.P. (1989) The ecology of *Heliothis* species in relation to agroecosystems. *Annual Review of Entomology* 34, 17–52.

Fitt, G.P. (1991) Host selection in Heliothinae. In: Bailey, W.J. and Ridsdill-Smith, J. (eds) *Reproductive Behaviour in Insects – Individuals and Populations*. Chapman and Hall, London, pp. 172–201.

Fitt, G.P. (1994) Cotton pest management: an Australia perspective. *Annual Review of Entomology* 39, 543–562.

Fitt, G.P. (1997) Risks, deployment and integration of insect resistant crops expressing genes from *Bacillus thuringiensis*i. In: McLean, G.D., Waterhouse, P.M., Evans, G. and Gibbs, M.J. (eds) *Commercialisation of Transgenic Crops: Risk, Benefit and Trade Considerations*. Cooperative Research Centre for Plant Science and Bureau of Resource Sciences, Canberra, ACT, pp. 273–284.

Fitt, G.P. (2000) An Australian approach to IPM in cotton: Integrating new technologies to minimise insecticide dependence. *Crop Protection* 19, 793–800.

Fitt, G.P. (2003) Deployment and impact of transgenic Bt cottons in Australia. In: Kalaitzandonakes, N.G. (ed.) *The Economic and Environmental Impacts of Agbiotech: A Global Perspective*. Kluwer, New York, pp. 141–164.

Fitt, G.P. and Cotter, S.C. (2004) The *Helicoverpa* problem in Australia. In: Sharma, H. (ed.) *Heliothis/Helicoverpa Management: Emerging Trends and Strategies for Future Research*. Oxford and IBH Publishing, New Delhi, India, pp. 45–62.

Fitt, G.P. and Daly, J.C. (1990) Abundance of overwintering pupae and the spring generation of *Helicoverpa* spp. (Lepidoptera: Noctuidae) in northern New South Wales, Australia: Implications for pest management. *Journal of Economic Entomology* 83, 1827–1836.

Fitt, G.P. and Wilson, L.J. (2000) Genetic engineering in IPM: Bt cotton. *Proceedings of the Conference on Emerging Technologies for Integrated Pest Management*, March 8–10 1999, Raleigh, N.C., pp. 108–125.

Forrester, N.W., Cahill, M., Bird, L.J. and Layland, J.K. (1993). Management of pyrethroid and endosulfan resistance in *Helicoverpa armigera* (Lepidoptera: Noctuidae) in Australia. *Bulletin of Entomological Research* Supplement 1, 1–132.

Fox, K.J. (1978) The transoceanic migration of Lepidotera to New Zealand – A history and a hypothesis on colonisation. *New Zealand Entomologist*. 6, 368–380.

Gonçalves, R.M, Mastrangelo, T., Rodrigues, J.C.V., Paulo, D.F., Omoto, C. *et al.* (2019) Invasion origin, rapid population expansion, and the lack of genetic structure of cotton bollworm *Helicoverpa armigera* in the Americas. *Ecology and Evolution* 9, 7378–7401.

Gregg, P.C. (1995) Migration of cotton pests: Patterns and implications for management. In: Constable, G.A. and Forrester, N.W. (eds) *Challenging the Future*. Proceedings of the First World Cotton Research Conference, Brisbane, February 13–17. CSIRO, Brisbane, Queensland, pp 423–433.

Gregg, P.C. and Wilson, A.G.L. (1991) Trapping methods for adults. In: Zalucki, M.P. (ed.) *Heliothis: Research Methods and Prospects*. Springer-Verlag, New York, pp. 30–48.

Gregg, P.C., Fitt, G.P., Zalucki, M.P. and Murray, D.A.H. (1995) Insect migration in an arid continent. 2. *Helicoverpa* spp in eastern Australia. In: Drake, V.A. and Gatehouse, A.G. (eds) *Insect Migration: Tracking Resources through Space and Time*. Cambridge University Press, Cambridge, pp. 151–172.

Gregg, P.C., del Socorro. A.P. and Rochester, W.A. (2001) Field test of a model of migration of moths (Lepidoptera: Noctuidae) in inland Australia. *Australian Journal of Entomology* 40, 249–256.

Gregg, P.C., Henderson, G.S., del Socorro, A.P., le Mottee, K. and Birchall, C. (2016) Polyphagy in an uncertain environment: *Helicoverpa punctigera* in inland Australia. *Austral Ecology* 41, 819–828.

Gregg, P.C., del Socorro, A.P. and Landolt, P.J. (2018) Advances in attract-and-kill for agricultural pests: Beyond pheromones. *Annual Review of Entomology* 63, 453–470.

Gregg, P.C., del Socorro, A.P., le Mottee, K., Tann, C.R., Fitt, G.P. and Zalucki, M.P. (2019) Host plants and habitats of *Helicoverpa punctigera* and *H. armigera* (Lepidoptera: Noctuidae) in inland Australia. *Austral Entomology* 58, 547–560.

Gregg, P.C., del Socorro, A.P., Wilson, S., Knight, K.M., Binns, M.R. and Armytage, P. (2022) Bisexual attract-and-kill: a novel component of resistance management for transgenic cotton in Australia. *Journal of Economic Entomology* 115, 826–834.

Gu, H.N., Fitt, G.P. and Baker, G.H. (2007) Invertebrate pests of canola and their management in Australia: a review. *Australian Journal of Entomology* 46, 231–243.

Gunning, R.V. and Easton, C.S. (1994) The response of *Helicoverpa punctigera* (Wallengren) (Lepidoptera, Noctuidae) to DDT, endosulfan, deltamethrin and fenvalerate. *Journal of the Australian Entomological Society* 33, 61–64.

Hardwick, D.F. (1965) The corn earworm complex. *Memoirs of the Entomological Society of Canada* 40, 1–247.

Hearn, A.B. and Fitt, G.P. (1992) Cotton cropping systems. In: Pearson, C.J. (ed.) *Field Crop Ecosystems*. Elsevier. Amsterdam, The Netherlands, pp. 85–142.

Hill, L. (1993) Colour in adult *Helicoverpa punctigera* (Wallengren) (Lepidoptera: Noctuidae) as an indicator of migratory origin. *Journal of the Australian Entomological Society* 32, 145–151.

Holloway, J.C., Furlong, M.J. and Bowden, P.I. (2008) Management of beneficial invertebrates and their potential role in integrated pest management for Australian grain systems. *Australian Journal of Experimental Agriculture* 48, 1531–1542.

Holloway, J.D. (1977) *Lepidoptera of Norfolk Island: Their Biogeography and Ecology*. Springer, Dordrecht, The Netherlands.

Hoque, Z., Farquharson, B., Dillon, M. and Kauter. G. (2000) Soft options can reduce costs and increase cotton profits. *Australian Cottongrower* 21, 33–37.

Hübner, J. (1805) *Sammlung Europaeische Schmetterlinge*, J. Hübner, Augsberg, Germany.

Jallow, M.F.A. and Zalucki, M.P. (1996) Within- and between-population variation in host-plant preference and specificity in Australian *Helicoverpa armigera* (Hubner) (Lepidoptera: Noctuidae). *Australian Journal of Zoology* 44, 503–519.

Jallow, M.F.A. and Zalucki, M.P. (1998) Effects of egg load on the host-selection behaviour of *Helicoverpa armigera* (Hübner) (Lepidoptera: Noctuidae). *Australian Journal of Zoology* 46, 291–299.

Jallow, M.F.A., Zalucki, M.P. and Fitt, G.P. (1999) Role of chemical cues from cotton in mediating host selection and oviposition behaviour in *Helicoverpa armigera* (Hübner) (Lepidoptera : Noctuidae). *Australian Journal of Entomology* 38, 359–366.

Johnson, M.L. and Zalucki, M.P. (2005) Foraging behaviour of *Helicoverpa armigera* first instar larvae on crop plants of different developmental stages. *Journal of Applied Entomology* 129, 239–245.

Jones, C.M., Parry, H., Tay, W.T., Reynolds, D. and Chapman, J.W. (2019) Movement ecology of pest *Helicoverpa*: Implications for ongoing spread. *Annual Review of Entomology* 64, 277–295.

Jones, L.C., Rafter, M.A. and Walter, G.H. (2021) Interactions of *Helicoverpa punctigera* (Lepidoptera: Noctuidae) larvae and adults with four native host plants relative to field use patterns. *Environmental Entomology* 50, 418–426.

Kirkpatrick, T.H. (1961) Comparative morphological studies of *Heliothis* species (Lepidoptera: Noctuidae) in Queensland. *Queensland Journal of Agricultural Science* 18, 179–194.

Kirkpatrick, T.H. (1962) Notes on the life-histories of species of *Heliothis* (Lepidoptera: Noctuidae) from Queensland. *Queensland Journal of Agricultural Science* 19, 567–570.

Kriticos, D.J., Ota, N., Hutchison, W.D, Beddow, D., Walsh, T. *et al.* (2015) The potential distribution of invading *Helicoverpa armigera* in North America: Is it just a matter of time? *PLOS One* 10(7), e0133224.

Luong, T.T.A., Cribb, B., Downes, S.J., Perkins, L.E. and Zalucki, M.P. (2019) Stay or move: How Bt-susceptible *Helicoverpa armigera* neonates behave on Bt cotton plants? *Entomologia Experimentalis et Applicata* 167, 868–887.

Luong, T.T.A., Downes, S.J., Perkins, L.E. and Zalucki, M.P. (2022) Drop-off behaviour of Bt-resistant and Bt-susceptible *Helicoverpa armigera* (Hübner) (Lepidoptera: Noctuidae) larvae on Bt-cotton and non-Bt cotton plants. *Bulletin of Entomological Research* 112, 604–612.

Maelzer, D.A. and Zalucki, M.P. (2000) Long range forecasts of the numbers of *Helicoverpa punctigera* and *H. armigera* (Lepidoptera: Noctuidae) in Australia using the Southern Oscillation Index and the sea surface temperature. *Bulletin of Entomological Research* 90, 133–146.

Matthews, M. (1999) *Heliothine Moths of Australia: A Guide to Pest Bollworms and Related Noctuid Groups*. Monographs on Australian Lepidoptrea, 7. CSIRO Publshing, Melbourne, Victoria.

McKechnie, S.W., Hoffmann, A.A., Kovacs, I.V., Cacoyianni, Z., Naughton, N.E. and Katsabanas, S. (1993) Genetic variation among Australian populations of native budworm, *Helicoverpa punctigera*. In: Corey, S., Dall, D.F.J. and Milne, W.M. (eds) *Pest Control and Sustainable Agriculture*. CSIRO, Melbourne, Victoria, pp. 428–431.

Mensah, R.K. (2002a) Development of an integrated pest management programme for cotton. Part 1: Establishing and utilizing natural enemies. *International Journal of Pest Management* 48, 87–94.

Mensah, R.K. (2002b) Development of an integrated pest management programme for cotton. Part 2: Integration of a lucerne/cotton interplant system, food supplement sprays with biological and synthetic insecticides. *International Journal of Pest Management* 48, 95–105.

Mitter, C., Poole, R.W. and Matthews, M. (1993). Biosystematics of the Heliothinae (Lepidoptera: Noctuidae). *Annual Review of Entomology* 38, 207–225.

Murray, D.A.H., Clarke, M.B. and Ronning, D.A. (2013) Estimating invertebrate pest losses in six major Australian grain crops. *Australian Journal of Entomology* 52, 227–241.

Oliveira, R.C.D., Zalucki, M.P., Pastori, P.L. and Kriticos, D.J. (2022) Current and future potential distributions of *Helicoverpa punctigera* (Lepidoptera: Noctuidae): Is this the next FAW? *Bulletin of Entomological Research* 112, 119–130.

Perkins, L.E., Cribb, B.W., Hanan, J. and Zalucki, M.P. (2009) The role of two plant-derived volatiles in the foraging movement of 1st instar *Helicoverpa armigera* (Hübner): Time to stop and smell the flowers. *Arthropod–Plant Interactions* 3, 173–179.

Perovic, D., Johnson, M.L., Scholz, B. and Zalucki, M.P. (2008) The mortality of *Helicoverpa armigera* (Hübner) (Lepidoptera: Noctuidae) neonate larvae in relation to drop-off and soil surface temperature: The dangers of bungy jumping. *Australian Journal of Entomology* 47, 289–296.

Qayyum, A. and Zalucki, M.P. (1987) Effects of high temperature on survival of eggs of *Heliothis armigera* (Hübner) and *Heliothis punctigera* Wallengren (Lepidoptera, Noctuidae). *Journal of the Australian Entomological Society* 26, 295–296.

Ridland, P.M., Smith, A.M., McDonald, G. and Cole, P.G. (1993) Pest management of *Helicoverpa punctigera* (Wallengren) (Lepidoptera: Noctuidae) in field peas. In: Corey, S., Dall, D.F.J. and Milne, W.M. (eds) *Pest Control and Sustainable Agriculture*. CSIRO, Melbourne, Victoria, pp. 229–232.

Rizvi, S.Z.M., Reynolds, O.L., Haque, A., Furlong, M.J., Mo, J. *et al.* (2022) Prospects for habitat management to suppress vegetable pests in Australia. *Austral Entomology* 61, 3–28.

Rothschild, G.H.L. (1978) Attractants for *Heliothis armigera* and *Heliothis punctigera*. *Journal of the Australian Entomological Society* 17, 389–390.

Rothschild, G.H.L., Nesbitt, B.F., Beevor, P.S., Cork, A., Hall, D.R. and Vickers, R.A. (1982) Studies of the female sex pheromone of the native budworm, *Heliothis punctigera*. *Entomologia Experimentalis et Applicata* 31, 395–401.
Roush, R.T., Fitt, G.P., Forrester, N.W. and Daly, J.C. (1998) Resistance management for insecticidal transgenic crops: Theory and practice. In: Constable, G.A. and Forrester, N.W. (eds) *Challenging the Future*. Proceedings of the First World Cotton Research Conference, Brisbane, February 13–17. CSIRO, Brisbane, Queensland, pp. 247–257.
Sadras, V.O. (1995) Compensatory growth in cotton after loss of reproductive organs. *Field Crops Research* 40, 1–18.
Sadras, V.O. and Fitt, G.P. (1997) Resistance to insect herbivory of cotton lines: quantification of recovery capacity after damage. *Field Crops Research* 52, 129–136.
Scott, K.D., Lange, C.L., Scott, L.J. and Graham, G.C. (2004) Isolation and characterization of microsatellite loci from *Helicoverpa armigera* Hübner (Lepidoptera: Noctuidae). *Molecular Ecology Notes* 4, 204–205.
Sequeira, R.V. (2001) Inter-seasonal population dynamics and cultural management of *Helicoverpa* spp. in a Central Queensland cropping system. *Australian Journal of Experimental Agriculture* 41, 249–259.
Sequeira, R.V. and Moore, A.D. (2003) Aggregative oviposition behaviour of *Helicoverpa* spp. Lepidoptera: Noctuidae) in contaminated chickpea crops. *Australian Journal of Entomology* 42, 29–34.
Tay, W.T., Soria, M.F., Walsh, T., Thomazoni, D., Silvie, P. *et al.* (2013) A Brave New World for an Old World Pest: *Helicoverpa armigera* (Lepidoptera: Noctuidae) in Brazil. *PLOS One* 8(11), e80134.
Tay, W.T., Walsh, T.K., Downes, S., Anderson, C., Jermiin, L.S. *et al.* (2017) Mitochondrial DNA and trade data support multiple origins of *Helicoverpa armigera* (Lepidoptera, Noctuidae) in Brazil. *Scientific Reports* 7, 45302.
Trowell, S.C., Forrester, N.W., Garsia, K.A., Lang, G.A., Bird, L.J. *et al.* (2000) Rapid antibody-based field test to distinguish between *Helicoverpa armigera* (Lepidoptera: Noctuidae) and *Helicoverpa punctigera* (Lepidoptera: Noctuidae). *Journal of Economic Entomology* 93, 878–891.
Walden, K.J. (1995) Insect migration in an arid continent. 3. The Australian plague locust *Chortoicetes terminifera* and the native budworm *Helicoverpa punctigera* in Western Australia. In: Drake, V.A. and Gatehouse, A.G. (eds) *Insect Migration: Tracking Resources through Space and Time*. Cambridge University Press, Cambridge, pp. 173–190.
Wallengren, H.D.F. (1860) Lepidopterologische Mittheilungen. *Wiener entomologische Monatschrift* 4, 162–176.
Walsh, T.K. (2016) Characterization of the complete mitochondrial genome of the Australian Heliothine moth, *Helicoverpa punctigera* (Lepidoptera: Noctuidae). *Mitochondrial DNA Part A* 27, 167–168.
Walsh, T.K., Perera, O., Anderson, C., Gordon, K., Czepak, C. *et al.* (2019) Mitochondrial DNA genomes of five major *Helicoverpa* pest species from the Old and New Worlds (Lepidoptera: Noctuidae). *Ecology and Evolution* 9, 2933–2944.
Walter, G.H. and Benfield, M.D. (1994) Temporal host plant use in three polyphagous Heliothinae, with special reference to *Helicoverpa punctigera* (Wallengren) (Noctuidae: Lepidoptera). *Australian Journal of Ecology* 19, 458–465.
Wardhaugh, K.G., Room, P.M. and Greenup, L.R. (1980) The incidence of *Heliothis armigera* (Hübner) and *Heliothis punctigera* (Wallengren) (Lepidoptera: Noctuidae) on cotton and other host plants in the Namoi valley of New South Wales. *Bulletin of Entomological Research* 70, 113–132.
Weeks, A.R., Endersby, N.M., Lange, C.L., Lowe, A., Zalucki, M.P. and Hoffmann, A.A. (2010) Genetic variation among *Helicoverpa armigera* populations as assessed by microsatellites: A cautionary tale about accurate allele scoring. *Bulletin of Entomological Research* 100, 445–450.
Wilson, A.G.L. (1983) Abundance and mortality of overwintering *Heliothis* spp. *Journal of the Australian Entomological Society* 22, 191–199.
Wilson, A.G.L. and Greenup, L.R. (1977) The relative injuriousness of insect pests of cotton in the Namoi valley, New South Wales. *Australian Journal of Ecology* 2, 319–328.
Wilson, L.J., Sadras, V.O., Heimoana, S.C. and Gibb, D. (2003) How to succeed by doing nothing: Cotton compensation after simulated early season pest damage. *Crop Science* 43, 2125–2134.
Wilson, L.J., Mensah, R.K. and Fitt, G.P. (2004) Implementing integrated pest management in Australian cotton. In: Horowitz, A.R. and Ishaaya, I. (eds) *Insect Pest Management: Field and Protected Crops*. Springer, New York, pp. 97–118.
Wilson, L., Downes, S., Khan, M., Whitehouse, M., Baker, G. *et al.* (2013) IPM in the transgenic era: A review of the challenges from emerging pests in Australian cotton systems. *Crop and Pasture Science* 64, 737–749.

Wilson, L.J., Whitehouse, M.E.A. and Herron, G.A. (2018) The management of insect pests in Australian cotton: an evolving story. *Annual Review of Entomology* 63, 215–237.

Yamasaki, A., Shimizu, K. and Fujisaki, K. (2009) Effect of host plant part on larval body color polymorphism in *Helicoverpa armigera* (Lepidoptera: Noctuidae). *Annals of the Entomological Society of America* 102, 76–84.

Zalucki, M.P., Daglish, G., Firempong, S. and Twine, P. (1986) The biology and ecology of *Heliothis armigera* (Hübner) and *Heliothis punctigera* Wallengren (Lepidoptera: Noctuidae) in Australia: what do we know? *Australian Journal of Zoology* 34, 779–814.

Zalucki, M.P., Murray, D.A.H., Gregg, P.C., Fitt, G.P., Twine, P.H. and Jones, C. (1994) Ecology of *Helicoverpa armigera* (Hübner) and *H. punctigera* (Wallengren) in the inland of Australia – larval sampling and host-plant relationships during winter and spring. *Australian Journal of Zoology* 42, 329–346.

33 Moths of Special Significance

Roger L. Kitching
School of Environment and Science, Griffith University, Nathan, Queensland, Australia

Abstract
Many moths attract attention by being particularly spectacular and visible. Early lepidopterists often showed a preference for the large and colourful in preference to the small and dull. A wide range of moths and knowledge concerning them are part of the culture and language of Australia's First Nations. These feature variously as food, as totems or as seasonal indicators, as well as being part of everyday life in the bush. Other species have become well known because they are of conservation concern: bogong moths, sun moths and the spectacular pink underwing are cases in point. A select set of other species impinge upon everyday life for other reasons. Some are simply so large, colourful or bizarre that they become widely known. Others, like the larvae of the cup moths, are indeed spectacular but also deliver a powerful sting. Yet other species attract attention through their products: webworms, bagworms and the familiar scribbly gum moths are cases in point. This overview provides directions to other chapters in the volume which expand upon some of these examples.

33.1 Introduction

There is no doubt that when moths enter general human awareness it is almost always as pests of one sort or another. Chapter 23 provides an overview of the veritable army of iniquitous species that trouble our furnishings, clothing, stored products, crop plants and pastures. It is no great surprise, then, that when research funds and efforts are being assigned, the vast majority has gone, over the years, to understanding the biology and consequent control of the most demanding of these pest species. Chapters 24–32 give accounts of some of the more troublesome of these species. Other species of moth, however, have attracted attention for a variety of other reasons.

From earliest times, some groups of moths have been a focus for, principally, amateur lepidopterists to a much greater extent than others. At its most fundamental, their choices of target species reflected the size, splendour and visibility of particular groups. Perhaps the most notable of these have been the Sphingidae, the hawkmoths. Pittaway (1993) gives a detailed account of this focus on the hawkmoths from their first appearance in Mouffet's *Insectorum sive Minimorum Aniimalium Theatrum* in 1634 through the many subsequent works. It is no surprise that Lord Walter Rothschild held a special enthusiasm for the hawkmoths (among many other things) and, together with his insect curator Karl Jordan, produced a (close to) 1000-page (1903) monograph on the group. This

*Corresponding author: r.kitching@griffith.edu.au

DOI: 10.1079/9781800627680.0033

enthusiasm for the spectacular, large and diverse hawkmoths persisted to modern times with Bernard d'Abrera's *Sphingidae Mundi* (1986) and, closer to home, Max Moulds and colleagues' monograph on the *Hawkmoths of Australia* (2020) (see also Chapter 19). A similar fascination with the large, spectacular and, in this case, utilitarian silkmoths and their relatives, the Saturniidae (poorly represented in Australia), no doubt led to the many monographs on the family (e.g. Bouvier, 1936; Ferguson, 1971–72; d'Abrera, 1995, 1998, 2012). In Australia, the giant cossids have attracted dilettante and professional attention over the years, no doubt in part at least because of their spectacular nature (see, for example, Oberthür *et al.*, 1916; Simonsen, 2018). Last but not least in this list of attention-grabbing groups is the day-flying moths, often grouped with the butterflies in regional treatments. Outside Australia, many lepidopterists have focused on the Zygaenidae, especially the zygaenine burnets, which exhibit both an overall diversity and a within-species variability that have attracted amateur and professional enthusiasm (Naumann *et al.*, 1999; Hofmann and Tremewan, 2023). It is perhaps not surprising that the Australian representatives are among the few Australian families of moths to have been monographed (Tarmann, 2004). The spectacular Aganainae (Erebidae) (Zwier, 2022) are also now receiving well-deserved attention, as are the Agaristinae (Noctuidae) (see Chapter 22).

When it comes to individual species, there are perhaps three categories of moths, other than pest species, that have attracted special attention in Australia, although these categories are far from isolated one from the other. First, particular moth species are of special significance in Indigenous culture and this prominence has been built upon by subsequent scientific study and public notice. Then, there are species which are of conservation concern as human activities have pushed their very existence into question. Finally, there is a small category which I would describe as being of general cultural interest because of their visibility and eminence especially to the non-lepidoptersist. Some species, of course, 'belong' to more than one of these categories and even trespass into the pest category covered in Part 3 of this volume.

33.2 Moths of Indigenous Importance

As has been noted in several of the contributions to this volume, many species of moths are of special significance to Australia's First Nations people (see Clarke, 2026). This significance has variously included moths as foodstuffs, as seasonal indicators, as symbols/characters in stories of country and, no doubt, as features of the environment in general. To a general, non-Indigenous lepidopterist, it would be fascinating to understand further this depth of knowledge and the underlying descriptive language and 'taxonomies' involved. Renewed interest in Indigenous languages may, to some extent, provide this opportunity although, for many regions and languages, the underlying knowledge may already be lost.

Perhaps the most familiar species of moths in this category is *Agrotis infusa*, the bogong moth; indeed, this species probably fits into all three of the artificial categories I have erected here. It is of undoubted Aboriginal significance; it is undergoing significant decline as a species; and it is (or was) known to most residents of south-eastern Australia as it invaded cities and homes by the hundreds of thousands as spin-off from its annual migrations. Who can forget its part in the Sydney Olympics or its widely reported impact on the air-conditioning of Parliament House? It is also an occasional pest of crops. The bogong moth may also be the only Australian moth species to be a central player in a children's book (although playing second fiddle – as a meal – to a very cute mountain pygmy-possum) (Murphy, 2021). At the other end of the spectrum, bogongs also featured recently in a whole two columns (including portrait) in the august pages of *The Economist* (Anonymous, 2025), perhaps the only moths ever to do so. The species is discussed at length in Chapter 37.

33.3 Moths of Conservation Concern

Tim New has explored moth conservation in Australia in his recent book *The Other Lepidoptera* (New, 2023) and discusses the issues at length in Chapter 9 of this volume. He identfies and discusses a few species that have received public (and research) attention in the same way that iconic mammals and birds of conservation concern have been a focus of attention. Indeed, the closest model for these examples may well be the equally iconic Richmond birdwing butterfly (Sands and New, 2013), probably the best known insect conservation 'story' in Australia. New (2023) also dwells on the whole-assemblage approach to environmental conservation. In this case, whole sets of species of locally occurring moths have great potential as reflections of ecosystem condition (both good and bad). This general approach is the subject of Chapter 8.

Returning to studies of individual threatened species, two outstanding examples are the subject of chapters in the accounts which follow.

Synemon plana, the golden sun moth, is a day-flying castniid moth specifically associated with native grasslands in south-eastern Australia. A suite of species of *Synemon* occur in Australia, all short-range endemics and all of conservation concern. In the south-east, areas of native grassland, never very extensive, have proved attractive for both agriculture and housing developments (examples include the national capital, Canberra) and surviving tracts are often degraded by exotic species of grass and other plants. *S. plana* is the best known of the sun moths and has been listed as 'endangered' in New South Wales and the ACT, 'threatened' in Victoria and as 'critically endangered' federally. These listings generated financial support for investigation of the species such that its status has now been upgraded to, simply, 'vulnerable'. The biology and conservation status of *S. plana* are discussed in Chapter 35 and a wider account of the family Castniidae is in Chapter 15.

Phyllodes imperialis smithersi, the southern subspecies of the spectacular pink underwing moth, is a rare denizen of subtropical rainforests from south-eastern Queensland to north-eastern New South Wales. Its larvae feed only on the vine *Carronia multisepalea*. Only a few blocks of subtropical rainforest remain, many of them restricted in size. The subspecies is listed as 'endangered' under both federal (2002) and NSW state (2003) legislations. Fortunately, these listings led to funding for further investigation of the species' distribution and biology. The outcomes and their significance are discussed at length in Chapter 36.

For both these examples, formal designation of their conservation status led to much increased knowledge of their biologies and distributions. That wider geographic distributions were discovered in this fashion reflects a pattern seen in other instances of designated species of concern. Both act as 'flagships', drawing attention to the broader conservation concerns for the threatened ecosystems in which they live.

Together with the bogong moth, discussed above, these three species are by far the best known examples of moths that have come to public and official attention for conservation reasons. There are undoubtedly many more species, perhaps less showy and publicly visible, which are deserving of just as much attention. Some may receive this attention in due course but the few well-worked examples that do exist underline the need for ecosystem-level conservation actions focusing on the well-known driving factors that lead to biodiversity loss: habitat degradation, invasive species and climate change. Chapter 9 elaborates on these generalizations.

33.4 Moths of General Cultural Interest

One final category of species is worthy of mention here. I struggled to find an over-arching descriptor for them: 'moths of social notice' was an early effort before I settled on 'moths of general cultural interest'. There are a few species of moths that draw public attention simply because they are spectacular, very large or otherwise obvious even to non-naturalists.

33.4.1 The day-flyers

Many of the day-flying moths attract popular attention perhaps because their diurnal habits, distastefulness and associated visual signalling (both among themselves and to potential predators) lend themselves to the evolution of brilliant colours and striking patterns.

The colourful vine-moth, *Phalaenoides glycinae*, common in southern gardens and often mistaken for a butterfly by the uninitiated, is a familiar case in point. The adults advertise themselves with bold black and white stripes. The striking black, yellow and orange larvae are indeed occasional pests of grape vines but occur on a range of familiar garden plants from Virginia creeper to fuchsias. Other Agaristinae offer similar brilliance as both larvae and adults. We suppose many of them are poisonous and their brilliant colours advertise this fact. They are discussed at length in Chapter 22.

Even more obvious and intrusive to the casual observer are the swarms of zodiac moths, *Alcides metaurus*, which can fill the air around their food trees in the tropical north. Again, brightly coloured caterpillars feed on a range of Euphorbiaceae and, like many other Uraniidae, sequester protective chemicals from their larval hosts. Indeed, in New Guinea *Alcides* spp. form a Batesian mimicry pair, with the distasteful moths acting as models for the mimicking papilionid butterfly, *Papilio laglaizei*, the larvae of which feed on the (presumably) non-toxic rainforest trees, *Litsea* spp. (Lauraceae).

33.4.2 Spectacular larvae

The adults of most moths are, of course, nocturnal and may escape general notice except by the equally nocturnal lepidopterist. This is not the case with many of their larvae, however, and some of these are large enough or spectacular enough to attract frequent popular attention.

Perhaps the most frequently noticed of the larger larvae are those of the emperor gum moth, *Opodiphthera eucalypti*, one of the few Australian saturniids. The very large larvae are blue-green with prominent lateral yellow stripes, the whole enhanced by orange and blue tubercles and white hairs. The habit of the larva, when disturbed, of hunching and expanding its front end before rearing up makes them even more visible and noticeable – and much photographed!

The posterior-horned larvae of hawkmoths also attract special attention especially when adorned with diagonal stripes and eye-like spots. For many species, the later larval instars reach substantial size and make themselves apparent to all. A browse through the superb colour illustrations of larvae in Moulds *et al*.'s (2020) excellent monograph on the Australian sphingids will show why they are so apparent. The Sphingidae are discussed with further illustrations of all stages in Chapter 19.

Finally among moths well known for their familiar larvae, we abandon sheer size and move to the brightly coloured and complex larvae of the cup moths, the Limacodidae. Most Australians will have wondered at the green, black and yellow colours of these eucalypt feeders with their fantastic arrays of tubercles and, not least, their extrusible arrays of stinging hairs, often themselves brightly coloured. The unwary hiker, lawnmower or naturalist brushing past succulent eucalypt saplings will know to their cost what acute pain these gorgeous larvae can elicit. The Australian cup moth fauna is discussed at length in Chapter 16.

33.4.3 The giants

Some moths demand popular attention just by being among the largest insects anyone will ever see.

Anywhere north of Cairns in Queensland, Australia's largest moth may be encountered. The giant Atlas or Hercules moth, *Coscinera hercules*, reaches up to 27 cm in wingspan. Its russet, yellow and white wing patterns and prominent wing 'tails' make it both unmistakeable and impressive.

Over southern Australia, the true giants are the wood borers, cossid moths of impressive heft that have claims to be the heaviest of all moths. Synchronized emergences of these giants are infrequent but always attract public attention, even awe. One of the largest, *Endoxyla cinereus*, is also an occasional pest of plantation eucalypts often through the depredations of one of its predators, black cockatoos, which rip apart growing trees to access the *Endoxyla* larvae. Chapter 26 gives an extended account of this species and cossid moths in general are discussed in Chapter 15.

33.4.4 The builders and artists

Among the moths that attract popular attention are those whose constructions and traces draw attention rather than the moths or larvae themselves. Some may cover trees in silken webbing, others construct caddis-like cases sometimes of impressive size, and yet others 'draw' their tracks on the white underbark of gum trees.

Communally living 'webworms' build highly visible webbing shelters enclosing parts of their host trees. A wide range of moths do this, often to the chagrin of gardeners; examples include the so-called tea-tree web moth, *Orthaga thyrisalis* (Pyralidae: Epipaschiinae) and the familiar processionary catepillar moth, *Ochrogaster* spp. (Notodontidae), among many others. The biology of *Ochrogaster* is discussed at length in Chapter 29.

One of the most familiar objects brought to entomologists for identification is the tightly sown and very solid shelter of the Psychidae, the 'bagworms'. Those created by the genus *Metura*, the 'giant bagworms', create cases anywhere from 10 to 30 cm in length and, especially when they wander away from the larval food plants as prepupae, attract attention and puzzle their finders. An extended account of the Psychidae is given in Chapter 14.

Finally there are the scribblers. The serpentine mines of, usually, micromoths elaborated on the leaves of host plants attract the eye of the lepidopterist. When these feeding tracks are written over large areas of the often blindingly white trunks of eucalypt trees, they attract everyone's attention. A whole class of such trees are commonly named 'scribbly gums' for exactly that feature. The scribblers themselves were long a mystery even to entomologists. Finally identified as the feeding tracks of moth larvae in the 1930s, it was not until 2012 that Marianne Horak and her colleagues redescribed the original *Ogmograptis* species. Horak *et al.* (2012) added 11 further species to the genus, all 'scribblers', and confirmed their placement in the rather obscure family Bucculatricidae. Chapter 34 gives an extended account of these *Ogmograptis* species and their distinctive artworks.

References

Anonymous (2025) Stars in their eyes. *The Economist*, June 21, pp. 69–70.

Bouvier, E.L. (1936) Etude des saturnioïdes normaux, Famille des Saturniides. *Mémoires de Musée d'Histoire Naturelle, Paris, N.S.* 3, 1–354.

Clarke, P.A. (2026) *Aboriginal People and Terrestrial Invertebrates in Australia: Historical and Cultural Relationships*. CSIRO Publishing, Melbourne, Victoria.

d'Abrera, B. (1986) *Sphingidae Mundi: Hawk Moths of the World*. Classey, Faringdon, UK.

d'Abrera, B. (1995) *Saturniidae Mundi: Saturniid Moths of the World Part 1*. Automeris Press, Keltern, Germany.

d'Abrera, B. (1998) *Saturniidae Mundi: Saturniid Moths of the World Part 3*. Goecke and Evans, Kelter, Germany.

d'Abrera, B. (2012) *Saturniidae Mundi: Saturniid Moths of the World Part 2*. Hill House, Melbourne, Victoria.

Ferguson, D.C. (1971–72) *The Moths of America North of Mexico* 20.2A, 20.2B. E.W. Classey, London.

Hofmann, A. and Tremewan, W.G. (2023) *A Systematic Catalogue of the Zygaenidae (Lepidoptera: Zygaenidae)*. Harley Books, Colchester, UK.

Horak, M., Day, M.F., Barlow, C., Edwards, E.D., Su, Y.N. and Cameron, S.L. (2012) Systematics and biology of the iconic Australian scribbly gum moths *Ogmograptis* Meyrick (Lepidoptera: Bucculatricidae) and their unique insect–plant interaction. *Invertebrate Systematics* 26, 357–398.

Mouffet, T. (1634) *Insectorum sive Minimorum Animalium Theatrum*. London.

Moulds, M.S., Tuttle, J.P. and Lane, D.A. (2020) *Hawkmoths of Australia: Identification, Biology and Distribution*. Monographs on Australian Lepidoptera, 13. CSIRO Publishing, Melbourne, Victoria.

Murphy, J. (2021) *Tiny Possum and the Migrating Moths*. CSIRO Publishing, Melbourne, Victoria.

Naumann, D.M., Tarmann, G.M. and Tremewan, W.G. (1999) *The Western Palaearctic Zygaenidae (Lepidoptera)*. Apollo Books, Stenstrup, Denmark.

New, T.R. (2023) *The Other Lepidoptera: Moth Conservation in Australia*. Springer, Cham, Switzerland.

Oberthür, C., Houlbert, C. and Dodd, F.P. (1916) *Contributions à l'étude des Grands Lépidoptéres d'Australia (Genres* Coscinera *et* Xyleutes*)*. Imprimerie Obethür, Rennes, France.

Pittaway, A.R. (1993) *The Hawkmoths of the Western Palaearctic*. Harley Books, Colchester, UK.

Rothschild, W. and Jordan, K. (1903) A revision of the lepidopterous family Sphingidae. *Novitates Zoologicae* 9, 1–972.

Sands, D.P.A. and New, T.R. (2013) *Conservation of the Richmond Birdwing Butterfly in Australia*. Springer, Dordrecht, The Netherlands.

Simonsen, T.J. (2018) *Splendid Ghost Moths and their Allies, A Revision of Australian* Abantiades, Oncopera, Aenetus, Archaeoaenetus *and* Zelotypia *(Hepialidae)*. Monographs on Australian Lepidoptera, 12. CSIRO Publishing, Melbourne, Victoria.

Tarmann, G.M. (2004) *Zygaenid Moths of Australia. A Revision of the Australian Zygaenidae (Procridinae: Artinini)*. Monographs on Australian Lepidoptera, 9. CSIRO Publishing, Melbourne, Victoria.

Zwier, J.H.H. (2022) *Aganainae of the World (Lepidoptera, Erebidae): Documentation and a Guide to their Identification*. Museum Witt, Munich, Germany.

34 The Scribbly Gum Moths: *Ogmograptis* Species

Julia Cooke[1]* and Matthew Brookhouse[2]
[1]*School of Environment, Earth and Ecosystem Sciences, The Open University, Milton Keynes, UK;* [2]*Fenner School of Environment and Society, Australian National University, Canberra, ACT, Australia*

Abstract
Unusually for an iconic animal, the scribbly gum moths (*Ogmograptis* spp. Bucculatricidae) are better known for the distinctive tracks they leave. Larvae feed in the phellogen, or cork cambium, of *Eucalyptus* bark. It is only months later, when a bark layer sheds, that their larval mines are revealed, appearing as if someone has scribbled on the trunk. The scribbles, an accurate record of larval behaviour of some species, have distinct attributes on different host tree species. Considered monotypic for over 75 years, there are now 12 *Ogmograptis* species described. Larval behaviour is best understood for the 'scribula' group which consists of *O. scribula, O. fraxinoides, O. racemosa* and *O. pilularis*. The potential distribution of the scribbly gum moths is indicated by records of the moths themselves and their eucalypt hosts. The life history of *Ogmograptis* species is described, where known, with an emphasis on bark traits which could explain why only some *Eucalyptus* are hosts for *Ogmograptis* species and how larvae identify the bark layer in which to mine. Increases in temperature and fire frequency/severity as climate shifts, together with habitat destruction, are concerning threats to scribbly gum moths.

34.1 Introduction

Bark 'scribbles' are a widely recognized feature of the Australian bush. The scribbles are zig-zagging marks on the smooth bark of some *Eucalyptus* trees, that look a bit like scrawled messages (Fig. 34.1, top row). The trees on which these marks occur, especially when characterized by otherwise entirely smooth bark, have long been known in English as scribbly gums. The Darug language name for the scribbly gum *E. haemastoma* is Tarrin'ny, and *E. pilularis* is Tarunde'a (as reported by Webb (1995) from records of when Daniel Moowattin was a guide to botanist George Caley in 1805). Other groups are rebuilding languages in conjunction with the Australian Institute of Aboriginal and Torres Strait Islander Studies (AIATSIS), such as the Ngunawal language, and may find or create the scribbly gums' names (Tyrone Bell, pers. commun.).

It was discovered in 1934 that moth larvae were responsible for the scribbles (Upton, 1997). The moths, with a wingspan of <1 cm (Fig. 34.1, bottom row), predominantly a brown and grey colouration and nocturnal habit, are unsurprisingly rarely seen. Hence scribbly gum moths are better known for the marks made when the larvae burrow within the bark, that are exposed as layers of bark slough off. Edward Meyrick (1935) described the first species of scribbly gum moth,

*Corresponding author: julia.cooke@open.ac.uk

DOI: 10.1079/9781800627680.0034

Fig. 34.1. Images of scribbles on four different *Eucalyptus* species, with the scribbly moths, the larvae of which are responsible, below. Scribbles on (A) *Eucalyptus pauciflora* made by (B) *Ogmograptis scribula* (male shown), on (C) *E. pilularis* by (D) *O. pilularis* (female shown), on (E) *E. rossii* by (F) *O. racemosa* (female shown), and on (G) *E. fraxinoides* by (H) *O. fraxinoides* (male shown). Scale bar is 2 mm and the same for all moths. Images A, C, E by Julia Cooke, images B, D, F, G, H reproduced from Horak *et al.* (2012), with permission from CSIRO Publishing.

naming it *Ogmograptis scribula*. The specimen was reared from a larva emerging from the trunk of a snow gum, *Eucalyptus pauciflora*. Meyrick found it difficult to determine the family to which it belonged but Nielsen and Common (1991) attributed it to the Bucculatricidae because the cocoon is very similar to those in the genus *Bucculatrix*. This taxonomic placement has since been confirmed by larval morphology and genetic analysis (Mutanen *et al.*, 2010; Horak *et al.*, 2012; Cameron, 2023). Although only one moth was described, Cooke and Edwards (2007) showed that the morphology of the scribbles varied significantly among host tree species and suggested that there were more species to be scientifically identified and described.

In 2012, Horak *et al.* (2012) published a new account of the genus, describing 11 new species and distinguishing three groups, the 'scribula', 'maxdayi' and 'triradiata' (Table 34.1). In allocating these three groups to the genus, Horak *et al.* (2012) presented much new information about the biology of the 'scribula' group and their associated host trees, but the larval biologies of the other two groups are not known. One species, *Ogmograptis notosema*, could not be assigned to a group as no holotype could be located. This chapter focuses on the 'scribula' group but includes the other scribbly gum moths where possible. There is also a related genus, *Tritymba*, the larvae of which also feed in the trunks of *Eucalyptus* species, albeit at the cambial interface, and produce 'ghost scribbles' in the phloem (Horak *et al.*, 2012).

The *Ogmograptis* genus is related to *Bucculatrix* which are leaf miners, also in the Bucculatricidae. With other leaf-mining families, they form a monophyletic superfamily Gracillarioidea and a clade of Gracillarioidea + Yponomeutoidea (Cameron, 2023). As noted, the larvae feed in the cork cambium but the terminal instar of *Ogmograptis*, which is the only one with legs, feeds on the callus tissue produced in this layer by the plant as a wound response to earlier larval damage (Horak *et al.*, 2012). This differential larval feeding and final instar morphology are consistent with the larval development ancestral to the clade Gracillarioidea + Yponomeutoidea (Cameron, 2023). Scribbly gum moth larvae probably gain similar benefits to leaf miners from their feeding environment, namely lower disease incidence, a microenvironment limiting desiccation due to a lower evaporative demand, protection from ultraviolet light and avoidance of plant

Table 34.1. The three groups of *Ogmograptis* described by Horak *et al.* (2012).

Group	Species	Notes
'scribula'	*O. scribula* Meyrick 1935 *O. fraxinoides* Horak 2012 *O. racemosa* Horak 2012 *O. pilularis* Horak 2012	Host trees known, some larval biology studies
'maxdayi'	*O. maxdayi* Horak 2012 *O. barloworum* Horak 2012 *O. paucidentatus* Horak 2012 *O. rodens* Horak 2012 *O. bignathifer* Horak 2012 *O. inornatus* Horak 2012	Host trees unknown, larval biology unknown
'triradiata'	*O. bipunctatus* Horak 2012 *O. pulcher* Horak 2012 *O. triradiata* Turner 1926 *O. centrospila* Turner 1923 *O. notosema* Meyrick 1935	Host trees unknown, larval biology unknown

defences (Connor and Taverner, 1997). These authors note that leaf miners have not achieved the diversity of their externally feeding relatives, and in a similar way, there are very few bark-mining species (Hoare and Dugdale, 2003).

The origin of the name of the first species described by Meyrick (1935), *Ogmograptis scribula*, is unclear. One suggestion for the generic taxon comes from Fennessy (2005) who wrote 'Ogma was a god of poetry and eloquence and was the supposed inventor of ogam writing which was an alphabetical script known as "ogam" or "ogham" and used by some Celtic races'. Indeed, this script, with letters marked on a central vertical line, does have similarities with the scribbles in the way they zig-zag up and down trees. Whitten (2012), however, suggested it was from the Greek *ogmos* meaning 'furrow' and *graptis* meaning 'inscribed or painted', followed by the Latin *scribula* meaning 'writing'. For subsequent species of scribbly gum moth, Horak *et al.* (2012) have named some species after the *Eucalyptus* species on which they occur and others after people involved in studying their biology.

The larval tracks of the scribbly gum moths are tactile and have both random and more predictable components. Gradual increases in width, characteristic of scribbles, make it possible to distinguish the beginnings and ends of feeding tracks. As permanent marks on the bark, their presence is not affected by daily weather conditions and hence they are a reliable phenomenon for bushwalkers and nature enthusiasts to see and for student study. They have inspired writers (Gibbs, 1946; Wright, 1955; Cooke and Crosby-Fairall, 2012) as well as painters, potters and other artists (for an example, see Fig. 34.2).

34.2 Occurrence and Distribution

Scribbly gum moths are native and endemic to south-eastern Australia. The animals are rarely seen as flying adults, and the larvae burrowing under the bark are not visible unless outer bark layers are forcibly removed. However, some pupae and moths have been collected from southern Queensland, eastern New South Wales, southern Victoria, Tasmania and far eastern South Australia and scribbles have more frequently been recorded across this distribution also (Fig. 34.3A).

The larvae feed in the bark of a narrow range of *Eucalyptus* species, so mapping these host trees can provide a maximum potential distribution for scribbly gum moths. Isolated tree populations or individual plants may not have scribbles, as the dispersal capacity of the *Ogmograptis* genus is unlikely to be fast or large given the size of the moths (as shown in another micro-moth in the Bucculatricidae; Vargas-Ortiz *et al.*, 2018). Euclid (2015), an identification key for *Eucalyptus*, which uses the taxonomy of Brooker (2000), records 'scribbles present' for 23 tree species. These include *Eucalyptus racemosa* subsp.

Fig. 34.2. A scribbly gum-inspired tea set by Shannon Garson. Photo: courtesy of Shannon Garson.

racemosa and *E. racemosa* subsp. *rossii* which have since been split into two species (https://id.biodiversity.org.au/instance/apni/613832), bringing this to 24 tree species. For 15 of these species (listed in Table 34.2, mapped in Fig. 34.3), scribbles are referred to in the species description in Euclid, indicating that their presence is sufficiently regular to be a diagnostic feature.

Scribbly gum moths leave scribbles on eucalypts described as gums (characterized by smooth bark across the entire bole and branches) and half-barks (possessing a rough, fibrous basal stocking on the lower part of the trunk and gum bark on the upper trunk and branches), and scribbles are also found on the smooth small branches of some rough-barked (peppermint) species (Table 34.2). In most years, only parts of the outer bark are shed in any season. The half-bark form is considered an adaptation to surface fires and semi-arid climates as it allows protection from fire on the lower trunk, while enabling photosynthesis on the smooth upper bark which, as an adaptation to water stress, allows a sparser leaf canopy with the equivalent photosynthetic gain (Lawes *et al.*, 2021). Common to both gum- and half-barked species is the presence of extensive smooth bark that may be essential for oviposition by *Ogmograptis*. It is not known, however, if *Ogmograptis* species also burrow in the lower trunks of half-barks and in the trunks of rough barks, but the scribbles simply never become visible.

For each of the scribbly gum moth species in the 'scribula' group, the *Eucalyptus* species on which its larvae feed is known (top of Table 34.2; Horak *et al.*, 2012) and includes both smooth and part-bark species. Hence the theoretical range of each of these five moth species is predictable by the bounds of their host species (Fig. 34.3B, C). Local climatic conditions, however, probably restrict the distribution of the associated *Ogmograptis* species to a narrower realized range than that offered by the host species. For example, where *Ogmograptis* species occur on *E. pauciflora* (snow gum), their occurrence is restricted to upper montane areas and they are not present in higher elevation subalpine regions (Costermans, 1981). It is noteworthy that *E. pauciflora* hosts two species of scribbly gum moths, and two moth species, *O. racemosa* and *O. fraxinoides*, occur on more than one species of tree (Table 34.2).

There are nine species of *Eucalyptus* on which scribbles routinely occur for which the associated species of *Ogmograptis* is not known (Table 34.2), and their distribution has been mapped (Fig. 34.3D). There are also 10 species of *Ogmograptis* in the 'maxdayi' and 'triradiata' groups (Table 34.1) about which little is known, including the host tree species. Research is needed to associate the other scribbly gum moth

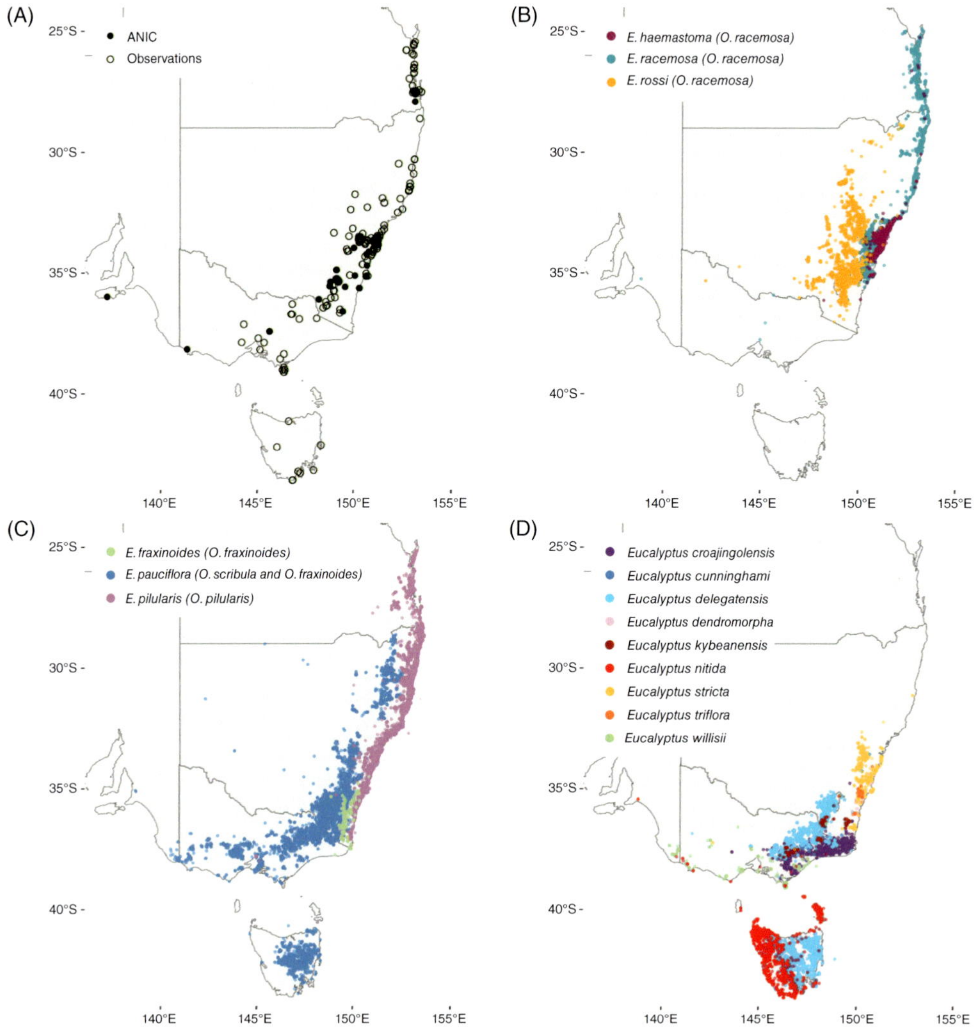

Fig. 34.3. Mapped locations of scribbly gum moths and host tree species records. (A) Data-based *Ogmograptis* records held at the Australian National Insect Collection (ANIC) and records of scribbles from iNaturalist (Observations). There were too few records of each *Ogmograptis* species from ANIC for presenting these separately to be meaningful. (B) The distribution of *E. haemastoma*, *E. racemosa* and *E. rossii* which host *O. racemosa*. (C) The distribution of *E. fraxioides*, *E. pauciflora* and *E. pilularis*, with the *Ogmograptis* species they host. (D) The distribution of the nine *Eucalyptus* species known regularly to have scribbles, but for which *Ogmograptis* species have not yet been identified. The *Eucalyptus* species occurrences were downloaded from the Atlas of Living Australia at https://biocache.ala.org.au/occurrences/search?&q=qid%3A1722256378227&disableAllQualityFilters=true on 29 July 2024.

species with their host plants and understand if there are more moth species yet to be described.

Of the 24 species which mention 'scribbles present' in the Euclid (2015) key, scribbles are not included in the tree species description for nine and it is presumed that although scribbles have been recorded, *Ogmograptis* species do not regularly inhabit these species. These include *Eucalyptus burgessiana*, *E. elata*, *E. fastigata*, *E. multicaulis*, *E nebulosa*, *E. olsenii*, *E. piperita*, *E. stellulata* and *E. stenostoma*. Scribbles occasionally appear on other species but these seem to be anomalies where the larvae

Table 34.2. The section, series, species and common name of the *Eucalyptus* host species from which scribbles have frequently been reported (Euclid, 2015), a description of their bark, the identity of scribbly gum moth species inhabiting these species (Horak *et al.*, 2012), where known, and notes about the specific biology of their moths. See text for descriptions of smooth/gum, half-bark and rough-barked species. Whether the scribbles have a single track or double track (see section 34.3 for more information) was determined from personal observations or photos uploaded to the Atlas of Living Australia.

Trees					Moths	
Section	Series	Eucalyptus species	English common name	Bark description	Species	Notes
Cineraceae	Psathyroxylon	*E. racemosa* (*E. scherophylla*, *E. micrantha*, *E. signata*)	Narrow-leaved scribbly gum	Smooth bark	*O. racemosa*	Single track scribble
		E. rossii	Inland scribbly gum	Smooth bark	*O. racemosa*	Single track scribble
		E. haemastoma	Broad-leaved scribbly gum	Smooth bark	*O. racemosa*	Single track scribble
	Pauciflorae	*E. pauciflora*	Snowgum	Smooth bark	*O. scribula* *O. fraxinoides*	Double track scribble. Only on sub-alpine trees (Costamans, 1981)
	Fraxinales	*E. fraxinoides*	White ash, white mountain ash	Half-bark, upper bark peels in strips	*O. fraxinoides*	Double track scribble
		E. delegatensis Baker	Alpine ash	Half-bark, upper bark peels in strips	Unknown	Double track scribble
Pseudophloius	(no series)	*E. pilularis*	Blackbutt	Half-bark, rough on lower half, smooth on upper.	*O. pilularis*	Double track scribble

Continued

Table 34.2. Continued.

Trees					Moths	
Section	Series	Eucalyptus species	English common name	Bark description	Species	Notes
Eucalyptus	Strictae	*E. triflora* (Maiden) Blakely	Pigeon House ash, three-flowered ash	Smooth to half-bark, with rough bark at base of larger trees	Unknown	
		E. stricta Sieber ex Spreng.	Blue Mountains mallee ash	Smooth bark	Unknown	Blue Mountains mainly, often with scribbles
		E. cunninghamii Sweet	Cliff mallee ash	Smooth bark	Unknown	Blue Mountains only, double track scribble
		E. dendromorpha (Blakely) Johnson and Blaxell	Budawang ash	Half-bark, rough bark on base of trunk	Unknown	Double track scribble
	Contiguae	*E. kybeanensis* Maiden and Cambage [vulnerable]	Kybean mallee ash	Smooth bark	Unknown	
Aromatica	Radiatae	*E. croajingolensis* Johnson and Hill	Gippsland peppermint	Rough bark throughout except branches <8 cm	Unknown	Often has scribbles on upper branches
		E. willisii Ladiges, Humphries and Brooker	Peppermint	Rough bark over whole or part of the trunk	Unknown	Sometimes has scribbles on upper branches
	Insulanae	*E. nitida* Hooker	Smithton peppermint	Rough bark except branches <5 cm	Unknown	Scribbles on upper branches

survive to, or near to, pupation but may not form a persistent population. Scribbles have, for example, been recorded on *E. polyanthemos* in Black Mountain Nature Reserve in 2023 and 2024, indicating that host tree species may either not be fully documented or expansion onto new hosts may be possible. The persistence of *O. racemosa* on *E. rossii* in the same stands, however, may facilitate intermittent colonization of *E. polyanthemos*.

Comparing the distribution of specimen records for *Ogmograptis* and observations of scribbles (Fig. 34.3A) with the distribution of the trees on which scribbles are considered sufficiently regular to be a diagnostic character (Fig. 34.3B–D), the distribution of *Ogmograptis* is probably broader than currently recorded. Notably, there are no moth records from south-east Victoria and much of Tasmania. The distribution limits of each species in the 'scribula' group can be predicted with some confidence (Fig. 34.3B, C) but identification of the host plants for the 'maxdayi' and 'triradiata' groups, and clarification of whether there are any more species to be described, are needed before further distributions can be predicted or confirmed.

34.3 Life Histories

The *Eucalyptus* species that are hosts for scribbly gum moths, described above, are key to the moth species survival as the tree bark provides sustenance, moisture and shelter for the majority of the moths' life cycle. Accordingly, to describe the life history of scribbly gum moths, an understanding of bark morphology and biology is helpful (Fig. 34.4, drawn here from Rosell *et al.*, 2015; Serra *et al.*, 2022 and Shtein *et al.*, 2023).

Bark replaces the epidermis and endodermis present in other plant parts and allows radial thickening of plant organs such as the stem and

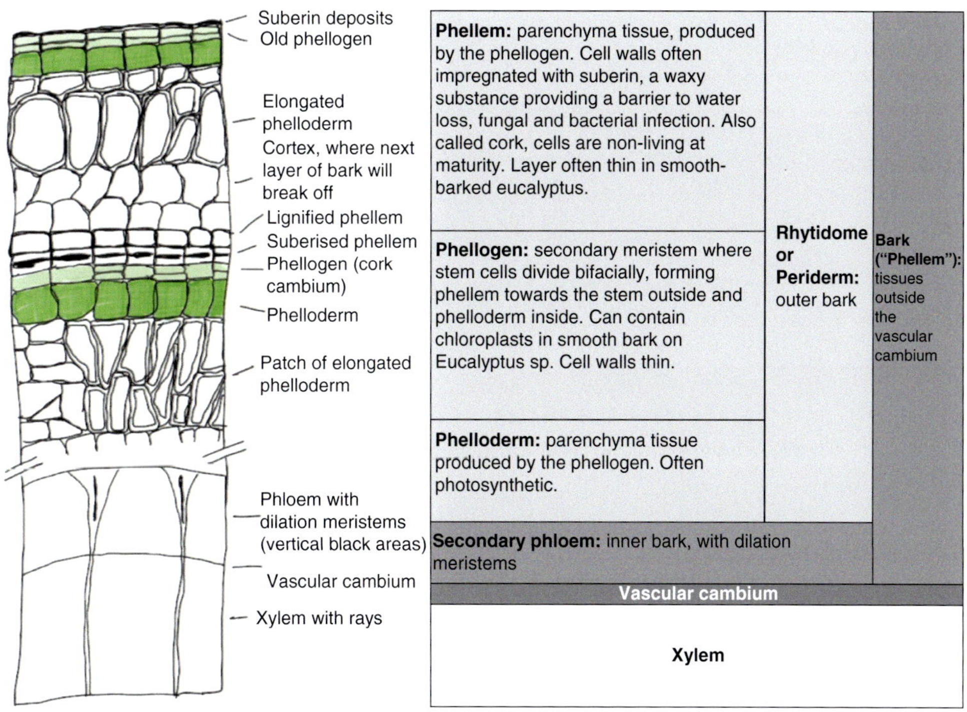

Fig. 34.4. A stylized depiction of bark and its relation to the rest of a woody stem (left) and a description of the tissues (right). The periderm component of the bark is shown at approximately 10× of that of the secondary phloem, vascular cambium and xylem, separated with //. The cells arrangement of the periderm is based on a *Eucalyptus* cross-section. The diagram is adapted from Chattaway (1953) and Shtein *et al.* (2023). Descriptions of tissue types are from Rosell *et al.* (2015), Serra *et al.* (2022) and Shtein *et al.* (2023).

trunk, while protecting inner tissues from a range of biotic and abiotic stresses. Bark describes all the layers external to the vascular cambium (Fig. 34.4). Bark is made up of the secondary phloem and layers of periderm, which itself consists of three layers: meristematic tissues known as the phellogen (or cork cambium) and its derivatives, the phellem which forms outwards and the phelloderm which forms inwards (properties of these tissues are described in more detail in Fig. 34.4). On the trunks and larger branches of *E. rossii*, photosynthesis occurs in layers immediately below the phellem, with some chloroplasts in the phellogen, but the highest density is in the phelloderm (Burrows and Connor, 2020) (coloured green in Fig. 34.4). During a growing season, new layers of periderm form in curving sheaths in the outer phloem, with new meristems developing in the parenchyma, close to the nearest layer of periderm (Chattaway, 1953). When the bark is shed, the break often occurs through the parenchyma with thin cell walls (shown in Fig. 34.4).

Horak *et al.* (2012) described the life history for the 'scribula' group of *Ogmograptis*, predominantly based on careful observations of *O. racemosa*, which are summarized here, using their lettering system to describe the scribble components (Fig. 34.5A). Eggs are laid on the surface of the eucalypt bark, often at the edge of an imperfection (Fig. 34.5B). The larva hatches, burrows through the egg and into the bark, then turns 90° once it reaches a phellogen layer. The larva feeds in this layer, thus staying parallel to the trunk of the tree, but the initial burrowing direction (left or right, up or down) is random (Cooke and Edwards, 2007). This larval instar follows a meandering path (Fig. 34.5A points a–b), generally forming a few loose zig-zags and always keeping within the confines of the bark patch to be shed. Then, probably after moulting, a more compact zig-zag path is made (Fig. 34.5A b–c), then the larva performs a U-turn (Fig. 34.5A c). The legless larva of *O. racemosa* then travels again along its previous track, substantially widening it (Fig. 34.5A points c–d, 34.5C) and sometimes forming another double

Fig. 34.5. A scribble and the stages in the lifecycle of *O. racemosa*, except (B) which is *O. scribula*. (A) A scribble with the beginning of different sections marked with lettering following Horak *et al.* (2012). (B) An egg laid in a bark crevice. (C) Penultimate instar. (D) Final instar feeding on callus tissue. (E) Exit hole in trunk. (F) Cocoon. (G) Adult moth. Image A the author. Images B, C, D, F, G reproduced from Horak *et al.* (2012), with permission from CSIRO Publishing. Image E courtesy of Marianne Horak.

route (Fig. 34.5A, near point d). Other species double back from 'c' in a track distinct from but parallel to their previous track (see, e.g., Fig. 34.1, the tracks of *O. scribula* and *O. pilularis*). Those species forming an enlarged single as opposed to a double (parallel) second track are noted in Table 34.2, where known. The larva then moults. The final instar, which unlike previous instars has legs and a spinneret, returns along the previous path, feeding on the callus, or wound, tissue that formed within the previous mine (Fig. 34.5D). The callus consists of large, nutritious cells. This final instar then burrows to the surface of the bark, leaving a small hole (Fig. 34.5E), and makes a ribbed cocoon at the base of the tree, attached to either the trunk or previously shed bark (Fig. 34.5F). Last, the moths hatch (Fig. 34.5G), mate, and oviposition occurs. It is assumed that the adult moths of *Ogmograptis* do not feed.

Scribbles from the other *Ogmograptis* species follow a very similar pattern in the different sections (Fig. 34.1A, C, E, G), with the variations in the double-thickness or parallel tracks occurring between points c and d noted. Cooke and Edwards (2007) showed that the lengths of the different parts of the scribble vary between subsequently described species. Amongst the tracks they measured, *O. racemosa* on *E. rossii* had a shorter track from a to b (Fig. 34.5A), but longer b–c than *O. scribula* on *E. pauciflora*. They showed that the scribbles on *E. delegatensis*, for which the species of moth responsible has not been determined, are very similar to those of *O. scribula*. The scribbles made by *O. scribula* and scribbles on *E. delegatensis* were also more compact than those made by *O. racemosa* (Cooke and Edwards, 2007).

34.4 Phenology

The specimens from the Australian National Insect Collection (ANIC) show that moths from the *Ogmograptis* 'scribula' group have been recorded from February to May (Fig. 34.6A, B, data compiled by

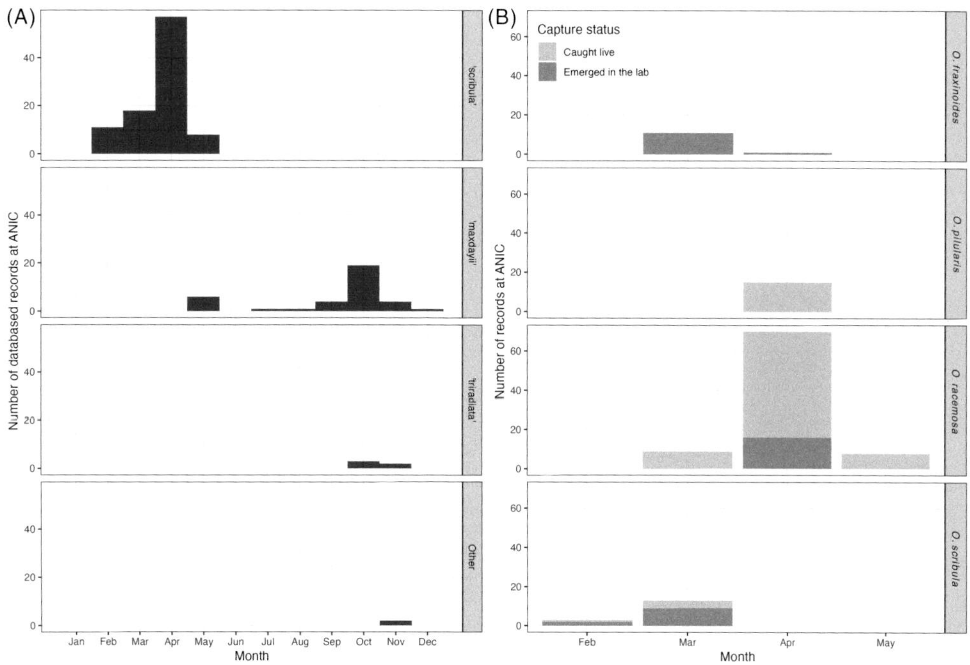

Fig. 34.6. (A) Using data-based records from ANIC, moths from the 'scribula' group appear to fly in the Australian autumn, while the other groups mainly fly in spring. (B) The number of specimens of the 'scribula' group held at ANIC (compiled directly from the collection) are shown plotted against the month they were caught live or emerged in the lab. The records for each species are from 1 to 7 years: *O. fraxinoides* (2009, 2010), *O. pilularis* (1975), *O. racemosa* (1963, 1964, 1974, 1975, 1977, 2006, 2007), *O. scribula* (1933, 1935, 2009, 2010).

the author from specimens). In contrast, the 'maxdayi' group have been recorded from May to December and the 'triradiata' group specimens from October and November, suggesting very different seasonal life histories (Fig. 34.6A). In the 'scribula' group, *O. scribula* moths fly earliest in the year in February and March, *O. fraxinoides* moths have been recorded predominantly in March, and *O. racemosa* from April to May and *O. pilularis* in May (Fig. 34.6B). The emergence could be associated with local meteorological conditions as the host trees of *O. scribula* and *O. fraxinoides* have a more southerly distribution than the host trees of *O. racemosa* and *O. pilularis* (Fig. 34.3B, C).

While the scribbles form a record of larval behaviour, these are only exposed after the larvae have burrowed to the trunk surface and pupated. Hence there is little information about larval phenology. However, in their investigations of *O. racemosa*, Horak et al. (2012) found that bark could be forcibly split from the cork cambium when the later instars were present. They concluded that the larval phenology is co-ordinated with bark development and larvae are well advanced by the time a discontinuous layer is forming as a first step towards decortication. Penultimate instars were found for *O. racemosa* in early December on the cooler, southern side of trunks, while final stage instars were found on the warmer, northern side of the same tree. Pupae of this species have been collected in December and January (Horak *et al.*, 2012). The moths have emerged or been trapped in March–May (Fig. 34.6B). Given decortication occurs during summer, oviposition in autumn enables egg laying when bark shedding is not imminent. Other species in the 'scribula' group may have similar phenology given similar emergence times (Fig. 34.6B). Those in the other groups, however, may be quite different based on the emergence times of the moth species (Fig. 34.6A) and because the *Eucalyptus* species regularly exhibiting scribbles but with unknown moth species have a more southerly distribution (Fig. 34.3D).

34.5 Scribbly Gum Moths and Bark Traits

There are two intriguing facts about scribbly gum moths, manifested in their scribbles. The first is that although there are many species of *Eucalyptus* with at least some smooth bark – 'wholly smooth' or 'partly rough' describes 675 of the 934 eucalypt taxa (Euclid, 2015) – scribbles routinely occur on only 15 species and, potentially, on nine others (Euclid, 2015). The scribbly gums themselves are spread across multiple subgeneric sections and series of the *Eucalyptus* genus and therefore are not closely related (Table 34.2). Why exactly these host species, and why are the other *Eucalyptus* species not suitable? The second curious thing is that as bark falls off in strips or patches, they reveal whole scribbles. Rather than the pattern of bark shedding responding to the scribble, scribbly gum moth larvae appear to confine their activity entirely to areas that will be shed during the subsequent season (Fig. 34.1E). Scribbles are often tightly fitted into these bark areas, with the track always turning back just before reaching the bark edge. The larvae must therefore be able to anticipate which areas of bark will be shed. But how do they know?

Chattaway (1953, 1955) examined the bark of 270 species of *Eucalyptus*. Of these, 70 had radially elongated cells in the phelloderm (illustrated in Fig. 34.4). All the tree species that routinely have scribbles (listed in Table 34.2), which also appear in Chattaway's list, have this phenomenon, which are *E. haemastoma*, *E. racemosa*, *E. pauciflora*, *E. rossii* and *E. piluaris*. Chattaway (1953) writes that in *E. haemastoma* and *E. rossii*:

> ... the smooth stems are abraded and peeled down to the innermost periderm, the cells of the phelloderm are radially elongated, and patches of fibres with greater cross-sectional diameter occur scattered among the rest of the phloem fibres

and

> ... most of them [in the redgums and snappy gums] have groups of fibres which have a greater diameter than the rest.

Does the presence of radially elongated phelloderm contribute to suitable habitat provision for *Ogmograptis* larvae, or does it, or the fibres, enable the first instar larvae to locate the appropriate layer of tissue, that which sheds all at once, in which to feed before layers have fully developed? In *E. cladocalyx* (which does not have radially elongated phelloderm) and in *Corymbia citriodora*, chloroplasts have been found in the inner phelloderm layers, which are some distance from the surface (Burrows and Connor, 2020). Deep chloroplasts are not necessarily photosynthetic but rather have a starch storage function (Pilarski, 1999). One hypothesis is that scribbly gum moth larvae are locating and burrowing in deeper phellogen layers, where the larvae benefit from the stored starches or nutrients supplied to the dividing layer from the adjacent phelloderm.

Questions about why scribbly gum moths use some *Eucalyptus* species and not others, where moths decide to lay eggs, and how the larvae decide where on the trunk to burrow, remain unanswered. The answers, however, could be related to specific bark traits, and the presence of elongated phelloderm cells is a promising avenue to explore. Bark traits remain poorly understood in general, but recent technological advances for better sampling and preparation of bark for anatomical studies (Shtein *et al.*, 2023) may enable new information to be uncovered.

34.6 Conservation

The density of scribbles varies around the faces of tree boles, suggesting that either the adult moths or their larvae have microclimatic preferences (Cooke and Edwards, 2007). Scribble density of *O. scribula* on *E. pauciflora* was roughly twice as abundant on southern and eastern trunk faces than north and west in the Orroral Valley, ACT (Cooke and Edwards, 2007; Fig. 34.7A). For *O. racemosa* on *E. rossii* in the same year (1996) on Black Mountain in Canberra, the southern faces had almost twice the density of scribbles as other faces (Cooke and Edwards, 2007), but this pattern was more muted in 2023 and 2024 (Fig. 34.7B). Cooke and Edwards (2007) suggested that the larvae may be susceptible to high temperatures, or that wind speed and direction may affect oviposition sites given the small size of the moths. The Ngunawal

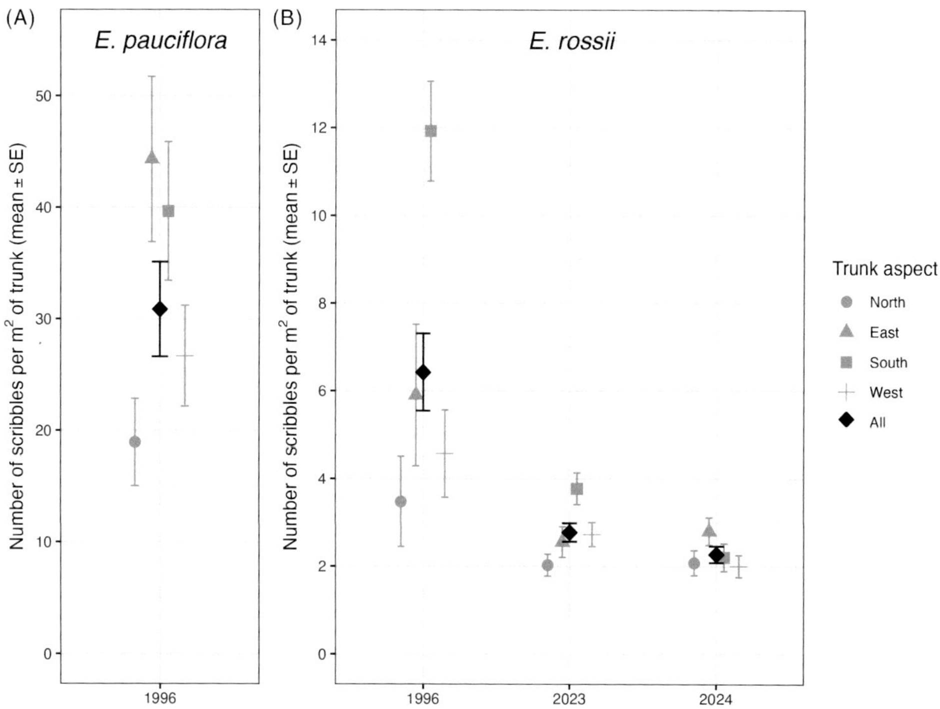

Fig. 34.7. Scribble density by trunk aspect (grey) and overall (black) on two host species. (A) *O. scribula* densities on *E. pauciflora* measured in 1996 recalculated from Cooke and Edwards (2007). (B) *O. racemosa* densities on *E. rossii* collected on Black Mountain in 1996 (Cooke and Edwards, 2007) and in 2023 and 2024 the data were collected by the students of ENVS1003/6103 'Introduction to Environmental and Social Research' (Fenner School of Environment and Society, Australian National University) and supplied by course convener Matthew Brookhouse.

people have a traditional story that when the scribbles disappear from the tree, it is a sign of drought (Tyrone Bell, pers. commun.). Taken together, this information suggests that climate change may affect local presence and abundance of the scribbly gum moths.

In the summer of 2019–20, the south-east of Australia experienced extreme hot and dry weather conditions following a prolonged period of drought (Squire *et al.*, 2021). We, for example, noted that on Black Mountain in Canberra, where scribbles had previously been abundant on *E. rossii*, there were almost none following decortication during the summer of 2020–21. During subsequent years, student-based monitoring assessed the presence and density of scribbles on Black Mountain, using 400 m^2 plots systematically located on north- and south-facing aspects along eight ridgelines. The proportion of trees on which scribbles occurred was, on average, 35% in 2021 and steadily increased to more than 75% by 2024 (Fig. 34.8A). In 2023 and 2024, when there appeared to have been some *O. racemosa* population recovery, scribble density was between 2 and 3 m^{-2}, less than half that recorded in 1996 which averaged more than 6 m^{-2} (Fig. 34.7B).

Although there are no data from immediately before the extreme conditions of 2019–20, the findings do suggest that annual conditions have a significant negative impact on the population of *O. racemosa* on Black Mountain. Earlier records of bark fall may provide insight, as the amount of bark shed by *E. rossii* varies substantially from year to year. Crockford and Richardson (1998) recorded bark fall of 5, 84.9, 0.0 and 177.2 g m^{-2} tree basal area, collected from June to June in 1980–81, 1981–82, 1982–83 and 1983–84 respectively from a site near Canberra. The year when there was no bark fall corresponded to a drought year, whereas the large bark fall the subsequent year corresponded to higher than average rainfall and occurred in autumn to early summer. The timing of the rainfall also appears important, and if rainfall between July to December is less than 250 mm, *E. rossii* will shed little or no bark (Crockford and Richardson, 1998). Chattaway (1953) notes that many *Eucalyptus* trees strip their bark, which can include multiple layers (phelloderm, phellogen and phellum; Fig. 34.4), after a period of unusually hot weather in summer. It is not clear, however, if it is the lack of bark fall or full decortication that affects scribbly gum moth populations, or if larvae simply don't survive because of lack of food or moisture associated with drought conditions. It may well be a combination of these factors as Horak *et al.* (2012) found dead last instar larvae in their tracks on *E. rossii* following a drought in 2006–7 at Gunning, NSW.

Extreme weather events, including drought and heatwaves, are predicted to increase in Australia with anthropogenic climate change (summarized in Hughes, 2011), and populations of scribbly gum moth species could be massively depleted, especially given that multiple years of more favourable conditions are needed for population recovery (Fig. 34.8A,B). The distribution of the *Ogmograptis* genus, from Queensland to Tasmania (Fig. 34.3A), and broad distribution of some of the host trees (e.g. *E. pauciflora*; Fig. 34.3C) suggest a range of temperature and rainfall tolerances. Accordingly, there could be scope for adaptation or species shifts, but the likely small dispersal range of *Ogmograptis* moths could be a limiting factor.

There are no specific data about how fire affects scribbly gum moths, but the timing and severity of the fire are probably important. Various species of *Ogmograptis* could be more susceptible to fire if they have distinctive distributions within a tree (Table 34.2), especially given the different phenology of *Ogmograptis* species (Fig. 34.6). Low, fast-moving fires may not be hot or high enough to kill larvae burrowing in the trunks or branches of their hosts. They could, however, destroy pupae in the bark litter or attached to the base of trees. Hotter fires probably kill the burrowing larvae. Of course, if the host tree is killed, the larvae perish. Historically, bushfires burn patchily, which enables some trees and their inhabitants to survive. As the vegetation recovers either through resprouting (common in *Eucalyptus* species through epicormic growth) or reseeding, populations of animals, such as scribbly gum moths, can then recolonize from these refugia.

The rate at which *Ogmograptis* can (re)colonize an area is unknown. In the Australian mega-fires of 2019–20, thousands of fires burnt more than 10 million hectares (Squire *et al.*, 2021; Driscoll *et al.*, 2024). The fires spanned much of the distribution range of scribbly gum species in southern Queensland, eastern New South Wales, the Australian Capital Territory and far eastern Victoria. At the time of writing,

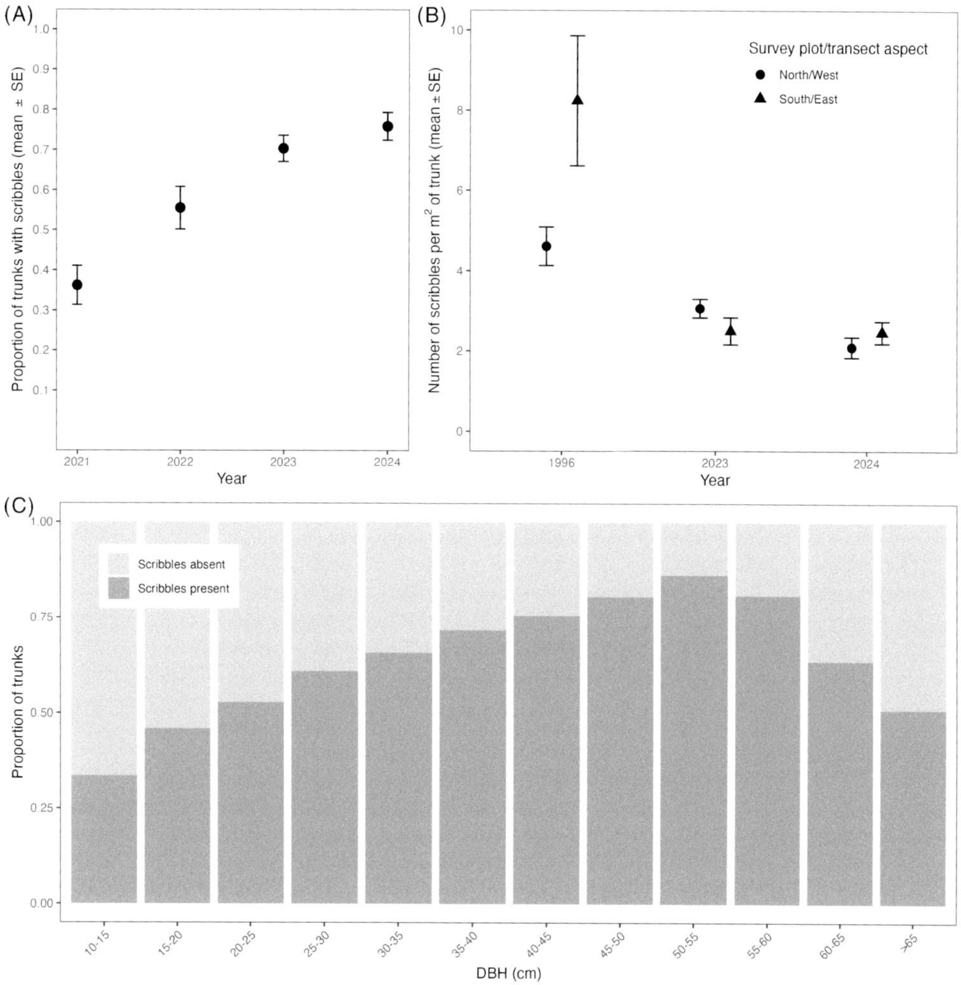

Fig. 34.8. Students from the Australian National University (ANU) surveyed scribbles on the trunks of trees on Black Mountain from 2021 to 2024, in plots distributed across the eastern flank of the mountain. While other *Eucalyptus* species occur here, scribbles are largely restricted to *E. rossii* and data for only this species are shown. Only *O. racemosa* has been described from this site. (A) The proportion of *E. rossii* trunks with scribbles within 400 m^2 plots over 4 years. (B) In 2023 and 2024 the density of scribbles on tree trunks (from ground level to 2 m) was remarkably consistent, but substantially lower than densities recorded in 1996 when there was also a more marked variation in density with mountain-side aspect. In these calculations, trees with no scribbles have been excluded. (C) The proportion of trees with scribbles versus tree size, given as diameter at breast height (DBH) from data pooled from 2021 to 2024. All data from Matthew Brookhouse (source as described in Fig. 34.7) except 1996 data are from Cooke and Edwards (2007).

no data on the impact of fires on *Ogmograptis* species populations had been published, and indeed little information is available on insect responses in general (Driscoll *et al.*, 2024). Climate predictions are that the number of 'very high' fire days could increase by 20–100% by 2050, and 'extreme' fire danger by 100–300% by 2050 under a high global warming scenario (0.4°C); 'catastrophic' fire days are predicted to become a possibility in more locations (Lucas *et al.*, 2007).

Tree host, or food plant, availability and health are obviously key to survival of scribbly gum moth species. Using all available data from 2021–24, it appears that trunk diameter affects either oviposition site selection or larval survival, because the proportion of trees hosting scribbles varies with tree size (Fig. 34.8C). The proportion of *E. rossii* with a diameter at breast height (DBH) of less than 20 cm hosting scribbles was less than 50%, whereas 50–80% of trees with a DBH of 20–65 cm had scribbles. Once DBH exceeds 65 cm, the proportion with scribbles declines, but these data are drawn from just a few trees and hence difficult to interpret. Possible reasons for this variation could be differences in bark structure as trees age or increase in size, or perhaps moths choose larger trees (similar to the way leaf size can affect oviposition site selection in some leaf miners; Kagata and Ohgushi, 2002) as a mechanism to reduce competition among larvae. The data, however, do indicate that scribbly gum moths require or prefer mature trees (DBH 20–55 cm) and hence it could take decades after an area is severely damaged by fire or land clearing before trees re-establishing in the area form suitable habitat for *Ogmograptis* populations again.

The protection of stable populations of the *Eucalyptus* hosts of the scribbly gum moths is key to the survival of these iconic moth species, whose tracks are a unique and defining feature of bushland in much of south-eastern Australia.

Acknowledgements

We extend much gratitude to Dr Marianne Horak for discussing the chapter, providing access to the collections at the Australian National Insect Collection, and generously supplying many of the photographs for this chapter.

References

Atlas of Living Australia. Species occurrence data: *E. croajingolensis* (accessed 29/7/2024) https://doi.ala.org.au/doi/10.26197/ala.faec067b-85d0-4973-a776-a46e7cab2c11; *E. cunninghamii* (accessed 29/7/2024) https://doi.ala.org.au/doi/10.26197/ala.2c532d82-d0f0-4157-a765-797169e0ebb5; *E. delegatensis* (accessed 29/7/2024) https://doi.ala.org.au/doi/10.26197/ala.964f7de8-5897-4b7e-ab61-4f1674e7ee76; *E. dendromorpha* (accessed 29/7/2024) https://doi.ala.org.au/doi/10.26197/ala.aa58cd69-750c-43e9-bb47-0a3cca243315; *E. haemastoma* (accessed 29/7/2024) https://doi.ala.org.au/doi/10.26197/ala.080d8079-25ed-4ec9-b81a-153b329d0582; *E. fraxioides* (accessed 29/7/2024) https://doi.ala.org.au/doi/10.26197/ala.7bb26a22-ccf8-4da8-ad2a-c95103757132; E. pauciflora (accessed 29/7/2024) https://doi.ala.org.au/doi/10.26197/ala.3fbf78f6-d16d-4d7b-b9f3-9b1b5df7efec; *E. pilularis* (accessed 29/7/2024) https://doi.ala.org.au/doi/10.26197/ala.eea29a49-4cc6-49fa-a229-149306d69ec3; *E. racemosa* (accessed 29/7/2024) https://doi.ala.org.au/doi/10.26197/ala.6c5dcebc-8fed-4d04-9ba9-ddf7c89ffd55; *E. rossii* (accessed 6/1/2025) https://doi.ala.org.au/doi/10.26197/ala.4f925b5e-b7ec-455c-ab91-f0fbb9b64b6d *E. stricta* (accessed 29/7/2024) https://doi.ala.org.au/doi/10.26197/ala.6291c465-6026-410e-8d9c-844e4f4760ae; *E. triflora* (accessed 29/7/2024) https://doi.ala.org.au/doi/10.26197/ala.d801ca57-4f01-41d8-afed-68bce88a1514; *E. kybeanensis* (accessed 29/7/2024) https://doi.ala.org.au/doi/10.26197/ala.e3a07528-d370-4adc-a83e-556f9fa6ee6f; *E. willisii* (accessed 29/7/2024) https://doi.ala.org.au/doi/10.26197/ala.52dcc88e-73ed-4cca-b79a-73d34a1ae5af; *E. nitida* (accessed 29/7/2024) https://doi.ala.org.au/doi/10.26197/ala.a43f0b11-0374-452d-8c51-9fa3e56a05ba; Ogmograptis sp. (accessed 29/7/2024) https://doi.ala.org.au/doi/10.26197/ala.b0cab42c-a5ff-48e9-8e85-70e13cedf501.

Burrows, G.E. and Connor, C. (2020) Chloroplast distribution in the stems of 23 eucalypt species. *Plants* 9, 1814.

Cameron, S.L. (2023) Mitochondrial phylogenomics of the Australian scribbly gum moth *Ogmograptis* (Lepidoptera: Bucculatricidae) and an examination of deep-level relationships within Lepidoptera. *Austral Entomology* 62, 449–463.

Chattaway, M.M. (1953) The anatomy of bark. I. The genus *Eucalyptus*. *Australian Journal of Botany* 1, 402–433.

Chattaway, M.M. (1955) The anatomy of bark. IV. Radially elongated cells in the phelloderm of species of Eucalyptus. *Australian Journal of Botany* 3, 39–47.

Connor, E.F. and Taverner, M.P. (1997) The evolution and adaptive significance of the leaf-mining habit. *Oikos* 79, 6–25.

Cooke, J. and Crosby-Fairall, M. (2012) *My Little World*. Omnibus Books, Gosford, NSW.

Cooke, J. and Edwards, T. (2007) The behaviour of scribbly gum moth larvae *Ogmograptis* sp. Meyrick (Lepidoptera: Bucculatricidae) in the Australian Capital Territory. *Australian Journal of Entomology* 46, 269–275.

Costermans, L.F. (1981) *Native Trees and Shrubs of South-eastern Australia*. Weldon, Sydney, NSW.

Crockford, R.H. and Richardson, D.P. (1998) Litterfall, litter and associated chemistry in a dry sclerophyll eucalypt forest and a pine plantation in south-eastern Australia: 1. Litterfall and litter. *Hydrological Processes* 12, 365–384.

Driscoll, D.A., Macdonald, K.J., Gibson, R.K., Doherty, T.S., Nimmo, D.G. *et al.* (2024) Biodiversity impacts of the 2019–2020 Australian megafires. *Nature* 635, 898–905.

EUCLID (2015) *Eucalypts of Australia*, 4th edn. Available at: https://apps.lucidcentral.org/euclid/text/intro/ (accessed 15 October 2025).

Fennessy, B. (2005) What's in a Name? Australian National Botanic Gardens. Available at: www.friendsanbg.org.au/sites/default/files/pdf/bernardpapers.pdf (accessed 15 October 2025).

Gibbs, M. (1946) *The Complete Adventures of Snugglepot and Cuddlepie*. HarperCollins (Australia), Sydney, NSW.

Hoare, R.J. and Dugdale, J.S. (2003) Description of the New Zealand incurvarioid *Xanadoses nielseni*, gen. nov., sp. nov. and placement in Cecidosidae (Lepidoptera). *Invertebrate Systematics* 17, 47–57.

Horak, M., Day, M.F., Barlow, C., Edwards, E.D., Su, Y.N. and Cameron, S.L. (2012) Systematics and biology of the iconic Australian scribbly gum moths *Ogmograptis* Meyrick (Lepidoptera: Bucculatricidae) and their unique insect–plant interaction. *Invertebrate Systematics* 26, 357–398.

Hughes, L. (2011) Climate change and Australia: Key vulnerable regions. *Regional Environmental Change* 11, 189–195.

Kagata, H. and Ohgushi, T. (2002) Effects of multiple oviposition on clutch size in a leaf-mining moth, *Paraleucoptera sinuella* (Lepidoptera: Lyonetiidae). *Entomological Science* 5, 407–410.

Lawes, M.J., Woolley, L.A., Van Holsbeeck, S., Murphy, B.P., Burrows, G.E. and Midgley, J.J. (2021) Bark functional ecology and its influence on the distribution of Australian half-butt eucalypts. *Austral Ecology* 46, 1097–1111.

Lucas, C., Hennessy, K., Mills, G. and Bathols, J. (2007) *Bushfire Weather in Southeast Australia: Recent Trends and Projected Climate Change Impacts*. Bushfire Cooperative Research Centre, Australian Bureau of Meteorology and CSIRO Marine and Atmospheric Research, Melbourne, Victoria.

Meyrick, E. (1935) *Exotic Microlepidoptera*, 4. Taylor and Francis, London, pp. 593–608.

Mutanen, M., Wahlberg, N. and Kaila, L. (2010) Comprehensive gene and taxon coverage elucidates radiation patterns in moths and butterflies. *Proceedings of the Royal Society B: Biological Sciences* 277, 2839–2848.

Nielsen, E.S. and Common, I.F.B. (1991) Lepidoptera (Moths and Butterflies). In: Naumann, I. (ed.) *The Insects of Australia*, *vol.* 2. Melbourne University Press, Melbourne, Victoria, pp. 817–915.

Pilarski, J. (1999) Gradient of photosynthetic pigments in the bark and leaves of lilac (Syringa vulgaris L.). *Acta Physiologiae Plantarum* 21, 365–373.

Rosell, J.A., Castorena, M., Laws, C.A. and Westoby, M. (2015) Bark ecology of twigs vs. main stems: Functional traits across eighty-five species of angiosperms. *Oecologia* 178, 1033–1043.

Serra, O., Mähönen, A.P., Hetherington, A.J. and Ragni, L. (2022) The making of plant armor: The periderm. *Annual Review of Plant Biology* 73, 405–432.

Shtein, I., Gričar, J., Lev-Yadun, S., Oskolski, A., Pace, M.R. *et al.* (2023) Priorities for bark anatomical research: Study venues and open questions. *Plants* 12, 1985.

Squire, D.T., Richardson, D., Risbey, J.S., Black, A.S., Kitsios, V. *et al.* (2021) Likelihood of unprecedented drought and fire weather during Australia's 2019 megafires. *npj Climate and Atmospheric Science* 4, 64.

Upton, M.S. (1997) *A Rich and Diverse Fauna: The History of the Australian National Insect Collection 1926–1991*. CSIRO Publishing, Collingwood, Victoria.

Vargas-Ortiz, M., Bobadilla, D., Huanca-Mamani, W. and Vargas, H.A. (2018) Genetic divergence of isolated populations of the native micromoth *Bucculatrix mirnae* (Lepidoptera: Bucculatricidae) in the arid environments of Northern Chile. *Mitochondrial DNA A*, 29, 1139–1147.

Webb, J. (1995) *George Caley: Nineteenth Century Naturalist*. Surrey Beatty and Sons, Chipping Norton, NSW.

Whitten, M. (2012) Deciphering nature's message stick. *Meanjin* 71, 30–38.

Wright, J. (1955) *The Two Fires*. Angus and Robertson, Sydney, NSW.

35 The Golden Sun Moth: *Synemon plana*

Tim R. New
Department of Ecological, Plant and Animal Sciences, La Trobe University, Melbourne, Victoria, Australia

Abstract

The threatened Australian endemic golden sun moth, *Synemon plana*, occurs on native grasslands on the south-east mainland. Many grassland habitats have been lost and the moth has become a significant flagship species for grassland conservation in the region. It is still unclear whether the moth is univoltine or has a longer larval development period of 2 or 3 years. The non-feeding adults are short-lived. The sexually dimorphic adults differ in activity: only males fly actively – under particular weather conditions and for a few hours in the middle of the day – and can be targeted in visual surveys for the moth. Females are relatively sedentary. At any site, the flight period extends over several months from spring to early summer, but the rapid turnover of individuals over this period ensures that repeated survey counts are needed to enumerate the seasonal cohort of the site population. Conservation interest flows from the losses of native grasslands and their associated biota. Development of conservation concern and awareness of the moth's biology are discussed in this chapter.

35.1 Introduction

The diurnal golden sun moth, *Synemon plana* (Fig. 35.1), is a castniid moth endemic to the south-east mainland of Australia, where it has become a significant flagship species for the conservation of increasingly fragmented lowland native grasslands which are recognized as among the most threatened of all the country's terrestrial ecosystems (Kirkpatrick *et al.*, 1995). Less than 1% of original grass cover remains and much of this is degraded by weed invasion. *S. plana* has been investigated more extensively, and over a longer period of awareness of vulnerability to habitat change, than any other moth of conservation concern in the region. Many gaps in biological understanding persist; for example, almost nothing is known of larval mortality.

The sexually dimorphic golden sun moth is by no means the only castniid signalled for conservation interest – some other locally endemic species have a far more tenuous existence and restricted range on grasslands in southern Australia, and new species of *Synemon* continue to be described from very small ranges as apparent narrow-range endemics (Chapter 15) – but its study has shown the difficulties of assessing population size, structure and vulnerability, and contributed significantly to approaches to resolving the conflicts involved with grassland conservation in the region. Until recently, the

Corresponding author: t.new@latrobe.edu.au

DOI: 10.1079/9781800627680.0035

Fig. 35.1. *Synemon plana,* adult moth showing wing colour contrast. Photo: Ethan Beaver.

moth was known from very few grasslands in the region but historically is suspected to have been far more widespread and to have declined as grasslands were lost and degraded. The recent known and suspected habitat range (Fig. 35.2) indicates the two disjunct areas of interest in Victoria and in New South Wales/Australian Capital Territory. The golden sun moth's notoriety flows from difficulties of study, with uncertainties over determining conservation status and needs for conservation in the face of continuing demand for subsuming periurban grassland areas for urban and/or industrial development, most notably around Melbourne and Canberra.

35.2 Life Cycle

Eggs of *Synemon plana* are laid singly near ground level around the bases of grass tussocks which also serve as the larval food plant (Fig. 35.3), in many cases inserted below the surface through the female's long ovipositor. Larvae are subterranean, feeding on the roots of grasses, and pupae protrude vertically from the ground at adult emergence. Adults do not feed and are short-lived, with published statements on longevity ranging over about 2–5 days. *S. plana* is sexually dimorphic in both appearance and activity. Males are active flyers and females fly little (in some cases, apparently, not at all) but can run or sit and expose their brightly coloured hindwings that attract patrolling males. At any site, the total flight period extends over several months from spring to early summer, over which emergences continue but with rapid turnover in the active adult population present. The short adult life renders mark–release–recapture attempts of limited value in measuring population size.

The onset and pattern of adult emergence are partially related to soil temperature, so that site topography and vegetation cover may be involved in seasonal timing. As Braby and Dunford (2006) found on sites near Canberra, the changing seasonal abundance of moths seen across sites is related to topography reflecting soil temperatures, so that 'hotspots' of moth abundance may simply be hotter spots. Cool wet spring and early summer weather may delay onset of the flight season for a month or more. It is also still uncertain whether the life cycle takes 1, 2 or even 3 years. Richter *et al.* (2013a) found larvae of three broad size classes but also suggested that under optimal conditions, development might be completed in 1 year. The apparent age cohorts of larvae may be isolated spatially so remain independent sectors of the site population. The total adults present at any site in one season thus represent an unknown cohort or proportion of the total resident population; very low visibility in one season may not necessarily imply a small population on that site.

Several reports summarized by Trust for Nature (2022) imply that the first instar larvae are free-living at and near the ground surface and around host grass tussocks. Larger larvae burrow and construct short silk-lined tunnels from which they burrow along grass rhizomes and produce a network of tunnels. Pupation occurs in a vertical silken tunnel at or very close to the soil surface.

35.3 Habitat and Larval Food

The major habitat of *Synemon plana* is treeless temperate C3 grassland dominated by wallaby grasses (*Rytidosperma* spp., formerly *Austrodanthonia*) and/or spear-grasses (*Austrostipa* spp.), with areas of low biomass and suitable intertussock

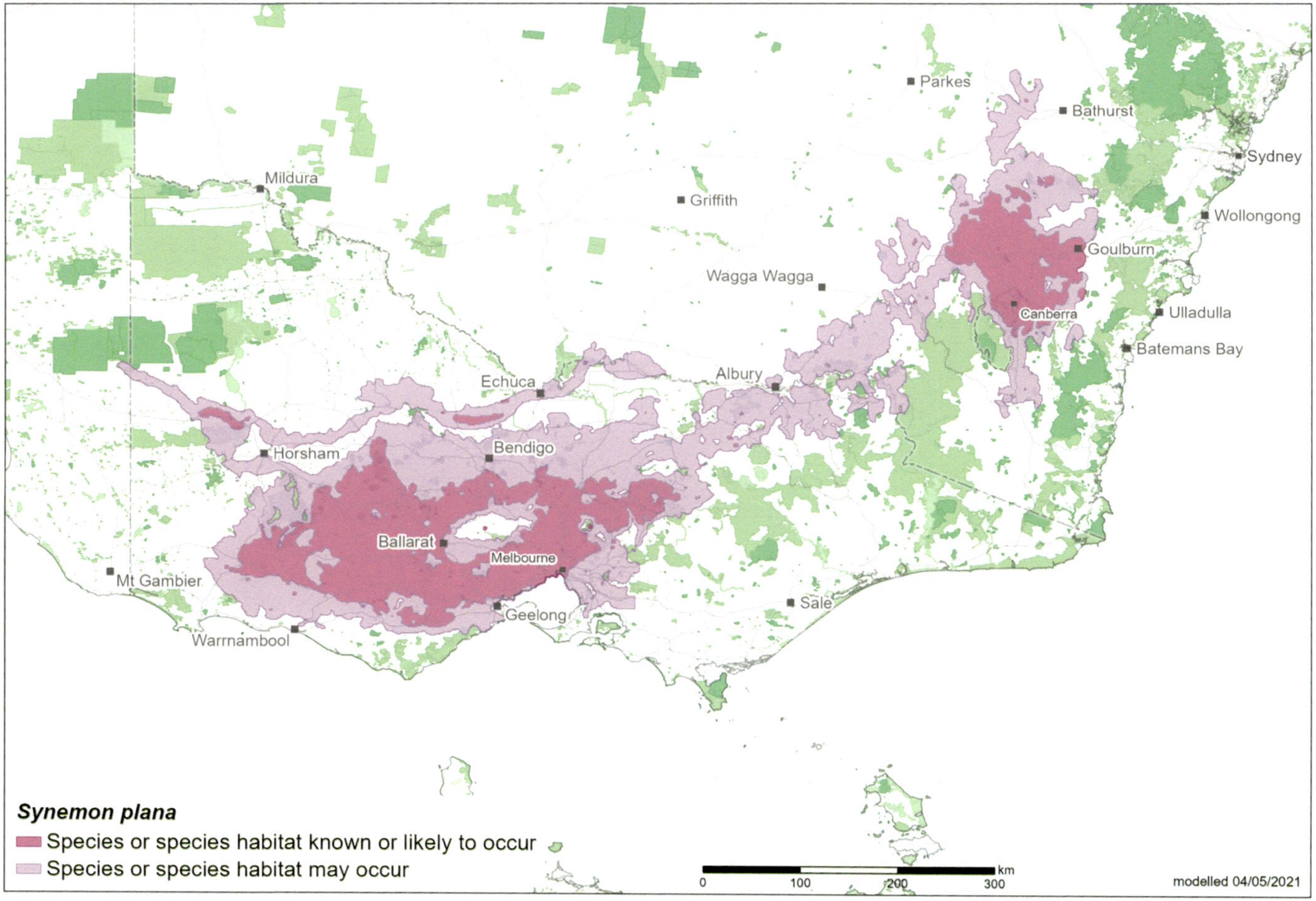

Fig. 35.2. Modelled distribution of *Synemon plana* in south-eastern Australia, as in DAWE (2021); deeper red: areas where species or species habitat known or likely to occur; purple: areas where species or habitat may occur. Map used with permission and under CC-BY.

Fig. 35.3. *Synemon plana,* female ovipositing. Photo: Ethan Beaver.

spaces (Kutt *et al.*, 2015). These grasslands show fluctuations in richness and abundance of these grass genera associated with fluctuations in moth abundance. Dense grass swards without bare spaces are largely avoided, and tussocks with a large root volume may be beneficial because larvae may then spend their whole life there without having to move through soil to find food elsewhere. Unlike some related species of *Synemon*, larvae have not been recorded feeding on sedges, such as *Gahnia* spp., or mat-rushes, *Lomandra* spp.

Moths also occur in larger clearings in open grassy woodlands. Most suitable grasslands in Victoria are at low elevations, whereas the inland sites in New South Wales and around Canberra are generally more elevated. Many grassland patches are now fragmented and increasingly isolated; others have been lost or severely degraded by weed invasion, overgrazing and trampling by stock, or fire. High tussock mortality from burns and losses of grazing-sensitive forbs contribute to site degradation.

Synemon plana, nevertheless, has been recorded from sites with low cover of the main larval food plants, *Rytidosperma* spp., with feeding records from a number of species and more inferential reports from *Austrostipa* spp. The invasive Chilean needle grass (*Nassella neesiana*), a declared noxious weed in Australia, may be an important food plant in altered grasslands (Richter *et al.*, 2013a) and, even, a more suitable food than some native hosts. Indeed, the sun moth has been reported from sites entirely lacking native grasses (Brown *et al.*, 2012). There are also suggestions that serrated tussock grass (*Nassella trichotoma*), also a declared noxious weed, may be used as food, because larvae have been found among root masses in the ACT (DAWE, 2021). The apparent value of such weeds, formal targets for eradication, for *S. plana* leads to conflicting management priorities, both in assuring larval food supply and the wider impacts of herbicides used to eliminate noxious weeds.

35.4 Assessment and Vulnerability

Counts of active male moths derived from standard techniques (such as transect walks) have been the traditional basis for detecting and enumerating moths at a site, with many reported absences contributing to conservation concern. Recognition that the moths have a very restricted activity pattern, however, has led to refined survey techniques and defined protocols that give a more realistic picture. Moths fly from mid-morning to mid-afternoon, but only in fine calm sunny weather with no precipitation and within a relatively narrow temperature range (Gibson and New, 2007); surveys at other times of day or under different conditions (such as in windy weather or cloud cover) do not detect the moth. Likewise, the rapid turnover of adults at a site means that any single survey (as the most common inspection method used), even under optimal conditions, cannot yield any meaningful estimate of abundance. Comparative surveys across sites may extend beyond the daily active period, or encounter different weather conditions whose significance is not recognized, so giving false negatives of incidence. Counts of emerged pupal cases protruding from the ground allow some cumulative estimate of numbers of adults but are laborious to undertake (Richter et al., 2013b). Low and uncertain detectability is a persistent hindrance to surveys.

Surveys for greater efficiency are now based on well-defined protocols: DEWHA (2009) advocated surveys at a site over at least 4 suitable days at approximately weekly intervals during the local flight season and, if presence is detected early, the effort should then switch to determining the relative distribution of sun moths on the site. Conditions for such surveys were also specified: surveys should be made on warm to hot days (above 20°C), during the warmest part of the day (1000–1400 hours), with clear or mostly cloudless sky, and still or relatively still wind conditions, and at least 2 days since rain. These surveys have revealed the moth's presence on many additional grasslands in the region, and cast doubt on some previously reported absences.

The resulting dilemmas over whether and how to protect those populations as they are discovered in the face of burgeoning pressures to develop sites and with consequent high financial implications – in addition to conservation ideals– persist. *S. plana* has become '... "the meat" in a very complex sandwich between developers and conservationists' (New, 2012). Project delays and prohibitions flowing from discovery of sun moths on sites where it was thought unlikely to occur have frustrated developers. Feelings ran high; for example, 'We've been frustrated by growling grass frogs, bandicoots, mouthless moths, and the golden sun moth in particular' (Tony Domenico, Chief Executive, Urban Development Institute of Australia, as quoted in the *Herald Sun* newspaper, 29 November 2011). In essence, reports of a moth population on a site cause stoppage pending fuller investigation and possible prohibition of any development.

35.5 Conservation Status and Concern

Synemon plana occurs in limited areas of Victoria, New South Wales and the Australian Capital Territory, in all of which it has declined as its habitats have been changed. The low detection levels of *S. plana* across most grasslands surveyed, coupled with the vulnerability of those grasslands to development, led to early designation of the moth as of concern under all three range jurisdictions (ACT, NSW – both 'Endangered' and Victoria – 'Threatened'), and eventually to national listing as 'Critically Endangered' under the EPBC Act in 2002. At that time the Area of Occupancy (AOO) was approximately 8.8 km^2 and the Extent of Occurrence (EOO) approximately 131,100 km^2; this has increased considerably to at least 1596 km^2 (AOO) and 145,322 km^2 (EOO) (DAWE, 2021). *S. plana* is now 'Vulnerable' (effective from December 2021), but lack of quantitative data on population size and trends meant that only Criterion 2 (dealing with geographical distribution) could be evaluated. Grounds for Vulnerable status (DAWE, 2021) are limited area of occupancy, severe fragmentation and continuing decline in area of occupancy, area, extent and quality of habitat, number of locations or subpopulations and number of mature individuals.

The earlier putative rarity of the moth coincided with losses and change to grasslands. Invasions by exotic pasture grasses have fundamentally altered the composition of many grasslands by replacing native grasses with non-palatable species. The list of threats to *S. plana*, however, extends well beyond habitat loss (Table 35.1).

35.6 Development of a Conservation Programme

Both site protection and site management are key to conservation planning, together with attention to single sites and their place within the wider landscape. Collectively, these reflect the presumed metapopulation structure of *S. plana* for which connectivity between occupied sites is critical. It is presumed that males may not be able to disperse more than a few hundred metres, whilst females are essentially site bound; in Victoria's guidelines, separation of sites by 200 m is considered sufficient to declare 'isolation' of sun moth populations. Sites on which the moth has become extinct are unlikely to be recolonized if separated more widely from existing populations.

Declines of native grassland have been severe and rapid across the moth's range, with most concern over areas subsumed or planned for urban expansion to the north and west of Melbourne and around Canberra, leading to urgent calls for quality assessment and conservation. These areas also harbour two threatened reptiles, so are recognized more widely for conservation significance. Many significant sites are very small, and many are also within areas of urban/industrial development. Surveys and monitoring of sun moths on each site

Table 35.1. Threats impacting the golden sun moth, as listed by DAWE 2021, abbreviated to the main headings used in report.

Threat
Habitat loss, degradation and fragmentation
Land clearing for urban development and agriculture
Soil disturbance
Lack of biomass removal
Intensification of grazing and/or mowing
Application of chemicals such as herbicides, pesticides and fertilizers
Planting and/or regeneration of shrubs/trees
Invasive species
Weed invasion
Habitat degradation by the introduced rabbit (*Oryctolagus cuniculus*)
Installation of artificial structures
Increasing predation by native and introduced birds
Fire
Inappropriate fire regimes
Climate change
Temperature increases, changes to rainfall patterns and hydrological flows, and more extreme weather events

proposed for development have led to continuing appearance of site-focused reports, advice and 'action planning', exemplified by the many listed in the national Conservation Advice issued by DAWE (2021), and the earlier sub-regional strategy for Victoria (DSE, 2011; DEPI, 2013). The proximity of some occupied sites to housing estates precludes some management approaches, notably use of fire, for fear of property damage.

Formal guidelines for assessing grassland site importance for the sun moth are obliged by its EPBC Act listing and were developed (DEWHA, 2008) to apply to grassland and open grassy woodland habitats across the range from western Victoria to south-eastern New South Wales and including the ACT. They recognize that disturbances to either the subterranean early stages or the adult can occur, and that habitat disturbance through removal of vegetation, inappropriate fire regimes, weed invasion, loss of intertussock spaces, overstocking, soil compaction and changes to agricultural practices can all be detrimental. An exercise in viability assessment involving many environmental factors (Mata *et al.*, 2017) suggested that adult survival, bare soil cover and cover of resource plants were especially important influences, but the workshop preceding that publication exposed many uncertainties and differences of opinion on how the moth 'works'.

'Habitat quality' of grassland can thus be assessed broadly, but the presence and abundance of the moth are the basis for formally categorizing three habitat classes in terms of their high, medium or low contribution to the species' persistence. Removal of each habitat in Victoria obliges designation of 'offsets' (DSE, 2010, for Victoria), and *S. plana* was the first insect to raise this issue in Australia as part of a wider duty to compensate for clearing of any native grassland or other native vegetation regarded as a 'patch'; formal prescriptions have been defined for need and extent, and can become complex. Habitat offset requirements differ across the three habitat classes, but offset areas must be defined, sourced and secured, and approved by the Department before any clearing is begun.

General conservation objectives are embraced by those expressed in the ACT Action Plan (ACT, 2017), superseding the initial plan published in 1998, and which place the moth in the context of its habitat needs. These are (i) conserve large populations in the ACT and protect other populations from unintended impacts; (ii) manage the species and its habitat to maintain the potential for evolutionary development in the wild; and (iii) enhance the long-term viability of populations through management of adjacent grassland to increase habitat area and connect populations.

In principle, conservation measures proposed across the moth's range have broadly embraced (i) protection of significant grasslands from despoliation or loss, where possible as 'reserves'; (ii) promoting connectivity among grassland patches, essentially preventing habitat fragmentation and isolation of moth populations; (iii) restoration and rehabilitation of degraded grasslands; (iv) prevention or reduction of current and future threats, including invasive species; (v) continued monitoring and documentation of *S. plana* populations; and (vi) increasing community awareness and citizen science interest and involvement in the moth as a 'flagship species' and the wider importance of grassland habitat conservation. The fuller listing by DAWE (2021) expands on these themes, as developed in several earlier documents; threat mitigation measures comprised three groups of measures (avoiding impacts, minimizing impacts, managing habitat) (DEWHA, 2008). Victoria's ambitious 'Subregional Strategy' proposed five main foci for conservation (DSE, 2011) within the major growth corridors for Melbourne (Table 35.2). Such documents have laid the foundations for practical conservation for *S. plana* undertaken since then.

A 'Draft National Recovery Plan' for *Synemon plana* was circulated for comment in May 2025 (DCCEEW, 2025). That draft, designed to operate over 10 years, lists four objectives as guidance for progress: (i) all known golden sun moth subpopulations are represented in secure conservation reserves or protected by conservation mechanisms; (ii) golden sun moth habitat quality is improved, fragmentation is reduced and connectivity is increased; (iii) threats to the golden sun moth and its habitat are stabilized and/or reversed; and (iv) the distribution and habitat occupancy of the golden sun moth are better understood. Series of 'recovery actions' are proposed for each of these.

Circumstances and priorities are largely site determined, and the specific measures promoted at each site contribute to the wider understanding of the moth's biology and guidance of management. The following examples illustrate the advances being made.

35.6.1 Reservation

A number of significant grassland patches are now protected far more conscientiously than previously. Some of the largest occupied sites, such as the Craigieburn Grasslands Flora and Fauna Reserve (Victoria), are thus acknowledged for their significance. At Nhill, western Victoria, a dedicated reserve for conservation of two co-occurring sun moths (the unique 'Nhill morph' of *S. selene* and *S. plana*) was the region's first 'moth reserve' (Douglas, 2004). The genetic variation shown for *S. plana* across its range (Clarke and O'Dwyer, 2000; Clarke and Whyte, 2003) emphasizes the importance of conserving possibly unique isolated populations and of avoiding any translocations that might threaten their integrity. *Synemon selene* presents another unusual biological feature among moths in that Victorian populations are parthenogenetic, although both sexes occur in recently discovered South Australian populations (Grund *et al.*, 2019).

35.6.2 Restoration

At Mt Piper, near Broadford in Victoria, patches occupied by *S. plana* had food plant cover greater than 40%, and methods to increase cover on other patches by a combination of weeding and

Table 35.2. The five main components of the 'Subregional strategy' proposed for conservation of *S. plana* in Melbourne's urban growth zones (DSE 2011, as summarized by New 2012).

1. Long-term target of 80% of the highest priority habitat for *Synemon plana* within the Victorian Volcanic Plains bioregion to be protected.
2. Prescription to guide day-to-day management, with clearing of habitat conditional on offsets designed to specified criteria.
3. Large areas of habitat permanently protected, including western grassland reserves, and to be acquired over 10 years.
4. Smaller reserves to manage risks and retain genetic diversity of *Synemon plana*.
5. Greatly improved information, through strategically placed surveys and habitat information to facilitate informed management decisions.

planted seedlings showed considerable promise in attaining this threshold (O'Dwyer and Attiwill, 1999, 2000). Other, less precise examples exist: at Salisbury Bushland Reserve, 6000 *Rytidopsperma* seedlings were planted in 2010–2011 into an area previously dominated by wild oats. About 80% survived. Additionally, restoration of *Rytidosperma* grasslands can be approached through biomass reduction – for example, by managed burning, but the relative effects of burning, slashing and grazing are unclear.

35.6.3 Adult predator control

Repeated observations of predation on adult moths by birds (notably willy wagtails, *Rhiphidura leucophrys*) and asilid flies have implied that this may be a significant contributor to mortality, but one difficult to control. Both these predators seek prey visually from perches, and fly to capture them. Suggestions have been made that using wire strand fencing around grasslands can reduce perching availability for predatory wagtails and other birds and may help reduce their impacts. Control of domestic cats on small reserves has also been recommended (Regan *et al.*, 2021).

35.6.4 Translocation

Translocation, the removal of individuals to a new site, is an extreme conservation measure used to 'rescue' insects from areas that are to be destroyed. A pioneering attempt in the ACT involved removal of soil slabs containing *S. plana* larvae from the road alignment of the proposed Majura Parkway and inserting them in a more secure area devoid of moths (Allen, 2014). Containment nets over the translocated slabs confirmed the success of this operation as large numbers of moths emerged in the following season. This technique has also been used in Victoria, notably as 'rescue' from a major water pipeline construction (Sugarloaf Pipeline Project), with slab removal completed in 2011 with monitoring for moths for two flight seasons following construction and three additional years to 2013–14 (Melbourne Water, 2022).

The broader approach pioneered in the ACT was later used in conjunction with direct relocation of larvae dug from the sacrificial sites and inserted into soil adjacent to food plants in the receptor sites (Rowell, 2019). Pupal cases were found subsequently in plots subject to each treatment.

35.6.5 Burning

Control burning is recognized widely in grassland management, but impacts on the sun moth require further research. Trust for Nature (2022) suggested that the 'best' time for burning is in cooler conditions in March and April, and that burns during the adult flight and oviposition period (October to January) should be avoided. As in other contexts, small-scale managed 'micromosaic burns' of sensitive sites are likely to sustain heterogeneity more than large-scale fires, with low-intensity fires less harmful to subterranean growth stages of the moth.

Each of these themes has augmented knowledge of the moth and how it may be managed based on increasingly sound ecological understanding and maintaining relatively low biomass. Grazing is sometimes preferred over burning for grassland maintenance because it is more flexible and can be done at higher frequencies; for the large Western Grasslands Reserve near Melbourne, high-intensity annual pulse grazing of no longer than 4 weeks, with perch control to reduce bird predation, was considered the 'best' management treatment (Regan *et al.*, 2021).

35.6.6 Citizen interest

Many volunteers have participated (as 'citizen scientists') in moth surveys or discovered additional moth populations, and helped also in exercises such as translocations, for all of which the level of 'official' support is generally inadequate. The unusual biology of *S. plana* has helped to garner and sustain public interest.

35.7 Continuing Ambiguity and Future Needs

Recent discoveries of further *S. plana* populations are reflected in reduction of its formal conservation status to 'Vulnerable' with correspondingly less stringent conservation obligations. Pressures for grassland development, however, do not lessen and the moth is still a potent conservation ambassador amidst continuing controversy, and frequent societal calls to preserve grassland remnants in urban and periurban areas (New, 2019). In view of its now-documented wide distribution and increased number of occupied sites, debate will continue over the real need for the moth's conservation. The complex behaviour and biology of *S. plana*, that it is diurnal (and hence visible), and that it occurs in grasslands close to urban developments have all helped to foster community interest and appreciation of grassland values, and aided in public education. Foci such as the sun moth children's playground, at Craigieburn, Victoria, accompanied by informative signage on the moth's biology, are both novel and important in periurban areas (New, 2015). Many sites historically or currently occupied by *S. plana* belong to a range of Traditional Owner groups (many of them listed by DAWE, 2021), whose participation in management should be solicited. The moth is known as Yurrnga Ganbina in the Bangerang language (DCCEEW, 2025).

The interest in *S. plana* generated over the last quarter century has been an important development in Australia's insect conservation, and provided a framework for wider conservation of grassland insects, as well as for other threatened species of Castniidae occupying similar open habitats, and for which conservation needs appear to be similar (Chapter 15). A joint Action Statement for five threatened Victorian species (other than *S. plana*) explored their common needs and framed three objectives as (i) achieve long-term protection and sympathetic management of key sun moth sites; (ii) improve current understanding of distribution, abundance, biology and habitat requirements of sun moths; and, (iii) increase public awareness of sun moths and other grassland-dependent invertebrates (Douglas, 2003). Wider interest is exemplified through the graceful sun moth, *Synemon gratiosa*, endemic to a small area of south-west Western Australia where its conservation has received considerable attention (Bishop *et al.*, 2010; WADEC, 2011) and its biology studied in some detail (Williams *et al.*, 2021). *S. gratiosa* has also become an important flagship for the vulnerable coastal heathland and *Banksia* woodland habitats it occupies. Mark–release–recapture studies of *S. gratiosa* (Williams *et al.*, 2018), as attempted for *S. plana* in the south-east (but using more conspicuous marks), implied that this species also is very short-lived. The maximum interval from mark to recapture of males was 5 days.

References

ACT (Australian Capital Territory) (2017) *Golden Sun Moth Synemon plana Action Plan*. ACT Government, Canberra, ACT.

Allen, W. (2014) Golden sun moth habitat translocation – an Australian first. *News of Friends of Grasslands* 2014, 9–10.

Bishop, C., Williams, M., Mitchell, D. and Gamblin, T. (2010) *Survey Guidelines for the Graceful Sun moth (Synemon gratiosa) and Site Habitat Assessments*. Department of Environment and Conservation, Kensington, Western Australia.

Braby, M.F. and Dunford, M. (2006) Field observations on the ecology of the golden sun moth *Synemon plana* Walker (Lepidoptera: Castniidae). *Austral Entomology* 33, 103–110.

Brown, G., Tolsma, A. and McNabb, E. (2012) Ecological aspects of new populations of the threatened Golden Sun moth *Synemon plana* on the Victorian Volcanic Plains. *Victorian Naturalist* 129, 77–85.

Clarke, G.M. and O'Dwyer, C. (2000) Genetic variability and population structure of the endangered golden sun moth (*Synemon plana*). *Biological Conservation* 92, 371–381.

Clarke, G.M. and Whyte, L.S. (2003) Phylogeography and population history of the endangered golden sun moth (*Synemon plana*) revealed by allozymes and mitochondrial DNA analysis. *Conservation Genetics* 4, 719–734.

DAWE (Department of Agriculture, Water and the Environment) (2021) *Conservation advice for Synemon plana (Golden Sun Moth)*. DAWE, Canberra, ACT.

DCCEEW (Department of Climate Change, Energy, the Environment and Water) (2025) *Draft National Recovery Plan for the golden sun moth* Synemon plana. DCCEEW, Canberra, ACT.

DEPI (Department of Environment and Primary Industries) (2013) *Sub-regional Species Strategy for the Golden Sun Moth*. DEPI, Melbourne, Victoria.

DEWHA (Department of the Environment, Water, Heritage and the Arts) (2008) *Approved Conservation Advice for the Natural Temperate Grassland of the Victorian Volcanic Plain*. DEWHA, Canberra, ACT.

DEWHA (Department of the Environment, Water, Heritage and the Arts) (2009) *Significant Impact Guidelines for the Critically Endangered Golden Sun Moth (Synemon plana)*. EPBC Act policy statement 3.12. DEWHA, Canberra, ACT.

Douglas, F. (2003) *Five Threatened Victorian Sun moths (Synemon species)*. Flora and Fauna Guarantee Act 1988, Action Statement No 146. Department of Sustainability and Environment, Melbourne, Victoria.

Douglas, F. (2004) A dedicated reserve for conservation of two species of *Synemon* (Lepidoptera: Castniidae) in Australia. *Journal of Insect Conservation* 8, 221–228.

DSE (Department of Sustainability and Environment) (2010) *Offset Requirements for Native Vegetation and Golden Sun Moth*. DSE, Melbourne, Victoria.

DSE (Department of Sustainability and Environment) (2011) *Sub-regional Species Strategy for the Golden Sun Moth – draft for public comment*. DSE, Melbourne, Victoria.

Gibson, L. and New, T.R. (2007) Problems in studying populations of the golden sun moth, *Synemon plana* (Lepidoptera: Castniidae) in south-eastern Australia. *Journal of Insect Conservation* 11, 309–313.

Grund, R., Stolarski, A.S. and Stevens, M. (2019) Rediscovery of *Synemon selene* (Lepidoptera: Castniidae) in the mid-north of South Australia. *Transactions of the Royal Society of South Australia* 143, 86–91.

Kirkpatrick, J.B., McDougall, K. and Hyde, M. (1995) *Australia's Most Threatened Ecosystems, the South-eastern Lowland Native Grasslands*. Surrey Beatty and Sons/World Wide Fund for Nature, Sydney, NSW.

Kutt, A.S., McKenzie, V.J., Wills, T.J., Retallick, R.W.R., Dalton, K. *et al.* (2015) Spatial and temporal determinants of golden sun moth *Synemon plana* distribution. *Austral Ecology* 40, 100–107.

Mata, L., Gerrard, G.E., Kutt, A.S., Wintle, B.C., En Chee, Y. *et al.* (2017) Eliciting and integrating expert knowledge to assess the viability of the critically endangered golden sun moth *Synemon plana*. *Austral Ecology* 42, 297–308.

Melbourne Water (2022) *Annual Report on Compliance with Conditions: Sugarloaf Pipeline Project, Goulburn River to Sugarloaf Reservoir, Victoria*. EPBC 2008/3960. Melbourne Water, Melbourne, Victoria.

New, T.R. (2012) The Golden sun moth, *Synemon plana* Walker (Castniidae): Continuing conservation ambiguity in Victoria. *Victorian Naturalist* 129, 109–113.

New, T.R. (2015) *Insect Conservation and Urban Environments*. Springer, Cham, Switzerland.

New, T.R. (2019) *Insect Conservation and Australia's Grasslands*. Springer, Cham, Switzerland.

O'Dwyer, C. and Attiwill, P. M. (1999) A comparative study of habitats of the Golden Sun Moth *Synemon plana* Walker (Lepidoptera: Castniidae): Implications for conservation. *Biological Conservation* 89, 131–141.

O'Dwyer, C. and Attiwill, P.M. (2000) Restoration of a native grassland as habitat for the Golden Sun Moth *Synemon plana* Walker (Lepidoptera: Castniidae) at Mount Piper, Australia. *Restoration Ecology* 8, 170–174.

Regan, T., Bruce, M., Batpurev, K., Farmilo, B., Scroggie, M. *et al.* (2021) *Melbourne Strategic Assessment. Population Viability Analysis Models for Threatened Species: Version 1.0*. Technical Report No 327. Arthur Rylah Institute for Environmental Research, Melbourne, Victoria.

Richter, A., Osborne, W., Hnatiuk, S. and Rowell, A. (2013a) Moths in fragments: Insights into the biology and ecology of the Australian endangered golden sun moth *Synemon plana* (Lepidoptera: Castniidae). *Journal of Insect Conservation* 17, 1093–1104.

Richter, A., Weinhold, D., Robertson, G., Young, M., Edwards, T. *et al.* (2013b) More than an empty case: A non-invasive technique for monitoring the Australian critically endangered golden sun moth, *Synemon plana* (Lepidoptera: Castniidae). *Journal of Insect Conservation* 17, 529–536.

Rowell, A. (2019) *Kinlyside Golden Sun Moth Translocation Monitoring 2018*. Prepared for the Environment and Planning Directorate, ACT Government, Canberra, ACT.

Trust for Nature (2022) *Golden Sun Moth (Synemon plana) Conservation Assessment Project*. Trust for Nature, Melbourne, Victoria.

WADEC (Western Australia Department of Environment and Conservation) (2011) *Conservation Advice for Synemon gratiosa (Graceful Sun moth) – January 2011*. WADEC, Perth, Western Australia.

Williams, A.A.E., Coppen, R.A.M. and Williams, M.R. (2018) A method for conducting mark-release-recapture studies of the graceful sun moth *Synemon gratiosa* (Lepidoptera: Castniidae). *Records of the Western Australian Museum* 33, 181–186.

Williams, A.A.E., Williams, M.R., Bishop, C.L., Gamblin, T., Fissioli, J. and Coppen, R.A.M. (2021) Notes on the biology, ecology, life history and conservation of the graceful sun moth, *Synemon gratiosa* Westwood, 1877 (Lepidoptera: Castniidae). *Austral Entomology* 48, 17–32.

36 The Southern Pink Underwing Moth: *Phyllodes imperialis smithersi* – A Flagship Species

Michael J. Andren[1]* and Donald P. A. Sands[2]
[1]*Biodiversity Conservation and Science, NSW Department of Climate Change, Energy, the Environment and Water, Coffs Harbour, NSW, Australia;* [2]*CSIRO Health and Biosecurity, Brisbane, Queensland, Australia, and School of the Environment, University of Queensland, St Lucia, Queensland, Australia*

Abstract

The pink underwing moth *Phyllodes imperialis* H. Druce (Erebidae, Calpinae) occurs in Indonesia, Australia and several tropical Pacific islands. The subspecies *P. i. meyricki* Olliff ranges from north-eastern Queensland to New Guinea, and the nationally endangered southern subspecies, *P. i. smithersi* Sands, is found in south-east Queensland and north-east New South Wales. Larvae of *P. i. smithersi* feed solely on a rainforest vine, *Carronia multisepalea*, that occurs from Gympie to Bellingen. Adult moths feed on soft, damaged or rotting fruit, withdrawing sugary fluids with their proboscis. *Phyllodes i. smithersi* is a 'fruit-sucking' moth, not a 'fruit-piercing' moth and lacks the thickened spines on the dorsal edge of the proboscis characteristic of *Eudocima* spp. It cannot pierce intact fruit and is not regarded as a pest of horticulture. Adults and larvae of *P. i. smithersi* avoid predation by camouflage and dramatic warning displays when disturbed. Intense droughts and fires are increasing threats and in 2019, severe drought caused widespread desiccation of the food plant and wildfires resulted in unprecedented burning of some rainforest habitat. *Phyllodes i. smithersi* was prioritized by the federal government as one of 11 Australian invertebrates requiring urgent conservation attention and the moth is recognized as a flagship species for insect conservation in subtropical eastern Australia.

36.1 Introduction

Twelve species of *Phyllodes* Boisduval, 1832, were referenced by Olliff (1889), 11 by Hampson (1913), 13 by Seitz (1923) and 11 by Poole (1989), from Borneo, Philippines, Indonesia, New Guinea and Australia. *Phyllodes imperialis* Druce and other species of the tribe Ophiderini were separated from Noctuidae for inclusion in the family Erebidae by Zahiri *et al.* (2012). Sands (2012a) reviewed the identity of the subspecies of *P. imperialis* from Indonesia, New Guinea, eastern Australia and several tropical Pacific islands including New Caledonia (Holloway 1979). In that review, Sands (2012a) described *P. i. smithersi* Sands from subtropical eastern Queensland and New South Wales, and noted *P. i. meyricki* Olliff occurred in northern Queensland (Hunter, 1939; Common, 1990) from near Paluma to Bamaga, as well as Papua New Guinea (Sands, 2012a).

The life history of *P. i. smithersi* was progressively outlined by Sands and House (1990), Clark and Spier-Ashcroft (2003), Sands (2012b) and Andren *et al.* (2021). Larvae feed on a subtropical rainforest vine *Carronia multisepalea*

*Corresponding author: mick.andren@environment.nsw.gov.au

DOI: 10.1079/9781800627680.0036

(Menispermaceae) that occurs from Gympie (Queensland) to Bellingen (New South Wales). Large larvae avoid predation through camouflage, with their shape and colouration resembling a dead leaf, and they have the ability to rear up to display large colourful eye spots and white bands that appear to mimic the teeth of vertebrate predators. Adult moths have similar strategies. They are camouflaged, with their forewing shape and colour mimicking dead leaves and, in males, a central spot resembling a leaf miner scar. They also display dramatic warning colours when disturbed, with the hindwings of both sexes possessing a central pink patch that can be flashed as a warning.

Sands (1999) considered *P. i. smithersi* to be threatened by loss of its rainforest habitat and Clark and Spier-Ashcroft (2003) thought it was potentially critically endangered. Although the subspecies was then known only by its common name, 'southern pink underwing moth', it was listed nationally as endangered in 2002 (Threatened Species Scientific Committee, 2002) and in New South Wales in 2003 (NSW Scientific Committee, 2003). These assessments were based on the species only being known from very few locations, such as the five confirmed locations used in the national determination (Threatened Species Scientific Committee, 2002). Later surveys, particularly in New South Wales (Andren *et al.*, 2021; Lloyd *et al.*, 2024), have shown the species to be more widespread and abundant than originally thought.

A conservation 'flagship' is a species with a high public profile that can act as an ambassador to gain support for conservation actions that will also assist other species (New, 2009; Taylor *et al.*, 2018). Often, they are spectacular species that have captured the public's imagination, but they may also be threatened species that have generated a groundswell of support for their recovery. They can be species of cultural or scientific importance (Taylor *et al.*, 2018). Sands (1999) and Clark and Spier-Ashcroft (2003) suggested that *P. i. smithersi* would make an excellent flagship species based on its large size and remarkable appearance as both adult and larva, as well as the importance of preserving the habitat it occupies. In recent years, despite being rarely seen by the public, the moth has become a flagship species for insect conservation.

36.2 Taxonomic Identity

The genus *Phyllodes*, previously referred to the subfamily Catocalinae, family Noctuidae (Kitching, 1984; Edwards, 1996), was placed in the family Erebidae following rearrangement of Noctuoidea by Zahiri *et al.* (2012). *Phyllodes imperialis* was observed breeding in rainforest areas near Maleny in south-east Queensland in the 1980s (Sands and House, 1990). Sands (2012a) later described the southern pink underwing moth *P. i. smithersi* after specimens from south-east Queensland and New South Wales were found to differ from specimens of *P. i. meyricki* from north Queensland. The wingspan and markings of adult moth specimens were different and the larvae of *P. i. smithersi* found at Maleny were feeding on a rainforest vine, *C. multisepalea* (Sands and House, 1990; Moss, 2002). This vine is restricted to subtropical rainforest and does not occur in north Queensland where the food plant of *P. i. meyricki* is another vine, *Pycnarrhena novoguineensis* (Menispermaceae) (Hunter 1939).

Adult *P. i. smithersi* (Fig. 36.1) are large moths with wingspan measures of 138 mm (Sands, 2012a), although 160–170 mm has also been cited. The forewings above are leaf-like with the apex tapered, mimicking a leaf petiole. The wings above are grey or grey-brown with a white spot on the costa, darker brown mottling and a postmedian dark brown leaf-like 'mid-rib'. In males, the leaf-like forewings are distinctive (for *P. imperialis* sspp.) with the very pale reinform mark resembling a leaf miner scar (Sands, 2012a); the mark is much darker in females. The hindwings are black

Fig. 36.1. *Phyllodes imperialis smithersi* adult male, upperside. Photo: D.P.A. Sands.

with outer margins chequered white at the vein ends and with a large central suboval pink patch. The dorsal edge of the proboscis of *P. i. smithersi* is finely spined, except for some broadened spines on the ventral edge near the apex. The moth appears to be incapable of penetrating firm fruit in the same way as related calpine species.

36.3 Distribution

Phyllodes i. smithersi occurs in rainforest patches from south of Gympie in south-east Queensland to Bellingen in north-east New South Wales (Fig. 36.2). In Queensland, breeding sites include Mary River, Kin Kin Creek, Conondale Range, Witta, Maleny, Blackall Range, Bellthorpe, D'Aguilar Range, Springbrook and Lamington National Park (NP). The species is known in New South Wales near the Queensland border at Tweed Heads, Numinbah Nature Reserve (NR) and Limpinwood NR, the rainforests of the Tweed caldera, Nightcap Range, remnants of the 'Big Scrub' east of Lismore and an isolated area further south that includes the Bellinger Valley and headwaters of the Orara River.

36.4 Life History and Ecology

36.4.1 Habitat and larval food plant

The habitat required by the immature stages of *P. i. smithersi* is delineated by the areas occupied by its food plant *Carronia multisepalea*. This dioecious vine occurs in wet, complex notophyll vine forests (Webb, 1978) up to 1100 m. It occurs from near Gympie to Bellingen, corresponding to the distribution of *P. i. smithersi*. *C. multisepalea* grows in rainforest on rich volcanic or alluvial soils in the mountainous hinterland and lowland riparian zones. Pollinators of the vine are not known. In mature rainforest, *C. multisepalea* grows as a large vine with most foliage in the rainforest canopy, although it is also a pioneer species that exploits forest edges and canopy gaps. The northern rivers of New South Wales provide a stronghold for *C. multisepalea*, and it is relatively common throughout rainforests of the region.

Andren *et al.* (2021) found that breeding habitat for the moth is characterized by vines with softer leaves (the current season growth) below a height of 2 m in shaded sites (with canopy cover greater than 60%). Foliage near ground level derives from four structural forms of *C. multisepalea*: seedlings, suckers from subterranean rhizomes, prolific low tangled vine found in canopy gaps and edges, and new growth at the base of mature stems (Andren *et al.*, 2021). The larvae of several other moths (mostly Calpinae), including *Eudocima phalonia* (Linnaeus), also feed on *C. multisepalea* (see Chapter 30).

36.4.2 Immature stages

The fecundity of *P. i. smithersi* has not been determined, but dissected females have been observed to carry up to 400 egg oocytes. Immature stages can generally be found from October to April with their abundance predominating in November and February–March (Lloyd *et al.*, 2024). Immature development rates depend on temperature and concentration of nitrogen in the food plant, except for pupal development times when in diapause.

Eggs (Table 36.1) are deposited singly or in small groups, less commonly more than 10, although 35 have been recorded on one leaf (Andren *et al.*, 2021). In 2023, Lloyd *et al.* (2024) recorded 1102 eggs over 4 days of survey in February and March in a relatively small area at Coolgardie (NSW). Eggs are laid on the underside of the subterminal leaves of *C. multisepalea*, on low (usually below 1.5 m) heavily shaded vines in rainforest. Eggs eclose in 8–10 days after deposition. There are five larval instars of *P. i. smithersi* (Table 36.1) or occasionally six when the food plant is low in nutrients. First instar larvae are green and feed on the new soft subterminal leaves. Second and third instars are greenish brown to dark grey with one, then two pairs of yellow and black eye spots (Fig. 36.3). Fourth and fifth instars are brown or black with dark eye spots, and may be highly ornamented (Fig. 36.4). Mature larvae measure up to about 11 cm when extended. Large larvae consume firm leaves as well as the young stems of vines and, as with many species of Calpinae, larvae of *P. i. smithersi* are probably distasteful to birds,

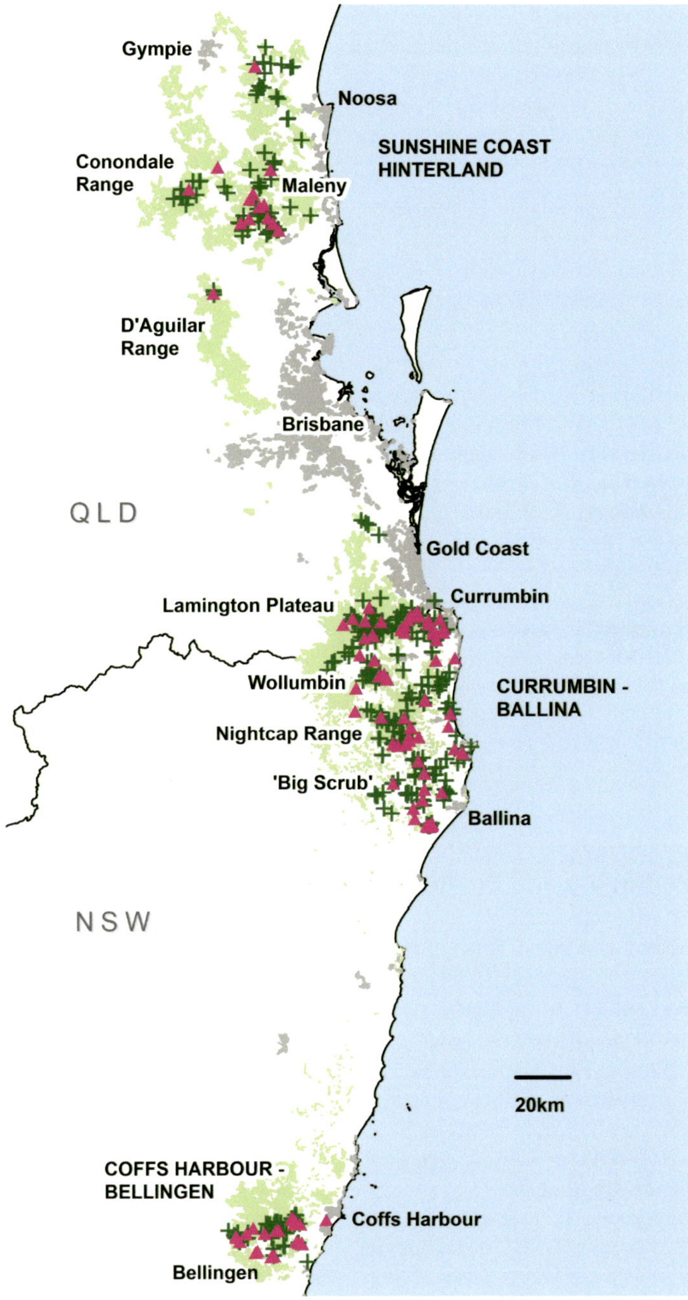

Fig. 36.2. *Phyllodes imperialis smithersi* current distribution (pink triangles) and that of host plant, *Carronia multisepalea* (green crosses), overlaid on the occurrence of rainforest within approximately 50 km of the coast (light green) and illustrating the three distinct regions occupied: the Sunshine Coast hinterland, the Currumbin - Ballina region and the Coffs Harbour - Bellingen region. *C. multisepalea* and *P. i. smithersi* data from the Atlas of Living Australia at https://doi.org/10.26197/ala.3a18c95c-6937-4bd0-a334-6d298578b56e and accessed on November 6 2024. Three incidental sites were added for completeness (Conondale Range, Kin Kin, nr Gympie and the D'Aguilar Range).

Table 36.1. Egg and larval instar measurements for *Phyllodes imperialis smithersi* (data from Don Sands). Note that head capsule widths of larval instars vary with concentrations of food plant nitrogen (amino acids).

	Head capsule width (mm, n = 5 for each instar)				
Egg (diam. mm, n = 5)	First instar	Second instar	Third instar	Fourth instar	Fifth instar
1.82–1.90	1.16–1.18	1.60–1.65	2.25–2.50	3.30–3.70	4.30–5.00

Fig. 36.3. *Phyllodes imperialis smithersi* second instar larva on the underside of a *Carronia multisepalea* leaf. Photo: John Turbill.

Fig. 36.4. *Phyllodes imperialis smithersi* warning display of a fifth instar larva. Photo: Peter Richards.

due to storage of toxic materials from the menisperm food plant (Blanchfield *et al.*, 1993).

Single larvae and groups of two to three occur commonly. Occasionally up to about 15 larvae have been observed (N. Clancy, pers. commun.). Lloyd *et al.* (2024) found almost 1000 larvae at Coolgardie (NSW) in 2023. When resting against a young stem, early instar larvae are camouflaged. Later instars are also well camouflaged, resembling dead leaves. Instars four and five, when disturbed, will rear up with a spectacular defensive display (Fig. 36.4), revealing large round patches resembling vertebrate eyes and white bands resembling teeth (Zborowski and Edwards, 2007).

Mature larvae leave the vine to find a site for pupation, sometimes dispersing some distance from the food plant before making a cocoon between the curled green leaves of low, heavily shaded shrubs. Pupae are approximately 44 mm in length and dark brown with reflective gold abdominal segments (Fig. 36.5). Pupal development occurs in 21–31 days in warmer months and 180–220 days when in diapause (D.P.A. Sands, unpubl.). Pupae generally overwinter in diapause from February to May until adults emerge in October to November (but see section 36.6 below). The very few pupae collected from the wild have completed development intact. In warmer months (October–March), egg to adult development, with non-diapausing larvae, is 69–88 days (D. Sands, unpubl.). For example, for five individuals raised at 25°C, average development times were: egg 8 days, larva 18 days, pupa 25 days and adult 30 days (Andren *et al.*, 2021 from data supplied by D.P.A. Sands).

36.4.3 Adults

Large adult *P. i. smithersi* can fly powerfully and are capable of foraging and dispersing well beyond

Fig. 36.5. *Phyllodes imperialis smithersi* pupa exposed within its coccoon. Photo: David Britton.

rainforest boundaries, including across rural and urban areas. An adult moth observed in suburban Coffs Harbour (NSW) was at least 5 km from rainforest habitat and the species is present in small, isolated remnants within rural landscapes up to 15 km from the nearest alternative habitat. Considering the large size of *P. i. smithersi* and reported movements of other moth species, much greater dispersal distances are possible. Adults rarely come to light but are very occasionally seen resting on foliage by day in rainforest. When disturbed, adults will arch their forewings to expose the pink hindwing areas as a warning pose.

Adults live for about 30 days and feed at night on fluids from soft, overripe, damaged, scarred (e.g. by fruit fly) or decomposing rainforest fruit from a variety of species including *Ficus watkinsiana* and *Syzygium* spp. They sometimes soften the firm pulp of injured fruit to feed on juice by inserting the apex of the proboscis into a cavity and then rotating it. They feed on fruit while attached to the host plant, not after it has fallen.

36.4.4 Natural enemies

Eggs are sometimes parasitized by a small wasp (*Telenomus* sp., Scelionidae) (Sands, 2012a). First instar larvae are heavily predated by ants, including *Tetramorium* spp., and the larger larvae are sometimes taken by bull-ants (*Myrmecia* spp.). Larvae appear to be susceptible to exposure to bright sunlight, possibly indicating an adaptation to heavily shaded habitats. Natural enemies of adults have not been observed but as large and mobile moths, they may be opportunistically targeted by microchiropteran bats.

36.5 Habitat Loss and Degradation

Rainforest in subtropical Australia has been dramatically reduced in area since European settlement (Threatened Species Scientific Committee, 2011). The impact has been particularly severe for lowland rainforest (below about 300 m elevation), most of which has been cleared. A remarkable example is the 'Big Scrub' lowland rainforest of the Lismore basalt region, once the most extensive lowland rainforest in south-eastern Australia, where Floyd (1990) estimated that the original 75,000 ha was reduced to only 300 ha (0.4%). Much of the lowland rainforest has also been lost from the developed valleys of the Mary, Tweed, Wilson, Bellinger and upper Orara rivers, as well as many of the smaller coastal catchments. The remaining lowland rainforest typically occurs as small remnants of variable quality that are often isolated within highly modified agricultural landscapes. Historically, therefore, *P. i. smithersi* is highly likely to have lost a large proportion of its original habitat.

This historical clearing has resulted in extreme fragmentation of lowland rainforest on the fertile soils that characterize habitat of *P. i. smithersi* and its food plant. Small and isolated remnants are particularly vulnerable to degradation, particularly by weeds and, in the more densely settled areas, direct human disturbance.

Lowland rainforest is listed under federal legislation as a Critically Endangered Ecological Community (Threatened Species Scientific Committee, 2011) across the range of *P. i. smithersi*. It is also listed in New South Wales as a Threatened Ecological Community (NSW Scientific Committee, 2006) and consequently it is becoming increasingly difficult to clear in that State. Although incremental loss is continuing, there are also widespread rehabilitation programmes that are trying to reverse the long-term declining trend. Unlike many insects of fragmented lowland rainforest habitat, *P. i. smithersi* appears to be capable of rapid dispersal across developed landscapes. If remnants can be rehabilitated to provide suitable habitat (including the food plant), the moth should be capable of exploiting these new opportunities.

Above about 300 m, rainforests in the region are more intact and there are extensive tracts remaining in the Conondale–Maleny region, Border Ranges, Wollumbin–Nightcap and Dorrigo NP. Significant large reserves include Conondale NP, Lamington NP, Springbrook NP, Numinbah NR, Limpinwood NR, Wollumbin NP, Mt Jerusalem NP, Nightcap NP and Dorrigo NP. Habitat in these reserves is in relatively good condition and more resilient to disturbance than smaller remnants. They are the core areas needed to sustain the regional population of *P. i. smithersi*.

36.6 Climate Change

The climatic extremes of drought and bushfire weather in south-eastern Australia are projected to increase, with a '1-in-20-year event' likely to at least double in frequency by the late 21st century (Herold *et al.*, 2021). Until relatively recently, fire was not considered a threat to *P. i. smithersi*, due to its wet rainforest habitat, and this was not mentioned in the early assessments by the Threatened Species Scientific Committee (2002), Clark and Spier-Ashcroft (2003) or NSW Scientific Committee (2003). The intense 2019 drought, however, demonstrated that the moth's rainforest habitat can become flammable under extreme conditions. Surveys during the height of that drought failed to locate any larvae and most leaves of *C. multisepalea* were observed to be wholly or partly desiccated. Many of the small understorey plants (i.e. the primary habitat of *P. i. smithersi*) appeared to be dead (M. Andren, pers. obs.). Effects of prolonged drought have been observed in Queensland, where leaves of the food plant become too tough for eclosing first instar larvae and, sometimes, for second instar larvae (D.P.A. Sands, pers. obs.). Although most *P. i. smithersi* habitat escaped the fires that followed the drought in late 2019, some of it burnt – in Binna Burra (Qld) and Mt Nardi (NSW). A repeat of the conditions experienced in 2019 appears inevitable and is a significant threat to *P. i. smithersi*.

Extreme rainfall events are also projected to increase in frequency (Herold *et al.*, 2021). Much of the floodplain and riparian lowland rainforest habitat occupied by *P. i. smithersi* may be inundated by floods. The severity of the impact on *P. i. smithersi* depends on the duration of inundation and the timing of the flood relative to the life cycle stage of the moth. Although there are no data on flood impacts, an example is the inundation of Boatharbour Nature Reserve (NSW), which was flooded for an extended period during the record Lismore flood of 2022. Subsequent larval surveys have not yet relocated the species in this remnant (G. Beyer, pers. commun.). The projected increase in flooding frequency is likely to be detrimental to the species, particularly in highly fragmented lowland rainforest habitat.

In contrast, there are indications that climatic conditions in 2020–24 have had a positive effect on *C. multisepalea* and *P. i. smithersi*. Following four seasons of generally favourable growing conditions and historically high temperatures, during early winter 2024 there was still new growth on *C. multisepalea* in the Big Scrub remnants in the Northern Rivers region of New South Wales, as well as Bellingen Island, the southern limit of the species. Larvae of all instars were found to be active in four coastal remnants of the Big Scrub up until 29 June (I. Stych, G. Beyer, pers. commun., M. Andren, pers. obs.). Breeding into winter is a newly observed behaviour that demonstrates an inherent capacity of the moth to exploit changing environmental conditions.

36.7 Conservation Status

The conservation status of *P. i. smithersi* has been consistently assessed as endangered (Threatened Species Scientific Committee, 2002; Clark and Spier-Ashcroft, 2003; NSW Scientific Committee, 2003). The most recent assessment, by the federal government, found *P. i. smithersi* to be in the top 100 priority species for recovery actions in Australia (Australian Government Threatened Species Action Plan, 2021–26). This assessment was based on a multicriteria decision analysis where species were scored against six principles: risk of extinction, multiple benefits, feasibility and effectiveness, importance to people, uniqueness and representativeness. We agree that *P. i. smithersi* would score highly on these criteria, with the possible exception of 'risk of extinction'.

Records from the Atlas of Living Australia indicate that there are now nearly 400 sites known for *P. i. smithersi* (including 28 reserves) and over 600 localities for *C. multisepalea*. The increased reporting of sightings is a result of the effective dissemination of knowledge of the species among local communities, particularly those involved in environmental management and rainforest restoration. Large areas of habitat, however, particularly in remote areas and on private property, are still poorly surveyed, so many more sites remain to be found. The stronghold of *P. i. smithersi* is the 100 km of coast and hinterland stretching from Currumbin (Qld) to Ballina (NSW), tracking the concentration of coastal subtropical rainforest in that region (Fig. 36.2). Within this stronghold, the species has been recorded in 17 national parks and nature reserves, six local council reserves and numerous private properties.

Comprehensive surveys for the immature stages of *P. i. smithersi* were conducted from 2016 to 2021 (Andren *et al.*, 2021). These larval surveys covered the geographic and environmental range of *C. multisepalea*. Larvae were found 5–790 m above sea level in extensive forests as well as small and isolated remnants. Although the population strucure of the moth has not been determined (most observations have been of eggs and larvae), the geographic and environmental distribution of larvae approximately matches that of *C. multisepalea*. This indicates that within the Currumbin to Ballina stronghold, where *C. multisepalea* is relatively common, *P. i. smithersi* is widespread and likely to be a relatively secure species. This explains our assessment above that 'risk of extinction' is not high for this species relative to many other endangered species.

The abundance of *P. i. smithersi* appears to be lower in the outlying northern (Sunshine Coast hinterland) and southern (Coffs Harbour to Bellingen) parts of its range. In the Sunshine Coast hinterland, this includes the Conondale and Blackall Ranges (Qld), where the lower abundance is likely to be due to the infrequency of soft leaves on the food plant on which the early instar larvae depend for their survival. Surveys in the Blackall Ranges between 2005 and 2009 identified 36 sites with the food plant *C. multisepalea* (N. Clancy, pers commun.). Larvae or eggs of *P. i. smithersi* were observed on 12 of these: five Municipal Reserves, four Nature Refuges and three Council Covenant areas (that is, privately owned land supporting 'old growth' rainforest). Five breeding sites were located between 2008 and 2009 in the Conondale Range where the food plant is uncommon (R. Thomas, pers. commun.). Until 2009 many of these areas were used for timber harvesting but the breeding sites for *P. i. smithersi* are now in reserves (R. Thomas, R. Patterson, pers. commun.). In the isolated southern (Coffs Harbour to Bellingen, NSW) part of its range, *P. i. smithersi* abundance is also reduced because suitable rainforest habitat and the food plant are both much less common than in the northern rivers. Here the species has been recorded predominantly on public land (two State reserves, Dorrigo NP and Bongil Bongil NP, two Council reserves and five State Forests, where rainforest is not permitted to be logged), which should benefit the conservation outcomes for *P. i. smithersi* in the area.

36.8 A Flagship Species

The striking warning display of late-stage *Phyllodes* larvae was reported more than eight decades ago by Hunter (1939). Following the national listing of *P. i. smithersi* in 2002, there was a resurgence of interest in this charismatic insect, and it became increasingly well known and publicized widely across media and the internet (e.g. Prigg, 2012; Jelinek, 2021; Runkovski, 2021; Georgiou, 2022; Beyer and Stych, 2023). A video about the species was produced by Natura Pacific (2021). Sands (1999) suggested that *P. i. smithersi* would be an excellent flagship species due to its 'magnificent form, size and colour' – an assessment endorsed by Clark and Spier-Ashcroft (2003).

Apart from its charismatic appearance, *P. i. smithersi* is an appropriate flagship because the charismatic larvae are sufficiently widespread and abundant that they are regularly found across most of its range – in small and isolated remnants that are often the target of local rehabilitation as well as in large national parks. It is also an outstanding flagship because the subtropical rainforest habitat of *P. i. smithersi* is one of the most biodiverse terrestrial ecosystems in Australia, maximizing the other species that derive benefit from any conservation actions. The national listing of *P. i. smithersi* as an endangered species also attracts conservation attention and funding.

The use of flagship species as a conservation approach can potentially lead to the inefficient allocation of resources if support gained for management benefits accrues to the flagship but not to biodiversity overall (Andelman and Fagan, 2000; McGowan *et al.*, 2020). Taylor *et al.* (2018) advocated for a strategic approach that entailed distributing insect flagships evenly across bioregions so that they represented biodiversity throughout Australia. From this perspective, *P. i. smithersi* is not a strategic flagship species because the subtropical region it occupies already receives more conservation attention than most bioregions in Australia. Other flagship species occur there, even another rainforest insect flagship (the Richmond birdwing butterfly, *Ornithoptera richmondia*). This is a valid criticism but the degree to which subtropical rainforest habitat has already been lost and the continued rapid development of the region (the Sunshine Coast – Brisbane – Gold Coast – Tweed Heads urban population is approaching 3.8 million) help justify the concentrated conservation focus in the area.

Conservation programmes for *P. i. smithersi* began with work undertaken in Mary Cairncross Scenic Reserve in the 1990s and Bellingen Island from 2006 (Britton, 2006), and have since extended through many organizations and landowners across the range of the species. The NSW Government Saving our Species programme funded comprehensive surveys (Andren *et al.*, 2021) and habitat regeneration by the Jaliigirr Biodiversity Alliance in national parks of the Bellingen region. The Foundation for Australia's Most Endangered Species undertook habitat regeneration in the northern rivers in 2019. Following the 2019 drought and bushfires, the federal government targeted *P. i. smithersi* for threatened species funding and financed several rehabilitation programmes, including the Watergum (Community for Environment) project at Binna Burra and Barung Landcare on the Sunshine Coast. In 2021, Natura Pacific and the Sunshine Coast University embarked on a habitat enhancement project involving the propagation and planting of 10,000 *C. multisepalea* vines. *Phyllodes i. smithersi* was one of only nine invertebrates selected in the top 100 national priority threatened species by the federal government in 2021–22, which resulted in funding for further habitat rehabilitation programmes (e.g. the southern pink underwing moth habitat restoration project by Envite Environment in the NSW northern rivers).

The benefits of these and other projects extend well beyond a single species and *P. i. smithersi* has become a key insect flagship for the protection and restoration of subtropical rainforest in eastern Australia.

Acknowledgements

We thank Georgia Beyer, Mark Bonner, Veronica Brancatini, David Britton, Rose Broe, Nick Clancy, the late Ian Common, Trevor and Carol Deane, the late Ted Edwards, the late Alex Floyd, Paul Grimshaw, Marianne Horak, Peter Richards, Shane Ruming, Martina Schotz, Iain Stych, Maree Thompson and the NSW Saving our Species programme.

References

Andelman, S.J. and Fagan, W.F. (2000) Umbrellas and flagships: Efficient conservation surrogates or expensive mistakes? *Proceedings of the National Academy of Sciences* 97, 5954–5959.

Andren, M., Richards, P.G. and Sands, D.P.A. (2021) The distribution and ecology of the Southern Pink Underwing Moth *Phyllodes imperialis smithersi* Sands (Lepidoptera: Erebidae) in New South Wales. *Australian Entomologist* 48, 133–148.

Beyer, G. and Stych, I. (2023) *Searching for the elusive Southern Pink Underwing Moth*. Big Scrub Rainforest Conservancy, Lismore. Available at: https://bigscrubrainforest.org/news/searching-for-the-elusive-southern-pink-underwing-moth/ (accessed 16 October 2025).

Blanchfield, J.T., Kitching, W., Sands, D.P.A., Thong, Y.H., Kennard, C.H.L. and Byrel, K.A. (1993) Alkaloids from Australian *Stephania* (Menispermaceae) species. *Natural Products Letters* 3, 305–312.

Britton, D. (2006) The hunt for the pink underwing moth, *Phyllodes imperialis* and some associated events, November 2006. *Society for Insect Studies Circular* 124, 88–90.

Clark, G.M. and Spier-Ashcroft, F. (2003) *A Review of the Conservation Status of selected Australian Non-marine Invertebrates*. Environment Australia, Canberra, ACT.

Common, I.F.B. (1990) *Moths of Australia*. Melbourne University Press, Carlton, Victoria.

Edwards, E.D. (1996) Noctuidae. In: Nielsen, E.S., Edwards, E.D. and Rangsi, T.V. (eds) *Checklist of the Lepidoptera of Australia*. Monographs on Australian Lepidoptera, 6. CSIRO Publishing, Collingwood, Victoria, pp. 291–333.

Floyd, A.G. (1990) *Australian Rainforests in New South Wales*. Surrey Beatty and Sons, Chipping Norton, NSW.

Georgiou, A. (2022) Giant moth with 'baby xenomorph' skull horrifies the internet. *Newsweek (US)*, 28 October 2022. Available at: www.newsweek.com/weird-moth-caterpillar-spooky-skull-markings-1754953 (accessed 16 October 2025).

Hampson, G.F. (1913) *Catalogue of the Lepidoptera Phalaenae in the British Museum, 12. Catalogue of the Noctuidae*. British Museum (Natural History), London, pp. 392–393.

Herold, N., Downes, S.M., Gross, M.H., Ji, F., Nishant, N. *et al.* (2021) Projected changes in the frequency of climate extremes over southeast Australia. *Environmental Research Communications* 3, 011001.

Holloway, J.D. (1979) *A Survey of the Lepidoptera, Biogeography and Ecology of New Caledonia*. Junk, The Hague, The Netherlands.

Hunter, R.L. (1939) The pink underwing moth. *Phyllodes meyricki* (Olliff). Family Noctuidae, Sub-family Catocalinae. *North Queensland Naturalist* 7, 1–2.

Jelinek, K. (2021) You've never seen a bug like this before. *Daily Mail (Australia)*, 4 May. Available at: www.dailymail.co.uk/news/article-9539927/Bizarre-caterpillar-pink-underwing-moth-shocks-internet-resembling-skull.html (accessed 16 October 2025).

Kitching, I.J. (1984) An historical review of the higher classification of the Noctuidae (Lepidoptera). *Bulletin of the British Museum (Entomology)* 54, 143–234.

Lloyd, P., Popple, L.W. and Thomson, C.N. (2024) Managing and monitoring the potential impacts of a new highway on southern pink underwing moth, a threatened insect. *Australian Entomologist* 51, 9–25.

McGowan, J., Beaumont, L.J., Smith, R.J., Chauvenet, A.L.M., Harcourt, R. *et al.* (2020) Conservation prioritization can resolve the flagship species conundrum. *Nature Communications* 24, 994.

Moss, J. (2002) Imperial fruit piercing moth (*Phyllodes imperialis*). *Butterfly and Other Invertebrates Club Newsletter* 25, 4–5.

Natura Pacific (2021) Back from the Brink, Season 3 Episode 5: Southern Pink Underwing Moth. Available at: www.natura-pacific.com/video-back-from-the-brink-southern-pink-underwing-moth/ (accessed 16 October 2025).

New, T.R. (2009) *Insect Species Conservation*. Cambridge University Press, Cambridge.

NSW Scientific Committee (2003) Phyllodes imperialis*, southern subspecies (a moth) – endangered species listing – final determination*. New South Wales Scientific Committee, Sydney, NSW.

NSW Scientific Committee (2006) *Lowland Rainforest in NSW North Coast and Sydney Basin Bioregion Endangered Ecological Community Listing*. Final determination. New South Wales Scientific Committee, Sydney, NSW.

Olliff, A.S. (1889) Description of a new moth of the genus *Phyllodes*. *Proceedings of the Linnean Society of New South Wales* 4, 113–116.

Poole, R.W. (1989). Noctuidae, Part 2. In: Heppner, J.B. (ed.) *Lepidoptorum Catalogus* (New Series), Fascicle 118. E.J. Brill, Flora and Fauna Publications, Leiden, The Netherlands.

Prigg, M. (2012) Moth larvae's bony face makes it look fierce (but it's just skullduggery to deter predators). *Daily Mail (Australia)*, 29 October 2012. Available at: www.dailymail.co.uk/sciencetech/article-2224727/The-incredible-skull-caterpiller-fighting-life-Australian-rainforest.html (accessed 16 October 2025).

Runkovski, M. (2021) The *Carronia* you want to get! *Land for Wildlife South East Queensland Newsletter* 15, 10–11.

Sands, D.P.A. (1999) Conservation status of Lepidoptera: Assessment, Threatening Processes and Recovery actions. In: Ponder, W. and Lunney, D. (eds) *The Other 99%. The Conservation and Biodiversity of Invertebrates*. Transactions of the Royal Zoological Society of New South Wales, Mosman, NSW, pp. 382–387.

Sands, D.P.A. (2012a) Review of Australian *Phyllodes imperialis* Druce (Lepidoptera: Erebidae) with description of a new subspecies from subtropical Australia. *Australian Entomologist* 39, 281–292.

Sands, D.P.A. (2012b) The southern pink underwing moth. In: Curtis, L.K., Dennis, A.J., McDonald, K.R.M., Kyne, P.M. and Debus, S.J.S. (eds) *Queensland's Threatened Animals*. CSIRO Publishing, Collingwood, Victoria, pp. 38–39.

Sands, D.P.A. and House, S. (1990) Plant – insect interactions: Food webs and breeding systems. In: Webb, L.J. and Kikkawa, J. (eds) *Australian Tropical Rainforests: Science – Values – Meaning*. CSIRO, Canberra, ACT, pp. 88–97.

Seitz, A. (1923) Noctuidae. In: Seitz, A. (ed.) *The Macrolepidoptera of the World II*. Alfred Kernan Verlag, Stuttgart, Germany, p. 470.

Taylor, G.S., Braby, M.F., Moir, M.L., Harvey, M.S., Sands, D.P.A. *et al.* (2018) Strategic national approach for improving the conservation management of insects and allied invertebrates in Australia. *Austral Entomology* 57, 124–149.

Threatened Species Scientific Committee (2002) *Advice to the Minister for the Environment and Heritage from the Threatened Species Scientific Committee (TSSC) on amendments to the list of Threatened Species under the Environment Protection and Biodiversity Conservation Act (EPBC Act): Phyllodes imperialis (southern subsp. – ANIC 3333) (a moth)*. Threatened Species Scientific Committee, Canberra, ACT.

Threatened Species Scientific Committee (2011) *Commonwealth Listing Advice on Lowland Rainforest of Subtropical Australia*. Department of Sustainability, Environment, Water, Population and Communities, Canberra, ACT.

Webb, L.J. (1978) A general classification of Australian rainforests. *Australian Plants* 9, 349–363.

Zahiri, R., Holloway, J.D., Kitching, I.J., Lafontaine, J.D., Mutanen, M. and Wahlberg, N. (2012) Molecular phylogenetics of Erebidae (Lepidoptera, Noctuoidea). *Systematic Entomology* 37, 102–124.

Zborowski, P. and Edwards, E.D. (2007) *A Guide to Australian Moths*. CSIRO Publishing, Collingwood, Victoria.

37 The Bogong Moth: *Agrotis infusa*

Rose M. Lownds[1], Joshua Waters[2], Connor Marsland[1], Eleanor Drinkwater[1] and Kate D.L. Umbers[1]*

[1]School of Science, Western Sydney University, Penrith, NSW, Australia; [2]Indigenous Knowledge Systems Laboratory, Deakin University, Burwood, Victoria, Australia

Abstract

Bogong moths are one of Australia's most iconic insect species. They hold great traditional importance to the First Nations people of Australia and their aestivating masses in the Australian Alps are (or were) one of the great biological wonders of Australia. They are likely Australia's most notable moth species, and their decline represents a shrinking opportunity for Australian people to learn about and connect with their native fauna in a world under threat. Bogong moths have key roles in the transport of nutrients and energy from their winter breeding grounds to their summer aestivation sites in the mountains. Their aestivating masses also provide essential food resouces for a wide range of specialist alpine animals, including the critically endangered mountain pygmy possum and the southern bent-wing bat. Bogongs are divisive economically, serving, on the one hand, in the ecological functioning of the Australian Alps and, on the other, as crop pests in their winter range. Bogong moth populations have undergone severe recent declines and are now considered as endangered. Threatening processes include climate change, agricultural intensification and non-native vertebrate predation. The chapter focuses on their conservation, identifying knowledge gaps and discussing appropriate habitat restoration and management together with surveying and monitoring programmes.

37.1 Introduction

The Bogong moth (*Bugung* (Ngarigu/Ngarigo); *Deberra* (Taungurung); *Agrotis infusa* Boisduval 1832) is a remarkable yet unassuming moth. It is a relatively small, nocturnal, rather plain-looking moth found across Australia, and occasionally in New Zealand (Fig. 37.1). This species has an identifiable black stripe and two lighter spots visible on each forewing and wingspans of up to 50 mm.

Bogong moths are remarkable for their migration and navigation abilities, known to travel distances greater than 1000 km in some instances between their breeding grounds and aestivation sites (Warrant *et al.*, 2016; Green *et al.*, 2021). During spring each year, bogong moths escape the heat of oncoming summer and fly towards the Australian Alps from their numerous breeding grounds situated across New South Wales, Victoria and Queensland. This journey may take several days, or even weeks, to complete (Warrant *et al.*, 2016). Upon arrival in the Australian Alps, the moths seek out cool caves and crevices, where they aggregate in their thousands staying in the caves

*Corresponding author: k.umbers@westernsydney.edu.au

DOI: 10.1079/9781800627680.0037

during the day (Fig. 37.2). During this period, bogong moths will occasionally form maelstroms, a brief period of activity where the moths swarm in large numbers outside their caves. The purpose of this behaviour is currently unknown but it has been suggested that the moths use this time to seek water or to orient themselves for migratory flight (Wallace *et al.*, 2022). Between February and March, the bogong moths will leave their aestivation sites and return to their breeding grounds. The moth is able to achieve this remarkable two-way migration by navigating using both the Earth's magnetic field and by star configurations in the night sky (Dreyer *et al.*, 2025).

Fig. 37.1. Bogong moth, *Agrotis infusa*. Photo: Ajay Narendra, with permission.

Although bogong moth populations have been recorded as far west as Perth in Western Australia, it is unknown whether these populations travel across the Nullarbor Plains to reach the Australian Alps in the east. Non-migratory populations of bogong moths have also been recorded, with known populations established south of Perth, WA, and Black Mountain, ACT (Common, 1954; Heinze and Warrant, 2016) (Fig. 37.3).

During migration, bogong moths are sexually immature and reach maturity upon return to their breeding grounds in autumn (Common, 1954). Although bogong moths are capable of producing up to three overlapping generations per year, most moths are incapable of breeding

Fig. 37.2. Aestivating bogong moths. Photo: Eric Warrant, with permission.

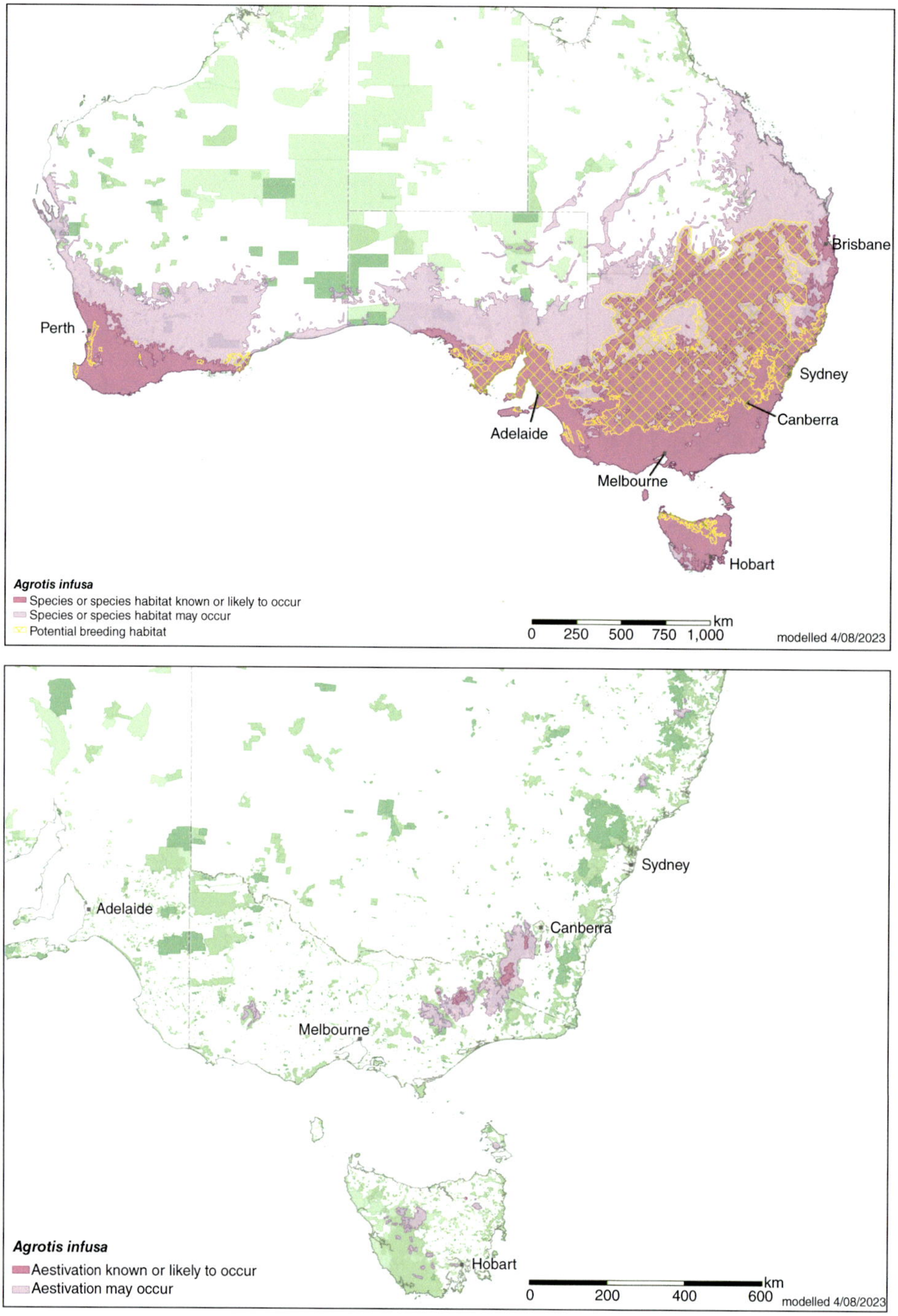

Fig. 37.3. Distribution maps for *Agrotis infusa* from draft Conservation Advice for EPBC Act modelling by ACT Govt. (A) Total distribution modelled for Conservation Advice (EOO). (B) Aestivation sites (AOO) modelled for Conservation Advice.

year round due to unfavourable conditions (Common, 1954). During breeding season, a female bogong moth can lay up to 2000 eggs in soil, or on foliage and plant stems if the soil conditions are not favourable (Common, 1981; Hill, 2007; McQuillan *et al.*, 2007).

Like other species of the *Agrotis* genus, the larvae of bogong moths are often referred to as 'cutworms' for their foraging behaviour, which involves cutting the base of low-growing plant stems to access food (Common, 1981; McQuillan *et al.*, 2007). Larvae grow through six instars, before pupating under the soil and emerging as adults.

37.2 Cultural Importance

37.2.1 Cultural importance to First Nations Australians

Bogong moths play a vital role in the lives of Indigenous Australians. The name 'Bogong' is derived from several Aboriginal languages from across the south-eastern parts of the continent (Love, 2010) and these small peanut-sized insects have served as a seasonal form of sustenance for First Nations peoples and communities in those areas for at least 1000 years but up to 2000 years, or more than 65 generations, according to newer research (Flood, 1983; Stephenson *et al.*, 2020).

During the summer aggregations of the moths, estimated to be in their millions with up to 14,000 per square metre across the various granite and basalt peaks of the Australian Alps, Aboriginal groups would gather in large numbers to collect, prepare, roast and consume the moths (Flood, 1983; Low, 1990; Pascoe, 2019). Once collected, they could be eaten immediately after roasting in hot coals or burning sand, as long as the scales, legs and wings were sufficiently removed, or they could be pounded and turned into cakes for later consumption or used in trade relations for several weeks (Waterhouse, 1971; Cherikoff and Isaacs, 1989; Cherry, 1991). Bogong moths are rich in fat, with the average fat content of adult males and females being 50–61% of their overall dry weight (Flood, 1980; Yen, 2005), and it is because of this and the high amounts of consumption (by the kilogram daily in many cases) that early settlers reported Aboriginal peoples returning from the mountains looking 'sleek and fat' (Low, 1990) as the skin of those consuming the moths 'literally shone, and their bodies showed the plumpness quite in contrast with the leanness of normal times' (MacDonald in Gale, 1927, pp. 57–58).

The main gatherings of the hundreds of Aboriginal people took place seasonally to celebrate the arrival of the moths into the Alps during early November to mid-December, and these occasions were marked with initiation ceremonies for young First Nations men and women, along with corroborees (dance rites), marriage arrangements across tribes and clan groups, and settling of intertribal disputes which often resulted in unsuccessful individuals or communities foregoing their supply of moths for the season (Flood, 1980). Aboriginal groups such as the Maneroo, Bidwell, Ngarigo, Yuin, Thawa, Diriganj, Walbanga and Ngunuwal travelled as far as 100 km to participate in the feast (Pascoe, 2019). There are oral accounts of bogong moth songlines which connect Australia's south-eastern alpine region (the Snowy Mountains) to inland tribes such as the Wiradjuri, Kamilaroi and Euahlayi, all the way up to the Kukatj, Gkuthaarn and Kurtjar peoples in the Gulf of Carpentaria (Fuller *et al.*, 2014; ABC Indigenous, 2019). The vastness of these connections across the continent perhaps emphasizes the species' overall social, cultural, spiritual and economic importance in Indigenous Australian contexts.

In recent times, changes in weather patterns, light pollution, agrochemicals and invasive pests such as feral pigs inhibit First Nations peoples' cultural and ceremonial connections to the bogong moth (Green *et al.*, 2001; Love, 2010; Colloff, 2020). However, stories, songs and dances continue to be shared and maintained across generations to allow for cultural participation and continuity (Troy and Barwick, 2020).

37.2.2 Cultural importance to mainstream Australia

Bogong moths are one of Australia's most iconic insect species. They have been seen at prominent events such as the 2000 Olympic games in Sydney (Rigby, 2011), the 2009 Australian Open finals, and occasionally at Parliament House in Canberra (McCormick, 2005). They are probably Australia's most notable moth species and their decline represents a shrinking opportunity for Australian people to learn about and connect with their native fauna in a world under threat.

37.3 Ecology

Bogong moths have key roles in the transport of nutrients and energy from their winter breeding grounds to their summer aestivation sites in the mountains. Alpine areas are often dependent on the introduction of nutrients from lower elevation areas, and elements such as nitrogen and phosphorus provided by bogong moths are essential to the ecological functioning of the Australian Alps (Green, 2011). It is estimated that approximately 1015 million moths die annually in elevations above 1500 m in the Australian Snowy Mountains, and that these moths are associated with contributing approximately 7.2 tonnes of nitrogen and 0.97 tonnes of phosphorus to this alpine ecosystem (Green, 2011).

Bogong moths are also an important source of food for many species within the Alps, acting as an essential source of protein for mammals such as wild pigs and foxes, birds, frogs and lizards (Wintle *et al.*, 2021). The critically endangered mountain pygmy possum (*Burramys parvus*) is particularly dependent on bogong moths as food, with their survival in the mountains inexorably linked to the continued presence of the moths (Wintle *et al.*, 2021). It has also been shown that the critically endangered southern bent-wing bat (*Miniopterus orianae bassanii*) predates upon bogong moths (Kuhne *et al.*, 2022).

An often overlooked further aspect of bogong moth ecology is that they are flower visitors. Coates *et al.* (2023) observed bogong moths foraging from local flowers around their aestivation sites in the Australian Alps, and suggested that they may have an unrecognized role in pollination within their alpine ecosystems during the summer. Evidence suggests that bogong moths are generalist herbivores (Coates *et al.*, 2023), so their role in pollination during their migration and while in the mountains may be significant.

37.4 Economic Importance

Bogongs are divisive economically – on the one hand, they are essential to the ecological functioning of areas of stunning natural beauty such as the Australian Alps; on the other, they can cause damage to crops in their larval stage as the common cutworm during the winter. Cutworm species are hard to differentiate, however, and so damage from other species may have been unknowingly attributed to bogong larvae (Wintle *et al.*, 2021). They are good sources for further research into economically important topics such as navigation and camouflage. They are also a potential species of interest to the growing edible insect industry (Reyes and Lessard, 2023).

37.5 Conservation Status

37.5.1 Listing status and issues

The bogong moth was listed as Endangered on the IUCN Red List in 2021 and in 2024 its conservation status was assessed for the Environmental Protection and Biodiversity Conservation (EPBC) Act. It was not listed under the EPBC Act, however, due to insufficient data. The bogong moth's listing as Endangered on the IUCN Red List triggered worldwide media attention and widespread concern in Australia (Fig. 37.4) along with some scepticism. How could such a numerous species be endangered?

For the IUCN Red List, the moth was listed under criterion B2ab(iii,v). Criterion B refers to the size of the moth's distribution and subcategory B2 refers to the Area of Occupancy (AOO) – the size in km^2 of the area that the moth inhabits at the most restricted part of its life cycle. For bogong moths, this is the total area in km^2 of the aestivation sites plus the area that is occupied by the 'non-migratory' populations. The AOO for bogong moths for the 2021 IUCN assessment was estimated at between 100 and 150 km^2 (and 172 km^2 for the EPBC Act assessment). The threshold for a species to be considered 'Endangered' is 500 km^2 or below; that is, a species with an AOO less than 500 km^2 can qualify as

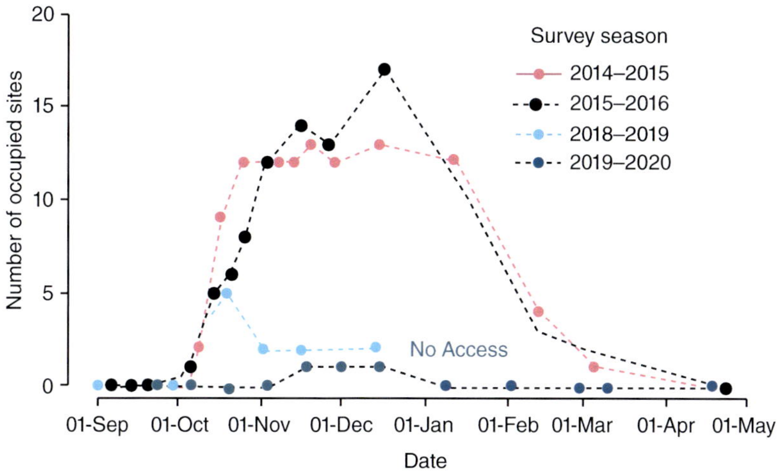

Fig. 37.4. Population numbers of *Agrotis infusa*. From Green *et al.* (2021), *Austral Entomology*, with permission.

Endangered, provided it meets at least two of three further criteria:

- the number of locations (defined by threat) is less than or equal to five
- there is a 'continuing decline observed, estimated, inferred or projected in any of: (i) extent of occurrence; (ii) area of occupancy; (iii) area, extent and/or quality of habitat; (iv) number of locations or subpopulations; (v) number of mature individuals'
- the species undergoes 'extreme fluctuations in any of: (i) extent of occurrence; (ii) area of occupancy; (iii) number of locations or subpopulations; (iv) number of mature individuals'.

The bogong moth is thought to have 3–5 populations for the IUCN and two locations for the EPBC assessment, triggering criterion B2a. In addition, there is a decline in the area of occupancy (B2biii) due to climate change reducing the available cool cave areas plus observed changes in the moth's use of aestivation sites in the north of the aestivation range (P. Caley, unpublished data). There has also been a decline in the number of mature individuals (Green *et al.*, 2021). The bogong moth is not thought to have extreme fluctuations in numbers as defined by the IUCN criteria.

37.5.2 Threats to bogong moths

There are multiple interacting threats which put pressure on bogong moth populations. The three threats currently of the highest conservation concern are climate change, agricultural intensification and predation pressures.

37.5.2.1 Climate change

Climate change is a major threat to the bogong moth, affecting both the adult and larval stages of this species' life history. For adult moths, aestivation is a key part of the life cycle which is disrupted by climate change. Under normal aestivation conditions, bogong moths show limited activity except at dawn and dusk. Laboratory studies have shown, however, that this crepuscular activity is disrupted at 12.5°C and lost at 15°C (Lownds *et al.*, 2023). This shift in behaviour has also been linked to increased dehydration in the moths (Lownds *et al.*, 2023). Given that modelling predictions suggest that by 2050 temperatures in the Australian Alps may increase by up to 2.9°C (Hennessy *et al.*, 2003), there are concerns that warming could lead to a critical loss of suitable aestivation sites.

For larvae, climate change and associated droughts are also a threat. Drought conditions in the agricultural areas of western New South Wales and southern Queensland in 2017–19, for example, were linked to the population crashes of bogong moths observed in 2017–18 (Green *et al.*, 2021). Although no direct studies have been carried out on larval survival under drought conditions, reduced availability of food coupled with hotter and drier soils are expected to lead to increased mortality of larvae (Green *et al.*, 2021). Further work is needed to understand fully the impacts that these changing conditions may have on this species.

37.5.2.2 Agricultural intensification

Agricultural change and intensification are further serious threats to the bogong moth, with the most significant impact thought to be on the larval stage (Caley, 2022). Farming practices affect available bogong moth habitat in different ways. Some farming practices are likely to reduce the suitability of soils for larvae. The production of rice, for example, involves irrigation which makes soil unsuitable for larvae (Green *et al.*, 2021), while the introduction of no-till soil management may reduce the availability of softer soils utilized by the larvae (Caley, 2022).

In addition to changes in soil structure, larvae are also threatened by a reduction in available food. Some farming practices may involve regimens which limit opportunities for herbivores. In some areas of cultivation within the bogong range, there has been a move towards more extensive use of herbicides in the late summer and autumn to reduce weed growth (Caley, 2022), which reduces weeds and crop pests and, by extension, limits feeding opportunities for the bogong moth larvae. Other methods for reducing opportunities for herbivorous pests involve choosing resistant crop types, and specific planting regimens. Within the bogong moth range, for example, genetically modified cotton is grown, which expresses a *bt* toxin. As well as having some pest resistance, this crop is grown in a monoculture without other plants growing among the crops (Green *et al.*, 2021). Methods like these may be effective in reducing crop pests, but also reduce feeding opportunities for the bogong moth larvae.

The impact of habitat loss due to farming practices like these is significant. It has been estimated that cotton production alone covers 483,890 ha of the bogong moth range, which has been estimated to have led to a loss of over 370 million moths (Green *et al.*, 2021), but the total impact of combined farming methods is likely to be much higher. The impacts are likely to be higher still when pesticide use is considered (Green *et al.*, 2021; Caley, 2022). A key group of pesticides used in areas where bogong moth are present are neonicotinoids (Green *et al.*, 2021) which affect the central nervous system of insects (through interference with insect nicotinic acetylcholine receptors). At high doses, neonicotinoids will kill or cause paralysis in insects; even sublethal doses have been found to disrupt navigation memory in well-studied insects like bees (Fischer *et al.*, 2014). Further study is needed to ascertain the impacts these and other pesticides have on the bogong moth (Green *et al.*, 2021).

37.5.2.3 Predation

Aestivating moths are predated upon by a range of native and non-native species (Caley, 2022). The greatest conservation concern, however, arises from non-native predators. European red foxes (*Vulpes vulpes*) arrived in the northern Australian Alps in the 1890s, and have become key predators of aestivating moths (Caley, 2022). Estimates suggest that foxes may consume as much as 56 tonnes of moths in the Snowy Mountains annually (Green, 2011). Introduced feral pigs (*Sus scrofa*) arrived later and have been in the northern Australian Alps since the late 1950s (Caley, 2022). It has been suggested that pigs have only learned to target aestivation sites more recently (Caley and Welvaert, 2018), and their impacts are most pronounced on more accessible aestivation sites (Caley, 2022).

Moths in aestivation sites are also important prey for native predators including bats, birds and reptiles (Green, 2011). One predator of particular conservation concern is the mountain pygmy possum (*Burramys parvus*), a critically endangered marsupial adapted to the alpine environment. Bogong moths are a crucial part of the diet of this possum (Broome, 2001). Faecal analysis has found that these moths can constitute up to 46% of the dietary items consumed by breeding female mountain pygmy possums (Mansergh *et al.*, 1990). Given the dependence of these possums on bogong moths, there are concerns that declines in these moths may have serious impacts on their survival (Gibson *et al.*, 2018).

37.5.3 Conservation actions

37.5.3.1 *Knowledge gaps*

Given the multitude of threats facing the already endangered bogong moth, it is vital that further research is carried out in order to develop our understanding of bogong moth ecology. Filling knowledge gaps will enable effective monitoring schemes and conservation actions to be designed and enacted to stabilize and protect bogong moth numbers in Australia. The development of conservation management schemes for insects in Australia is said to be generally challenging due to limited taxonomic and background ecological knowledge of many insect species, low funding and relatively low levels of public engagement and education. Effective bogong moth monitoring schemes and conservation actions have the potential to become a flagship example for the promotion of further data collection and conservation of other Australian insect species.

Very little is known about the bogong moth's migratory flyways between their breeding sites and their aestivation sites in the Australian Alps, including the knowledge of which plants they feed upon while migrating and which areas are favoured as temporary resting sites en route to the mountains. Knowledge of the migratory flyways of monarch butterflies (*Danaus plexippus*) is well developed thanks to successful citizen science programmes, and has been used to propose conservation actions such as nectar resource enhancement in threatened areas (Tracy *et al.*, 2019). If the geographical locations of bogong migratory flyways were better known with the knowledge of where moths are at every point of their migration, it would enable the design of targeted conservation actions which could improve the survival of bogongs while migrating.

Whole-genome sequencing has been performed on bogong moths (Wallace *et al.*, 2022), and the availability of this genome potentially will allow further insight into particular genes associated with migratory navigation and potential evolutionary adaptability in the face of global change. Developed understanding of population structure in bogong moths will also be helpful in understanding the interaction of individuals while migrating and the choices of migratory flyways and breeding/aestivation habitats.

In the context of climate change and elevated global temperatures, it is important that we understand how bogong moth physiology is responding to warmer temperatures and what is their physiological tolerance and susceptibility to climate change. It is thought that insects are particularly sensitive to climate change – elevated temperatures, for example, may increase the rate of metabolic activity while simultaneously affecting the availability of suitable food and water resources (Huey and Kingsolver, 2019). This may be particularly true for aestivating bogong moths, as elevated temperatures and increased metabolism in their aestivation sites may reduce survival. Developing this knowledge may include investigating thermal melanism in bogong moths.

Further information is also needed to better understand the distribution of the moths at both their breeding and aestivation sites. The spread of the breeding sites across south-eastern and southern Australia may vary from year to year in accordance with precipitation levels across the country during autumn (Wintle *et al.*, 2021). It will also be important to monitor the distribution of aestivation sites within the Alps as the number of climatically suitable sites may shrink due to climate change-associated elevated alpine temperatures (Green *et al.*, 2021). Better understanding of the bogong moths' choice and use of habitat both in their breeding sites and within their alpine aestivation sites will aid in conservation efforts.

37.5.3.2 *Habitat restoration and management*

Habitat restoration in partnership with traditional custodians has been identified as a priority conservation action for bogong moths (K. Umbers and others, unpublished submission). Part of these habitat restoration efforts centres on predator control. At Mt Gingera an approved barrier to aestivation sites has been installed to protect the sites from feral pigs, and fox numbers near aestivation sites are being reduced through targeted baiting (Green *et al.*, 2021). In addition to predator control, restoration of host plants in larval areas and nectar-rich floral resources in migration areas have been proposed to support survival at larval and adult stages of the bogong life cycle (Coates *et al.*, 2023). Despite steps which can be taken to support populations, however, the most significant habitat threats to bogong moths are driven by climate change, which is an issue which needs to be urgently addressed through government and industry action.

37.5.4 Surveying and monitoring

The continued monitoring of bogong moth populations is vital to their conservation. Currently, bogong moth aestivation sites are monitored throughout summer each year to keep track of populations (Green, 2010; Green *et al.*, 2021). Over the last several decades, their absence has been noted in several previously visited aestivation sites at lower elevations of the Australian Alps, providing evidence for their decline (Green *et al.*, 2021). This monitoring is essential for informing conservation efforts and must be maintained.

In recent years, citizen science projects have been introduced to involve the community in bogong moth conservation. Several projects, such as Moth Tracker (Zoos Victoria, 2025), ask members of the general public to upload images of sighted bogong moths to their websites. For each image, the location and date of the sighting are recorded. The data acquired from these websites aids in the identification of bogong moth migratory flyways, and potential locations for breeding grounds. Over time, repeated observations may provide a clear indication of the paths taken by bogong moths during their yearly migrations.

To provide further information regarding bogong moth migratory flyways, citizen science projects could be expanded to include mark–recapture methods. In the United States, the programme MonarchWatch has been successful in determining the migratory paths of monarch butterflies (*Danaus plexippus*) using community-driven mark–recapture projects (Ries and Oberhauser, 2015; Taylor *et al.*, 2019). In this programme, the community is provided with small paper tags, which can be applied to the wings of monarch butterflies using a heat-activated adhesive (Davis, 2015; Taylor *et al.*, 2019). Each butterfly is provided with a unique code printed on the tag, and the date and location of tag application are recorded. Once the tag is applied, the butterfly is released. If a tagged butterfly is encountered by a member of the public, then the time and location of recapture are recorded. This data can then be used to track the direction and distance of dispersal for each recaptured butterfly (Davis, 2015). Monarch butterflies and bogong moths demonstrate similar migratory behaviours, capable of migrating distances greater than 1000 km (Warrant *et al.*, 2016). With the success of MonarchWatch, these methods may also prove useful for tracking the dispersal of bogong moths.

References

ABC Indigenous (2019) The journey of the Bogong moths – this place. Available at: www.youtube.com/watch?v=NPI-YIOW6-A (accessed 16 October 2025).

Broome, L.S. (2001) Density, home range, seasonal movements and habitat use of the mountain pygmy-possum *Burramys parvus* (Marsupialia: Burramyidae) at Mount Blue Cow, Kosciuszko National Park. *Austral Ecology* 26, 275–292.

Caley, P. (2022) Identifying hypotheses for drivers of decline of the bogong moth (*Agrotis infusa*). *Pacific Conservation Biology* 29, 429–444.

Caley, P. and Welvaert, M. (2018) Aestivation dynamics of bogong moths (*Agrotis infusa*) in the Australian Alps and predation by wild pigs (*Sus scrofa*). *Pacific Conservation Biology* 24, 178–182.

Cherikoff, V. and Isaacs, J. (1989) *Bush Food Handbook*. Ti Tree Press, Sydney, NSW.

Cherry, R.H. (1991) Use of insects by Australian aborigines. *American Entomologist* 37, 8–13.

Coates, J.M., Keaney, B., Scheele, B.C. and Cunningham, S.A. (2023) Endangered Bogong moths (*Agrotis infusa*) forage from local flowers after annual mass migration to alpine sites. *Global Ecology and Conservation* 44, e02482.

Colloff, M. (2020) *Landscapes of Our Hearts: Reconciling People and Environment*. Thames & Hudson Australia, Cremorne, Victoria.

Common, I.F.B. (1954) A study of the ecology of the adult bogong moth, *Agrotis infusa* (Boisd) (Lepidoptera: Noctuidae), with special reference to its behaviour during migration and aestivation. *Australian Journal of Zoology* 2, 223–263.

Common, I.F.B. (1981) The Bogong moth. *Bogong* 2, 4–5.

Davis, A.K. (2015) Opinion: Conservation of monarch butterflies (*Danaus plexippus*) could be enhanced with analyses and publication of citizen science tagging data. *Insect Conservation and Diversity* 8, 103–106.

Dreyer, D., Adden, A., Chen, H., Frost, B., Mouritsen, H. *et al.* (2025) Bogong moths use a stellar compass for long-distance navigation at night. *Nature* 643, 994–1000.

Fischer, J., Müller, T., Spatz, A.K., Greggers, U., Grünewald, B. and Menzel, R. (2014) Neonicotinoids interfere with specific components of navigation in honeybees. *PLOS One* 9, e91364.

Flood, J. (1983) *Archaeology of the Dreamtime*. University of Hawaii Press, Honolulu, Hawaii.

Fuller, R.S., Norris, R.P. and Trudgett, M. (2014) The astronomy of the Kamilaroi and Euahlayi peoples and their neighbours. *Australian Aboriginal Studies* 2014, 3–27.

Gale, J. (1927) *Canberra: History of and Legends Relating to the Federal Capital Territory of the Commonwealth of Australia*. A.M. Pallick & Sons, Queanbeyan, NSW.

Gibson, R.K., Broome, L. and Hutchinson, M.F. (2018) Susceptibility to climate change via effects on food resources: The feeding ecology of the endangered mountain pygmy-possum (*Burramys parvus*). *Wildlife Research* 45, 539–550.

Green, K. (2010) The aestivation sites of bogong moths, *Agrotis infusa* (Boisduval) (Lepidoptera: Noctuidae), in the snowy mountains and the projected effects of climate change. *Australian Entomologist* 37, 93–104.

Green, K. (2011) The transport of nutrients and energy into the Australian snowy mountains by migrating bogong moths *Agrotis infusa*. *Austral Ecology*, 36, 25–34.

Green, K., Broome, L., Heinze, D. and Johnston, S. (2001) Long distance transport of arsenic by migrating Bogong moths from agricultural lowlands to mountain ecosystems. *Victorian Naturalist* 118, 112–116.

Green, K., Caley, P., Baker, M., Dreyer, D., Wallace, J. and Warrant, E. (2021) Australian Bogong moths *Agrotis infusa* (Lepidoptera: Noctuidae), 1951–2020: Decline and crash. *Austral Entomology* 60, 66–81.

Heinze, S. and Warrant, E. (2016) Bogong moths. *Current Biology* 26, R263–R265.

Hennessy, K., Whetton, P., Smith, I., Bathols, J., Hutchinson, M. and Sharples, J. (2003) *The Impact of Climate Change on Snow Conditions in Mainland Australia*. CSIRO Atmospheric Research, Aspendale, Victoria.

Hill, L. (2007) *Agrotis* (Lepidoptera: Noctuidae) species in Tasmania including montane, summer aestivation of the bogong moth, *Agrotis infusa* (Boisduval, 1832). *Victorian Entomologist* 37, 3–9.

Huey, R.B. and Kingsolver, J.G. (2019) Climate warming, resource availability, and the metabolic meltdown of ectotherms. *American Naturalist* 194, E140–E150.

Kuhne, J.G., Austin, J.J., Reardon, T.B. and Prowse, T.A. (2022) Diverse moth prey identified in the diet of the critically endangered southern bent-wing bat (*Miniopterus orianae bassanii*) using DNA metabarcoding of scats. *Wildlife Research* 49, 571–582.

Love, P. (2010) Spatial and temporal characteristics of arsenic in the bogong moth (*Agrotis infusa*). PhD thesis, LaTrobe University, Bundoora, Victoria.

Low, T. (1990) Bush tucker: Australia's harvest of the wild, adapted from Bush Tucker: Australia's Wild Food Harvest by Tim Low. *Habitat Australia*, 18, 20–23.

Lownds, R.M., Turbill, C., White, T.E. and Umbers, K.D. (2023) The impact of elevated aestivation temperatures on the behaviour of bogong moths (*Agrotis infusa*). *Journal of Thermal Biology* 113, 103538.

Mansergh, I., Baxter, B., Scotts, D., Brady, T. and Jolley, D. (1990) Diet of the Mountain Pygmy-possum, *Burramys parvus* (Marsupialia: Burramyidae) and other small mammals in the alpine environment at Mt Higginbotham, Victoria. *Australian Mammalogy* 13, 167–177.

McCormick, B. (2005) *Bogong Moths and Parliament House*. Parliament of Australia Research Brief 5, Canberra, ACT.

McQuillan, P., Ireson, J., Hill, L. and Young, C. (2007) *Tasmanian Pasture and Forage Pests: Identification, Biology and Control*. Department of Primary Industries and Water, Hobart, Tasmania.

Pascoe, B. (2018) *Dark Emu: Aboriginal Australia and the Birth of Agriculture*. Magabala Books, Broome, Western Australia.

Reyes, R.P. and Lessard, B.D. (2023) Towards an inclusive edible insect industry: Perspectives from Australia. *Journal of Insects as Food and Feed* 10, 363–368.

Ries, L. and Oberhauser, K. (2015) A citizen army for science: Quantifying the contributions of citizen scientists to our understanding of monarch butterfly biology. *BioScience* 65, 419–430.

Rigby, K. (2011) Getting a taste for the bogong moth. *Australian Humanities Review* 49, 1–18.

Stephenson, B., David, B., Fresløv, J., Arnold, L.J., Gunai Kurnai Land and Waters Aboriginal Corporation, *et al.* (2020) 2000 year-old Bogong moth (*Agrotis infusa*) Aboriginal food remains, Australia. *Scientific Reports*, 10, 22151.

Taylor, O.R. Jr, Lovett, J.P., Gibo, D.L., Weiser, E.L., Thogmartin, W.E. *et al.* (2019) Is the timing, pace, and success of the monarch migration associated with sun angle? *Frontiers in Ecology and Evolution* 7, 442.

Tracy, J.L., Kantola, T., Baum, K.A. and Coulson, R.N. (2019) Modeling fall migration pathways and spatially identifying potential migratory hazards for the eastern monarch butterfly. *Landscape Ecology* 34, 443–458.

Troy, J. and Barwick, L. (2020) Claiming the 'song of the women of the Menero tribe'. *Musicology Australia*, 42, 85–107.

Wallace, J.R., Maleszka, R. and Warrant, E.J. (2022) Large-scale whole-genome sequencing of migratory Bogong moths, *Agrotis infusa*, reveals genetic variants associated with migratory direction in a panmictic population. *bioRxiv*. DOI: 10.1101/2022.05.27.493801.

Warrant, E., Frost, B., Green, K., Mouritsen, H., Dreyer, D. *et al.* (2016) The Australian Bogong moth *Agrotis infusa*: A long-distance nocturnal navigator. *Frontiers in Behavioral Neuroscience* 10, 77.

Waterhouse, D.F. (1971). Insects and Australia. *Australian Journal of Entomology* 10, 145–160.

Wintle, B., Tunney, O., Cadenhead, N., Visintin, C., Mackay, E. *et al.* (2021) *The Bogong Moth, Agrotis infusa: Cultural Context, Knowledge Gaps, Conservation and Monitoring Options*. Interim Report 4(1). NESP Threatened Species Recovery Hub Project, Brisbane, Queensland.

Yen, A.L. (2005). Insect and other invertebrate foods of the Australian aborigines. In: Paoletti, M.G. (ed.) *Ecological Implications of Minilivestock: Potential of Insects, Rodents, Frogs and Snails*. Routledge, Abingdon, UK, pp. 367–387.

Zoos Victoria (2025) *Moth tracker*. Available at: www.zoo.org.au/moth-tracker/ (accessed 19 May 2025).

38 Knowledge Gaps and Opportunities: A Future for Moth Studies in Australia

The Authors[1]

Abstract

This final chapter summarizes the responses to a questionnaire sent to all authors soliciting opinions on knowledge gaps, future directions and technological opportunities for students of moth studies in Australia going forward. Almost all higher taxa demand more work but particular groups such as the Oecophoridae, the basal families, Heliozelidae and several others are obvious near-term targets. *Spodoptera frugiperda* and *Heliothis punctifera* were highlighted as demanding particular attention from applied researchers. Geographical areas requiring further attention are legion. Further work on the drier ecosystems west of the eastern ranges and on hard-to-access habitat islands such as mountain tops and mangroves is suggested. Biogeographical links between Australia, New Zealand and New Guinea all require further elucidation. Technological developments in molecular, artificial intelligence and other fields will assist future studies, as will ongoing engagement with amateur and citizen scientists. The role of moths in providing ecosystem services needs to be promoted and their public image as drab and pestiferous needs to be revised. There are clear future roles for professional and amateur societies in achieving these goals.

38.1 Introduction

Each of the preceding chapters was constructed by inviting authors or teams of authors to submit drafts which then proceeded through the processes of reviewing and editing like any other scientific article. For this concluding chapter, a different approach has been taken.

A few weeks before going to press, a questionnaire was circulated to all authors or teams of authors asking a set of general questions about current knowledge gaps and future needs for moth studies in Australia. These questions form the section headings that follow. A majority responded (see author list), some at great length, and this chapter is a collation and summary of their responses. Some of these knowledge gaps have also been canvassed within individual chapters in more specific contexts. There was a great deal of convergence in responses, as was to be expected, but some individual ideas of great originality emerged. In the account that follows, contributions are not ascribed to individual respondees.

Corresponding author: r.kitching@griffith.edu.au

[1]Roger Kitching, Louise Ashton, Ethan Beaver, Georgina Binns, Michael Braby, David Britton, Catherine Byrne, Julia Cooke, Alastair Drake, Eleanor Drinkwater, Mark Elgar, Marc Epstein, Gary Fitt, Peter Gregg, Peter Hendry, Robert Hoare, Axel Kallies, Ying Luo, Peter McQuillan, Scott Miller, Michael Moore, Tim New, Erica Odell, David Rees, Don Sands, Alan Wills, Myron Zalucki

DOI: 10.1079/9781800627680.0038

38.2 Which Taxa are Most Worthy of Future Attention Either Based on Earlier Neglect, Especially Interesting Biology, Applied Importance and/or Any Other Particular Interest for Australian Science?

38.2.1 The fauna as a whole

From a systematics point of view, most moth families in Australia still need major work. The late Ebbe Nielsen proposed a taxonomic triage approach: that is, in Australia (if not beyond) it was most important to use the limited resources available to publish robust generic classifications. Organizing the species into recognizable units ('morphospecies') against such robust genera in major museum collections would then allow their rapid recognition and description when needed for economic, conservation or ecological reasons. In the case of genera, it is important that they be rigorously defined, monophyletic lineages, but also that correct names be applied (many Australian moths are still lumped into misapplied generic names derived from the European or Asian fauna). DNA barcoding within such an approach allows species-level information to be accessed without investment in full taxonomic treatments. Detailed monographs and less extensive taxonomic revisions can then folllow at their own slower pace reflecting available expertise and resources.

The high levels of endemism and diversity within the Australian fauna principally reflect a dominance of the micromoths, as traditionally defined. Not only have particular families radiated in Australia, but they have done so, in many cases, by evolving unique associations, co-radiations and co-evolutionary systems with the equally special Australian flora. These co-radiations, in general, are poorly known and offer future research opportunities of global scientific significance. Specific examples are explored below.

These general needs notwithstanding, some Australian taxa stand out as obvious candidates for further research building on previous efforts and their uniqueness in global terms.

38.2.2 Future target taxa

The **Oecophoridae** (Chapter 12) are by far Australia's most diverse moth family, representing perhaps 60% of the moth fauna. Most of these species are endemic and, at current count, about half remain undescribed. Although Ian Common's three volumes (1994, 1997, 2000) defining genera are foundational, we await work-in-progress towards a definitive phylogeny using phylogenomic methods. At the species level, both taxonomy and biology present daunting tasks for future researchers. The biology of oecophorine species, in general, is poorly known but 1000+ species are likely to feed on dead leaves of eucalypts. This makes them ecologically important in decomposition processing and of general conservation importance as ecosystem fire regimes change under global warming. Known species with exceptional and unique dependence on specific vertebrate hosts (breeding in bird nests, marsupial droppings, etc.) present opportunities for studies of novel life cycles.

The Oecophoridae are placed within the superfamily Gelechioidea which has a strong claim to be the largest superfamily of Lepidoptera globally (if estimates of undescribed species are taken into account). Heikkilä *et al.* (2013) have gone some way to resolving the family structure within the larger taxon. Gelechioidea are massively diverse and ecologically important in Australia, but many of the larger families in the fauna are in a state of taxonomic neglect, apart from the Elachistinae (Kaila, 2011), Oecophoridae (see above) and, to some extent, the Xyloryctidae (see Chapter 13). Many genera and species are undoubtedly assigned to the wrong family in the current classification (L. Kaila, pers. commun.). There is the potential for extraordinary discoveries and taxonomic advances in this group.

Australia is home to a suite of **primitive moth families** (Chapter 3) about which little is known but which have the potential to contribute to our understanding of the evolution of the Lepidoptera as a whole. Almost all the species have very restricted known distributions and are, potentially, of considerable conservation concern. Recent discoveries of new Micropterigidae in Western Australia have modified global understanding of the evolution and biogeography of that ancient family (Gibbs, 2016). What little we can hypothesize about their life histories is based on what is known about New Zealand species. The Agathiphagidae are probably the second most basal of all surviving Lepidoptera: they are closely associated with old kauri trees and have only recently been rediscovered. Their biology remains poorly known. The Lophocoronidae are basal Glossata restricted to Australia with six known species in sclerophyllous ecosystems yet nothing is known of their life histories. Within the basal clades of the Heteroneura, the Nepticulidae and the Incurvariidae are diverse in Australia but largely undescribed and understudied. Each of these families offers opportunities for future work of global significance. The Nepticulidae probably have a strong association with the Australian Myrtaceae and deserve more attention for that reason alone. The recent description of the 'basal' Aenigmatineidae is clear evidence that whole families may still await discovery in Australia.

A set of little researched families offer opportunities in understanding moth–plant co-evolution. Most recent attention has been devoted to the **Heliozelidae** (Chapter 11) and, in particular, a group of taxa that have apparently developed pollination-based mutualisms with their *Boronia* (Rutaceae) host plants. The family is highly diverse in Australia but the vast majority of species and genera remain unnamed. The diverse Australian fauna is likely to throw light on the evolution of the family globally. Other taxa exhibiting tight pollination relationships with native plants are also worthy of increased attention (e.g. *Epicephala* spp. (Gracillariidae) with *Glochidion* and other Phyllanthaceae).

A wide range of other Australian taxa have their strong advocates for greater attention based on existing preliminary studies; examples include the parasitic endemic Cyclotornidae, the endemic Cossidae, the Psychidae, the aquatic Acentropinae, the enigmatic Heliocosmidae and the highly diverse but little studed Tineidae. Other little worked groups have the potential to be flagship species for conservation and increasing public awareness. These include highly visible day-flying groups within the Agaristinae, Zygaenidae, Castniidae, Larentiinae and Syntominae. Others, like the bogong moth, *Agrotis infusa*, are important indicator species for monitoring the health and well-being of threatened habitats (Green *et al.*, 2021).

38.2.3 Potential future economic species

The full negative economic potential of some noctuid species demands more precautionary research on their general biology, phenology, migration and life cycle biology.

The fall armyworm, *Spodoptera frugiperda*, has greatly expanded its global range in recent years, and arrived in Australia in 2020. It has the potential to become a much more important pest in Australia, though at present damage is mostly confined to maize and sorghum. It is highly polyphagous even though it has some obvious host preferences. Although there is research happening, it is narrowly focused on the usual aspects: pheromones, economic thresholds, which insecticides work best and insecticide resistance. A broader view of its behaviour and much more field work are needed. Key questions relate to its adaptability to the vagaries of the Australian landscape; what non-crop vegetation, particularly native grasses, the species uses; the source populations from which seasonal migrations occur, and so forth.

A species that might reward some broad scale ecological work is *Heliothis punctifera*, which occupies similar habitats and has a host range similar to *Helicoverpa punctigera*. There are anecdotal reports of 'clouds' of this moth, almost like locust swarms, in some cases apparently following a long-distance migration. Although it never seems to cause problems in cropping areas, a study of its ecology, contrasting it with *H. punctigera*, would be interesting and informative. However, at present its population seems still unrecovered from the Millennium drought.

In addition to these species-level studies, there is great potential for expanding studies on higher taxa that contain important actual or potential pest species. The monographic study of the Heliothinae by Marcus Matthews (1999) offers a model for placing major pest species into their wider taxonomic and faunistic context. Taxa such as the stem-boring Crambinae or sections of the Tortricidae, among others, would benefit from such an approach.

38.3 Which Areas of General Interest in Evolution, Ecology and/or Biogeography Would You Most Like to See Developed Further?

38.3.1 The evolution of the Australian fauna

Although higher level phylogenies of many lepidopteran groups are being produced and regularly updated, many of these include only poor representation of Australian species. There is need for much more granular molecular phylogenies with wider Australian taxon sampling aimed at better understanding the local and global context of the Australian fauna. Local clades need to be identified and understood and Australian endemics uncovered at the family and subfamily level. In other words, we need a better understanding of Lepidoptera evolution and diversification in the specific Australian, and wider Gondwanan, contexts. We know a lot about speciation events in the context of vicariance or within tropical rainforest niches. These concepts do not necessarily apply easily to most of Australia which has been geologically stable and much more homogeneous in most parts than tropical regions to the north – yet we have super-diverse groups and many local endemics.

Once again, the Oecophoridae feature in proposed ways forward. Resolution of the systematics of this, our largest family, will provide fundamental insights into the evolution, ecology and biogeography of the Australian Lepidoptera in general. The dominance of micromoths and their radiation in the predominantly arid environment of Australia are hallmarks of the Australian fauna.

38.3.2 Biogeographic issues

As is apparent in Chapter 4, understanding the biogeography of particular Australian faunal elements leads to hypotheses which can only serve as temporary challenges for future testing. Explicit hypotheses can now be tested using modern phylogenetic approaches but only when we have both reliable phylogenies and taxonomies, and extensive data on distributions.

The relationships between the Australian fauna and those of nearby (or not so nearby) land masses need much more work. Understanding of many groups, especially in the tropics, demands knowledge of the faunas to our north which are often less well known even than those of Australia. Reconciling taxonomy across these faunas is a challenging but essential task. Australia and the great tropical island of New Guinea are fundamentally a single biogeographic unit separated intermittently by the vagaries of quatenary ice ages. The overlap in faunas between Australia and New Guinea, accordingly, is worthy of much more analysis. Many species have been given names in both places that have not been connected and, in other cases, the same name has been applied to different species. Elucidating faunistic connections between Australia and New Zealand is also needed and may be more 'do-able' in the near term than it is with the super-rich and lesser known tropical faunas in New Guinea and elsewhere. Jeremy Holloway's work on the faunas of Norfolk Island and New Caledonia (and elsewhere) are exemplary in this respect (Holloway, 1977, 1979, 1996a,b).

Pest species (Chapter 23) may be endemic or more global in distribution. In research terms, understanding the emergence, impact and ecology of the endemic species will likely generate the most interesting results. Of course, the widespread more global species cannot be ignored and demand attention from a management viewpoint. Indeed, many of these supposedly widespread species may well turn out to be complexes of cryptic species upon closer examination.

38.3.3 Migration and movement

Although a lot of work has been done on migration, the amount of firm, species-specific knowledge is quite limited. The simple north–south spring–autumn rhythm seen in much of the northern hemisphere hardly applies here, though there is a spring burst towards higher latitudes in the southern half of the continent. As well as studying single, economically important species, some broader ecological questions could be addressed. How predictable and consistent is migration in a continent dominated by highly variable rainfall, temporally and spatially, with its consequent effects on host resources? How do patterns of migration increase fitness and ensure population survival? How are the migratory adaptations maintained through continual contemporary natural selection? Do species that breed sympatrically migrate in the same way? Are there genetic differences between migratory and non-migratory populations?

Taxa other than the major pests are also worthy of attention. *H. punctifera* has already been mentioned. A recent study indicates that breeding and movement by *Agrotis infusa* may be more varied than the long-recognized annual exchange between winter inland breeding areas and alpine aestivation sites (McDonald *et al.*, 2024). Sphingids have been caught at light in large numbers in inland New South Wales along with massive catches of noctuids of known migratory species, and have also been recorded as vagrants on offshore islands and even New Zealand, but almost nothing is known of their spatial population processes within the continent. Given our advanced knowledge of their taxonomy and life histories in Australia, and their known migratory status elsewhere in the world, more attention to their long-distance movements here seems overdue.

38.3.4 General ecology

38.3.4.1 Life history studies

There is a fundamental and oft-repeated need for more life history information on Australian moths, many of which are known only from the adults. Filling this knowledge gap involves rearing species from eggs or larvae and carefully documenting the changes that occur as the animal matures. Herbison-Evans and Crossley's website and iNaturalist entries are excellent starting points for would-be rearers. Of course, rearing is an arduous task requiring dedication and a tolerance for failure. We do, however, have excellent models for highly useful and usable outcomes. At the continental level, the Moulds *et al.* (2020) monograph on the sphingids is a perfect case in point. More localized models are those of McFarland (1988) for South Australian geometrids, and McQuillan *et al.* (2019) for 'southern' Australia. Successful understanding of almost all aspects of moth biology depends on an understanding of life histories, particularly with respect to their larval resources.

Combined with expert photography and local publication (readily available), such contributions can be valuable to the entire community of lepidopterists. There is also considerable unrealized opportunity for citizen science engagement in rearing programmes.

The possibility of co-ordinating and managing such efforts through organizations like Moths and Butterflies of Australasia (MABA) is worthy of investigation. One option is the creation of local 'rearing groups' drawing on assistance from the more experienced. Producing an electronic (or hard-copy) guide to 'best practice' would be a useful first step.

Complementing traditional rearing techniques, there is an increasing literature describing molecular approaches to detecting host plant signals from larval gut contents. Generally, we suppose a gut will likely contain a single species of plant residue. In such cases, a single plant barcoding approach will be appropriate. Effective use of such an approach of course has technical demands more appropriate to a molecular laboratory than an amateur workroom. It also requires a barcode library for the flora in question; there are few such regional libraries currently, but we may expect their availability to increase.

38.3.4.2 Climate change impacts

Beyond basic life history ecology, the most frequently raised issues relate to the likely impacts of ongoing climate change on moth populations. These begin with basic needs for monitoring change through time of abundances and distributions (see, for example, Chapter 37). Indeed, this encompasses the key question 'are moth numbers declining?'. Such declines, where they occur, may reflect climate change, habitat degradation, pesticide use and so on, but understanding and untangling any of these impacts require answering the primary question 'is decline happening?'. To answer this question, we require data from repeated sampling of the moth community carried out in standard fashion at regular intervals. Such data are rare globally and, apart from a few modest efforts for pest noctuids, are largely absent in Australia.

Other climate-related issues worthy of further investigation relate to changes in food plants, both their quality in chemical terms and their phenologies inasmuch as they determine availability to herbivores such as moth larvae. There will also be top-down effects upon moths from their more climate-sensitive natural enemies.

Under climate change there will be, naïvely, winners and losers: we fear the 'winners' will be the abundant polyphages and the 'losers', the local endemic specialists. These remain speculations, however, and we urgently need careful comparative studies on selected species for both conservation and pest management purposes.

38.3.5 Behavioural ecology

Our knowledge of the natural history of moth reproductive behaviour is limited compared with other insect groups. The most extensive knowledge relates to the chemical nature of sex pheromones, largely driven by the possibility of using synthetic sex pheromones for pest control. There are two outcomes of this very strong focus – a concentration of knowledge around 'pest' moths, and a remarkably poor understanding of the nature of female calling (pheromone release) behaviour. The former is important practically because lesser known species may have traits that could offer considerable benefits. The latter is important because conventional wisdom holds that females exercise little control over their calling behaviour, and yet studies of *Uraba lugens* (Chapter 31) indicate that this view is misplaced, with important implications for more effective use of pheromones in pest control. In a broader context, and following earlier remarks of Charles Darwin, many of the moths with feathery antennae are poorly documented – these moths are likely, *a priori*, to have interesting pheromone-driven reproductive behaviour.

38.3.6 Chemical ecology

The unique Australian moth fauna interacts and has evolved with the equally unique Australian flora. We need to better understand how the one has shaped the other historically. In this context the chemical ecology of the Australian moths, about which we know little, is likely to be both interesting and 'different', particularly with respect to defensive chemicals and their consequences in phenomena such as aposematism and mimicry. Of course, initally we need to know a great deal more about the food plant relationships of our moths, larval feeding behaviour and, in particular, the gut and leaf surface microbiomes (see below). Syntomine moths (Chapter 21) may be useful models for partially rectifying this omission. These moths are visually striking and charismatic and appear to be a part of a large Müllerian mimicry ring. *Amata* and *Eressa* in particular seem to be successful generalists found in varied habitats. There is no information on what these moths are using for chemical defences, only that it is effective, as there is little evidence of other taxa predating upon them. Similar patterns occur in the Australian Agaristinae (Noctuidae) (Chapter 22). The larvae (and adults) of these moths are highly aposematic and thus assumed to be unpalatable to most predators, yet the roles of secondary plant compounds (e.g. sequestration of alkaloids and other metabolites) in their chemical defence are unknown.

38.3.7 Indigenous knowledge

The full appreciation of Indigenous knowledge of moths (and other insects) has been little explored beyond a few celebrity species such as bogong moths and widgetty grubs (but see Clarke, 2026). Collaborative work combining field knowledge, linguistic tools and cross-cultural comparisons would be fascinating and socially valuable, perhaps leading to some intersection of western science and traditional knowledge, and ongoing exchanges and collaborations between these two knowledge communities.

38.4 Which Geographical Areas of Australia are Least Well Known in Terms of Moth Diversity and Biology?

In general, the more remote areas of Australia are probably the least known biologically. Broadly, that means that everywhere west of the eastern ranges and, perhaps, the south-west corner of the continent needs more attention in terms of moth surveys. That having been said, anywhere that has not had a *resident* lepidopterist interested in moths is likely to be undersampled; indeed, even resident lepidopterists turn up new things on a regular basis. However, some areas are regularly identified as needing much more attention: the arid and semi-arid interior, the Top End of the Northern Territory, Cape York, virtually all of Western Australia including the vast Kimberley and Pilbara, and the offshore islands.

Other neglected areas are simply hard to access regardless of their geographical location. These include mountain tops, mangroves and saltmarshes, extensive wetlands, remote uplands, gorge complexes, artesian springs and salt lake margins. All will likely yield local endemics. Tiny remnants of bushland on stock routes within otherwise cleared agricultural 'steppes' would also reward survey work.

A significant proportion of our moth fauna was collected and described many decades ago and targeted collecting areas near growing cities were visited repeatedly; many type localities reflect a relatively small selection of these periurban locations. Some of these localities still exist as reserves, others have long been lost to development and urbanization. Targeted resampling possibly by citizen science-driven 'bioblitzes' would generate valuable information on the impacts of urbanization and the resilience (or otherwise) of moth populations.

Returning to our earlier point, expedition-style visits even to promising remote or hard-to-access locations will yield only snapshots of the target biodiversity and will store up yet more ecological and taxonomic questions for the future. Of course, they should continue to happen but so should more intensive studies closer to home. These are likely to produce more ecological insights and more intimate knowledge of our moths than the more intrepid excursions. Again, properly prepared citizen scientists will be essential if these intensive local studies are to happen. A brainstorming session among existing lepidopterists to develop key questions and standard protocols for such local studies would be a useful starting point.

38.5 What Technological Developments Do You See as Holding Most Promise for the Future of Our Work?

There is no doubt that new technologies can have a profound impact on the questions we can ask, and the quality of the data provided by empirical studies. However, it would be a profound mistake for research questions in biology to be driven entirely by technological advances, rather than by those sparked by natural history observations that translate into clever and rigorous observations and experiments. Our understanding of the natural history and biology of moths, like other insects, will be enhanced by technologies but key insights and understanding will be, and must be, intellectual.

Nevertheless, technical advances will affect what natural historical questions, in the broadest sense, can be tackled.

38.5.1 Molecular studies

Without doubt, the most profound technical advance impacting our understanding of the Lepidoptera over the last 30 years has been our ability to study identities and relationships through DNA analysis. Various accounts in this volume examine advances in moth research since the publication of Common's *Moths of Australia* (1990). That work makes no mention of DNA studies.

38.5.1.1 Phylogenies

Molecular systematics is very useful for resolving the higher-level classification of the Australian Lepidoptera, but taxonomic work at the species and genus level is still pivotal for describing new species and understanding biology. A substantial investment in taxonomy by orders of magnitude is required to achieve this goal (Taxonomy Decadal Plan Working Group, 2018).

Next-generation and whole-genome sequencing are getting cheaper and easier to access. In terms of evolutionary relationships, 'museomic' techniques, such as low coverage whole-genome sequencing, are going to be pivotal in understanding Australian moth phylogenies and diversity going forward. One of the biggest gaps for Australia is a phylogenetic framework for endemic species; without this, it is often difficult to proceed with species discovery and conservation. Life history, ecological and morphological traits can then be mapped onto these phylogenies.

38.5.1.2 Barcoding

Barcoding using short nucleotide sequences such as the CO1 sequence is hardly new technology now. Although large libraries for the Australian Lepidoptera do exist, compiling geographically limited subsets will set the scene for underpinning local bioblitzes, metabarcoding of local mixed samples and other point surveys. The software for extracting Australia-only results from the BOLD website is already available. Extracting DNA, amplifying it and then getting it sequenced, even for short-locus barcoding, remain time-consuming and technically demanding processes. Any new technologies for streamlining, cheapening and/or otherwise opening up this process would change the scene for moth students dramatically.

38.5.1.3 Metabarcoding

The mechanics and early applications of metabarcoding are discussed in Chapter 8. It demands access to the appropriate technology and remains relatively expensive. There is no doubt, however, that it can be complementary to the traditional collect, set and identify approach.

38.5.2 Imaging, automated detection, identification and species description

The promise of artificial intelligence (AI) approaches awaits testing. Much-discussed AI approaches are beginning to transform our abilities to identify species. One prospect for moth studies is automated image recognition linked to databases of standard images. This approach is beginning to be realized in automated trapping and recognition particularly of known pest species (see, e.g., Sun *et al.*, 2022; Zhang *et al.*, 2022; Roy *et al.*, 2024). The availability of high-resolution cameras in mobile phones, electronic connectivity and cloud computing puts such automated identification well within the near-future realms of possibility. Appropriate smart apps are central to such a tool. Key precursor steps are needed.

38.5.2.1 Standard imaging

Moths generally have a characteristic resting posture on light sheets and elsewhere (as illustrated by many of the photographic 'portraits' in this volume). This means that standard images can be taken from above or the side to become part of the requisite AI training database. An associated approach can encompass set specimens. Existing internet sites such as the CSIRO's Moths of Australia and Herbison-Evans and Crossley's 'Butterfly House' already contain many appropriate images. Existing images of larvae are more variable, and an appropriate set of general standards should be developed and published. Role models for such standardization exist in works such as Moulds *et al.*'s (2020) monograph of the Sphingidae.

38.5.2.2 Baseline identification

Any AI-based identification tools need to be, above all else, authoritative. This means that any images used must be correctly identified in the first place. For currently known and named species and even some documented but unnamed, this is achievable using available resources and expertise. For Australia's large undescribed fauna, this is much more challenging.

38.5.2.3 Species description

We need a discussion on how we deal with our large undescribed fauna (at the species level); if we insist on only publishing wordy monographs, we will never be able to describe the thousands and thousands of unnamed species. Indeed, an insistence on the superiority of 'complete' monographs has led to an unfortunate reluctance to publish shorter genus-level revisions or even single species descriptions. An emphasis on high-quality standard images and associated DNA barcodes could short-cut the present cumbersome processes of species descriptions. The discussion paper by Forum Herbulot (2014), principally dealing with the Geometridae, points ways forward that would be foundational to such a discussion.

38.5.3 Traditional identification tools: the role of books

The era of automated AI-based identification is not far away but the place of traditional hard-copy books and their electronic avatars will remain. Of course, these may assist, indeed in some cases be central to, species identification but they serve many more purposes. Although an AI-based tool may pinpoint a name for a given moth, only books allow one to leaf through their contents and get a general image of a fauna and its underlying biology. In this way enthusiasts are created and ideas generated.

The appearance of major works on moths engenders whole further waves of professional research and amateur interest. Ian Common's (1990) *Moths of Australia* inspired a whole generation of moth workers and provided foundational information for their future efforts. Indeed, the very existence of the current volume can be traced to that earlier work. Regionally based works act in conjunction with the more grandiose efforts to inspire and underpin local efforts: the *Moths of Victoria* series (Marriott *et al.*, 2008–2020), *Moths in the A.C.T.* (Cocking *et al.*, 2022) and *Caterpillars, Moths and their Plants* (McQuillan *et al.*, 2019) are all excellent cases in point and role models for others.

Last, it may be too much to hope for illustrated species-level guidebooks to the moths of Australia, as we have for the much less diverse butterflies (at *ca* 460 species). Nevertheless, some families could conceivably be dealt with in this fashion. Not only would such volumes be resources of immense general value, they would also be available to assist applied entomologists. Several candidate taxa have been suggested, perhaps the most persuasive arguments being for the Australian Geometridae. The superb series *Geometrid Moths of Europe* (Hausmann *et al.*, 2001–2019) deals with a similar sized known fauna and shows what can be achieved with dedication and an appropriate 'champion'.

38.6 How Do You Feel We Can Best Promote the Study of Moths (and Other Insects) for Both the General Public and the Next Generation of Professionals?

For knowledge of moth biology to advance, both amateurs and professionals must work together. It is often the unpaid hobbyist who has the time and dedication for rearing and collecting, but perhaps lacks the resources or knowledge to describe a species formally. Professionals must be willing not only to embrace the work of amateurs but to foster their interest and even upskill them. As with most communities in the modern day, amateurs (and professionals) can be found in special interest groups online, such as in Facebook groups and online platforms such as iNaturalist and NatureMapr. Increasingly, these are great places to search for interested minds and share factual information.

38.6.1 Public engagement

It goes without saying that you need to be passionate about any group of animals to pursue a lifetime of learning and engagement. Moths in general have had a poor press (unlike butterflies), largely reflecting misplaced assumptions about their pest status, drabness and general 'sameness'. It behoves current moth enthusiasts to engage through readily available media, targeted short courses and constant promotion. Again, there is a role here for the Moths and Butterflies of Australasia Association to look outwards and organize appropriate public events.

Public engagement with biodiversity assessment is essential if both information-based and conservation goals are to be met. Supporting internet-based citizen science is an important tool in this respect, generating better data on distribution and phenology than has been available up to now. Intensive 'bioblitzes' complement these activities and, properly planned and managed, can generate useful information while being fun events. Students and others can be introduced to moth studies through public lectures and seminars. Light trapping events, particularly when hosted by universities, land care, catchment and conservation community groups, and entomological societies, are exciting as well as instructive.

38.6.2 Higher education

An early passion for the Lepidoptera has frequently led to subsequent professional engagement. To encourage more people into professional fields of entomology, we need more and better courses at the university level. There are very few entomology courses being taught at undergraduate level across Australian universities and, as senior entomologists retire, these get even fewer. This loss is opportunistic on the part of cash-strapped institutions and in no way reflects any reduction in the importance of the subject matter. Indeed, while the need for informed pest management remains, entomology as a discipline has expanded its applications into environmental management, conservation and human health. If we want to educate globally competitive entomologists, we must have good university courses to spark and maintain interest in students. Having enough jobs for them is another matter, but just ensuring some ongoing expertise in the taxa that form such a significant part of global biodiversity is very important.

38.6.3 Moths and ecosystem services

There is need to promote a sense of urgency around the study of moths as key parts of our environment and the services they provide within it. Their roles as pollinators, as vital parts of food chains (not just as food for birds) and as indicators of environmental quality need to be promoted. Alongside these economic and ecological dimensions, we need to stress the uniqueness of our fauna, build national pride and a drive to better understand and protect our fauna in general and moths specifically.

38.7 What Do You See As the Key Nexus Between Moth Science (in the Broadest Sense) and Human Welfare Going Forward?

Lepidoptera are one of the most diverse animal groups in the world and play a vital role in maintaining healthy ecosystems. In this time of increasing threats to the natural environment, the protection of biodiversity is very important to human wellbeing and survival. More than ever, governments are assigning a value to biodiversity, recognizing its role in mitigating the effects of climate change. Lepidoptera are a very important part of that process. According to reinsurance giant Swiss Re, $44 trillion of economic value generation – over half the world's GDP – is moderately or highly dependent on biodiversity and ecosystem services that 'underpin all economic activity in our societies globally' (World Economic Forum, 2022). Such services are critically dependent on functionally diverse invertebrate groups like insects that contribute a wide range of ecosystem services (Losey and Vaughan, 2006), comprise over half of the described species (Chapman, 2009) and are hosts for millions of unique bacterial species, nematodes and mites (Larsen *et al.*, 2017).

For Australia, the Deloitte Access Economics (2020) cost–benefit analysis of a mission to discover and document Australia's species, prepared for the Australian Academy of Sciences, estimates net benefits ranging between $3.7 and $28.9 billion in a generation.

The former Minister for the Environment and Water, Tanya Plibersek, released the *Threatened Species Action Plan: Towards Zero Extinctions* in 2022. This Plan sets out a pathway for threatened species conservation and recovery over the next 10 years.

The Action Plan has ambitious targets, which include preventing any new extinctions of plants and animals, and protecting and conserving at least 30% of Australia's land mass: https://minister.dcceew.gov.au/plibersek/media-releases/minister-launches-threatened-species-action-plan-toward-zero-extinctions (accessed 16 October 2025).

Some principal points are as follows.

- The value of biodiversity to humans has now become more recognized than ever to governments and people in general.
- The best estimate is that 70% (or 400,000) of all Australian species of plants, animals, fungi and other organisms have yet to be discovered, documented, named and classified (Cassis *et al.*, 2017). https://soe.dcceew.gov.au/overview/environment/biodiversity.
- It is estimated that Australia has approximately 320,465 invertebrate species, of which about 35% have been described.
- The vast majority of this diversity remains undescribed (Stork, 2018), leaving humanity dangerously ignorant, with most species largely inaccessible for ecological studies and biodiversity inventories.
- The bulk of the planet's unknown diversity is in hyperdiverse, taxonomically neglected groups.
- The documentation of the distribution of named species is vitally important in generating knowledge of Australia's biodiversity and for conservation, land management and biosecurity.

The sheer diversity of moths in Australia highlights the rich biodiversity our country supports. Moth biodiversity is a natural resource that is vastly underappreciated by the public (and decision makers), yet it performs vital services (e.g. nutrient cycling, herbivory, pollination and energy flow in terms of providing food for other animals) that maintain healthy ecosystem function of natural habitats. Human welfare is dependent on the retention and maintenance of these natural habitats, and moth biodiversity is just one component of nature.

Moths are so diverse and ubiquitous that when the general public are given opportunities to learn about them, they notice them more and take more interest in understanding or finding them. We can build awareness of the impacts of issues such as pesticide use and land clearing, and of planting natives in backyards to support biodiversity. Community science events are programmes that everyone can participate in and which have a direct impact on mental health and welfare of humans.

Fundamentally, then, we need to pursue some or all of the following.

- We need to inspire the public and improve the reputation of 'moths' as a vital part of the living world.
- We need to include 'biodiversity security' as a path to maintaining ecosystem services on which we all depend.
- We can promote moths as indicators of habitat restoration.
- We can encourage the development and promotion of the idea of 'caterpillar gardens' where food plants are encouraged in suburban gardens and newly planted urban mini-forests – make 'moth-eaten' a badge of pride!
- As a community we can agitate for a biodiversity dividend from projects which clear native habitat, to be pooled as a fund for invertebrate research and conservation.

38.8 Any Other Suggestions?

Three unrelated issues have been raised under this heading.

38.8.1 Moth microbiota

The microbiomes of moth larvae have been little studied. So many species of moths spend part of their lives in leaf litter and soil that there may be secrets awaiting unveiling about their gut health and the impacts of symbiotic relationships with fungi and microbes on their development.

38.8.2 Cost–benefit analyses

Cost–benefit predictions and postproject analysis are needed by agencies to accompany research projects on moths as indicators for public spending and its justification.

38.8.3 State moths

Appoint, promote and legislate a 'state moth' for each state.

38.9 Concluding Remarks

The knowledge gaps, new approaches and socio-scientific issues raised by the authors clearly offer endless future research agendas for experienced professionals, doctoral, masters and honours projects, amateur enthusiasts and citizen scientists. There are opportunities for participants across a wide range of subfields and interests from the laboratory 'nerds' to the booted field naturalists to the casual observers. Some issues are achievable in short time frames while others are truly 'blue sky'.

Progress on any or all of these suggestions will be enhanced by existing lepidopterists – that is us – engaging with each other, with funding bodies and the public at large. Scientific societies have a large role to play in this connection. Motivated groups within those organizations can, and should, set standards, promote information dissemination and advocate for public and financial support.

Finally, the growing pre-eminence of AI images and other representations across electronic media can certainly facilitate progress in moth studies. If, however, this becomes a barrier to and substitute for first-hand engagement with nature itself then the game is lost – in terms of science, conservation and society.

References

Cassis, G., Laffan, S. and Ebach, M. (2017) Biodiversity and bioregionalisation perspectives on the historical biogeography of Australia. In: Ebach, M. (ed.) *Handbook of Australasian Biogeography*. CRC Press, Boca Raton, Florida, pp. 11–26.

Chapman, A.D. (2009) *Numbers of Living Species in Australia and the World*, 2nd edn. Australian Biological Resources Study, Canberra, ACT.

Clarke, P.A. (2026) *Aboriginal People and Terrestrial Invertebrates in Australia: Historical and Cultural Relationships*. CSIRO Publishing, Melbourne, Victoria.

Cocking, G., Bond, S. and Edwards, T. (2022) *Moths in the A.C.T.* Glenn Cocking, Canberra, ACT.

Common, I.F.B. (1990) *Moths of Australia*. Melbourne University Press, Collingwood, Victoria.

Common, I.F.B. (1994) *Oecophorine Genera of Australia I. The Wingia Group (Lepidoptera: Oecophoridae)*. Monographs on Australian Lepidoptera 3. CSIRO Publishing, Collingwood, Victoria.

Common, I.F.B. (1997) *Oecophorine Genera of Australia II. The* Chezala, Philobota *and* Eulechria *groups (Lepidoptera: Oecophoridae)*. Monographs on Australian Lepidoptera 5. CSIRO Publishing, Collingwood, Victoria.

Common, I.F.B. (2000) *Oecophorine Genera of Australia III. The Barea Group and Unplaced Genera*. Monographs on Australian Lepidoptera 8. CSIRO Publishing, Collingwood, Victoria.

DeLoitte Access Economics (2020) *Cost Benefit Analysis of a Mission to Discover and Document Australia's Species*. Deloitte Access Economics, prepared for the Australian Academy of Science, Perth, Western Australia.

Forum Herbulot (2014) Statement on Accelerated Biodiversity Assessment (Community Consensus Position). *Spixiana* 37, 241–242.

Gibbs, G. (2016) *Ghosts of Gondwana: The History of Life in New Zealand*, 2nd edn. Potton and Burton, Nelson, New Zealand.

Green, K., Caley, P., Baker, M., Dreyer, D., Wallace, J. *et al.* (2021) Australian Bogong moths *Agrotis infusa* (Lepidoptera: Noctuidae), 1951–2020: Decline and crash. *Austral Entomology* 60, 66–81.

Hausmann, A. (and others) (2001–2019) *The Geometrid Moths of Europe* (7 volumes). Apollo, Stenstrup, Denmark/Brill, Leiden, The Netherlands.

Heikkilä, M., Mutanen, M., Kekkonen, M. and Kaila, L. (2014) Morphology reinforces proposed molecular phylogenetic affinities: A revised classification for Gelechioidea (Lepidoptera). *Cladistics* 30, 563–589.

Holloway, J.D. (1977) *The Lepidoptera of Norfolk. Island, their Biogeography and Ecology*. Series Entomologica 13. W. Junk, The Hague, The Netherlands.

Holloway, J.D. (1979) *A Survey of the Lepidoptera, Biogeography and Ecology of New Caledonia*. Series Entomologica 15. W. Junk, The Hague, The Netherlands.

Holloway, J.D. (1996a) Geological signal and dispersal noise in two contrasting insect groups in the Indo-Australian tropics: R-mode analysis of pattern in Lepidoptera and Cicadas. In: Hall, R. and Holloway, J.D. (eds) *Biogeography and Geological Evolution of SE Asia*. Backhuys, Leiden, The Netherlands, pp. 291–314.

Holloway, J.D. (1996b) The Lepidoptera of Norfolk Island, actual and potential, their origins and dynamics. In: Keast, A. and Miller, S.E. (eds) *The Origin and Evolution of Pacific Island Biotas, New Guinea to eastern Polynesia: Patterns and Processes*. SPB Academic Publishing, Amsterdam, The Netherlands, pp. 123–151.

Kaila, L. (2011) *Elachistine Moths of Australia (Lepidoptera: Gelechioidea: Elachistidae)*. Monographs on Australian Lepidoptera 11. CSIRO Publishing, Collingwood, Victoria.

Larsen, B.B., Miller, E.C., Rhodes, M.K. and Wiens, J.J. (2017) Inordinate fondness multiplied and redistributed: the number of species on Earth and the new pie of life. *Quarterly Review of Biology* 92, 229–265.

Losey, J.E. and Vaughan, M. (2006) The economic value of ecological services provided by insects. *BioScience* 56, 311–323.

Marriott, P., Hewish, M., Edwards, E.D., Kallies, A., Williams, S. *et al.* (2008–2020) *Moths of Victoria* (9 parts with various combinations of authors). Entomological Society of Victoria, Melbourne, Victoria.

Matthews, M. (1999) *Heliothine Moths of Australia: A Guide to Pest Bollworms and Related Noctuid Groups*. Monographs on Australian Lepidoptera 7. CSIRO Publishing, Collingwood, Victoria.

McDonald, G.J., Maino, J.L., Smith, A.M., Gregg, P.C. and Ridland, P.M. (2024) New insights from old data: the complex migration and breeding patterns of the bogong moth, *Agrotis infusa* (Boisduval) (Lepidoptera: Noctuidae), in eastern Australia. *Austral Entomology* 63, 466–484.

McFarland, N. (1988) *Portraits of South Australian Geometrid Moths*. Allen Press, Lawrence, Kansas.

McQuillan, P., Forrest, J., Keane, D. and Grund, R. (2019) *Caterpillars, Moths and their Plants of Southern Australia*. Butterfly Conservation South Australia, Adelaide, South Australia.

Moulds, M.S., Tuttle, J.P. and Lane, D.A. (2020). *Hawkmoths of Australia: Identification, Biology and Distribution*. Monographs on Australian Lepidoptera 13. CSIRO Publishing, Collingwood, Victoria.

Roy, D.B., Alison, J., August, T.A., Bélisle, M., Bjerge, K. *et al.* (2024) Towards a standardized framework for AI-assisted, image-based monitoring of nocturnal insects. *Philosophical Transactions of the Royal Society of London. B Biological Sciences* 379, 20230108.

Stork, N.E. (2018) How many species of insects and other terrestrial arthropods are there on Earth? *Annual Review of Entomology* 63, 31–45.

Sun, G., Liu, S., Luo, H., Feng, Z., Yang, B. *et al.* (2022) Intelligent monitoring system of migratory pests based on searchlght trap and machine vision. *Frontiers in Plant Science* 13, 897739.

Taxonomy Decadal Plan Working Group (2018) *Discovering Diversity: a Decadal Plan for Taxonomy and Biosystematics in Australia and New Zealand 2018–2028*. Australian Academy of Science, Canberra, ACT, and Royal Society Te Apārangi, Wellington, New Zealand.

World Economic Forum (2022) *Scaling Investments in Nature: the Next Critical Frontier of Private Sector Leadership*. World Economic Forum, Geneva, Switzerland.

Zhang, H., Zhao, S., Song, Y., Ge, S., Liu, D. *et al.* (2022) A deep learning and Grad-Cam-based approach to accurate identification of fall armyworm (*Spodoptera frugiperda*) in maize fields. *Computers and Electronics in Agriculture* 202, 107440.

Taxonomic Index

Note: Page numbers in **bold** type refer to figures; page numbers in *italic* type refer to tables.

General Index

Note:- Page numbers in bold type refer to figures; page numbers in italic type refer to tables.